ALLE ZEIT WACH 1842

Hermann Lautensach

KOREA

A Geography Based on the Author's Travels and Literature

Translated from the German,
Supplemented with a Thoroughly Revised and Expanded Index
and Edited by

Katherine and Eckart Dege

With 42 Photographs, 95 Diagrams, 46 Tables
and a Map

Springer-Verlag
Berlin Heidelberg New York
London Paris Tokyo

Prof. Dr. Hermann Lautensach (1886–1971)

Published by
Dr. Wolf Tietze GeoWissenschaftliche GmbH
POB 1647, D-3330 Helmstedt, FR Germany

Translated by
Prof. Dr. Eckart Dege · Katherine Dege
Institute of Geography, University of Kiel
Olshausenstraße 40
D-2300 Kiel, FR Germany

German Edition: Lautensach, H., KOREA, eine Landeskunde auf Grund eigener Reisen und der Literatur. 542 Seiten, K. F. Koehler-Verlag, Leipzig 1945

ISBN-13: 978-3-642-73580-6 **e-ISBN-13: 978-3-642-73578-3**
DOI: 10.1007/978-3-642-73578-3

Softcover reprint of the hardcover 1st edition 1988

Typesetting, printing and binding: Waisenhaus-Druckerei GmbH, D-3300 Braunschweig, FR Germany

2131/3130-543210

CONTENTS

LIST OF FIGURES

LIST OF TABLES

LIST OF PHOTOGRAPHS

Foreword

Why translate Lautensach, 40 years after publication of the book and 50 years after the field work that led to it? And why read Lautensach, with the tremendous volume of literature that has since appeared on Korea? The fact is that to this date no geography of Korea has been written in any Western language that anywhere near approaches Lautensach in thoroughness or sheer volume. His book remains the most comprehensive geography of the entire country of Korea. It is also a historical document that shows Korea at a time when the Japanese endeavors to develop the country as part of their empire had reached their zenith. In addition, Lautensach's Korea became known as a classical example of regional geography for its methodological approach. Unfortunately, the book was published at a most inauspicious moment, a few months before the end of the Second World War, and most copies were destroyed by bombs. Its readership is thus not only limited to persons who understand German; it is practically impossible to get hold of.

Of course, if one were to write a geography of Korea today, one would tackle the problem from a different standpoint. New developments in the field of geography would have to find their way into such a book, and the results of the great volume of more recent research by Korean colleagues would have to be incorporated. For this reason, we have not attempted to update the book to any extent. What we have tried to do is to render a faithful translation, using modern terminology as far as possible, only correcting a few obvious errors.

The one big change we have made is in the geographical names. For reasons explained in the third chapter of Part One, Lautensach chose to use a transcription of the Japanese pronunciation of all Korean place names. This may have seemed the most logical procedure when Lautensach was working on the book; unfortunately it meant that very shortly after the book appeared all place names had become obsolete, so that the regional descriptions in this book were basically as useless as the many earlier travel accounts and other books on Korea that he so criticizes because of their use of unidentifiable place names. Therefore we have attempted to identify every place mentioned in the book on maps showing the Chinese characters, the ultimate authority, and have used the currently approved romanization system of the *han-gŭl* characters for the Korean pronunciation, a tedious task involving consulting and comparing many detailed maps. Our most sincere thanks go to Mr. Kim Chong-kyu and Mr. Kang, Kyoung-won, Department of Geography, University of Kiel, for the countless hours they spent pouring over old maps, searching for the correct Chinese characters. Where they were unsuccessful, we turned to colleagues in Korea, Japan and China for assistance. We should especially like to mention Profs. Cho Dong-kyu and Park Dong-won (†), Seoul, Prof. H. Morikawa, Hiroshima, Profs. M. Ishii and T. Chiba, Tōkyō, and Mr. Tian Jing, Yanji. The few places names that we were still unable to identify appear in the text with their Japanese pronunciation in angle brackets.

Translating a handbook such as this requires a volume of knowledge that is more than one person alone can have. It would be impossible to mention by name all of the books and textbooks we have consulted for details of specific areas of geography. Our sincere thanks also go to the many colleagues in related fields who kindly and patiently answered our many questions on the correct English terminology for phenomena pertaining to their fields of study. Having noticed that the names of birds, fish and other fauna had changed in some cases during the past decades, and were inaccurate in other cases, we attempted to identify and include the correct scientific names and the common English equivalent. The chapter on fauna lists the sources for these names. For information on names of fish we are greatly indebted to Dr. A. Post, Division of Ichthyology, Federal Fisheries Research Institute, Hamburg. Unfortunately, we were not able to get hold of more recent material on East Asian flora: therefore the names used in the translation are for the most part those used in the English language sources listed by Lautensach in his bibliography.

The original book was accompanied by four colored maps, which it has unfortunately not been possible to reproduce (cf. Editors' note on p. 7). As an aid in orientation we are including a modern map, which contains a large number of place names. We should like to express our particular thanks to Prof. Cho Dongkyu, president of the Korean Geographical Society, for his efforts in procuring these maps. Nor has it been possible to reproduce all of the original photographs, as some unfortunately no longer seem to exist. The photographs that are included belong to the collection of the Sektion Sprache und Kultur Koreas, Fakultät für Ostasienwissenschaften, Ruhr-Universität Bochum, and are reproduced with their kind permission. The many maps and drawings in the figures have however been included, with the Korean place names, thanks to the tireless efforts of Miss P. Scheu, who redrew many of them.

Without the incessant endeavors and initiative of Dr. W. Tietze, this translation would never have come about; to him we owe a particularly great debt of gratitude. The German Research Council provided a grant for a research project, in the course of which the place names were identified and the maps redrawn; Inter Nationes supported the translation. For his interest in the project and his constant encouragement we thank S. McCune, Professor Emeritus of the University of Florida. From him we had a rough translation of extracts of the book done in 1949 as part of a research project on Korea by G. Waas, at that time Associate Professor of German at Colgate Unversity. This was a useful aid in the difficult first stages of translation. Last but not least, we thank our children, who had to share their parents with Lautensach for 2 years.

Wielen, April 1988

Eckart and Katherine Dege

PREFACE

Time and again the irresistible urge to master distant, unknown climes, disregarding all hardships, leads us geographers to undertake at least one long expedition to investigate the regional problems of a non-European area. In so doing we spread knowledge of the countries of the earth, which is based on profound personal experience, among our fellow countrymen. At the same time nothing increases one's awareness of the responsibilities of geography at home more than the investigation of an environment that totally forces the researcher to apply different standards from the ones to which he is accustomed at home. As such a goal I chose Korea, for reasons mentioned in the chapter outlining the objectives of the present book at the end of Part One.

A geographical study of an East Asian country entails particular difficulties, because of the great significance of the ancient cultural development and because of the necessity of also considering the sources published in the native languages. In the case of Korea these difficulties are even compounded, for here various foreign cultures overlie each other, and the Japanese literature on the country is often not easy to come by. I owe a great debt of gratitude to the Japanese and Korean lectors who assisted me in translating this literature. Alone the correct Japanese or Korean pronunciation of place names, which are written in Chinese in the official maps, was often a matter requiring great effort. In difficult cases I owe the Japanese pronunciation to the Japan Institute in Berlin. I am particularly indebted to the director of this institution, Prof. M. Ramming. For reasons explained in Chapter 3 of Part One, the place names are generally given with the Japanese pronunciation of the Chinese characters, which are the sole authority. At the end of the book, however, an extensive list of place names with their Korean pronunciation is appended. (Editors' note: cf. Chapter 3 of Part One for the transcription of place names used in the translation. The list of place names appended to the translation has been greatly expanded to include all place names mentioned in the book.) Occasionally the sources of information from the literature flowed very sparsely and if in these cases my own observations could not help out, the depiction was of necessity scant and was closely limited in its methodology. In such cases I also consulted reports and articles in Japanese newspapers, particularly on the most recent development. The Institute of World Economy and World Trade in Kiel and the Archives of World Economy in Hamburg have large numbers of these, and I am very grateful to these institutions for their hospitality. It was not possible to include such newspaper reports in the already extensive bibliography, however. For the rest, all remarks that are not verified by references to the literature are based entirely or for the most part on the author's observations.

Since recent works on Korea in European languages are rare, and there is no geographical handbook thus far in any of these languages, I was not rigid in how I drew the lines between geography and related sciences, but rather included my own observations in the fields of geological structure, vegetation, ethnology, etc., as far as at all compatible with the geographical objectives. In the method of organizing the material I sought new ways, which are closely adapted to the peculiarities of the area in question. In this manner it was possible to trace the geographical effects of Japanese activities in Korea in a special section.

A handbook is not generally read from beginning to end non-stop. Therefore I have not bothered to avoid brief repetitions. They spare the reader the nuisance of having to follow up on all too many page references. Arabic numerals in brackets in the text refer to the books and articles in the bibliography, Roman numerals to periodicals, capital letters to bibliographies and small letters to map series. The values in yen were not reduced to gold parity, because there has been only very slight inflation in the country despite the international devaluation of the Japanese currency. Production and investment values for different years can thus only be compared with each other in unreduced yen. In contrast, the measures, which are given in the sources in Japanese or Anglo-American units of measurement, were transposed into metric units. The data on population figures refer, unless otherwise noted, to 1936 or 1937.

During my journeys in Korea the Japanese authorities supported me to a degree that was most gratifying. I am particularly grateful to the Government–General of Chōsen, the Supreme Command of the Imperial Japanese Army in said place, the Japanese Embassy in Berlin, the Geological Survey of Chōsen in Keijō, the central Forest Experiment Station in Keijō–Seijōri, the Agricultural Experiment Stations in Suigen and Shariin and numerous scholars in Keijō, Tōkyō and Kyōtō. Likewise I am full of gratitude to my intrepid, skillful chauffeur and my knowledgeable interpreter. Without the constant commitment of these two Koreans, Mr. Oh Hung Kun and Mr. Ji Djong Sun, it would not have been possible to carry out the trips to the extent that we did. In addition many other Koreans supported me as valiant local guides, porters or pony drivers. My native country also extended its assistance into this remote country. The German Embassy in Tōkyō, the German–Japanese Cultural Institute in said place and my fellow countrymen in Korea gave me competent advice and granted me comforting hospitality.

The expenses of the trip were shared by the German Research Council, the Gießen University Society and the author. The maps (cf. Editors' note on p. 7) and figures were drawn with tireless zeal by the cartographers A. Schott, F. Siepmann and H. Ehser. A few further maps were drawn by students during a laboratory course of the department of geography in Greifswald. The drafts were done by the author, where no other source is mentioned in the caption. The same holds for the photographs, with the exception of Nos. 1, 2, 5, 10, 11, 20, 21, 22, 24, 33, 34, 36, 39 and 41, which are reproductions of photographs acquired in Korea. The costs of drawing the maps were paid by the Colonial Science Division of the National Research Council. From the Society of Friends and Supporters of the Ernst Moritz Arndt University in Greifswald came a sizable grant for printing. The Geographical Service of the Foreign Ministry and the German–Japanese Society in Berlin also gave their support for the book. My wife labored unceasingly to type the manuscript and helped me daily with her expert advice. Prof. Krebs and Dr. Zorell were kind enough to help me read the proofs. Finally, I would like to thank the publisher, K.F. Koehler Verlag, for the sympathetic, always helpful cooperation.

Greifswald, November 1942 Hermann Lautensach

PART ONE:

GEOGRAPHICAL AND HISTORICAL BACKGROUND

1 Geographical Location and Boundaries

If one tries to imagine the sea level being lowered by only 90 m, the entire Yellow Sea and the adjacent portion of the East China Sea would run dry. Korea would lose its peninsular nature and disappear into the continental mass of Asia. The eastern coast of northern China would move approximately 800 km eastward to coincide with the present eastern coast of Korea. The eastern coast of Korea would be dislocated only minimally by such a lowering of the sea level, however, for the Sea of Japan is relatively deep, its floor dropping to more than 3000 m below sea level not far from the coast of the continent. The steeply sloping coastline continues its sharp decline under the water level for the entire length of the eastern coast of Korea. The 200–m isobath lies less than 4 km from the coast, and the 1000–m isobath lies 8 km from the coast.

During the Ice Age the sea level was in fact approximately that much lower than today due to the eustatic lowering of the sea level resulting from the large amount of water bound up in the continental ice masses. At that time Korea was actually a part of the East Asian mainland and was strongly influenced in both its flora and its climate from the west. As the paleogeographical maps of Y. Otuka show [327, p. 453], this is also true of most of the Tertiary period. The island of Cheju to the south of the peninsula was probably also connected to the continent during the Ice Age [883]. In the geological present the only frequent influence on Korea from the west is climatic. This influence is particularly strong in the winter, when the western coast is much colder than the eastern coast with its maritime influence. There are also relationships between the geological structure of Shandong and the part of Korea lying across from Shandong. Anthropogeographical relationships are, however, much less frequent. The western coast of Korea is very dangerous, and both Chinese and Koreans are basically land–based peoples who for thousands of years confined their navigation primarily to their coastal waters and to the purposes of fishing and piracy. These were practically the only areas in which historical connections existed across the Yellow Sea between the two peoples. There are very few reports of Chinese war fleets being sent to Korea.

The Chinese cultural influence, which lasted for thousands of years and had a profound effect, did not reach Korea by sea as much as by land, i.e, not from the west, but from the north or northwest. The same is true of most of the Chinese military intrusions into Korean life. From the north by land extensive racial and cultural influences from Central Manchuria also reached Korea in both prehistorical and historical times. Similar influence came from the south from the southwestern end of the Japanese islands, which is separated from the Korean peninsula only by the 180–km–wide Korea Strait. In contrast, Japan almost never reached out from the east across the Sea of Japan toward Korea, for the "Western

Country of Treasures" – an honorary title given to Korea by the Japanese during the golden age of Buddhism – turns its defensive side toward the Sea of Japan. The souvenirs of former cultural and diplomatic relations with Korea found by W.E. Griffis [45] in 1871 in Tsuruga, the most important gateway to Japan on the Sea of Japan, most certainly entered the empire of the Mikado via the south coast rather than the east coast. Thus, in the most important spheres it is not Korea's position between northern China in the west and the Japanese islands in the east that is decisive, but rather its position between Manchuria in the north and the end of the Japanese islands in the south. Korea is a land bridge that juts out north–northwest to south–southeast from the mainland of Asia. It is also a transition zone between the purely continental and the purely maritime portions of extratropical Monsoon Asia.

The actual Korean peninsula begins between the mouths of the Ch'ŏngch'ŏn–gang and the Sŏngch'ŏn–gang (Fig. 1). The indentations forming Sŏhanman (Korea Bay) and Tonghanman (East Korea Gulf) constrict the land here to a width of 170 km. A short distance farther south, however, the Hwanghae peninsula juts out far westward into the Yellow Sea toward Shandong. The cape of Changsan–got approaches to within 180 km of Shandong, and the peninsula widens to a width of 320 km. At the latitude of the capital city, Seoul, it again narrows to 190 km, only to widen out again to 320 km at its broad southern end between Mokp'o and the Changgi peninsula. All of these changes in width are caused by bays and peninsulas along the western coast, whereas the eastern coast forms a symmetrical, almost continuously convex arc. Its entire length amounts to 580 km. Both geographically and politically the thousands of islands off the western and southern coast naturally belong to Korea. After all, they are nothing but former spurs of the Korean mainland that were cut off by the postglacial rise in the sea level that inundated the Yellow Sea and the Korea Strait. We also consider Chejudo as belonging to Korea. Politically, linguistically and culturally it belongs to Korea, whereas from the standpoint of physical geography it is clearly related to Kyūshū. The island of Ullŭngdo has a similar relationship to Korea. It is a volcanic island rising from a depth of more than 2000 m in the Sea of Japan 130 km from the Korean coast but still in sight of the mainland. The two islands of Tsushima in the middle of the Korea Strait, on the other hand, have belonged culturally and politically to Japan for centuries and are not considered to belong to Korea.

The former Kingdom, or Empire, and present Government–General of Korea is not limited to the peninsula. It extends northward onto the continent of Asia, which bulges out southward toward the neck of the peninsula. For more than 500 years the political boundary has been formed by the Yalu (Kor. Amnokkang) and Tumen (Kor. Tuman–gang) rivers, which rise on the slopes of Paektusan (2744 m, Chin. Baitou Shan), by far the highest mountain in the entire region. This continental northern Korea is 570 km wide between the mouths of the Yalu and the Tumen. Its maximum width measured perpendicular to this line is 230 km.

Geologically and morphologically the Yalu–Tumen line is not a definite boundary, except in a few places. From the standpoint of climate and vegetation it is even less of a boundary. The entire immense upland area extending from the northern end of the peninsula to the Manchurian plain is a decidedly transitional

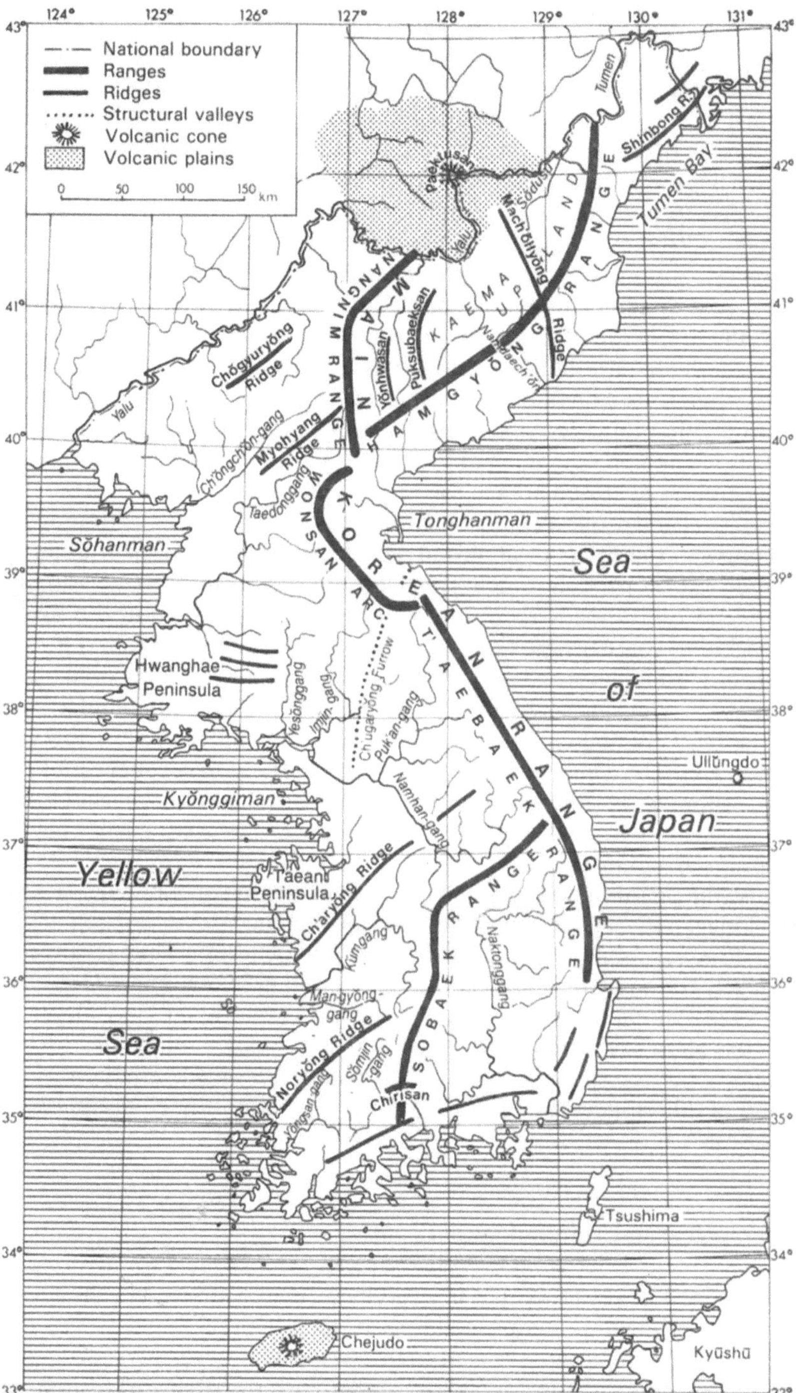

Fig. 1. Generalized orographical map

zone. Although in earlier millennia the cultural influence from the north was stronger than from any other direction, the function of the Yalu–Tumen line as a geographical boundary is much clearer from the viewpoint of cultural geography, due to its long existence as a political boundary and the pronounced isolationist tendency that dominated both Chinese and Korean policy during the last centuries. It is true that in recent times the Korean population has spread across this boundary, and 80% of the population of the Manchurian province of K'ientao (corresponding approximately to the present–day Yanbian Korean Autonomous Prefecture, Jilin Province) directly north of the Tumen is Korean. Nevertheless, even in a geographical investigation the Yalu–Tumen line is the relatively best dividing line for defining Korea's northern boundary. Its central portion, the area around Paektusan and the surrounding basalt plateau, is the only part in which the political boundary dissects a landscape of pronounced individuality in such a manner that only the southeastern sector belongs to Korea, whereas the other three sectors belong to Manchuria. In this area the geographical boundary cannot coincide with the political boundary. We prefer here to include the entire Paektusan area in the geographical unit of Korea.

The political and geographical unit Korea extends from the northernmost point of the course of the Tumen (43°02') to the islet of Marado on the southern coast of Chejudo (33°06') and thus measures 1100 km north–south. The latitude at its southern end corresponds to northern Kyūshū or central Morocco and at its northern end to southern Hokkaidō or the northern coast of Spain. Of this considerable longitudinal spread only a bit more than one–half is covered by the actual peninsula. The land boundary measures 1329 km, the sea boundary without the islands 8637. With the islands included, the length of the coast is now given as 18,204 km [XLIII]. This is considerably more than Schröter measured in 1904 using maps that were still insufficient [354]. The total number of islands is 3479, 1164 of which are more than 1 km in circumference. The coastal length of the latter measures 8334 km.

Because the peninsula is so narrow, Korea has an area of only 220,741 km^2, despite its long north–south extension. If we include the Paektusan area, the entire country of Korea comprises approximately 237,000 km^2. Thus it corresponds fairly closely to the main island of Japan, Honshū (223,520 km^2), Great Britain (228,275 km^2) or mainland Italy before the First World War (238,900 km^2), with which it has often been compared since Carl Ritter [133]. Compared with the vast land areas of the rest of the East Asian mainland, it is only a minor entity, therefore, and in this respect it definitely belongs to the maritime–peripheral outer zone of East Asia.

2 Topographical and Morphological Features

As is the case in Japan, Korea is not only a small country, but the individual landscapes of which it is composed are also small. Korea is an extremely mountainous country, and because the ultimate base level is never very far, it is highly dissected and therefore almost has the appearance of patchwork. For an initial overview we can turn to Richthofen's concept of the East Asian step fault structure [351]. Korea belongs to the "Manchurian fault block". Directly north of the neck of the peninsula is a junction zone, in which the Tungusic and the Korean arcs of this step fault meet. The peninsula forms the northern leg of the Korean arc, which trends north–south. Roughly speaking we can consider it to be a tilted fault block, whose curved crest runs very close to the coast of the Sea of Japan. The floor of the Yellow Sea consists of the depressed parts of the dip slope, which are now submerged. From the coast of the Yellow Sea the dip slope gradually rises eastward toward its crest, where it drops off abruptly to the coast of the Sea of Japan. The westward flowing rivers arise on the crest very close to the east coast. The longest of them, with the most branches, is the Han–gang, the river of the capital city. It begins 20 km from the Sea of Japan and flows 482 km before it reaches its mouth south of the Hwanghae peninsula. Very nearly the same is true of the Imjin–gang and the Taedonggang (Fig. 1).

Richthofen was of the opinion that the western leg of the Tungusic arc, with its southward trend, overlapped the northern leg of the Korean arc and that therefore its crest formed a mountain range along the entire southern border of continental Korea. The only evidence of the crest of the Korean arc in continental Korea was to be seen in north–south fractures that revealed how thoroughly linked these arcs were. Richthofen believed at the time that he could actually present this relationship between the two arcs as a paradigm [351, Vol. IV, p. 9]. In reality, the reverse is true, at least if we are speaking of the surface forms. The Korean arc is "dominant" in the junction zone,to use the more recent nomenclature of Motizuki [396]. The Tungusic arc is "recessive". The peninsular block extends north into continental Korea at least as far as the Yalu, perfectly distinct, and, accordingly, its crest runs northward from the peninsula in the form of a huge mountain ridge that is an almost uninterrupted extension of the crest of the peninsula. Thus in its surface forms, the western half of continental northern Korea resembles the peninsula. The Yalu downriver from Chunggangjin and the parallel Ch'ŏngch'ŏn–gang belong to the same series of west–southwest flowing rivers.

The crest with its surroundings forms a mountain chain running north–south that is the actual backbone of all of Korea, with the exception of the northeast. I have given this the name Main Korean Range. The Koreans call the continental part the Nangnim Range (Kor. Nangnimsanmaek) and the peninsular part T'aebaek Range (Kor. T'aebaeksanmaek). It is, however, advisable to have a uniform name. The Nangnim and T'aebaek Ranges must be considered parts of the Main Korean Range. The former rises to a height of 2262 m (Wagalbong). The passes crossing it lie at an altitude of 1300–1500 m (Sŏrhallyŏng 1427 m). The Main Range arches around the innermost portion of Tonghanman, Yŏnghŭngman (the bay at Wonsan). In this area only relatively few peaks reach a height of more than 1000 m (Haramsan 1486 m, Maebongsan 1665 m). Here the Main Range is crossed by a diagonal depression running from Wonsan to Yŏnch'ŏn, which Kotō [352] has entitled the "Tungarion graben" (Kor. Ch'ugaryŏng rift valley). Since it is not, as we will observe later on (p. 73ff), a true tectonic graben, we prefer to call it the Ch'ugaryŏng furrow. Here the drainage divide between the Yellow Sea and the Sea of Japan lies at an altitude of just under 600 m; thus it forms the most important passageway from the east coast to the capital. The T'aebaek Range follows the curvilinear coast of the Sea of Japan, with its very small tidal range, at a distance of 20–30 km. The rivers on its eastern side are therefore short, flowing into narrow coastal plains with lagoons blocked by bars. The short, narrow alluvial plains are separated from each other by spurs ending in rounded rocky capes. At its northern end the T'aebaek Range reaches the height of 1638 m in the spectacular Diamond Mountains (Kor. Kŭmgangsan), and a bit farther south in Sŏraksan even 1708 m. On the other side of the pass of Taegwallyŏng (865 m), which connects the central eastern coast with the Han–gang basin, it again rises to a height of 1573 m in the tangle of mountains at T'aebaeksan, then slowly dips, ending at the broad tectonic east–west valley of P'ohang–Kyŏngju–Taegu.

In the southeastern part of the peninsula a secondary feature of the step fault blocks takes predominance. Here we find mountain ranges and valleys trending north–northeast to south–southwest, which were included by Tokuda [362] and Motizuki [395 to 397] in their concept of an echelon pattern in which the individual elements of the East Asian island and peninsular arcs are arranged. The coastal range running from the Changgi peninsula to Pusan is followed to the west by the structural valley of Kyŏngju–Yangsan and then by a mighty mountain range that reaches an altitude of 1240 m in Kajisan. Above all, however, the Sobaek Range branches off to the southwest from the Main Range at the tangle of mountains at T'aebaeksan. Initially this range gradually dips until it reaches the pass of Ch'up'ungnyŏng, with a height of only 254 m, but then it rises again sharply until it attains an altitude of no less than 1915 m in Chirisan, only 40 km from the south coast. Between the Sobaek Range and the above mentioned mountain ranges lies the basin of the Naktonggang, an extraordinarily characteristic feature of southeastern Korea. The big Naktonggang flows in several arms through a substantial delta just west of Pusan, discharging into the eastern end of the Korea Strait, where the tidal range is small.

The westward flowing rivers are joined farther south by the Kŭmgang and the Yŏngsan–gang, which in contrast to the former, rise in the Sobaek Range. All of these rivers that flow into the Yellow Sea, from the Yalu to the Yŏngsan–gang, have formed small alluvial plains in their lower reaches. These are separated from each other by low hills, the spurs of the ridges branching off from the Main Range. Due to the mighty tidal range of the Yellow Sea, they all have tidal estuaries, which are enclosed by low, but usually rocky peninsulas and islands. Thus one low–lying, fertile basin follows the next along the western coast from the mouth of the Yalu to the southwesternmost tip of the peninsula, facing an exceptionally stormy sea, which abounds in dangerous tidal currents, sandbanks, cliffs and islands. The western side of the peninsula, together with the Naktonggang basin, has always been its cultivated side. It turns its back on the Sea of Japan with its coast lacking in bays (Fig. 1).

As we emphasized above, the eastern half of continental northern Korea can be considered to be the western leg of the Tungusic arc, in Richthofen's sense. Thus it also forms a tilted fault block, which, however, tilts north to northwest, in contrast to the main Korean tilt block. The crest of this block also runs parallel to the coast of the Sea of Japan, first northeast, then, on the other side of Namdaech'ŏn, north–northeast. In contrast to the crest of the main Korean tilt block, it lies at the considerable distance of 40–60 km from the coast. It rises in a generally northeasterly direction and reaches 2541 m in Kwanmobong, the highest all Korean peak. Its name, Hamgyŏng Range, refers to the province. The southern tributaries of the lower reaches of the Yalu and the Tumen follow the direction of the tilt of this block: Changjin–gang, Pujŏn–gang, Hŏch'ŏn–gang, Sŏdusu. Since the ultimate base level is far away, these rivers are much less deeply entrenched than the rivers of western Korea. The valley floors lie at 500–1400 m, the passes crossing the crest at 1300–1600 m. Going back to a name that occurs as early as the Imperial Chinese Geography of 1744, and even the annals of the later Han dynasty (25–220 A.D.), Japanese geologists refer to the dip slope as the Kaema plateau. I prefer the term Kaema upland, because above the valleys mentioned before the relief is still quite strong. The ridge that runs along the eastern side of the Pujŏn–gang parallel to the Nangnim Range towers even higher than the latter (Puksubaeksan 2522 m), and the drainage divide between the Yalu and the Tumen is formed by the Mach'ŏlly'ŏng Range, which rises toward the northwest, culminating at 2435 m in Namp'ot'aesan.

West of the crest of the main tilt block, there is a ridge that can be considered an extension of the Hamgyŏng Range penetrating the crest of the main block. It is the Myohyang Ridge (Myohyangsan 1909 m). Farther to the north and parallel to this ridge, not far from the Yalu, lies the Chŏgyuryŏng Range (Sungjŏksan 1994 m). The two are separated by the basin of the Ch'ŏngch'ŏn–gang.

Continental northern Korea has by far the greatest altitudes of all Korea, both on an average and in its peaks. As Map 1 shows (editor's note: Maps 1–4 refer to the colored maps accompanying the original German version of the book, which it unfortunately has not been possible to reproduce. We have translated the references to these maps to give the reader an idea of their role in the original work. The reader who is truly interested in consulting these maps will have to refer to one of the rare extant volumes of the original German book.), this is the only part of the country with extensive areas lying over 1500 m, and whereas in the rest of Korea no peak reaches an altitude of 2000 m, many here are more than 2300 m high. Still, the only places where the relief is Alpine in character are where recent headward erosion has cut directly into the areas with the highest peaks. Almost the only area in which this is the case is the side of the Hamgyŏng Range facing the Sea of Japan.

Nevertheless, the slope of the Hamgyŏng Range down to the Sea of Japan is broader and more highly dissected than the scarp of the T'aebaek Range on the peninsula. This is particularly true of the northeastern section. Here a hill country with its own morphology and geology has developed in front of the high mountains. It begins between Sŏngjin and the cape of Musudan and extends from there into the extreme northeastern corner of Korea, which is encircled by the Tumen. Still there are many peaks here exceeding an altitude of 1000 m.

Between the Korean cape Musudan and the Ussurian Cape Povorotnyj, in the northeastern extension of the Kaema tilt block, lies a bay that is 340 km across and 115 km deep, at the maximum. Since the Tumen flows into the center of this bay, it is best named after that river. Richthofen [351, Vol. II, p. 8] considers it to be a collapse within the Tungusic arc and compares it with the bay of Beijing, which it, however, exceeds in size by far (Fig. 1).

The northward sloping Kaema upland ends north of the upper Yalu in a depression that is filled with tremendous masses of basalt. These incline outward from Paektusan in all directions and form an extensive plateau. In its center rises the mountain ring composed of acidic extrusive material that forms Korea's highest mountain. The subsidence caldera in its interior is filled by a fair sized lake, whose surface lies 500 m below the highest peaks.

3 Place Names and Their Transcription

Editors' note: At the time that Lautensach wrote his Geography, Korea was a colony of Japan, and most scientific literature on Korea was written in Japanese. For this reason, and because he was unable to find a uniform transcription of the Korean pronunciation of place names, he chose to use the Japanese pronunciation and spelling of all place names. Since Korea's independence from Japan in 1945, there is no reason to use Japanese names for Korean places, and in his publications after Korea's independence Lautensach himself turned to the Korean pronunciation of place names. The modern reader will most certainly refer to modern maps with Korean place names. Therefore, we have looked up all place names mentioned by Lautensach on maps and transcribed them according to their Korean pronunciation, using the modified version of the McCune–Reischauer system that was approved on 13 January 1984 by the Ministry of Education of the Republic of Korea. In the following chapter we have not bothered to translate the portions that referred directly to the Japanese place names, but only the more general portions that will also be of interest to readers of this translated version.

Almost all names mentioned in the previous sections and in the chapters to come may be found in Map 1 (cf. editor's note on p. 7). The spelling of Korean place names is to date a matter of almost utter confusion. A scientific evaluation of many of the available travel reports is difficult or even impossible, because the spelling of the names of villages, rivers or mountains makes it impossible to identify them. The only really exact method of writing Korean place names is by means of Chinese characters. Two, three, less often four, characters used in combination represent the name of a given place, exactly the same as in China, Japan and the other countries in which Chinese characters are used. All of these characters are still alive and in use today. Thus, in Korea, as in the rest of East Asia, all place names have a meaning, and it is not normally necessary to go into the etymology of such names, as is so often the case in regions with alphabetical and syllabic writing systems. A few examples are: Mountain of Great Virtue, Peak of the Thousand Buddhas, Mountain that Touches Heaven, Seven Peak Mountain, Yellow Dragon Mountain, White Cloud Dome, Diamond Mountains, Valley of Cool Shadow, Nine Dragon Fall, Flying Phoenix, Great Southern Stream, Yellow Sea Province, New County Seat of Peace, Great South Gate, Temple of Eternal Peace. Many of these names are related to the Koreans' strong feeling for nature, to superstition, Buddhism or to the Confucian system of ethics.

Naturally it would be just as possible to write all place names in Korea's own national alphabet, the *ŏnmun* (p. 30, Editors' note: known in South Korea today as *han–gŭl*). But only a few old, relatively small scale maps use this writing system. Moreover, the old Korean maps use Chinese characters, as do the newer maps and atlases published by the Japanese administration, as long as these are intended mainly for use in the country, which is almost always the case when the scale is larger than 1:1 million.

The Chinese characters on the maps of Korea can be read in any language. Practically, of course, only Chinese, Korean and Japanese come into question. The Chinese pronunciation is used in older works in European languages, which are generally translations of Chinese sources or were written on the basis of Chinese sources, such as the works of du Halde [125], d'Anville [126], Klaproth [131], Carl Ritter [133] and others. To the present day the Korean pronunciation is used preferentially in books and articles written by missionaries and strangely enough in some of the official sea charts of European countries. Naturally the same is true of the few books written by Koreans in European languages that are of use in a geography of the country. The localization of place names according to their Korean pronunciation is simplified by one gazetteer published by the Japanese and two by the British [U, V, W]. By far the greater part of the scientific literature on Korea today is written in Japanese or in English by Japanese. Thus the only possible way to straighten out the terrible confusion inherent to the spelling of Korean place names today is to use the Japanese pronunciation of the Chinese characters (cf. Editors' note above.)

To make the confusion even greater, each of the three possible pronunciations has been transcribed into the European phonetic writing systems in many different ways. The French missionaries prefer a French transcription of the Korean pronunciation, and the English-speaking missionaries an English transcription of the same pronunciation, although the many inconsistencies of English spelling are often the cause of ambiguity. Books written in English in China about Korea use an English transcription of the Chinese pronunciation, etc. Particularly unfortunate are the Russian transcriptions, because the Russian alphabet lacks a number of letters that are necessary for a phonetically correct pronunciation of the Korean language. The translators of French or Russian books about Korea into German never take the trouble to attempt to write the names such that a phonetically correct German pronunciation is possible. Instead they retain the original spelling or transcribe the Cyrillic transcription a second time. Thus Paektusan becomes **Бай Тай** in Russian for M.A. Pogio, and in his German translation v. Ursyn–Pruszyński [156, p. 3] transcribes this as Baj–Taj. The reader must have an extremely good knowledge of the language to be able to identify such garbled names at all. Later German authors then retain these garbled transcriptions from other languages. Thus Ritter [133], for instance, writes some of his Korean names as they were written in his French sources (e.g., [125], [129], [131]).

For the two Chinese characters that symbolize the name of the second largest city of Korea, and its oldest capital, I have found in non-Russian, European books and the corresponding maps and globes alone no less than 56 different spellings:

Feanjo, Heijō, Heizyō, Hejo, Hēnzyō, Hpieng–Yang, Hpyeng yang, Peng Yang, Phieung–iang, Phingjang, Phionjan, Phjöng Jang, P–hjöng–jang, Phŏngyang, Phöngyang, P–hyeng–yang, Phyong An, Phyongjang, Phyŏngyang, Phyöng–an, Phyöngyang, Piengan, Pieng–'an, Piengiang P'iéng–iang, Pieng–Jang, Pien–yang, Pieun–Yang, Pin–an, Pingan, Pinggan, Ping–jang, P'ing–jang, Pingyan, Ping–yang, P'ing–yang, Pinjam, Pinjang, Pinyan, Pinyang, Pion–Anj, Piongyang, P'jöng–jang, Pjöng–jang, Pjöngyang, Pöngyang, Pyangan, Pyeng–yang, Pyeung–yang, Pyongyang, P'yongyang, Pyöng–jang, Pyöngyang, P'yöngyang, P'yŭngyang. This does not even include minor differences, such as the hyphen between the two syllables and whether or not the second syllable is capitalized. And this list, which is probably not even complete, would be much longer if we wished to include the transcriptions into the Cyrillic alphabet. Finally, we have ignored the fact that the city, or an earlier city at the same location, had a different name in the first centuries A.D., i.e., a name that was written with different Chinese characters. These 56 spellings all refer to the same two Chinese characters.

As we mentioned previously, Korean place names consist of two–four Chinese characters, each of which symbolizes a certain concept. Thus each place name has a meaning. To help the reader understand this meaning, I have made a list of the most common terms used in Korean place names.

Korean	Japanese	English
ak	gaku	peak
an	an	peace
chang,jang	chō	long (space or time)
ch'il	shichi	seven
chin,jin	–chin	fortress
chin,jin	–shin	harbor, port
cho,jo	shio	current
ch'ŏl	tetsu–	iron
chŏn,jŏn	–den	field
ch'ŏn	–sen,–kawa,–gawa	small river
chu,ju	–shū	trad. provincial capital
chuk,chung,juk,jung	chiku–	bamboo
chung,jung	chū,naka–	middle
–do	–dō	province
–dong	–dō	central village, city precinct
–gun	–gun	county
ha	ka–	lower
hae	–kai	sea
haehyŏp	–kaikyō	strait, sound
ho	–ko	lake
hong	aka	red
hŭk	kuro	black
hŭng	–kō	port
hwa	ka	fire
hwang	kō	yellow
i	ni	two
il	ichi	one
il	nichi	day
kang,gang	–kō	large river
ki,gi	–zaki	cape
ko,go	kyū–	old
kot,got	–kan	cape
ku,gu	kyū	nine
kuk,guk	koku,goku	land, area, country
kŭm,gŭm	kin	gold
kundo,gundo	–guntō	group of islands
kyŏng,gyŏng	kei–	capital
ma	–ma–	–touching–
ma	ba–,ma	horse
man	–wan	bay
man	man	ten thousand
mun	mon	gate
–myŏn	–men	rural community
nam	nan–,minami–	south
nyŏn,yŏn	nen	year
o	go	five
onch'ŏn	onsen	hot spring
paek,baek	haku	white
paek,baek	hyaku	hundred
p'al	hachi	eight
pando,bando	–hantō	peninsula
p'o	–ho	bay, beach
pong,bong	–hō	peak
pu,bu	–fu	city
puk,buk	hoku,kita	north
p'yŏng	hei	peace
–ri	–ri	village
rim,lim,nim	–rin	woods
ryŏng,lyŏng,nyŏng	–rei	pass
sa	–ji	Buddhist temple
sa	shi	four

sam	san	three
san	–san, –zan	mountain
sang	jō–, kami	upper
sanmaek	–sammyaku	mountain range
shin	shin–	new
ship	jū	ten
so	shō–	small
sŏ	sai–, sei–, nishi–	west
song	matsu	pine
sŏng	–jō	fortress, fortified city
su	–sui	water, creek
tae, dae	dai, tai	big
tae, dae	tei	terrace
tan, dan	–tan	cape
to, do	–tō, shima	island
tong, dong	tō, higashi	east
wŏl	gatsu, getsu	moon, month
yŏlto	rettō	chain of islands
yong, ryong, lyong, nyong	ryū–	dragon
yuk, yung	roku	six

As the reader can see from this brief list, the same syllable can have more than one meaning (Editors' note: i.e., different Chinese characters may have the same pronunciation). Thus the name Paeksan means White Mountain and Paektusan, written with the same Chinese character, means Whitehead Mountain, whereas Paeksan written with a different Chinese character but the same Korean letters means Cypress Mountain.

To increase the confusion even more, European naval expeditions around the end of the 18th century and the first three quarters of the 19th century, i.e., during Korea's period of isolation, gave mountains, bays and capes along the Korean coast names of their own invention. On the east coast Russian names predominate, and on the west and south coast English and French names. Thus there was a Prince Jerôme Gulf, a Prince Impérial Archipelago, a Broughton Bay, a Sir James Hall's Groupe [named by Basil Hall (cf. p. 45) after his father, the president of the Royal Society of Edinburgh], a Mt. Auckland, a Port Hamilton, a Port Lazarev, an Unkovsky Bay, a Cape Peshchurov, a Gontcharov Island, etc. These names had a right to exist only as long as the native Korean names were unknown, but somehow they still manage to exist in modern atlases and charts of European origin. Certainly we owe our respect to the great European sailors who gave these names. We shall deal with their accomplishments further on (p. 45ff). Still the impossible situation resulting from this confusion about the spelling of Korean place names calls for a radical solution (Editors' note: i.e., for Lautensach at the time of writing, to use the Japanese pronunciation of the Korean place names. For us today, as for Lautensach after Korea's independence, the solution is quite naturally to use the best possible transcription of the Korean pronunciation of Korean place names). The explorers who gave foreign names to Korean places also considered this to be a rather makeshift solution. Broughton, e.g., did not hesitate to call the Bay of Pusan "Chosan Harbor", because he heard this term from the natives and was under the impression that it referred to the bay. Such cases are, however, rare, since foreign sailors seldom had a chance to go on land during that period of isolation, and even when they did have a chance, it was often impossible to communicate with the natives. Therefore, I avoid all foreign names given to Korean places during the Age of Exploration, with one exception.

The name Quelpart has become so well known everywhere in Europe through Hamel's tale of woe (p. 31, 42f), that it would not be justifiable to replace it with the name Saishūtō. (Editors' note: Today it of course would be unjustifiable not to use the Korean name Chejudo.)

Names of places in Manchuria are written according to their Chinese pronunciation using Pinyin, the Chinese phonetic alphabet officially used in the People's Republic of China today. The names of the two rivers forming the boundary with China are given as they are generally known in Europe, with their Chinese names Yalu and Tumen, rather than their Korean names Amnokkang and Tumangang.

4 Provincial Boundaries

In the late 14th and early 15th centuries Korea was divided up into eight provinces. The boundaries of these provinces have remained unchanged through all political upheavals until the present day. Only the two northernmost and the three southernmost provinces, in each of which a "right" and a "left" half had previously been distinguished, were each divided into two separate provinces in 1896. The three central provinces have remained undivided. The new provinces resulting from this division retained the old names and are distinguished from each other by the prefixes "north" (Kor. puk,buk) and south (Kor. nam). It is worth mentioning, however, that the two parts of the old northeastern province Hamgyŏng lie northeast–southwest of each other, rather than north–south, and the two Ch'ungch'ŏng provinces even east–west instead of north–south (Fig. 2). Korea's division into provinces was based on the Chinese system, and it played and plays as important a role in Korea as in China. Thus we cannot begin a geography of Korea without an evaluation of this system from the standpoint of geography.

The central province on the western side contains the capital city, Seoul. Therefore it is called Capital City Province (Kyŏnggi-do). The names of the other seven provinces were formed on the model of the Chinese provinces Gansu, Jiangsu and Anhui, i.e., the first syllables of the names of two important cities in a given province were combined to form a new word. Since each of these syllables represents a Chinese character, each has its own meaning, and in combination they have a new combined meaning. The resulting names of the provinces – from north to south – are listed in Table 1.

The location of the cities after which the provinces are named can be seen in Fig. 2. In general it would be beside the point to attempt to explain these names objectively. On the contrary, they bear witness to the lofty philosophical spirit that prevailed among Korean government officials at the time that the names were chosen. The fact that the name "Peaceful Quiet" was chosen for the province that was the northwestern entrance to Korea, which before and after the names were given saw more warlike activity than any other province, is certainly a paradox that did not escape the persons who chose the name. In the case of Hwanghae-do it was a matter of chance that there were two cities in this province with initial syllables meaning "yellow" and "sea", thus making it easy to choose the geographically very pertinent name Yellow Sea Province. It is a name that de-

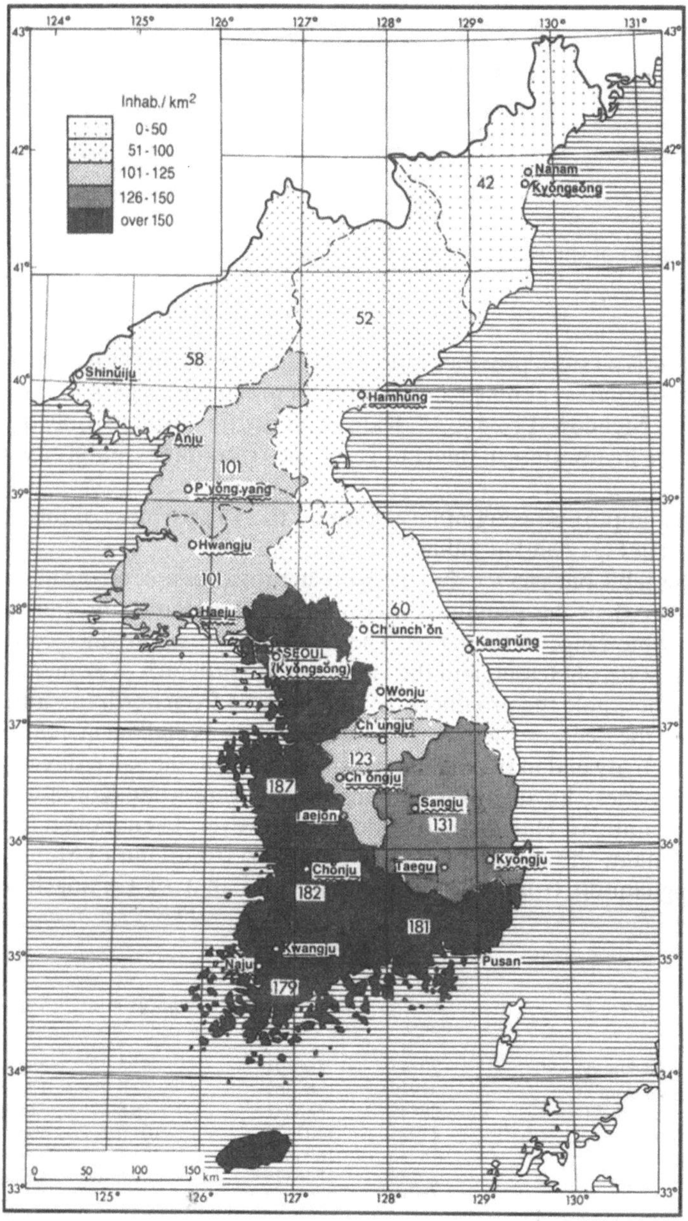

Fig. 2. Provinces and capitals. The population density of the provinces (end of 1938) is represented by *hatching*. Provincial capitals are underlined with *straight* lines, the cities from which the names of the provinces were taken, with *wavy* lines. *I* Hamgyŏngbuk-do; *II* Hamgyŏngnam-do; *III* P'yŏng-anbuk-do; *IV* P'yŏng-annam-do; *V* Hwanghae-do; *VI* Kang-won-do; *VII* Kyŏnggi-do; *VIII* Ch'ungch'ŏngnam-do; *IX* Ch'ungch'ŏngbuk-do; *X* Kyŏngsangbuk-do; *XI* Kyŏngsangnam-do; *XII* Chŏllabuk-do; *XIII* Chŏllanam-do.

Name of province	Names of cities after which they are named	Meaning of province's name
Hamgyŏng-do	Hamhŭng, Kyŏngsŏng	Perfect Mirror
P'yŏng-an-do	P'yŏng-yang, Anju	Peaceful Quiet
Hwanghae-do	Hwangju,Haeju	Yellow Sea
Kang-won-do	Kangnŭng,Wonju	River Meadow
Ch'ungch'ŏng-do	Ch'ungju, Ch'ŏngju	Serene Loyalty
Kyŏngsang-do	Kyŏngju,Sangju	Respectful Congratulation
Chŏlla-do	Chŏnju, Naju	Completed Network

Table 1. Origin of the names of the Korean provinces

finitely fits this western province more than any other, lying as it does on the most prominent secondary peninsula. The attempt to establish a geographical basis for the "River Meadow Province," Kang-won-do, is justified only by the fact that all of the headwaters of the Han-gang, the river of the capital city, rise in this province.

The most important geographical data on the 13 provinces of the present day can be seen in Fig. 2 and in Table 2.

From this table one can see that the Korean provinces are, on the average, somewhat smaller than the provinces of Prussia. The variation in their size lies in about the same range as the latter. The smallest Korean province corresponds in area to the state of Hessia, and the largest to the province of Pomerania (before 1938). On the whole the provinces decrease in area from north to south. This corresponds to the decrease in size of discrete landscapes and the increase in agricultural productivity and in population density running in the same direction. In general, the Japanese have left the capitals as they were when Korea was independent. The capital of Hamgyŏngbuk-do has, however, been moved from Kyŏngsŏng to Nanam, the capital of P'yŏng-anbuk-do from Ŭiju to Shinŭiju, that of Kyŏngsangnam-do from Chinju to Pusan and that of Ch'ungch'ŏngnam-do

	Name of province		Area (km^2)	Inhabitants and of 1938	Population density	Capital
I.	Hamgyŏngbuk-do		20347	860191	42.3	Nanam
II.	Hamgyŏngnam-do		31979	1662369	52.0	Hamhŭng
III.	P'yŏng-anbuk-do		28445	1648041	58.0	Shinŭiju
IV.	P'yŏng-annam-do		14925	1507579	101.0	P'yŏng-yang
V.	Hwanghae-do		16732	1695858	101.4	Haeju
VI.	Kang-won-do		26263	1566375	59.6	Ch'unch'ŏn
VII.	Kyŏnggi-do		12814	2528829	197.4	Seoul
VIII.	Ch'ungch'ŏngnam-do		8106	1518552	187.2	Taejŏn
IX.	Ch'ungch'ŏngbuk-do		7418	905284	122.2	Ch'ŏngju
X.	Kyŏngsangbuk-do		18989	2479662	130.6	Taegu
XI.	Kyŏngsangnam-do		12305	2225467	180.9	Pusan
XII.	Chŏllabuk-do		8531	1553106	182.1	Chŏnju
XIII.	Chŏllanam-do		13887	2482438	178.9	Kwangju
	Korea total	1938	220741	22633751	102.5	Seoul
		1940		24326327	110.2	

Table 2. The most important data on the provinces

from Kongju to Taejŏn, for political reasons. After all, the new capitals are among the cities with the largest proportions of Japanese residents (Fig. 83).

Hamgyŏngbuk–do is the northeasternmost province. It is embraced by the Tumen River and consists of the narrow hilly strip mentioned in Chapter 2, beyond which rise the high mountains of the Hamgyŏng Range and the flat upland areas of the Kaema upland with its virgin forest cover. The border between this least densely settled province and Hamgyŏngnam–do follows the drainage divide between the Tumen and the Yalu and thus runs along the above mentioned Mach'ŏllyŏng Range.

The heart of the largest province, Hamgyŏngnam–do, consists of most of the Kaema upland and its southern escarpment along the coast of the Sea of Japan. The bulge on its southwestern end, however, also includes the fertile lowlands around Yŏnghŭngman. Its entire western boundary follows the crest of the Nangnim Range.

The northwestern provinces P'yŏng–anbuk–do and P'yŏng–annam–do extend from the crest of the Nangnim Range southwest to Sŏhanman following the general inclination of the west Korean dip slope. The border between the two provinces is initially a ridge (Myohyang Ridge) and then a river (Ch'ŏngch'ŏngang). The southern border with Hwanghae–do is of a similar nature, eventually following the tidal estuary of the Taedonggang. Like all of the provinces on the main Korean tilt block, the eastern portion is mountainous, and the population is crowded into the fertile basins close to the Yellow Sea coast.

Hwanghae–do fills the peninsula that is named after the province and extends eastward to the depressed central section of the Main Range between the Nangnim and T'aebaek Ranges. By far the largest part has an elevation below 200 m, so that the province already has a medium population density.

Kang–won–do, in contrast, encloses the entire southern branch of the Main Range, whose axis it follows, and the headwater region of the Han–gang system. No other province has as few areas with an elevation under 200 m. Kang-won-do is the province that is the backbone and the back side of Korea, and its population density is extremely low, by central Korean standards.

Kyŏnggi–do frames the capital and profits from all the advantages of its location on the west coast. It even surpasses the other western provinces due to its central location and its relatively good access to the east coast (Ch'ugaryŏng furrow, Taegwallyŏng). It coincides with the western portion of the Han–gang basin. It is the province with the highest population density, but only because of the capital, where almost three tenths of the population live.

The location of Ch'ungch'ŏngnam–do corresponds to that of Hwanghae–do. Similar to Hwanghae–do, its boundary on the side away from Kyŏnggi–do is a deep tidal estuary, that of the Kŭmgang. It has a very high population density, the second highest of any province, due to its much more favorable climate, compared with Hwanghae–do.

The only province that does not have a sea boundary is also the smallest, Ch'ungch'ŏngbuk–do, which extends eastward to the crest of the Sobaek Range. Its mountainous location far from the sea helps to keep the population density low.

The two Kyŏngsang provinces exactly fill the Naktonggang basin and are thus a perfect example of a political unit with boundaries dictated by physical geography. The border between the two runs east–west cutting across the step fault blocks (cf. p. 6) along diagonal ribs that rise up as high as 1490 m in Kayasan. The southern province contains considerably more low–lying paddy land than the northern province does. Additionally, it has the impulses that arise from its character as the southeastern entrance to Korea. Thus its population density is higher than that of the northern province. The island of Ullŭngdo belongs to Kyŏngsangbuk–do and was settled primarily from there.

The Chŏlla provinces, finally, make up the southwestern part of Korea, with its mild winters and hot summers. The wide band of islands and Chejudo also belong to them. They too are separated from each other by a lateral ridge that reaches a height of 620 m in Munjusan. The very high population density can be explained by their predominantly low elevation, the fertile soil, the favorable climate and good communications.

The provincial boundaries of Korea represent a very sensitive adaptation to the natural conditions of relief and climate, which has proved its worth over a period of 500 years and, apart from minor details, no really fundamental improvements are either possible or necessary. For most of their length the boundaries follow important drainage divides.

The provinces are divided into counties. At present there are 220 counties (*gun*) and 18 cities (*pu*, now called *shi*). The large rural communities (townships) within the counties are called *myŏn*. There are currently 2365 *myŏn*.

5 National, Political and Cultural History

The history of Korea is closely related to the geographical characteristics outlined above. Over and over again Korea played the role of a dead–end street. Into it streamed the peoples of East Asia, who were land–living peoples unaccustomed to the sea. They influenced the Koreans racially and culturally in the early centuries. At the same time Korea played the role of a "land bridge" that was used by southern peoples, particularly the Japanese, to gain a foothold on the Asian continent (Richthofen [253, p. 466]; Trautz [57]). The Japanese historian Kuno is quite correct when he stresses ([87, p. 23]) that "China, Japan and Korea are so closely connected that the rise of either China or Japan regularly affects the other two and that small, weak Korea, the path of military intervention between the two stronger powers, always has to suffer, when the supremacy in the Orient passes from one to the other." We shall take only a brief look at Korean history to see where in the course of this history the geopolitical themes we have just touched on are visible. More importantly, we are interested in how the history of Korea can help us to understand its present geography, i.e., what effects it has had that are visible in the Korean landscape today. I will base my account primarily on the studies by Hulbert [42, 43, 70], Ross [44], Griffis [45–47], Courant [50], Parker [48], Longford [53], Franke [56], Eberhard [52], Kuno [87], v. Siebold-Hoffmann [132] and the translations of Klaproth [129, 131], Pfizmaier [134, 136] and Nocentini [157].

There are as yet no certain traces of the Paleolithic age in Korea. Neolithic and Eneolithic remains have, however, been found [11–13]. Their distribution is limited to the coast. If further research confirms this, the interior of Korea must have been uninhabited at that time. Thus far remains have been found chiefly in five areas, in the extreme northeast, around the mouth of the Naktonggang, near Seoul, in the southern half of the Hwanghae peninsula and in the basin of the Taedonggang near P'yŏng–yang (Fig. 3). Most of the remains consist of shell mounds. According to their contents, S. Yokoyama distinguishes between a "coastal culture" and a "hill culture". The coastal culture, which is found at sites directly along the present–day coast, is predominantly Neolithic and is closely related to the Japanese Jōmon culture. This culture was based primarily on fishing. But the people living at these sites were also hunters and gatherers. The fishermen of Yup'an in the northeast gathered walnuts, and the fishermen of Tongsam–dong hunted several kinds of deer [12]. In Sadong and Changjin–ri, opposite P'yŏng–yang, deer and wild boars were hunted [22]. This shows that the now almost barren surroundings of these sites must have been wooded during Neolithic times.

The hill culture preferred the hills near the coast. Agriculture had begun to develop. This culture resembles the Japanese Yayoi culture, which is more recent

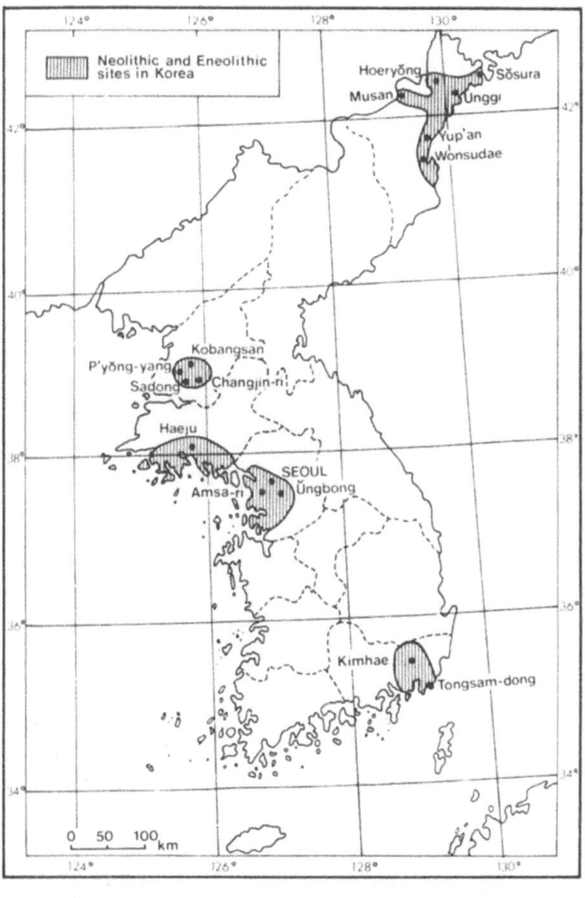

Fig. 3. Prehistoric sites in Korea

than the Jōmon culture. An Eneolithic Yayoi culture appears to have developed in an area that included both western Japan and southern Korea. Ohyama [10] believes that the precursors of the Japanese Yayoi culture came from Korea and that the metal implements of the Japanese Yayoi sphere were also imported from Korea. R. Torii [1,2] is of a similar conviction. The Yayoi immigrants were probably Tungusic tribes. Thus, the cultural relations between southern Korea and Japan that are seen time and again during the historical period began in prehistoric times. In the area around the mouth of the Naktonggang, which is shown in Fig. 4, there are four sites each of the coastal and the hill cultures. The famous Kimhae shell mound belongs to the latter [13]. On the basis of a Chinese coin found there, it can be dated to the 2nd century B.C. The Japanese Yayoi culture had already developed into a rice–growing culture, and grains of rice were also found in the shell mound in Kimhae. Perhaps rice cultivation spread from the area of the East Chinese coastal culture (Lungshan culture) (W. Eberhard [7, 8]; A. Herrmann [73]) to southern Korea and western Japan. This is the opinion of Z. Nakaya [3, 4]. The coin found in Kimhae would appear to be evidence of this. The pottery of the Korean hill culture has traits that have been preserved up till the present day, as Yokoyama points out.

Presumably the Tungusic–Korean stone cist coffins [21] and, above all, the dolmens belong to the same period as that revealed by these prehistorical remains. Figure 5 shows the distribution of dolmens. With the exception of conti-

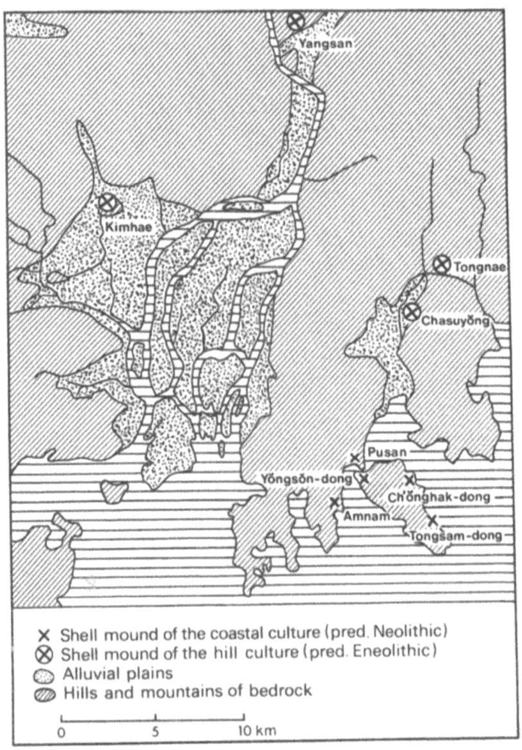

Fig. 4. Neolithic and Eneolithic sites at the mouth of the Naktonggang in southeastern Korea (after S. Yokoyama)

nental northern Korea, dolmens have been found much farther in thc interior and at higher elevations than the previously mentioned sites of the coastal and hill cultures. Korean dolmens are tremendous, detached structures topped by flat, stone slabs up to 9 m in length. Korea's wealth in hard, exfoliating rocks (phyllite, gneiss, granite) facilitated their construction. The investigations to date [15–19, 22] show that they resemble the northwestern European megalithic graves very closely. In this context it is perhaps interesting to note that there are blue-eyed, light-haired Koreans. Moreover, the studies of Koppelmann [67] and Günther, most of whose results Jensen [68], however, rejects, have shown that the Korean language appears to have some strange affiliations with Indogermanic languages. H.F.K. Günther has collected the various indications of a Nordic admixture in East Asia *(Die Nordische Rasse bei den Indogermanen Asiens* = The Nordic Race Among the Indogermanic Tribes of Asia, Munich 1934, p. 194–207). Baelz [17] already considered it possible that there were migratory streams from the western to the eastern edge of the earth's largest continent, and v. Eickstedt, in his *Rassenkunde und Rassengeschichte der Menschheit* (Study of Races and Racial History of Man, Stuttgart 1934), developed an extensive view of the racial dynamics of Central and North Asia, in which these areas figure as a source from

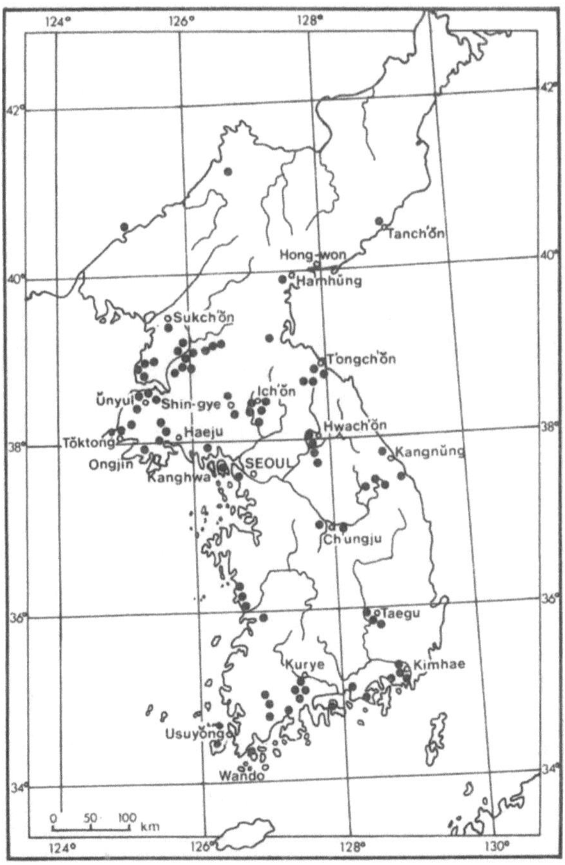

Fig. 5. Distribution of dolmens (after R. Torii)

which mighty racial movements emanated. On the other hand, the question whether there may be a connection between Korean dolmens and the Southeast Asian megalithic culture [20] does not appear to have been seriously studied.

Tradition, as handed down by Chinese and, above all, Korean annalists and historians, begins long before the period that these prehistorical remains have helped us to understand. In the beginning, of course, it is rather legendary in character, though undoubtedly with a factual basis. Figure 6 shows the rather complicated Korean history and the parallel epochs in Chinese and Japanese history.

According to tradition, Korean history begins with Tan–gun. He is said to have been the offspring of the son of the divine creator, who came down to earth, and a bear who was transformed into a woman. The savage people found him under a sandalwood tree (*Berberis*) on top of Myohyangsan (p. 7) and made him their king (Tan–gun = sandalwood prince). This is said to have happened in 2332 B.C. A cave on Myohyangsan is still named after Tan–gun today, and Myohyangsan is one of the five holy mountains of Korea. Additionally, in 1904 a Korean founded a religious sect that is devoted to the worship of Tan–gun [582, p. 203]. Certain racial traits of the Korean people and ancient features of their tradition make it appear likely that in this earliest period people of the same type of physique as the present–day Tungus had already migrated into northern Korca from central Manchuria (Wedemeyer [86], Haloun [76]), and this is probably the origin of this legend. The shamanistic features that play an important role in the religious concepts of Koreans even today and have made their impression on the landscape probably also entered the country from the north at this very early date (Clark [580], Umehara [65], Hentze [66]). It is also probable that the feudal system that influenced the social structure of the Korean people until the end of the country's national independence entered the country from the north at a very early date, it being an equally characteristic feature of the early organization form of the Manchus. Political, racial and cultural influences from central Manchuria on Korea are seen again and again until the 11th century A.D., because the borderland between Korea and Manchuria does not present any real obstacles to migration.

The other area from which pervasive influences on northern Korea emanated was northern China. After the fall of the Shang dynasty Kija (Chin. Chi–tzu) is said to have migrated to Korea by land in 1122 B.C. with 5–10,000 followers and to have settled in P'yŏng–yang, which was known by the Chinese in ancient times as Wangxian. Kija is said to have introduced the cultivation of rice and barley, sericulture and weaving to northern Korea from China. Prior to this the only grain that had been cultivated had been millet. In fact, according to Vavilov foxtail millet (*Setaria italica* Beauv.) and broomcorn millet (*Panicum miliaceum* L.) are among the cultivated plants that originated in the East Asian grain gene pool. Vavilov also classes naked barley and soybeans among these plants and includes Korea in the gene pool. Thus it is quite possible that these and other plants were domesticated in Korea itself. Wheat, however, does not appear to be one of these plants. It is not certain whether wheat was brought to Korea from China. As Hulbert stresses, its Korean name is not of Chinese origin [557]. Hulbert includes among the non–Chinese, mostly native crops not only wheat but the black soybean, two of the three kinds of kaoliang, wild sesame (*Perilla*) and buckwheat. As

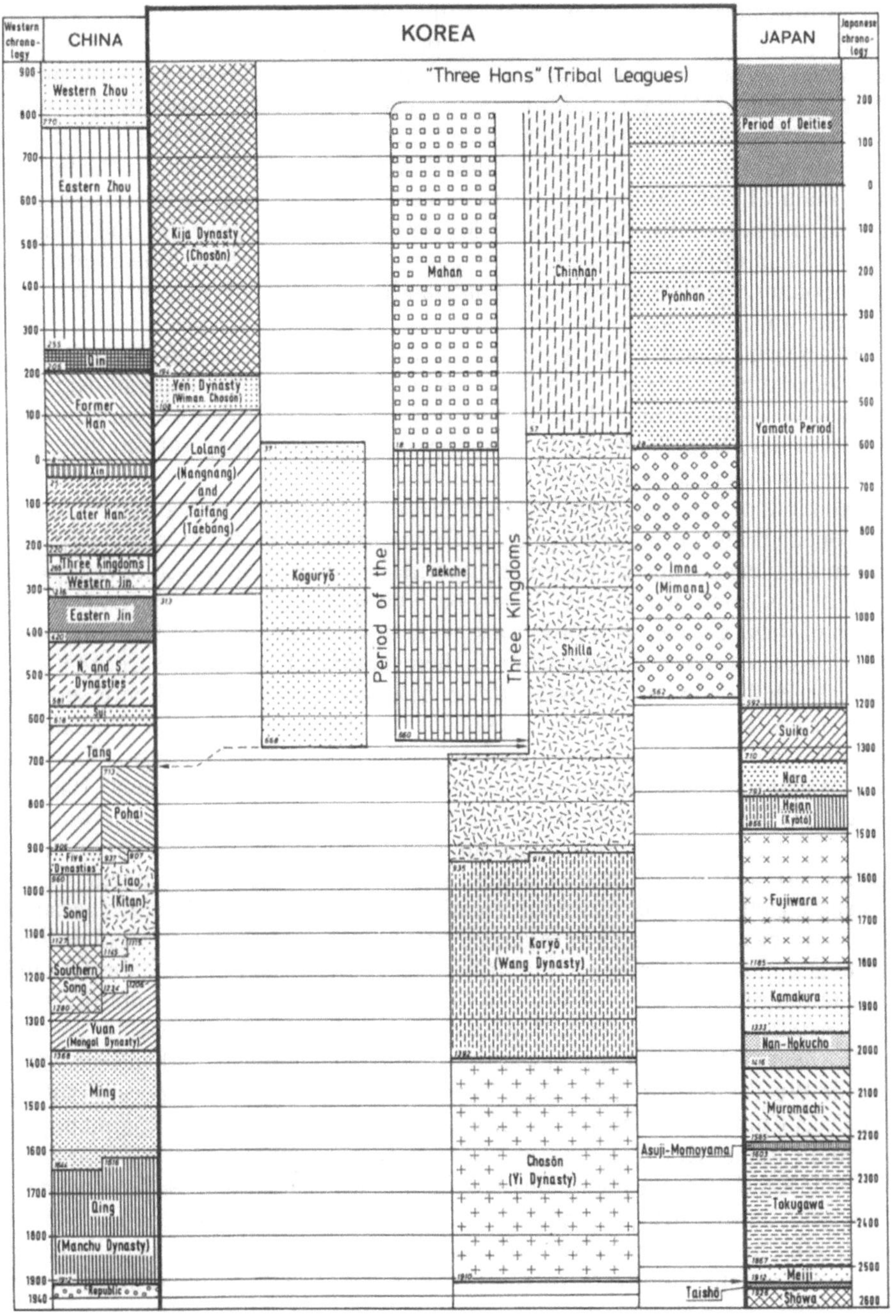

Fig. 6. Chronological table of Korean history. Comparison with Chinese and Japanese history (after a compilation by the government–general)

far as barley and rice are concerned, it is quite possible that they were introduced to northwestern Korea from China at the end of the second millenium B.C., according to the legend of Kija. Rice was probably introduced to southern Korea, however, by sea from the south. Other crops that were brought from China to Korea were probably the third kind of kaoliang, which is one of the oldest Chinese cultivated plants [73], and several kinds of beans. Perhaps hemp was also already cultivated in Korea in this period. After all, it also belongs to the famous five main Chinese crops. Kija is said to have ruled the half–wild native people wisely and peacefully. He and his followers learned their language, and he introduced them to the merits of Chinese culture. Today Koreans still venerate the grave of this mythical personality of outstanding significance outside of the northern city wall of P'yŏng–yang.

It is he who is said to have given his country the name Chaoxian (Kor. Chosŏn) and thus to have created the name by which Korea is known again all over East Asia today. This ancient Chosŏn is not, however, spatially identical with the Chosŏn of today. It included the Liaodong peninsula and extended from there into the northwestern part of the Korean peninsula.

All kinds of incorrect translations of the word Chosŏn have been circulating in European books, even today (e.g., [227]): morning calm, dawn, morning peace, morning silence, sunrise, morning brightness. Attempts at a factual interpretation of this name, which may already be 3000 years old, have been made on the basis of these erroneous translations, with more or less taste. And many books about Korea use one of these names in their subtitles. Klaproth [131, p. 145, footnote 1], Hulbert [43, p. 70], Baelz [51, p. 5] and others have already stressed time and again that the two Chinese characters that are pronounced Chaoxian in Chinese and Chosŏn in Korean can only mean "morning freshness". Thus one can at best translate Chosŏn–guk as the "Land of Morning Freshness". It is very doubtful, though, whether this meaning was intended when the name was chosen. Franke is probably correct in assuming that the name Chaoxian originally referred to a tribe of non–Chinese origin and name living north of the Taedonggang [56, Vol. I, p. 326]. According to him, the two Chinese characters for morning and freshness were chosen only because their Chinese pronunciation sounds similar to the name of that tribe. This is the normal Chinese way of handling new terms for which there are no Chinese characters. Franke's interpretation is supported by a passage in the Imperial Chinese Geography of 1744 [131, p. 93].

In 108 B.C. the state Chosŏn was destroyed in a combined land and sea attack by the Wu Ti emperor. Liaodong and the northern half of Korea became subject to direct Chinese rule. Northwestern Korea was divided into two commanderies named Taifang (Kor. Taebang) and Lolang (Kor. Nangnang) (Fig. 7). Lolang has become famous for the high state of its culture, which the Japanese have been uncovering during the past decades in extensive and exemplary excavations [23–32]. This culture is entirely Chinese and reveals to us the high standard of art at the time of the Chinese Han dynasty. The heart of Lolang also lay in the vicinity of P'yŏng–yang. Just south of the city, on a terrace approx. 30 m high on the other side of the Taedonggang, there is an enormous field with semicircular tumuli from this period up to 6 m in height (Fig. 8). It extends 8 km from east to west and 4 km from north to south. By 1926 the Japanese excavations had uncovered no less than 1386 mounds, and many had been destroyed previously. Indeed I found beautifully decorated tiles from these mounds being used as bricks in farmhouses in the neighborhood. Each of the grave mounds contains a rectangular chamber with walls covered with glazed tiles. The floor is covered with thick planks and holds at least two coffins. The samples I took, which were kindly identified by the Forest Experiment Station, showed that the wood of the planks is of the common oak, Manchurian walnut and the elm known as *Abelicea serrata*. Since large quantities of these planks were used, the now denuded vicinity of P'yŏng–yang must have been covered with trees of these species during the Lolang period. The boards used for the ceilings of the grave chambers were made of the wood of the Korean nut–pine and the Korean yew, which was probably rafted down the Taedonggang. These boards are arranged in a peculiar diagonal manner that is also found in Afghanistan. Thus there may possibly have been influences from that direction on Chinese–Korean architecture, or vice versa [75, p. 26f, Fig. 38 and 47]. Some of the coffins are made of cedar (*Chamaecyparis*) from southern China [27]. Splendid graves of this type continued to be built after Chinese rule ended and the Lolang commandery had been disbanded (313 A.D.), during the native Koguryŏ period. The decorative tiles on the inner walls disappear and are replaced by colorful murals showing, among other things, the race and clothing of the 4th to 7th century [571]. Most of the Koguryŏ graves lie west of P'yŏng–yang.

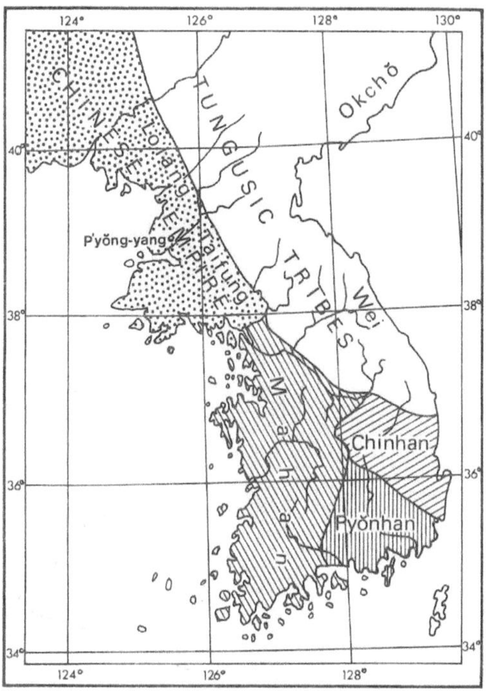

Fig. 7. Political divisions of Korea at the time of the "Three Hans" during the last century B.C. (after Wedemeyer)

While northern Korea's development was being strongly influenced by central Manchuria and, above all, northern China, as we have seen above, in the southern half of the country there were three tribal leagues, who were distinguished from each other and from the inhabitants of northern Korea by the languages they spoke. The Mahan occupied the western part of central and southern Korea and the Chinhan the eastern part of southern Korea, while the Pyŏnhan shoved themselves northward like a wedge between the two (Fig. 7). A total of 78 tribes with 1000 to 10,000 families each are said to have belonged to these leagues [131]. That would mean that approx. 2 million persons lived in an area where now 11 million live. Chinese sources emphasize that these tribes of the "Three Hans" (Kor. Samhan) differed completely from the inhabitants of northern Korea in their language, customs, house types and many other ways. The old descriptions give the impression that particularly the Mahan had connections with southern peoples, such as the inhabitants of Kyūshū, and the prehistorical record also indicates such connections. The "hill culture" mentioned above belongs to the period of the Three Hans. According to the Chinese reports the Mahan tilled the ground and lived in villages but they did not use either oxen or horses as draft or pack animals or for riding. Their houses were pits in the earth, like graves, with the door in the roof. From the fact that they had pearls we can assume that the famous

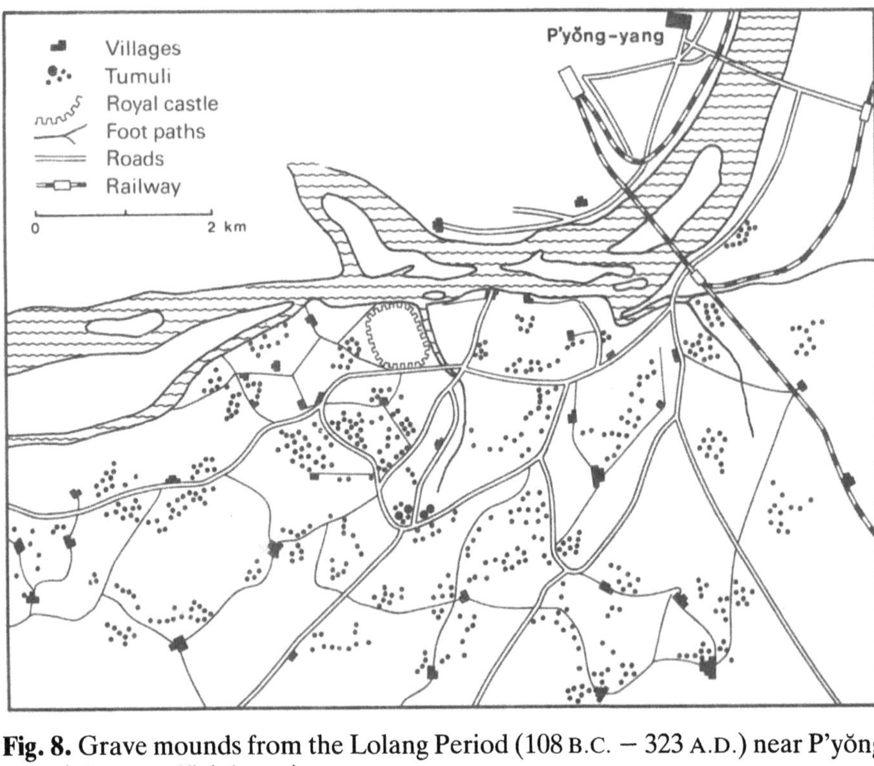

Fig. 8. Grave mounds from the Lolang Period (108 B.C. – 323 A.D.) near P'yŏng-yang (after an official map)

pearl fishing off the coasts of the Hwanghae peninsula in later periods had already begun at this time. The pheasant and the rather insipid Korean pear, which has the form of an apple, are already mentioned as culinary delights of the Mahan. Around the time of the dynastic change from the Qin to the Han dynasty (206 B.C.) Chinese refugees joined the Chinhan, and under their influence the level of Chinhan civilization increased considerably. They lived in palisaded cities, cultivated the mulberry tree, practiced sericulture and wove silk. They also possessed wagons with teams of oxen and draft horses and knew how to make iron. The cultural objects of the Pyŏnhan were similar to those of the Chinhan. The Chinese reports, however, expressly mention rice as a crop, quite in accordance with the remains found in the shell mound in Kimhae (p. 18).

Under the influence of the Chinese colonists, the Chinhan were the first to form a unified kingdom, which soon expanded its territory. Its foundation is dated to 57 B.C. From this time on Korean history loses its legendary and mythological character. From 504 A.D. on this kingdom was known as Shilla (Jap. Shiragi, Chin. Xinlo). The fertile Naktonggang basin formed its core, and in it the capital Kyŏngju developed. Two decades later than Shilla, in 37 B.C., the above mentioned kingdom of Koguryŏ (Jap. Kokuri, Chin. Gaogouli) was founded north of the Yalu. It spread slowly southward, gaining in strength, and eventually annexed the commanderies of Lolang and Taifang. Finally in 18 B.C. a refugee from Koguryŏ founded the kingdom of Paekche (Jap. Kudara or Hyakusai, Chin. Baiji) in the western half of southern Korea. From 371 to 475 its capital was what was later to become Seoul, thereafter for several centuries Puyŏ on the Kŭmgang. This period of Korean history, which began shortly before the birth of Christ and lasted until the 7th century, is known as the "Three Kingdoms" (Kor. Samguk) (Fig. 9).

Chinese sources stress that in the two southernmost of the three kingdoms and in the state of Okchŏ in northeastern Korea the five types of grain, the fruits and the vegetables were all the same as in the Middle Kingdom. Thus China's civilizing influence was already considerable in all of Korea by that time. This is particularly true of Shilla, which developed into a peaceful, flourishing, densely populated state. The temples, pagodas, royal graves and monuments in and around Kyŏngju are still witnesses to its wealth. Shilla fostered the arts and sciences, especially astronomy [84] and thus continued to develop the cultural achievements of Chinhan. From the 6th century on Buddhism contributed to its cultural development, there are said to have been 808 Buddhist temples in and around Kyŏngju. The Chinese writing system probably came into common usage in Shilla at an earlier date than in other parts of Korea. In 512 Shilla conquered the island of Ullŭngdo, which has been under Korean rule continuously since then.

In contrast to Shilla, the more mountainous Koguryŏ with its harsher climate was short on fertile soil and productive fields. As a result it had a more martial character and was often engaged in feuds with its two southern neighbors or with China. It held back the frequent Chinese attempts at conquest, and for a while it kept the political pressure from China off its southern neighbors. From 313 A.D. on its heartland again lay in the lower Taedong basin around P'yŏng-yang. From then on the entire northern half of Korea belonged to Koguryŏ, including, temporarily, the present-day province of Hwanghae-do and the northern part of Kang-won-do. It extended far into southwestern Manchuria, for a while even as far as the Great Wall at Shanhaiguan. As a result particularly large numbers of Tungusic tribes again migrated into northern Korea during the time of Koguryŏ. Its capital was originally Kungnaesŏng (Hwando), which lies north of the Yalu, but was moved to P'yŏng–yang in 427 A.D. Despite its frequent clashes with China, Koguryŏ, lying as it did so much closer to China, absorbed important Chinese cultural assets at an earlier date than Shilla and Paekche. These included Confucianism, which appeared in Koguryŏ shortly after the birth of Christ, and Buddhism, which was introduced in 372 A.D.

Paekche spent its time in complacency and developed both agriculture and crafts to a high degree. It tended more strongly than its neighbors to ally itself politically with Japan, in order to preserve its national identity in the face of these neighbors. Confucianism won more ground in Paekche than in the other two of the Three Kingdoms.

Korea has two politically and strategically very important gateways, which lie diagonally opposite from each other. The one in the northwest provides the connection with southern Manchuria and northern China. It lies at the mounth of the Yalu. Time and again the history of northwestern Korea and the fate of its capital, P'yŏung–yang, have borne eloquent witness to the importance of this gateway.

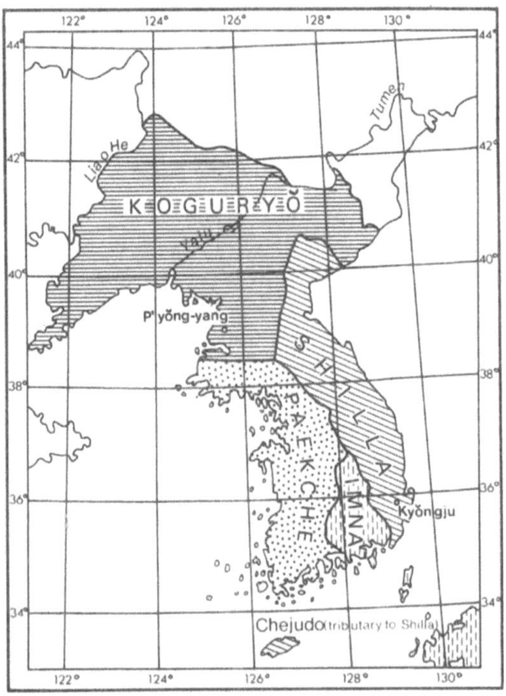

Fig. 9. Political divisions of Korea at the time of the "Three Kingdoms" around 550 A.D. (after Kuno)

The southeastern entrance – and exit – lie around the mouth of the Naktonggang. The narrow Korea Strait with Tsushima in the middle, like the pile of a bridge, leads over to the islands of Japan. Through this gateway the above mentioned racial and cultural relations with southwestern Japan came into being. During the Three Kingdoms period these relations were intensified. Trade flourished, and not only the region around the mouth of the Naktonggang participated in this trade. Paekche did as well, and for a time Paekche even had the closest relations with Japan. In these centuries Korea played the role that geography destined it to play as an intermediary between China and Japan. Via this route Korea transmitted Buddhism to Japan in 552, and 2 years later the Chinese calendar. It also brought sericulture, porcelain, weaving and papermaking to the island nation. As v. Siebold describes these relations [132, Vol. I, p. 14], "colonists now often came from Kooraï (i.e., Korea); although interrupted by bloody wars, the traffic between the two countries...became more and more lively and Kooraï, from whence the teachings of Confucius were conveyed to Japan, around the year 285 A.D. – according to the newer chronology 405 – now became a school of moral refinement for the nation. In the Japanese annals those years are exactly noted in which arts and sciences were bestowed upon them by this neighboring country, with which they associated even more frequently than with China." Thus southern Korea became for the Japanese the "Western Land of Treasures" (cf. p. 1 f).

The south coast, however, and particularly the mouth of the Naktonggang, also became the gateway for intrusions by Japanese pirates and warriors. The source materials on these earliest hostile contacts between southern Korea and the neighboring island nation have been carefully studied by A. Wedemeyer [86] and Y.S. Kuno [87]. According to their sources, in one of the first decades A.D., after the Japanese state had consolidated its hold on the islands of Japan, the district of Kaya, which until that point had belonged to Shilla, became a Japanese protectorate. Similar to Pyŏnhan previously, Kaya extended northward like a wedge from the center of the south coast of Korea. This protectorate was known as Imna (Jap. Mimana). The Japanese sources report further that in 362 (according to the newer chronology) the Japanese empress Yingō Kōgō forced Shilla to capitulate without a struggle. Paekche, too, indeed even Koguryŏ, supposedly then accepted Japanese suzerainty. Be that as it may, it is certain that during the Three Kingdoms period Japan often intervened militarily in their internal disputes and that from the 4th century on, at the latest, it had gained a firm footing on southern Korean soil. The Japanese influence on Paekche was particularly strong. In 562, however, Japan's bridgehead Imna was lost to Shilla. From then on Japan's political influence in Korea sank, despite repeated efforts, for a long time.

In the first half of the 7th century the Chinese emperors repeatedly attacked Koguryŏ with tremendous armies. These attacks were foiled by the Koreans time and again with admirable bravery. After these attacks by land from the northwest had failed, China attempted to gain a foothold in southern Korea by sea. In 660 China, with Shilla as its ally, attacked Paekche and sent a fleet eastward across the Yellow Sea. The Japanese attempt to turn the tables in favor of Paekche was repulsed in a bitter land and sea battle along the western coast of Korea. In 668 Koguryŏ also succumbed to the combined forces of China and Shilla. In the following years fighting again broke out, this time between China and Shilla. At the end of this period China had established its sovereignty over Korea. "No country was subjugated by China with more stubborn perseverance, more vigorous efforts and more oppressive sacrifices, but none offered more bitter resistance than Korea in this...period of the Sui and Tang dynasties" (Franke [56, Vol. II, p. 410]). But the victory actually belonged not to China, but to Shilla. The Tang dynasty had to be satisfied with Shilla as a tributary but autonomous state. From this point on Shilla ruled over the entire peninsula. Thus territorial unity was established for the first time. It has never been lost since. In the following centuries Japan was completely out of the picture. Thousands of Korean refugees, however, particularly from Paekche, emigrated to settle in Japan. In both western and eastern Japan there was from then on a "Kudara (Paekche) and Kōrai Region". E. Baelz [522, p. 443f] reports, evidently on the basis of the Japanese source Shinsen shōjiroku of 815, that at that time no less than 170 of the 1100 noble families of Japan traced their ancestry to Korea.

Thus the first major period of Korean history ends. It is characterized by internal fighting between the different tribal leagues and kingdoms and by great diversity in the degree of civilization, and presumably also in race and language. This diversity is caused by the very difficult conditions facing travel and communication in the narrow, mountainous country. And it is constantly being regenerated by influences from without that reach this land bridge from the north, from central Manchuria, the home of so many peoples, from the northwest, the cultural center of northern China, and the south, the Japanese islands. The two principal gateways, the continental gate at the mouth of the Yalu and the maritime gate at the mouth of the Naktonggang, play a decisive role again and again.

In the following period the continental northern portion of Korea belonged to different independent states, some of which were highly civilized, such as the state Pohai (Matveeff [77]). Territorial unity was initially restricted to the actual peninsula. The change in political forms in northern Korea resulted in renewed immigration from southeastern Manchuria into the sparsely settled and partly devastated valleys in the mountains south of the Yalu and Tumen.

In the new kingdom of Shilla peace and order within the borders formed the basis upon which culture developed. The language, laws and civilization of the former state of Shilla spread through the entire peninsula, preparing the birth of a unified Korean people. Buddhism spread widely. A tremendous number of temples were built. The knowledge of Chinese characters became common among the privileged classes. The development of a syllabic writing system, known as "*idu*" is an independent Korean cultural achievement [78,79]. It can be traced back to a learned statesmen by the name of Sŏl Ch'ong (late 7th, early 8th century), the reason for its development being that Korean is an agglutinative language and quite unrelated to the uninflected, monosyllabic Chinese. In vast areas, however, the transcendent Chinese culture became dominant at exactly this time. "There can be no question as to the great debt which Korea owes to China, but, on the other hand, this was not accomplished without a certain amount of harm to the Korean people. They were still in a formative period. They were just beginning to feel their own powers, and at this very moment they were flooded with the finished products of an older civilization, which took away all incentive for personal effort. The genius of the people was smothered at the start, and never have they recovered from the intellectual stagnation which resulted from the overloading of their minds with Chinese ideals" (Hulbert [43, p. 76]).

This overloading finally so enervated Shilla that the kingdom fell apart. Wang Kŏn, a nobleman of the former kingdom of Koguryŏ, was made king of a part of the country. The tired king of Shilla abdicated in 935 in favor of this energetic, intrepid, steadfast sovereign. Shortly afterward Paekche, which had also seceded from Shilla and waged war with Wang Kŏn, was also defeated. After a period of upheaval that lasted basically only from 918 to 938 Korea was again united politically, though under new leadership. The rather loose tributary relationship with China continued to exist. From 933 on the king of Korea received the Chinese calendar annually as a sign of Korea's tributary status. Wang Kŏn made Kaesŏng (formerly Songdo) his capital and named his kingdom Koryŏ (Jap. Kōrai, Chin. Gaoli), a shortened version of the name of the former northern kingdom of Koguryŏ. To interpret this name as meaning "Beauty of the Mountains", which is sometimes done, is rather risky, as Kibat–Gotha informs me. Probably this is also a case of an ancient, non–Chinese tribal name being written phonetically with Chinese characters, for the Koreans had long since been called by this name by the Chinese and the Japanese [131, p. 93]. The peoples who flowed across the Yalu into Koguryŏ from Puyŏ in central Manchuria around the birth of Christ were already called Kōrai or Gaoli. Both people and country later became known to Europeans under this name (p. 41 f).

Apart from raids by the Khitan, against whom a palisade was constructed in northern Korea from around 1033 on, the three centuries after 938 were basically a time of peace within and without. The two great East Asian powers were occupied with themselves. For the first time the state of Korea pushed its land boundary north to the Yalu, where it lies today. Chejudo was annexed in 1150. The administration was rearranged more and more along Chinese lines. Korea developed a bureaucratic hierarchy, and Buddhism gained such might that the king himself eventually had to don monk's clothing. A large number of the 1336 temples that still exist in Korea date back to the Koryŏ dynasty. The same is true of the giant Buddha statues, as much as 26 m in height, hewn out of cliffs or carved in stone and whitewashed (*Mirŭk*), which can still be found here and there in places that were centers of the cult at that time [41].

In the middle of the 13th century, i.e., during the Koryŏ dynasty, the Chinese inspector of foreign trade in the province of Fujian, Zhao Rugua wrote some remarks on Korea, which he still called Xinlo [74]. Zhao Rugua emphasizes the similarity between many elements of the cultural landscape of Korea and that of China. He remarks that the soil is very well suited to rice cultivation. The vessels used in the household and other implements are all made of copper. Among the products he mentions are ginseng, quicksilver, musk, pine nuts, hazelnuts, shellfish, medicinal roots, cotton clothing, porcelain and straw mats.

The court eventually became smothered in intrigues, plots and murders. As a result Korea was powerless to repel the Mongol invasion that burst over the country in 1231. Despite much brave resistence, the whole country, including Chejudo, was overrun in a very brief period. As a result of the perpetually repeated invasions of the Mongols Korea became one of those "lands of ruins" that Ratzel studies in his anthropogeography (2nd edition, p. 334) in the borderland "between settled and nomadic cultures from Morocco to Korea." Khubilai Khan finally changed to a more friendly attitude toward Korea. Thus he assured himself of the assistance of the land bridge in his plans to conquer the Japanese empire. The attempts made in 1274 and 1281 with the help of a Korean fleet failed completely, however, and many thousands of Koreans lost their lives.

With the collapse of the Mongol dynasty in China, Korea automatically became free again. But simultaneously the Koryŏ dynasty faced its end due to internal disintegration. When the last ruler gave general Yi the order to march against the rising Ming dynasty of China, the general turned back at the Yalu and deposed his lord. The circumstances impelled him to assume the throne in 1392. With great energy and political foresight he set out, under the name T'aejo, to reorganize the state. Thus the last period of Korean political independence begins.

The Ming awarded the new state the name Chaoxian (Chosŏn), which its northwestern quarter had already borne at an earlier date. With great geopolitical tact T'aejo moved his capital to the lower reaches of the Han–gang at Hanyang. It has not been changed again since then. The name Hanyang, however, was soon abandoned in favor of the vernacular Seoul.

It was the founder of the Chosŏn dynasty (Editors' note: It has been more common to use the term Yi dynasty in Western writings on Korea, but Chosŏn dynasty is more correct, Yi having being the family name of the founder of the dynasty) who divided the country into eight provinces and gave them their names (p. 12). Buddhism, which had fallen into disrepute due to the depravity of the court at Kaesŏng, was strongly repressed under the Chosŏn dynasty. Most of the temples that are still extant today lie in mountainous areas with difficult access. Confucianism was made the political creed, and its principles gradually became established among the common people. The cultural relations with China were intensified. The clothing of the Koreans today is that of the Chinese during the Ming dynasty. The white color that predominates in the everyday clothing of the Koreans also comes from China. It is the Chinese color of mourning, which has lost this significance in Korea (Editors' note: Lautensach was mistaken, white is every bit as much the color of mourning in Korea as it is in China.)

The Ming dynasty maintained very close, friendly relations with the Chosŏn dynasty. The tributary relations were very mild. Between 1393 and 1450 the tribute sent by the king of Korea consisted mainly of linen, hemp and cotton cloth and horses. It appears that considerable quantities of cotton were already grown. How highly developed horse breeding was can be seen from the fact that the tribute for the one year 1423 consisted of 10,000 horses [72, p. 8f].

From 1404 on censuses were held, the results of which are still extant. Evidently it was the fourth king, named Sejong (1418–50), who invented the Korean phonetic alphabet *ŏnmun* (known today as *han–gŭl*) with its 25 characters (Courant [H]). This invention, which was unique in East Asia, made it possible to print books with movable metal type during the rule of Sejong, and thus a decade earlier than in Germany. Beginning in 1403 Chinese classics had already been printed with metal type in Korea [82]. In 1425 the same king had the amount of rainwater that had seeped into the earth measured, and in 1441 he had a rain gauge made out of cast iron. The precipitation measurements made with this gauge are thus around 200 years earlier than the earliest measurements in Europe [401]. Sejong also had a book about agriculture compiled in 1442 dealing with the most suitable crops for the different regions of the country and with agricultural methods. It was during the reign of Sejong as well that the boundary in the northeast was moved up to the Tumen River. Thus for more than 500 years the land boundary has lain where it is today. The seventh king, Sejo (1455–68), also ranks with his worthy ancestors as a patron of agriculture, arboriculture, art and science. This period of calm, which lasted for 200 years, brought Korea a new and last apogee.

With the year 1592, however, a new period of almost 50 years began in which the land bridge was to play the role that was so disastrous for its inhabitants. The generalissimo of the Japanese emperor, Hideyoshi, also known by the honorific name Taikōsama, had conceived the plan to establish a tremendous empire in East, South and Central Asia under Japanese leadership, and in his intention to wage war with China, he first turned to the Korean land bridge. This monstrous plan failed completely, but for 7 years Korea was the scene of terrible battles and tremendous devastation. When Hideyoshi died in 1598, the campaign was abandoned. Many Japanese remained in southern Korea, but on the other hand, during the Japanese invasion large numbers of Korean prisoners were taken to Japan. Above all, all Korean makers of porcelain and faience appear to have been taken to Japan; as a result the art of porcelain making, which had flourished in Korea until then, died out [85]. In the next decades Korea sent annual tribute missions to the Tokugawa shogunate. Apparently one of these missions around 1615 brought back tobacco cultivation from Japan, where it had been introduced by the Portuguese. Thus, as was appropriate to its geographical location, from 1598 on and for the next centuries Korea had to pay tribute to two powers.

The country was nowhere near having recovered from the destruction of this 7–year war, when in 1627 the Manchus began to take advantage of the weakened condition of their neighbor and to attack the country from the north, bringing renewed devastation. From then on Korea also paid tribute to the Manchu kingdom. This tribute was, however, reduced when the Manchus took over the imperial throne of China. Later it was lowered so much that it signified no more than a formal recognition of Chinese suzerainty. From this time on the political relations with China were always friendly.

After the bitter experiences of these terrible times Korea began a 250 year period of complete isolation. With the exception of the imperial mission, no foreigner was allowed to enter the country, and no Korean was allowed, on pain of death, to leave the country or to trade with foreigners. Korea became the "hermit nation" or "forbidden land", as it was described in numerous European books at the end of this period [45, 169, 213]. Shipwrecked sailors who became stranded on Korea's coasts were held there for the rest of their lives. This happened mainly to Dutch sailors en route to the island of Deshima in the harbor of Nagasaki, the only foreign settlement allowed by the Japanese in their island empire after they had begun the same isolation policy as the Koreans. Well known is the story of the shipwreck of the Dutch yacht "Sperwer" (Sparrow Hawk) in 1653 on the southwest coast of Chejudo and the 13 years its secretary Hendrik Hamel and his surviving companions spent in captivity [171–173].

Even before the Manchu (Qing) dynasty was established in China, the Manchus and the king of Korea reached an agreement that a strip of land on the northwestern bank of the Yalu 100 Chinese li (approx. one–half of a degree of latitude, i.e., about 55 km) wide, at the maximum, and 500 km long should remain no–man's–land. According to a calculation in Petermanns Mitteilungen (1881, p. 72) this strip of land measures 13,882 km^2. To create this strip of neutral territory four cities and a large number of villages were razed. It was forbidden on pain of death to cross this strip. Along the northwestern boundary of this neutral territory there was a wooden palisade that has been shown on European atlases up till the present day (cf. d'Anville [126, Plate XXXI]). Richthofen, who gave a vivid description of this neutral territory in 1869 [140, 141, 253], found instead of a palisade only an insignificant ditch when he visited Gaolimen. The Chinese customs office, however, still lay at the northwestern boundary of the neutral zone at that time. This neutral territory did not fulfill its purpose of safeguarding the boundary areas on both sides, though; instead it became a refuge not only for wild animals but also for refugees and thieves. In 1875 the king of Korea complained to the court in Beijing about the misdeeds of these robbers. Li Hongzhang (Li Hung–chang), the governor–general of Zhili (Chihli), sent gunboats with troops up the Yalu in response. They discovered that in the meantime quite a number of peaceful farmers had settled down in this area. Thereupon the region was occupied by the Chinese, and the customs boundary was moved forward to the Yalu ([44, p. 270], [45, p. 182], [204, p. 199ff]). A similar neutral area was created in 1715 along the Tumen [131, p. 119]; consequently the nationality of the K'ientao region (Kor. Kando) was a repeated object of dispute until 1909 (p. 394).

Korea continued this policy of absolute seclusion even after China found itself forced by the Treaty of Beijing (1860) to relinquish the Ussuri region to Russia. This made czarist Russia Korea's neighbor for a stretch of 25 km along the lower Tumen. Thereafter it was forbidden on pain of death to cross the river. The villages along the Tumen were moved inland, and the market at Kyŏng–won was closed down.

Along the sea coasts as well, the settlements were moved inland where possible, partly in the 15th and 16th centuries because of the danger from Japanese pirates, partly in the 17th and 18th centuries in compliance with the isolation policy. This partially explains the impression of forbidding desolation that many European explorers gained from their ships of the "Land of Morning Freshness."

Almost the only contact that Korea had with the outside world during these two and a half centuries was the annual mission sent to Beijing by the king to bring tribute and receive the lunar calendar from the Manchu emperor and the return mission that was sent annually from Beijing to Seoul. These missions travelled via Ŭiju and Gaolimen on the best road in the country, which was suited for pack animals at the best. Although it was a much longer route, these missions travelled by land. The presents that the Korean mission took along with them to Beijing give an impression of the economic strength of the "hermit kingdom": white linen cloth made of the fibers of the chu plant (probably a kind of ramie), embroidered satins, cotton cloth, woven mats, white satin paper, white rice, white rice straw, deer hides, wolves' pelts, swords. This list is taken from the Imperial Chinese Geography of 1744, which Klaproth translated [131, p. 167f]. As corresponding gifts from the king of Korea to the emperor of Japan, Rinsife lists in 1786 ([131, p. 21], similarly Courant [900, p. 14]): ginseng roots, tiger and leopard skins, fine green leather, fish skins, pieces of satin, very fine white cotton cloth, horses, falcons. A few times a year Korean merchants were allowed to sell other goods at the markets in Gaolimen and Hunchun and to the Japanese in Tsushima: gold dust, lead, raw silk, silk cloth, medicinal powder made of tiger bones, tiger hair, tiger, leopard, fox, marder and wildcat furs, deer antlers, cowhides, fresh and dried fish, trepang, seaweed (*Fucus saccharinus*), human hair, beans, millet, hemp, oiled paper, bamboo and paper fans ([141, Vol. II, p. 163–65], [156, p. 241f]). The export of ginseng, which was extraordinarily highly prized in China, was a monopoly of the king himself. These lists of exports give us a picture of the self–sufficient economy of an East Asian country with extensive forests where valuable animals can be hunted, in which edible products from the sea play an important role, the most important food and textile plants are grown and mining is just beginning to be known, but where some crafts have reached a high degree of excellence.

Whereas Christianity had already gained a foothold in China and Japan in the 16th century, the small "Land of Morning Freshness" was initially overlooked by the mission. The first traces of Christianity are not authenticated until 1777 A.D. They were brought to Korea in the beginning by native believers. The movement increased rapidly, as did the persecution by Korean officials. In 1836 the first French missionary entered Korea across the Yalu, dressed in Korean mourning dress, which preserves the wearer's privacy. Others followed, some by the same route, and some by boat across the Yellow Sea. They perished in the persecutions of 1839. The most fanatical persecution was that beginning in 1866, to which nine French bishops and missionaries and 20,000 Koreans fell victim. After the opening of the country the French Catholic mission thus found a field that was fertilized by the sufferings of untold martyrs, and this field they proceeded to till with increasing success. Later the German Benedictine mission joined them, and since 1927 this mission has been assigned a special apostolic vicariate in the form of the two Hamgyŏng provinces [94, 99]. Immediately after the end of the seclusion policy numerous Protestant churches began to missionize. These Protestant missionaries are exclusively British and, above all, American [95–98].

The centuries of complete isolation resulted in a terrible decline in Korean culture. No stimulation from without, no chances of comparison, no rivalry between peoples were possible. The king reigned as an absolute monarch with the help of the official class, which was recruited from the ranks of the nobility, the *yangban*. This official class was extremely avaricious and exploited the peasants with arbitrarily levied taxes. The class system, which is probably very ancient and was not abolished until 1894 (p. 20), now had disastrous effects. There were two classes in addition to the *yangban*, the commoners (peasants, merchants, craftsmen) and the low–born (slaves, butchers, monks, etc.). The two lower classes became more and more impoverished. This system stifled every impetus toward cultural progress before it began, and material impoverishment went hand in hand with spiritual impoverishment. The two and a half centuries of stagnation that followed a period of the most dreadful trials kept the country in approximately the same condition it had been in at the end of the Japanese and Manchu invasions. In many ways Korea in 1870 resembled the China of the Ming dynasty (1368–1644). The mummified state of civilization and culture contrasted sharply with the national arrogance that was customary in government circles and among the *yangban*, as numerous decrees issued during the first two thirds of the 19th century amply testify. By secluding itself, Korea had robbed itself of all chances to rise again of its own accord. When East Asia was torn out of centuries of contemplative self–content by the European nations, this little country was more than ever at the mercy of great–power politics.

After China in 1842 and Japan in 1854 had been forced by the European powers and the United States of America to open their boundaries, it could only be a question of a short period of time before Korea was forced to do the same. That this did not happen until a few decades later was due to its small size, which made it politically less interesting. In addition, the policies of the energetic and cruel regent Taewon–gun were able to prolong the state of things for a while, though this only made the situation more difficult later on.

During the expedition of the Russian admiral Putiatin, the frigate "Pallas" sailed along the entire east coast of Korea from Pusan to the mouth of the Tumen in 1854. In the bay near Wonsan it found a very favorable ice–free port, which was named "Port Lazarev" [236]. This evoked Russian desires to annex territory. These desires increased greatly after the acquisition of the Maritime Province and the establishment of Vladivostok (1860), but were never fulfilled. In 1866 a French fleet attempted to take the mouth of the Han–gang and advance to Seoul to obtain redress for the murder of French missionaries. After a slight mishap this attempt was dropped by Admiral Roze, for some inexplicable reason. In the same year the xenophobia artificially kindled by the government triumphed again when the American schooner "General Sherman" ran aground in the Taedonggang just below P'yŏng–yang and was unable to maneuver. The American expedition sent to retaliate in 1871 was not particularly successful either [92]. But when the king (Kojong), who had been underage until that point, took the throne (1873), it meant the end of the isolation policy. It can be seen as a symbol of the development of all of East Asia since then, that the first beneficiary of the opening of Korea was not the European powers, but the country they had just opened, Japan. Japan had not begun its spectacular restoration until 1867, the beginning of the Meiji era, and in 1876 it was already able to play an active role in Korea. It could not allow a European power to dominate the land bridge that is practically an invitation to leap across to the island empire. When the crew of a Japanese ship was fired upon on the coast of Kanghwa island in 1876, Japan forced a treaty upon the Korean government opening one port on the west coast and one on the east coast to Japanese trade, in addition to Pusan, where a Japanese trade settlement had long existed. Accordingly, Wonsan was opened in 1880, and Inch'ŏn (then known to Europeans as Chemulp'o) in 1883.

This was the beginning of the third period of Korean history, the Japanese period. Between 1882 and 1892 the king of Korea was forced to sign similar treaties with the United States of America, England, Germany, Russia, Italy, France and Austria–Hungary. From then on foreigners began to flow into the country, Europeans in small numbers, and Japanese in swarms. Of the foreigners living in Korea at the turn of the century 19,200 were Japanese, 3500 Chinese, 230 American, 110 English, 95 French, 70 Russian and 50 German [260, p. 253]. Trade with Japan developed rapidly, after the initial reluctance of the Korean government had been abandoned. A Korean Customs Service patterned after the Chinese Maritime Customs was established, and its first director was the German P.G. v. Moellendorff (1883–85). At the court in Seoul a bloody rivalry developed in 1884 between pro-Japanese and pro-Chinese groups. These two great powers then agreed in the Treaty of Tianjin (Tientsin) to both withdraw their troops from Korea. The overwhelming influence of Western civilization and inventions spread rapidly in Seoul, Chemulp'o (Inchŏn) and the other treaty ports, but this did not improve the living standard of the broad masses of the people. In this situation the xenophobic, philosophic-religious cult of Tonghak (= Eastern Learning) gathered influence. During the Tonghak rebellion of 1893 the king appealed to China for help. In response Chinese troops were sent back to Korea, thus serving as the ostensible motive for the Sino-Japanese War, which was fought over the control of Korea (1894/95). The military developments on the land bridge again showed the strategic significance of the Taedonggang basin around P'yŏng-yang. The short, bloody war ended with an overwhelming victory for Japan. Russia and France, joined by Germany, intervened, and Japan was robbed of much of the fruits of its labors, but it did succeed in having Korea's nominal independence from China proclaimed. Obscuring the true situation, the king of Korea thereupon took the title of emperor in 1897 and changed the name of the country from Chosŏn to Taehan-guk (= Great Han Country, referring back to the tribal leagues of the "Three Hans").

The actual winner of the war of 1894/95 soon turned out to be Russia, which increased its influence in Manchuria tremendously. In 1898 it gained an ice-free port on the Pacific in the Chinese Port Arthur, which the Japanese had just captured. And it also received permission from China not only to build the Trans-Siberian Railway through Manchurian territory, but also to build a branch line from Harbin to Port Arthur. The Korean court again became the scene of incessant intrigues, this time between the pro-Japanese and the pro-Russian factions. The weak king was a helpless pawn. For more than a year he lived as a fugitive in the Russian legation of his own capital. Two treaties signed by Russia and Japan in 1896 and 1898 (the Yamagata-Lobanov and Nishi-Rosen conventions), again affirming the independence of Korea, did not help to stabilize the situation. After the total defeat of China in the Boxer war, Russian predominance in Korea increased even more. The Russian concessions, particularly around the port of Masan near Pusan and in the forests of the Yalu region and at the mouth of the Yalu near Yong-amp'o, raised storms of protest in Japan. Finally Russian troops appeared on Korean soil, and Russian munition and provision depots were set up there.

To prevent the land bridge from falling into the hands of what was now the largest continental power on the earth, the youthful Japan took the risk of the war of 1904/05, which was fought mainly in Manchuria, though the final decision in Japan's favor fell at the battle of Tsushima. Korea now became a Japanese protectorate. The first resident-general was the eminent Marquis Ito. After he was assassinated by a Korean in Harbin in 1909, Japan went a step further and "annexed" Korea in 1910. The country became a Japanese Government-General. Its name was changed back to Chosŏn. The capital Seoul was given the name Keijō (Kor. Kyŏngsŏng). "At the beginning of the new era the Japanese did not intend to take up the burden of remolding the neighboring kingdom, with whose many difficulties they were familiar. Their desire was that Korea should develop of itself and reform itself to become a small, autonomous power, a friend of Japan, and strong enough to exist in its own right. It was the senility of the peninsula, the internal difficulties rising out of the government and out of the royal family, European ambitions, in a word, the conditions, that gradually and quite naturally led to a change in Japan's plans" (A. Halot, Japanese consul in Brussels, [714, p. 71f]).

Its annexation by the Japanese empire is for the Land of Morning Freshness the beginning of the geographical present. Certainly the bitter national fate that Korea has endured time and again because of its small size and unfortunate geographical location and its weak social organization deserves our deep sympathy. But if we take a realistic view of the political situation we cannot help but recognize that in the period from 1876 to 1910 it was no longer a question whether Korea should develop of its own will as a free and independent state. The only question was whether it would become a part of the Russian or the Japanese empire. The decision that was reached at that time was for the Koreans the better of the two choices, for Korea was now developed by a nation that is related to the Koreans racially, linguistically and spatially. And Japan was able to concentrate its energy much more fully on the tasks that needed fulfilling in the neighboring country than Russia could have. As early as 13 October 1894, during the first phase of the Sino–Japanese War, Richthofen uttered the following prophetic words [253, p. 474f]: "Korea will not be able to rise of itself; the talent for commerce and the spirit of enterprise are lacking, and the country lies helpless in the network of foreign relationships that has been being spun. The Chinese yoke, which could be renewed, is...light, it is true, but it would not help the inhabitants on. It is, however, conceivable that the influence of a country that is as active as Japan, if exerted with wisdom and fairness, could lead to the development of resources, the restoration of industries and to the material and spiritual improvement of the population. This conjecture is supported by the glowing report of a British customs official on the onset of Japanese colonization in southern Korea. If Japan should play a role in the administration of the country, it would not be the Japan of Taikōsama, but a new Japan that, as the conduct of the current war shows, has adopted not only the material advantages, but also the humanitarian ideals of European civilization."

In Korea's development we can clearly recognize the political effects of the geographical conditions. A small country, it lies between China, the gigantic continental power that has for millennia been far superior militarily and culturally, and Japan, the maritime power that is certainly not extraordinarily large, but is protected by its island nature and has a larger population. It is torn back and forth between these two poles, and its only chance to develop according to its own precepts is when these two great powers are occupied with themselves. Although northern China, Korea and Japan are oriented east–west of each other, the historical movement flows through the country from north to south, for Korea forms a land bridge projecting out in this direction. From the north Tungusic–Manchurian and Chinese influence penetrates the country, and from the south Japanese influence. The Tungusic–Manchurian influence has had a profound effect on the racial constitution and is still ingrained in the religious concepts of the ordinary people. The Chinese influence has formed the cultural attitudes of the entire country, while the Japanese influence has formed the geographical landscape more strongly from year to year in the past two generations since 1876. In view of these facts it is an impressive achievement that the inhabitants of this land bridge were able to overcome the division into individual tribes and kingdoms characterizing the first millennia of their history. And it is every bit as impressive that they were able to form a nation with a unified language, writing system, life style and national identity, a nation that has formed the cultural geography of its country in its own manner and quite differently from its neighbors.

6 History of Discovery and Exploration

One can begin the history of exploration of a country such as Korea, with an ancient culture unknown to Europeans, from various standpoints. Naturally for us the development of geographical knowledge of Korea in the Occident is by far the most important point. But it would be wrong to overlook the geographical and cartographical work that the Koreans themselves accomplished in their own country, just because the Occident had no knowledge of this work until the last decades of the last century or because even since then modern European science has failed to evaluate these books, written as they are in Chinese characters. Finally, it is also profitable to study the development of geographical knowledge about Korea in its two larger neighbors, who were the medium by which almost all information on Korea reached Europe until the end of the isolation policy. We shall attempt to do this within the framework of the history of European exploration. But first we must pay special attention to the development of Korean geography in Korea itself.

Since the 12th century A.D. the Koreans have produced a respectable number of important historical works on their country. Hulbert ([42] and Kor. Rev. 1, 1901, p. 30f) lists eight of them [102, 103, 106, cf. 101 and 836]. There is evidently much less geographical literature. Of course we must consider the possibility that valuable works have completely disappeared in the course of the many wars and invasions Korea suffered, especially in the period before the invention of printing, when only a few copies of a given work could have existed at one time.

By far the most important work is the 55 volume *Tongguk Yŏji Sŭngnam* (Survey of Korean Geography). In his treatise *Corean Books and their Authors* [101, p. 98], a work displaying profound professional knowledge, Trollope calls it "Korea's only geographical work". According to Hulbert (42) it was begun by No Sa–sin around 1420. It was completed by Sŏ Kŏ–jŏng (1420–88), who began his work on it about 1470 (Editors' note: there appears to be some confusion here between the *Tongguk yŏji sŭngnam* and the *Tongguk t'onggam*). It was printed in 1537. Volumes 1–3 describe Seoul, Vols. 4 and 5 the former capital Songdo (Kaesŏng), the other 50 volumes describe the eight provinces in 3 to 12 volumes per province.

B. Kotō [352, p. 6] also mentions a Korean geographer Yi Chung–hwan. He lived in the late 18th, early 19th century (Editors' note: 1690–ca. 1750). The passages that Kotō translates from his book *P'aryŏk chi* (Accounts of the Eight Provinces) show that this geographer was already reflecting on both the relationships between the mountain ranges on the peninsula and those between them and the mountain ranges of Manchuria and China.

Korea's most famous historical work, the *Samguk Sagi* (Records of the Three Kingdoms) by Kim Pu–sik [102],was completed in 1145. In Volumes 34–36 it contains a geography of the kingdom of Shilla and in Volume 37 one of Koguryŏ and Paekche. Trollope [101, p. 101] also cites a three–volume *Tonggyŏng Chi* (History of the Eastern Capital) that describes the ancient capital of Shilla, Kyŏngju, not only historically but also geographically. A similar history of P'yŏng–yang, the first part of which was printed in 1590 and the later part not until 1837, also provides information on the buildings, fortifications, products, etc. of this city. There are numerous Korean books on the Diamond Mountains (Kŭmgangsan) and their scenic beauties. Gale has translated some impressive passages from them [864]. The most important appears to have been that by Hong Ŭng–gil (16th century).

In the commentary to his translation of Chinese writings on the history and geography of Korea, Nocentini [157] used not only an old Korean map but also a text supplementing it, which he refers to as a "Geographical Compendium of Korea". He does not mention the author or the date of publication of these sources. The excerpts he cites, however, reveal that the text contains valuable geographical information. Finally, in his Bibliographie Corénne [H] M. Courant mentions several unprinted manuscripts with geographical content, such as the *P'al–to yuk–taero ko* (Examen des six grandes routes de la Coré) and the *P'al–ro p'yo* (Itinéraires des huit provinces), and Trollope refers to a "Text–book of Korean Geography" published in one of the last years of the last century by the Ministry of Education of the Korean government [829, p. 7].

With the exception of the work by Yi Chung–hwan the geographical contents of these works are confined basically to enumerations of mountains, rivers, prefectural cities of grades one to four and their curiosities. Often they also contain historical details, just as the accounts of the history of Korea contain a fair amount of chorographical information. The methodology of these old books still does not distinguish clearly between geography and history. This is of course particularly true of encyclopedias, such as that of Hong Pong–han [107].

In modern times Korean scholars have not been very prominent thus far. Still it should be mentioned that many of them have diligently assisted in the publication of Japanese works. Thus, for instance, the Korean T.T. Keng, together with the Japanese J. Kokumen, prepared an excellent *Survey of the species of trees in Korean forests* [486]. Independent, geographically useful works have been written primarily by the American–educated Korean Hoon K. Lee, professor of agricultural science at Union Christian College in P'yŏng–yang. His most important work, on Korean immigrants in Manchuria, which is also useful for studies of their country of origin [700], is written in *ŏnmun*. His comprehensive study on land utilization and rural economy in Korea [605] is in English.

There are many more old maps compiled by Koreans of their country. In his *Catalog of Old Korean Maps,* which was issued in 1933, Ōtani mentioned 42 maps prepared by foreign scholars and 78 by Koreans, 3 of which he reproduced [109]. The most famous of all is the map of Korea that hung in the palace of the king of Korea during the reign of the Kangxi (K'anghsi) emperor (p. 43). In 1846 the Korean Christian André Kim drew a map for French missionaries with French legends based on native sources. It was reproduced in a smaller scale in 1849 in the Bulletin de la Société géographique. In 1884 C. Gottsche obtained a map of Korea in 23 sheets on a scale of approx. 1:150,000 that had been published in Seoul in 1861 [188, p. 249]. When the German geologist B.

Knochenhauer traveled through Korea from 1897–99 in search of exploitable gold deposits, he used an atlas which he described as follows [214, p. 85]: "For orientation I had the use of a large old Korean map, which represented the country in some 20 sheets on the approximate scale of 1:240,000. As childishly simple as the presentation is, the roads being represented by straight lines, the mountains pictorially as if viewed from below and greatly enlarged, still the distances stated therein agreed surprising well with the actual distances, at least in the southern provinces." It is likely that both geologists had the same atlas and that it was the *Taedong yŏjido* described by Courant [H, Vol. II, p. 485]. This is obviously by far the largest scale and most modern cartographical work compiled during the time of Korea's independence. C. Chaillé–Long published the sheet Chejudo of this atlas [195, 196].

All other old Korean maps that I have become acquainted with show the country on a single sheet and are accordingly small in scale (cf. v. Siebold [132, Vol. VII, Plates XI and XV]). Knochenhauer's description fits them to an even greater degree. The outlines are almost always incorrect and it is obvious that astronomical bearings were taken in only a few cases. The same can be said of a Korean atlas consisting of 14 sheets from the 18th century described by H. Cordier, part of which he reproduces in facsimile [111]. It contains a map of Seoul, another of the environs of the capital and one each of the eight provinces, among others. It would lead us too far afield at this point to enter into a critical discussion of the old maps.

Arab traders followed the coasts of Asia as far as China. Through them the first knowledge of Korea reached the Occident. In the merchant Sulaiman's report on India and China, written in 851 A.D., Korea appears under the name Sīla [112]. The last sentences of the report read (p. 72f): "On its maritime side China borders on the islands of Sīla. Their inhabitants are white. They exchange gifts with the emperor of China. They believe that it would not rain in their country, if they did not exchange gifts with him. No Arab has traveled to this country to bring us an account of these people."

A few decades later Ibn Khordadhbeh, who was employed at the court of Caliph Motamid from 870–892, writes in his *Book of Roads and Provinces* the following: "What lies on the other side of China is unknown land. But high mountains rise up densely across from Kantu. These lie over in the land of Sila, which is rich in gold. Mohammedans who visit this country are often persuaded by its fine qualities to settle down there. Ghorraib, kino resin, aloe, camphor, nails, saddles, porcelain, atlas, cinnamon and galanga are exported from there." It was Richthofen who made this report known, and I have cited his translation into German [141, Vol. I, p. 575f]. He probably quoted it from a translation of the works of Khordadhbeh that had appeared shortly before in the Journal Asiatique. He identifies ghorraib as ginseng, kino resin as the highly prized yellow varnish of the Koreans, which gives the appearance of gilt (from the tree *Rhus vernicifera* Decne.) and was often praised by Chinese authors writing later on Korea, and galanga as the medicinal root *Alpinia galanga*. He notes, however, quite correctly that not all of the products mentioned by Khordadhbeh can be from Korea.

The Arab scholar Abu Zaid Hasan also mentions Korea around 916 in his remarks on the Sulaiman report [112, p. 93], and in 947 the Arab historian and geographer Masudi, who traveled as far as the China Sea, wrote his book *Meadows of Gold and Mines of Precious Stones,* which is incorrectly referred to as *Les prairies d'or* in the French translation [113]. In it he reports on Korea [I, p. 346] "Beyond China, on the maritime side, there are no more known kingdoms nor regions that have been described, with the exception of the region of es Sila and the islands that are subject to it. Seldom has a stranger who has come there from Iraq or another country left it afterwards. So healthy is the air there, so pure the water, so fertile the soil and so plentiful all good things. The inhabitants live in good relations with the peoples of China as with their rulers, to whom they constantly send gifts. They form, it is said, a part of the large family of the Amur descendents and have established themselves in this country in the same manner as the Chinese occupied theirs."

Above all, the world map of Idrisi of 1154 and its text contain references that are worthy of mention. The map [115] shows in its southeastern corner a group of islands, five of which bear the name al sīla, lying off the continent, at the point where the coast of the continent makes a bend from an easterly to a northerly direction and the city maidalū (= Shanghai, according to K. Miller) lies. From what we have said above, there can hardly be a doubt that the name is a memory of the preceding period of Arab traders and refers to Korea. The text by Idrisi [114] reports: "From this island (Almaid) one can travel to the Sīla islands, which are very numerous and lie very close together. There lies a city Ankuah (Anju, according to Campbell, [252]) whose area is so fertile and so rich in goods of all kinds that strangers who come to visit settle down and are reluctant to leave. There is gold in such large amounts that the inhabitants even fashion dog chains and collars for their monkeys from this metal. They also fashion woven clothing of gold and sell it." Some of the features described here may fit the Korea of the previous centuries. This applies particularly to the mining and manufacturing of gold. But simultaneously Idrisi's map shows an egg–shaped peninsula with a narrow connection to the coast farther north on the east coast of Asia, at about the latitude of Korea, which bears the name gezīre al jakūt (Hyacinth Island). The text says the following about it: "The city where the king of the Khirkhir lives (according to Miller: Beijing)...lies in the vicinity of the Hyacinth Peninsula, which is separated from the continent by an isthmus and surrounded by a round mountain range that is so difficult of access that its summit can be reached only with unspeakable efforts; as far as the lower ground of the peninsula is concerned, it is impossible to get there. It is said that there are snakes whose bite is deadly and a large number of hyacinths. The inhabitants apply special trades and crafts to obtain these precious stones." Thus hyacinth refers to the familiar mineral zircon. I do not know if it really is common in Korea. Many of the features in this description fit Korea better than the Liaodong peninsula, with which Miller would like to identify the Hyacinth Peninsula. It is not surprising that Idrisi's description of the eastern side of Asia is much more vaguely related to reality than the reports of Sulaiman, Khordadhbeh and Masudi and that it shows two separate localities, both of which could be interpreted as being Korea.

It took another 100 years until Europeans could themselves report on the country of Korea. The first such information comes from the Franciscan monk Friar William of Rubruck, who traveled to Tartary in 1253–55 in the name of Louis IX of France and reached the court of the Mongol khan Mangu in Karakorum. There he met Master William Buchier, a goldsmith from Paris, and quotes from a conversation with him [116, p. 328f]: "Et narravit michi magister W. quod ipse vidit nuntios quorundam hominum qui dicuntur Caule et Manse, qui habitant in insulis quorum mare congelatur in hyeme, ita quod tunc possunt Tartari currere ad eos, qui offerebant triginta duo milia tumeniascot (guilders) annuatim dum modo dimitterent eos in pace." There is no doubt that Caule refers to the Gaoli of the Chinese, i.e., the Koreans. Probably the basis of this report is the fact that the Yalu freezes over in winter (p. 120). The lower Yalu with its many islands is not unlike an inlet; however, there are no true sounds that freeze over in winter along Korea's coasts.

Marco Polo, who made his famous voyage between the years of 1271 and 1295, also mentions Korea. He writes about the annexation of four provinces by the Chinese empire under Khubilai Khan [117, Vol. II, Chap. V]: "Et tunc quidem quatuor provinciae accesserunt ditioni eius, quarum nomina sunt Funotia, Cauli, Barscol & Sinchintingui."

Then 3 centuries passed before Korea again appeared within the horizon of the Occident, this time never to completely disappear again. This occurred in connection with the beginning of missionary activity in China and Japan. These missionaries brought back tidings of the country of Coria or Coray to Europe. Until 1650 it is rare that much more than this name is mentioned [118]), and in the maps of this period Korea appears now as an island – perhaps because of Rubruck's report – and now as a peninsula with an outline conforming to reality only in that it extends from north to south.

Father Luis Froës, one of the Portuguese missionaries sent to Japan, directed attention to Hideyoshi's expedition to "Corai, altrimente detta Koria" [118, 252] in the annual letter to the Superior of the Jesuit order in Rome dated 1590 (correctly 1592). In the same year of 1592 the famous Italian Jesuit missionary Matteo Ricci heard of Coria in Nanking. This fact was transmitted to us in the travel report by Ricci's companion Trigault, who writes the following about the impression made by Hideyoshi's invasion of Korea [122, p. 316f]: "Nanchinum ubi ventum, omnia formidinis plena repererunt. Nam Japones suis finibus egressi regnum Coria infestis armis impetiuerant." Thus all at once because of Hideyoshi's invasion the name of Korea reached Europe several times. Along with one of the armies of Taikōsama the first European came to Korea. This was Father Gregório de Céspedes [88], who was born in Madrid and was also a member of the Portuguese Jesuit mission. His letters written in the winter of 1594/95 from Korea contain, however, almost nothing but unimportant news of the country (cf. p. 82).

Since the Hideyoshi expedition the name Korea has also not disappeared again from maps made in Europe. In the map "Japoniae insulae descriptio" compiled by the Portuguese Ludoico Teisera in 1595, which was incorporated into the Theatrum Orbis Terrarum of 1612 as Plate 110, a "Corea Insula" is recorded. And a cartographical representation of East Asia from the same year, 1595, which is included in the travel account by the famous Dutch hydrographer Jan Huygen van Linschoten [120, map V] shows Korea as an almost circular island near the continent bearing the name Ilha de Corea. In the atlases of Mercator (1607 to 1633) and Blaeuw (1635), indeed even in those of Jansson (1650), Sanson (1652) and Duval d'Abbeville (1681), the name Corey or Coria is assigned to an island, which is exceedingly long from north to south, lying between China and Japan. Linschoten, however, has his doubts as to the island nature of Korea. He writes in 1599 in his Japoniae Nova Descriptio: "Corea, indigenis Caoli, Japonensibus Coras dicitur, quae aut sit insula aut pars continentis nondum constat", and as A. Pawlowski was the first to notice [159], some European maps from the first half of the 16th century already show a peninsula on the east coast of Asia at the latitude that comes into question for Korea. Among these are the maps that are included in the cosmography of Sebastian Münster. On Petrus Plancius' world map of 1594, which is also appended to Linschoten's itinerary [120, map III], Korea is shown with this name as an elongated peninsula. Hakluyt, too, in the map included in his Principal Navigations (1589–1600, [119]), shows Korea as a peninsular extension of Manchuria, to which he gives the name Coray regnum. A few years later the German Mathias Quade [121] also recognized Korea as a peninsula. In the atlas of the Jesuit M. Martini [123], which was first published in 1653, Korea appears on Plates 5 and 17 as a good–sized peninsula, though still too narrow ("Corea Peninsula"), and in the accompanying text we read "De Corea, insulane sit aut continens, varia est ac dubia apud Europaeos fama; at mihi certum omnino est, peninsulam esse." The atlases of Sanson (1697) and Homann (1716ff) are in this respect copies of Martini.

The text to Martini's atlas of China contains a very commendable first historical and geographical outline of Korea. With it a new epoch of European knowledge of Korea begins. It is the first epoch of active European participation. It lasts until the end of the isolation policy, in other words, from about 1650 to 1880. During this entire period foreigners are still forbidden on pain of death to set foot in the country. But Korea has loose relationships with its larger neighbors: through the exchange of envoys, through trade across the land and sea boundaries, as slight as it is in scale, through refugees, prisoners and shipwrecked sailors. In these neighboring countries, however, live Europeans who are eager to penetrate the veil of secrecy that lies over the Hermit Kingdom. To be sure, their number is very small until 1842 and 1854, respectively, but among the Jesuit missionaries in Beijing and the Dutch traders in Deshima–Nagasaki are some highly interested scholars, who know how to take advantage of the slightest opportunity. For instance, one meets Koreans living temporarily in China and Japan and obtains information on their country and its inhabitants from them. One translates Chinese and Japanese writings on Korea and publishes maps one has obtained there with European inscriptions.

At the beginning of this epoch we have, in addition to the work of Martini, who was a missionary to China, one book that does not belong to this category of scholarly publications based on secondary sources. It is an eyewitness report from Korea itself, the likes of which was not seen again in the next 200 years. In 1653, the same year in which Martini's atlas and text were first published, the Dutch yacht "Sparrow Hawk" stranded on the coast of Chejudo, as described on p. 31, and in 1668 in Rotterdam its secretary, Hendrik Hamel, published the story of his 13 years of captivity in Korea (171–173). The book was published in 1670 in French, and in 1672 in English and German [124]. The doubts that were sometimes entertained as to the truth of Hamel's tale of woe have proven to be unfounded. It has been possible to trace the route by which the prisoners were transported from Chejudo to the capital with some certainty (v. Siebold [132 Vol. I,

Hist. Overview, p. 159f]). This route has accordingly been included in the map of routes of explorers (Fig. 10). Of course, Hamel was inwardly too occupied with his fate and that of his comrades and outwardly too restricted in his freedom of movement for his book to be a treasure trove of geographical and historical details. But it contributed greatly to the interest in more information on this inhospitable country and its inhabitants.

A tremendous cartographical improvement is the Jesuit atlas of 1718. The credit for the Korean portions of this map is principally due to Father Régis. At his instigation the Kangxi emperor sent envoys to this tributary country in 1709 to obtain topographical data [158]. These envoys copied the map that hung in the palace of the Korean king in Seoul (p. 38) and measured the latitude of Seoul by astronomical means as 37°39'. The Jesuit fathers did not come along to Korea themselves. Régis, Jartoux and Fredelli did, however, reach the mouths of the Yalu and Tumen in 1709 and measured their positions. The Korean map was corrected by Fathers Régis and Gaubil on the basis of the observations of the lunar eclipse of 1612 in Ingolstadt and Nagasaki, the geodetic observations made en route by the envoys and geographical information obtained by the envoys in Korea. Such was the origin of the map of Korea that was included in the Chinese Jesuit atlas of 1718. The same map appeared as Plate XXXI with inscriptions transcribed in French in the Atlas of d'Anville [126] that was published in 1735. The river network and the cities are reproduced surprisingly well. The peninsula also has the correct length from north to south. The eastern coast lies at approximately the correct longitude, but the characteristic convex curvature is lacking. Above all, the western coast is displaced far into the Yellow Sea, and little is to be seen of the multitudes of islands along the western and southern coast. Thus the peninsula is much too wide and plump, particularly its southern half. On d'Anville's map it extends across 5.5° of longitude instead of 3°. Father Régis estimated the degree of cartographical accuracy he was able to obtain correctly when he observed [127, Vol. IV, p. 70]: "We are lacking in some observations on the noon and evening sides; then we would have everything we could desire on Corea for our design to provide a general geographical description of Asia." It was to take another 300 years before the Yellow Sea coast of Korea with its tremendous wealth of forms was recorded relatively accurately.

The Chinese envoys mentioned above had access to a geographical description of Korea that was in the hands of Korean mandarins. Father Régis used the thus gained knowledge of the country not only for his map, but also for a written description. This is included in du Halde's famous Description of China [125, 127]. The section on Korea in this work was the first detailed and reliable geographical and historical description of Korea by a European author. It is one of the main sources that Carl Ritter used for his comprehensive and, for its period, excellent geographical work on the "mountainous peninsula of Korea" [133].

In the 19th century we must first of all mention J. Klaproth, who translated Chinese, Japanese and Manchurian sources on Korea into French and added critical annotations. The most abundant information on historical geography is provided by his Aperçu *Général des Trois Royaumes* [131]. This contains a translation of the description of Korea published by the Japanese Rinsife in 1786 in Edo and one of Chapter 353 of the Imperial Chinese Geography (*Tai t'sing i t'ung chi*), edition of 1744. In the same manner Pfizmaier [134, 136], Scherzer

[142, 154], Nocentini [157], Dimitrewski [148, 149] and others later made the effort to translate Far Eastern sources on Korea. From what we have said above on the nature of these sources, it is understandable that they do not clearly distinguish between historical and geographical topics.

At approximately the same time as Klaproth P. v. Siebold and, at his suggestion, J. Hoffmann concerned themselves with the Japanese sources on Korea. The two scholars did not, however, publish translations of these sources, but combined them into an original new presentation. This is included in Part VII, the part dealing with Korea, of Siebold's gigantic work [132]. By the same method, W.E. Griffis, an American working at the University of Tokyo, wrote around 1880 for that period, an excellent, comprehensive work on the history and ethnology of Korea [45], without ever having set foot on Korean soil. In a similar manner and at approximately the same time, the Scottish minister J. Ross, who lived in Manchuria for 7 years, studied the Chinese sources [44].

Appended to many of these works, such as those by Klaproth, v. Siebold and Griffis, are maps of Korea that came from Japan and were revised by the authors. Superior to these maps is the one that was prepared by the Japanese Ministry of War in 1875 under the title "Chōsen zen dsu" on a scale of 1:960,000. It was published in 1883 in Petermanns Mitteilungen [150], using the well–founded transcription system of E. Satow [V]. The exactitude of this map can be estimated from the fact that planimetric measurements showed the value of 218,192 km^2, which very closely approximates reality (Pet. Mitt. 1883, p. 434, cf. p. 4).

Siebold was not content, however, with merely processing the Japanese sources. Being stationed in Deshima, he had the opportunity to meet shipwrecked Koreans, who were tended to by the Japanese in Nagasaki until a chance was found to send them back to Korea. He obtained further information from the Japanese officers and officials who were employed in Tsushima and in the Japanese settlement in Pusan. He was able to recognize and describe the two racial types that actually do predominate in Korea and reproduced fine drawings of them. Independent of Siebold, Richthofen recognized the same racial types during his visit to Gaolimen in 1869, and he himself gathered observations and information on the people and the economy of Korea [140, 141, 253]. He also attempted to extrapolate from his geological–morphological studies in Shandong and Liaodong, to draw inferences as to the structure of the mountains of northern Korea [141]. Among the other works of this period based on material gathered in the neighboring countries we should mention that of the Russian chargé d'affaires in Beijing for many years, M.A. Pogio [240]. It was not published until after the end of the isolation policy, and although it was translated into German [156], it is not very well known. Certainly it is not always reliable, but it does contain a fair amount of geographically useful data that are missing in the other books from this period.

The conclusion of this period can be considered to be pére C. Dallet's colossal work on the history of the church in Korea [137], which was published in 1874. It is based entirely on the reports of French missionaries. The main portion of the book is, as the title indicates, devoted to a very detailed description of the atrocious persecutions of Christians, which had reached their climax a bit earlier. An introduction of almost 200 pages, however, is concerned with the geographical conditions in the Land of Morning Freshness. This introduction suffers from the missionaries' lack of scientific knowledge, and the topographical data are dry and schematic. Still Reclus used it as the main basis for the chapter on Korea in his Nouvelle Géographie Universelle [146], so that its contents were widely disseminated.

The second half of this epoch, lasting until the opening of the country, is additionally characterized by European naval expeditions, which surveyed the outline of the coast of the "Land of 10,000 Islands" and measured the depth of the coastal waters. It was through these expeditions that the mathematical position, which had been quite inaccurate on maps of Korean origin, was finally fixed by astronomical means. In this manner the map of Korea received, for Koreans and foreigners, the characteristic form that is familiar to everyone today. This surveying activity was very time-consuming among the swarms of islands along the southern and western coasts. It was not completed until the end of this epoch by European warships, among them German ones, and particularly by the Japanese. Many of these expeditions also made brief landings, during which they gained an impression of the nature of the country. The first expeditions had purely scientific goals. Beginning with the middle of the last century, however, there were parallel economic and political motives. It was these expeditions that gave English, French and Russian names to geographical objects along the coasts, as was mentioned earlier (p. 11f, cf. Fig. 10).

The first in the long line of daring mariners was the ill-fated La Pérouse. During the third year of his great voyage of discovery through the Pacific he passed through Korean waters [160]. On 21 May 1787 he sailed along the southern coast of Chejudo with his ships "Bussole" and "Astrolabe" and then through the Korea Strait between Tsushima and the continent close to the latter. On 27 May he discovered the island of Ullŭngdo, which until then had been known only to Koreans and Japanese and had not been shown on any European map. He named it Dagelet after the astronomer who accompanied the expedition. Ten years later Captain W.R. Broughton, on the British sloop of war "Providence", sailed in the opposite direction along the entire eastern coast from the mouth of the Tumen to Pusan [161]. In the process he discovered Tonghanman and Broughton Bay, which was named after him. He landed in Pusan and then took the dangerous course through the multitudes of islands off the southwestern tip of the peninsula and past Chejudo.

In 1816 Lord Amherst was sent with an embassy to China. On this occasion the captains Murray Maxwell and Basil Hall with the ships "Alceste" and "Lyra" were sent on a cruise in the Gulf of Zhili (Chihli) and along the west coast of Korea [162 – 164]. On the basis of their measurements the coast of Korea was moved by 2° of longitude in relation to the map of d'Anville. In the northern perimeter they discovered several groups of islands far from the coast and in the southern perimeter an archipelago through which fierce tidal currents rushed. The ship's doctor on board the "Alceste" was McLeod, who also wrote a report on this voyage [165].

In 1832 another British ship, the "Lord Amherst," which had been equipped by the British East India Company in Canton, followed in the same waters. On board were Captain H.H. Lindsay and the missionary Karl Gützlaff from the Brandenburg Marches. The "Lord Amherst" anchored for a long time on the west coast near Inch'ŏn, in the vain hope of taking up contact with the Korean

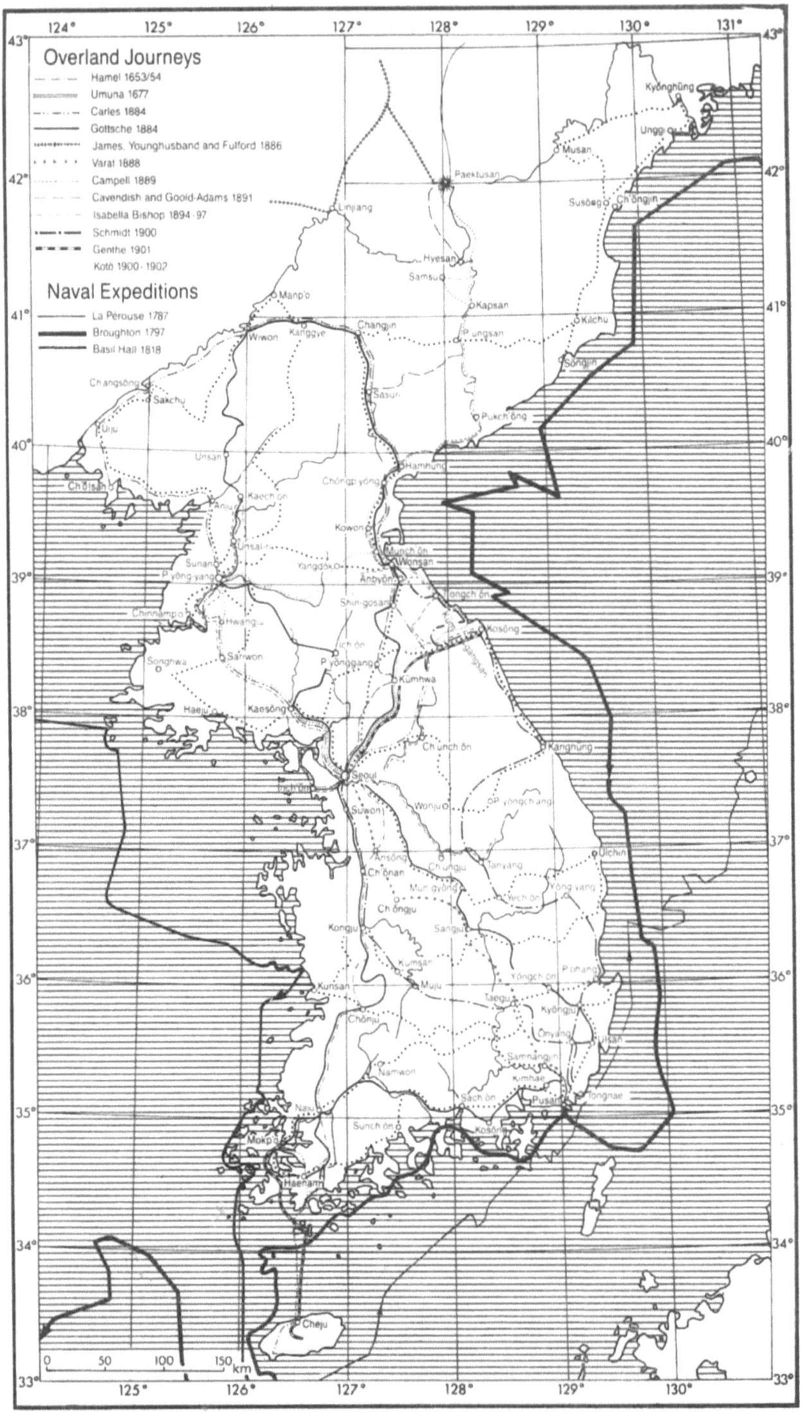

Fig. 10. Voyages of discovery in and around Korea, 1653–1902

government and working out a trade treaty. During this time Gützlaff taught the Koreans how to plant potatoes. The contributions this expedition made to our knowledge of Korea were slight [166, 167].

In 1845 Sir Edward Belcher was able to travel around Chejudo by land directly along the coast. He accomplished this within a period of 3 weeks and was constantly guarded by the British warship "Samarang". In the course he compiled the first serviceable map of the outline of the coast. It was published in his travel account [168] on a scale of 1:390,000. Although the attitude of the inhabitants was often hostile, so that it was almost impossible to enter the settlements, his report contains some very valuable information on the geography of the island. In autumn of the same year Belcher went on to investigate the swarms of islands festooned along the coast of the peninsula to the north. There he discovered a safe harbor surrounded by three small islands, which he named Port Hamilton. With his boats he advanced as far as 34°40', i.e., very close to the coast of the peninsula.

In the next year French warships visited the area west of Korea. The "Sabine", under the command of Captain Guérin, sailed along the western fringe of the southwest Korean archipelago from 34° to 37° N, surveying the islands. In response to the execution of French priests by the Taewon–gun, French warships visited the Korean coast three more times before the end of the seclusion policy. The frigates "La Gloire" and "La Victorieuse" were completely wrecked in Kyŏnggiman with its many cliffs and strong tides. The crew safely reached an island, where they needed the friendly aid of just the persons whom they intended to call to account. Thus scientific observations were also impossible. The last of these visitors, the above mentioned unfortunate expedition of Admiral Roze in 1866, managed to complete a hydrographical survey of the mouth of the Han–gang. The French naval officer H. Zuber also had the opportunity to make observations on the country and its people, which he published in a vivid, reliable article [174]. Partly before and partly after the expedition of Admiral Roze, the Jewish merchant from Hamburg in British services, Ernst Oppert [169], made three futile attempts to establish trade relations with the Korean government. The third time, with the assistance of the American Jenkins and the French missionary Féron, he combined this with an unsuccessful attempt to appropriate royal relics from a grave near the coast, an affair that triggered a lot of indignation, especially in America ("international body snatching expedition"). The three attempts did, however, result in improved sea charts. The Russian frigate "Pallas", which we have already mentioned, also surveyed the coast from Pusan to the mouth of the Tumen [236].

At the end of this epoch the topography of Korea was known in broad outline, approximately to the extent that can be represented on a map of the scale 1:1.5 million. Quite a bit was also known about the Korean administration, history, culture, political divisions, manners and customs and a general outline of the economy was known. Several attempts were made at a geographical summary of this knowledge, of which Carl Ritter's great synthesis [133] is impressively superior to Reclus' presentation [146]. Using all of the literature that was available in 1834, Ritter brings out masterfully the main characteristics of the country. His work does have a tendency toward verbosity, though, because the contents of many of the sources are reproduced in great detail.

Around 1880 there was, however, still no information on Korean regions and how they look, on the physical geography of the country, except for a rough idea of the climate, or on rural life. Reclus [146, p. 662], for instance, could still maintain in 1882 that there were no oak trees in Korea. Today the "Survey of the species of trees in Korean forests", which makes no pretensions at being complete, lists no fewer than 22 different types of oak, and the practical significance of each type is known in detail. There was a strange discrepancy in the state of geographical knowledge, which can only be explained by the centuries of seclusion. The topographical knowledge of Korea was no worse than that of the European portions of Turkey, but the knowledge of the cultural geography of the Land of Morning Freshness was about as accurate as that of tropical Central Africa.

With the opening of the country cultivated and interested Europeans streamed into the country all at once: officials of the legations and consulates, merchants, newspaper reporters, Catholic and Protestant missionaries, naval officers and later Russian staff officers and globetrotters. Scientists, in contrast, were few in number, and there were practically no geographers. Many of these people undertook journeys around this country, which had been kept under a mysterious veil of secrecy for so long and thus was observed from outside with great curiosity and high expectations. They generally wrote down their travel impressions in some literary form. Travel accounts, and literature on Korea in general suddenly increased tremendously. Underwood's bibliography [B], which is still incomplete, lists 152 publications on Korea in Western languages, not including Russian, for the period from 1595 to 1880. For the period from 1881 to 1931, in contrast, it mentions 2730. Unfortunately, most of the travel accounts and printed versions of travel lectures have little to offer that is of use in a modern geography of the country. They usually restrict themselves all too much to listing one superficial impression after the other. And other chapters repeat familiar ethnographical information from earlier epochs. There are dozens of travel accounts, all written according to this same recipe, and many of the lectures published in scientific periodicals are not much better. As long as the trips being described were the first expeditions to particular areas, they at least contribute topographical information, unless this is useless because of the above mentioned confusion with regard to place names. Nevertheless, among this great volume of travel literature, which has been decreasing in the course of several decades, there are some excellent examples to which we shall call appropriate attention in the following. The books written by Protestant and Catholic missionaries about the country and its people are generally not travel accounts in the strict sense of the word. Although most of them were written by men who spent many years in the country and were in daily contact with the common people, it is often surprising how little geographical information they impart. The descriptions are quite often superficial and lacking in life. At times they take on a moralizing tone and present a one–sided, anti–Japanese point of view of the social and political situation. I know of only two books providing excellent geographical information and rising way above the standard that were written by missionaries. One is the *Geschichte der koreanischen Kunst* (History of Korean Art) by the then German Benedictine priest A. Eckardt [571]. The other is C.A. Clark's *Religions of Old Korea* [580]. In the English–language journals [I–IV, XLIX, L, LIV] there are also some articles by missionaries that are quite useful from the viewpoint of geography.

The first trade treaties had hardly been signed between the Korean government and the United States and Great Britain in the early summer of 1882, when the first Europeans began to make exploratory trips and write about them. Initially their trips were restricted to the coastal area of Inch'ŏn – Suwon – Seoul (J.C. Hall Oct. 1882, H.A.C. Bonar March/April 1883, P. Lowell winter 1883/84, [175 to 179]). In July 1884 S.B. Bernerston of the American navy traveled from Seoul to P'yŏng–yang, where he was the first Westerner to enter the city, aside from a few French missionaries in hiding [182]. The first European traveler of standing in Korea was the British Consul W.R. Carles [186, 187]. After a brief trip in the direction of the Diamond Mountains (Kŭmgangsan) in December 1883, he undertook the first big expedition through northern Korea, which led him as far as the Yalu and to Changjin in the autumn of 1884 (route in Fig. 10). In the Ch'ŭgaryŏng furrow Carles discovered the tremendous lava fields embedded in the relief and wrote a vivid description of the morphological conditions of this area. Unfortunately, its influence was later undermined by the authority of the Japanese geologist Kotō. Everywhere Carles confines himself to a brief, objective description of what he himself observed and experienced.

At almost the same time as Carles, the German geologist Dr. Carl Gottsche traveled in Korea. He had previously worked in Japan and was invited to Korea by P.G. von Moellendorff (p. 34). A brief visit in 1883 was followed the next year by a stay of 8 months. From 11 June to 15 August 1884 Gottsche undertook a trip all the way down to the south coast, and from 18 September to 28 November another one to the north following essentially the same route as Carles. Gottsche traveled a total of 2550 km in Korea. For decades he remained the only European traveler who had seen both the northern and the southern halves of the country. He was a scientist and a skilled observer, and to him we owe the first overview of the geology of Korea and the country's mineral deposits [309, 330]. For one and a half decades he reviewed the travel literature on Korea for Petermanns Mitteilungen, and his brilliant knowledge of land and literature strikes the reader time and again. It is very regrettable that he published almost nothing on the geography of Korea, except for a lecture at the Geographical Society of Berlin [188]. He would have had the capacity to write the definitive geography of Korea before the turn of the century that the author is only now attempting with this book. His contributions to the exploration of Korea have met with much too little attention thus far, presumably because he did not write a voluminous book on his travels, as others did. The first comprehensive book on Korea by an independent observer is that of P. Lowell [179].

In 1884 the Englishman Gowland also traveled from Seoul to Pusan [184], and the British Consul General W.G. Aston traveled from the capital to Kaesŏng [178]. Due to the political unrest in December 1884 the number of travelers decreased, and the only ones worth mentioning in the next years are the American G.C. Foulk [183] and the German F. Kraus [191, 192]. Neither wrote very much, but what they wrote is worth reading. In 1888 the Frenchman C. Varat traveled from Seoul to Pusan, following basically the same route as Gottsche and Gowland and wrote a rousing report that was published with a large number of illustrations in a popular travel magazine [194]. In autumn of the same year the American consul general C. Chaillé–Long visited Chejudo, but this visit was no great success because of the negative attitude of the governor [195, 196].

Our knowledge of Korea was brought a great step forward again in 1888 by the trip taken by the British consular official C.W. Campbell, who was the first European to visit Kŭmgangsan with its many Buddhist temples and simultaneously the first non–Russian to travel to northeastern Korea [197]. His goal was Paektusan. It had been climbed in 1886 from the Manchurian side by the Englishmen James, Younghusband and Fulford, the first Europeans to do so [190]. Since he was not able to leave Seoul until rather late in the year, he was caught unexpectedly by the first snowfalls of the winter and had to turn back shortly before reaching his goal. From a geographical viewpoint his trip was still an outstanding achievement. For his cartographical surveys the Royal Geographical Society of London gave him a medal of honor.

Where he failed in the last minute, his fellow countryman H.E. Goold–Adams, an officer of the British army in Hongkong, succeeded 3 years later [202]. Starting again at Seoul, Goold–Adams and his comrade A.E. Cavendish more or less followed Campbell's route. They did not visit Kŭmgangsan, but did make a detour via Changjin – Kapsan (Kaema upland). Cavendish turned back in the vicinity of Hyesan, but Goold–Adams reached the rim of the crater on 16 October and looked down onto the "Heavenly Lake." Their report on their trip and the hunting interests that they simultaneously pursued was written primarily by Cavendish and was published as a book. It is full of original descriptions of new facts [203]. In 1891, the German Consul E. Brass traveled in Korea with the owner of the large German export firm in Inch'ŏn, C. Wolter, one of the greatest living experts on Korea [259], but his only publication on his abundant observations was a long lecture [201].

There now follow three travel accounts that are radically different in character. A.H. Savage–Landor is an amiable globetrotter, who writes attractively about Korea but makes no pretensions at presenting new information [205]. E. v. Hesse–Wartegg did not undertake any pioneering trips in Korea either and only visited places that were already known. His book is, however, the first detailed description of Korea in the German language and it is based on older books, information gathered from Koreans and from the Europeans living in Korea and on reports in the government news bulletins, etc. In particular, the author takes pains to portray Korea as a battle field of Chinese and Japanese interests [204]. By far the most profound and absorbing observations are those of G.N. Curzon, the later viceroy of India, who traveled from Wonsan to Seoul in 1893. His *Problems of the Far East,* in which he treats Korea in detail, is certainly no travel account in the style of the others. It is an eminently political book [206, 865].

In January 1894 Isabella Bird Bishop entered the country for the first time. She was a traveler–author who had already made herself a name in the field of travel accounts. She made four trips to Korea between 1894 and March 1897. The routes of her trips were not unusual, aside from a trip on a sampan up the Han–gang. Nevertheless, her book *Korea and Her Neighbors* [207], which was published shortly thereafter in two volumes, attracted more attention than any other book on Korea. It truly is the most comprehensive, most thorough and most reliable book on the country that we have from this period of travel accounts. The chapter on her visit to Kŭmgangsan and the lives of the monks there has become famous. Her own impressions from her travels and splendid descriptions of the forms and colors of the landscape alternate with general passages on ethnology, culture and current politics. The book tells the history of the agitated years from 1894–97 from firsthand experience. She was the first author to examine Korean shamanism closely and also the first to provide detailed information on vegetation. Thus her book deserves the attention it received.

In the same and the following years the German geologist B. Knochenhauer [214], the famous German doctor working in Japan, E. Baelz [218, 522], the Englishmen Willis and L.O. Warner [199] and the Frenchmen D. Liévre [212] and M. Monnier, who wrote sensitive descriptions of Kŭmgangsan [215], all traveled in Korea.

In 1901 Dr. Siegfried Genthe came to Korea as a correspondent for the "Kölnische Zeitung" (Cologne newspaper). He had received his Ph.D. in Marburg under Theobald Fischer with a thesis on the Persian Gulf and before and after this he got around quite a lot in the world. Of the many trips that he took in Korea, his visit to Kŭmgangsan is worthy of special mention, as is the fact that he was the first to climb Hallasan, the volcano on Chejudo. A few years later, after Genthe had been assassinated in Morocco, Georg Wegener published some of the reports he had written during his approximately 1–year stay in Korea along with a detailed biographical introduction. He dedicated it to Fischer, under whom they had studied together [217]. Genthe depicts his adventures in a particularly gripping manner and his descriptions of landscapes are geographically well–founded and to the point. At the same time, with the penetrating view of a trained reporter he gives a clear and versatile portrayal of the political development of the past years and of the cultural and social conditions in Korea. Genthe's book is one of the best travel accounts that the literature on Korea has to offer.

With the works of Bishop and Genthe we reach the acme. The Russo–Japanese War soon led to a new flood of books and articles about the sorely tried country. Particularly the versatile portrayal by the Englishman A. Hamilton, which was published in three languages [219], the work by É. Bourdaret [261] with its many ethnological details, the voluminous book by the Italian chargé d'affaires, C. Rosetti [216] and the vivid descriptions by the Pole W. Sieroszewski [220] are worthy of mention. In 1937 Sieroszewski also published a two volume work on Korea, but only in Polish [290]. The book that is the most comprehensive and at the same time most profound portrayal of Korean culture is *The Passing of Korea* by the American missionary–educator H.B. Hulbert. It is based on many years of daily contact with Korean life [43]. The book by the abbot Norbert Weber [226], which has been widely read in Germany, contains little new geographical information, especially since his travels covered only a very small portion of the country. Excellent modern descriptions can be found in the book by Sten Bergman, who roamed through the country in 1935/36 on paths that often had not yet been described, studying the life and distribution of birds in northern Korea and gathering specimens for the Swedish Museum of Natural History. Stories about his expeditions and hunting trips alternate with perceptive descriptions of various Korean animals and sketches of daily life in Korea [227–230].

In view of such titles as J.M. Velter: *Dämonen und Masken. Als Jäger und Forscher durch das unbekannte Korea und die Mongolei.* (Demons and Masks. Travels as hunter and scientist in unknown parts of Korea and Mongolia, Cologne

1934, 156 p.), we must stress that books that intentionally fail to draw a line between fact and fiction are absolutely useless for serious research.

The rest of the Western literature on Korea in the past 60 years can be omitted here (cf. references). But we must make special mention of the only European scientific journal in Korea, the Transactions of the Royal Asiatic Society, Korea Branch, which has been being published for 40 years without interruption and contains a number of valuable geographical articles [I]. How incomplete our knowledge of the geography of Korea as a whole is can be seen by the unsatisfying attempts to summarize this knowledge recently made in the *Handbuch der Geographischen Wissenschaft* [289], the *Géographie Universelle* [278] and the Italian *Geografia Universale Illustrata* [288]. Our knowledge of the spheres of economy, trade and commerce is better, as K. Haushofer's chapter in *Geographie des Welthandels* (Geography of World Trade) demonstrates [275].

It is easy to understand this disappointing situation, if we consider the following. The travel accounts were not written by scientists who had the intention of solving the complex geography of Korea systematically bit by bit. On the contrary, each of these travelers had the desire to tell interested readers of his adventures while traveling and at the same time to provide as much information as possible on this country, which the course of political events so often hurled into the focus of attention. Neither the choice of routes nor the contents of the books follows any constructive scientific plan. Each author begins anew from the beginning, so to speak.

The Russian literature on Korea from the period of 1885 to 1904, which E. Bretschneider competently reviews [234], reveals quite a different trend. Russia is the only European country that is a neighbor of Korea. During this period Korea became one of the objects of czarist expansion policy. Therefore, Korea was systematically explored by Russian scholars and officers, who soon produced remarkable results. Whereas travelers from Western Europe generally started out from Seoul, the Russians often entered the peninsula by land from the Ussuri region, crossing the Tumen at Kyŏnghŭng.

The first one, however, the trader P.M. Delotkevitch, still took the opposite route. He arrived at Inch'ŏn by ship and traveled from December 1885 to February 1886 via Seoul – Wonsan and along the coast of the Sea of Japan back to Vladivostok [237]. From 1885 to 1887 the Polish naturalist J. Kalinovski collected zoological specimens in northeastern Korea for the University of Warsaw. In May 1889 lieutenant colonel Webel traveled from Vladivostok via Kyŏnghŭng – Wonsan – P'yŏng–yang to Seoul [238]. In 1895/96 lieutenant colonel Alftan followed the same route as far as Wonsan [239], and in the same year lieutenant colonel Strelbitzki traveled from Kyŏnghŭng through the northern part of Hamgyŏngbuk–do via Musan and the headwater region of the Tumen to Kapsan and via Changjin – Kanggye to Manchuria. He was the first European to climb Paektusan from the east [235]. In 1896 the Imperial Russian Geographical Society sent an expedition to explore Manchuria and northeastern Korea. It consisted of the German mining engineer E. Ahnert, the botanist W.L. Komarov, the zoologist A. Jankovski and a topographer. They first explored Manchuria and did not enter Korea until 1897, following the route Hunchun – Kyŏng–won – Hoeryŏng – Musan – source of the Tumen – Paektusan – Samsu. Komarov returned to Manchuria from there. Jankovski went to Wonsan [244]. In 1900 A.J. Zweginzov undertook trips through all of northern Korea [242], and simultaneously the zoologist P.I. Schmidt traveled from Wonsan through Kŭmgangsan and along the coast of the Sea of Japan to Kangnŭng, looping around through the interior via Tamyang – Yŏngju – P'ohang to Pusan and finally via Kongju – Suwon to Seoul (route in Fig. 10, [243]), again on behalf of the Imperial Russian Geographical Society.

From this enumeration it is clear that the Russians systematically studied those areas of Korea that were almost completely neglected by Western Europeans, particularly in the north and east. They were simultaneously the areas that Russia would have considered occupying if the outcome of a war or some such belligerent activity had been favorable to her. As the conclusion of these efforts, the Russian Ministry of Finance published a large three volume compendium on Korea in 1900. It includes the Western European research and is accompanied by a map on a scale of 1:1,355,000 with Russian inscriptions [245]. This work is a political geography written from an objective point of view and aiming particularly at practical use. It is a reliable summary of the knowledge of Korea at the turn of the century. It does not attempt to be a definitive geography of Korea based on genetic–causal relationships and regional description. In 1904 the above mentioned A.J. Zweginzov published a 20–verst map of northern Korea (1 Russ. inch = 20 Russ. verst, i.e., 1:840,000), together with N.A. Korf. Primarily on the basis of these Russian sources, the Prussian major general Krahmer published a military geography of Korea that remained almost unnoticed, but is very reliable and contains a lot of valuable information. From the title *Die Beziehungen Rußlands zu Japan* (Russia's Relationships with Japan), one would hardly guess that it is the best compendium on Korea in the German language [263].

With the, for Russia, unfortunate outcome of the war of 1904/05 and the takeover of Korea by Japan, Russian exploration in Korea naturally became impossible, and Russian publications on Korea become rare (cf., however, [246 and 247]). With this the last period in the exploration of Korea begins, the Japanese period. Modern Japanese research had already begun before 1905. For instance, between 1900 and 1902 the Japanese geologist B. Kotō investigated large portions of both southern and northern Korea on foot, laboriously covering 6300 km in 14 months ([352, 312], route in Fig. 10).

From the outset Japanese research tackled the problems of the newly acquired country from a standpoint that did not become fully established in Germany until the last decade. It is that of regional planning based on thorough specialized investigations. This is not to say that the Japanese neglected other fields of research. With their typical energy they very soon began to study the old Korean literature and to carry out archeological research. But research was oriented primarily toward practical goals related to the regional planning policies of the government–general (Sōtokofu).

The initial basis for this research of course had to be a sufficiently exact cartographical survey [292]. During World War I the government–general had this done without attracting much attention. A set of "Topographical Maps 1:50,000" was published for all of Korea except for a few fortified zones (grid map; 723 sheets with a width of 15' of longitude and a height of 10' of latitude; watercourses are shown in blue, on most sheets everything else is shown in black; normal distance between contour lines 20 m; in the vicinity of the fortified zones the terrain is indicated by shading in brown; topographical names in Chinese characters with the Korean pronunciation given in the Japanese syllabic script). In these excellent maps the terrain is represented with an accuracy that I tested thousands of times. The road network, in contrast, is shown according to a plan that was later revised, so that the maps are often useless from this point of view. A series of General Maps on a scale of 1:200,000 was also published for all of Korea (65 sheets with a width of 60' of longitude and a height of 40' of latitude; thus one of these sheets covers the same area as 16 of the sheets of the 1:50,000 scale map; details as above). Ten areas of particular cultural significance are additionally represented in contiguous series of sheets on a scale of 1:25,000 (a total of 97 sheets with a width of 7½' of longitude and a height of 5' of latitude; distance between contour lines 10 m). And finally there are 45 official city maps on a scale of 1:10,000 printed in four colors, most of which are outdated today. Each of the 13 provinces is shown in a map on a scale of 1:500,000 in four colors, with brownish–gray contour lines at a distance of 100 m. The Japanese maps of Manchuria on scales of 1:100,000, 1:200,000 and 1:500,000 extend into northern Korea [292].

In the International Map of the World 1:1 Million, Korea is shown on four sheets for which the Japanese government assumed responsibility; three of them have been published. Korea is shown much better on the excellent Map of Japan and Adjacent Regions on a scale of 1:2 million that was recently published by the "Society for International Cultural Relations". Map 1, which is appended to this book (cf. editors' note on p. 7), is based primarily on this map, and it is the authority for the spelling of place names in this book (Editors' note: this of course referred to the Japanese spelling, which we are not using in this translation. It may, however, interest some readers to know the basis for the Japanese spelling listed in the index of place names in this edition). With the exception of the world map 1:1 million the official maps do not use any European writing. Japan, and particularly the Government–General of Korea, accomplished an outstanding feat with the surveying and publication of these maps.

This was the basis upon which the research in regional planning of the past 25 years was founded. The recent geological research is based on the investigations of B. Kotō [312]. It was he who established the foundations of the stratigraphical development, which T. Kobayashi [319–324] and others have been expanding recently. Most of Kotō's concepts of tectonics and morphology, however, which carried good ideas of Richthofen's to an extreme in a schematic manner [352], have had to be abandoned. T. Kobayashi has also made progress in the explanation of the morphology [380]. The director of the Geological Survey of the Government–General of Chōsen, S. Kawasaki, and his current successor, I. Tateiwa, have energetically pursued the official geological survey of the country. There is now a general geological map in two sheets on a scale of 1:1 million of all of Korea [j]. Nineteen issues with a total of 63 sheets of the Geological Atlas of Chōsen 1:50,000 have appeared thus far on the basis of the topographical maps on the same scale and with an ample number of supplements [l]. In addition, a large number of articles on tectonics, stratigraphy and mineral deposits have been published, primarily in the Bulletin of the Geological Survey of Chōsen, the Report on the Geological Survey of Coalfields in Chōsen and the Journal of the Geological Society of Tōkyō.

Immediately after the protectorate was set up the Japanese established a network of meteorological stations. They had already begun this in the treaty ports, and have gradually extended it with persistent energy. In 1933 it consisted of one central observatory (Inch'ŏn), 15 first-order stations and 234 auxiliary stations, in other words, a total of 250 observation points. The results have been published annually since 1907 in the "Annual Report of the Meteorological Observatory of the Government-General of Tyōsen" [XIV].

In the past 10 years extensive oceanographical investigations have been made of the surrounding seas and the biological conditions of the fish living in them. The study of the botany and plant geography of Korea have been very greatly advanced by the indefatigable T. Nakai. He has not only written six regional monographs on plant geography [766, 826, 860, 879, 902, 935], but he is also working on a gigantic *Flora silvatica Koreana.* During the period from 1909 to 1936 21 volumes were published [480-484].

Recently these studies on plant geography have been carried out under the auspices of the Forest Experiment Station of the Government-General in Seoul-Ch'ŏngnyang-ri. It has already published a number of monographs on forestry research that are of use to a geographer [486, 488] and an atlas of woody plants and bamboos [t]. The director of the Forest Experiment Station is the most worthy T. Kaburaki. Forestry research is carried out primarily by H. Uyeki (488-492).

Corresponding to the Forest Research Station is the Agricultural Experiment Station, founded in 1906 in Suwon, south of Seoul, and its many branches in different parts of the country. The most important men in this institution are I. Nagai, T. Nakagawa, T. Takasaki and N. Takahashi. It runs experimental stations testing different varieties of crops, breeding, use of fertilizer, irrigation, agricultural implements, tillage, etc. and schools teaching plant and animal breeding, and particularly sericulture. It has also published standard works on the most important varieties of paddy rice and their distribution [620] and on the distribution and breeding of the most important crops [624], etc.

The last group of regional analyses that we shall mention are those of Korean rivers (water levels, rate of flow, floods, water regulation and utilization, [451-453]).

All of these studies are, necessarily, analytical in nature, and they almost always deal with Korea as a whole. They are valuable aids, but not ready-made building blocks, for a regional geography. In accordance with the objectives they pursue, they are all written in Japanese. They are published by the government-general in Seoul, generally in small editions for official use only and cannot be purchased in normal bookstores. Therefore, I consider it quite likely that I have missed some important publications of this type, especially if they were printed after my stay in Korea.

Important articles on Korea with geographical content appear now and then, although not too frequently, in journals in Japan. The most important of these are listed among the references. There is no journal thus far that is concerned solely with the geography of Korea. To be sure, Jimbun Chiri is an anthropogeographical journal that appears in Seoul, but its contents are not restricted to the country.

There are also Japanese publications on Korea in English. Above all the *Annual Report on Administration of Chōsen,* compiled by the Government-General of Chōsen should be mentioned. It has been published annually since 1908, until 1923 under the title "Annual Report of Reforms and Progress in Chōsen" [VI]. This attractive yearbook is a type of illustrated statistical report dressed up with a text tailored to propaganda. The form of the yearbook and the contents of the individual chapters do not vary much from year to year, unless necessitated by special events. From time to time, particularly for festive occasions, surveys of the Japanese accomplishments in various areas of Korea appear in English. As we shall point out later, the Japanese are quite proud of this development, and occasionally, to bolster their reputation, they attempt to call the attention of Westerners to their achievements. One of these abundantly illustrated summaries is "Thriving Chōsen. A Survey of Twenty–five Year's Administration" [304]. Another is T. Hoshino's "Economic History of Chōsen" [716], which was published for the tenth anniversary of the Chōsen Bank. The Japanese Journal of Geology and Geography is also written almost entirely in English. It reviews a large number of articles that have appeared in Japanese in other publications [XXVII]. Although it deals with the entire Japanese Empire, it is still indispensable for a geography of Korea. It is a pity that geography is almost entirely restricted to geomorphology in it.

Japanese have attempted several times to summarize the most important geographical information on Korea in the Japanese language [297–299, 306]. The 19–volume Nihon Chiri Fūzoku Taikei [299] devotes two volumes with a total of more than 800 pages to this largest Japanese colony. One covers the entire country, the other the 13 provinces. This amply and tastefully illustrated work is written for the general public, and accordingly it consists primarily of an unpretentious description of beautiful landscapes, customs and cultural institutions of the Koreans and of the successful Japanese efforts at colonization.

7 The Author's Objectives and Travels

The scientific situation today is as follows. On the one hand we have the popular European travel accounts with their tendency toward graphic and sometimes quite vivid descriptions, and on the other hand the exact scientific research of the Japanese with its orientation toward regional planning, which is naturally rather dry and analytical. There are very few regional monographs on different parts of Korea. Thus it is a very stimulating idea to attempt a synthesis of the older sources of geographical knowledge and the more recent ones and simultaneously to initiate a kind of geographical research that is based on empirical observation and in–depth analysis, applying this research to the country as a whole and to its principal areas. That is the objective of this book.

To pursue this objective I traveled through Korea from March to November 1933. Initially I spent a good month in the capital city, Seoul, and its immediate environs. After that I undertook four separate trips, each time starting from Seoul (Fig. 11). The first lasted from 22 April to 15 June. On this trip I covered 4300 km. Its objective was an investigation of the half of the peninsula that lies south of 37°45'. This we crossed four times from east to west and back, traveling along stretches of both the east and the west coast, and I spent 5 days on Ullŭngdo. The second trip was devoted to the northwestern part of the country (28 June to 8 August, 3200 km). It led us first to the former capital Kaesŏng, then around the Hwanghae peninsula to the mouth of the Taedonggang and P'yŏng–yang. Then I investigated the upper portions of the Taedonggang basin in the Ch'ŏrongsan and Myohyangsan areas. From there we continued on via Hŭich'ŏn and Kanggye into the big bend of the Yalu at Chunggangjin and then downriver as far as Pyŏktong, most of the way across high passes and through side valleys, since there generally is no road along the river itself. At Pyŏktong we looped south to Pakch'ŏn, returning to the Yalu at Sakchu, and then followed the river to its mouth at Yong–amp'o. From there my journey took me southeastward along the coast of the Yellow Sea and through the center of the peninsula, across the lava fields in the region of the Imjin–gang, and back to Seoul.

The third trip (21 August to 18 October, 4600 km) began by railway to the mouth of the Yalu and from there 700 km upriver to Shin–galp'ajin on a boat equipped with an airplane propeller. Starting from Hyesan on the southeastern bend of the Yalu we climbed Paektusan in the period from 30 August to 7 September. In the uninhabited virgin forest in this region my pony caravan and I were attacked and robbed by Chinese bandits twice. From Hyesan we headed south to the coast of the Sea of Japan at Pukch'ŏng, and from there we began a particularly difficult stretch along the coast toward the northeast. On a side trip we climbed Kwanmobong and then investigated the extreme northeastern tip of Korea. Then I returned southward down the coast. From Hamhŭng we took another long side trip northward to Changjin and visited Maengbusan in the Nangnim Range. Then we continued south around Yŏnghŭngman to Kŭmgangsan, and I had the chance to hike through it when the autumn foliage was at its best. We then followed the coast of the Sea of Japan southward as far as Kansŏng and crossed the peninsula once more to Seoul. After a stay of only 2 days, the fourth trip, devoted to Chejudo (1400 km), began. I traveled by train from Seoul to Mokp'o and by ship to Cheju, the island's northern port. From there we climbed the extinct central volcano Hallasan and made a round trip of 210 km around the island with a number of side trips by foot.

I traveled a total of 15,000 km in Korea, 8900 of them in an old Ford car, 1600 by foot and 4500 by train or boat. When traveling by foot, at least if the trip lasted for several days, I employed a caravan of porters in the southern part of the country and a pony caravan in the north. The trips by foot were often extremely difficult. Since base level is not far, the slopes are frequently very steep, even if the altitude is not very great, and the virgin forest and secondary scrub formations are often completely devoid of paths. Most of the peaks I climbed had not been climbed by anyone since the Japanese survey 15 to 20 years ago. On many of them I was the first European. Climbing mountains even of an altitude of less than 1000 m sometimes required 10 to 12 hours of uninterrupted climbing and crawling. The trips in the old car were also beset with difficulties. The roads often did not allow speeds above 10 to 20 km/h, and frequently they had to be prepared for passage with hard work. More than once we barely avoided fatal plunges into infathomable depths. Especially difficult were the river crossings, because the monsoon floods destroy the bridges.

As Fig. 11 indicates, it was my endeavor to gain as complete an insight as possible into the entire country. If all of my routes were arranged in the form of a

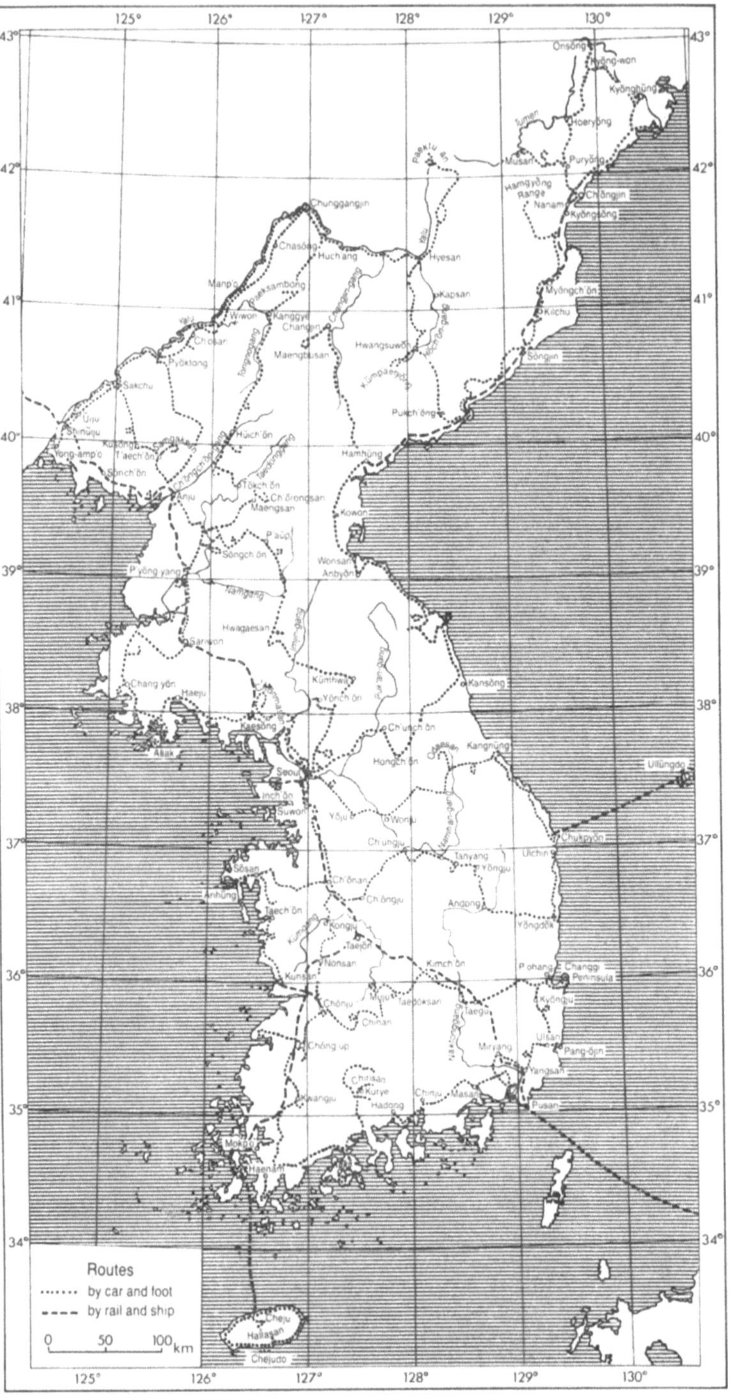

Fig. 11. Routes traveled by the author

square grid, the width of a mesh, and thus the distance between adjacent routes, would be 30 km.

On these trips I collected more than 1100 plants that are of importance for the physiognomy of the country, and Director T. Kaburaki was so kind as to have them identified in the Forest Experiment Station immediately upon arrival, enabling me to return home with completely finished lists of flora. In addition, I brought home not only a large number of photographs, but also a fine collection of rocks and soil samples, most of the official topographical and geological maps and hundreds of Japanese books and brochures.

I believe, therefore, that I can attempt to write a geography of Korea according to modern methodological principles. In a number of articles I have already summarized the main results of my trip [231–233], edited a review of the geographical literature [A], investigated certain problems pertaining to the whole country [14, 100, 292, 398, 559, 583, 595, 601, 618, 697] and written monographs on some of the most important areas of Korea [775, 778, 843, 876, 883].

PART TWO:

NATURE AND TRADITIONAL CULTURE

1 Geological Structure and Surface Forms

In its geological structure Korea is predominantly a part of the East Asian continent that has few features in common with the island arcs festooned along the coast of the continent. The prevalence of crystalline schists and granites, which together cover much more than half of its surface, and the almost complete lack of active volcanoes link it with southeastern Manchuria and Shandong, but distinguish it from the Japanese islands. There are close relationships between northern Korea and southeastern Manchuria in the trend of the crystalline schists, the stratigraphy of the surface rocks and the significance of recent basalt flows. The stratigraphy of the extreme northeast, which is encircled by the Tumen, is related to that of eastern Manchuria and the southern Ussuri region. The Hwanghae peninsula is correspondingly related in part to the Liaodong peninsula, and in part to Shandong. Only the southeastern portion of the country, especially the Naktonggang basin, which consists primarily of Cretaceous rocks, went through the same development as the Tsushima islands, northwestern Kyūshū and western Chūgoku in this period. More closely related to the Japanese islands are Ullŭngdo and Chejudo, the Korean islands that lie closest to Japan.

T. Kobayashi recently presented an extremely useful English language survey of the tectonics and structure of the Japanese empire [327] and had previously written several important articles on the geology of Korea [319–324]. He divides Korea into a series of main stratigraphical structural regions, partly on the basis of earlier studies by Kotō [312, 352] and Yamanari [885]. These regions are as follows, with slight modifications made by me (Fig. 12): (1) Tumen region; (2) Kilchu–Myŏngch'ŏn region; (3) Eastern Kaema and Paektusan region; (4) Western Kaema and P'yŏngbuk massif; (5) P'yŏng–won massif; (6) P'yŏngnam fold belt; (7) Kyŏnggi massif; (8) Okch'ŏn fold belt; (9) Yŏngnam massif; (10) Tsushima basin; (11) Chejudo and Ullŭngdo.

Because of the wide distribution of lignite–bearing Tertiary strata and Carbo–Permian formations in which sandstone, shale and marine limestone predominate, the Tumen region occupies a special position relating it to eastern Manchuria, as we have pointed out. This structural region forms the northern flank of the hill country of Hamgyŏngbuk–do (cf. p. 7). The southern flank of this area, a peninsula jutting out into the Sea of Japan between Ŏdaejin and Musudan, comprises the structural region of Kilchu–Myŏngch'ŏn. This most likely represents the extreme southwestern extension of the Alpidic orogenic belt of northeastern Asia. It is remarkable for its thick and varied Tertiary strata, which distinguish it from the eastern Kaema region adjoining it to the west. It is a piece of the mainland that remained standing when the Tumen bay collapsed (p. 7). Prior to this the Sea of Japan had also inundated it and deposited the strata of the Myŏngch'ŏn group. The Kilchu–Myŏngch'ŏn graben runs north–northeast to south–southwest through the middle of it, with the Ch'ilbosan horst form-

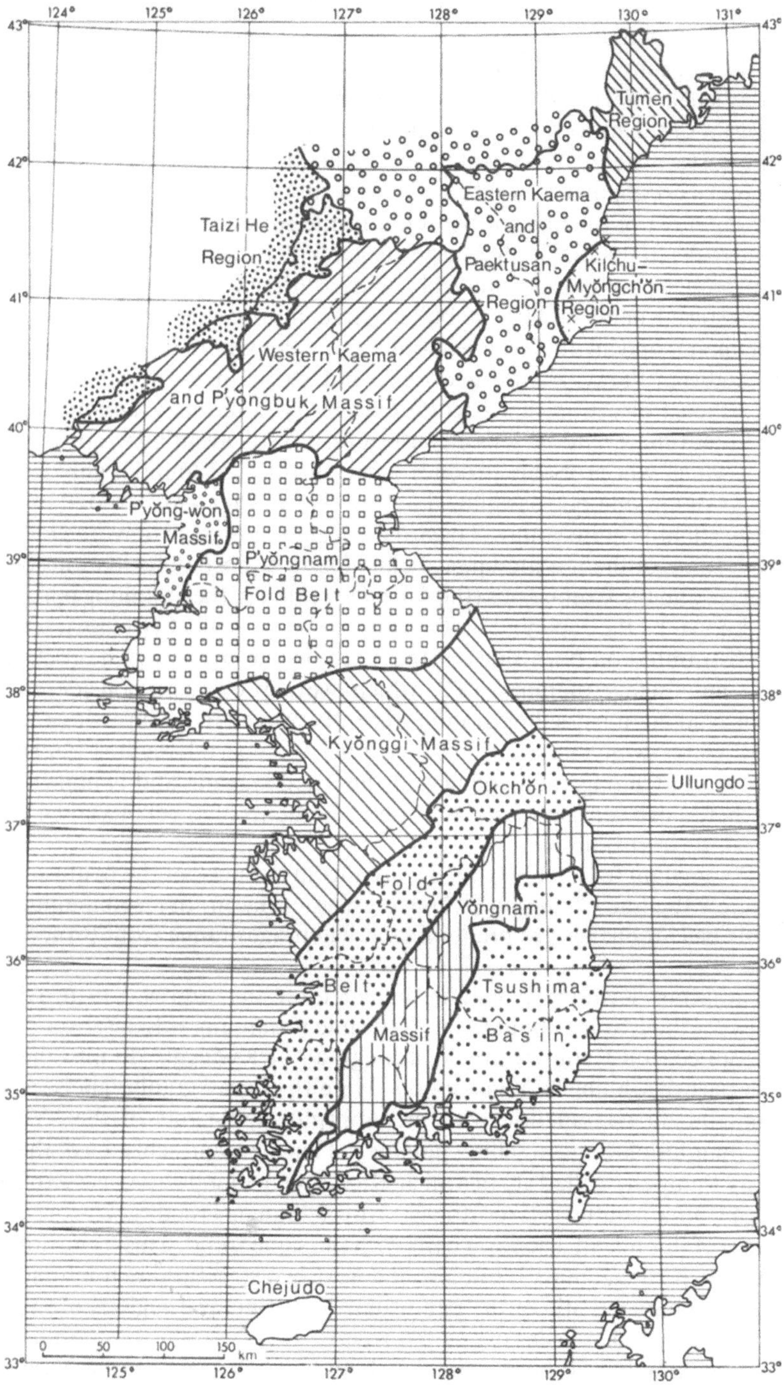

Fig. 12. Korea's main stratigraphical structural regions (primarily after T. Kobayashi)

ing its seaward end. On the basis of the precise geological surveys by I. Tateiwa [744] and F. Yamanari [743], the age of the graben can be dated to approximately the Early Miocene. The Tumen bay presumably is of the same age. The present–day surface forms of the Kilchu–Myŏngch'ŏn region are not, however, the direct result of the process of graben and horst formation.

In the eastern Kaema and Paektusan region late Tertiary and Quaternary extrusive rocks predominate. Acidic and basic eruptions alternated many times and covered up most of the ancient bedrock. In the southern part of this region Archeozoic rocks of the Mach'ŏllyŏng group are widely distributed. Here metamorphic limestones play an important role, along with mica schist and amphibolite.

The western Kaema and P'yŏngbuk massif comprises the western half of Hamgyŏngnam–do and all of P'yŏng–anbuk–do. It consists of tremendous expanses of Precambrian "gray granite gneisses" that strike in the west–southwest to east–northeast direction that Richthofen observed on the Liaodong peninsula (Kotō: "Liaotung Trend"). This gigantic massif is still awaiting a thorough investigation, which will probably produce some surprises. I found highly metamorphic rocks of varying petrographical nature there. They are intruded by granite that is presumably quite young. Toward the Yalu this massif is covered by patches of strata, some of them quite large, belonging to almost all formations between the Silurian and the Cretaceous (the Taizi He region in the adjoining part of Manchuria, known by the Japanese as Taitzuhō). The so–called Ŭiju series, which is widely distributed on both sides of the Yalu, is one of them. The rocks of this series, some highly metamorphic, which were previously considered to be Paleozoic and are intruded by granite, recently proved to be Jurassic (325, 785).

The P'yŏng–won massif, which occupies the coastal regions between the mouths of the Ch'ŏngch'ŏn–gang and the Taedonggang and is named after a county there, also consists of metamorphic sediments of as yet unknown age with small granitic intrusions.

The P'yŏngnam fold belt covers the area from the neck of the peninsula to approximately the line Changjŏn (Kŭmgangsan) – Haeju. In it sedimentary rocks of mainly calcareous nature predominate. Their ages range from Proterozoic to Triassic.

The adjacent Kyŏnggi massif is named after the capital city province. It stretches southeastward about as far as the line Kangnŭng – Kunsan and thus extends from central Kang–won–do through all of Kyŏnggi–do into the northwestern part of Ch'ungch'ŏngnam–do. Here the "gray granite gneisses" again predominate, in conjunction with metamorphic sediments of the "Yŏnch'ŏn group" (amphibolite, mica schist, phyllite). The gneis and schist have the Sinitic trend, as Pumpelly called it, of which Richthofen found immense expanses in southern China (southwest–northeast; Kotō: "Sinian Trend"). The great granitic masses of this massif also form belts with the same trend. This trend dominates not only the Kyŏnggi massif, but also the next two structural regions and is thus particularly characteristic of the peninsula. The structural regions of the peninsula are arranged in belts following this trend.

The approximately 50–km–wide Okch'ŏn fold belt again consists primarily of nonmetamorphic sediments from the Cambrian to Cretaceous periods. It is named after the county of Okch'ŏn and is bounded to the southeast by the line T'aebaeksan – Chindo. Thus it extends from southern Kang–won–do through

Ch'ungch'ŏngbuk–do and the western half of the two Chŏlla provinces to the southwestern tip of the peninsula.

The Yŏngnam massif has the same features as the Kyŏnggi massif. It ends in the southeast at a lobular line, where it is covered by the Cretaceous sediments of the Tsushima basin. This line runs basically southwest from Ulchin to the mouth of the Sŏmjin–gang and thus approximately follows the boundary between the provinces of Chŏlla–do and Ch'ungch'ŏngbuk–do on the one side and Kyŏngsang–do on the other side.

The eastern two–thirds of the Kyŏngsang provinces consist of the Korean portions of the Tsushima basin. As indicated above, in the Cretaceous period this basin and the western end of the Japanese islands formed a stratigraphical region with the Tsushima islands in its center ([327, p. 371], [402]). The rock formations of this period, which are interspersed with "young" granites, give this depression a uniform character. The eastern margin is characterized by Tertiary rocks extending into the Late Neogene (Yŏnil series).

Chejudo and Ullŭngdo, finally, consist almost exclusively of very young volcanic rocks.

By far the greatest part of Korea is characterized by the alternation of three massifs and two fold belts. Kobayashi characterizes the relationship between them as follows [324, p. 589]: "The two negative elements, the Heinan (P'yŏngnam) between the positive Heihoku (P'yŏngbuk) and Keiki (Kyŏnggi) areas and the Yokusen (Okch'ŏn) between the Keiki (Kyŏnggi) and Reinan (Yŏngnam) positive tracts took the rôle of geosynclines. In these geosynclines or troughs thick sedimentary strata, extending from the Proterozoic to the Ordovician and from the Carboniferous to the Triassic were laid down, while the positive Keiki (Kyŏnggi) and Reinan (Yŏngnam) lands, together with most part of the Heihoku (P'yŏngbuk) region, remained practically without sediments on the Archean metamorphics, but being intruded to a large extent by the neo–granite. The few patches of Jurassic sediments deposited on the positive areas have escaped folding of both the Liaotung and the Sinian trends."

The preceding structural details already include some of the essential features of the stratigraphy. For a better understanding of the detailed morphological descriptions in Part Three, the appended Table of Geological Formations (Table 46) gives a complete list of the stratigraphical formations. This table shows the following facts, which were determined for the most part by the indefatigable I. Tateiwa [318, 744, 750, 794, 886].

The basic foundation upon which all of Korea rests is the "basement complex", an extraordinarily thick mass of gray gneiss, likewise ancient "Kogurian" granite and correspondingly thick stratigraphical sequences of highly metamorphic sediments. This basement is exposed primarily in the massifs and is truncated by the post–Kogurian erosional plane. It is overlain by the masses of mainly limestone and shale of the Nangnang complex, which is divided into the Proterozoic Sang–won system and the Lower Paleozoic Chosŏn system. These, like the younger sediments, were deposited primarily in the fold belts, which accordingly must also have begun to develop at a very early period in time. Only very recently have strata from the period between the Late Ordovician and the Early Carboniferous been discovered in Korea, and they are too limited in extent to be of geographical significance. Almost everywhere the sandstone, shale and anthracite of the P'yŏng–an system lie directly on top of the strata of the Chosŏn system. They range in age from the Middle Carboniferous to the Middle Triassic. The subsequent Late Triassic "Songnim disturbance" created the basins in which the thick conglomerates, sandstones and coal–bearing shales of the Lower and Middle Jurassic Taedong system were deposited.

The Late Jurassic was the time of the great Taebo orogeny. It was the most important tectonic upheaval after the archaic folding and corresponds to the Yenshanian orogeny in northeastern China (363) and the Hida phase of the Sakawa orogeny on the Japanese islands (327). In the massifs the Taedong sediments were left almost undisturbed by the orogeny, but the thick sediments in the two geosynclines underwent intense folding and faulting. The folds of the P'yŏngnam belt followed the Liaodong trend, and those of the Okch'ŏn belt the Sinian trend. In the former folding predominates, leading in places to pronounced overthrusts. The structure of the P'yŏngnam fold belt is shown schematically in Fig. 13, after Kobayashi [327, p. 517]. In the latter the pressure from the northwest produced an imbricate structure that was investigated particularly by I. Yoshimura in the Yŏng–wol district [846, 847]. The individual folds are highly compressed and crumpled. The typical pattern of this structure is shown in Fig. 14 after I. Yoshimura. The Taebo folding altered the strata in the geosyncline dynamometamorphically in places. This alteration does not, however, by any means have the same extent or intensity as in the mountain ranges of Alpidic folding.

With the Taebo orogeny the principal features of the stratigraphical sequence and the structure of most of Korea are completed. Large areas of intensive sedimentation occurring after this period are found only in the Tsushima basin and the southwestern half of the Okch'ŏn fold belt. In these two regions the extremely thick strata of the Cretaceous Kyŏngsang system, also known as the upper Taedong system, were deposited. In the lower Naktong series sandstone, conglomerates and shale, which is bituminous in places, predominate. The Middle Shilla series is dominated by a thick succession of eruptive rocks. Porphyritic flood basalts and their associated tuffs are especially widely distributed. The upper series of the Kyŏngsang system, which is named after the famous temple of Pulguksa near Kyŏngju, consists almost exclusively of eruptive rocks. Along with very heterogeneous effusive rocks, among which quartz porphyry is particularly prominent, this series also contains intrusive rocks, including diorite and above

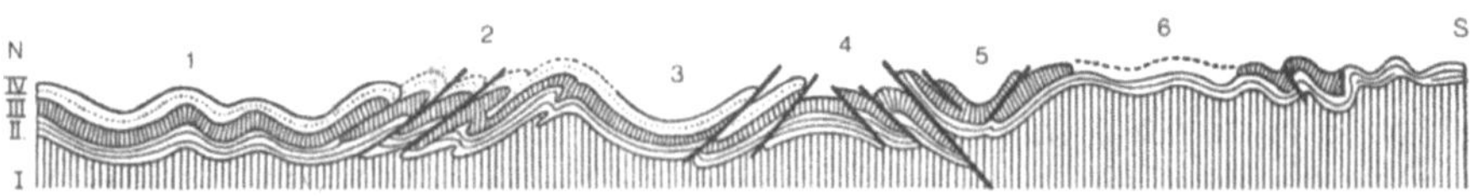

Fig. 13. Diagrammatic north–south section through the P'yŏngnam fold belt after the Taebo orogeny (after T. Kobayashi). **I** Archaic basement complex; **II** Sang-won system; **III** Chosŏn system; **IV** P'yŏngnam system; *1* Northern P'yŏng-annam-do coalfield; *2* Sŏngch'ŏn area; *3* Southern P'yŏng-annam-do coalfield; *4* Chunghwa–Sang-won area; *5* Kwangju syncline; *6* Chŏngbangsan area

all "young" granite. These "young" granites permeate not only the strata of the Kyŏngsang system, but all of the massifs and fold belts of Korea and dominate much larger areas than the ancient Kogurian granites do. Therefore the ratio of the total area with granite to the area of the individual provinces is very high. It is highest in Ch'ungch'ŏngnam–do (35.0%), Chŏllabuk–do (32.7%), Hamgyŏngbuk–do (32.3%) and Kang–won–do (30.8%). The intrusion of "young" granites presupposes that in all structural regions of Korea there were still very thick strata during the Late Cretaceous period, which have subsequently eroded away. This erosion probably occurred primarily during the Eocene.

The interior of Korea is free of Tertiary deposits. These are distributed only along the margins and are mainly terrigenous Paleogene and Miocene formations, some of which are extremely thick. They consist primarily of sandstones, conglomerates and shale, sometimes containing lignite. Most lie near the coast of the Sea of Japan. On the Yellow Sea coast the only deposits are those of Shinanju and Pongsan. The Lower Miocene deposits in the Pongsan coalfields east of Sariwon in Hwanghae–do are partially of marine origin and prove that at this time the peninsula was inundated from the east by the Sea of Japan. Once during the Oligocene period and a second time during the Late Tertiary and the Pleistocene large masses of volcanic material of extremely varied petrographical nature were extruded in the vicinity of the Kilchu–Myŏngch'ŏn graben and on and around the Changgi peninsula. The last–named area also contains a 600–m–thick Neogene sedimentary deposit, the above mentioned Yŏnil group.

Korea is rich in mineral deposits, some of which are of considerable economic significance. Mining will be discussed in Chapter 7 of Part Four. In the present context we shall only be concerned with a brief summary of the geological conditions of the deposits, based primarily on the surveys by Inouye [310] and Kawasaki [315].

Gold deposits are found in quartz veins, contact zones and in alluvial deposits. Gold–bearing quartz veins are found predominantly in the archaic "gray gneisses". They are the most important deposits. Everywhere these rocks are rich in sulfurous minerals, of which pyrite and arsenopyrite are very often auriferous.

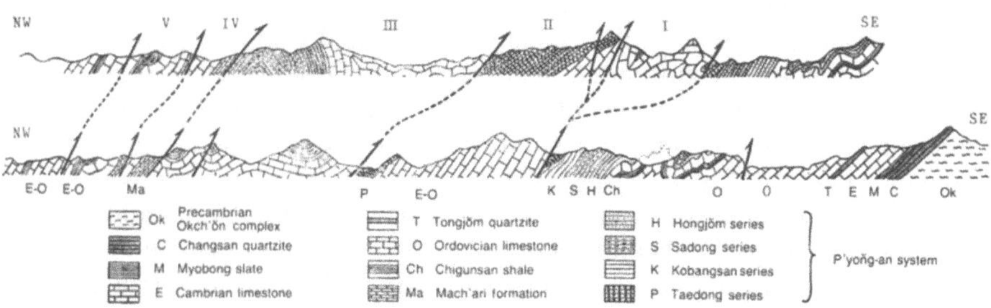

Fig. 14. Imbricate structure of the Okch'ŏn fold belt in the Yŏng-wol district (after I. Yoshimura). **I** Kongsuwon shingle; **II** Tŏkp'ori shingle; **III** Mach'ari shingle; **IV** Mohari shingle; **V** Namaeri shingle

The contact deposits are found in the limestones of the P'yŏngnam fold belt, specifically where they are intruded by granite. They either have infiltrated into interstices or are deposited in irregular masses in fissures. Placer gold deposits are generally found in the river gravel in the vicinity of the primary deposits, often not far from their base. At times the gold of the primary deposits is concentrated in lower layers of the residual soil even without fluvial transport. As examples of the three types of gold deposits we may mention the mines of Unsan (Pukchin, P'yŏng–anbuk–do), Suan (Holgol, 60 km east–southeast of P'yŏng–yang) and Kŭmsŏng (Tanggogae, near Hoeyang in the headwater region of the Puk'an–gang), which was operated by Germans from 1897 to 1901.

Iron ores are found in the form of magnetite, hematite and limonite. Titaniferous magnetite occurs in or at the margin of basic eruptive rocks; nontitaniferous magnetite, in ancient metamorphic rocks as precipitated sediments. Such magnetite–bearing quartz schists are found in the districts of Ch'ungju (Ch'ungch'ŏngbuk–do), Musan (Hamgyŏngbuk–do) and Tanch'ŏn (coast of Hamgyŏngnam–do), among others. Hematite and limonite deposits occur primarily as lenses or layers in the "Great Limestone Series" of the Chosŏn system.

Copper deposits occur in three forms: as metalliferous veins, as contact deposits and as metasomatic deposits. The most important deposit, that of Kapsan in Hamgyŏngbuk–do, belongs to the third group. It is intercalated with limestone and consists of extensive, irregular masses of sulfide that replace the limestone along the bedding planes. An example of the first group is the deposit of Ch'ang–won (near Masan in Kyŏngsangnam–do), which is superjacent to a layer of green porphyry. The two mines of Tongjŏm on the Chasŏnggang (P'yŏng–anbuk–do) belong to the second group. They lie in the contact between the limestone and shale of the Chosŏn series and granite.

Recently nickel ores have also been found and mined in Korea. According to Kato [343] they are of two different types. In the mine at Kŭmsŏng (northern Kang–won–do) the ore lies in a dyke of basic eruptive material that penetrates the metamorphic sediments of the Sang–won system. The other type is represented by the mine at Unsong–ri in the district of Tanch'ŏn (Hamgyŏngnam–do). Here a hydrothermal solution originating from the underlying, solidifying basic magma metasomatically displaced the metamorphic dolomitic limestone of the Mach'ŏllyŏng group, forming the ore.

Graphite occurs in two forms, as crystalline flakes and in an amorphous form. The former is found in veins or seams, or finely distributed in the "gray gneisses"; the latter in ancient metamorphic sediments or in the P'yŏng–an strata.

The most important anthracite deposits are found embedded in shale and sandstone in the upper horizons of the Carbo–Permian Sadong series and the lower horizons of the Permo–Triassic Kobangsan series of the P'yŏng–an system in the vicinity of P'yŏng–yang. The Taedong system also contains some anthracite seams. Generally the anthracite is of a powdery consistency and must be made into briquettes for heating purposes. Here and there hard lump coal is found. The total deposits of anthracite and bituminous coal in Korea are estimated at 1 billion tons.

With the exception of the above mentioned Pongsan coalfield and that of Shinanju, lignite deposits are limited to Tertiary deposits close to the Sea of Japan. The most important are near Hoeryŏng (Hamgyŏngbuk–do), in the Kilchu–Myŏngch'ŏn graben, near Kowon (north of Wonsan), T'ongch'ŏn (northern Kang–won–do) and on the Changgi peninsula. The dark–colored lignite usually contains a lot of water and is very friable. In zones of contact with intrusive basalt it is of better quality. The thickness of the individual strata seldom appears to exceed 2 m.

Other exploitable minerals that occur in Korea are silver, zinc, lead, bismuth, tungsten and molybdenum ores and kaolin, alunite, fluorite, barite and mica deposits.

On the basis of Kato's observations [339] three major metallogenic epochs can be identified in Korea. The first is the Precambrian, during which gold quartz veins and sedimentary iron ore deposits developed. The second is during the Paleozoic Chosŏn period. Its "Great Limestone Series" contains hematite and limonite deposits. The third is Cretaceous in age. During this period iron, nickel and copper ores developed in connection with effusions of porphyry and the formation of dykes of basic material, on the one hand, and gold and copper contact deposits in the periphery of the "young" granitic intrusions, on the other hand. The main orogenic phase, the Taebo orogeny, is not a major metallogenic epoch, because it evidently was not connected with intrusions.

The time has not yet come for a division of Korea into levels of deposits, such as I attempted for Portugal (Pet. Mitt. Suppl. 213, 1932, p. 35–37), since most of the ore deposits cannot as yet be classified according to Niggli's system. The results would presumably not be as simple as in Portugal, since we have here not one, but several metallogenic epochs, which certainly lead to a complicated interference pattern. E. Ahnert's study of the distribution of ore deposits in southern Manchuria also shows that the results are anything but simple [344].

B. Kotō, to whom much credit is due for his research on the structure and stratigraphy of Korea (p. 54), also made a preliminary attempt to interpret the surface forms of the country [352]. Out of the confusing maze of mountains he picks out a large number of ridges, which he combines into systems on the basis of their trend, and gives them Korean names. The T'aebaek Range runs north–northwest to south–southeast and consists of three main ridges and three secondary ones. The Sobaek Range forms a fan of ridges that trend basically southwest. The T'aebaek and the Sobaek Range combined form the Korean system. In northern Korea he finds a Liaodong range, whose three main ridges trend in the well–known Liaodong direction. In the eastern part of continental Korea he recognizes the ridge of the "little Changbaeksan". He divides the P'yŏng–an structural region into a grid of ridges that cross each other at approximately right angles. From Wonsan the Ch'ugaryŏng graben supposedly runs south–southwest and, according to Kotō, it widens out like a funnel in this direction, including at the west coast the entire mouth of the Han–gang as far as Kunsan. In the Ch'ungch'ŏng and Cholla provinces he recognizes two axes of the Sinian fold system, which he calls the Ch'aryŏng and Noryŏng ridges. In the extreme south, finally, he traces the east–west ridges of the Hansan system. The southernmost of these extends along the peninsulas and islands on the Korea Strait, which are bent in the shape of a hammer.

For him the two axes of the Sinian system are anticlinal ridges that were lowered by erosion. The other ridges he considers to be the crests of tilt blocks, which were separated from each other by faults. He does not demonstrate these faults geologically, but deduces them from the escarpments that these ridges supposedly have. He extends these orographical escarpments into the crust of the earth and simply considers them to be faults. According to the direction in which these escarpments face, he evaluates the movements that supposedly took place along these faults. The question whether or to what extent the ridges were formed or modified by subaerial erosion is not mentioned in a single line.

Immediately after the appearance of Kotō's book in 1903, Pervinquière [355] and Lorenz [356] protested against this undifferentiated presentation of the orography of Korea and its unproven and equally undiscriminating interpretation, and in a later publication [312] Kotō energetically replied. The objections of Pervinquière and Lorenz were disregarded. In the next 3 decades no geomorphological research was carried out in Korea and after that only very little. Consequently even today in East Asia Kotō's results are often considered to be the key to the understanding of Korean morphology, as shown, e.g., by Nihon Chiri Fuzōku Taikei [299, Vol. 16, p. 26, and the color diagram in Vol. 17, p. 88/89]. Meanwhile, geological research in Korea has revealed faults here and there that strike in a direction corresponding to Kotō's concepts. In the individual issues of the Geological Atlas of Chōsen these concepts are referred to in each case, but without stressing that the faults are not manifest in the present morphology in any of these cases. This is true, for instance, of the faults that strike in the direction of the T'aebaek Range in southern Kang–won–do, causing bands of Shilla strata to lie next to granite, and are truncated by the surface [888]. The geomorphological research initiated in the last 10 years in Korea by T. Kobayashi and I. Tateiwa is tacitly based on quite different principles and of Kotō's concepts it retains only the few real facts, which were emphasized simultaneously by Richthofen [351] at the beginning of the century. Only Kotō's deplorable Ch'ugaryŏng graben is considered by Kobayashi to be a reality even in his most recent book of 1941 [327].

In Part One I outlined the main morphological features (p. 4ff). Our primary purpose here will be to explain them. As Kobayashi has stressed several times [324, 327, 380, 383], the main body of Korea was so highly reinforced by the Taebo orogeny and the intrusions of "young" granite that the more recent uplift that formed the main features of the present–day relief is unrelated to the ancient structure. This is particularly true of the actual peninsula, where the ancient structural lines follow the Liadong and Sinian trend, while the Main Korean Range and with it the coast of the Sea of Japan, indeed the entire body of the peninsula, trends north–northwest to south–southeast. The main features of the present–day relief are thus unrelated to the structure, and the coastal outlines cut through the old structural regions, inconsistently carving out pieces that remained dry land, as Fig. 12 clearly shows. Richthofen already recognized this [351], and Motizuki recently termed mountain arcs of the type represented by the Main Korean Range discordant arcs [395].

In the publications we have just quoted, Kobayashi considers the initial basis of the recent morphological development of the peninsula to have been a peneplain that must have developed prior to the Early Miocene transgression to which the Pongsan coalfield attests. In fact, Tateiwa [384] discovered level erosional planes forming the base of both the Ch'ilbosan granite porphyry in the Kilch–Myŏngch'ŏn region and the Neogene sediments around the Changgi peninsula. Based on the age of the overlying strata, these planes must be dated as

Miocene. In the T'aebaek Range Kobayashi found erosion surfaces at 800–1200 m in the region where the main divide lies, with peaks such as Odaesan (1563 m) towering as much as 650 m above them (Fig. 15). He names this peneplain the Yukpaeksan plain after a mountain in the southern part of this range, whose summit consists of a remnant of such a surface. He traces remnants of this plain throughout the entire peninsula and determines that today it is warped upward in the form of an asymmetric "geanticline". Thus this uplift occurred after the Miocene. Contrasting with this geanticline is the geosyncline of the Sea of Japan. According to Kobayashi the Kaema upland was also asymmetrically uplifted in the same manner. Whereas the axis of warping of the Yukpaeksan Plain coincides with the crest line of the Main Korean Range and thus trends north–northwest to south–southeast, the axis of warping of the Kaema upland runs east–northeast.

Kobayashi also traces an extensive peneplain in the western half of the peninsula. This he studied particularly in the vicinity of Ch'ungju, Changhowon, Yŏju, Ich'ŏn and on the lower reaches of the Taedonggang. In this last area Nakamura [361] had already identified this peneplain and named it the Nangnang plain after the field of tumuli south of P'yŏng–yang (p. 22ff). At the bottom of the basin near the Yellow Sea coast this plain is very wide. It extends along the main river valleys into the interior of the peninsula, accompanied by isolated monadnocks and narrowing down as it proceeds. In the same direction it increases considerably in height, showing that the geanticline continued to form after it had developed. Corresponding to the Nangnang plain on the Yellow Sea coast is the narrow Yŏdo plain along the Sea of Japan. Between the Nangnang plain and the Yukpaeksan plain and along the Han–gang and other rivers, Kobayashi traces three levels of terraces. The present–day river valleys with their broad alluvial plains and sharply undercut banks are entrenched into the Nangnang plain. According to Kobayashi the Korean peninsula is thus an asymmetrical, large–scale fold, which has been changed by erosion into a two–step and two–sided peneplain "staircase" (Fig. 15).

There can be no doubt that the Nangnang plain actually exists. As the recent geological surveys have shown, in the P'yŏng–an fold belt south of P'yŏng–yang

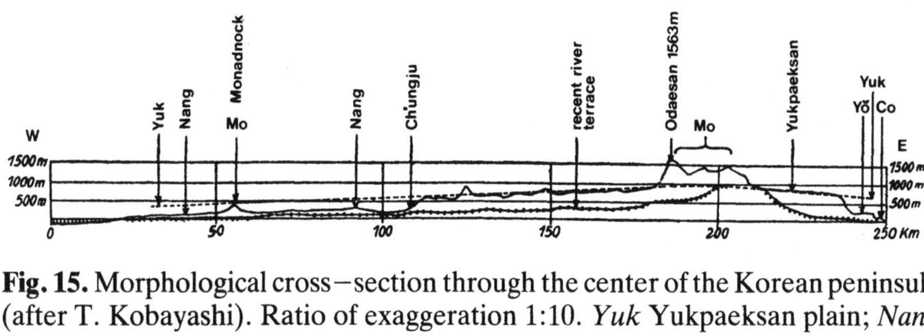

Fig. 15. Morphological cross–section through the center of the Korean peninsula (after T. Kobayashi). Ratio of exaggeration 1:10. *Yuk* Yukpaeksan plain; *Nang* Nangnang plain; *Yŏ* Yŏdo plain; *Co* coastal terraces; *Mo* monadnocks above the Yukpaeksan and Nangnang plains. *Dotted lines*: longitudinal profile of the present-day westward flowing rivers: Han-gang

it truncates a wildly agitated structure and is covered by a thick layer of terra rossa. It is also true that monadnocks rise very abruptly from the Nangnang plain, for reasons that we shall explain later. The erosion surfaces lying above the Nangnang plain are, however, often quite small and vary greatly in altitude. Without knowing Kobayashi's investigations, which were not published in English until after my stay in Korea, I myself observed several such surfaces at much higher altitudes even than Kobayashi's Yukpaeksan plain (e.g., the summit plateau of Haramsan at the border between P'yŏng–annam–do and Hwanghae–do at 1450 m). That a number of them combine to form a continuous, asymmetrically warped surface of the type represented by the Yukpaeksan plain is a matter that has not yet been definitely proven, as Tateiwa [384] points out. Kobayashi's principle is, however, certainly correct. The peninsula was uplifted in the form of an asymmetrical geanticline or of a tilt block with a downwarped eastern side. As the veins of Pulguksa granite porphyry trending in the T'aebaeksan direction in southern Kang–won–do demonstrate, this uplifting began during the Late Cretaceous period. It is possible, though, that the present–day relief actually did originate in an Early Miocene peneplain that was briefly inundated by the Sea of Japan prior to the uplift, so that the torso mountains rising above it became islands. The subsequent uplift occurred on the steep eastern side as a slight flexural fold that appears to merge into step faults in places. Thus the rivers on the eastern side are short and steep and are shifting the continental divide westward by headward erosion at the expense of the rivers flowing toward the Yellow Sea.

It is conceivable that the Kaema upland developed in the same manner, except that the differences in the relief must have already been much more pronounced during the Miocene period. Nor is anything known of flooding by the Miocene seas there. Thus the main characteristics of Korea's relief have developed essentially as Richthofen assumed [351]. Only we must substitute large–scale folding for his step fault arcs with stretching faults. Richthofen himself uses the term "swelling" several times in the context of the Korean coastal arc [351, Vol. II, p. 10f]. As emphasized earlier (p. 5), we must also abandon his concept of the overlapping Tungusic arc (Kaema upland).

In the Early Miocene the northwestern and western coast of the Sea of Japan already lay approximately where it does today. The obviously well–founded maps of Otuka clearly show this (reproduced in [327, p. 453]). If Kobayashi's concepts of geanticline and geosyncline, probably with corresponding subcrustal flows of sima [cf. 371, 387], are correct, the Sea of Japan must not have reached its quite considerable present depth of 2000 m close to the Korean coast and 3712 m in its northeastern portion until after the Early Miocene. Off the coast of Hamgyŏngbuk–do near Kyŏngsŏng there is a drowned valley branching toward the land that was first pointed out by Yabe and Tayama [369, 386]. As Fig. 16 shows, it is not a shelf channel, but a groove in the edge of the shelf cutting into the continental slope. Without going into the tremendous volume of recent literature on "submarine valleys", we can at least say that in this case it is most likely a subaerial formation that has moved from the region of the geanticline to that of the geosyncline due to a landward shift in the apsis of the fold line. It is not yet known whether the submarine valley of Kyŏngsŏng is an isolated phenomenon. On the maps mentioned above [cf. also 376], Y. Otuka shows the Pliocene coastline much farther eastward than the Miocene or the present–day coastline. I too have conjectured that Ullŭngdo must have been connected to the continent again at the end of the Tertiary, on the basis of the vegetation [883, p. 192]. An indication of this is the fact that the closest relatives of the beech *Fagus multinervis* Nakai, which is indigenous to Ullŭngdo, are to be found in the Neogene flora of southern Korea and Japan. From the highly endemic nature of the flora of this island we can conjecture that after that period it was never again connected to the mainland.

The axis of the large–scale peninsular fold, and with it the coast of the Sea of Japan, takes the form of an arc that is convex toward the east, as is characteristic of the marginal East Asian mountain ranges. The projection of this "Korean coastal arc" probably extends across the Tsushima and Gōtō islands, or through western Kyūshū, and across the Ryūkyū islands. In the center of this arc lies the East China Sea (Tunghai = Dong Hai). For this reason Japanese geologists speak of a Peri–Tunghai arc (Motizuki [396], Fig. 17). It is likely that horizontal move-

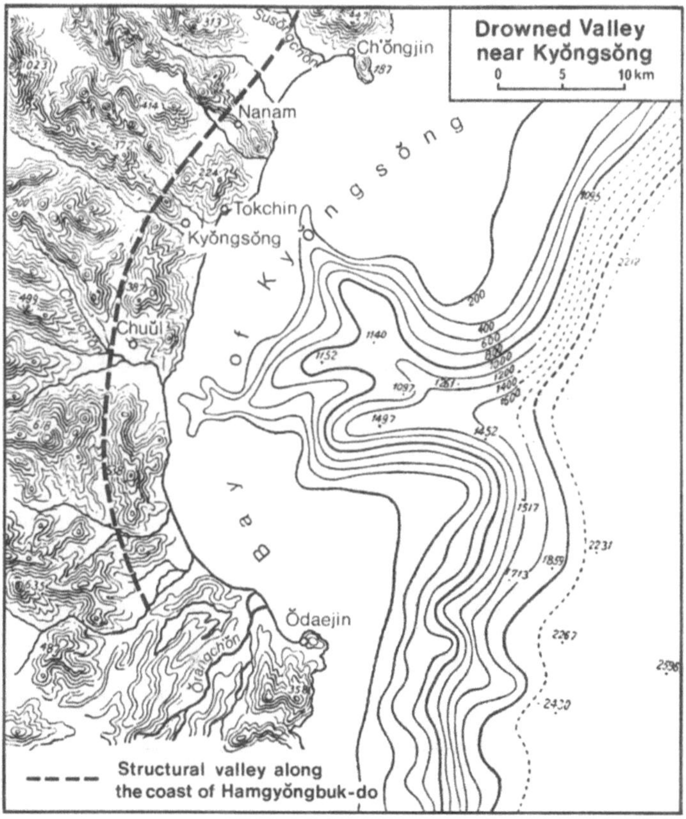

Fig. 16. The drowned valley near Kyŏngsŏng (northeastern Korea) (after Mukoyama, with modifications)

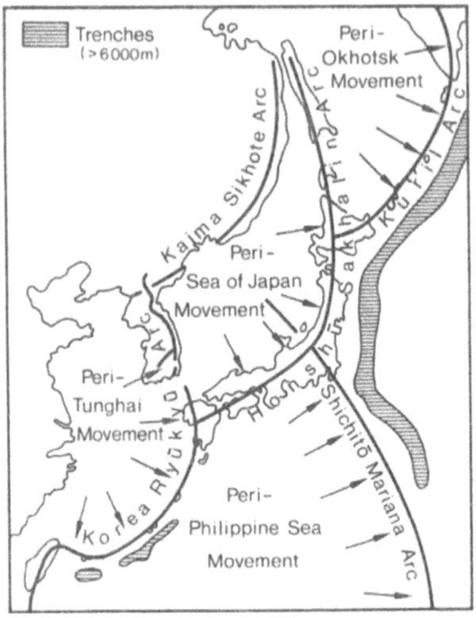

Fig. 17. The marginal mountain arcs of East Asia (after Motizuki and Kobayashi)

ments from the inside outward contributed to the formation of this and the other arcs.

The outer East Asian mountain arcs, especially those that take the forms of island chains, have an echelon pattern that was first investigated by Tokuda [362]. They consist of individual small blocks whose axes run at sharp angles to the curvature of the arc. The blocks are arranged in the form of an echelon, such that each block is shoved slightly forward and to the side of the one behind it, in the direction of the curvature of the arc. This model of Tokuda's, which is represented most clearly by the Kuriles, is reproduced in Fig. 18. As Tokuda demonstrated, the Ryūkyū arc has this echelon pattern (Fig. 19). It also exists on the western edge of Kyūshū. The Gōtō and Tsushima islands are presumably further blocks lying to the back of such an echelon. Both Tokuda and Motizuki [396] trace this echelon type structure into the Korean peninsula. According to my observations, it does actually exist here. As I already emphasized on p. 5, it can be seen particularly clearly in the extreme southeast, in the region of the Tsushima basin. Here linear mountain ranges trend north–northeast to south–southwest, separated by equally linear valleys. Yamanari pointed out this

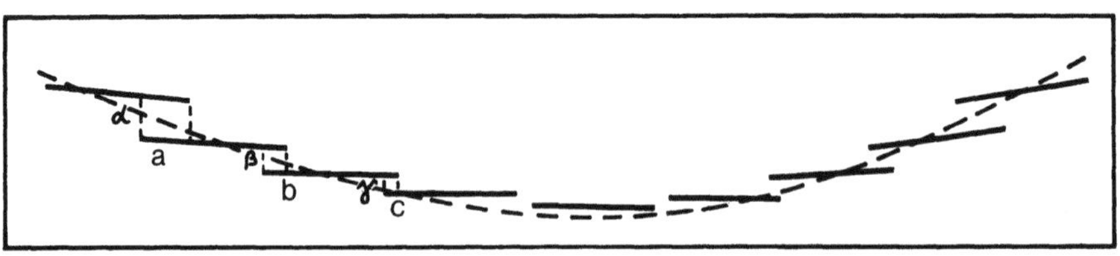

Fig. 18. Model of the echelon pattern of the East Asian island arcs (after S. Tokuda). *Dotted line*: curvature of the arc; *Thick solid lines*: the individual blocks; *a*, *b*, *c*: overlap between blocks; α, β, γ: distance between blocks

structure as early as 1924 [885]. According to Yamanari, whose description I can confirm, this is not a matter of tectonic movements that are directly expressed in the relief. Rather, these are older fault systems along which the rocks were crushed, so that they were later predisposed toward valley formation. To put it briefly: The linear alignments of valleys are structural valleys and not grabens. Perhaps the strange structure of the Sobaek Range is somehow related to this structural trait of the southeastern part of the Korean peninsula. In the T'aebaek Range there are many straight valleys flowing into the Sea of Japan that approach the coast from the south–southwest at an acute angle instead of following the steep eastern slope consequently, and thus at a right angle, to the coast. It is likely that these valleys are also structural valleys belonging to the echelon pattern. Some of these valleys even extend south–southwestward across the continental divide into the drainage basins of the rivers that flow into the Yellow Sea (Fig. 19). This is most conspicuous in the case of the Ch'ugaryŏng furrow, which we

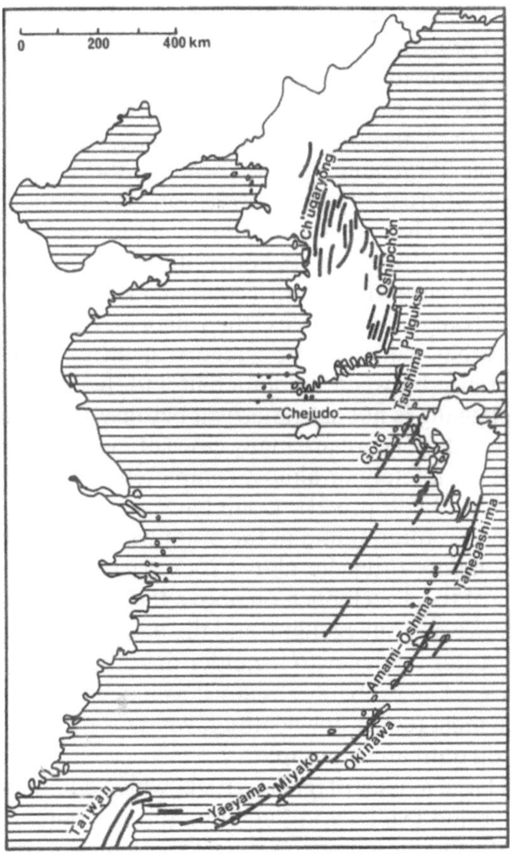

Fig. 19. The echelon pattern of the Peri–Tunghai arc (after S. Tokuda, in Korea after the author)

have mentioned several times. It forms a linear valley crossing the continental divide, which extends from a point east of Wonsan 200 km south–southwest to Ttuksŏm near Seoul. It passes through "gray gneiss", granite, metamorphic sediments and Chosŏn strata. Nowhere did I find recent faults along its edges. Thus it is presumably also a structural valley of the same type as the valleys in the southeast. The northwesternmost of these valleys lie in the drainage basin of the Yonghŭnggang, which flows into Yŏnghŭngman. In the Kaema region there is no echelon pattern. It already belongs to the inland type of East Asian step fault arcs, which also appear to lack this pattern [cf. 397]. When approaching from the northwest, one therefore encounters this echelon pattern, in the form of structural valleys, for the first time on the Korean peninsula. Thus, from the standpoint of tectonics as well, the peninsula is a transition zone between the inland arcs and the coastal arcs.

A striking contrast to the straight structural valleys is seen in the meandering valleys, which are extremely common in Korea. Along the steep consequent rivers flowing into the Sea of Japan valley meanders are lacking, due to the steep gradient. But they are a conspicuous feature of many of the rivers that flow into the Yellow Sea, particularly in their central reaches. That they are not necessarily lacking in the lower reaches of even the larger rivers is shown by the tremendous loop of the Taedonggang just above its mouth. Valley meanders are found in all rock formations in Korea. In granite they are rare, but they are common in basalt flows, shale and in the limestones of the Nangnang complex. In this last mentioned formation the Piryugang and its tributary the P'yŏng–wonch'ŏn form splendid meanders, whereas slightly upstream in granite and gneiss the valley is straight and wide [794]. Here too the amplitude of the bends has the well–known relationship to the size of the rivers. In horizontal strata that tend to break off vertically, particularly in basalt flows with a columnar structure, but also in the horizontal strata of the Nangnang complex, the valley meanders have a symmetrical cross section. Thus they belong to the category of entrenched meanders with no recognizable difference between the undercut and the slipoff slopes. The great loop of the Taedonggang is not asymmetrical because it is completely subject to the influence of the tides. There are many cases of meanders in which the meander neck has been completely or almost completely cut through, in contrast to China (Schmitthenner [379, p. 92]). Completely or almost completely isolated meander spurs are also often found at the point where two meandering streams meet. Solitary meanders often lie above rock barriers (cf., e.g., [905], sheet Namp'o: meander of the Taech'ŏnch'ŏn above the Kaehwari conglomerate). In the Yalu region between Chunggangjin and the mouth of the Yalu the larger rivers have a much greater tendency to meander than the smaller ones. The meanders of the large rivers had probably already developed prior to the last strong uplift that caused the most recent downcutting, whereas the smaller rivers presumably did not develop until after this uplift.

The eruption of large masses of igneous rock, which dominated the Middle and Late Cretaceous period (Shilla and Pulguksa Series) continued throughout the Tertiary and into the Quaternary (cf. Table of Geological Formations, Table 46). The Late Tertiary and Quaternary formations are not only structural components of the earth's crust in Korea. They simultaneously form more or less undisturbed accretive forms that dominate the relief of Korea over vast areas. Japanese geologists are generally of the opinion that the acidic or alkaline extrusive rocks belong to the Neogene, whereas the basaltic rocks are Quaternary in age (e.g., Tateiwa [318]). Such acidic eruptive rocks are found particularly in the mountain ring of Paektusan, on the top of the Mach'ŏllyŏng Ridge between Pukp'ot'aesan (2239 m) and Turyusan (2309 m) and in the projection of the latter on Ullŭngdo. Kawasaki [768] and Tateiwa [318] seek to establish a connection between these occurrences by including the Oki islands off the northwestern coast of Japan and reflect on a series of eruptions along a rectilinear fissure ["Hakuto = (Paektu) Volcanic Chain"]. As I pointed out in another publication [778], I consider the acidic rhyolites and trachytes of Paektusan to be definitely more recent than the basalts of the tremendous plateau surrounding it, which again are more recent than the rhyolites of the Mach'ŏllyŏng Ridge. Acidic and basic extrusions appear to have alternated again and again in Korea since the Neogene. The exact surveys of the few sheets of the geological map that have appeared of the area in question also show this ([762], in particular sheet Shinbokchang).

The recent eruptions have occurred in two forms, as stratovolcanoes and as flood basalts. There are only three of the first type: Paektusan, the volcano on Ullŭngdo and Hallasan on Chejudo. To construct a peri–Korean arc connecting them, as Kobayashi does [327, p. 457, 460], appears to me a bit daring. The lava flow on the ridge between Pukp'ot'aesan and Turyusan consists of rhyolites; otherwise all of the lava flows are basalt. Most belong to several different geological periods. The formation of magnificent columns promotes permeability and thus the preservation of the flows. The flood basalts fill present–day valleys and basins, proving their recent origin. Often they have diverted the rivers in a characteristic manner. Generally, though, the rivers have cut as much as 100 m deep into them again. The gorge of the Yalu between Hyesan and Chunggangjin is entrenched as much as 500 m deep into the basalt plateau of the Paektusan area. This shows that the formation of these flood basalts was not the most recent geomorphological event. The most recent of the four flood basalts in the Tumen region that are distinguished near Musan in the geological survey of Manchuria [758] lies way down in the steep–walled floodplain valley with its surface only 10 to 20 m above the level of the river. Thus the age of the flood basalts varies greatly, and their relationship to the present–day relief is accordingly varied.

Almost the entire length of the Ch'ugaryŏng furrow is filled by basalt flows. Above their broad, level surface the old valley walls rise abruptly; in it the continental divide lies at an altitude of barely 600 m (cf. p. 5). This gives the superficial impression of a grabenlike form, which may have led Kotō to his erroneous opinion [352, Pl. I, Fig. 3].

Korean historical records show that on Chejudo the most recent eruptions of two parasitic cones occurred in 1002 and 1007 A.D. According to Chinese records, Paektusan emitted masses of pumice as recently as 1702. Thus it appears that volcanic activity continued on into historical times. Nevertheless, in Korea it has more or less come to a standstill, in contrast to the Japanese islands, and on the whole Volz' [358] statement is correct that since the Tertiary intensive volcanic activity, along with the formation of step fault arcs, has moved from the interior to the periphery.

Independently of each other, T. Kano [391] and Y. Sasa and K. Tanaka [393] discovered traces of Pleistocene glaciation northwest of the summit of Kwanmobong. Most of the cirques face eastward along a single line below a highly dissected ancient peneplain (Fig. 20). During the last Ice Age the winds that brought the snow were thus evidently also westerly winds. Moraine ridges up to 30 m in height at the open sides of the cirques define the ends of the cirque glaciers and allow us to set the climatic snowline of that period at 2100 m. There is a less well–developed stage of valley glaciers extending farther down the valleys belonging to a snowline that was approximately 100 m lower. A snowline at only 2000 m is an abnormally low altitude for the East Asian continent. In Wutai Shan in northern China, only 3° farther south, the climatic snowline of the last Ice Age lay at approx. 3000 m [399]. It appears that the snowline of the last Ice Age dropped very steeply toward the north in this part of the East Asian continent.

In the northern Hinggan Plaetschke even discovered a cirque indicating a snowline of 1200 m at the most, and I myself saw a cirque on a mountain that was barely 900 m high 10 km northeast of Mogocha on the Amur railway. It would probably be worthwhile to look for cirques on other peaks in northern Korea that

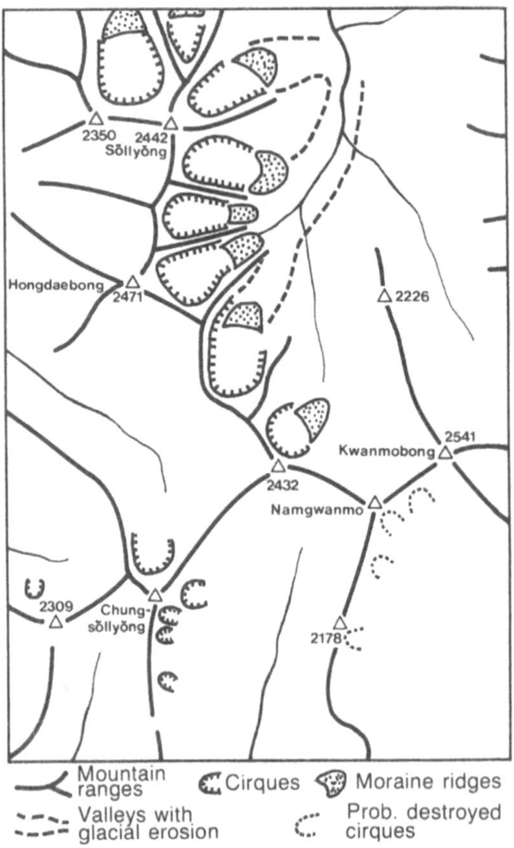

Fig. 20. Cirques and other Pleistocene forms northwest of Kwanmobong (after T. Kano)

have an altitude of more than 2200 m. To be sure, the orographical conditions are probably nowhere as favorable for the formation of cirques as they are to the northwest of Kwanmobong. On the northern slope of Pukp'ot'aesan (2435 m) there is a niche with a floor level at approx. 2000 m that also appears to be a cirque. Whether the three niches inside the caldera of Paektusan are cirques too is doubtful, in my opinion [778]. If they do indicate that Paektusan was also glaciated, they would require a snowline at approx. 2400 m and would thus belong to a period when the last Ice Age was receding.

The east coast of Korea from the mouth of the Tumen to Pang–ŏjin, 55 km north–northeast of Pusan, is a smooth, downwarped coast with faulting here and there, particularly along the last stretch between the cape of Changgi and Pang–ŏjin. On the eastern side of the Changgi peninsula I was able to demonstrate such faulting. Despite its smoothness the east coast does have recent ingression bays on a small scale, which probably developed as a result of the eustatic rise in the sea level since the climax of the last glaciation. The largest by far is the bay at Wonsan, Yŏnghŭngman, at the northern end of the Korean arc, where a tectonic basin was presumably inundated by the sea. The process of straightening that began immediately after the ingression is very far advanced on the coast of the Sea of Japan. The current that flows southward along the coast and the very small tidal range have contributed to this process of straightening, which consists in the formation of cliffs, sandbars, lagoons and delta plains. The largest of these deltas were deposited in the vicinity of Yŏnghŭngman by the relatively young rivers Yonghŭnggang and Sŏngch'ŏn–gang. The Neolithic sites at Yup'an and Wonsudae (Fig. 3) are proof that the ingressive forms were already fully developed during the Neolithic period and that the formation of sandbars and barrier beaches had already begun before the Eneolithic period. Abrasion platforms at a height of 20–150 m above sea level, their lower levels covered with marine pebbles and shells, attest to the rise of the land that occurred prior to the eustatic rise in the sea level. These platforms continue up the rivers in the form of alluvial terraces. Southeast of Kilchu the Yangdo islands lie 4 km off the coast. According to F. Yamanari [743] two of the three islands consist of mid–Pleistocene melanite basalt that is truncated by an abrasion surface at a height of 45 m above steep cliffs. The marine terrace at a height of 40 to 50 m that is so common in northern Korea thus developed after the middle of the Pleistocene.

As indicated previously (p. 2), the south and west coasts have a completely different character. Here the ingression resulting from the postglacial eustatic rise in the sea level did not affect a steep, mountainous coast. Instead it inundated an area of lowlands and hills that had probably previously occupied the entire region of the Yellow Sea (p. 1) and extended southward at least as far as Chejudo. Since the flora of Chejudo is much less highly endemic than that of Ullŭngdo, and since many of the species that it has in common with the continent have seeds that cannot be transported by wind or birds, e.g., oaks, the connection between Chejudo and the mainland must have existed longer than that between Ullŭngdo and the mainland. The offshore waters between Chejudo, the Ch'uja islands and the southwestern tip of the mainland are 100 m deep at the most, so that the eustatic lowering of the sea level during the Ice Age probably was sufficient to connect Chejudo to the mainland.

The bays between the islands and the protruding peninsulas, which have taken the place of the broad valleys of the Würm glacial period, do not really represent a ria shoreline of the Spanish type. One can very well gain the superficial impression of a Spanish ria when the high tide has just filled all of these bays, and the slopes of the thousands of islands and peninsulas dip more or less steeply under the surface of the sea, even where there are no active cliffs. In one's mind's eye one involuntarily projects the steep slopes under the surface of the water and gains the impression of deep bays. How great is the surprise 6 hours later when the entire bay has run dry except for a few tidal inlets and is filled with shiny brown mud flats. At the bottom of one of these tidal inlets near Inch'ŏn I found an outcrop of schist in the middle of a bay. The layer of mud in these tidal flats is thus rather shallow. The coastline of the Würm glacial period was too far away for the lowering of the sea level during that period to have caused deep erosion in the bays.

In contrast to the eastern coast, there are almost no spits and sandbars on the western and southern coasts of Korea. Neither have any large expanses of marsh begun to develop, although the bays are very shallow. The only exceptions are the mouths of the large rivers Kŭmgang, Han–gang, Ch'ŏngch'ŏn–gang and Yalu. Otherwise, the solid rocks of the mainland dip under the mud flats with no fringe of land forming in between. The enormous floods coming down the rivers do carry large amounts of sand and silt into the Yellow Sea, but the very strong tidal currents transport the material away from the land and distribute it over the bottom of the sea. Off the mouth of the Ch'ŏngch'ŏn–gang there are dangerous sandbanks that lie dry at low tide. They extend south–southwest in parallel rows for as much as 100 km (cf. the German sea charts). And off the mouth of the Han–gang the relief of the sea floor shows the effects of deposition by the tidal currents for the same distance. In the narrow channels between the islands the same tidal currents appear to have an erosive effect. At least, Ogura discovered "many submarine valleys" off the Korean Yellow Sea coast [470, 471]. Thus the principal area of sedimentation of the westward flowing rivers does not lie in the region of the permanent mainland, but in the tidal flats. In contrast to northeastern China on the other side of the Yellow Sea, no extensive alluvial plains have formed in Korea.

The conditions are quite different at the mouth of the Naktonggang, where the tidal range has dropped to 2 m. Here this large river has deposited a delta 10 km wide and more than 15 km long into the former bay (Fig. 4). The objects found in the shell mound at Kimhae, which lies on a granitic hill at the head of this bay, indicate that during the Yayoi period in the 2nd century B.C. the coastline still lay very near, or that the hill was even an island at that time.

We have just explained the contrasts between the different types of Korean coasts. Let us now demonstrate them with morphometrical figures calculated by Schröter in 1904 on the basis of maps that were admittedly still quite inadequate [354]. According to Schröter the west coast, including the islands off it, is 5500 km long, the south coast 5189 km. Although it extends so far from north to south, the east coast is only 1800 km long. By "articulation coefficient" Schröter means the quotient of the actual length of a stretch of coastline (including the islands) and the straight line connecting its endpoints. He calculates the articulation coefficient of the west coast between the mouth of the Yalu and 37°40' N as 4, that of the south coast as 11, and that of the east coast as 2.25.

Soil formation is closely related to climate and to the state of preservation of the natural vegetation. Where the mixed forest of the lower and medium elevations has been preserved, a brown forest soil predominates. The steeper the slope is, the more pieces of bedrock are found in the topsoil. In the deforested lower elevations, which have a climate with very hot and moist summers, brightly and often glaringly colored red soils predominate. On limestone the already mentioned terra rossa has developed at these elevations. At higher elevations the limestone mountains are covered by a humus carbonate soil (rendzina). In the extreme northeast podzolic soils are found instead of brown forest soils. Nowhere in Korea do we find loess deposits.

The so widely distributed granites do not have as strong a tendency to form red soils [231, 377]. This occurs only in locations where chemical weathering can continue to its end stage, i.e., where the slope angle is very low and at particularly low elevations. Only on the southern coast with its mild climate do we find granitic red soils on moderately steep slopes as well. Otherwise yellow grus soils predominate in the granitic regions. At the northern end of the peninsula such soils extend up to an elevation of 400 m; in southern Korea they are found at 800 m and above and on slopes of up to 50°. The chemical decomposition of the granite has proceeded in situ to such a high degree that even at great depths (10 m and more) it can be dug up with a spade (saprolite). Starting from the surface and from joints the decomposition has advanced deep into the interior. In the innermost core of what used to be blocks bounded by joints there generally are undecomposed remnants with the shape of a giant potato. The chemically decomposed granitic masses are subject to the erosive effect of summer's downpours, whereas in winter ice wedging loosens the grain structure. Deep gullies cut their way down into the deforested granite slopes and eventually dissect each other, so that the last traces of vegetation disappear. Because the half decomposed granite is so extremely porous, the beds of the gullies and the sandy beds of the rivers flowing through these granitic regions are often still dry after 24 hours of heavy rain. Once they do start to flow, though, they transport enormous amounts of debris and sand. The highly aggrading rivers build up their beds, endangering the rice paddies in the floodplains downriver. In places the river bed lies 2 to 5 m higher than the rice fields, contained by anxiously maintained sand dams. The tremendous erosion and transportation activity of these rivers is restricted entirely to the summer floods. At normal water levels the autochthonous rivers of these granitic regions carry very little water and are very shallow. They often divide up into many arms in their broad, sandy beds.

In the granitic areas a region of barren rocky slopes and peaks lies above the region in which chemical decomposition and mechanical erosion prevail. These barren rocks usually have a covering of algae, giving them a grayish–black color that contrasts quite strikingly with the yellow, orange and red of the chemically decomposed lower portions of the slopes. There are four types of slopes and peaks in granitic areas. The most common type is a kind of rounded rocky surface, which can form completely smooth slopes up to 500 m in height. In places they merge from all sides to form sugarloaf or dome–shaped mountains. For climatic reasons they are lacking in northeastern Korea. The only place here with a slight tendency toward such forms is along the coast of the Sea of Japan, where the winter is milder. In the rest of Korea their elevation has a limit, lying at around 800 m at Kŭmgangsan and 1000 m at Myohyangsan. In granite they occur only in coarse to medium coarse–grained rocks, but they are also found in other unstratified rocks, such as porphyry, quartz, conglomerate and breccia. In Korea they are never found in fine–grained granite, clearly stratified rocks or in gneiss or mica schist. They are also lacking in rocks that are highly interlaced with joints and clefts. On these rounded slopes physical weathering predominates, particularly "exfoliation from below" (Schmidthenner [364, 379]). This exfoliation can lead to overhangs measuring as much as 10 m perpendicular to the arch of the plate. In Samgaksan near Taech'ŏn (P'yŏng–anbuk–do) I even observed an overhang of 30 m. I did not find so–called "exfoliation from above" anywhere in Korea. Where two rounded slopes lie opposite each other, they are capped by a keystone that is sometimes split by a radial crack, or they end in a ridge a few meters wide that is covered by detritus and rock fragments. The origin of these rounded slopes and dome–shaped mountains has nothing to do with the artificial deforestation. In some mountain ranges, for instance, Kŭmgangsan or Ch'ŏnmasan near Kaesŏng, the rounded slopes jut up out of untouched alpine forests, and occasionally they disappear higher up, where the slope angle is more gentle, under a covering of brown soil and forest.

The sugarloaf and dome–shaped mountains often jut up out of a low surrounding area that consists of exactly the same rock, both petrographically and in its joint structure. It is thus impossible to explain the steep slopes on the basis of the structure of the rock or its joints. On the contrary, the structure of the joints explains only why one granite mountain has a rounded shape and another one nearby does not. Thus these dome–shaped mountains are closely related to tropical inselbergs (cf. N. Krebs, *Über Wesen und Verbreitung der tropischen Inselberge* = On the Nature and Distribution of Tropical Inselbergs. Abh. Pr. Ak. d. Wiss. 1942. Math.–naturw. Kl. No. 6).

Whereas the rounded slopes have a convex arch, the deep valleys between them often have a concave arch and also consist of solid rock. Their longitudinal profiles have oversteepened sections lying in areas with particularly coarse grains or particularly large spaces between joints. These oversteepened sections sometimes take the form of broadly projecting steps, which again show exfoliation from below in or beside the river. The side valleys sometimes also end with steps of this kind. All of these forms are strongly reminiscent of glacial forms in crystalline rocks.

At higher elevations the rounded forms give way to jagged ribs, pillars, slabs and sharp–edged ridges. This change in forms with altitude can be explained only by climatic influences, especially by the increasing intensity of frost action during the cold winters. The contrast in forms is simultaneously a contrast in colors. The black color of the algae yields to a grayish white color produced by lichens. The edges of the pillars and columns are rounded, except where shattering along a joint has occurred recently. This rounding is the result of the solvent action of the warm summer rain. It has produced forms that are often quite fantastic and may resemble mushrooms or animals.

Solvent action has also produced solution pans up to 1 m deep. They are found in both regions with jagged forms and ones with rounded forms and in both coarse–grained granite and quartzite. As I observed several times, the water in the pans is whipped out by storms, and the new rainwater filling the pan decreases the concentration of the solution each time. Solvent action is also the explanation for the grooves that sometimes furrow the surface of the rounded slopes in the direction of the steepest gradient and that can be up to 0.5 m deep.

In areas where the granite is very fine–grained and highly jointed the rounded forms do not develop. In these areas the jagged forms extend down to the region in which chemical weathering predominates. Where there is very regular cleavage in three directions that are approximately perpendicular to each other, we find the rectangular rock walls and towers we are familiar with in Germany as chimney rocks. Such peaks are often called *nojŏkpong* (= grain heap peak in Korean), because East Asian rice bags have the shape of a block with rounded edges when filled.

Finally, in regions with gentle or moderate slopes, in which the blocks are affected by chemical weathering, they are replaced by heaps of rounded boulders looking like gigantic potatoes that jut up out of slopes of grus.

In the Main Korean Range, which, as we have discussed, has been subjected to intensive uplifting since the Miocene, straight slope profiles predominate. In the widely distributed crystalline slates and shales these slopes are highly dissected by side valleys, and the ridges are often very sharp–edged, when seen in cross section. The forms in the region of the Kaema plateau are of quite similar, exhausting uniformity, particularly along its southeastern escarpment. In the western half of the country, on the contrary, as in the immediate vicinity of the Sea of Japan, there is only a narrow band of slopes with moderate slope angles, or such slopes are completely lacking, just as Schmitthenner described in China [379] and Credner in Southeast Asia [378]. The high elevations are dominated by slopes with very large slope angles, and the low elevations by slopes with small ones. Occasionally these two form systems abut directly on each other at a sharp angle. In granitic areas the lower form system is often found in the region in which chemical weathering predominates, and the upper form system in the region of rocky slopes. The lower form system often drops off abruptly to the broad alluvial valleys, in which case it forms a zone of foothills. Toward the Yellow Sea these foothill zones merge into the broad, moderately elevated and dissected regions of the Nangnang plain with no sharp angle between. I tend to agree with Schmithenner that regions with moderate slope angles have been consumed by the peculiar form of erosion induced by the climate in the extratropical monsoon regions.

In regions in which a dense, original forest cover extends into the the valley bottoms, such as Kŭmgangsan and Odaesan (central Kang–won–do), the transport of debris is not remarkable, despite the paroxysmal changes in stream flow. Quite often the concavely arched rocky bed, shining in the natural color of the unweathered rock up to the high watermark, has been swept completely free of detritus. Nevertheless, considerable amounts of material from the creeping forest soil are swept through these rocky channels, as the masses of large boulders farther downstream attest. The amount of transported debris is incomparably greater in areas that have been artificially deforested or ones that tower above the natural timberline. The climate in northern Monsoon Asia is especially favorable to such record transports of debris. In winter the bare, steep slopes, robbed of their topsoil in artificially deforested areas, are subject to frost shattering, and in summer to thermal shattering. Simultaneously, in summer the terrible floods provide tremendous forces for transporting loads. Gigantic rock slides move the broken down masses of blocks into the mountain torrents. Boulders as much as 4 m in diameter can often be found in such streams. Downstream the trenches of the mountain torrents very quickly turn into floodplain valleys filled from one side to the other with tremendous angular blocks or boulders with only slightly rounded edges. As their fresh color shows, these masses of boulders are moved on by each flood. They destroy the concrete bridges spanning these rivers time after time and pile up against the ruined piers in heaps several meters high. At first these active boulder–filled beds occupy the full width of the valley. The layer of boulders is, however, only a few meters thick. In the region of the active low water bed or in abandoned low water beds the bedrock is often visible in the midst of this rubble. The moving masses of debris cause a very effective lateral erosion, which leads to the very sharp angle at the foot of slopes that we have mentioned several times. This type of abnormally wide, shallow valley choked with debris continues on seaward, and on the western and southern coasts it empties into the shallow bays filled with mud flats that we have also already described. Naturally the size of the debris decreases rapidly as we approach the sea, and soon the still wide boulder–covered bed continues on within an alluvial valley that is not active at present and is therefore covered with fields. But the abrupt angle at the edge of the valley remains right up to the mouth of the river.

2 Climate

Beginning in 1442 precipitation was measured regularly in several Korean cities at the order of King Sejong (p. 30). These measurements were continued for more than 100 years, but were then abandoned during the period of invasions. In 1770 they began anew, and precipitation data are available for Seoul without interruption from 1776 to the present day [reproduced in 453, Vol. I, p. 210–15]. These data were evaluated by Wada [400] in 1917 and Tada [401] in 1938. By means of harmonic analysis Tada identified periods of 2, 6, 13, 35 and 37 years. No data on the other climatic elements appear to exist for the Korean dynastic period.

The letters written by the first European to set foot on Korean soil, the Jesuit priest Gregório de Céspedes (p. 41), already mention the main differences between the Korean and Japanese climates. In the winter of 1594/95 Father de Céspedes writes [88, p. 44]: "The cold in Korea is very severe and without comparison with that of Japan. All day long my limbs are half benumbed, and in the morning I can hardly move my hands to say mass." Hamel's report [171–173] is also full of complaints about the annual winter cold and the large amount of snow. It also contains some information on the storms in the Korea Strait. More exact data on the climatic elements commence only after the country was opened to Westerners. They are based on measurements of temperature and precipitation in the first treaty ports, Inch'ŏn, Pusan and Wonsan. From these data A. Woeikow determined as early as 1887 that the east coast of the peninsula has higher temperatures in winter and lower ones in summer than the west coast, and he gave an explanation that is still valid today [402]. Even more detailed are the data that Dinklage reported in 1891 on the basis of the observations of the German port official Mörsel. His brief paper [403] outlines many of the traits of the Korean climate accurately. Essentially the same is true of the reports of the Russian Minister to the court at Seoul, C. Waeber, from the same year [404]. Five years later A. Supan published the first data on mean precipitation in Korea [405].

As pointed out earlier (p. 55), when the Japanese established their protectorate, they set up a network of meteorological stations, and in the course of the generation that has since passed, they have continued to expand it with perspicacity and energy [406–410]. A first attempt to summarize these data in a study of the climate of Korea was made by T. Hirata [411] in 1916, and in 1919 the American medical doctor and missionary J.D. van Buskirk published a survey of the annual variation in the climatic elements at the meteorological stations. It was based on the Japanese data and contained isoline maps. Stimulated by Huntington, he included comments on the physiological effects of the Korean climate [412]. Okada's big climatic atlas [m] and the climatography of Japan by the same author [416] include Korea. It is quite clear that these studies cannot be comprehensive enough or on a large enough scale for our purposes. Okada uses only the means from the central observatory and the 15 meteorological stations, but not those from the 234 auxiliary stations. In addition, his means are based on observation periods of varying length, i.e., they are not reduced to a standard period. Using Okada's published means, Biel and Moese have attempted for the first time to apply the principal of duration to the Japanese climate [418]. They too include Korea in their text and maps.

To be able to utilize the abundance of material published in the "Annual Reports of the Meteorological Observatory of the Government–General of Tyōsen" as fully as possible, I proposed two climatographies of Korea, one for the south and one for the north. Both of them calculate the means of all stations for the period from 1914–1933. The dissertation on the climate of southern Korea by G. Schumacher has appeared [413]. It also evaluates the official Japanese weather maps in an attempt at a physiological interpretation of the processes upon which the means are based. The dissertation on the climate of northern Korea, on which E. Trojan is working [414], has been interrupted by the present war. But all of the means are available.

Almost all of the climatic charts included in this book were drawn by me and are based on the laborious calculations of these two authors. The tables on the climate of each individual region in Part Three of this volume use the same 20–year means. These tables were drawn up from the following point of view. The annual cycle of most climatic elements in Korea is extremely periodic and is characterized by a single maximum and a single minimum. There are only isolated local exceptions to this rule. The tables contain the maximum and minimum values of the annual cycle. The maximum values are almost always associated with a summer month, and the minimum values with a winter month. These months are not, however, the same for all elements and all stations. The headings to the tables name the months in which the maxima and minima of the columns in question usually fall. Months that deviate from the mean are mentioned in the explanatory notes. The Korean lighthouse stations do not record the maximum and minimum temperatures; consequently, data on these elements are occasionally missing in the tables. The location of almost all meteorological stations mentioned in the tables and in the text can be found in Map 1 (cf. editors' note on p. 7). The more subtle regional details of the Korean climate are dealt with in Part Three, using the tables as references. In this chapter I shall limit myself to the principle features.

In connection with the general air pressure conditions in East Asia, the average annual cycle of pressure in Korea displays a strong simple periodicity almost without secondary extremes (Fig. 21). The maximum occurs in January, the minimum in June or August. In January the pressure in northern Korea is even higher than in southern Korea. In midsummer the opposite is the case. These facts coincide well with the pressure distribution in East Asia as a whole, as described by Okada [m, 416] or earlier, though with some major errors, by Froc [415]. The difference in air pressure between Chunggangjin and Pusan is +4.3 mm in January and −2.6 mm in July. Thus in midwinter there is a very strong pressure gradient toward the south−southwest, and the prevailing winds come from north−northwest to northerly directions. This is the well−known winter land monsoon. In midsummer the pressure gradient is directed toward the northwest and is less pronounced. The sea monsoon then blows from south−southeast to south−southwest. Thus twice in the course of a year the prevailing winds make a turn of 180°. On the whole, the map in Fig. 30 confirms this rule. It shows the prevailing wind direction in January at the central observatory and the 15 meteorological stations, represented by arrows with solid lines. The wind directions at sea are recorded according to the new Dutch atlas [s]. The prevailing wind direction is shown as that direction of the 16−sector wind rose that was recorded most frequently on the average of the period from 1914 to 1933. For some locations the second most frequent direction is also shown. The length of the arrows corresponds to this frequency. The same is true of the prevailing wind direction in July, which is rep-

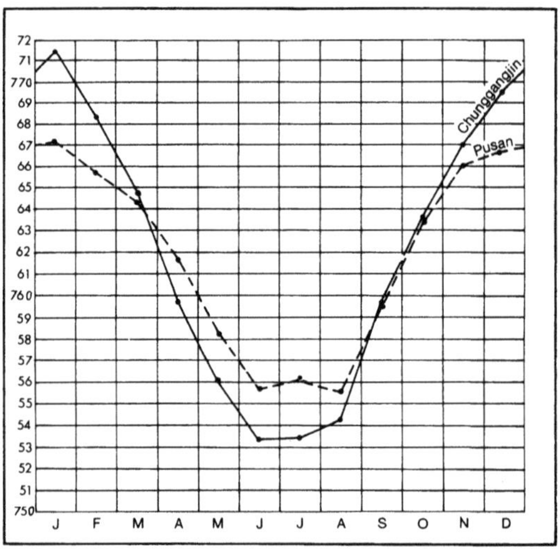

Fig. 21. Annual cycle of air pressure in Chungangjin and Pusan

resented by arrows with dotted lines. The map shows that at most of the coastal stations northerly winds strongly prevail in January, and southerly winds in July. An important exception is the two stations on the central eastern coast, Wonsan and Kangnŭng. There west to southwesterly winds prevail in January. The reason for this is evidently that in winter, and autumn, a local low pressure area forms over the warm Sea of Japan. Okada [416, p. 203] also supports this view. Everywhere in the interior the arrows are much shorter than along the coasts. There periods of calm constitute a large percentage of the observations in both January and July, and the wind direction is influenced to quite an extent by the mountainous relief. The principles of monsoon circulation apply there only in attenuated form or not at all. The influence of the relief is seen particularly clearly in the case of Cheju, which lies on the northern coast of Chejudo, protected to the south by the sprawling mass of Hallasan (1950 m). In winter north–northwesterly winds develop their full potential. The summer monsoon, however, flows around the mountain and comes from the east. With 8.6%, wind from the south–southeast is only the fourth most frequent wind in July in Cheju.

The average wind velocity is also highly influenced by the relief. At the coastal stations it is much greater than in the interior. Inch'ŏn has an annual average of 4.5 m/sec. In contrast, Seoul, which lies 35 km away in the interior, has only 2.2 m/sec. The lowest wind velocity is recorded at the stations farthest from the coast (Ch'osan 1.1). Since the pressure gradient is greater in midwinter than in midsummer, the average wind velocity at the coastal stations is greater in winter than in summer (e.g., Unggi January 7.9, July and August 3.5). At the stations in the interior the average wind velocity is always rather low, and often enough it is greater in summer than in winter. This is presumably due to the influence of the local relief.

The monsoons bring allochtonous air masses into the country with great frequency. In winter these are cold air masses originating from the north; in summer they are warm and come from the south. For this reason Korea has extraordinarily great seasonal temperature contrasts. They are much greater than the average for the geographical latitude. The mean January temperature at the southern end of Korea is about 5.5° C lower than the mean for that latitude, and at the northern end it is even as much as 22° C lower. Conversely, the July temperature is only approx. 0.5° C higher in the south and about 0–2° C in the north than the mean temperature. Thus the negative anomaly in midwinter is much greater than the positive anomaly in midsummer, and both are greater in the north than in the south.

This last mentioned fact shows that in its climate as well Korea is definitely a transition zone between the continental and the maritime spheres. Its transitional character with regard to temperature is expressed most strongly in midwinter. At all meteorological stations January is the coldest month in the average annual temperature cycle. The diagrams in Fig. 27 show examples of this from all parts of the country. In most cases this is also true of the individual years. Only in the more strongly maritime islands in the south and west is February often the coldest month in the individual annual cycles [413]. In the north December is fairly frequently the coldest month [414]. Figure 22 shows the isotherms of the mean January temperature reduced to sea level. From Chunggangjin to the lighthouse island of Marado south of Chejudo this mean temperature increases from −19.4°

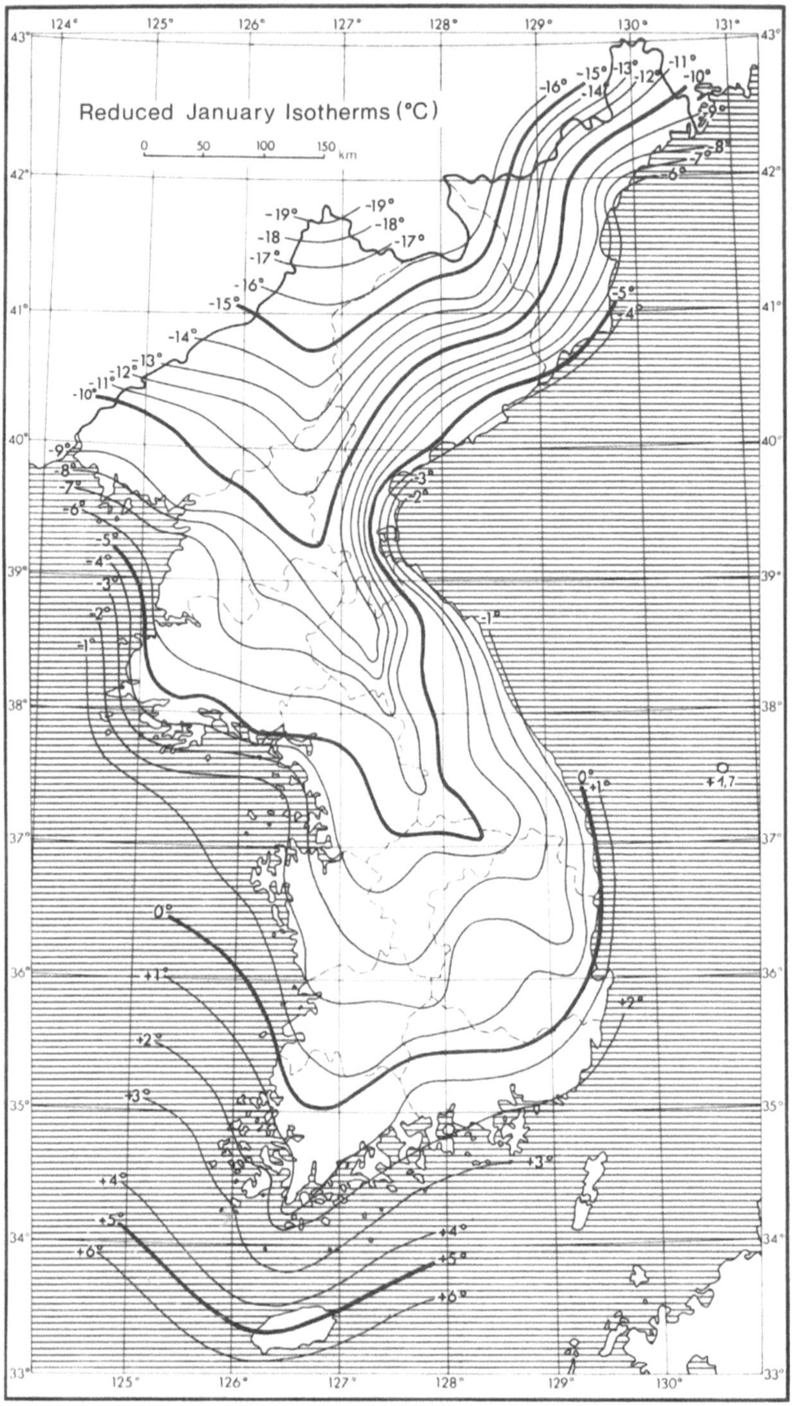

Fig. 22. Reduced January isotherms

to +6.0° C. These two stations also have the lowest and highest unreduced mean January temperatures in Korea (−21.0° and +5.8° C). In by far the greater part of the country the mean January temperature is below 0° C. The 0° C isotherm passes through the southern fringe of the country. In Korea the January isotherms are crowded more closely together than in almost any other place on earth. Another characteristic trait that stands out on this map is the fact that the interior is always colder than the coast at the same latitude. The islands are also always much warmer than the adjacent stations on the mainland. This characteristic feature can be traced all the way down to the southwestern tip of the peninsula. The coldest parts of Korea are the parts that are farthest from the sea. The greater the distance between a station and the coast, the less often is it affected by the seas on both sides of the country, neither of which freezes over. The third characteristic trait, finally, is that the stations on the coast of the Sea of Japan are approx. 1−6° C warmer than those on the Yellow Sea coast. The Main Korean Range largely protects the former from surges of cold air from the northwest, whereas the latter is defenseless against them. Moreover, as Woeikow [402] already recognized, the prevailing west to southwest air currents along the central eastern coast are of necessity warmed adiabatically when they drop from the Main Korean Range to the coast of the Sea of Japan. It is certain that there are foehn winds in Wonsan (p. 309), and it is likely that they also exist in Kangnŭng (p. 314f). In addition, the surface of the shallow Yellow Sea cools off in midwinter much more than that of the deep Sea of Japan (p. 124), and winds from the eastern quadrant occasionally bring the relatively high surface temperatures of the latter to the adjacent coastal margin. In January, however, winds from north−northeasterly to southeasterly directions occur with a frequency of only 10.7% in Wonsan and 6.4% in Kangnŭng.

The maximum monthly temperatures are not as highly concentrated in one single month as the minimum temperatures are in January. At all stations both July and August have been the months with the highest monthly mean temperature during the course of the period from 1914−1933. In the extreme north July predominates, and the farther south one proceeds, the more often are the cases in which August predominates. If we observe the 20−year means, we find a clearly defined contiguous area with a maximum mean temperature in July. The border between it and the main area with a maximum in August is shown in Fig. 23. It forms the continental northern edge of Korea. Nevertheless, the difference between the July and August mean temperatures is always quite small, as shown by the temperature curves in Fig. 27. They are always flattened off in midsummer. The fact that most of Korea has an August maximum demonstrates that the country is subject primarily to maritime influence in midsummer. In Fig. 23 the reduced August isotherms are given. It shows that in this month the differences in temperature between north and south are only very slight, as the path followed by the 25° C isotherm demonstrates particularly clearly. The reduced August temperature at Hyesan on the Yalu, at 25° C, is even higher than at the lighthouse island of Ŏryongdo off the southwestern tip of the peninsula (24.1° C). By then the upper Yalu and Tumen region has been strongly heated up continentally, whereas cool upwelling water off the southwestern coast (p. 126) lowers the air temperature. In midsummer the principal temperature gradient runs from southwest to northeast, the northeastern tip having by far the lowest temperatures of

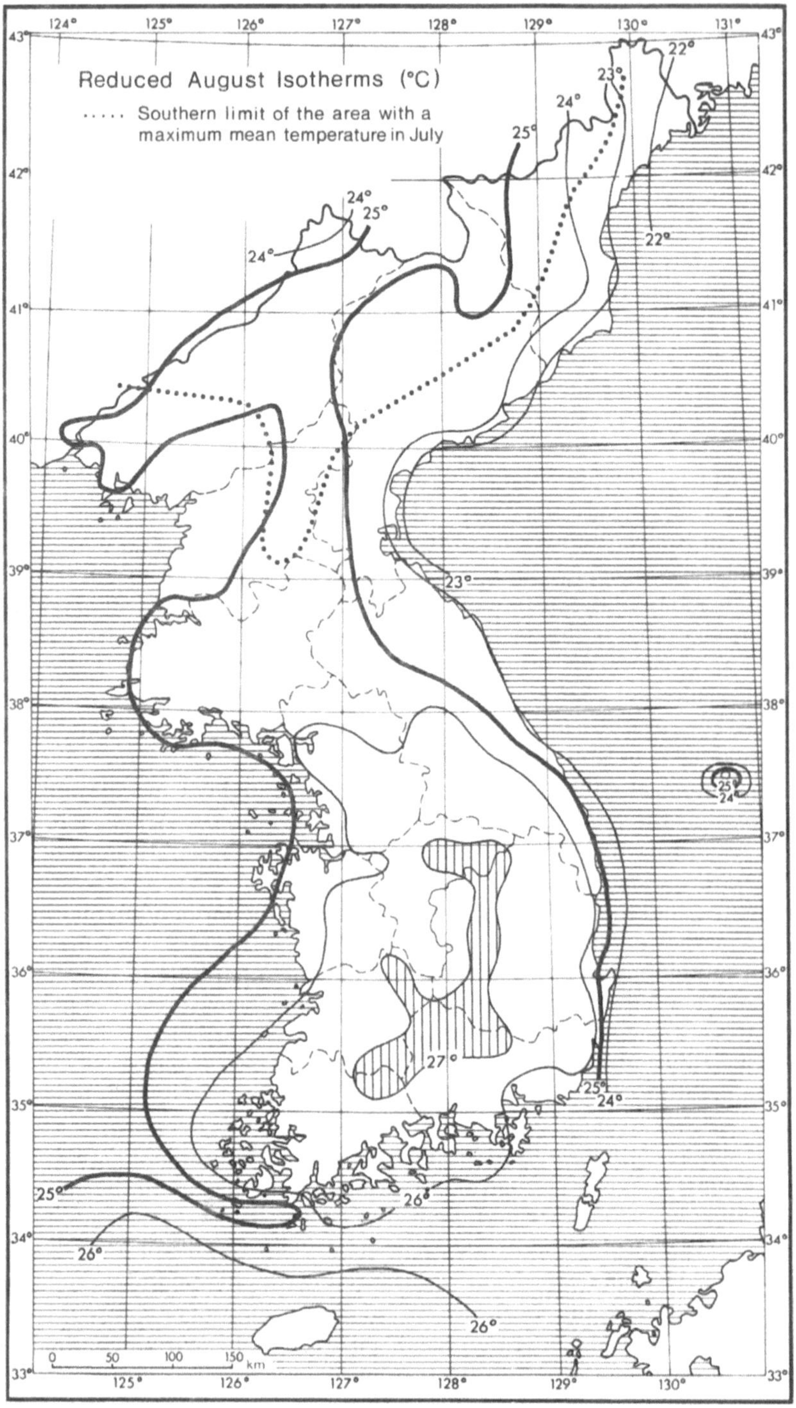

Fig. 23. Reduced August isotherms

this season. The contrast between the coastal margins and the interior is highly manifest in August as well, though in the opposite sense. The coastal margins and particularly the islands are about 1–3° C cooler than the interior. In the broad southern part of the peninsula there is an inland area with a reduced average temperature of more than 27° C. The highest unreduced August temperature (26.9° C) is found, however, in Ch'ungmu on the south coast. The coast of the Sea of Japan is now about 1–2° C cooler than the Yellow Sea coast, since in midsummer the water temperature of the Sea of Japan is lower than that of the Yellow Sea (p. 124), and onshore winds have plenty of opportunity to spread these cooler temperatures across the coastal margin. In August winds from north–northeast to southeast have a frequency of 31.3% in Kangnŭng, and even 43.7% in Wonsan.

Since the regional differences in temperature are slight in August, the map of the annual range of temperature (Fig. 24) closely resembles the isothermal map for January. The annual range was calculated as the difference between the mean temperatures for August and January. This was also the case for the northern fringe, where the July temperature is even somewhat higher. The annual range increases from 20.6° C in the south (Marado) to 42.7° C in the north (Chunggangjin). It decreases from the interior toward the coast and the islands. Along the coast of the Sea of Japan it is about 2–7° C lower than on the Yellow Sea coast. This last mentioned fact demonstrates more clearly than any other that the east coast is the maritime edge of Korea from a climatic point of view as well.

The temperature charts (Fig. 27) show the entire annual cycle. At all stations the temperature cycle reveals a simple, clear periodicity completely lacking in secondary extremes. Generally the descending side is somewhat steeper than the ascending side. The stations Chongsŏng and Chunggangjin show the tremendous annual contrasts in the extreme north and the July maximum. The stations Kanghwa – Ch'unch'ŏn – Kangnŭng form a cross section through the peninsula at a latitude of approx. 37°45' and demonstrate what we have said above about the thermal characteristics of such a cross section. Sŏdo exhibits the small temperature range of the island stations in the Yellow Sea. In Ullŭngdo and Chejudo we see the very mild winters and extremely small range of the remote islands.

In their spatial distribution and their annual cycle the average extreme temperatures closely resemble the average monthly mean temperatures. The minimum January temperatures are extremely low in the north. Everywhere in the Kaema upland and in the big bend of the Yalu they are below −20° C. Changjin has the lowest figures (−29.1° C). Toward the south they of course increase greatly. But everywhere on the mainland they remain below freezing. The span between the lowest and the highest average minimum January temperature is 31.2° C and thus larger than that between the January mean temperatures (25.5° C). In January Cheju is the only station with an average minimum temperature above freezing (+2.1° C). The mean maximum January temperature is still below −10° C in the extreme north. With −12.2° C Chunggangjin has the lowest figure. In the interior mean maximum temperatures that are below freezing extend far south, the southernmost station of this type being Chech'ŏn (37°8' N). Along the coast of the Sea of Japan the area with mean maximum temperatures above freezing begins far to the north, namely near Tanch'ŏn (40°28' N). Along the Yellow Sea coast, in contrast, it does not begin until the mouth of the Han–gang (37°50' N), in ac-

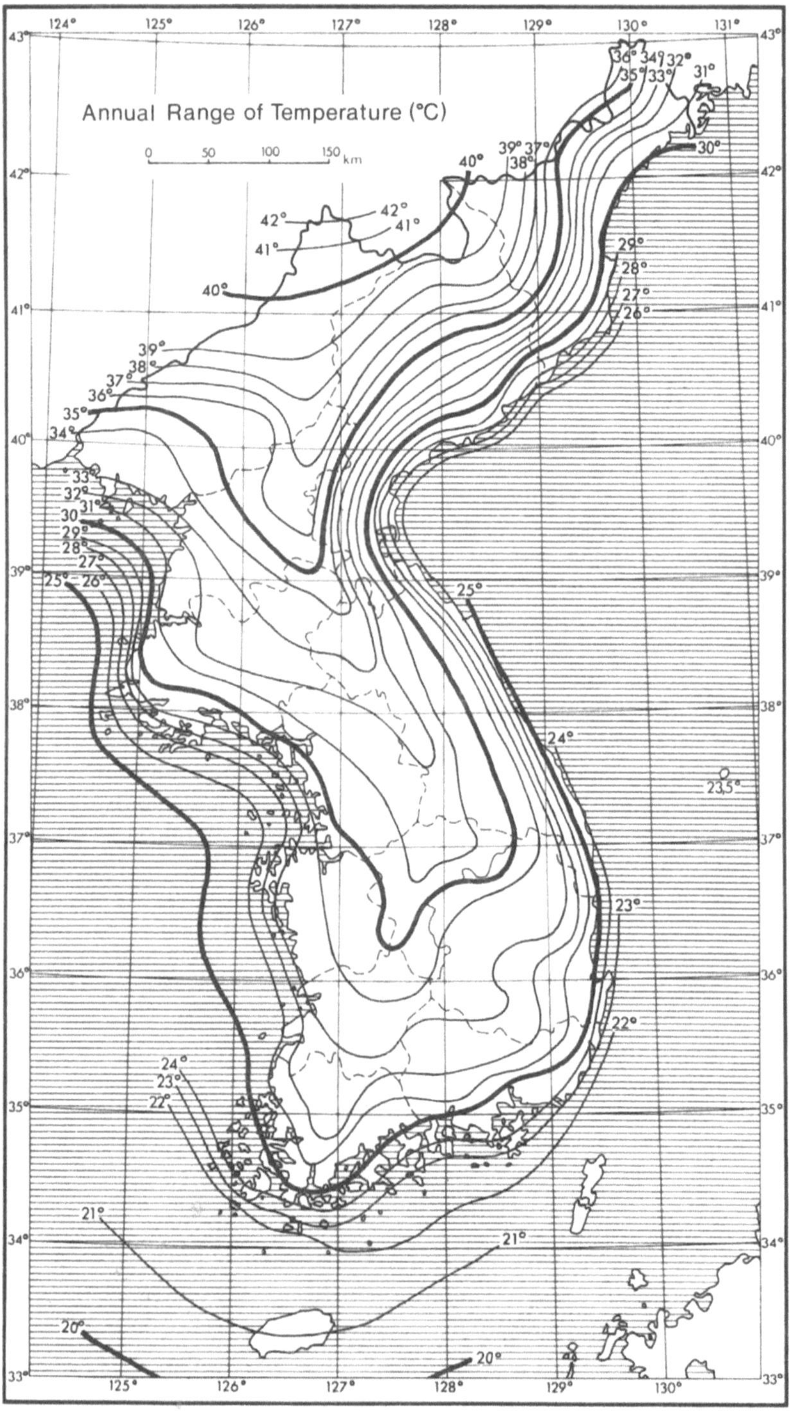

Fig. 24. Annual range of temperature

cordance with what we have said about the winter there. Again Cheju has the highest figure (+7.5° C). Thus the span between the lowest and the highest mean maximum January temperatures is only 19.9° C.

The highest mean extreme temperatures occur in July in approximately the same area as the highest monthly mean temperatures do; otherwise they are generally found in August. If we eliminate the influence of altitude, we see that in midsummer there is, similar to the mean temperatures, only a slight difference in the mean extreme temperatures between the north and the south. In August P'ungsan in the Kaema upland, at 1150 m the highest station in Korea, has a maximum temperature of 30.9° C, reduced to sea level. Cheju, in contrast, has a maximum temperature of 29.7° C. The mean minimum August temperatures reduced to sea level lie between 18° and 23° C, the corresponding maximum temperatures at 25°–31° C. The highest unreduced mean minimum temperature is 23.8° C (Ch'ungmu), the highest maximum temperature 31.7° C (Kimch'ŏn, in the center of southern Korea).

The absolute minimum temperatures are reached in the months of December to February. They are about −12° to −14° C in the south and −30° to −40° C in the north. The lowest temperature measured in Korea in the period between 1914 and 1939 was in Chunggangjin (−43.6° C). The temperature fell the least in Cheju (−5.7° C). The absolute maximum temperatures are measured in July or August, occasionally also in June or September. They are around 35° to 40° C almost everywhere. It is particularly worth mentioning that the stations situated high up in the Kaema upland also lie in the same range. Chunggangjin has an absolute maximum temperature of 38.0° C; thus during the period from 1914 to 1933 it experienced a temperature range of 81.6° C. Korea's highest absolute maximum temperature was recorded in Kyŏngju at 42.0° C, the lowest in Samsu (1025 m) at 34.0° C.

The average daily temperature range of a given month can be calculated, as the reader will know, as the difference between the mean maximum and minimum temperatures. In all seasons it is small near the coast and large in the interior. In summer it is generally smaller than in winter because of the many clouds. Therefore the Kaema upland has the largest range, and it is found in January (Changjin 18.6° C). The southern coast has the smallest range, which is found in August (Pusan 6.7° C).

Months with a mean minimum temperature below 0° C are termed frost months. Cheju is the only station in Korea that has no frost month. In the extreme southern part of the mainland some stations have two frost months. Otherwise southern Korea has three to five, the Kaema upland six to seven. On the coasts there are fewer frost months than in the interior at the same latitude.

Ice months, i.e., months in which the mean maximum temperature is also below 0° C, are lacking in southern Korea. In the Kaema upland the number increases to four (December–March). In most of Korea the five months from May to September have mean maximum temperatures above 20° C. In the south October joins them as the sixth month. In the center of southern Korea the number increases to seven, because April is also added. In northeastern Korea, in contrast, the number decreases to four and three, due to the influence of the cool coastal current or the altitude. In the high interior they are the months of June to August, on the coast, July to September.

Figure 25 shows the duration of the period in which the daily temperature is below 0° C on the average in weeks, according to Biel and Moese. On the islands and on the mainland in the extreme south the daily average remains above 0° C. Toward the north the number of such weeks increases rapidly, in the interior more rapidly than on the coasts. In the northern part of the Kaema upland it finally amounts to 20–24.

On the basis of the 20-year means of the 250 meteorological stations I have drawn up the map of annual precipitation (Map 2, cf. editors' note on p. 7). Since stations are lacking on slopes and mountain tops, the influence of the relief had to be evaluated on an exclusively deductive basis. Like any map of precipitation for a country with relatively few stations this is one of the possible interpretations of the figures. Naturally it is the one that appears most likely to me. As befits the highly diversified surface structure, the pattern of spatial distribution of

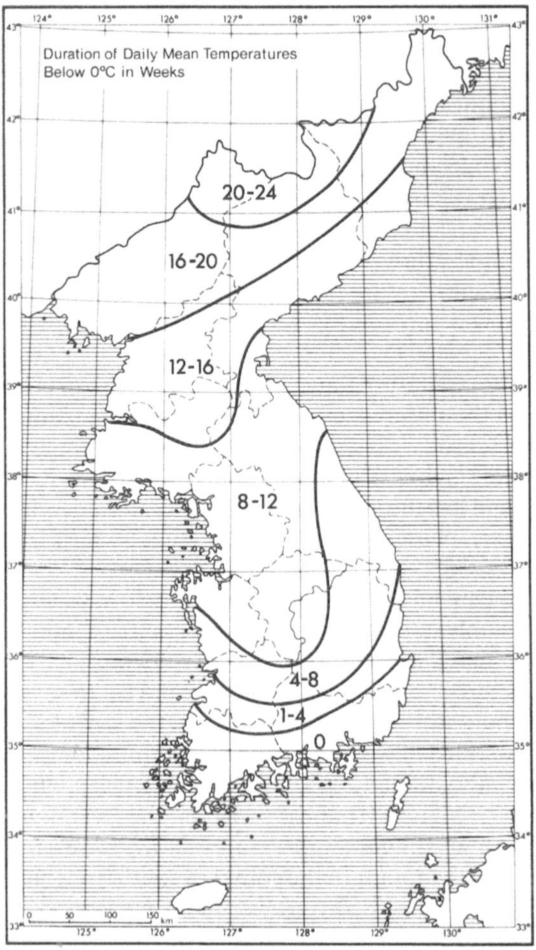

Fig. 25. Duration of daily mean temperatures below 0° C (after Biel and Moese)

precipitation is quite variegated. The following basic characteristics can be recognized. (1) The northeastern part of the country up to the Nangnim Range is an area with conspicuously low precipitation; no station has a mean above 900 mm. Particularly dry are the coastal zone and the inland region along the upper Yalu and the upper and central Tumen, which is blocked off from the sea by mountain ranges. (2) Another dry area is the Hwanghae peninsula and the adjoining southwestern portion of P'yŏng–an–namdo. (3) The islands are dryer than the adjacent regions of the mainland; the islands off the coast of the Hwanghae peninsula register the smallest annual amounts in all of Korea (480 and 490 mm). J. Lee discovered that the same rule holds for the islands off the Chinese coast [426, 427]. He relates this phenomenon to the low air temperatures in summer due to the cool surface of the sea. In Korea, however, this does not hold for the high islands such as Sohŭksando (639 m, 1122 mm), Ullŭngdo (984 m, 1510 mm) and Chejudo (1950 m, 1380 mm). Therefore, it appears to me that the most important point is that most of the islands are so low that the only precipitation they receive is that which is not orographical. (4) Precipitation increases from the coast toward the interior. This is particularly true of the coast of the Yellow Sea. (5) Following the longitudinal axis of the peninsula there is a zone of high precipitation and beyond it lies a zone of diminished precipitation, which is divided up into separate cores. It commences in the north in the longitudinal valley of P'aŭp – Yangdŏk. The second area follows the upper Puk'an–gang, and the third the upper Namhan–gang. The largest and driest of these areas lies east of the Sobaek Range in the northern and central part of the Naktonggang basin, where the amount of precipitation decreases to 852 mm. (6) The Main Korean Range and the Sobaek Range are most probably zones with large amounts of precipitation. An indication of this is the fact that despite its low elevation (104 m), Onjŏng–ri in Kŭmgangsan has an annual average of 1755 mm and is thus the station with the most precipitation in Korea. (7) The southern fringe of the mainland from Changhang to Pusan is an area with very high precipitation.

The variations in the annual amount of precipitation, which are combined in the means shown in the map, are quite extreme, a fact that will confront us again in the economic sphere. In overly wet years the amount of precipitation can exceed the mean by as much as 90%. In overly dry years it can sink as low as 57% below average. A deficit of more than 40% has a devastating effect on the rice harvest. In southern Korea 1915, 1924, 1932 and particularly 1939 were very dry years, and 1916 and 1933 were very wet years. In the disastrous year of 1939 precipitation dropped by two thirds in the decisive months of April to June. The mean deviation of the annual amounts from the average is greater than that given in an article in the Scott. Geogr. Mag. of 1936 [421]. Nor does the regional distribution of the mean deviation show the simple increase from east to west that v. Wissmann portrays [420]. It is greatest on the islands off the west coast (up to 28%), on the one hand, and in the northeastern tip on the other hand (up to 31%). In the center of the peninsula and on the east and south coast it sinks below 18%.

What is true of the years is also true of the individual months. Their figures also vary greatly at every station. At any rate each year shows one principle fact: except for a few island stations, in particular Ullŭngdo (p. 327) and Chejudo (p. 376), much less precipitation falls in the winter half–year than in the summer half –year. The precipitation minimum tends to be in December, January, or February, the maximum in July or August. At any rate, the seasonal deviation is rather large [413, p. 18f]. In the annual cycle of mean precipitation the maximum is in July at most stations. The only exception to this rule is northeastern Korea (maximum in August). The minimum lies in January or February, occasionally even in December or March. This is shown by the diagrams in Fig. 27.

As we have said, almost everywhere in Korea very small amounts of precipitation fall in January. The January average for the extreme northeast is less than 5 mm. Kyŏng–won, with 2.7 mm, is the station with the lowest monthly mean. In the region of the main Korean tilt block the January precipitation increases from north to south. The Yellow Sea islands and coasts also have a lower January average than the interior. Along the northeastern Korean fringe of the Sea of Japan, in contrast, precipitation decreases inland. In the interior of the peninsula we again find cores with particularly low figures. The Main Korean Range appears also to get more precipitation in midwinter than the western slope of the peninsula, an indication of this being the fact that Onjŏng–ri records the highest January mean on the mainland (93 mm). Relatively high averages are also found along the central Korean coast of the Sea of Japan. The absolute largest January amount of precipitation is measured on the island of Ullŭngdo to the east (116 mm).

The distribution of precipitation in July presents a much more complicated picture. Its main characteristics resemble those of the annual sum, so that we can refer the reader to the explanation of the map of yearly precipitation. The cape of Musudan in northeastern Korea has the least July precipitation (91 mm), P'och'ŏn north of Seoul has the most (468 mm).

In the average cycle the transition from the midwinter precipitation minimum to the midsummer maximum takes place gradually at most stations (diagrams in Fig. 27). After a noticeable increase in early spring, precipitation remains constant or sinks slightly in May at most stations. Then it rises to the midsummer maximum and is still appreciable in September, sinking thereafter to the low October and winter figures. In the south the entire period from June to September is the main rainy season and from October to March the dry season. Toward the north the main rainy season becomes shortened to the midsummer period. The midsummer rainy season (Kor. *changma)*, however, does not stand out as sharphly as the "plum rains" (Jap. *bai–u*) in the Japanese islands [416, p. 257–63].

In a regional classification of Korea according to the seasonal precipitation cycle we must of course disregard the absolute figures for the average monthly precipitation and restrict ourselves to a comparison of the ratios of the monthly amounts to the annual amount. Let us initially divide the stations into three groups, depending on whether the proportion of the month with the least precipitation is ‹ 20 per 1000, 20–35 per 1000 or › 35 per 1000 of the annual precipitation. In Fig. 26 the regions with relatively slight amounts (s) of rain in winter are designated by horizontal hatching, those with moderate amounts (m) by vertical hatching, and those with relatively large amounts (l) by cross–hatching. s dominates almost all of northern Korea and central Korea apart from the eastern coastal margin and is also found in a narrow band not far from the southern coast. There the very high annual total has the effect of lowering the January percentage, although the absolute amount is not that small. m fills all of the rest of southern Korea except for a small fringe along the east coast and extends across the

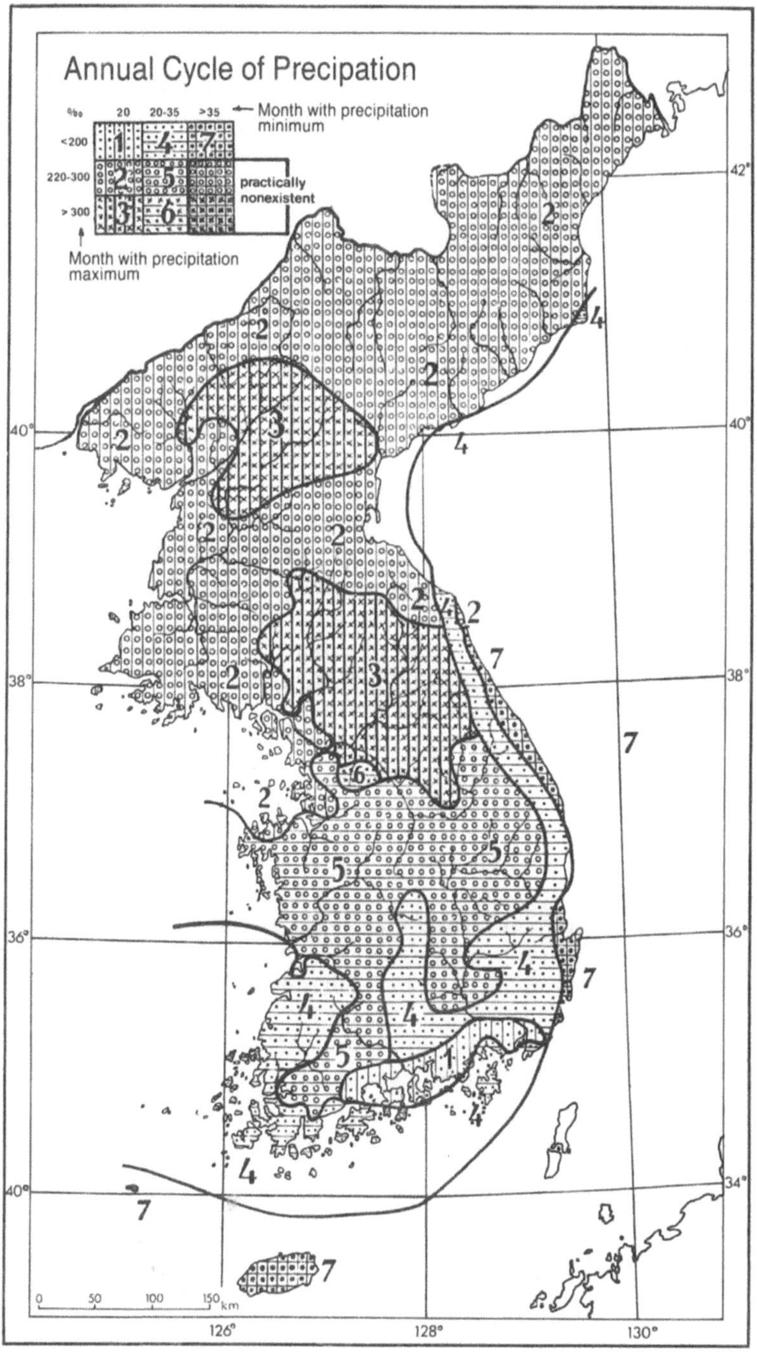

Fig. 26. Annual cycle of precipitation

T'aebaek Range as far as the cape of Musudan. l, finally, is restricted to the remote islands, Sohŭksando, Chejudo and Ullŭngdo, and a strip of coast along the Sea of Japan in southern and central Korea.

In the same manner we shall divide the stations into three other groups according to whether the proportion of the month with the most precipitation is ‹ 220 per 1000, 220–300 per 1000 or › 300 per 1000 of the annual precipitation. On the same map the areas with relatively slight amounts (S) of rain in summer are represented by dots, those with moderate amounts (M) by circles and those with relatively large amounts (L) by crosses. S dominates the islands far from the coast, the southern Korean band of islands and peninsulas, portions of the southern half of the southern Korean mainland and a strip along the Sea of Japan extending as far as the cape of Musudan. In these areas precipitation is distributed rather widely across the summer months, so that the month with the most precipitation does not stand out in a peak. M is found in wide areas of northern and southern Korea. In central Korea it is restricted to the west coast. L, finally, has two separate areas of distribution north and south of the neck of the peninsula. It is a continental type that does not extend out to the sea.

It is now possible to combine s, m and l with S, M and L. This has been done in the map by linking the symbols. Theoretically this results in nine combination types; however, two of them, lM and lL, practically do not exist. We are therefore left with the following seven types (cf. diagram, Fig. 27):

1. sS (rel. small min. and max.). South coast.
2. sM (small min. and moderate max.). Most of northern Korea and northern part of central Korea, entire central Korean west coast
3. sL (small min. and large max., the largest contrast in the annual cycle). North and south of the neck of the peninsula.
4. mS (moderate min. and small max.). Much of the southern half of southern Korea. Narrow strip along the T'aebaek Range northward as far as Musudan.
5. mM (moderate min. and moderate max.). The remaining southern Korean mainland.
6. mL (moderate min. and large max.). Only a small area in southwestern central Korea around Ich'ŏn – Yong–in.
7. lS (large min. and small max., i.e., fairly even annual cycle). The remote islands and the eastern fringe of southern and central Korea.

Thus the overall picture reveals both continental influences (3) and the effects of the Sea of Japan and the Korea Strait (1, 4, 7).

Everywhere in Korea some of the precipitation falls in the form of snow. With the tremendous regional differences in temperature in the winter climate, however, the proportion of the total precipitation that falls as snow varies greatly, as does the duration of the period during which snow falls. Between 1914 and 1933 Pusan had an average of 8 days of snowfall, and Chunggangjin 72 days. Those are the extremes. At 13 of the meteorological stations and at the central observatory, which are the only stations for which snowfall data are published, the annual number of days of snowfall is around 30. Snow of course falls most frequently in the three winter months. In Chunggangjin snow falls on one–half of the days in December, on the average. In Cheju and Pusan the snowfall period lasts from December to March; at almost all of the stations in the two Hamgyŏng provinces,

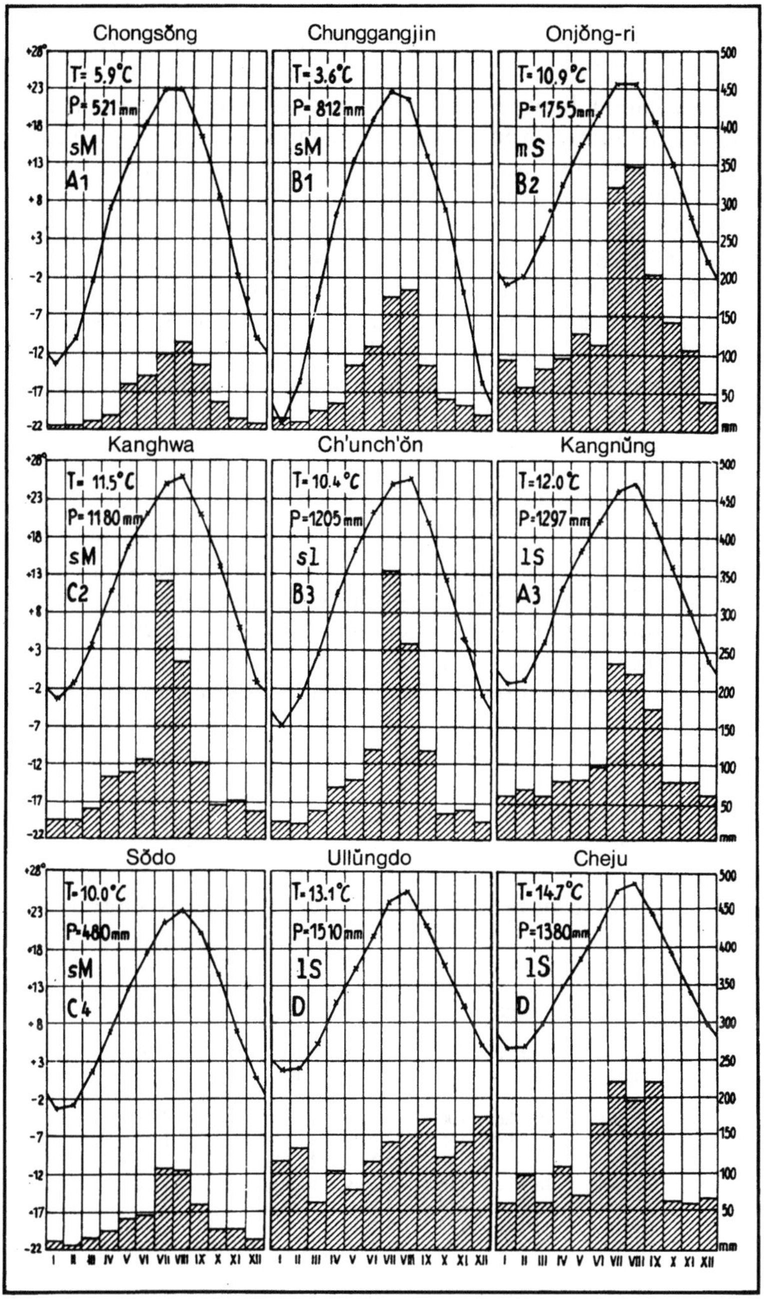

Fig. 27. Temperature–precipitation graphs of selected stations. The *curves* represent the average temperature cycle, the *columns* the average precipitation cycle. *T* mean annual temperature; *P* average annual precipitation. The letters *s*, *m*, *l*, *S*, *M*, *L* refer to the annual precipitation cycle (Fig. 26); the letters A_1, etc., to the climatic regions (Fig. 30)

in contrast, it lasts from October to May. In the winter of 1931/32, a normal year, it snowed in Chunggangjin on 67 days between 25 Oct. and 24 April. The snow cover lasted from 27 Oct. to 24 April, though with some interruptions. In the same winter in Cheju, however, it only snowed on 18 days, between 12 Dec. and 15 March, and the snow stayed on the ground only on 7 days between 13 Dec. and 22 Feb. There are no systematic observations on the depth of the snow cover. Table 3 provides information on the average date of the first and last occurrence of snow, hoarfrost and ice cover on standing fresh water surfaces, on the average of the years of 1909–28 (according to Okada [416, p. 354–56]).

Almost everywhere in Korea hoarfrost has the earliest date of onset and the latest final date and thus the longest period. The opposite is true of snowfall, while the dates and period for ice cover lie in the middle. The stations are arranged in order of increasing latitude. One can see how the date of onset and the final date generally diverge more and more with increasing latitude. The exceptions to this rule demonstrate the contrast between the coastal margins and the interior especially well. In this respect a comparison between Inch'ŏn and nearby Seoul is particularly instructive. The hoarfrost period is 27 days longer in Seoul. Above all, the very serious effect of an inland location on vegetation is apparent in Chunggangjin, which lies more than half a degree of latitude farther south than Unggi. The periods in Chunggangjin are as much as 73 days longer than in Unggi. Finally, we can establish that in this sphere too the Sea of Japan moderates the severity of the winter along the coastal margin more strongly than the Yellow Sea. This is shown by a comparison of Sŏngjin and Yong–amp'o. Although the former lies almost 1° of latitude farther north, the period of ice cover is about 8 days shorter and the hoarfrost period about 18 days shorter. Kangnŭng is equally favored, compared with Inch'ŏn.

Table 4 presents data on the relative humidity, percentage of possible sunshine (according to Jordan), cloudiness (on a ten–division scale) and number of days with fog. For the figures on days of fog I again refer to Okada's work ([416, p. 352]; unfortunately means of very unequal and, in part, very short periods). The

	Snowfall (average)			Ice cover (average)			Hoarfrost (average)		
	Earliest date	Latest date	Days	Earliest date	Latest date	Days	Earliest date	Latest date	Days
Cheju	7 Dec.	16 Mar.	100	–	–	–	24 Dec.	20 Mar.	87
Mokp'o	22 Nov.	16 Mar.	115	22 Nov.	28 Mar.	127	12 Nov.	7 Apr.	147
Pusan	20 Dec.	9 Mar.	80	19 Nov.	25 Mar.	127	7 Nov.	29 Mar.	143
Chŏnju	23 Nov.	18 Mar.	116	2 Nov.	14 Apr.	164	17 Oct.	24 Apr.	190
Taegu	27 Nov.	19 Mar.	113	3 Nov.	4 Apr.	153	23 Oct.	11 Apr.	171
Inchŏn	19 Nov.	23 Mar.	125	5 Nov.	3 Apr.	150	5 Nov.	6 Apr.	153
Seoul	18 Nov.	27 Mar.	130	23 Oct.	11 Apr.	162	14 Oct.	21 Apr.	190
Kangnŭng	6 Dec.	2 Apr.	118	12 Nov.	4 Apr.	144	11 Nov.	8 Apr.	149
P'yŏng-yang	14 Nov.	28 Mar.	135	19 Oct.	15 Apr.	178	13 Oct.	23 Apr.	193
Wonsan	17 Nov.	4 Apr.	139	2 Nov.	7 Apr.	157	24 Oct.	15 Apr.	174
Yong-amp'o	12 Nov.	26 Mar.	135	18 Oct.	16 Apr.	180	11 Oct.	23 Apr.	195
Sŏngjin	8 Nov.	31 Mar.	144	25 Oct.	15 Apr.	172	23 Oct.	17 Apr.	177
Chunggangjin	21 Oct.	26 Apr.	188	26 Nov.	1 Apr.	249	25 Nov.	15 May	233
Unggi	2 Nov.	16 Apr.	166	21 Oct.	15 Apr.	176	14 Oct.	26 Apr.	195

Table 3. Average dates of the first and last occurrence of snow, ice and hoarfrost

other figures are from the dissertations by Schumacher [413] and Trojan [414]. To make the annual cycle apparent, the minimum and the maximum values are given along with the month to which they belong.

The average annual cycle of these climatic elements generally does not have the simple periodicity that air pressure, temperature and precipitation customarily have. The location on the windward or leeward side, which in turn changes with the monsoon, introduces considerable disturbances into the simple cycle. The simple periodicity is least affected in the case of relative humidity, which con-

	Relative humidity (%)				Percentage of possible sunshine			
	Minimum		Maximum		Minimum		Maximum	
	Amount	Month	Amount	Month	Amount	Month	Amount	Month
Cheju	69.0	Nov.	81.1	July	27	Jan.	63	Aug.
Mokp'o	70.0	Mar.	85.3	July	46	Dec.	68	Oct.
Pusan	49.5	Jan.	83.7	July	47	July	68	Nov.
Chŏnju	70.0	Mar.	79.5	July	47	Dec.	67	Oct.
Taegu	57.2	Mar.	74.7	Sep.	49	July	65	Oct.
Inch'ŏn	66.0	Feb.	85.9	July	44	July	68	Oct.
Seoul	63.0	Mar.	78.5	July	42	July	66	Oct.
Kangnŭng	54.9	Jan.	81.7	Aug.	40	July	66	Jan.
P'yŏng-yang	62.9	Apr.	79.9	July	48	July	71	Oct.
Wonsan	50.0	Jan.	81.7	Aug.	41	July	68	Jan.
Yong-amp'o	68.0	Mar.	87.9	July	45	July	72	Oct.
Shinŭiju	60.6	Feb.	84.9	July	39	July	74	Oct.
Sŏngjin	61.3	Dec.	89.0	July	40	July	67	Oct.
Ch'osan	61.2	Apr.	81.6	Sep.	46	Sep.	61	Oct.
Chunggangjin	60.8	Apr.	81.2	Aug.	44	May	63	Feb.
Unggi	49.2	Dec.	88.3	July	35	July	69	Jan.

[a]also Jan., [b]also Nov, [c]also Dec. and Jan., [d]also Sep., [e]also Dec.,

	Cloudiness				Days with fog			
	Minimum		Maximum		Minimum		Maximum	
	Amount	Month	Amount	Month	Amount	Month	Amount	Month
Cheju	6.3	Apr.	8.7	Jan.	0.0	Dec.[a]	2.1	July
Mokp'o	5.1	Oct.	7.7	July	0.3	Oct.	4.2	July
Pusan	3.9	Dec.[a]	7.4	July[f]	0.1	Oct.	2.3	July
Chŏnju	4.8	Oct.	7.7	July	0.4	Jan.	1.3	May
Taegu	4.2	Jan.	7.3	July	0.4	Mar.	3.1	Nov.
Inchŏn	4.3	Oct.	7.4	July	0.8	Oct.[b]	10.3	July
Seoul	4.3	Jan.	7.7	July	0.9	Mar.	1.8	May
Kangnŭng	3.7	Jan.	7.5	July	0.0	Nov.[c]	1.7	July
P'yŏng-yang	3.8	Jan.	7.7	July	0.9	Feb.	4.2	Aug.
Wonsan	3.3	Jan.	7.6	July	0.1	Oct.[d]	2.3	May
Yong-amp'o	3.0	Jan.	7.3	July	0.5	Jan.	4.4	July
Shinŭiju	3.7	Jan.[g]	7.9	July	–	–	–	–
Sŏngjin	4.1	Jan.	7.7	July	0.0	Nov.[e]	7.5	July
Ch'osan	3.2	Dec.	6.6	July[f]	0.0	Jan.	4.5	Sep.
Chunggangjin	3.8	Jan.	7.1	July	0.0	May[f]	3.3	Jan.
Unggi	3.3	Jan.	8.1	July	0.2	Dec.[a]	10.9	June

[f]also June, [g]also Oct. and Dec.

Table 4. Relative humidity and percentage of possible sunshine, cloudiness and days with fog forms directly to the monsoon rhythm. In correspondence with the conditions of the two main monsoon air masses, it naturally has a tendency toward a summer maximum and a winter minimum. This can be clearly seen in Table 4.

The same is true of cloudiness. The maximum is in July, the minimum primarily in January; in some cases it is in December or in the calm autumn, when in other cases there is a secondary minimum. The cycle is totally different in Cheju, which is completely exposed to the northern monsoon after it has just blown across wide expanses of sea. Therefore cloudiness is at a maximum in January. This is the highest maximum in Korea.

Cloudiness and percentage of possible sunshine naturally have the tendency to follow inverse annual cycles. Cheju's absolute maximum cloudiness in January corresponds to an absolute minimum of possible sunshine in the same month. The sun shines only a good quarter of the possible time. The otherwise predominant July maximum cloudiness corresponds to a July minimum of sunshine. That Mokp'o and Chŏnju, like Cheju, have their minimum amount of sunshine in midwinter is again because the northwest monsoon reaches them after having crossed the Yellow Sea. In July these stations have a second minimum of possible sunshine that is only about 2% higher. The contrast between the summer minimum and the winter maximum of possible sunshine is particularly great around Unggi, which, due to its location, is generally reached by both of the monsoons in unmodified form. The parallel seasonal contrast in cloudiness and relative humidity is just as great in Unggi. Almost exactly the same is true of Pusan, which is similarly located. At most of the other stations the maximum possible amount of sunshine is found during one of the winter months. The windless month of October is particularly favored. Cheju, in contrast, has its maximum in August, because it then lies in the lee of the sea monsoon.

There are generally no great regional differences in the maximum and minimum figures for these three elements. Thus it is not worthwhile to present them in map form. But we should point out the strong midwinter decrease in cloudiness from south to north and in relative humidity from west to east (foehn). It is quite a different case with the days with fog. Here the regional differences are great as well. The interior is generally less foggy than the coastal zones. Along the coasts two main areas of fog can be distinguished. One lies in northeastern Korea (Sŏngjin 25.7 days with fog on an annual average, Unggi 40.6); the other is off the southern west coast (Inch'ŏn 40.9, Mokp'o 18.8) and in the southwestern archipelago with its cold upwelling water in the summer, where there are as many as 70 days with fog. Extremely little fog is found along the east coast from Wonsan to Pusan, which is under the influence of falling westerly winds for much of the year. Along the coasts midsummer is the period of greatest frequency, a result of the general atmospheric conditions in this season. In addition, the seawater is then cooler than the air (p. 126). In the southwestern archipelago fog sometimes lasts for 3 days and nights without interruption in June and July. In the interior, which cools off extremely during the winter half of the year, the maximum number of days of fog often occurs in autumn (Taegu, Ch'osan) or winter (Chunggangjin).

The regional distribution and annual cycle of the climatic elements described above and the interrelationships between them can only be understood fully by means of a physiological characterization of the weather processes. Such a characterization follows, based on the studies on air masses, cyclones and typhoons by Japanese and Chinese meteorologists [428–450] and the climatological analysis of the isobaric maps in the "Daily Weather Charts of the North Pacific Ocean" (Central Meteorological Observatory Tōkyō, Vols. 1926–33) by G. Schumacher [413].

What we are accustomed to calling the winter monsoon (Mw) is not in reality a stationary air motion over a period of several months. Rather it is a continuous or discontinuous series of cold waves of varying origin and accordingly distinct properties. Sjan–zsi Li, to whom we owe a thorough study of cold waves in East Asia [436], distinguishes three East Asian types. They correspond to the types of Korean winter monsoon air pressure conditions identified independently by Schumacher [413]. The first type (Schumacher: Mwc) originates, according to Li, over the Kara Sea and western Siberia, the second (Mwa) in eastern Siberia, the third (Mwb) over the Sea of Okhotsk. Other authors assume that the source regions of the first two types are nearer (Arakawa [429], [428], Gherzi [431], Chang–Wang Tu [433]). The first two consist of continental polar air masses with large vertical temperature gradients in the lowest atmospheric layer. The third consists of maritime polar air masses. Cold waves of the first type are associated with northwesterly winds in Korea. They reach Korea when the high pressure center lies over central China and the low pressure center over Kamchatka. A cold wave of this type on 15 May 1928 caused the temperature in Chefoo (= Yantai, Shandong) to drop by 15.8° C in 15 min [439]. In Korea they often arrive in a somewhat modified form, i.e., they have warmed up somewhat and gained some moisture. The winter precipitation in southwestern Korea and Cheju can be explained in part by these cold waves. Under the influence of the first two types a slight amount of precipitation and cloudiness develops outside of this area as well, when a new cold wave shoves itself over a stagnating old one. From time to time cold waves of the first type bring yellow loess dust to the west coast of Korea as far south as Mokp'o [438]. Cold waves of the second type manifest themselves as northerly or northwesterly winds that bring almost unmodified dry, very cold, mighty air masses. Under the influence of these Mwa air masses the sky is usually completely clear. In this case the high pressure center lies in Mongolia and the low pressure center around the Kuriles. Cold waves of the third type follow the sea corridor toward the southwest. The center of the high then lies near Lake Baikal, the low again around the Kuriles and the Aleutian Islands. They contribute to the formation of the relatively large amounts of winter precipitation in the T'aebaek Range and its coastal margin. Schumacher divides each of these types into two subgroups (1 and 2), according to whether the pressure gradient, and with it the wind velocity, is large or small. He uses the term Mwx_1 to describe all three types when they are strongly developed and Mwx_2 when they are weakly developed. On an annual average Mwx_1 types prevail in Korea on 40.9 days, Mwx_2 types on 80.5 days.

In the three summer months Mw types are completely lacking. Korea is then more or less often under the influence of maritime tropical air masses that flow in from the southern quadrant. They originate in the oceanic high pressure areas and flow toward the continental low pressure areas. In their lower layers they are stable and very moist. Arakawa [428] says of this type of weather that it is "generally fine and hot." All other recent scholars also agree that this summer monsoon (Ms) itself generally brings only little precipitation to East Asia [441, 426]. It is a different situation when this warm, moist air mass must stream up over stagnating, cooler and heavier air along a stationary front. This is the origin of the plum rains in Japan and a small amount of the midsummer rain in Korea. The cooler and heavier air appears to lie over the Sea of Japan in this case. Incidentally, East Asian meteorologists reject the widespread view that the summer monsoon is the most important or even the only bringer of summer rain in their countries, and for good reasons. On clear, windless summer afternoons the stratification of the modified maritime tropical air becomes moist and unstable and local thunderstorms develop.

Schumacher also distinguishes three Ms types, according to the location of the centers of action, which now consist of the continental low and the subtropical Pacific high. In Msa the low pressure center lies over Manchuria, the high pressure center has advanced as far as the Pacific side of the Japanese islands. In Msb the high pressure center has moved farther to the northeast, the low pressure center is in Mongolia. In Msc the low pressure center dominates the region around Lake Baikal, and the high pressure center lies far to the east of Japan. Again two degrees of intensity are distinguished; however, the higher one occurs only as Msa_1. All three types are associated in Korea with south–southwesterly to south–southeasterly winds. Types a and b bring clouds. Schumacher also stresses for each of these types that they confer very little precipitation on Korea. According to his count, Ms types occur only in the months of May to September. The type Msa_1 prevails on an average of 10.9 days in a year, the type Msx_2 on 49.2 days. The frequency is thus significantly lower than that of the Mw types (60.1 compared with 121.4 days). All in all monsoon weather types prevail in Korea for somewhat less than half of the year.

It has long been known that the weather in East Asia is controlled not only by the monsoons but also by cyclonic weather types. Typhoons have been studied primarily for practical reasons. They consist of cyclonic eddies with extremely large gradients and high wind velocities but rather small diameters. In them only tropical air masses are moved. Most East Asian typhoons develop in the South Pacific along the equatorial front (trade wind inversion). According to Li [443] surges of cold air from the southern hemisphere into the northern hemisphere during the southern winter help to trigger them. They move along the equatorial front toward the continent of Asia (Deppermann). As soon as they reach the triple point where the southern monsoon, subtropical and northern air masses meet, they either continue to move in approximately the same direction over the continent, where they soon disperse, or their track veers off sharply to the northeast and then follows the front between the subtropical and the northern air masses [415, 442]. When it passes close to Korea, the typhoons influence the weather there. This occurs most frequently in late summer. Everywhere in East Asia this is the season of the greatest typhoon frequency, which is in complete accordance with Li's results.

According to Schumacher's count, Korea is influenced by typhoons on 14.2 days a year. These are concentrated in the months of July to October, particularly in August and September. When these eddies hit upon the saturated air masses previously brought in by the summer monsoon, this can cause tremendous amounts of rain to precipitate within a short period, particularly in southern Korea. For instance, the typhoon of 4 August 1927 with its southeasterly storm winds with a velocity of 30.9 m/sec brought 282 mm of precipitation to Cheju, and that of 11 Sept. 1927 brought 301 mm to the same station. The typhoon on 20 Sept. 1933 even brought 353 mm to the cape of Ulgi on the southeastern coast. The maximum amount of precipitation within a period of 24 h always falls between the months of July and September in Korea and it is often brought by typhoons. Typhoons occur more often in southern Korea than in northern Korea and their effects are much more destructive there. On the basis of the weather charts Schumacher identified five different typhoon tracks that strongly influence Korea. He designated them with the letters a–e. Figure 28 shows these tracks, as he calculated them. He calls the corresponding weather types La_1 ... Le_1. The map shows that tracks b, d, e influence only southern Korea, whereas a and c also affect northern Korea.

The characteristics of Korea's weather are influenced much more often by the second group of cyclonic weather types, the extratropical cyclones. Only recently has much attention been paid to their great importance for the weather and climate of East Asia. The reader is referred to the studies by Ooma [449], Tokuyama [445], Gherzi [446], Chang–Wang Tu [433] and particularly Shio Wang Sung [447]. In the extratropical cyclones tropical air enters into a reaction with polar air, so that they consist of a warm and a cold sector, as in Europe. According to Gherzi many of them develop in China at the line of contact between warm moist air and cold dry air. The northern ones come from southern Siberia or the lower Amur region. Tokuyama investigated 21 cases of cyclones that developed over the Sea of Japan. A large proportion of them occur in winter, when this sea is warmer than the surrounding area. Another important region of origin is the East China Sea. Mitchell [444] discovered that of the 83 primary cyclones that developed in the northern hemisphere between January and April 1925, nine had their center of origin in the 10° field 20–30° N, 120–130° E. This is by far the highest figure of all 10° fields. The cyclones travel from their centers of origin toward the eastern quadrant. In so doing they have a definite tendency to follow the polar front. Over the sea this front moves northeastward with the Kuroshio Current.

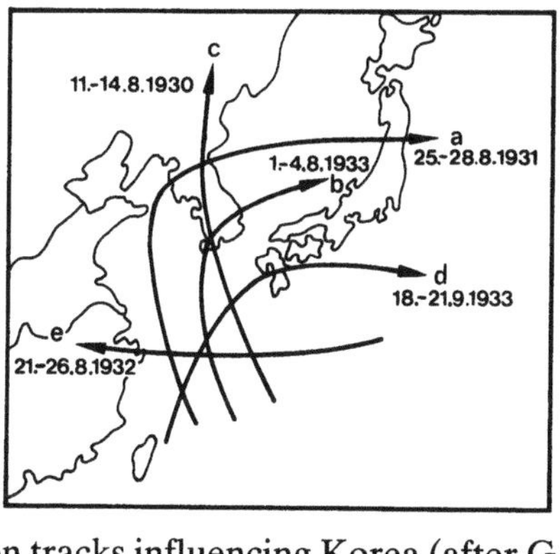

Fig. 28. Typhoon tracks influencing Korea (after G. Schumacher)

The Chinese meteorologist Shio Wang Sung meticulously studied all extratropical cyclones in eastern China and vicinity in the period from 1921–1930 [447]. There were 841 of them. He divided them into eight types according to center of origin and track followed. Six of them correspond to the types (a–f) that Schumacher independently established for Korea on the basis of the Japanese weather charts for 1926–1932. Figure 29 shows the tracks of Sung's eight types for the four seasons. All of his types except III and VII, which are lacking in Schumacher's classification, influence Korea's weather. A comparison of the four small maps shows the seasonal shifts in the tracks quite vividly. According to Sung the long axes of the cyclones measure as much as 1900 km, the short axes as much as 1100 km. Their moving speed reaches its maximum in winter (up to 62

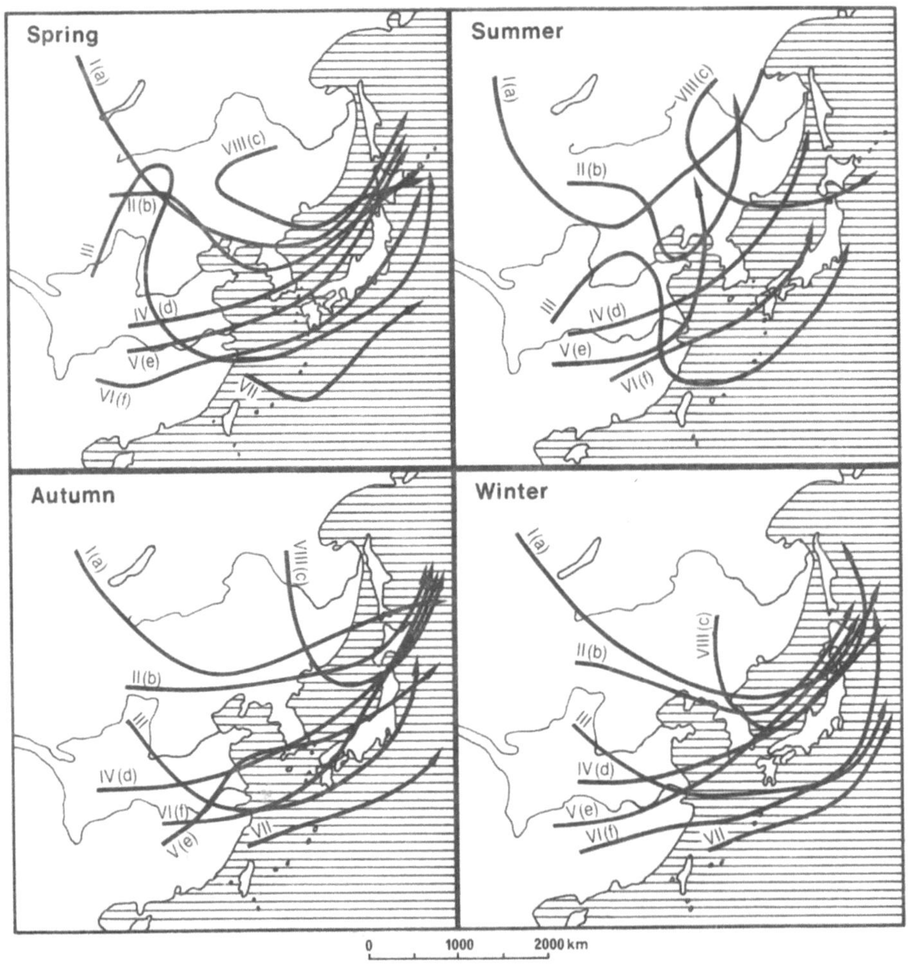

Fig. 29. East Asian cyclone tracks (I–VIII) and their seasonal shift (after Shio Wang Sung) The letters *a–f* refer to G. Schumacher's classification

km/h) and its minimum in summer (more than 27 km/h). Type I occasionally reaches speeds of more than 80 km/h.

Cyclones influence Korea most easily when weak monsoon types (Mwx_2 or Msx_2) have previously prevailed, because the gradient is then smallest and the air over Korea is not subject to strong wind action. Therefore cyclone activity is greatest in spring and autumn. This is the result unanimously reached by Sung [447], Chu [448], Ooma [449] and Schumacher [413]. In winter cyclones following Schumacher's track a (Sung's track I) generally influence only northern Korea. In summer, however, when they are strongly developed, they affect all of Korea. Even then precipitation usually does not fall in southern Korea. Pretty much the same is true of b (II). Track c (VIII) brings large amounts of precipitation everywhere with westerly winds, as the cyclone crosses the Sea of Japan. Track d (IV) proceeds across southern or central Korea and brings heavy rainfall to the western half of the country. Cyclones following track e (V) bring not only typhoons; they also confer upon south Korean stations the maximum amount of rainfall in a given month. And when the center lies around the axis of the peninsula, all stations register very strong winds. Finally, cyclones on track f (VI) produce maximum amounts of rain with northerly storms in the southwest and on Chejudo. Central Korea receives very little precipitation, and north Korea none.

In winter it is not uncommon for cyclones on track a (I) to repeat themselves in periodic succession. This is the basis of the Korean weather maxim "*sam han sa on*" (= three cold four warm, i.e., three cold days follow four warm ones). By harmonic analysis of the temperatures for the period from November 1929 to March 1930, Okada confirmed this rule for southern Manchuria and Korea [423]. Takahashi and Harada attribute the periodicity of cyclone recurrence to a free 7–day oscillation in the atmosphere [424, 440].

Extratropical cyclones are for Korea the most important bringers of rain. According to Schumacher they govern the weather in the country on an average of 105.9 days per year, and the largest proportion of the precipitation is of cyclonic origin. Naturally, the amount of rain that a cyclone causes to precipitate depends on the air temperature and is thus larger in summer than in winter. The record amounts of precipitation ever measured within 24 h (P'yŏngsan, 490 mm, 19 June 1924, Kwangju near Seoul, 486 mm, 1 August 1920) came from summer cyclones. The cyclones coming from the west develop the greatest precipitation intensity along the western slope of the main Korean tilt block. Northeastern Korea, which lies in the rain shadow of the Main Korean Range, receives relatively little rain in all seasons. In southern Korea typhoons in addition to the extratropical cyclones contribute noticeably to the increased amounts of rain in midsummer and early autumn.

Finally, we must mention one last type of weather type influencing Korea. Above Korea itself a high pressure area (H) can develop. Then gentle winds flow out of the center. Generally the center lies over central Korea, so that the northern stations experience winds from the southern quadrant, the western stations from the eastern quadrant, etc. Often it is calm. The sky is generally clear. Both insolation and radiation are great. There is no precipitation. Korea then has an autochtonous weather type. These types exist on an average of 34.8 days per year.

Schumacher has calculated the mean monthly frequency of the different weather types described above for the period from 1926–33 (cf. Table 5).

By analyzing the frequency data in this table and applying them to the above mentioned monthly averages for the climatic elements we arrive at a physiological view of the annual cycle of the Korean climate.

In January the prevailing winter wind direction from north–northwest to north–northeast is particularly pronounced. Only along the east coast does a low over the Sea of Japan draw in westerly winds. At all stations January is, on the average, the coldest month. The temperature contrasts between north and south are greatest then. Simultaneously this month records the lowest average extreme temperatures. The daily temperature range is great, particularly in northern Korea. Cloudiness and relative humidity have a primary or secondary minimum, and sunshine a primary or secondary maximum. Precipitation often has a minimum, whereas the number of days with snowfall often has a maximum. A relatively large amount of precipitation falls on the remote islands and on the east coast.

The winter continental high is most stable and most strongly developed in this extreme winter month. Korea lies between it and the northern Pacific low. Mwx_1 and Mwx_2 weather types reach their maximum in this month. Together they prevail on 22.6 of the 31 days of January. This fact explains a large number of the characteristic features that we have just enumerated. The lowest temperatures and the greatest wind velocities occur in connection with Mwa_1 types. The sky is then crystal–clear, and the radiation is enormous. Ground fog develops only locally. When Mwx_2 and H weather types prevail there is a pronounced temperature inversion in northern Korea. Since the existence of a strong pressure gradient is an obstacle to the development of cyclones, they occur on only 5.7 days. That is the lowest monthly rate, except for December. They bring most of the little precipitation that does fall. Among these cyclone days are 1.5 with Lc_2 types, which, along with the Mwb types, bring precipitation to the east coast, and 1.9 with Lf_2 types, which bring moisture to the coasts and the islands in the extreme south.

The examples in Table 6 may suffice to show the contrasts between different months of January. They show very clearly in what manner the character of a

Weather conditions[a]	Jan.	Feb.	Mar.	Apr.	May	June	July	Aug.	Sep.	Oct.	Nov.	Dec.	Year total	Year (%)
Mwx_1	9.2	6.1	4.6	1.2	0.0	–	–	–	0.8	4.0	6.2	8.8	40.9	11.2
Mwx_2	13.4	11.1	9.8	6.1	2.3	–	–	–	3.5	10.9	10.8	12.6	80.5	22.1
Msx_1	–	–	–	–	1.4	2.9	3.4	2.9	0.3	–	–	–	10.9	3.0
Msx_2	–	–	–	–	3.0	14.7	17.6	12.5	1.4	–	–	–	49.2	13.5
Lx_1	–	–	–	–	–	–	1.6	6.4	5.6	0.6	–	–	14.2	3.9
Lx_2	5.7	7.5	9.9	16.6	17.9	7.2	5.5	4.7	11.4	8.1	6.2	5.2	105.9	28.9
H	1.4	2.0	4.0	3.9	3.5	1.6	0.3	0.6	3.4	5.8	5.0	3.3	34.8	9.6
Unclassifiable	1.3	1.3	2.7	2.2	2.9	3.6	2.6	3.9	3.6	1.6	1.8	1.1	28.6	7.8
Sum	31.0	28.0	31.0	30.0	31.0	30.0	31.0	31.0	30.0	31.0	30.0	31	365.0	100.0

[a] Mwx_1 = winter monsoon, strongly developed. Mwx_2 = winter monsoon, weakly developed. Msx_1 = summer monsoon, strongly developed. Msx_2 = summer monsoon, weakly developed. Lx_1 = typhoon. Lx_2 = extratropical cyclones. H = high pressure in Korea itself.

Table 5. Mean monthly frequency of the main weather types in days given year is determined by the frequency of the severe and mild Mw weather types and the different types of cyclones. The year 1916 had a particularly warm January; in it the entire southern half of Korea had average temperatures above 0° C. January 1917 was extremely cold; in it the −5° C isotherm lay at the extreme southern tip of the peninsula [409, next to the last map].

In some years the lowest mean temperatures and the lowest mean extreme temperatures are not reached until February. Everywhere the average February temperatures are only slightly higher than the January temperatures. At many stations, particularly in the north, precipitation has its minimum in February, even under average conditions, and otherwise the amount of precipitation is not much greater than in January. Snow falls almost as often as in January. On the whole, February is very similar in character to January. The frequency of the weather types is in accordance with this. Mw types decrease, cyclonic types increase. Particularly great is the decrease in Mwx_1 types. This explains the moderation of the mean temperatures and wind velocities. The increase in cyclonic types is mainly in Lf_2, thus explaining the increase in precipitation at some stations in the extreme south.

During the month of March the mean temperature rises above +5° C in southern Korea and above 0° C in central Korea. In southern Korea March is the last month with snowfall. In the north, however, the snowfall frequency often equals that of January, and at some stations even surpasses it (e.g., Shinŭiju, Jan. 5.0 days, March 6.7). At some stations in western and southern Korea March is the month with the absolute minimum of relative humidity. Generally, however, the mean amount of precipitation is quite a bit higher than in the midwinter months. In many of its averages March appears to be the first transitional month. In some years, e.g., 1933, it can still be a cold and dry winter month. In others, e.g., 1930, it is already a warm and moist spring month. It depends on whether the continental high is still strongly developed, or whether it has begun to weaken, allowing cyclones to take their course through the Korean sphere. On the average the frequency of Mw weather types decreases, accompanied by a corresponding increase in Lx_2 and H types. The frequency of La_2 types explains the considerable increase in precipitation in northern Korea.

In April the monthly mean temperature has again risen noticeably. The temperature span between March and April is the greatest of the whole year. In central and northern Korea the average temperatures are now also above +5° C, in fact they are generally considerably higher. In most of Korea April brings the last

	Mwx_1	Mwx_2	H	La_2	Lb_2	Ld_2	Le_2	Lf_2	T = temperature P = precipitation
Jan. 1933	19	6	1	0	0	2	0	2	T: very cold, P: average
Jan. 1932	7	18	4	1	1	0	0	0	T: relatively mild, P: in the south very little, in the north somewhat more
Jan. 1928	4	16	2	1	0	1	2	5	T: even milder, P: in the south rather heavy, in the north little

Table 6. Frequency of January weather types in days

frost. Cloudiness and precipitation increase greatly. In the plains of central Korea, April is the last month with occasional snowfall. Almost everywhere hoarfrost occurs for the last time at low elevations.

The barometric gradient between Mongolia and the surrounding seas has dropped to almost zero. Thus cyclonic weather types increase noticeably; with 16.6 days they prevail during more than half of the month. Particularly frequent are La_2 types on the one hand and Lf_2 types on the other hand, so that both the extreme north and the extreme south receive moisture. The b and d cyclones bring abundant rain to central Korea. Severe Mw types are now encountered very infrequently and then do not have such drastic effects, since the continent north of Korea has also warmed up considerably meanwhile. Mw_2 types now make their appearance in the form of fine weather, generally with cloudless skies and weak, cool, dry winds from northerly directions.

In May monthly temperatures above 15° C begin in southern and central Korea. In the north the monthly temperatures are already 11–13° C. The regional similarity in the summer climate begins to show itself. At many stations in the south there is less precipitation than during the previous month. In the north, in contrast, precipitation continues to increase noticeably. There the last snowfalls and the last frost are recorded. Everywhere relative humidity and cloudiness increase strongly.

May is the only month in the first half of the year in which both Mw and Ms types occur. Mw types, however, occur only in the weak form, which brings good weather with cool northerly winds at this season, and their effects are thus not very different from H types. The absolute temperature minima are found in connection with H types. More frequent are Ms types, resulting in cloudiness and high humidity. They already occur in the strong form, which is associated with strong winds from the southern quadrant. Above all, May stands out due to its maximum number of days with cyclonic weather types. The 10.9 La_2 days are responsible for the increase in the amount of precipitation in northern and central Korea. With their southerly and westerly winds they also cause the noticeable increase in temperature. The occasional Lf_2 types already bring large amounts of precipitation of great intensity to southern Korea.

In June the monthly mean temperatures rise above 20° C in southern and central Korea, in northern Korea to 17–19° C. The mean maximum temperatures are around 24–29° C; the absolute temperatures rise above 37° C. The amount of precipitation continues to increase and reaches its maximum in some years. The air is damp and sultry, and the sky is very cloudy. Midsummer has arrived.

The subtropical Pacific high has now moved far to the north and skirts the Pacific edge of Japan. It is the origin of the summer monsoon, which blows on an average of 17.6 days. Only 2.9 days have Msx_1 types, but they bring the highest temperatures to Korea. The large amounts of precipitation are not brought by monsoon weather types, but by cyclones. If cyclonic types are lacking, as in June 1931, the month is relatively dry despite continuous Ms types.

July is in many respects the antipode to January. The prevailing winds come from south to south–southeast. At the northern edge of Korea the highest monthly mean temperatures and mean maximum temperatures are registered. With the exception of northeastern Korea the average precipitation cycle reaches its maximum. Almost everywhere the highest figures for precipitation intensity, relative humidity and cloudiness are reached and the lowest for percentage of possible sunshine. On the coasts fog has its maximum frequency. Midsummer is in its prime.

Compared with the previous month the average pressure gradient from sea to land has become even stronger. The summer monsoon reaches its maximum frequency with 21.0 days. For the first time the extratropical cyclones are joined by typhoons. Both cause tremendous amounts of precipitation to fall from the saturated monsoon air mass. Of the 378 mm of rain that fell on Mokp'o in the wet July of 1926, 338.5 mm fell on the 6 days with cyclones. The same is also true of the other southern Korean stations in the same month. In July 1929, in contrast, stable monsoon weather prevailed from the 10th on, and the precipitation from this day on till the end of the month did not measure more than 1 mm at any southern Korean station. Thus the summer monsoon itself is not the basis of rice cultivation, as one generally reads. It is its enemy.

With the exception of the northern edge, August has the highest monthly mean temperatures. The mean extreme temperatures also reach their maximum. The differences in temperature between north and south have almost disappeared. Only northeastern Korea has relatively low temperatures, due to the Sea of Japan. There the maximum amount of precipitation in the annual cycle is also not reached until August. The differences between the July and August figures for all climatic elements are, however, only slight. Like January and February, July and August are very similar.

The sea monsoon is already less frequent. Still, it prevails on half of the days in August. The temperature of the monsoon air mass is often even higher now than in July, which explains the August maximum in the average temperature cycle. Only at the northern edge is the continental heating no longer quite as effective, so that on the average a slight drop in temperature sets in here. Typhoon activity has increased significantly compared with July and now reaches its maximum. Days with maximum wind forces or maximum precipitation are now almost always days with typhoons, at least in southern Korea.

The drop in temperature from the average August figures to the September values is suddenly very great everywhere. In the north the mean temperatures begin to sink below 15° C toward the end of the month, and the first early frosts commence. By Central European standards, however, September must be considered a summer month, as far as its temperature characteristics are concerned. On the average a lot of rain still falls, but the month of September varies considerably from year to year in this respect. The amount of sunshine increases noticeably, and cloudiness and humidity decrease accordingly. The average wind forces have a primary or secondary minimum at many stations, with the exception of the northeast.

The continental high around Lake Baikal begins to gain in influence again. In addition to 1.7 Ms days, 4.3 Mw days also occur in September, although generally still in the weaker form. The Msx_2 and H weather types give September a quiet, sunny character at times. The number of cyclones and typhoons together reaches its maximum for the second half of the year, however, which explains the large amounts of precipitation that still fall. In some Septembers Mw and H types predominate, and in other L types, and this is the explanation for the fact that both the maximum and the minimum amount of precipitation in the annual cycle can fall in this month.

From September to October the temperatures drop even more rapidly. In most of the country the first frosts begin. Precipitation also decreases considerably, on the average. Quite often the primary precipitation minimum in a given year falls in this month. In northern Korea hoarfrost sets in and an ice cover begins to develop on standing fresh water surfaces. In many stations in northern and central Korea snow falls occasionally. The average temperatures drop below +5° C again in northern Korea. At quite a number of stations the percentage of possible sunshine has its primary maximum, and in many years October is just as quiet and sunny as September. In others it is a stormy wet autumn month. October 1927 belonged to the first type, October 1931 to the second type.

H weather types reach their average maximum value with 5.8 days. The lowest temperatures are brought by Mw types, which already prevail on almost half of the days in October again. The maximum temperatures are brought by cyclonic weather types, because the seas are still warm. Ld_1 typhoons influence southern Korea's weather for the last time. Much more common are Lx_2 types.

The drop in temperature from October to November is the steepest of all at many stations. Thus in much of the country the temperatures fall below the +5° C mark. In northern Korea below 0° C monthly mean temperatures are sometimes recorded. Even in the extreme south frosts begin to occur, and at all stations in the country, except for the southernmost ones, snowfall and hoarfrost begin and an ice cover starts to develop on standing water surfaces. The precipitation figures approach their winter minimum. In northern Korea November is already a winter month.

The winter continental high has become more stable, and Mw weather types already prevail on 17.0 days. Mwx_1 types now again bring the maximum wind forces and the lowest temperatures. H and Lx_2 types are on the decrease. With the latter the maritime influence on the country also diminishes.

In December the mean temperatures are below freezing in almost all of northern and central Korea. Only at the southern fringe do they still linger around and above +5° C. Hoarfrost and snowfall, however, now occur there, too. In many years December is already the coldest month, and in many the month with the least precipitation. Simultaneously the precipitation intensity reaches its minimum, for the most part.

The continental high has increased in strength. The number of Mwx_1 days with maximum cold and wind velocity has risen to 8.8. Winter monsoon weather types even prevail on a total of 21.4 days. Otherwise, as in January, only cyclonic types (5.2) are of any importance. They bring milder temperatures and precipitation. In December, however, they reach their minimum for the entire annual cycle.

To summarize, we can say that in the climatic sphere Korea fully reveals its character as a transition zone, from the seasonal and, particularly in winter, regional point of view. The surges of cold air that we customarily know as the land monsoon bring extremely low temperatures and continental dryness to the country during the winter. This cold air reaches northern Korea in a more or less unmodified form. Winter precipitation is brought primarily by the infrequent cyclones, which also bring in mild maritime temperatures. On its way across the peninsula and the adjacent open seas, the dry and cold land monsoon is modified. Cyclone activity is also greater in the south than in the north. Therefore, in the winter months temperature and precipitation increase considerably toward the south. In summer the sea monsoon blows, bringing warm, moist air from more southerly climes to the country. It envelops all of Korea in its uniformly warm mantle. For many weeks the entire country resembles a hothouse. The mountains hide behide a thick layer of low–lying gray clouds, and fog hangs over the coasts. But the sea monsoon does not bring much more precipitation than the land monsoon modified by the sea. The very heavy precipitation that falls in the summer is brought by cyclones and typhoons, which penetrate into the air mass that the sea monsoon has saturated with moisture. As a result, the summer precipitation is not highly dependent on the exposure to the monsoon, as in India and Indochina. It is true, though, that those parts of the country that lie in the lee of the cyclones receive little rain, for instance, northeastern Korea and the northern portion of the Naktonggang basin. Korea has a summer corresponding to that of the perennial climate of the tropical rainforests and a winter that is similar to Ireland in the extreme south and to the Pechora basin in the extreme north.

Thus far no attempt has been made at a special synthetic examination of the climate of Korea. Okada's work [416] is limited to a climatographical description of the means of the meteorological elements in the Japanese empire. On the other hand, Korea of course also appears on the maps of climatic divisions of the earth. I shall restrict myself to Köppen's famous system and v. Wissmann's modifications thereof for Eurasia [419]. Köppen's maps have of course changed somewhat in the course of the many years that this tireless scholar has continued to work on relevant problems, partly because more recent meteorological data have made it possible to draw the boundaries more exactly and partly because Köppen's definitions have changed somewhat. Therefore I refer only to Köppen's last climatic map (Köppen–Geiger, Climatic Map of the World. Wall map by Justus Perthes, Gotha, 1929) and the definitions upon which it is based.

Korea lies in the border region between the C and D climate groups. In continental Korea most of the stations, in Hamgyŏngbuk–do even the coastal stations, have average temperatures below +1° C for at least 4 months and thus belong to the D group. Figure 30 shows the boundary, based on all meteorological data. For the most part, the mountains of central and even southern Korea certainly also belong to the D group, but there are no stations at high elevations to provide the necessary evidence. Therefore I have not attempted to show these islands of D climates inside the bounds of the C group. The boundary drawn between D and C by Köppen–Geiger themselves, though, lies too far south. All stations in Hwanghae–do and southwestern P'yŏng–annam–do and the stations near the coast in Hamgyŏngnam–do are C stations.

The boundary between the regions with a dry season in winter (w) and those that are moist in all seasons (f) in Köppen's temperate climates passes through Korea. Köppen classifies a region as winter dry if the month with the least precipitation falls in the winter half–year and if this month has less than one tenth of the amount of precipitation that the month with the most rain has. Most of northern and central Korea lies in the w region. The f region is restricted to the southern two–thirds of southern Korea and a small strip stretching north along the coast of the Sea of Japan as far as the cape of Musudan. After what we have said above about the annual cycle of precipitation this is easy to understand. Thus the w/f

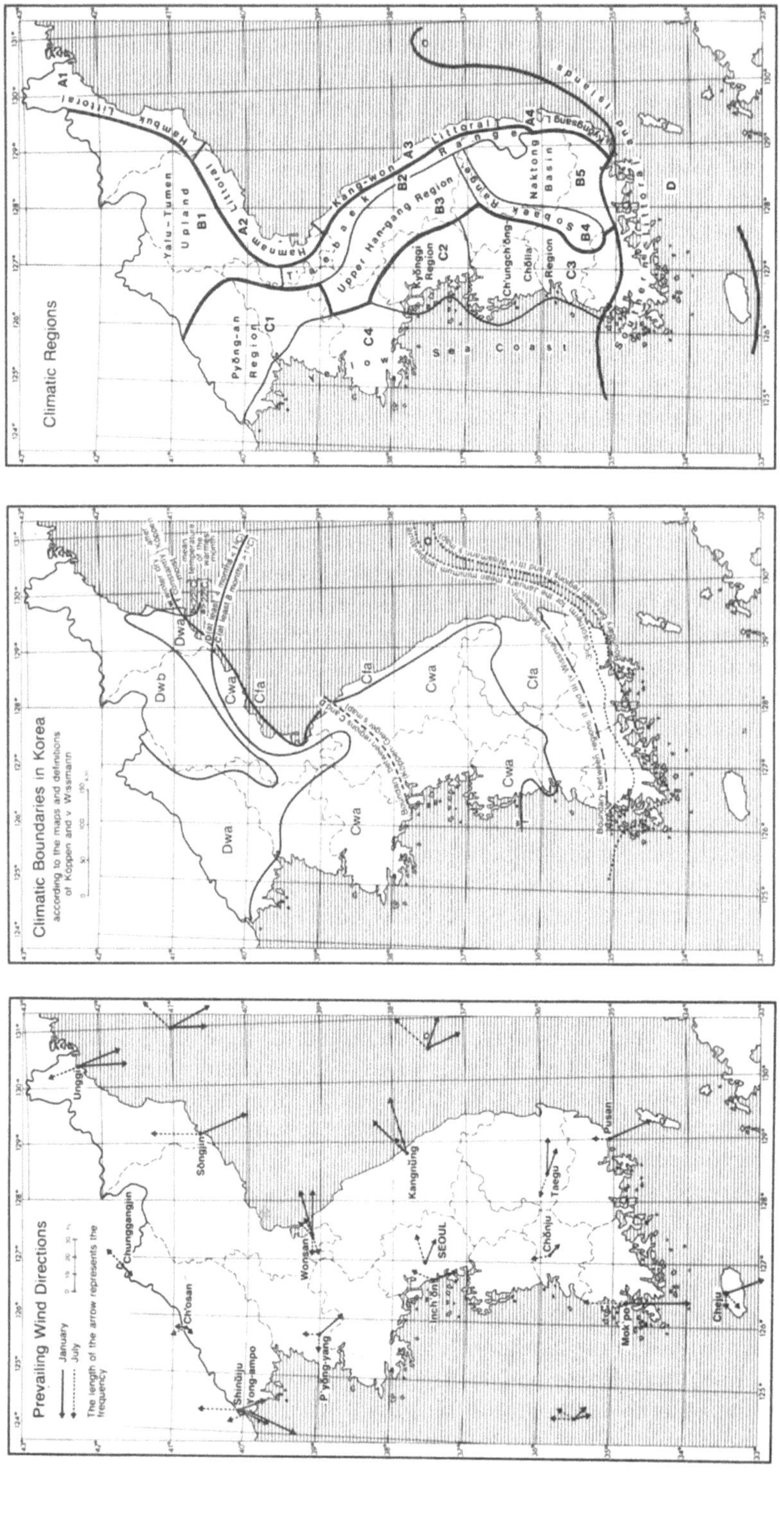

Fig. 30. Average wind directions and climatic divisions

boundary does not coincide with the D/C boundary in Korea. Between the Cf climate of southern Korea and the Dw climate of northern Korea there is a large area with Cw climate. Köppen–Geiger's map, however, shows Cw climate only on the southern coast of Hwanghae–do.

Finally, Köppen subdivides his climates again according to the temperatures in the warmest and coldest months. The a regions have a mean temperature above +22° C in the warmest month, the b regions less than 22° C. The unreduced 22° C isotherm of the warmest month includes the Kaema upland and the coast of northeastern Korea. Actually, the high mountains in central and southern Korea also belong to the b region. Even the top of Hallasan on Chejudo probably only has an average temperature of approximately 16° C in the warmest month.

From what we have said above, theoretically the following eight types of climate should occur in Korea: Cfa, Cfb, Cwa, Cwb, Dfa, Dfb, Dwa, Dwb. Since the D/C boundary always runs south of the b/a boundary, the combinations between C and b, i.e., Cfb and Cwb, are virtually nonexistent. Nor is there a station representing the Dfa climate. Perhaps there is a small Dfa region on the southern coast of Hamgyŏngbuk–do. If the station Sŏngjin had only 4 mm more precipitation in February, it would represent this type. The Dfb climate is represented by the neighboring lighthouse station Musudan. Basically only the Dwb, Dwa, Cwa and Cfa climates exist in Korea. Dwb prevails in the Kaema upland and the northeastern tip of the country, as well as in the mountains of central Korea. Dwa is found in most of P'yŏng–anbuk–do, in the Wonsan arc of the Main Korean Range and in the southern part of Hamgyŏngbuk–do. Cwa is the climate of most of central Korea and the northern third of southern Korea. Cfa is found in the southern two– thirds of southern Korea and on the east coast as far north as the island Mayangdo (Fig. 30).

There is no clear relationship between these climatic regions and the distribution of vegetation in Korea. Previously and again in his most recent brief publication in the *Handbuch der Klimatologie,* Köppen has his C/D boundary coincide with the January −3° C isotherm. In Korea this isotherm most likely does indeed include the region with an annual snow cover, as Köppen wishes.

Of course we do not wish to disparage Köppen's impressive achievement. But in spite of the great attention that his classifications have rightfully received, it is after all important to examine their repercussions in regions the size of Korea, as well. Von Wissmann [419] takes the same standpoint and classifies all of Eurasia according to viewpoints that deviate from Köppen's in some decisive points. He uses the +2° C isotherm of the coldest month to separate his II climates from the III climates. Figure 30 shows how this line runs according to v. Wissmann's illustration, on the one hand, and to Schumacher's climatological averages [413], on the other hand. The difference is in fact very slight, and both lines approximate the northern limit of evergreen plant formations quite closely. It would certainly not be worth wasting any words on this difference, at least from the standpoint of Eurasia as a whole, if one basic point did not seem important to me. Von Wissmann wishes to draw the boundary between primarily evergreen and primarily summergreen plant formations at the +2° C isotherm of the coldest month. However, the determining factor for the northern limit of evergreen plant species, which are sensitive to cold, is low night temperatures or even individual cold waves, which only barely appear in the mean temperatures. Therefore, the northern limit of primarily evergreen plant formations must be correlated with the mean or absolute minimum temperatures. In southern Korea this limit coincides quite well with the −3° C isotherm for the January mean minimum temperature. As Fig. 30 shows, this isotherm lies between the II/III boundary according to v. Wissmann's illustration and that according to his definition. South of it even the absolute minima of the period from 1914–1933 only very occasionally reached −15° C (p. 91).

Von Wissmann includes the Kaema upland and the adjacent portion of the Main Korean Range in his climatic region IVT, which occurs here only at high elevations (h) as an alpine belt. He has the III/IV boundary following the 4° C annual isotherm. The region v. Wissmann indicates does actually coincide with the area of distribution of stations with a mean annual temperature below +4° C. What I do not understand is that this region is assigned to the type T (trocken = dry). In regions with the rainy season in summer the boundary between T and F (feucht = moist) is considered to lie at the line where six times the monthly precipitation, measured in cm, is less than the monthly temperature plus 7 in more than 4 months (in more than 4 months: $6\varrho < \tau + 7$). In the region in question 6ϱ is greater than $\tau + 7$ in all months. After all, in the months with a temperature below −7° C, and those are the driest months, 6ϱ can never be less than $\tau + 7$, for 6ϱ is always positive.

The elevation of the 4° C annual isotherm varies tremendously in northern Korea. In extremely continental regions, with their exceedingly low winter temperatures, it is very low. The cold winter lowers its level to 400–500 m. On the outer edge along the coastal margin of the Sea of Japan, with its milder winters, it probably lies at approximately 1100–1200 m, as far as we can calculate by extrapolating from the figures for the stations in the valleys. This isotherm does not have any relationship whatsoever to any upper limits of plant formations, such as v. Wissmann intends to establish. To summergreen trees and shrubs the intensity of the cold in winter is more or less immaterial. The important thing is the length of the growing season and the amount of warmth they receive from the sun during this period. Thus only the principle of duration allows us a thermal delineation of the transition zone from primarily mixed forests to predominantly coniferous forests, which lies at 1400–1500 m in northern Korea (p. 138f). Nothing more can be said today, because only one station, P'ungsan (1150 m), lies close to the level of this transition zone. There are no higher stations.

In a regional geography it hardly seems necessary to provide a regional classification of climate with exactly defined dividing lines based on threshold values, as is certainly advantageous for specific climatic purposes and for large regions. In the following I shall therefore content myself with outlining a purely qualitative classification of the climate of Korea.

A small fringe along the coast of the Sea of Japan (A) and a much wider strip on the western side of the country (C) stand out clearly against a central region that follows the axis of the peninsula (B). All three regions extend over almost the entire length of the country from north to south and, with decreasing latitude, are subdivided into separate smaller regions. At the end of the peninsula they give way to the southern coast and island fringe (D), to which Ullŭngdo also belongs climatically. Thus we arrive at the following regional classification of the climate (Fig. 30).

A. Sea of Japan Littoral. Maritime outer edge of Korea. Rel. mild winters and cool summers. Rel. little precipitation.

1. Hambuk Littoral (Climatic chart Chongsŏng, Fig. 27). Very little precipitation, particularly in winter. Cold winter, cool summer. Much summer fog.

2. Hamnam Littoral. Precipitation and winter temp. increasing southward. Foehn. Rel. warm spring and autumn.

3. Kang–won Littoral (Kangnŭng). Rel. abundant winter precipitation. Foehn. Very mild winter, rel. cool summer.

4. Kyŏngsang Littoral. Rel. abundant winter precipitation. Very mild winter, hot summer. Little fog.

B. Central Interior Region. Tendency toward continental climatic features.

1. Yalu–Tumen Upland (Chunggangjin). In midwinter extremely cold radiation climate with temperature inversion and cold air pockets in the foggy basins and frequent snowfall. Highest monthly temp. in July. Extremely large annual temperature range. Little precipitation.

2. T'aebaek Range (Onjŏng–ri). Mountainous climate, in winter much snow, in summer abundant fog and rain. Winter cold decreases with latitude.

3. Upper Han–gang Region (Ch'unch'ŏn). Continental precipitation cycle and continental decrease in precipitation. Tendency toward large annual temperature range.

4. Sobaek Range. Mountainous climate similar to southern portion of B 2. Still rel. cold winter.

5. Naktong Basin. Basin region with little precipitation and hot summer. Continental climatic elements. Rel. large number of days with snowfall. Large daily amplitude.

C. Western Region. Outer edge of Korea facing the continent.

1. P'yŏng–an Region. Abundant annual precipitation. Cold winter, hot summer.

2. Kyŏnggi Region (Kanghwa). Mod. cold winter, hot summer. Abundant precipitation, in summer of very high intensity.

3. Ch'ungch'ŏng–Chŏlla Region. Milder winter, hot summer. Abundant precipitation, even in winter rel. large amount.

4. Yellow Sea Coast (Sŏdo). Rel. small annual temperature range. Very little annual precipitation. Much summer fog.

D. Southern Littoral and Islands (Ullŭngdo, Cheju). Average January minimum temperature above −3° C, January average above 0° C. Infrequent snowfall. Very abundant precipitation, on the islands even in winter. High precipitation intensity. High rel. humidity and cloudiness in all seasons.

3 Rivers and Seas

Korea's principal rivers are treated in detail in Part Three. Thus it will suffice if we mention only a few main traits here.

The size of the peninsula (p. 2ff) obviously limits the extent to which large river systems can evolve. This would be even more the case if the main drainage divide followed the axis of the peninsula. As we pointed out in the introduction, however, because Korea is essentially a tilted fault block, the main divide lies only a short distance from the coast of the Sea of Japan. For this reason all of the rivers flowing into the Sea of Japan from the peninsula are very short. Only very few exceed 50 km in length and a drainage basin of 1000 km^2. The river systems that develop on the dip slope of the tilt block are all the more imposing. The Han–gang system spreads across the middle of the peninsula. If the Imjin–gang and the Yesŏnggang, which join the Han–gang in its estuary, are counted as part of the system, we have a drainage basin of 38,522 km^2, i.e., not much smaller than that of the Weser (approx. 46,000 km^2). In contrast to the Weser, however, this large area is only due to the extremely large number of branches, and the main branch of the Han–gang (482 km) is much shorter than that of the Weser (710 km). In the southeastern part of the peninsula the orographical features of the T'aebaek Range and the Sobaek Range provide conditions in which another large river system can develop. It is the very regular Naktonggang system, which is completely adapted to the form of its basin. The Naktonggang, with its 526 km, is even the longest river of the peninsula, but its system (23,860 km^2) is much less extensive than that of the Han–gang. This is truer still of the Taedonggang, which flows in the very narrow neck of the peninsula. In addition to these there are a number of good–sized rivers flowing into the Yellow Sea, which are listed in Table 7. They all have in common that their drainage basins narrow down noticeably near their mouths, leaving wider or narrower strips between their mouths that are drained by very short coastal rivers or creeks.

The rivers in the more spacious continental portion of the country have much more impressive proportions. The most important ones are the Tumen and the

Name of river	Length (km)	Drainage Area (km^2)	Name of river	Length km	Drainage Area (km^2)
YELLOW SEA RIVERS			KOREA STRAIT RIVERS		
Yalu	813	62,639[a]	T"amjin-gang	45	1034
Taeryŏnggang	150	3635	Sŏmjin-gang	212	4897
Ch'ŏngch'ŏn-gang	213	9778	Naktonggang	526	23,860
Taedonggang	431	16,465			
Chaeryŏnggang	129	3671	SEA OF JAPAN RIVERS		
Yesŏnggang	174	4049			
Imjin-gang	254	8118	Hyŏngsan-gang	62	1168
Han-gang	482	26,355	Namdaech'ŏn (Anbyŏn)	82	1162
Ansŏngch'ŏn	76	1722	Yonghŭnggang	135	3397
Sapkyoch'ŏn	61	1619	Sŏngch'ŏn-gang	99	2338
Kŭmgang	402	9886	Namdaech'ŏn (Pukch'ŏng)	67	2056
Man-gyŏnggang	99	1602	Namdaech'ŏn (Tanch'ŏn)	161	2405
Yŏngsan-gang	116	2798	Puktaech'on	118	1373
			Namdaech'ŏn (Kilchu)	99	1371
			Ŏrangch'ŏn	103	1898
			Tumen	516	41,243[b]

[a] In Korea 32,064 km^2. [b] In Korea 10,513 km^2.

Table 7. Length and drainage area of Korean rivers

Yalu. With its 813 km, the Yalu is the longest Korean river, and the Tumen, with 516 km, is not much shorter than the Naktonggang. Both of them form the border, however, so that only half of the drainage basin of the Yalu and only a quarter of that of the Tumen belong to Korea. In continental Korea the main divide between the rivers flowing into the Yellow Sea and those flowing into the Sea of Japan no longer lies on top of the Main Korean Range. The Yalu has broken through and moved its sources back to the Mach'ŏllyŏng Ridge (Fig. 1). Because of this most of the Kaema upland belongs to the Yalu drainage basin, and the main divide between the Yellow Sea and the Sea of Japan generally lies close to the latter in northeastern Korea as well.

Table 7, which lists all Korean rivers with a drainage basin of at least 1000 km^2, gives a vivid picture of the tremendous differences in size between the rivers on the western and eastern slopes. They would be even more conspicuous if the small coastal rivers with less than 1000 km^2 along the fringe of the Sea of Japan were also listed in the table.

From the large-scale work on Korean rivers by the government-general [453] we can gain an overview of their water balance [Vol. I, p. 276–286], although it unfortunately bogs down in details much too often. It contains the means of all measurements of stream flow up to and including 1927. Thus the means refer to periods of different lengths and can be compared only with certain reservations. The amount of precipitation in the drainage basin is calculated simply as the mean of the figures for the stations. Because stations at high elevations are lacking, these figures are presumably less and less correct the higher the drainage basin of a given river extends. Therefore the drainage ratios are probably in reality lower. Still, a number of important facts can be drawn from a table of the water balance (cf. Table 8).

Korea's water balance is governed almost entirely by the summer. In winter the amount of precipitation and evapotranspiration is so slight that it has very little influence on the water balance. In summer the large amount of precipitation greatly exceeds evapotranspiration everywhere, since, although the temperatures are very high, the likewise high humidity holds down evapotranspiration. The extensive deforestation facilitates runoff. In Korea, therefore, everywhere and every year precipitation exceeds evapotranspiration, so that a considerable amount remains as runoff. All of Korea belongs to Penck's "humid regime". The drainage ratio, i.e., drainage expressed as percentage of precipitation, is high everywhere. According to the Japanese calculations it often lies between 70 and 75% in the upper and middle reaches. For the reasons mentioned above these figures are probably too high. Nevertheless, there is no doubt that in Korea, too, the mountains have particularly high drainage ratios due to the steep slopes and the relatively slight evapotranspiration. At the gauging stations on the lower reaches of rivers and on rivers that flow only in less mountainous terrain the drainage ratio is generally lower. The south, which is warmer in summer on the average, often has lower ratios too.

Discharge is at its minimum in winter, for precipitation is low in this season. At higher elevations some of the precipitation remains on the ground as snow, and due to the severe frost detrital springs dry up. In spring a moderate amount of snowmelt flooding occurs, particularly in the upper reaches, which rise in the high mountains. This often occurs in April in the north; in the south it frequently blends in with the generally increased spring discharge. The maximum lies, with the precipitation maximum, in midsummer. Even the average maximum is relatively limited. Floods disrupt the normal course of summer discharge, causing a paroxysmal increase. Figure 31 shows three examples. These show how the summer discharge increases to gigantic proportions in such cases. The Han–gang at Koan, for instance, had a discharge rate of 100 m^3/sec on 16 July 1926 and on the 22nd of the same month of 14,400 m^3/sec. As discussed in the chapter on climate, the summer precipitation is brought primarily by cyclones and typhoons, which cause enormous amounts of water to precipitate in only a few days and often only locally. These cloudbursts inevitably lead to floods, especially since the country has lost its forest cover in many centuries of reckless exploitation. The Taedonggang disaster of 28 July to 1 Aug. 1923 was the result of an Lb_2 cyclone that passed across the neck of the peninsula, the Yalu disaster of 13 Aug. 1923 that of an Lc_1 typhoon, the terrible flood damage in Kyŏngsangnam–do on 6/7 Sept. 1925 that of an Lb_1 typhoon, the Naktonggang flood of 19/20 July 1920 that of a cyclone on track c [452].

River System	Gauge station	Observation period	Precipitation (mm)	Discharge (mm)	Drainage ratio
Taeryŏnggang	Yongt'an	1916–27	1388	896	64.6
Ch'ŏngch'ŏn-gang	Pugwon	1923–27	988	626	63.4
	Puksong-ri	1923–27	1187	893	75.2
Taedonggang	Mujindae	1915–27	1194	719	60.2
	Sŏngch'ŏn	1922–27	1072	545	50.8
	Samdŭng	1917–27	1057	629	59.5
Chaeryŏnggang	Naedae-dong	1924–27	807	565	70.0
Yesŏnggang	Kŭmch'ŏn	1918–27	1173	633	54.0
Imjin-gang	Yŏnch'ŏn	1919–27	1268	950	74.9
	Chŏn-gok	1923–27	1312	941	71.7
Han-gang	Tongnyang	1923–26	1187	886	74.6
	Koan	1915–27	1319	882	66.8
	Homyŏng	1926–27	1417	736	51.9
Kŭmgang	Annam	1923–26	1168	523	44.8
Yŏngsan-gang	Marŭng-ri	1916–23	1233	489	39.7
	Naju	1915–27	1328	692	52.1
Sŏmjin-gang	Okchŏng-ri	1923–26	1169	511	43.7
	Songjŏng-ri	1917–26	1270	676	53.2
Naktonggang	Naktong-ri	1925–27	806	596	73.9
	Waegwan	1924–27	1068	486	45.5
	Hyŏnp'ung	1918–27	971	474	48.8
Yonghŭnggang	Yongban	1923–25	998	582	58.3

Table 8. Water balance of Korean rivers at selected gauging stations

Along with the main precipitation, these flood disasters occur primarily in midsummer. The Korean annals list 168 flood disasters for the Han–gang around Seoul during the period from 1407–1859. Of these 2 occurred in April, 5 in May, 15 in June, 76 in July, 44 in August, 17 in September, 4 in October, 2 in November, 1 in December, 2 in January [453, Vol. I, p. 357–65]. The dates on which the large rivers reach their highest water level are likewise concentrated in the months of July and August. The Yalu had risen 9.61 m above its lowest level for the same month on 27 Aug. 1926, the Naktonggang 12.55 m on 13 July 1925, the Sŏmjin–gang 14.21 m on 23 July 1924, the Imjin–gang 16.80 m on 21 July 1924 and the Han–gang 19.32 m on 18 July 1925. Although these rivers are so short, their flood levels are among the highest in the world. In the estuaries of the Yellow Sea rivers the normal rhythm of the tides is completely lost when such floods occur.

These floods regularly wreak havoc on the fields on the valley bottoms. The proportion of fields that are endangered by floodwaters is 8.6% of the total fields in the country, but in some river basins it is much higher (Han–gang, 13.5%; Ch'ŏngch'ŏn–gang, 21.5%; Yesŏnggang, 24.9%; Yonghŭnggang, 28.0%). The total damage wrought by the branches of the Taedonggang system within the period from 1916–27 is estimated at 7.2 million yen, that by the Kŭmgang at 8.5 million, by the Naktonggang at 33.9 million and by the Han–gang at 46.2 million. A detailed study [824] deals with the terrible damage resulting from the Han-gang and Imjin–gang flood of the second half of July 1925, which was caused by two successive typhoons on tracks b and c. In the typhoon during the second third of

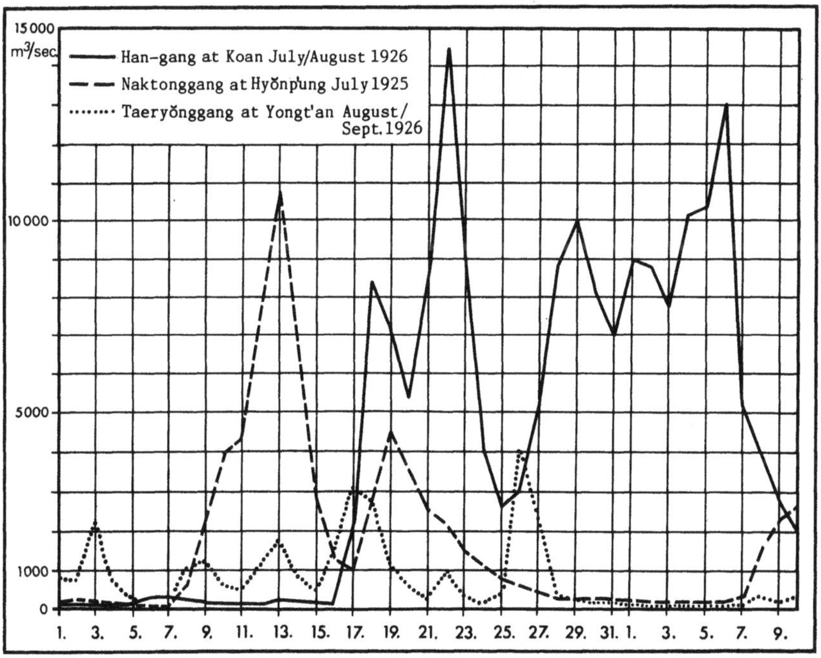

Fig. 31. Flood discharge of selected Korean rivers

September 1938, 1309 people died, 346 were reported as missing, and 666 were injured. 6413 houses were washed away, 2288 knocked down, 3776 partially damaged and 42,335 inundated. Most of this damage occurred in southern Korea.

Due to the intensive transport of large debris and the concurrent lateral erosion described on p. 81 f, the beds of Korean rivers are remarkably shallow and wide. Because of this and because of the recent uplifting, navigation is extremely difficult. If the Koreans did not patiently exhaust even the slightest possibilities with their flat-bottomed boats, the navigable stretches would be shorter still. In many small rivers, such as the Chaeryŏnggang, the Yesŏnggang, the Man-gyŏnggang and the Yonghŭnggang, navigability thus extends only as far as or slightly farther than the tidal portion of the river. In most of them only a fourth to a third of the total length of the river is navigable. As the example of the Yonghŭnggang shows, this proportion is even smaller in the short, steep rivers that flow into the Sea of Japan, especially since the tidal range there is much smaller. The same is true of such rivers as the Sŏmjin-gang, which antecedently crosses a zone of recent upheaval shortly above its mouth. The navigable length of the large rivers is appreciable, both absolutely and relatively speaking. In the Naktonggang it rises to almost two thirds. In this mountainous country, however, this is only possible by exploiting every last opportunity.

Practically only the main branch of the short rivers is navigable, particularly if no tributaries join it within the tidal portion. On the large rivers with many branches, navigation penetrates far up into the tributaries as well. The navigable length of the entire system is then more than twice as great as that of the main branch. This is particularly true of the Taedonggang and the Han-gang. All of this is demonstrated in Table 9.

During a long period in winter in the north, decreasing in length toward the south, navigation comes to a standstill because the rivers are frozen over, whereas the sea freezes only locally for short periods (p. 126). In the Yalu at Chunggangjin the period of freezeup lasts on an average from 17 Dec. to 5 April, in other words, 110 days, at Yong-amp'o at the mouth of the Yalu from 9 Dec. to 26 Mar., i.e., 108 days. The reason that the Yalu freezes later up at Chunggangjin than down at Yong-amp'o is the higher velocity in the middle section of the stream. For the

	A (km)	B (%)	C (km)	D (km)	E (%)		A (km)	B (%)	C (km)	D (km)	E (%)
Taeryŏnggang	67.5	45	69.2	25.6	17	Kŭmgang	120.9	30	144.7	70.0	17
Ch'ŏngch'ŏn-gang	87.5	41	87.5	18.3	8	Man-gyŏnggang	37.0	37	37.0	48.0	48
Taedonggang	243.0	56	513.6	98.0	23	Yŏngsan-gang	55.8	48	60.3	39.0	34
Chaeryŏnggang	31.2	24	81.4	42.0	33	Sŏmjin-gang	59.8	28	62.4	14.8	7
Yesŏnggang	39.0	23	53.4	51.9	30	Naktonggang	340.1	65	409.6	42.4	8
Imjin-gang	121.3	48	138.1	41.4	16	Yonghŭnggang	18.7	14	20.5	15.0	11
Han-gang	298.4	62	883.4	57.6	12	Sŏngch'ŏn-gang	–	–	–	2.0	2

[a] A = Navigable length of main river. B = A in % of total length of the river. C = Total navigable length of all branches of the river system. D = Length of tidal portion. E = D in % of total length of river.

Table 9. Navigable length of Korean rivers and length of tidal portion[a]

Taedonggang at P'yŏng–yang the average days are 17 Dec. and 8 Mar. (82 days), for the Han–gang at Seoul 17 Dec. and 4 Mar. (78 days). The Tumen has the longest period of freezeup. In the vicinity of Kyŏng–won it is frozen from 14 Nov. to 31 Mar., i.e., 138 days (454). A Manchurian source quoted by G. Fochler–Hauke in his work on Manchuria (p. 16) names the period from 28 Nov. to 7. Apr., i.e., 131 days, for the Tumen at Musan. I am not familiar with the dates for the southern rivers. Presumably the period of freezeup decreases rapidly.

As we have already indicated, the tidal range contributes greatly to navigation. Indeed, the section of the river that is navigated by steamships generally does not extend beyond the tidal portion. Table 9 also includes the length of this tidal portion, both absolute and in percent of the total length. It is small in the rivers flowing into the Sea of Japan and the Korea Strait, as the tidal range is small there. The Yellow Sea rivers, in contrast, have quite long tidal portions, especially the Taedonggang, whose lower part lies in an area of local subsidence (p. 280). In small Yellow Sea rivers such as the Man–gyŏnggang, which flow through broad alluvial plains for much of their length, almost half of this length is under tidal influence.

Figure 32 shows the progress of the river tide in the lower reaches of the Han–gang. The line of stations begins at Inch'ŏn, runs up the strait between the

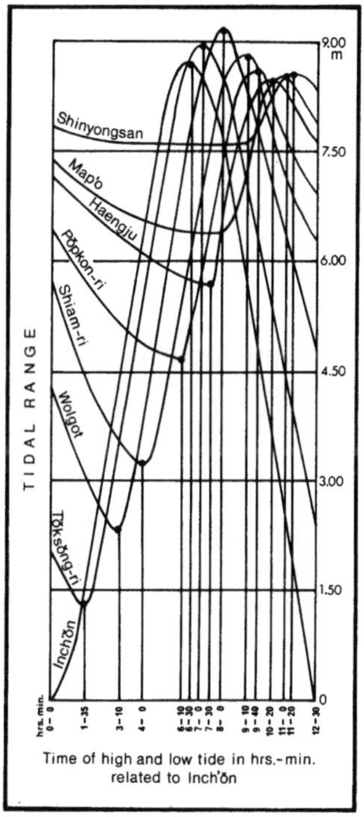

Fig. 32. The progress of the river tide in the lower Han-gang (after Chōsen kasen chōsa sho). Shown are the tide curves from 8 gauging stations, related to Inch'ŏn. These stations lie on a line running from Inch'ŏn through the strait between Kanghwa and the mainland and up the lower Han-gang to the vicinity of Seoul.

island of Kanghwa and the mainland and then follows the actual Han–gang to Shinyongsan near Seoul. It shows how the height of the wave at first increases as it moves upstream, although it is already extremely high at Inch'ŏn. To be sure, the ebb current also increases upstream, in fact much more strongly, so that the range decreases rapidly in this direction. The illustration shows quite clearly how the river tide moves upstream. It does not arrive in Shinyongsan until 11 h and 20 min later. In the other estuaries of the Yellow Sea rivers there are similar river tides. In the very deep, narrow, gradually converging estuary of the Taedonggang, however, the range at first increases upstream. It does not begin to decrease until the confluence of the Chaeryŏnggang.

This is shown by two of the maps in Fig. 33. They present results of studies by S. Ogura, who has made a very great contribution to the research on tides in the seas surrounding Korea [474, 476, 478, 479]. The Yellow Sea is less than 80 m deep, except for very few places. Sŏhanman barely reaches 50 m, Kyŏnggiman just under 20 m. Both are subject to sedimentation from the large Korean rivers in their full expanse (p. 252, 286). The Korea Strait drops to just under 100 m below sea level in its northern half and locally to 200 m in the channel between Pusan and Tsushima. But at the northeastern end of the Korea Strait the depth increases rapidly, and the Sea of Japan is a deep sea. Very close to the Korean coast its floor already lies way below 1000 m, and 65 km southeast of the mouth of the Tumen it reaches a maximum depth of 3658 m.

In the Yellow Sea and the Korea Strait tidal behavior is thus closely related to the horizontal structure of the coasts and the vertical structure of the sea floor. The tidal range on the west coast is tremendous. The mean spring range is approx. 4 m at the southwestern tip of the peninsula (Fig. 33). It increases northward and reaches its maximum values in the estuaries of Kyŏnggiman. In the estuary of the Ansŏngch'ŏn near Asan almost 10 m have been measured. In Sŏhanman there is a second maximum. The high tides are higher in summer and autumn than in winter and spring. Between June and September, particularly during southwesterly storms, tidal ranges up to 11 m have been observed in Kyŏnggiman [832].

In the Yellow Sea tides are predominantly semidaily. According to Ogura [479], the ratio of the tidal ranges of daily to semidaily tides is 0.36 at the southwestern tip, 0.19 in the middle of the west coast and 0.25 in Kyŏnggiman. Lacking more details on the overall phenomenon, Ogura's map with the lines of equal spring range for the semidaily tide $M_2 + S_2$ (Fig. 33) provides useful information for evaluating the latter. The rising tide enters the Yellow Sea from the south and is deflected to the right by the rotation of the earth, so that the tidal range is much greater along the Korean coasts than along the Chinese coasts. As Fig. 33 shows, the M_2 tide develops two counterclockwise gyres in the western half of the Yellow Sea.

The M_2 tide enters the Korea Strait from the south, heading for the Yellow Sea, on the one hand, and the Sea of Japan on the other hand. At the southwestern entrance to the latter there is a third counterclockwise gyre that is formed not only by the M_2 tide, but also by the M_1 tide. It dominates almost all of the Sea of Japan. The cotidal lines lie so far apart that on the eastern coast high water occurs almost everywhere at the same time. The semidaily tide decreases greatly in range as it enters the Sea of Japan, where the mean spring range is less than 0.25 m. The total tidal range is, however, much higher. It is determined primarily by daily

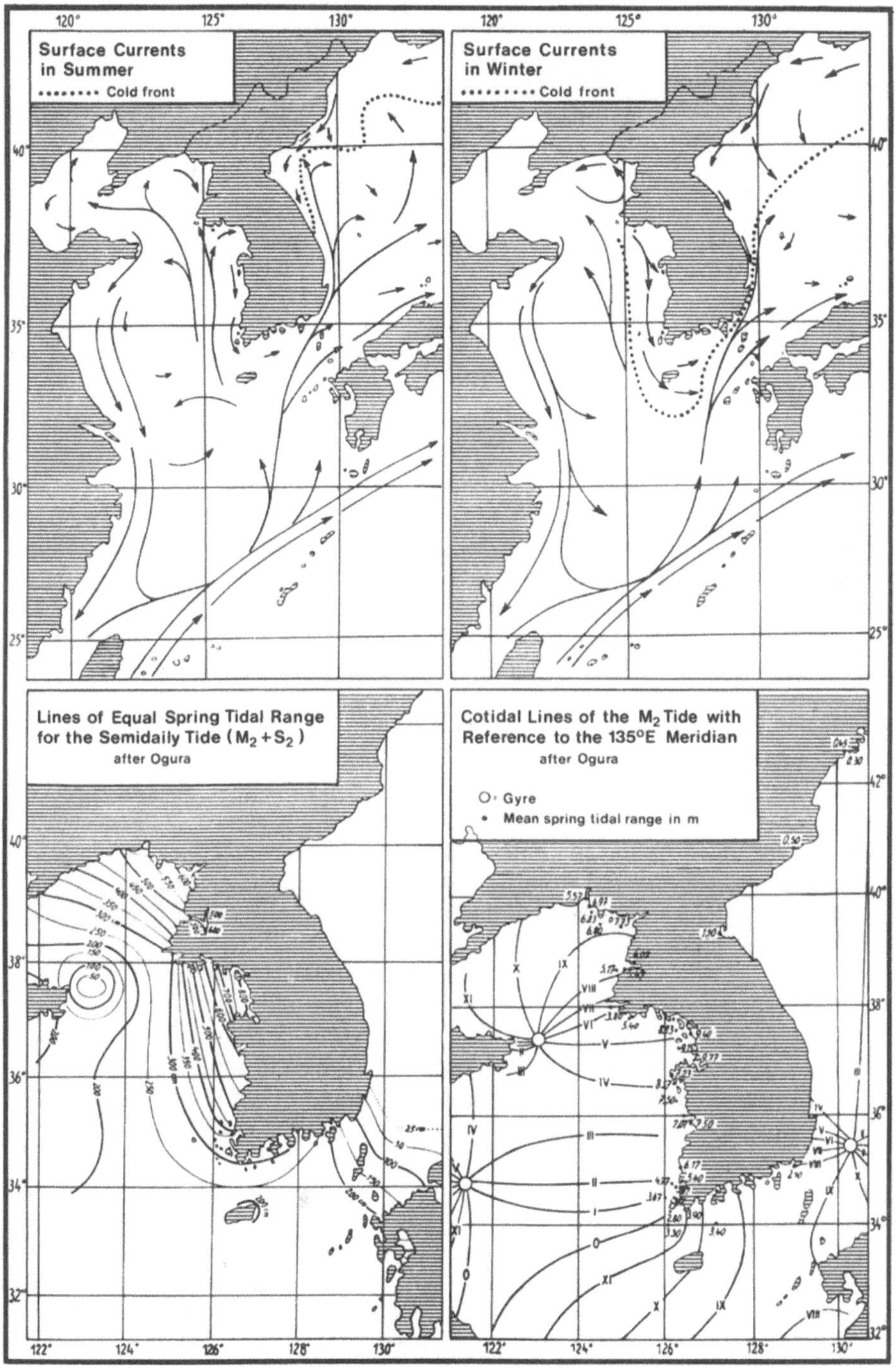

Fig. 33. Ocean currents and tidal behavior in the Korean seas (after Ogura and Suisan Shikenjō)

tides. Still, even it is very much lower than in the Yellow Sea; between Pusan and the mouth of the Tumen it sinks from 2.10 to 0.30 m.

In the Sea of Japan tidal currents play a very limited role, but they are that much stronger in the narrow waterways between the islands and peninsulas in the south and west. At rising tide the tidal current in the Yellow Sea generally moves northward, at falling tide southward, in the western half of the Korea Strait it moves west or eastward, accordingly. In the narrow straits the tidal currents currents develop immense velocities (p. 285, 366), preventing sedimentation, so that they are often much deeper than adjacent, more open parts of the sea.

In the Sea of Japan with its weak tidal influence, ocean currents play a much more important role than in the Yellow Sea with its strong tidal influence (Fig. 33). Von Schrenck [455] named a cold current, one branch of which enters the Tatar Strait, the Liman Current. He was of the opinion that this current, known by the Japanese as Oyashio, follows the coast of the Ussuri region and then the Korean coast southward. In fact there is a cold southwesterly current off the coast of northern Korea. Off the mouth of the Tumen, though, it is very weak and inconstant and does not appear to be connected with the Oyashio. Thus M. Uda distinguishes a separate "North Chōsen Cold Current System" [467, 468]. From the southwest a branch of the warm, high–salinity Kuroshio, the "Tsushima Warm Current System," enters the Sea of Japan along a wide front. The two currents converge and overlay each other to some extent. The uppermost layer of the Tsushima system (down to approx. 25 m) consists of water of low salinity from the Yellow Sea and the East China Sea. Along the peninsular coast of the Sea of Japan a narrow band of the cold current, deflected toward the right by the rotation of the earth, extends more or less far southward, depending on the season. The boundary between the two currents forms the cold front. As a comparison of the two pertinent maps in Fig. 33 shows, this cold front shifts seasonally. In summer a branch of the warm current flows far to the north at a short distance from the east coast of Korea and combines with the cold current to form a gyre in Tonghanman. In winter the cold front flows so close to the stretch of coast between the cape of Changgi and Pusan that the cold water is almost completely displaced at the surface even then.

Temperature and salinity are of course closely related to these currents. February and August are the extremes in the annual cycle of these two oceanographical elements and were thus chosen for the maps (Fig. 34). Surface salinity is higher in both seas in winter than in summer. It is lowest in September. The probable reason for this is the dilutive effect of river and rainwater. This again is higher in the Yellow Sea than in the Sea of Japan. As a result the surface salinity is lower in the Yellow Sea at all seasons than in the Sea of Japan. Under a thermocline at a depth of 15–30 m, which is particularly pronounced in summer, the Yellow Sea bottom water is water of much higher salinity flowing in from the south [466].

The surface water temperatures show much stronger seasonal and regional contrasts. In summer the Sea of Japan is somewhat cooler than the Yellow Sea and in winter much warmer. This is particularly true of the southern halves of these seas. The annual range of surface temperatures amounts to 16–23° C for the Yellow Sea and 14–20° for the Sea of Japan. Thus the surface temperatures have only a weak share in the tremendous temperature contrasts on land. Still the ranges we have just mentioned are extremely unusual for seas at a Mediterranean

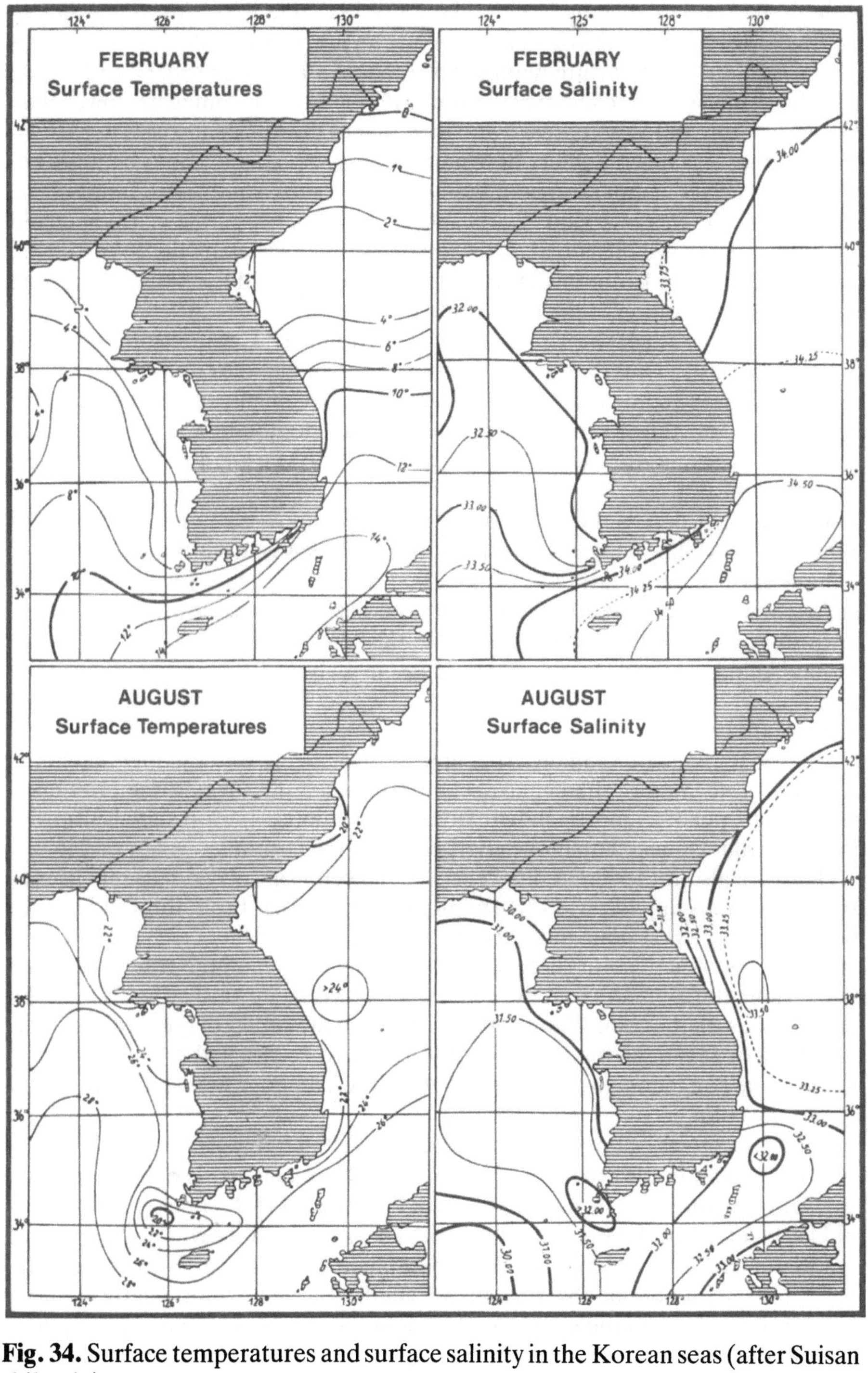

Fig. 34. Surface temperatures and surface salinity in the Korean seas (after Suisan shikenjō)

latitude (G. Schott, *Geographie des Indischen und Stillen Ozeans* = Geography of the Indian and the Pacific Ocean. Hamburg 1935, Plate XXIV). According to the isoanomalous maps drawn up by G. Wüst on the basis of Uda's monthly maps [469, p. 15], the temperature anomaly is −5° to −10° C in February and +5° to +7° C in August. In midsummer the water temperature is somewhat lower than the temperature of the air above it and than that of the adjacent coasts in particular; in winter it is much higher. In winter there is also an enormous contrast between the surface temperatures in the north and in the south. The trend of the isotherms in the Sea of Japan clearly shows the influence of the warm and cold currents; one can see the cooler water flowing southward along the coasts. From the crowding of the isotherms in winter and spring around the 38th parallel one can deduce the position of the cold front.

Only at the mouth of the Tumen does the February 0° C water isotherm touch on Korea's coasts. Thus they are basically ice free, unlike the coasts of the Ussuri region. The bays and straits in the north are only occasionally blocked by ice floes from the rivers, this occurring especially in the mouths of the Yalu and the Ch'ŏngch'ŏn–gang, the bay of Haeju, the strait of Mayangdo and the bays of Unggi and Sŏsura.

At only slight depths the water is already very cold year round. At a depth of 100 m the prevailing temperature in the Sea of Japan is more than 22° C lower than the surface temperature of 24° C. The water flowing into the bottom of the Yellow Sea has a temperature of 8° C. In the Sea of Japan warm water extends down to greater depths at a distance from the coast than it does near the coast. Around Ullŭngdo the water at a depth of 100 m is about 10° C warmer than at the nearby coast of the mainland. Thus the interface between the two bodies of water rises toward land.

A particular kind of anomaly is the appearance of cold water of high salinity at the southwestern tip of the peninsula in midsummer. This water is approx. 6° C cooler and 1 per 1000 more saline than its surroundings. It is presumably upwelling water that is sucked up in the angle between the currents.

There are also rather great differences in the color of the two seas that border on Korea (maps in [468, p. 117], and in map series [p]). From three sides sediment–bearing rivers pour into the Yellow Sea, some of them gigantic streams, and the predominant colors here are those on the far side of the scale from deep blue. Along the Korean coast the color sequence runs upward as far as 9 on the Forel scale, with the color 4 being observed only in the center of the southern half. In the Sea of Japan the highest level of the scale to be reached is 7, in the innermost part of Tonghanman and at the mouth of the Tumen, whereas all of the open sea has the color 3.

Ten years ago the above remarks would have been impossible, for it is only since then that the study of the seas surrounding Korea has made noticeable progress. Older reports and studies are those of v. Schrenk [455], Glocker [456], Mörsel [457] and Schott [458]. Since 1925 these seas are also shown on the maps of monthly means of the Japanese oceanographical atlases and journals. But the scales and the number of observations are small. We should particularly mention the map series [n] and the studies by Uda and Okamoto, on the one hand [464], and Suda and Seiki, on the other hand [473]. Since 1932, however, large–scale investigations are being carried out simultaneously by a large number of ships under the auspices of the Imperial Fisheries Experiment Station in Tōkyō. These we owe to the initiative of M. Uda [466–468]. In 1933 S. Ogura's standard work on the tides in the seas adjacent to Japan was published [479]. It deals with the Yellow Sea and the Sea of Japan in detail. Since 1924 the Fisheries Experiment Station of the Government–General in Pusan also publishes annual and other reports [XV, XVI], with an appendix containing monthly oceanographical maps for the individual years [map series o]. Above all, in 1936 this fisheries experiment station published an excellent little oceanographical atlas of the seas surrounding Korea [p]. It shows the averages of all observations up to 1935 in monthly maps of temperature (0, 25, 100 m), salinity (as above), currents, etc. The maps shown in Figures 33 and 34, with the exception of the tidal maps, were taken from this atlas. The non–Japanese ship observations are compiled in a Dutch atlas [s]. Its usefulness in the Korean seas is limited by the very small number of observations per 1° field of the graticule. The two Pilots of the British Admiralty [459, 460] contain a wealth of useful data, as do the occasional instructions in the "Seewart" [738, 832], the charts of the British Admiralty [q] and the German sea charts [r].

4 Flora

The growing season coincides approximately with the period in which the average daily temperatures exceed +5° C. On Chejudo this is the case at lower elevations for 48–44 weeks per year. On the southern coast this period lasts for 44–40 weeks, in the rest of southern Korea and along the central east coast 40–36 weeks, in the rest of central Korea 36–32 weeks and in northern Korea 32–28 weeks [418]. Thus the growing season on Chejudo comprises almost the whole year, whereas in northern Korea it is only somewhat more than half of the year. The thermal conditions for plant growth rapidly become less favorable from south to north; in the interior they are poorer than along the coasts, and in this mountainous country they are worse at higher elevations than in the lowlands. But many of the difficulties arising from the briefness of the growing season are compensated for by the heat of midsummer, which is much the same everywhere at lower elevations, indeed even in the Kaema upland. Of particular importance is the fact that the temperature maximum coincides with the precipitation maximum. Eighty to 90% of the precipitation falls during the growing season, in the extreme south even 90–100%. Thus the Korean climate provides optimum conditions for luxuriant plant life in several respects.

Outside of the growing season the temperatures drop to extreme lows, especially in central and northern Korea. In spite of the tropical heat of midsummer, summergreen elements of vegetation consequently prevail in most of the country, and the only important evergreen elements are conifers. Only in the extreme south, south of the –3° C isotherm for the January average minimum temperature (p. 114), are broadleaf evergreen trees an important element in the landscape. Only this narrow fringe belongs to the subtropics. All of the rest of Korea lies in the temperate zone and does not differ fundamentally in its flora and vegetation from the original conditions in Central Europe. Considering that the western and the eastern fringes of East Asia are connected by land, it is not surprising that a number of genera of both higher and lower plants are identical. This is not true of the species except in a few cases, most of them subalpine plants, for the distance between the center of Korea and that of Central Europe is 9500 km as the crow flies, and along the way there are numerous climatic and orographical obstacles to plant migration. Nevertheless, both Korea and Central Europe belong to the Eurasian woodland region of the Holarctic plant kingdom.

In one respect, though, the flora of Korea is fundamentally different from that of Central Europe; it consists of a much larger number of species. According to the most recent data in the Japan Yearbook 1940/41 [XLI] this relatively small country has 160 families with 883 genera, 3070 species and 550 varieties. Only 161 species with 5 varieties are considered to have been imported. Of the species no less than 800 belong to the physiognomically important woody plants. This basic difference is partially due to the fact that the climatic span encompassed by the country is much greater than in Central Europe. What is probably more important, however, is the fact that Korea was free of inland ice during the Ice Age, as Hofmann [495, p. 23] and Wilson [496] have already stressed. Certainly there too the climate was colder during the Ice Age than today, and the upper timberline probably lay at approx. 1300 m in northern Korea, i.e., 700 m lower than today (cf. p. 138). But the plants had plenty of room to retreat, especially since the continental shelf to the west and south was dry (p. 1). Many plants, such as the gingko tree and some magnolias, have thus been preserved unchanged from the Tertiary to the present day, and others have undergone only slight changes. No great extinction of species appears to have occurred during the Ice Age.

To give an idea of the abundance of Korea's flora and simultaneously to list the most important genera that are identical with Central Europe let me cite the genera that have a particularly large number of species according to Mori [485], Uyeki [488], Kokumen–Keng [486] and others. The number of species and varieties represented in Korea follows the name:

Yew (*Taxus*) 1; pine (*Pinus*) 8; larch (*Larix*) 3; spruce (*Picea*) 3; fir (*Abies*) 3; juniper (*Juniperus*) 6 – willow (*Salix*) 38; poplar (*Populus*) 8; hornbeam (*Carpinus*) 7; birch (*Betula*) 15; alder (*Alnus*) 13; edible chestnut (*Castanea*) 2; oak (*Quercus*) 22; elm (*Ulmus*) 12; mountain ash (*Sorbus*) 7; hawthorn (*Crataegus*) 3; wild pear (*Pyrus*) 14; wild cherry (*Prunus*) 41; maple (*Acer*) 28; linden (*Tilia*) 12; ash (*Fraxinus*) 3 – hazel (*Corylus*) 6; clematis (*Clematis*) 20; hydrangea (*Hydrangea*) 9; spirea (*Spiraea*) 12; rose (*Rosa*) 10; blackberry (*Rubus*) 25; lespedeza (*Lespedeza*) 14; spindle tree (*Euonymus*) 15; azalea (*Rhododendron*) 18; blueberry (*Vaccinium*) 7; lilac (*Syringa*) 14; honeysuckle (*Lonicera*) 21 – monkshood (*Aconitum*) 25; anemone (*Anemone*) 13; corydalis (*Corydalis*) 19; buttercup (*Ranunculus*) 14; violet (*Viola*) 44; cranesbill (*Geranium*) 20; cinquefoil (*Potentilla*) 30; vetch (*Vicia*) 29; speedwell (*Veronica*) 23; aster (*Aster*) 24; daisy (*Chrysanthemum*) 14; ragwort (*Senecio*) 11; leek (*Allium*) 22; lily (*Lilium*) 14; saussurea (*Saussurea*) 45 – woodrush (*Luzula*) 8; rush (*Juncus*) 14; wild millet (*Panicum*) 14; bluegrass (*Poa*); nut grass (*Cyperus*) 12; sedge (*Carex*) 130; pondweed (*Potamogeton*) 19; polypody (*Polypodium*) 19; club moss (*Lycopodium*) 18.

This very incomplete selection contains 6 genera of coniferous trees with 24 species, 15 genera of broadleaf trees with 225 species or varieties, almost all of them deciduous, 12 other genera of woody plants with 171 species, 15 genera of herbs, etc. with 347 species and varieties, 9 species of grass and nonflowering plants with 249 species and varieties. These 57 genera alone have no less than 1016 species and varieties in Korea. Almost all of these genera can be designated with common German names, showing how deeply rooted they are in Central European soil.

Besides these there are also many genera that are lacking in Germany and for which there are at best Germanized names. Let us mention the following as occurring particularly frequently: *Cephalotaxus*, *Torreya* (coniferous trees), *Phyllostachys*, *Sasa* (bamboo grasses), *Lithocarpus*, *Aphananthe*, *Zelkova*, *Morus*, *Schizandra*, *Machilus*, *Stephenandra*, *Maackia*, *Phellodendron*, *Zanthoxyllum*, *Daphniphyllum*, *Micromeles*, *Elaeagnus*, *Kalopanax*, *Acanthopanax*, *Aralia*, *Benzoin*, *Styrax*, *Paulownia*. With the exception of bamboo, all of the species of these genera found in Korea are trees or tall shrubs and thus play an important role physiognomically. Most of them are East Asian genera, and among these the number of evergreen plants is substantially greater.

Korea has a long border linking it to the East Asian continent. During the Ice Age it was connected with the continent to the west as well. It is separated from Japan by a fairly narrow strait that was probably much narrower or which may have even disappeared during the Ice Age. As a consequence, there is little floral endemism. It is greatest on Ullŭngdo, which was already separated from the mainland prior to the Ice Age (p. 327f). Among its endemic species the island harbors Korea's only beech, which is lacking on the mainland. On the mainland there is only one endemic genus of trees, the *Chosenia*, which is related to the willows, and there are very few endemic tree species. Their range is usually closely limited, and their significance for the geography of the country is accordingly slight.

Korea thus has most of its plants in common with Manchuria, the Ussuri region, northern China or the Japanese islands; some it has in common with one of these countries and others with several or even all of them. Of the Korean species 61% also occur in Japan. From this figure, however, we must not derive any exaggerated ideas about the degree of relationship between the Korean flora and the Japanese. A very large number of these species also occur in continental East Asia. Korea's relationship to the continent is in fact substantially closer than its relationship to the islands, for it has its cold winters in common with the continent.

A large number of physiognomically important plants extend into the country from the north. Those that require considerable cold in the winter find their limits along the coasts of the peninsula and do not occur on the islands, except in some higher altitudinal zones. Listed according to how far they have advanced southward, the important trees among these are Korea's tallest and most widely distributed tree, the poplar *Populus maximowiczii* Henry (up to 50 m in height), the small–leaved linden (*Tilia amurensis* Kom.), the Amur corktree (*Phellodendron amurense* Rupr.), the mountain ash (*Sorbus amurensis* Koehne), the tall birch that yields an exceptionally prized hardwood (*Betula schmidtii* Regel), Korea's most common oak, a powerful, tall tree not unlike our common oak (*Quercus mongolica* Fisch.), the maples *Acer manchuricum* Max. and *Acer triflorum* Kom., the big ash tree (*Fraxinus manchurica* Rupr.), the Manchurian walnut (*Juglans manchurica* Max.), the red birch (*Betula dahurica* Pallas) and the alders *Alnus sibirica* Fisch. and *Alnus hirsuta* Turcz. (Fig. 35, No. 7).

The antipodes to these are the trees that are sensitive to cold and extend more or less far into Korea from the south from Kyūshū and western Honshū. Of these the ones that reach only the southern fringe of the mainland, or even only Chejudo, are evergreen (ev). This group includes the camphor tree (*Cinnamomum camphora* Nees. and Eberm., ev), the tall, spreading shiino tree

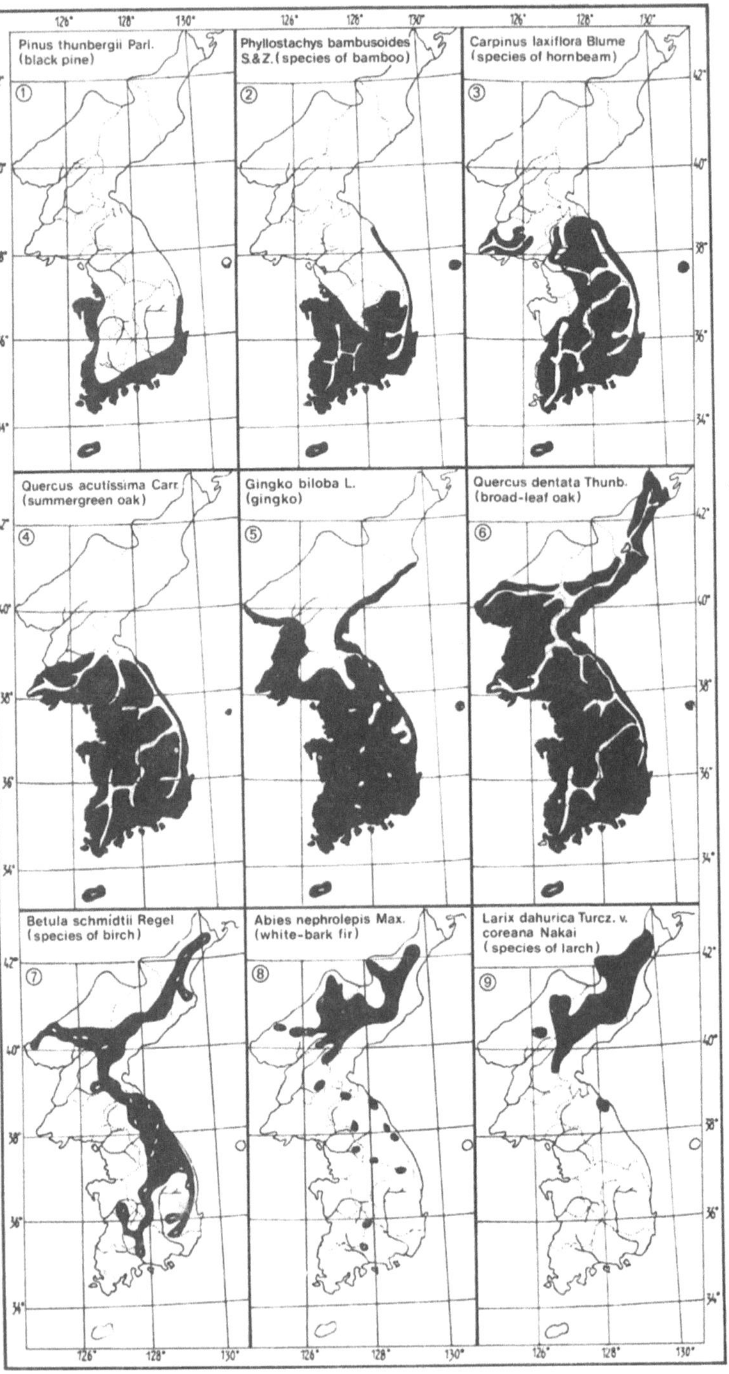

Fig. 35. Examples of the ranges of Korean plants (after Uyeki and Chōsen sōtokufu ringyō shikenjō yō ran)

(*Lithocarpus cuspidata* Nakai, ev), the oaks *Quercus glauca* Thunb. (ev) and *Quercus acuta* Thunb. (ev), the camellia (*Camellia japonica* L. (ev), a striking tree with its large, pink blossoms that is very characteristic of the southern fringe of the mainland and the remote islands), the conifer *Torreya nucifera* S. and Z., the Korean black pine (*Pinus thunbergii* Parl.), which has thick, dark–colored needles, the hornbeam *Carpinus tschonoskii* Max., with its brilliant orange catkins, the bamboo genera *Phyllostachys* and *Sasa* (ev), the large–leaf silverbells *Styrax japonica* S. and Z. and *Styrax obassia* S. and Z. with their clusters of magnificent white, bell–shaped blossoms, the hornbeam *Carpinus laxiflora* Blume with yellow catkins and the Korean benzoin (*Benzoin obtusilobum* O. Kuntze), which has small yellow blossoms and seeds containing oil (Fig. 35, Nos. 1–5).

A third group is eurythermic in both its winter and its summer requirements and thus extends from the Japanese islands through the lower elevations in Korea and into Manchuria. These include many of the wild cherry and pear trees, such as *Prunus serrulata* var. *pubescens* Wils. and *Pyrus ussuriensis* Max., the nut–gall tree (*Rhus javanica* L.), the broad–leaf oak (*Quercus dentata* Thunb.), the highly serrated narrow–leaved oak (*Quercus serrata* Thunb.), Korea's most famous maple, *Acer pseudosieboldianum* Kom. var. *coreanum* Nakai, whose finely articulated leaves turn fiery red and orange in autumn, the giant common ash (*Fraxinus rhynchophylla* Hauce) and the most common tree, the red pine (*Pinus densiflora* S. and Z). The last mentioned tree has long, thin needles in clusters of two and often assumes the parasol shape so fondly reproduced by East Asian painters. Its upper limit lies between 400 and 1150 m (Fig. 35, No. 6).

Finally there is another group of trees that is geographically important. These trees make slight demands on the temperature but do not tolerate great heat and dryness and are therefore found only in the higher altitudinal zones of this extremely dissected country. Thus in Korea their ranges always have an upper and a lower limit, the limits of the individual species lying at differing altitudes. These include the maples *Acer tegmentosum* Max., *Acer ukurunduense* Trautv. and Mey. and *Acer barbinerve* Max., which extend from 600 m to approx. 1500 m. The limits usually lie even lower for the fir *Abies holophylla* Max., which has spreading, wide–spaced branches and broad, prickly, dark green leaves. It is much beloved by the Buddhists and is planted in rows along the paths of the temples. The lower limit of the Korean nut–pine (Korean white pine, *Pinus koraiensis* S. and Z.) drops from 1200 m on Kajisan in the south to 700 m at Paektusan in the north [488], while the upper limit lies at approx. 1900 m in the south (Chirisan) and 1600–1700 m in the north (Kwanmobong). This magnificent tree has long, bluish green needles in clusters of five ("blue pine") and edible seeds the size of beechnuts ("edible nuts"). Its bark is dark gray. The stately white–trunked Korean white–bark fir (*Abies nephrolepis* Max. with broad needles cleft at the top) has about the same lower limit but extends upward to higher altitudes. On Maengbusan one can find it up to 2050 m. The same is true of the limits of the small–needled spruces *Picea jezoensis* Carr. and *Picea koraiensis* Nakai, which are often associated with the white–bark fir. In October Korea's only yew, *Taxus cuspidata* S. and Z., completely adorns itself with sweet red berries, which, in contrast to those of our native yew, can be eaten without harm. South of 35°30' its lower limit lies at 1400 m, at 40°30' it lies at 700 m, and on Ullŭngdo it even extends down to sea level. On the volcano Hallasan on Chejudo stunted specimens extend up almost to the edge of the crater (1950 m). The Korean conifer that extends up to the highest altitude is the larch *Larix dahurica* Turcz. var. *coreana* Nakai. On Paektusan it forms the timberline at approx. 1950 m, and on the southern side of Kwanmobong it is found up to 2200 m. Except for Paektusan it is the white–trunked birch, (*Betula ermanii* Cham.) that forms the timberline on northern Korean mountains, covering a belt above the conifers spanning another 100 m in altitude. It is found up to 2300 m on the southern side of Kwanmobong. Finally, on the crystalline peaks of northern and central Korea there is a dwarf pine region consisting of the pine *Pinus pumila* Regel, which becomes more and more stunted toward the top, only to merge with the subalpine plant associations of the highest elevations (Fig. 35, Nos. 8 and 9).

Many of the trees in the first group, those extending into Korea from the north, and many of the trees that are common only at higher altitudes have the tendency to advance southward or downward, according to Wilson [497] ("northern aggressive types"). Among these are the aspen *Populus tremula* L. var.*davidiana* Schneid., *Alnus hirsuta*, *Betula dahurica*, *Betula ermanii*, *Pinus pumila*, *Abies nephrolepis*, *Picea jezoensis* and *Picea koraiensis*. Their counterparts, which strive to advance northward or upward ("southern aggressive types"), are: *Pinus densiflora*, *Carpinus laxiflora* and *Fraxinus rhynchophylla*. Between them stand the persistent species ("types in possession"): *Abies holophylla*, *Pinus koraiensis*, *Betula schmidtii*, *Acer pseudosieboldianum* and *Acer manchuricum*, *Pyrus ussuriensis* and *Prunus manchurica* Koehne. Korea's floral character is thus slowly changing.

In its primeval condition, Korea was most probably entirely woodland [14]. Many of the reasons supporting this conclusion were mentioned in Part One (p. 17, 23). Only the active dunes and the floodplains of the rivers were naturally unwooded, along with the very limited areas that extend above the upper timberline.

Korea's present–day forest cover is listed as 161,670 km^2 (1938). That is 73% of the national area. But from the viewpoint of plant geography only one–third of this area can be considered forest. Map 4 (cf. editors' note on p. 7), which portrays the forest according to the only available draft, still shows stands of tall timber as covering too large an area, but it clearly demonstrates the most important fact, namely, that the mountain ranges of northern and central Korea bear most of the remaining forest cover. In the south only tall, steeply towering mountains, such as Chirisan and Hallasan are still densely wooded. At all elevations and in all parts of the country there are still more or less well preserved remnants of forests on the grounds of the 1336 Buddhist temples. These temples go back to the 4th century (p. 25). The Korean kings usually endowed them with the surrounding forests. The temples often had the power to prevent the land and wood–hungry inhabitants from destroying their forests. The kings themselves also took measures to protect the woods, these having favorable consequences especially for the vicinities of the royal graves around Seoul and Kaesŏng. Taken together there are thus sufficient clues to enable us to reconstruct the floral and sociological traits of Korea's original forest landscape. An excellent aid in this is the atlas of plant geography of the government–general [t]. This work, based on plant records most of which we owe to T. Nakai's decades of untiring research [480–484], shows the possible range of 128 woody plants and bamboos, based on their ecological requirements. Using this and my own investigations as a basis, the ranges of plant formations are depicted in Fig. 36.

The extreme southern part of the mainland, along with the peninsulas and adjacent islands and the lower elevations of the remote islands Ullŭngdo and Chejudo form the predominantly broadleaf evergreen forest region. We shall deal with this plant formation in detail in Part Three (p. 369f). In Chejudo it extends up to 400 m, on the mainland to 300 m, on Ullŭngdo only to 200 m.

Above and north of it begins the summergreen mixed forest zone. This covers by far the greatest part of the country. At lower elevations it even extends far beyond the northern boundary, and it does not give way to primarily coniferous forests until very high altitudes. This mixed forest is composed of the extremely large number of species characterized above. The splendid monograph on the forest of Kwangnŭng, northeast of Seoul (p. 289f and [826]), which has a maximum length of only 10 km and a width of 5 km extending in elevation from 50 m to 600 m, lists 649 species. They include 71 species of trees. At lower elevations broadleaf deciduous trees predominate by far in Korea's untouched virgin forests. Proceeding toward higher elevations, the proportion of coniferous trees gradually increases. On dry slopes with southern exposure conifers sometimes outnumber deciduous trees, whereas on the northern slopes deciduous trees often prevail alone.

In contrast to Central Europe, many of the latter are distinguished by a luxuriant array of blossoms. In early spring the hornbeams unfold an abundance of orange, brown to greenish yellow catkins. Soon afterwards the wild apples, cherries and pears blossom in white, pink and red. The blossoms of *Prunus manchurica*, *Prunus serrulata* var. *pubescens*, *Pyrus ussuriensis* and *Malus baccata* Borkh. var. *manchurica* Schneid. are particularly magnificent. In midsummer the ashes, such as the stately *Fraxinus sieboldiana* Blume, the mountain ashes, such as *Sorbus commixta* Hedl. and *Sorbaria stellipila* Schneid., *Maackia amurensis* Rupr. and Max. and *Kalopanax ricinifolium* Miquel develop large clusters of white blossoms. Sweet is the smell of the large–blossomed lindens *Tilia amurensis* and *Tilia manchurica* Rupr. and Max.

The abundance of blossoms is augmented by a goodly number of shrubs, which often give off an intoxicating fragrance. These include the *Philadelphus* species, which are relatives of our mock orange, the many species of snowball, the south Korean *Gardenia jasminoides* Ellis, *Weigela florida* S. and Z. and *Weigela subsessilis* Nakai, the roses, the *Lespedeza* species with white to violet blossoms, the rhododendrons with their gorgeous pink, red and mauve blossoms, the magnolias, the lilacs, the two species of *Styrax* and the benzoin tree.

In addition to the showy blossoming shrubs there are many woody lianas, which often make these primeval forests difficult to penetrate. These include the many species of clematis, which have a wonderful fragrance when they bloom, such as *Clematis patens* Morr. and Decne. with star–shaped blossoms 12 cm in diameter, the honeysuckles, of which *Lonicera japonica* Thunb. features sweet–smelling white blossoms that gradually turn yellow, and kudzu (*Pueraria hirsuta* Mats.) with erect 20–cm–long spikes covered with purple pealike blossoms. Other lianas are more remarkable for their vividly colored fruits, such as the greenbrier *Smilax china* L., which develops cherry red clusters of round berries in late autumn, the strangler *Celastrus orbiculatus* Thunb., which is covered

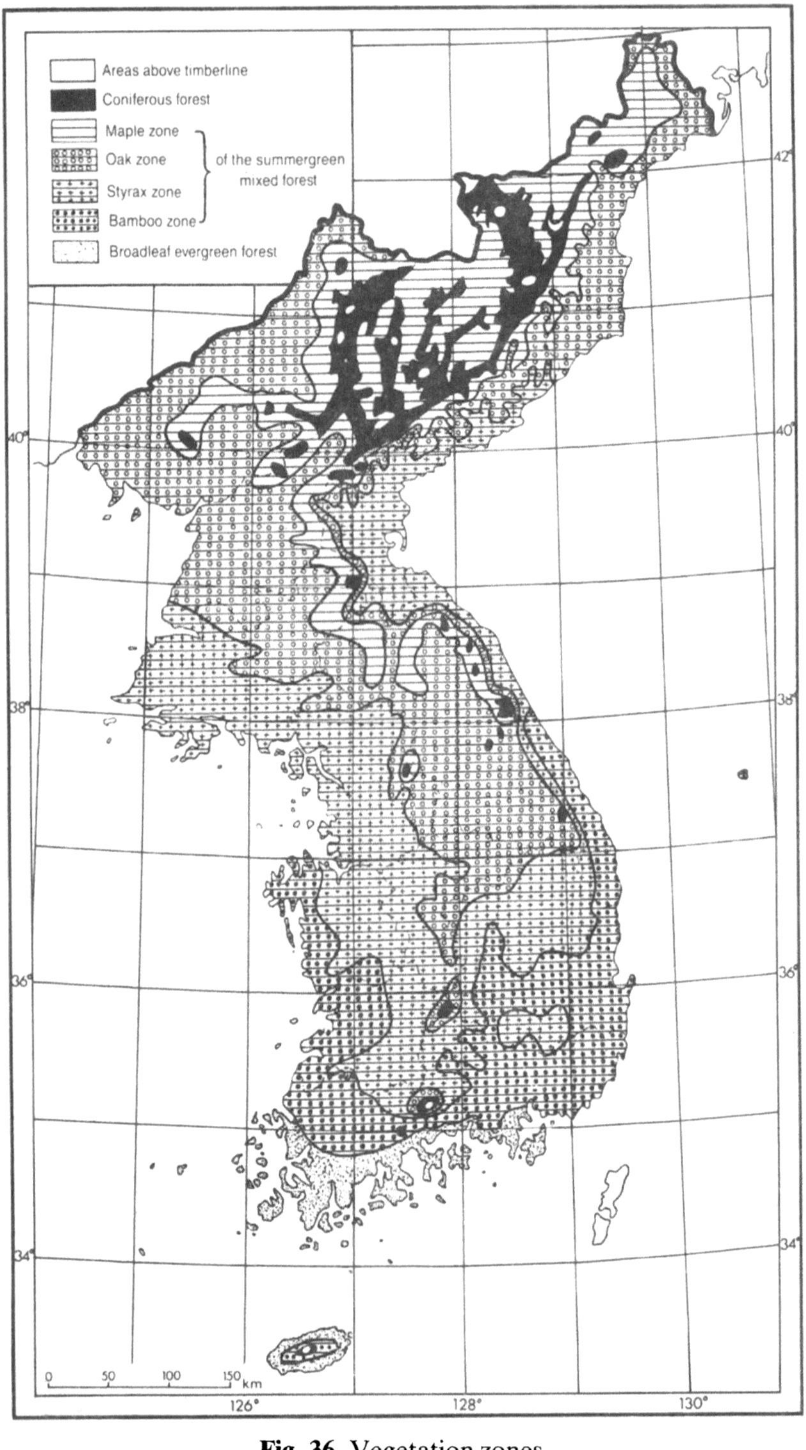

Fig. 36. Vegetation zones

with masses of yellow fruits in autumn and early winter, *Actinidia arguta* Planch. with greenish brown edible fruits and the wild grapevine *Vitis amurensis* Rupr., very common in the north, whose purplish black grapes are very tasty. The climbing bush *Tripterygium regelii* Sprague and Takeda, which is particularly common at high elevations, has large clusters of very white blossoms in late summer later followed by pink, balloon–shaped fruits. The climber *Actinidia kolokmikta* Rupr. has leaves that are totally or partially white and is therefore called "snow on the mountains" by the Koreans. In autumn this white turns to pink and red. The leaves of the relative of our Virginia creeper, *Ampelopsis brevipedunculata* Koehne, which climbs mainly on rocks, also assume fiery tints in autumn. Ivy is missing among the lianas in Korea's mixed forests. There is only one species of ivy in Korea, *Hedera tobleri* Nakai, and it is restricted to the evergreen southern fringe. Parasitic mistletoes of the genera *Viscum* and *Loranthus* are, however, common in the trees of the mixed forest, and here, too, they stand out particularly in winter because of their evergreen nature. Their blossoms and berries are inconspicuous, however.

The broad crowns of the deciduous trees in this summergreen mixed forest rise 40 m and more above the ground. Frequently they are arranged in tiers. Below them are the lower layers of large and small shrubs, and the ground is covered by a host of herbaceous plants blooming in many colors in early spring. Among the ground plants are such decided shade lovers as the balsams, for example *Impatiens textori* Miquel. The wealth of forms and colors of Korean blossoms is impressively depicted by the lovely book by Florence Heddlestone Crane [501], in which 150 watercolors are reproduced in color and in natural size. In spring the leaves come in all shades from the palest to the darkest of greens. The sporadic sprinkling of coniferous trees only adds to the abundance of forms and colors. As a whole this forest displays the great vigor and luxuriance befitting the tropical climate of summer. This is also manifested in the stately height of as much as 4 m achieved by annuals and in the gigantic size of the leaves that grow in one summer. Woody plants whose leaves can become particularly huge are *Quercus dentata*, *Quercus aliena* Blume and *Quercus mongolica*, *Tilia manchurica* Rupr. and Max., *Paulownia tomentosa* Koch, the climber *Pueraria hirsuta*, *Magnolia parviflora* S. and Z. and *Populus suaveolens* Fisch. I collected leaves of *Pueraria hirsuta* measuring 25 by 25 cm and of *Quercus dentata* measuring 17 by 35 cm. The most colossal leaves borne by Korean plants are those of *Rodgersia tabularis*, a plant restricted to northern Korea, whose leaves become as big as small tables. In each chapter in Part Three there are regional descriptions of the composition and appearance of the summergreen mixed forest.

In the calm autumn the leaves of a large number of the deciduous trees and shrubs in this forest take on the brilliant coloring that is famous all over East Asia. Central Europe has nothing that can compare with it in variety and vigor; it is equaled at best by central North America. It begins earlier in the north and at higher elevations than in the south or at lower elevations and progresses in these directions from the end of September until the beginning of November. The succession of colors runs, on the one hand, from the green of summer to greenish–yellow, bright yellow, beige, yellowish–brown and the dull brown of the fallen leaves. Plants belonging to this color succession are lindens, e.g., *Tilia amurensis*, elms, e.g., *Ulmus japonica* Sarg., willows, e.g., *Salix integra* Thunb., hornbeams, e.g., *Carpinus erosa* Blume, and birches, e.g., *Betula schmidtii* and *Betula latifolia* Kern. var. *japonica* Nakai. The second color succession runs from green to pink, in such a manner that both colors are sometimes combined in one leaf, and then to scarlet, crimson, wine red, russet and finally reddish–brown and brown. Best known for this succession is the maple *Acer pseudosieboldianum* var. *coreanum*. But many other species of maple also belong to it, such as *Acer tschonoskii* Max. var. *rubipes* Kom., *Acer mono* Max., *Acer mono* var. *savatieri* Nakai and *Acer ginnala* Max. In addition there are the common oak, which, due to its frequency, plays an especially important role in the autumnal coloring of the slopes, the bird cherry *Prunus padus* L. var. *rufo–ferruginea* Nakai, the spindle tree *Euonymus alatus* Sieb. and the sumac *Rhus trichocarpa* Miquel. Bushes such as the red–stemmed blackberry, *Rubus crataegifolius* Bunge, also belong to this color sequence. Some of them, such as the Daurian rose, show their red coloring only on the underside of the leaves.

In all parts of Korea these primeval forests have often been replaced at lower elevations by red pine forests consisting of only this one species. Many such forests are also found in the evergreen zone. In many cases there is a close spatial relationship between them and the more or less untouched mixed forests, and one can observe all stages of the transition. In these cases they developed because the Koreans gradually cut out the other trees. Others of these red pine forests were planted on previously deforested land during the Korean dynastic period (p. (p. 31). In the years between 1789–1803, 5.5 million trees were planted in the districts around Suwon [605, p. 182]. Since the red pine lives for 400–500 years, a historical act of this sort may have occurred far in the past. Still others grew up semispontaneously on land that had been burned over and then left to itself. The most recent ones developed during the last generation as a result of the reforestation measures of the Japanese forest authorities, which are discussed in Part Four (p. 429ff). Many of these open forests have a lower layer of gorgeously blooming shrubs. In most of them, however, the soil becomes impoverished because the inhabitants constantly glean them for every bit of usable material, scratching away every last little plant. These single species stands are often severely damaged by the pine bug, *Gastropacha pini* L. The thin needles of this pine are destroyed in an unbelievably short time. According to my observations, a caterpillar completely devours a fully developed needle longer than itself in a minute and a half. The thus infested red pine forests stand stiffly in a ghostly pale green. The pure black pine forests found only on the coastal fringe of southern Korea are hardly affected because their needles are so thick and hard. Therefore, in this part of the country protective strips of black pine are planted between the stands of red pine. Nowhere do the Korean nut–pine and *Abies holophylla* form stands of a single species, but they are quite often found scattered about in the summergreen forest.

Plant communities with special compositions are found in the valley bottoms. In southern Korea, to be sure, the valley bottoms are completely filled with paddies and dry fields today, where they are not entirely occupied by active beds of pebbles and sand. In northern Korea, however, there is an abundance of such communities, especially at higher elevations. Nowhere are natural meadows to be found. On the contrary, above the high water beds the valley floors are covered with lush, proliferous woods in which the above mentioned lianas are particularly common. The special character of this bottomland forest is due to certain trees that favor a high water table and are lacking or are rare on the adjacent slopes. These include the willows *Salix gracilistyla* Miquel, whose leaves have a silvery sheen on the underside, and *Chosenia bracteosa* Nakai, the elms *Ulmus davidiana* Planch. var. *japonica* Nakai and *Ulmus macrocarpa* Hance and the poplar *Populus suaveolens*. It should be emphasized, though, that willows, alders and poplars are not as strictly limited to the bottomland forests as in Central Europe. Many of their species occur just as well in the forests on the slopes.

A special kind of plant community is found on dunes, which are confined to the coast of the Sea of Japan and Chejudo. They are covered by a low plant community whose elements have the familiar characteristics of our native dune plants. On Chejudo the first series in the succession consists only of *Vitex rotundifolia* L. fil.; the second of the same species, the long runners of *Rosa polyantha* S. and Z. spreading across inactive dunes, and a species of aster.

The tremendous range of the natural summergreen mixed forest changes its characteristics from south to north and from lower to higher elevations. To date it is not possible to give a classification based on ecological and sociological viewpoints. For the time being floral viewpoints must suffice. Uyeki has prepared the way for such a classification [490–492]. I shall elaborate on his suggestions here on the basis of the atlas of plant geography [t] and my own collection. Figure 36 shows the classification. Four regions are distinguished, each of which begins in the south as an altitudinal zone. The first three descend toward the north into the lower elevations and as far down as the coasts. Each extends farther north along the coast of the Sea of Japan than along the Yellow Sea coast. The fourth region exists only as an altitudinal zone among the predominantly coniferous forests and does not reach the coast in Korea.

The range of the four regions depends of course on the climate and conforms to its basic characteristics, which were expounded in Chapter 2. In the south the plant species that extend across from Kyūshū are more prominent, in the north those that extend in from Manchuria and the Ussuri region. Let us call the first region the bamboo zone; although it is not their exclusive range, it is where the wild bamboo species most commonly occur. On the east coast it extends northward as far as the 38th parallel, and on the west coast to 37°. In the interior the mountains cause it to recede far southward. Along with a large number of uncharacteristic species the following trees and shrubs are important: *Pinus thunbergii*, *Torreya nucifera, Cephalotaxus drupacea* S. and Z., *Carpinus tschonoskii* Max., *Zanthoxyllum piperitum* D. C. and *Zanthoxyllum planispinum* S. and Z., *Aphanante aspera* Planch., *Acer palmatum* Thunb. var. *coreanum* Nakai, *Quercus stenophylla* Makino. The same species also occur in the adjacent exclusively evergreen zone to the south and at lower elevations. Similarly, the plants that are characteristic of the following region usually are also found in the bamboo zone, but in smaller numbers.

Let us call the following region the styrax zone after the two silverbells *Styrax japonica* and *Styrax obassia*. On the coast of the Sea of Japan it extends to north of 40°, on the Yellow Sea coast it does not quite reach the 39th parallel, while in the interior it falls back to Chirisan. Characteristic of it, besides the two species of styrax, are: *Carpinus laxiflora*, *Quercus acutissima* Carr., *Quercus serrata*, *Quer-*

cus dentata, *Zelkova serrata* Makino, *Pyrus faurei* Schneid., *Benzoin obtusilobum*, *Ulmus sieboldii* Dav. var. *coreana* Nakai and *Kalopanax pictum* Nakai.

Let us call the third region the oak zone, because on the basis of the number of individual trees, oaks are most prominent in its virgin forests. In northern Korea the upper limit for oaks lies at approx. 1100 m, in southern Korea at 1400 to 1500 m. In the south the oak zone exists only as a belt around the high mountains. From about the middle of the Sobaek Range northward it forms an uninterrupted altitudinal zone in the mountains. Everywhere in Hamgyŏngbuk–do and the two P'yŏng–an provinces it extends down to the coasts. The following can be considered characteristic of this zone: *Quercus mongolica*, *Quercus variabilis* Blume, *Juniperus chinensis* L., *Juglans manchurica*, *Fraxinus rhynchophylla*, *Fraxinus manchurica*, *Carpinus erosa*, *Aralia elata* Seemann, *Pyrus ussuriensis*, *Tilia manchurica*, *Betula dahurica*, *Betula schmidtii*, *Acer pseudosieboldianum* var. *coreanum*, *Malus baccata* var. *manchurica* and *Phellodendron amurense*.

Let us call the uppermost zone of the summergreen mixed forest the maple zone, because the red–stemmed *Acer tschonoskii* var. *rubipes*, *Acer tegmentosum*, *Acer ukurunduense*, *Acer manchuricum* Max., *Acer barbinerve*, *Acer mono* and *Acer mono* var. *savatieri* are heavily represented in it. It also includes *Sorbus amurensis*, *Sorbaria stellipila*, *Tilia amurensis*, *Prunus padus*, *Prunus maackii* Rupr., *Betula costata* Trautv. and *Syringa robusta* Nakai. For the rest, conifers, particularly the nut–pine and firs and spruces are already quite frequently found. In southern Korea the maple zone is absent. Here oaks extend upward to the lower limit of the coniferous forest zone. It is most extensive in the Nangnim and Hamgyŏng Ranges and in the Kaema upland. In the latter, though, the primeval forests of this altitudinal zone have largely been destroyed.

Since the transition from mixed forest to predominantly coniferous forest generally takes place very gradually, and since the altitude at which this transition takes place can vary by several hundred meters in one and the same mountain group, depending on the exposure and the soil, it is difficult to express this important altitudinal limit in figures. Nevertheless, we can say that on the north slope of Hallasan on Chejudo it lies at 1600 m (p. 377), on Chirisan at 1400 m (p. 348), on Odaesan at 1450 m (p. 317), on Haramsan at 1100–1400 m (p. 264), on Paeksambong in the bend of the Yalu at 1400 m (p. 254) and on the southern slopes of Kwanmobong, Myohyangsan and Maengbusan (p. 224f) at 1400–1500 m. Thus on the mainland it is only here and there that it sinks to lower elevations toward the north. On southern slopes the upper timberline lies at 2200 m on Kwanmobong, at 2150 m on Maengbusan, at 1950 m on Paektusan. On Chirisan and Hallasan in southern Korea it has a lower elevation than on these northern Korean peaks, probably due to the larger amount of snow there. On the whole, therefore, it lies no higher than in the Alps, despite the lower latitude. The vertical span between the upper limit of mixed forest growth and the absolute timberline is thus 550 to 800 m in northern Korea. In the primarily coniferous forest of this zone the trees described previously (p. 131) predominate: yew, nut–pine, white–bark fir (which gives way to the Korean fir in the south), flat–leaf spruce, Korean spruce and the three larches *Larix gmelini* Ledeb. var. *coreana* Nakai, *Larix dahurica* Turcz. var. *coreana* Nakai and *Larix dahurica* Turcz. var. *principis ruprechtii* Rehd. The deciduous trees found in the coniferous forest zone include especially: the birches *Betula japonica* Sieb., *Betula costata*, *Betula manchurica* Nakai, *Be-*

tula schmidtii and *Betula ermanii*, *Ulmus laciniata* Schneid., *Tilia amurensis* and *Populus maximowiczii*. Many of the conifers suffer from growths of the giant lichen *Usnea longosissima* Ach.

Within the coniferous forest zone we can distinguish various plant associations, each of which has its own undergrowth. Uyeki [492] mentions the fir–spruce type, the larch–fir type and the pure larch type, among others. The first two types prefer old crystalline rocks, the latter the permeable basalt and pumice soils of recent volcanic deposits. In other locations, however, they stand in a direct successional order. When, for example, a fire caused by lightning or negligence sweeps through a larch–spruce–fir forest, the very resinous spruces and firs with their low–hanging branches burn more easily than the larches, which lose their needles in winter. In their resistance to fire the Korean larches resemble their relatives, *Larix sibirica* Ledeb. (R. Pohle in Z.G.E. Berlin 1917, p. 209). But even when the larches are also destroyed, they are the first to grow again, along with birches and aspens, since the light seeds of these plants are brought by the wind from afar. At first the larch seedlings develop in the shadow of the birches and aspens, but they outgrow them after 20 to 30 years and eventually kill them. Then a pure, open larch forest grows up. After 60 to 100 years nut–pines, firs and spruces can again develop in the shade of the larches, which are by then as much as 40 m tall, and after approximately 150 years the old larch–spruce–fir type has been restored (Wilson [496]). The course of events is quite similar when a larchless fir–spruce forest falls victim to a fire.

In continental northern Korea, where elevations above 2000 m are not uncommon, considerable expanses reach up into the region that is barren of forests. Nevertheless, here too the subalpine zone is limited to a few isolated spots (Fig. 36). The largest of these surrounds the Paektusan caldera and has a diameter of about 22 km. All of the others are a great deal smaller. In this very low–growing plant association woody plants are still relatively numerous. They include the Japanese stone pine *Pinus pumila*, *Rhododendron chrysanthum* Pallas, *Rhododendron parvifolium* Adams and *Rhododendron redowskianum* Max., the blue mountain-heath *Phyllodoce coerulea* Bab., the black crowberry *Empetrum nigrum* L. and the bog huckleberry *Vaccinium uliginosum* L., the last two of which are, like many others, relics of the Ice Age here as well. In addition there are two species of edelweiss (*Leontopodium coreanum* Nakai and *Leontopodium leiolepis* Nakai) several species of gentian, the stonecrop *Sedum telephium* L., white mountain avens (*Dryas octopetala* L. var. *asiatica* Nakai) and other low–growing alpine plants with beautiful flowers, reindeer moss, grasses, mosses and many others.

As we have already said, only one third of the official forest area today consists of plant associations that can lay claim to the designation forest. And many of them are no longer untouched virgin forests. We have just given an example of the effect of fires in the region of predominantly coniferous forests. In this region, however, fires are only rarely set systematically. In the two upper zones of the mixed forest this is the case, though, even today. Here fire–field farming is widespread. This and the resulting secondary and tertiary plant formations will be discussed in detail in Chapter 2, Part Three, on the Kaema upland, and I shall come back to it when evaluating each of the northern and central Korean landscapes that extend upward into high altitudes. In the higher elevations of the Kaema upland, open oak, birch and larch forests generally develop as the climax of the secondary plant formations. Oaks, liking warm soil, advance much farther upward in the north Korean second growths than in the primary forests. There they extend up to 1500 m, i.e., to the upper limit of the maple zone of the natural forests. The secondary forests of the oak zone can become much more varied and luxuriant in southern Korea; Part Three contains some examples (p. 316, 345). Here they are distinguished from the primary forests in that conifers are much less prevalent. Often, however, secondary forests do not develop at all, but only open fields of scrub and weeds.

Whereas in the higher zones of the mixed forest fire–field farming is the chief cause of forest destruction, in the lower zones and in the region of evergreen forests it is due to other human practices. At lower elevations the fields are mostly restricted to the alluvial valley bottoms and floodplains. Only where dry field cultivation is very extensive do the fields in the first two regions extend upward onto gently sloping hills. As Hofmann [494] reports, formerly even the semispontaneous vegetation of these low elevations was burned over, in order to use the ash to obtain tender grass and leaves for green manure for the rice fields. Today this is prohibited by the Japanese, and I did not observe any violations against this regulation, but large amounts of grass and leaves are still being picked today for this purpose. Above all the vegetation of the lower elevations must serve as sources of fuel. Because of the intensive farming, these two regions are densely populated and the per capita fuel requirement is very large due to the cold winters and the bad habit of lighting the universal floor heating system (p. 157) even on summer nights. As a result the surrounding hills and mountains, with the exception of temple grounds, are usually completely deforested. Although measures taken by the Korean kings to protect the forests can be traced back for almost 1000 years, to the reign of Sŏngjong (982–997), they were generally not very successful, especially during the last decades of the Chosŏn dynasty. The reckless deforestation thus made tremendous strides, particularly in the 1870s and 1880s, and even today many trees are being cut secretly, in spite of the strict forestry measures of the government–general. Therefore, seen from the railways, which tend to follow the valley floors, Korea gives the appearance of a barren country. The hills and mountains are covered by low shrub formations. Everywhere the farmers and the children scrape the ground between the bushes down to the last leaf with their fan–shaped rakes, pick up the twigs and branches, pick the leaves and cut down the bushes. Every day one can see groups of white–clad women and children wandering around on the green bush–covered slopes, and the men carry off to their farms tremendous amounts of leaves, needles, twigs and branches in their clumsy A–frames (Kor. *chige*).

The state of preservation and the composition of the shrub formations varies with the soil, the intensity of the summer precipitation and particularly with the intensity with which they are exploited. They are scrawniest in the vicinity of the large cities, especially Seoul, and in the northern portion of the Naktonggang basin and the adjoining region to the east. The limestone regions are generally quite uniformly green, because the residual loam resulting from the weathering of the limestone fully absorbs the moisture. In the slate and gneiss regions, in contrast, the shrub vegetation is often full of gaps, so that the bare, boulder–covered ground is visible. Grasses and herbs grow in between. Here the shrub formations are composed primarily of the red pine, various oaks, in particular, *Quercus mongolica*, *Quercus mongolica* var. *manchurica* Nakai, *Quercus dentata* and *Quercus serrata*, and juniper bushes. Characteristic of the more vigorous of these formations is the especially luxuriant growth of lianas, making them almost impenetrable. In addition there are a striking number of shrubs that bloom magnificently in spring; they are much more common in these thickets than in the forests. They include the forsythias, e.g., the canary yellow *Forsythia viridissima* Lindl., and most especially the rhododendrons, whose variety and glorious colors are for the plant geographer a virtual characteristic of the Korean spring. The three most striking ones are: *Rhododendron mucronulatum* Turcz. with mauve to pink blossoms that appear before the leaves do, *Rhododendron pukhanense* Lév. and Van. with big wine–red blossoms and the beautiful pale pink *Rhododendron schlippenbachii* Max. Another group of shrubs that plays an important role in these second growths on gneiss and slate is the lespedeza species. In Korea and Japan they are so common that Nakai has devoted a special study to them [484, cf. also 503]. They have a very large number of small, lanceolate, orbicular or spatulate leaves and are completely covered in summer with white or violet pealike blossoms. It is not uncommon for the bushes to completely disappear from these second growths and be replaced by fields of grass and herbs, as on the grave fields. These meadows adorn themselves in spring with beautiful flowers, for instance, "one of the glories of Korea," the orange–colored lily *Hemerocallis disticha* Donne [501]. The shrub formations on granitic soils are usually much sparser still, since the sand and grus that are the products of weathering are highly subject to erosion (p. 79). Thus the Japanese reforestation, which will be discussed later, has begun primarily on them.

On the whole it can be said of the second growths that the less often they are subject to human influence, the more they resemble the original formations of the same locality. But even the first interference alters the ecological conditions fundamentally for a long time to come. The location has become drier, warmer and sunnier. As a result, firs and spruces are missing. Pines, oaks and lianas, rhododendrons and lespedezas are, in comparison, particularly common.

The relationships between the Korean population and the flora of their immediate surroundings are extremely close and varied in other respects as well. Its use forms an integral part of farm life. We shall discuss this use in the context of fire-field farming later, but here we shall sketch the corresponding relationships in permanent agriculture. The amount of red pine wood, branches and needles used by the Koreans annually for heating alone is estimated by Uyeki [492] at 4 million cubic meters of solid timber. Whereas *Abies holophylla* is used very little because its wood is so soft, the wood of *Gingko biloba* is used for making dishes, tables and the like; the yew yields floor boards, chopsticks and combs; the very durable wood of the nut–pine, coffins and beams; the red pine, building material, jugs and agricultural implements; the larch, telephone poles and bridge timbers; the white–bark fir, shingles, furniture and matches; the corktree, soft sticks for fastening the packsaddles on the backs of oxen; the extraordinarily hard wood of *Betula schmidtii*, wagon axles. Brooms and fences for the courtyards are made of switches cut from lespedeza. The bark of *Quercus dentata* is used in tanning. The Koreans make their famous cabinets and chests of the wood of *Zelkova serrata*. The climber *Pueraria hirsuta* furnishes tough ropes for shoes and baskets and for use in house building. Cosmetic oils are obtained from the fruits of the camellia and the benzoin tree. The seeds of the nut–pine are zealously gathered in October and are a delicacy, stuck together with honey. An edible oil is also pressed from them. With great enthusiasm other forest products are also gathered to serve as food, such as mushrooms, berries, herbs, roots and nuts. The Neolithic site at Yup'an (p. 17) proves that the nuts of *Juglans manchurica* were already highly prized during the time of the coastal culture. The medicinal plants yield a further group of forest products. Their number is immense. By far the most important of them since olden times is the ginseng root (*Panax ginseng* C.A. Meyer). Ginseng is a plant growing high up in the mountain forests whose white root bears a resemblance to the human figure (Alraune = mandrake, [505]). The root of the wild ginseng is very much in demand in Korea and especially in China as a tonic and an aphrodisiac, and its effects surpass those of the cultivated ginseng by far (p. 182). One last group of forest products is used in dying, such as the seed cases of *Gardenia jasminoides*, which produce a rich yellow, the sap of *Rhus vernicifera* Decne., which is an excellent varnish (p. 39), the bark of *Fraxinus rhynchophylla*, which yields a black dye, and the leaves of *Acer ginnala*, from which blue dyes are made.

In this exploitation of the ultimate possibilities that nature offers, the Koreans stand alongside the Chinese, from whom many impulses came [69]. Most of the imported plant species also come from China. In many cases this is verified by the Chinese and Korean classics.

5 Fauna

Editors' note: It is not always easy to identify the names of animals. In the case of birds we have used the English names used by Cumming [508], upon which Lautensach evidently based most of his German names. It appears that a number of names have changed in the last 40 years, particularly the scientific names. To make it easier for the reader to identify the animals in question, where there appears to be a change, we have added the name used in *The Birds of Korea* by M.E.J. Gore and Won, Pyong–Oh (Royal Asiatic Society, Korea Branch, 1971) and in *Checklist of the Mammals of Korea* by Won Pyong–Oh (Institute of Ornithology, Kyung Hee University, Seoul 1976) in brackets.

Since approximately three–quarters of Korea is covered by forests and shrub formations, there is still a lot of living space for wild animals. The list of fauna in the small forest of Kwangnŭng (p. 289) includes 504 species of higher animals [826]. In addition to woods and shrubland, the beaches and cliffs along the coasts afford a wide variety of possibilities. Finally, there are also the cultivated areas, particularly the paddies and dry fields and the areas surrounding the villages. Wild animals are much more evident in the landscape both physiognomically and acoustically than in Central Europe, so that a geography of Korea cannot avoid examining the fauna and its life. The groundwork for such an examination is, however, particularly meager. Aside from Cumming's thorough study on Korean birds [508], the zoological literature consists only of lists of fauna, which are of little value to the geographer. And of the travel accounts hardly any but Sten Bergman's magnificent books [227–229] contain useful observations. I shall thus have to rely heavily on my own investigations, which of course could only be very incomplete. The following remarks will have to be accordingly brief. In Part

Three I shall go into the fauna of certain regions at appropriate places (p. 239, 317, 328, 349). In this chapter I shall dispense with a description of the marine fauna; it will be described in Part Four in connection with fisheries.

In its fauna Korea is just as closely related to Central Europe as in its flora (p. 127f). The number of completely identical species is even much larger, because animals, being themselves able to move, were more easily capable of spreading across the largest landmass. Like Central Europe, Korea belongs to the Paleoboreal Realm. Along with northern China, Manchuria and the Japanese islands, it forms the Manchurian Province of the same. Of the regions forming this province, the southeast Manchurian and the southwest Manchurian extend into Korea. The Kaema upland and the rest of the northeast belong to the former, the rest of the country to the latter. As the following remarks will often show, there are fundamental differences in the fauna of these two parts of Korea.

Until a few decades ago the Manchurian tiger (*Tigris tigris longipilis* Fitzinger) was king of the forests and shrublands. It is a colossal animal with a coat that closely resembles that of its near relative, the Bengal tiger, but is even more beautiful in winter. Folk traditions and Korean fairy tales are full of this figure. One often sees it portrayed, in clouds of smoke and fire, on screens, ceilings and gates. During the dynastic period there was a special guild of tiger hunters, and as late as 1866 its members rendered the state valuable services as intrepid warriors. Tiger skins and hair were important items in the scant foreign trade of ancient Korea (p. 32f). In 1627 the Korean king sent 100 tiger skins, among other things, as compensation to the Manchu troops that had invaded the country [72, p. 13]. The older travel accounts contain many reports of tigers attacking houses, animals and people and on how tigers were hunted. After the end of the isolation period tigers still appeared now and then even in the streets of Seoul in winter. I myself was often warned of the alleged danger prior to ascending high mountains, but when checked more carefully, these warnings almost always proved to be unfounded. The Manchurian tiger is well–nigh exterminated in Korea today. Only the most remote mountain refuges, particularly in the Kaema upland, still provide shelter here and there for roaming pairs of tigers.

The smaller relative of the Manchurian tiger, the Korean tiger (*Tigris tigris coreensis* Brass. [*Panthera tigris coreensis* Brass.]), which has a black coat with yellow stripes, has probably been completely eradicated in the wild. In contrast, the East Asian leopard (*Pardus pardus orientalis* Schlegel [Siberian leopard, *Panthera pardus orientalis* Schlegel]) is still relatively common in the mountains of northern Korea and on Chirisan. The final peace treaty between the Korean king and the Manchus (1637) contained a provision that 142 leopard skins be sent annually to the Manchu court, and until 1712 this number appears to have been frequently achieved [72, p. 25f]. Even today leopard hunts are worthwhile in winter. Much more frequent, though, are lynxes (*Lynx lynx* [East Siberian lynx, *Felis lynx cervaria* Temminck]) and wildcats (*Felis catus*). All of these members of the cat family are shy forest creatures that live on other mammals or birds but are driven to the proximity of human settlements by necessity in the bitter cold winters.

The Korean bears are also forest dwellers. According to Underwood [506], there are two species, the black bear (*Ursus tibetanus* [Ussurian black bear, *Selenarctos tibetanus ussuricus* Heude]), which I saw in the forests of Paektusan, and the oriental brown bear (*Ursus arctos orientalis* [Manchurian brown bear, *Ursus arctos lasiotus* Gray]). Not only bears but other inhabitants of the cold forests of the north too are forced to hibernate in winter. They include the badger (*Meles melanogenys* Allen and Andrews [Korean badger, *Meles meles melanogenys* Allen and Andrews]) and, according to Sten Bergman, a small tailless rodent called the pika (*Ochotona coreana* [Korean pika, *Ochotona hyperborea coreana* Allen and Andrews]), large colonies of which live in the Kaema upland in holes they dig themselves [229, p. 59].

Wolves come into Korea from Manchuria in the winter. Two species are also known of them, the Korean variety of the European wolf (*Canis lupus coreanus* sp. *novus* Abe, [507]) and the doglike *Nyctereutes coreensis* Mori [Korean raccoon dog, *Nyctereutes procyonoides koreensis* Mori], which appears in packs, in contrast to the former.

A real nuisance for the farmers living close to virgin forests, especially fire-field farmers, is the wild boar (*Sus coreanus* Heude [Korean wild boar, *Sus scrofa coreanus* Heude]). A single one of these black, iron–gray to grayish-brown animals with a weight of up to 160 kg is capable of ruining the crops in a day's plowing in one night. In summer and autumn they live separately and feed primarily on wild roots and the potatoes and root crops of the neighboring fields. In late autumn, however, herds of as many as 15 animals gather together and head for the oak forest zone, to root for acorns under the thin layer of snow.

According to Underwood there are five species of *Cervus* in forest and scrub-land: the Chinese water deer (*Hydrelaphus inermis* [Korean water deer, *Hydropotes inermis argyropus* Heude]), both sexes of which are hornless, and which prefers the vicinity of the rivers, the grayish–brown musk deer (*Capreolus manchuricus* Noak [*Moschus moschiferus parvipes* Hollister]), the small, fawn–colored to dark grayish–brown, tailless sika deer (*Sika coreanus* [*Cervus nippon hortulorum* Swinhoe]), our native red deer (*Cervus elaphus*) and the Manchurian sika deer (*Sika manchuriens*), a large, dark animal that is spotted in summer. A real mountain dweller is the goral (*Naemorhedus raddeanus* Heude [Amur goral, *Naemorhedus goral raddeanus* Heude]), which is found from the Himalayas to northern Korea. In summer it climbs up above the timberline, where I encountered it on Kwanmobong. In winter it searches for food on rocks and scree slopes at lower elevations. The stockily built, powerful animal is similar to the chamois, but has short, straight horns.

The Korean representatives of the squirrel family are extremely well adapted to forest life. There are three of them. In addition to the Korean variety of our native animal (*Sciurus vulgaris coreae* Sowerby) there is a delightful chipmunk (*Eutamias asiaticus orientalis* Bonhote [Asiatic chipmunk, *Tamias sibiricus asiaticus* Gmelin]) that romps about, jumping gracefully on the shady rocks close to streams, and is thus a ground animal. Its fur is grayish–brown with four dark lengthwise stripes on its back. Smaller still, i.e., only about half as big as our squirrel, is the north Korean flying squirrel that Sten Bergman describes so vividly [229, p. 193ff]. It is a nocturnal animal with very big, round, protruding black eyes and sleeps curled up in hollow tree trunks during the day. Its silken fur is silvery gray on the back and white on the underside. Folds of skin between the forelegs and the hind legs allow it to glide from tree to tree. The broad, yellowish brown tail with bushy hairs on two sides functions as a balancing pole. It lives mainly in deciduous forests and feeds on young birch, aspen and hazel shoots.

Whereas the forest mammals are only seldom heard, many of its feathered inhabitants can be heard over a great distance. In spring, especially in southern Korea, the call of the cuckoo reverberates through the forest. According to Cumming [508] there are five species, one of them our native cuckoo (*Cuculus canorus telephonus* Heine). The little cuckoo *Cuculus poliocephalus* utters an extremely strange six note call. In the deciduous forests the hollow, two note calls of the wild doves sound. Not all of the four species are woodland birds. The very common eastern turtledove (*Streptopelia orientalis orientalis* Latham) is, however. Flying slowly and silently in the deep forest, with its olive–gray back and pinkish–dun breast, this bird nearly one–third of a meter long is an almost unearthly apparition. Through the upper levels of the forest resound the hoarse cries of the jays: Eastern Brandt's jay (*Garrulus glandarius taczanowski* Lonnberg [jay, *Garrulus* glandarius brandtii Eversmann]) and the long–billed nutcracker (*Nucifraga macrorhynchus* Brehm [nutcracker, *Nucifraga caryocatactes macrorhynchos* C.L. Brehm]). At night these calls give way to the cries of the owls, of which there are no fewer than 13 species. Among them is the most imposing of the Korean nocturnal birds of prey, the great eagle owl (*Bubo bubo tenuipes* Clark [eagle owl, *Bubo bubo kiautschensis* Reichenow]), which is closely related to our eagle owl, and the secretive dwarf owl (*Otus japonicus*), whose metallic call bopp–po–so is heard from the forest every summer night. During the day the knocking of woodpeckers is heard over a great distance. Of the 15 species, our native great black woodpecker (*Dryocopus martius martius* L.) and the Korean green–backed woodpecker (*Picus canus jessoensis* Stejneger [Eurasian green woodpecker, *Picus canus griseoviridis* Clark]) deserve mention.

Of the remaining woodland birds, which live mainly on fruits, seeds and young shoots, let us mention the rare Ussurian black–cock ([black grouse] *Lyrurus tetrix ussuriensis* Kothe), which is limited to the Kaema forests and spends the winter night in holes in the snow, the Korean hazel grouse (*Tetrastes bonasia coreensis* Kuroda and Mori [hazel grouse, *Tetrastes bonasia vicinitas* Riley]), the grayish blue Korean gray nuthatch (*Sitta canadensis coreae* Grant [Chinese nuthatch, *Sitta villosa* Verréaux]) and the widely represented families of tits and finches.

In many ways there is a gradual but not fundamental difference between the habitats in the primeval forests and the semispontaneous shrub formations. As a result, the woodland animals for whom tall stands and absolute solitude are not an unconditional must are also found in the shrub formations closer to the villages, through which people often roam. On the other hand, animals that prefer thickets and warm summer soil and ones that need plentiful supplies of herbs and grass are more common here than in the forests. Others again switch back and forth between the two habitats at will. The second group includes hares, mice and many birds, and among the last group are quite a few migratory birds. Of the nonmigratory birds we should mention particularly the pheasant, whose cries fill the thickets. There are two of them, both of which have a pure white ring of feathers around the neck, the Manchurian ring–necked pheasant and the even more common Korean ring–necked pheasant (*Phasianus colchicus pallasi* Rothschild and *Phasianus colchicus karpowi* Buturlin). Both are hunted by the Koreans with falcons [229, p. 110ff]. These are tame specimens of the Siberian peregrine falcon (*Falco peregrinus calidus* Latham [peregrine falcon, *Falco peregrinus pealei* Ridgway and *Falco peregrinus japonensis* Gmelin]), which is also found wild in Korea.

The art of falconry is very old in Korea. In 355 it was brought to Japan by a Paekche prince. Among the shrubland birds are many songbirds, such as Korea's three most famous songsters, the black-naped oriole (*Oriolus chinensis diffusus* Sharpe), whose loud clear fluting tones are heard unceasingly, the so-called Japanese nightingale (Manchurian bush warbler, *Horeites cantans borealis* Campbell [bush warbler, *Cettia diphone borealis* Campbell]), which resembles our native nightingale only in the nature of its song, and the ruby-throat (*Luscinia calliope* Pallas [*Erithacus calliope calliope* Pallas]). They also include Korea's most colorful bird, the paradise flycatcher (*Tchitrea paradisi incei* Gould, [Japanese paradise flycatcher, *Terpsiphone atrocaudata atrocaudata* Eyton]), with its truly tropical plumage. The beak and a ring around the eyes are sky blue, the head a shining blue-black, the back is purple and the breast white. The adult males have two purple-colored tail feathers of an incredible length. Among the ground-living animals of the bushlands there are a large number of snakes in summer, nor are they lacking in the virgin forests, even at higher elevations. In addition to several venomous snakes, such as *Ancistrodon blomhoffi brevicaudus* Stejneger, which belongs to the Old World pit vipers, there are many nonvenomous snakes that often exceed 2 m in length, such as the tiger water snake (*Natrix tigrina* Boie). The foes of these animals are primarily birds of prey, such as buzzards (*Buteo*, *Pernis*), kites (*Milvus*), harriers (*Circus*), goshawks (*Astur*), sparrow hawks (*Accipiter*), falcons (*Falco*), vultures (*Aegypius*, *Gypaetus*) and eagles. Most of the last mentioned are tremendous birds, such as the golden eagle (*Aquila chrysaetus japonica* Severtzov) and the Chinese imperial eagle (*Aquila heliaca ricketti* Swann and Wetmore [imperial eagle, *Aquila heliaca heliaca* Savigny]).

Many of the animals that are associated with man's habitation have chosen the fields as their habitat. A decorative element of the flooded nursery paddies and rice paddies is often a pair of herons, which live on the masses of frogs in these shallow ponds and breed in a tall pine tree nearby. Of the four species, the most common are the Eastern gray heron (*Ardea cinerea rescitostris* Gould [gray heron, *Ardea cinerea jouyi* Clark]), a close relative of our heron, and the Amur green heron [green-backed heron] (*Butorides striatus amurensis* Schrenk). On the dry fields one can see the Siberian bustard [great bustard] (*Otis tarda dybowskii* Taczanoski), the Tibetan hoopoe [hoopoe] (*Upupa epops saturata* Lonnberg) and many of our other species of wildlife of the fields, such as mice, moles, hedgehogs, crows, quail, larks and particularly magpies, the latter also being woodland birds. There are two representatives of the same: the magpie (*Pica pica sericea* Gould), with black and white plumage, and the magnificent Chinese blue magpie (*Cyanopica cyanus interposita* Hartert [blue magpie, *Cyanopica cyanus koreensis* Yamashina]). Particularly favored habitats of many of these animals that associate themselves with man are the bamboo groves in the vicinity of the southern Korean villages. Birds of these habitats are, e.g., Henson's brown-eared bulbul (*Microscelis amaurotis hensoni* Stejneger [brown-eared bulbul, *Hypsipetes amaurotis hensoni* Steiniger]), the eastern turtledove and the crowtit (*Suthora webbiana fulvicauda* Campbell [*Paradoxornia webbiana fulvicauda* Campbell]). In Korea, too, swallows and martins (*Hirundo*, *Delichon*, *Riparia*), sparrows (*Passer*), thrushes (*Turdus*), larks (*Alauda*) and many other songbirds are associated with man's dwellings in the strict sense of the word.

Along the coasts wildlife is particularly abundant. The coasts of the Sea of Japan are favored regions because of their abundance of fish, those of the Yellow Sea because of the wide fringe of mud flats with their highly varied grounds, and in the southwest thousands of cliffs provide undisturbed living and breeding grounds. An aquatic mammal that occurs is the harbor seal (*Phoca vitulina* [Western Pacific harbor seal, *Phoca vitulina largha* Pallas]), and occasionally the fur seal (*Callorhinus ursinus* L.) strays onto the Korean Sea of Japan coast. More abundant still is the birdlife of the coasts. The sea eagles include the biggest of all of Korea's birds, among them the Steller's sea eagle (*Haliaeetus pelagicus pelagicus* Pallas) and the black sea eagle ([Steller's sea eagle, subspecies] *Haliaeetus pelagicus niger* Heude). The beautiful blue or red kingfishers (*Halcyon*, *Alcedo*) plunge skillfully down to the water and return with wriggling fish. They dig nests in the sandy and gravelly banks. In addition there are cormorants, particularly the blackish Chinese cormorant (*Phalacrocorax carbo hanedae* Kuroda [common cormorant, *Phalacrocorax carbo sinensis* Blumenbach]), sandpipers (*Tringa*, *Glottis*, *Terekia*), plovers (*Squatarola*, *Charadrius*), gulls (*Larus*), murrelets (*Synthliboramphus*) and guillemots (*Uria*). The last two reach the southern limit of their western Pacific range in Korea.

A similar type of habitat is provided by fresh water coastal lakes, swamps and watercourses on dry land. We have already mentioned the water deer, the chipmunk and herons as being animals that prefer the vicinity of water. Others are: the pied harrier (*Circus melanoleucus* Forster [*Circus melanoleucus* Pannant]), bitterns (*Botaurus*, *Ixobrychus*), kingfishers, divers (*Colymbus*), wagtails (*Motacilla*), buntings (*Emberiza*), snipes (*Capella*), curlews (*Numenius*) and waterfowl such as geese (*Anser*, *Cygnopia*), swans (*Cygnus*), ducks (*Anas*, *Aix*, *Nyroca*) and water cocks (*Gallinula*, *Gallicrex*).

The eastern front of Asia, which separates earth's largest landmass from its largest expanse of ocean, forms the busiest migration route for birds in the world. Many of the previously mentioned birds are migratory birds that spend only part of the year in the Korean habitats described here. Cumming [508] distinguishes between permanent residents, summer residents, winter visitors and migrating visitors. The latter stay in Korea for only a few days or weeks in spring during their flight north and in autumn during their flight south. The permanent residents, according to Cumming, include the gray heron, the carrion crow (*Corvus corone corone* L. [*Corvus corone orientalis* Evermann]), the two species of magpie, the turtledoves, the two species of pheasant, the Japanese common kingfisher (*Alcedo athis japonica* Bonapart [common kingfisher, *Alcedo athis bengalensis* Gmelin]), the Korean great tit (*Parus major takahashii* Momiyama [great tit, *Parus major wladiwostokensis* Kleinschmidt]), the white–faced wagtail [pied wagtail] (*Motacilla alba leucopis* Gould), the Korean crested lark [crested lark] (*Galerida cristata coreensis* Taczanowski) and the brown–breasted meadow bunting [meadow bunting] (*Emberiza cioides castaneiceps* Moore). Summer residents are the black–naped oriole, the swallows and cuckoos, the Phillipine redtailed shrike [red–tailed shrike] (*Lanius cristatus lucionensis* L.), some species of plover and the green–backed heron. The Arctic willow warbler (*Acanthopneuste borealis borealis* Blasius [*Phylloscopus borealis borealis* Blasius]) is a summer resident only in the high mountains. At lower elevations, it is a passage migrant on its way to the north, as its name indicates. The same is true of the white-rumped swift (*Micropus pacificus pacificus* Latham [*Apus pacificus pacificus* Latham]), which spends the winter in southern China, Formosa, India and Aus tralia. Among the winter visitors Cumming includes: various wild goose species and many members of the duck family, the Eastern common crane [common crane] (*Grus grus lilfordi* Sharpe), the Daurian jackdaw (*Corvus dauricus dauricus* Pallas [jackdaw, *Corvus monedula dauricus* Pallas]), several species of robin, the Australian curlew (*Numenius cyanopus* Vieillot [*Numenius madagascariensis* L.]), which, however, remains in southern Korea only in mild winters, and several species of bunting. The migrating visitors, finally, consist of two main groups. The first includes the wading birds, particularly plovers, sandpipers and snipes. The Eastern gray plover (*Squatarola squatarola hypomelaena* Pallas [gray plover, *Squatarola squatarola* L.]) and the terek sandpiper (*Terekia cinerea* Guldenstaedt [*Xenus cinerea* Guldenstaedt]) breed in eastern Siberia, even in Alaska, and spend the winter in India, southern Africa and Australia. The second group consists of smaller insectivores, particularly flycatchers (*Tchitrea*, *Cyanoptila*, etc.), swallows, songbirds (*Acanthopneuste*, *Phylloscopus*, etc.) and wagtails and pippits (*Motacilla*, *Anthus*). Swinhoe's redtailed robin (*Larvivora sibilans* Swinhoe [Swinhoe's bushrobin, *Erithacus sibilans* Swinhoe]) spends the summer in southern Siberia and the winter in southern China. In spring and autumn Sten Bergmann saw tens of thousands of ducks and swarms of geese, swans, water cocks, sandpipers, cranes and snipes on the coastal lakes southwest of the mouth of the Tumen. They were followed by birds of prey: sea eagles, vultures, buzzards, goshawks, falcons and sparrow hawks [229, p. 45f, 96f]. During my own visit to these lakes on 29 September I made the same observations.

The Koreans use all manner of parts of wild animals just as versatilely as they do wild plants. Naturally, in a country in which livestock breeding is not very developed, by European standards, the meat of hunted animals is very highly prized. Particularly sought after is bear and leopard meat. Above all, however, parts of animals serve as medicine, and superstition plays just as great a role as with medicinal plants. A tiger in its winter coat is worth at least 2000 yen, and a good half of this is for the meat, the bones and the entrails. They are used for making medicines, which are exported especially to China (p. 33). The gallbladders of bears, gorals and wild boars are dried with their contents, pulverized and taken in rice brandy. This medicine is supposed to protect women against complications during and after childbirth. The raw liver of many animals is considered to be particularly effective. The different species of deer, especially the Manchurian sika deer, are hunted primarily for the horns while they are still in the velvet. These are cut off with a saw, boiled, dried and pulverized. Powdered deer antler has a very high reputation in China as a rejuvenating potion. Even the broth from boiled snakes is considered to be an excellent medicine.

In this sphere as well the old and strong influence of China is apparent. Here as well as there all wild animals are ruthlessly caught and used in the most varied manners. Revulsion is something the Korean does not know. As a result wildlife is already quite meager, as regards the number of individual animals, particularly in the densely populated regions of southern and central Korea. It would be even more so, were it not for the strong influence that Buddhism and its ban on killing and eating animals has had for 1000 years. This still continues to have its effect, supported by superstitious awe. In the temple forests hunting is still often strictly forbidden and the obligation to refrain from eating any meat is still in force. In 1916, 129 Koreans were still killed by wild animals (tiger, leopards, bears, wolves, wild boars), and 105 were injured. In the same manner 2839 head of livestock were lost (China Medical Journ. 32, 1918, p. 176). Since then the danger from wild animals has diminished greatly.

6 The Anthropogeographical Character of Ancient Korea

On 1 October 1940 Korea had 24,326,327 inhabitants (p. 14); 97% of them were Koreans. Thus even today the anthropogeographical character of the country is determined primarily by the Korean element of the population. It pervades the entire country, whereas the Japanese element is restricted mainly to the larger settlements. As the remarks at the end of the chapter on flora and fauna showed, it has many and close contacts to its natural environment. This broad contact has its roots in a history of several 1000 years, and therefore the basic features of Korea's cultural geography are still of Korean origin today. These features will be the subject of this chapter. In it we shall articulate all of the anthropogeographical traits that developed without Japanese influence, most of which thus extend back into the Korean dynastic period. We shall depict the anthropogeographical character of traditional Korea around the turn of the last century, particularly noting the phenomena that have preserved their geographical significance from that time into the present.

Korea has one anthropogeographical feature of unmistakable originality that is unique in East Asia: the white clothing of most of its inhabitants and the traditional hairstyle and headwear of its men. Thus it has often been asserted that it was easy to distinguish the Koreans racially from all other inhabitants of East Asia, especially the Chinese and Japanese. In reality this is not the case. Of late there are many Korean men dressed in Japanese uniforms. It is usually impossible to distinguish them from Japanese, and this is not only impossible for Europeans. A Korean policeman who speaks Japanese well and is transferred to a new place can pass himself off to his own countrymen as a Japanese for as long as he chooses. There is no fundamental racial difference between the individual East Asian peoples. Most of the racial types occur among all of them, only in varying frequency.

Most Koreans are brachycephalic, and the back of their head is very flat. According to Chantre and Bourdaret [525] the average cephalic index is 83.6. The forehead is low, the hair generally black to bluish–black and straight. There is very little hair on the body. Beards are limited mainly to the chin, that of the old men often being graced by a thin goatee. The color of the skin is a very pale yellowish brown, usually somewhat darker in farmers than in city–dwellers. The cheeks are often quite reddish, particularly among the women. The color of the eyes is almost always brown. The edges of the eyelids have the shape of a Greek circumflex with the steep downward curve on the inner side of the eye. The fully developed Mongolian fold is, however, relatively rare. On the average the Korean is considerably taller than the Japanese. Baelz gives the difference as 3 cm [523]. Kubo calculates the average height as 164 cm, though on the basis of only 3719 measurements [528]. Despite the consumption of tremendous amounts of food, obesity is rare.

Von Siebold already recognized two racial types among the Koreans [132]. Independently of him, Richthofen identified the same types [140, 141]. Baelz studied both in more detail and mapped their distribution [520–523]. He calls them the Manchu–Korean and the Mongolo–Malayan types. Baelz rightly opposes Hulbert's opinion that the Koreans are related to the Dravidian peoples [518, 519]. Rather, many features of Baelz' Mongolo–Malayan type correspond to v. Eickstedt's Paleomongoloids [*Rassenkunde und Rassengeschichte der Menschheit* (= Study of Races and Racial History of Man), Stuttgart 1934], whose main area of distribution lies in southern China and Indochina. Von Eickstedt does not single out the Manchu–Korean type, but includes them among his Sinitic peoples. Yet many of the features in his description of the latter do not fit that type. My own notes, written down before I knew Baelz' results, distinguish the same two types.

The Manchu–Korean type is taller than the Mongolo–Malayan type, very slender, orthognathous, with delicate limbs, narrow hips and a long face. The bridge of the nose is high, the ridge of the nose narrow and straight; even aquiline noses occur. The round nostrils are not visible from either the front or the side when the head is held normally. The lips are thin. The cheekbones protrude only moderately. The Mongolian fold is rare. Arms and legs are relatively long. "It is a stately race of men, aristocratic compared with other East Asians, that brings forth quite splendid figures, both in size and stature and in facial features" [Baelz 520, p. 183]. The Manchu–Korean type is considered an ideal of beauty, and most of the Korean geishas, the *kisaeng,* belong to it.

The Mongolo–Malayan type is shorter. The bridge of the nose is low; the profile of the nose is concave. From the front the viewer looks into the nostrils of the broad, upturned nose. The face is round, the upper jaw broad, and the cheek bones protrude strongly. The entire face lies in a single plane. The alveolar prognathism is fairly pronounced. The mouth is wide, the lips somewhat thick. The Mongolian fold is found more frequently than in the former type. The eyes are only slightly recessed, so that the line of the forehead continues on over the eyes to the lids. The physique is more thickset and sturdier. The powerful legs are very short.

In the historical section of Part One we described Korea as a land bridge into which movements have flowed from both the north and the south. Those from the north brought the Manchu–Korean type into Korea. This type is thus more common in northern Korea than in southern Korea, whereas the Mongolo–Malayan type reached Korea from the south, with the Kuroshio perhaps acting as a vehicle of race and culture, according to Baelz. Still, it is worth mentioning that v. Eickstedt's Tungusic (North Mongolian) type, which shares many characteristics with the Paleomongoloids, predominates in the Amur region. It is thus possible that Baelz' Mongolo–Malayan type includes two different individual types. Krahmer actually does already distinguish two types with protruding cheekbones, thick lips and slanting eyes [263]. Only one would have migrated into Korea from the south then, while the other must have come from the north. The number of pure representatives of the Mongolo–Malayan type is much larger than that of the Manchu-Koreans. I had the impression that conspicuous characteristics of the former are dominant, so that even in persons of mixed descent the Mongolo–Malayan strain predominates phenotypically. This is particularly true of the shape of the face. On the other hand, the greater height of the Manchu–Korean type appears to be inherited preferentially, resulting in the tall average stature of the Korean.

Occasionally there are individuals who possess certain Nordic racial traits. As Oppert already observed [169, p. 8f], they appear more often in children than in adults. These features include, above all, brown, or even blond hair. A few times I observed men with decidedly blue eyes. Krahmer [263] already points out the occasional occurrence of blue eyes. Regarding the question as to how these Nordic racial vestiges came to Korea, we refer to the passage on p. 19f and Günther's [*Die Nordische Rasse bei den Indogermanen Asiens* (= The Nordic Race Among the Indogermanic Tribes of Asia), München 1934] and v. Eickstedt's (p. 259–333) discussions.

Finally, Baelz mentions [520, p. 172] that in northern Korea there are very slight traces of Ainu blood, which he, like v. Eickstedt, ascribes to the Caucasian racial stock. Koganei [524] comes to the same conclusion.

On the whole the Koreans are a powerfully built and vigorous breed of men. Soon after the Chosŏn dynasty came into power in 1392, censuses of households and individuals began. The Korean economist Hoon K. Lee has published them for the eight provinces of ancient Korea [605, p. 40f]. The first two of these censuses, those of 1404 and 1648, are so incomplete, though, that they must be disregarded. But from the censuses of 1678, 1717, 1753, 1807, 1852 and 1904 we can get an approximate idea of the distribution and movement of population for two and a quarter centuries. Of course we must not attach too much importance to the absolute figures even here, since they are obviously still much too low. But the degree of inaccuracy should be approximately the same in the different periods and provinces, so that by comparing the differences from region to region and from time to time we can obtain certain results. Beginning in 1678 there was at first a considerable increase, which slowed down between the later census years until 1807. After 1807 it changed to a decrease that was almost twice as rapid in the second half of the 19th century as in the first. Even Lee admits that this unsatisfactory situation was the result of governmental maladministration prior to the Japanese takeover. The percentages that I calculated for the provinces show that there was a shift during the 226 years. The three southern provinces, which initially have the main portion of the increase, already begin to decrease in 1753 and then bear the main share of the decline. The two northern provinces show this decline only mildly, in fact, Hamgyŏng–do is even the site of a vigorous increase between 1852 and 1904 (cf. p. 393f). The main center of population thus gradually shifts from the south toward the north. The three southern provinces, in which 52% of the population still lived in 1678, sink to 47% in 1904, while the two northern provinces rise from 21 to 26%. This gives the latter the character of a pioneer zone to a great extent. At the end of the period of isolation the population of the country probably amounted to about 10 million.

The unsatisfactory characteristics of the population shifts are probably partially the consequence of extensive epidemics. Formerly such epidemics were brought into Korea mainly by land. Since the opening of the treaty ports in 1876 epidemics have often been carried into Korea by this route as well. In 1895 more than 600,000 persons are said to have fallen victim to cholera in the border districts, and as late as 1920 there were about 13,000 victims in Seoul alone. A large proportion of the Koreans, about one fifth of the men, have pockmarks. Among

Korea's ubiquitous deadly diseases are also tuberculosis and typhoid fever [537]. Other diseases are ones that are found in warm countries, providing evidence of the fact that Korea has a tropical climate in summer. Beriberi is, however, rare, although polished rice is the main item of the diet in the densely populated parts of the country. Probably the vitamins that are lost during the polishing process are made up for by the large amounts of raw vegetables that are eaten. Pellagra is common, though [534, 535], as are malaria, hookworm disease [536] and above all leprosy, which occurs primarily in southern Korea [530–533]. Taro, whose sapotoxin content Oberdörfer considers to be one of the environmental conditions for contracting leprosy, is, however, lacking in Korea, except for a minor crop on Chejudo. Nor does the corn cockle occur as a weed in the fields. As late as 1937 13,786 advanced cases were registered in southern Korea. At the instigation of R.M. Wilson, the American mission has rendered important services by establishing leprosy colonies in southern Korea. In 1917 the Japanese government created an excellently equipped home for lepers on the isolated island of Sorokto (Chŏllanam–do), which accomodates several thousand affected persons today. Since the takeover the Japanese accomplishments in the struggle against other contagious diseases have also been exemplary.

It has not been possible thus far to relate the above racial classification of the Korean people to any corresponding differences in its mentality. On the contrary, in its psychological structure the nation gives the impression of a uniform whole, whose individuality has been decisively influenced by the historical fate of the past three and a half centuries. Opinions on the character of the Koreans are extremely varied, and A.J. Brown is correct when he asserts that they are considered tormented saints by some and the most despicable people in the world by others [704, p. VI]. The first attitude, which contains an indictment of Japan, is assumed primarily by American missionaries, while the other is common in other circles. The detailed description of the Korean character and its consequences, for example, which is identical in all editions of Terry's *Japanese Empire* (the English "Baedeker" [276, p. 719ff]), is nothing but malevolent, sarcastically formulated criticism, and the opinion of such an unprejudiced traveler as Isabella Bishop is hardly more cautious except in its form [207, Vol. I, p. 4, 185, Vol. II, p. 148, 189, 278ff].

In general the Korean is good–natured, helpful and hospitable; he is affectionate toward parents and children, intelligent and very talented with languages. But the maladministration of two and a half centuries has left him lazy and apathetic. During this entire period there was no public justice. Beneath the eyes of a weak government, avaricious officials robbed the industrious of what they had laboriously acquired. Fawning servility and poverty were the only means of avoiding this system of exploitation. This subservience to the caste of officials, the *yangban*, and the tenacious adherence to conventions are outgrowths of Confucianism that were the inevitable results of seclusion. The consequences of this period have not been overcome to the present day. Even today most Koreans work only as much as necessitated by their most urgent needs, and these are incredibly slight. All foresight and all effort are lacking. The conditions of the untidily kept fields, of the forests and brush, exploited to the point of ruin, and of many houses and their evil–smelling surroundings bear eloquent witness to this mentality. The Korean men spend as much time as possible in idle bliss on the streets or in front of their huts, squatting close to the ground, in their mouths the meter–long pipe with the tiny bowl, which many lay aside only to sleep or eat. As much work as possible is unloaded unto the shoulders of the much more delicate women, who work from morning to night without a holiday. The disposition toward idleness is combined with the inclination to acquire money without exertion. Quickly acquired money is wasted on immoderate eating and drinking. The clan system, still universally in effect today, in which the head of the family is the master of the fate of all its members, hampers all activity, as does a very primitive form of demonism. The entire mentality has become ossified in a naive childishness. Even Koreans who have come into close contact with Western culture only vaguely recognize the failings of their people and the need to do away with these failings. A haughty pride in the worth of their own culture hinders them from recognizing ways toward improvement. Hulbert, who is one of the most benevolent critics [43, p. 27–44], sees Buddhist influences in many of the traits of the Koreans [70, p. 39]. He recognizes in them the four foundations of this religion: mysticism, fatalism, pessimism and quietism. It is quite possible that certain racial dispositions coinciding with the 1000-year-old dominance of Buddhism and the oppressive historical fate are the cause of this present mentality. But Hulbert also stresses that the Koreans possess a very healthy mixture of Chinese rationalism and Japanese idealism. This justifies the hope that they will gradually find the way out of their present stagnation. The first signs of such a development are visible.

"Just as China could, in many respects, be called a big Korea, Korea appears to the observer as a small China. Not the China of the present, but rather as it must have been at the time of the Ming dynasty in the 17th century." These words, with which E. v. Hesse–Wartegg begins his travel account of Korea in 1895 [204], still deserve to be placed at the beginning of a cultural geography of ancient Korea. In the first volume of the Transactions of the Korea Branch of the Royal Asiatic Society (1900) there is a dispute between Gale [69] and Hulbert [70] on the origin of the basic features of Korean culture. Whereas Hulbert emphasized the native aspects excessively, Gale wanted to interpret absolutely all phenomena of cultural life as Chinese influences and summarized his arguments with the following sentence: "Korea has to–day no life, literature or thought that is not of Chinese origin" (p. 14). One is as little correct as the other. In the chapter on history in Part One we often had the opportunity to differentiate between what was originally Korean and what was transmitted from China. But there can no longer be any doubt today that in very many spheres of life that are expressed geographically the Chinese features predominate, although partially altered, to a greater or lesser extent. The widely traveled v. Hesse–Wartegg is also correct, however, with the second part of his assertion. At the time that the Manchus seized power in China as well, Korea entered its period of absolute seclusion and thus of cultural stagnation. As a result the essential features of the geography of traditional Korea developed at an early period, and as far as they were imported from China, they date back to at least the Ming period (1368 to 1644, Fig. 6).

This is true first of all of the clothing, which is very striking physiognomically because of its white color. I shall describe only the costume of the mainland here. Due to its insular isolation Chejudo has preserved still older traits (p. 379). The main article of the man's clothing is a very wide pair of pants, tied around the waist and at the ankles with bands. The feet are covered with straw sandals or cloth shoes with felt soles. They have been being replaced in the last few decades by variously colored rubber shoes, which, like the cloth shoes, are Chinese in shape. The upper part of the body is covered by a jacket, which is tied on the right side of the breast with a bow. Korean clothing knows no buttons or buttonholes. When going out, a coat tied in the same manner is put on over this outfit. Among the better off, jacket and coat have a collar that is glued on with rice paste. Older children of both sexes wear the same clothing, and it is often difficult to distinguish boys from girls, especially since both wear their hair parted and in the same long braid. Men lay aside this hairdo on the day of their marriage, which usually occurs very early, and exchange it for the topknot. This 1000 - year-old symbol of masculine dignity consists of a knot of hair the length of a finger, looking like a horn, held together at the top of the head with a pin of bone. Over it is placed the main adornment, a broad–rimmed high–crowned hat that is much too tight by our standards. This is the most expensive article of clothing. It is woven in such a manner that it is transparent of horsehair or bamboo fibers and lacquered black. It rests on a band of the same material encircling the whole head, which is often not taken off for decades. It is kept on this band by means of a string tied under the chin. When it rains a funnel made of oiled paper, which the Korean always carries with him folded up, is inverted over the hat. Farmers also have large, bowl–shaped rain hats and straw coats. A string is wrapped around the waist, and on it hang the tobacco pouch and other utensils, since there are no pockets.

The clothing is made for the most part of cotton materials. In the hot summer hempen and thin silk materials are also worn. In the cities people often wear thin frames of bamboo or willow rods on the bare skin of the torso and around the wrists, thus improving the ventilation. In the cold winter many garments are worn on top of each other, or they are exchanged for special, padded winter clothing. The feet are then protected by socks made of pieces of cloth sewn together. Woolen clothing is nonexistent, as is sheep raising. All of these materials have the preferred color white. It is therefore inevitable that the farmers normally look dirty. Nevertheless, a good part of the women's work is spent on washing these clothes. They are beaten in the cold water of the wells and creeks, starched with rice starch and finally ironed by beating them with round wooden sticks. Half the night through the monotonous sound of this chore reverberates from the women's rooms of the farms. More recently bluish gray trousers or vests are frequently seen. Since time immemorial the men have worn special mourning clothes. A son mourns for his parents for 3 years. The mourning coat is a wide sack-like garment of brownish–yellow hemp. Of the same color is the gigantic mourning hat of coarse wickerwork covered with hemp, which extends far over the shoulders and has the shape of an inverted bowl with a curved edge.

The women's costume is similar to that of the men. Over the trousers the female sex wears a cloth that is wrapped around the body one and a half times and reaches to the ankles like a skirt. Town women part their black hair and braid it to form a knot at the nape of the neck, while farm women form it into a ring that they wind around their head to use as a base for their heavy burdens, for women do not possess hats. In the female dress, too, white predominates, and for women over 30 it is prescribed by tradition. Female persons between 20 and 30 often wear a light purple, those under 20 a pale green or pink. Most brightly colored are the clothes of the small children, whose gaiety contrasts strangely with the seriousness of the adults. In summer the smallest ones often romp about completely naked in the dirt.

In the last few decades European garments, sometimes with Korean touches, have appeared, and they are rapidly spreading in the cities. Even the school children there quite often wear something like a uniform of gray cotton cloth. The topknot and the braid are also giving way to short haircuts in young men and boys, although riots broke out only a few decades ago to see them kept. Thus the typical features of the Korean costume are on the decline. The Japanese administration favors the abandonment of white clothing, because it is uneconomical, and cutting off the topknot, because it indicates a certain receptiveness for the problems of modern times.

Life in the cities and in the countryside is naturally highly linked to the seasons. We can divide the civil year according to such customs. Winter can be characterized by the length of the period during which the ondol heating of living rooms is kept going (p. 157), summer by the length of the period during which thin clothing is worn. Civil spring thus begins with the end of the winter heating period, summer with the beginning of the period for wearing thin clothing, autumn with the end of this period, and winter with the beginning of the heating period. The Nihon Chiri Fūzoku Taikei [299, Vol. 15, p. 62] lists the average dates for these seasons, classified according to these customs, for the cities that have meteorological stations. The situation shown in Table 10 is the result.

As can be seen from this table, civil spring advances in uniform fronts from south to north, the difference in the time of onset within the country amounting to one and one–third months. Summer starts first in the southeast (Taegu) and advances northward very rapidly. Between Cheju and Yong–amp'o at the mouth of the Yalu there is a time difference of only 5 days. In the extreme northeast, however, it begins very late. Between Wonsan and Unggi the time difference is one full month. The beginning of autumn, again, advances in a uniform front from north to south, and the time difference between the extreme ends of the country is just as great as in spring. The onset of winter clearly shows how the in-

	Beginn of civil				Length of civil			
	spring	summer	autumn	winter	spring	summer	autumn	winter
Cheju	1 Mar.	15 June	30 Sep.	20 Dec.	106	107	81	71
Mokp'o	15 Mar.	5 June	25 Sep.	5 Dec.	82	112	71	100
Pusan	5 Mar.	15 June	25 Sep.	10 Dec.	102	102	76	85
Chŏnju	15 Mar.	5 June	15 Sep.	20 Nov.	82	102	66	115
Taegu	15 Mar.	1 June	20 Sep.	15 Nov.	78	111	56	120
Inch'ŏn	25 Mar.	15 June	15 Sep.	20 Nov.	82	92	66	125
Seoul	25 Mar.	10 June	15 Sep.	15 Nov.	77	97	61	130
Kangnŭng	20 Mar.	10 June	15 Sep.	30 Nov.	82	97	76	110
P'yŏng-yang	1 Apr.	15 June	10 Sep.	5 Nov.	75	87	56	147
Wonsan	1 Apr.	20 June	10 Sep.	20 Nov.	80	82	71	132
Yong-amp'o	5 Apr.	20 June	5 Sep.	5 Nov.	76	77	61	151
Sŏngjin	5 Apr.	15 July	5 Sep.	10 Nov.	101	52	66	146
Chunggangjin	10 Apr.	25 June	25 Aug.	20 Oct.	76	61	56	172
Unggi	10 Apr.	20 July	31 Aug.	30 Oct.	101	42	60	162

Table 10. Beginning and length of the civil seasons

terior cools off earlier; even in Taegu it comes as early as 15 Nov. The time difference between the ends of the country, however, is a full 2 months. On the warmer Sea of Japan coast the onset is 15 days later than on the Yellow Sea coast. The interior is correspondingly disadvantaged with regard to the length of spring. On the coasts of the Sea of Japan and the Korea Strait it is as much as 30% longer than in the interior and on the Yellow Sea coast. The length of summer is tremendously variable, depending on the latitude. It is particularly short in the extreme northeast. Cheju's summer is more than 2.5 times as long as Unggi's. Autumn is again shorter in the north than in the south and in the interior than on the coasts and longer on the Sea of Japan coast than on the Yellow Sea coast. The length of winter decreases from north to south quite uniformly to less than one half. In the extreme south spring and summer are long seasons, and autumn and winter are short seasons. In the extreme north winter stifles all other seasons.

As in clothing there are close relationships to China in house construction. But while the Korean costume does not exist in China today, the forms of buildings that were brought to Korea are still alive in the Middle Kingdom today too. One need only compare A. Eckardt's History of Korean Art with the magnificently illustrated works of E. Boerschmann to be convinced of these relationships. To be sure, these relationships are not very pronounced physiognomically in the most common and most primitive Korean dwelling, the straw–roofed hut. They do exist here as well, though, for the Chinese house, too, is often basically a wooden construction. The resemblance, indeed the derivation, becomes very apparent, as soon as tile roofs appear, and the more magnificent the palace, temple or monastery buildings are, the more amazing the Chinese influence is. It extends to the details of the forms of tiles, wooden pillars and beams and the type of technical construction. Geographically these relationships are less prominent only because in China stone buildings predominate, while in Korea wooden and half–timbered buildings do. In spite of this relationship higher Korean architecture does exhibit many basic features of its own, so that here and there one can speak of a Korean national style.

Korean settlements consist almost entirely of one–story buildings, which even in the cities are more deserving of the name hut than house [574]. As already mentioned, Korea has a half–timbered or wooden architecture. The Imperial Chinese Geography of 1744 already reports on Ch'ungch'ŏng and Hwanghae–do: "There are no stonework buildings. The houses are constructed of earth and have the form of graves" [131, p. 92]. Exclusively wooden architecture is closely connected with fire–field farming and is found only in areas with extensive forests, i.e., in the higher mountain ranges, on Ullŭngdo and in the Kaema upland. It will be discussed in connection with the latter (p. 231f). Likewise I shall not include the houses and settlements on Chejudo (p. 379) in this chapter.

The houses in most of Korea are half–timbered constructions. The timbers are usually cut from red pine logs and like these are often somewhat crooked. The main uprights rest on foundation stones dug into the ground and are connected by cross beams and tie beams, the latter of which bear the roof. The spaces in between are filled with sticks tied together with straw ropes; wickerwork is spread between the sticks, and finally the thus constructed outer walls are plastered with mud. Generally they are left the natural color of mud; only wealthier persons have the outer walls whitewashed. Occasionally the lower portions of the walls are covered on the outside by a stone surface to protect them from the influence of the weather. The stones are pressed into the mud of the walls while it is still moist and then daubed with mud. Straw ropes hold this dilapidated cover to the walls. On the outside the stones stick out of the mud coating.

The houses consist of rooms called *kan*. The *kan* have standard sizes varying between 3 x 3 and 5 x 5 m in different parts of the country. In addition to these square measures there are some slightly rectangular ones (for instance, 3 x 4 m). The distance between the main posts, which stick out of the clay walls on the outside, is determined by these measures. Inside several *kan* are sometimes joined to form a single room by leaving out the partitions between them. The simplest houses consist of two to three *kan* lined up next to each other, resulting in a longish rectangular floor plan. This is the basic form (Fig. 37b); it has many modifications. At the ends of the rectangle additional *kan* can be attached on one or both sides, so that shapes with angles, T–shapes or shapes with two wings arise (Fig. 37a and d), or for the entire length of the house two *kan* may be arranged behind one another in pairs, forming a double *kan* house. In northern Korea the floor plans are appreciably more diverse than in southern Korea.

The individual *kan* have different uses. One always serves as kitchen and heating source, one or several others as accomodations for the male and female members of the family. The floor of the kitchen lies at ground level, those of the rooms about knee high above it. Under the floors of the rooms run the flues for the floor heating system (Kor. *ondol*), which is operated from the kitchen. The fireplace for this universal, very practical Korean national heating system is built into the large, low kitchen stove. In larger houses rooms lying off to the side have separate holes for heating along an outer wall (Fig. 37d). The twigs and leaves collected daily in the nearby shrub formations serve as fuel (p. 140). The smoke comes out of a chimney hole at the end of the flues and envelops the villages and towns morning and evening in acrid fumes. In regions with wooden architecture, actual chimneys made of hollowed out logs are placed on top of the holes (p. 231f). From there the use of chimneys has spread to the parts of northern Korea that have little wood. The chimneys then consist of boards nailed together or simply of clay pipes and are usually covered with straw as a protection against the weather. They rise like factory smokestacks at some distance from the house. Since the *ondol* heating system is started up every time the kitchen stove is used, the rooms connected to it are unbearably hot in midsummer. The Koreans then prefer to sleep in the open on their verandas.

Even the simplest houses have a veranda extending along outside the rooms at the level of the floors, and like them, it is never stepped on with shoes. In the simplest cases it consists of earth piled up and paved on top. Usually, however, it is made of wooden boards that are kept clean, like the mud floors covered with oiled paper in the rooms. In larger houses the veranda is supplanted or supplemented by a hall (Kor. *maru*) that occupies one or more *kan* (Fig. 37d and e). Veranda and hall are both covered by the roof. From the veranda and the hall latticed wooden doors lead into the rooms. They turn on iron hinges and are thus not sliding doors, as in Japan. They are covered with rice paper so that light shines into the rooms. Usually there are no separate windows. The thresholds of the doors are, however, about 2 ft. high; thus one could just as well regard the doors as low windows through which one enters.

Most Korean houses have a flat hipped roof. In the northeast, in contrast, houses with gabled roofs are more common, presumably due to Chinese influence. Forming a transition between the two is the half–hipped roof, very common in the north, which has a triangular gable approx. 1 m high above the end slope. In regions with irrigated paddies the roofs are covered with rice straw. At times it hangs over the eaves in disorderly fringes, so that the whole house looks like a big heap of straw. That is probably the origin of the comparison with – East Asian – graves made by the Imperial Chinese Geography. Usually, however, the layers of straw are neatly cut off at the eaves, and indeed the individual layers of

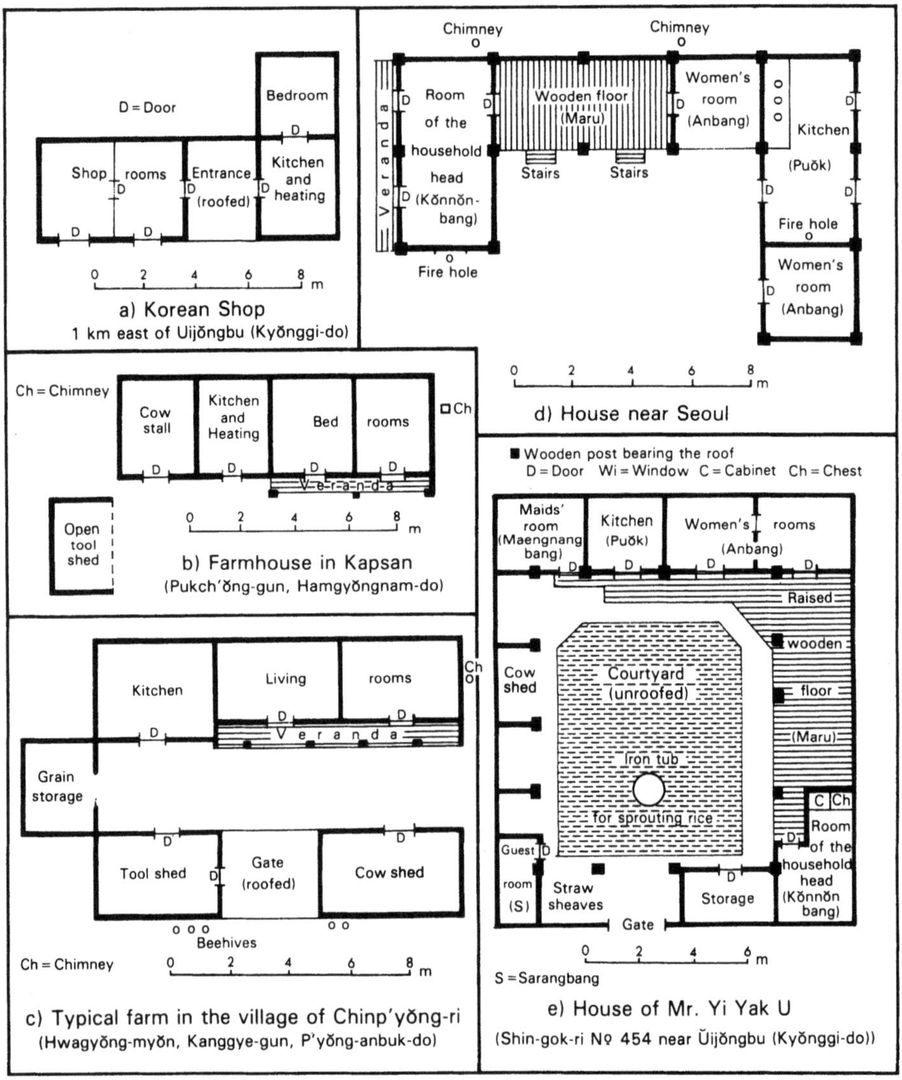

Fig. 37. Floor plans of typical houses and farmsteads (**a–c** and **e** by the author, **d** after A. Eckardt)

straw occasionally lie like tiles, one above the other, in rows parallel to the ridge. At the ridge the opposing layers of straw are covered with a wide braid of straw for protection against the rain, the braid usually extending down somewhat farther than the sloping ends. On gabled roofs it hangs way down and its two ends are weighted down with heavy stones to give the necessary tension. The top layers of straw need to be renewed every 3–5 years, because they decay, and the new layers are generally thicker than the ones that were removed, so that the thickness of the straw roofs increases. The sloping ends disappear, and the roofs look like overturned boats. In summer the roofs of farm houses are often half overgrown by climbing gourds and cucumbers. In autumn they are red from the pepper pods spread out to dry on them. Since the straw roof needs frequent repair, the inhabitants of areas where rice farming and forests lie close together like to replace it with the shingle roof that is the rule in the areas where there is wooden architecture. In P'yŏng–annam–do the outcropping limestone shale is also used as roofing material. Otherwise, outside of the areas where there is rice farming and forest the roof consists of grass, kaoliang straw or reeds. Near the stormy coasts the roofing is often fastened down specially, in the south with bamboo poles, in the north with ropes of straw or raffia.

Wealthier persons choose a tile roof. Slightly curved "female" tiles, lying with their concave side upward, are covered at the interstitial lines by strongly curved "male" ones with their convex side upward, which lie like tubes in the direction of the slope. Ridge and hips are covered with curved tiles, both with the outward concave sweep so familiar from Chinese architecture. Never, however, does one find the inverted uplift of the ends (ting roof) that is so common in China. Traditional Korean public buildings are always tile roofed. But thatched roofs still predominate today in the Korean sections even of the cities. Recently many people have unfortunately begun to use ugly roofs of galvanized sheet iron. In the county (gun) seat Hŭich'ŏn (P'yŏng–anbuk–do), for example, about one–third of the houses have straw roofs, another third tile roofs, and finally a good third galvanized roofing. In northern Korea roof shapes and roofing material are much more varied than in southern Korea, where flat hipped roofs made of rice straw are the monotonous rule.

In their architecture the pavilions closely resemble tile roofed houses. Picturesquely set into nature at particularly attractive scenic spots in the vicinity of cities, their serene and stately presence dominates a cliff, a river valley or an isolated hill. They have no side walls, however, so that the eye can roam freely to all sides, and the main supporting pillars are very thick tree trunks, suited to the weight of the mighty roof. Some of these pavilions have two stories.

The Koreans tend to shut their houses off from the outside world, mainly to hide the women of the house from sight. Thus in the simplest houses the verandas are occasionally concealed from the outside by hanging up straw mats. Generally, though, the house is surrounded by a high fence, making it a courtyard farmstead. Fences are made of lespedeza switches, hemp stems from which the fibers have been removed, or kaoliang stalks. Among the wealthy dirt or stone walls take their place. Inside the thus created enclosure the farm buildings are often set off from the house (Fig. 37b). They almost never have tile roofs. In very humble homes the ox and the few chickens and pigs sleep in a corner of the kitchen, which often is not even separated from the living *kan* by a wall. In wealthier establishments the house and the farm buildings enclose a courtyard that is accessible through an entrance gate (Fig. 37c). The enclosure can be so complete as to result in a square courtyard (Fig. 37e). In some corner or other of the courtyard stand the huge brown earthenware jars, in which soy sauce (Jap. *shoyu*, Kor. *kanjang*) is prepared and stored.

In southern Korea densely crowded irregular clustered farm villages are the rule. In northern Korea difficulties with the water supply in areas with limestone shale and recent lava flows are often the reason for living just as closely together. From a distance, the low, closely packed huts with their gray, rounded straw roofs often look like piles of big rounded boulders. In northern Korea more frequently than in southern Korea we find villages aligned along previously existing roads. In mountainous country the relief is often the reason for such a linear arrangement. In the centers of large villages one–story shops with gabled roofs line the street, house on house. Outside of the centers the farmsteads are less closely crowded. Even in densely populated areas the villages are frequently no larger than hamlets, but they are then located only short distances from each other. In many of the regions with rice farming along the western side, the dotlike distribution of the cultivated area makes hamlets appear the natural form of settlement. Dispersed settlements are confined primarily to the pioneer region of the north. They too are usually lined up along the creeks and rivers. Because of the flood danger, settlements in valleys seldom lie directly along the river. They favor low terraces, talus cones or slipoff slopes. Even where there is no danger of flooding, the settlements often lie at the foot of the slopes, and the fertile river bottoms are entirely given over to cultivation. The fishing villages on the coast of the Sea of Japan tend to crowd together on bars, where the boats can easily be pulled up on the shore.

The Neolithic and Eneolithic settlement of Korea was still closely linked to the coast (p. 16f). In the dolmen period we see that the settled area had already expanded. The settlement of Korea proceeded from lower to higher elevations. Even today four–fifths of the settlement is agrarian, and it has accordingly close ties to the fertile areas. The basins and plains on the west side, in the Naktong basin and in the northeastern Sea of Japan borderland, are therefore the only places where an extensive area can be settled. In the mountains the settlement sticks closely to the valleys, whose floors provide the relatively best soils, and is thus linear. How frequently settlements are related to valleys can be seen from the place name endings –ch'ŏn (= river) and –dong or –kol. The place name ending –dong or –kol, which is lacking in Japan, denotes a small village lying in a valley [W]. The dot map of population distribution (Map 3, cf. editors' note on p. 7) shows the contrast between extensive and linear settlement exceptionally clearly. Since the permanent settlements in the mountains keep to the deeply entrenched valleys, they generally do not extend very far upward. The Buddhist monasteries form one exception (p. 166ff). The upper limit of permanent agricultural settlements lies at approx. 600 m on Chejudo, 700 m in Chirisan, 800 m in Kŭmosan near Kimch'ŏn (p. 337), 700 m again in Kŭmgangsan, 650 m on Haramsan, i.e., very low for this low latitude. Recently it has even gone down somewhat in places, because here and there the government has refused to recognize persons who had settled without authorization. The upper limit of permanent agricultural settlements lies much higher in the Kaema upland, where deeply entrenched valleys are lacking. On it even county seats such as Pongsan lie at 1150 m, and the highest permanent agricultural settlements reach 1300 m. Thus, under the influence of the relief, the upper limit of permanent settlements rises poleward.

In the mountains, and even in the Kaema upland, the shifting settlements of the fire–field farmers (Kor. *hwajŏnmin*, Jap. *kadenmin*, [555–564] extend higher upward. In the Sobaek Range I found them up to around 1200 m, in the Kaema upland up to 1550 m, since they do not cultivate the valleys, but the steep slopes above them. Thus there is extensive settlement in the zone between the upper limit of permanent settlements and that of fire–field settlement. It is, however so sparse that it is not manifest as such in Map 3 (cf. editors' note on p. 7). Presumably the "fire–field folk" have always formed the uppermost layer in the gradual conquest of the currently settled area of Korea. Indeed, according to Oda [560], the oldest written record of fire–field farming dates back to the Shilla period. The fire–field farmers form the lowest stratum of the Korean population, having moved up from the lowlands because they had to break off with their previous existence there for some reason or other. In contrast to the peoples who practice slash–and–burn agriculture in Southeast Asia, the Korean fire–field folk are thus ethnically the same as the people who inhabit the basins and valleys of the country, and there is a continuous chain of gradations from the true, migratory fire–field farmer to the sedentary farmer who farms only officially registered permanent fields in the Korean mountains. The true or "wild" fire–field farmer high–handedly preempts the land that he cultivates. Somewhere in the virgin forest he clears a plot of land by burning it over and cultivates it with primitive methods for the few years that it yields a crop. In the vicinity of the field he builds a log cabin. When the soil is exhausted he looks for a new place that is suitable for his shifting cultivation and likewise moves his homestead, perhaps returning to the first field after quite a number of years. It is meanwhile covered with second growths, or at least weeds and bushes. Somewhat closer to the sedentary farmers are the fire–field farmers who clear only second–growth stands of trees and shrubs, because in their vicinity the natural forest is restricted to particularly inaccessible spots at the most, where it ekes out its existence in seclusion. A third group has abandoned the nomadic life and lives in permanent settlements, where they also have livestock and thus a manure supply. The fire fields can then be reused more often. A fourth group, finally, cultivates not only the preempted fire fields but also registered, leased permanent fields, which lie in the valleys or on the lower portions of the slopes, in contrast to the fire fields.

This continuous transition naturally makes it more difficult to make clear distinctions. The area of the fire fields under cultivation in 1930 still amounted to about 2000 km^2, which is less than 1% of the total area or 4.3% of the land under cultivation. But the effects of fire–field farming on the landscape and its economically deleterious influence go much further than these figures indicate (p. 431 f). If we include the fourth group we have just distinguished, the number of families engaged in fire–field farming comes to 137,678 [605, p. 158], according to the statistics of the government–general. The number of persons who are members of these families is presumably at least 0.8 million, and thus about one thirtieth of the population. The most important fire–field areas, according to my observations, are shown in Fig. 38. In central and southern Korea they include only the higher elevations of the mountains. In northern Korea, in contrast, they include all of the Kaema upland and the area bordering it to the southwest. Settlement and economy of the fire–field farmers will therefore be treated in detail in Part Three, in Chapter 2, on the Kaema upland.

Of the permanent settlements the old Korean cities deserve special attention. They include all settlements with the ending –chu or –ju and most with the endings –sŏng and –won (p. 10f). All travel accounts from the end of the last century agree that in their overall impression they do not differ much from large villages, and the parts of towns that have remained unchanged up to the present fully confirm those descriptions. In every city there are extensive areas that have remained so, except that the houses are even more closely packed together than in the villages, because the old cities were surrounded by walls. The percentage of houses with tile roofs is also higher (p. 293). This maze of houses is crisscrossed by narrow, crooked streets. The only impressive buildings, some of which are still standing today, were the seats of the county and provincial governors and the temples to Confucius. One of the latter was founded in each county seat by the Chosŏn dynasty. Buddhist temples are, however, lacking in the cities due to the hostile attitude of this dynasty toward Buddhism. In the cities that were former capitals, such as Seoul, Kaesŏng, P'yŏng–yang, Kyŏngju and Puyŏ there were additionally the old royal palaces and other historical monuments. Since they were not constructed of stone, however, many of these old palaces have fallen victim to fires. Nevertheless here and there in the regional descriptions (Part Three) we shall be able to point out cases in which such buildings contribute to the overall impression of the cities in which they are located. For the same reason private houses are generally of very recent date, even though their style does date back much earlier.

The historical survey in Part One showed how often the land bridge Korea was at the mercy of invading conquerors. These vicissitudes were the reason that all cities were surrounded by fortifications, and imposing remnants of these often still exist. Where they are missing, the routing of the streets or the contrast between a densely built–up inner city and more open suburbs often still reveals their location. Frequently they consisted of high earthen embankments ending in a vertical wall on the outer side only. In these cases little has been preserved, generally speaking. There are, however, some small towns that still have such walls today. They include Haemi, Hongsŏng and Namp'o north of the mouth of the Kŭmgang and Yŏng–am, Usuyŏng and Kohŭng in the southern coastal margin. In northern Korea the defences often consist of ramparts made of unhewn rocks. Larger cities were surrounded by high walls of rectangular squared stones, and long stretches of these stone ramparts often still remain. The most solidly built were the corner towers and the gates, particularly large numbers of which are still extant today. Because of their artistic value they are now generally classified as historical monuments. There are always at least four of them, named after the four cardinal points of the compass. Nowadays they often stand isolated in the middle of the modern city. The lower structure consists of large squared stones vaulting a high arch. On top of this is a massive wooden superstructure, sometimes two stories high, with a sweeping tile roof not unlike the pavilions described earlier. The superstructures to these gates were the tallest and thus most conspicuous buildings in traditional Korean cities and still are today, except where multistoried Europeanized buildings have been constructed in their immediate vicinity. The South Gate of Kaesŏng, which dates back to the Koryŏ dynasty, is the oldest extant secular building in Korea.

In cities lying in the open plain, the city wall forms a square closely surrounding the old town. Examples are (from north to south): Hoeryŏng, Kyŏngsŏng, Haeju, Myŏngch'ŏn, Sangju, Chŏnju, Kyŏngju and Taejŏng (Chejudo). If a nearby mountain towers above a city situated on a plain, it is often included in the fortifications. In this case the city never completely filled the encircling wall. Cities of this type are Kanggye, Hamhŭng, Chŏngju, Yŏnan and Kaesŏng. If the city lies in a basin, the wall runs along the surrounding peaks, so that the only places where the built–up area extends up to the encircling wall are at a few passes. The most splendid example of this type is Seoul (p. 293). Finally, some towns are located on summits, and the fortifications surround the summit plateau, in which case its size determines how narrow or wide the enclosure is. Since the inhabitants of the smaller towns are also predominantly occupied with farming, such locations are, however, quite unsuitable from the standpoint of economics and transportation, and therefore these cases are rare. The only example of a still existing county seat of this type is Yŏngbyŏn (p. 258f). The county seat Kwangju, 18 km southeast of Seoul, lies on top of a mountain, but the Japanese have moved the county government down to the village of Kyŏng–an on the main road from Seoul to Ch'ungju.

These fortified mountain towns lead over to the mountain fortresses (Kor. *sansŏng*), which are common in all parts of the country. They differ from the fortified mountain towns only in that they do not contain any permanent settlements, or if they do only hamlets. Some of them, such as the famous Puk'ansansŏng and Ch'ŏnmasan, lie in the vicinity of cities (Seoul, Kaesŏng) and were intended as a retreat for the population in times of danger. Some lie on important main roads, such as Kŭmosan, which flanks one of the Seoul – Pusan roads, and Naejangsan, not far from the pass of Noryŏng (Seoul – Mokp'o). Others are on peninsulas and islands to protect the approach against attacks from the sea, such as the island of Kanghwa and the promontories of Ponggang–ri and Anhŭng. Porphyritic flood basalts and granite intrusions have the tendency to form summit plateaus, which in the case of granite are often surrounded by sugarloaf or dome–shaped mountains (p. 79f). Such locations were a virtual invitation to construct a stronghold. The rims were fortified with walls of enormous squared stones that run along them looking not unlike the Great Wall of China. Sturdy gates mark the few entrances into the interior. Inside, cisterns or wells were built for the garrison and for refugees. Frequently one also finds the ruins of palaces, temples or hamlets in them even today, the latter occasionally bearing the name Sansŏng or Sansŏng–ri. These mountain fortresses reach an elevation of up to 1000 m.

In a discussion of Korean settlements one cannot ignore the Buddhist temples. They are not, however, the only features in which the religious sphere has left its imprint on the Korean landscape. There is a host of such features, differing completely in their form of expression and in their religious background [580–599, 299, Vol. 16, p. 94ff]. According to statistics published in *The Young East,* 2.8 million persons in Korea profess to believe in a particular religion [581]. That would be only about one–eighth of the total population, and even this figure is probably too high. Religious life is not very highly developed in Korea, due to the gradual cultural stagnation during the Chosŏn dynasty. Under the influence of the political unrest and of Christianity a large number of sects have sprung up during the past 100 years, only adding to the religious chaos of the preceding period. Thus far, however, this sectarianism has found only slight expression in the landscape. On the contrary, the landscape is influenced for one by the widespread popular religion, a spirit worship that is Shamanist in nature, and for the other by the two religions that entered the country from China but meet with only slight fervor among the broad masses today, Confucius' philosophy of life and Buddha's transcendent mysticism. All three have influenced each other mutually in Korea, so that the average Korean does homage to the faith and customs of the various spheres simultaneously.

Shamanism entered the country from Manchuria in the oldest period of Korean ethnic formation (p. 20) and is still closely related to Siberian Shamanism today (Clark [580, p. 173ff]). It is a belief in an innumerable host of spirits in water, air and earth, in house, yard, field and woods, indeed in a sick person himself. In the Koreans' attitude toward these spirits "feelings of veneration and loyalty are subordinate to those of fear. The cult does not consist so much in adoration and care as in appeasement, deception, submission" (Gundert [582, p. 189]). In more difficult cases this is the business of sorcerers. *Pansu* are blind men who drive the bad spirits out of houses. More effective and influential are the *mudang*, women who perform exorcising dances. Korean geomancy also has its roots in the belief in spirits. It was of decisive importance in the choice of burial sites, which depended on determining the *p'ungsu* (Chin. *fengshui*), i.e., the spiritual character of the sites in question.

Of the Shamanist features finding expression in the landscape one of the most important is the belief in the "God–Generals of the Five Quarters of the Sky" (*O–Pang Changgun*, [580, 586]). The eastern sky is ruled by the green general, the western sky by the white one, the northern sky by the black one, the southern sky by the red one and the zenith by the yellow one. The associations between the colors and the different points of the compass reveal a profound nature symbolism. Posts are set up in honor of these generals at the appropriate entrances to the villages. Every year sacrifices are offered to them, so that the generals will keep the bad spirits of their quarter of the sky away from the village. These *changgun* posts are tree trunks up to 3 m in height, their upper ends carved into grinning faces resembling humans with the hats of mandarins. Under the face there is an inscription to the effect, e.g., "West (east, etc.) side. White (green, etc.) king. Chase away spirits. Male great hero." Every few years a new post with the same shape and inscription is set up, without the others being removed, so that one occasionally finds five or six similar ones standing in a row, the older ones more or less weather–beaten. Such *changgun* posts are even found here and there by paths leading up to Buddhist temples, a sign of the amalgamation of totally different religious spheres. The externally quite similar *changsŭng* posts are of a different symbolic nature. They are supposed to keep alive the repulsive memory of a minister who was executed because of his reprehensible conduct, but serve simultaneously as milestones (Scott. Geogr. Mag. 1894, p. 572 and Kor. Rep. 1895, p. 143). They too are renewed in the same manner. On the newest post in such a group I found the inscription: "Woman general under earth. Erected in the year of *mujin* (= 1928) 4th month 5th day. To the county seat 20 *li*"

(1 Kor. *li* = approx. 400 m). On the middle post on the opposite side one could read: "Great general under heaven. Erected in the year of *imjin* (= 1892) 5th month 1st day. To Seoul 80 *li*."

Korea is a mountainous country and the Koreans have a profound, intimate relationship to their mountains. This is expressed in the important role that the Mountain Spirit (*Sanshin*) plays in this pantheon of demons. The gatherers of berries, mushrooms, herbs and roots and the ginseng hunters, who spend entire weeks in the virgin forest in late summer, sacrifice to him. It is not uncommon to find a small wooden shrine to him with a heavy, sweeping roof of tile or limestone shale at the entrance to a forest. Inside he is represented as an old man in the robes of a mandarin riding on a tiger.

At crossroads, on passes, even on mountain peaks one frequently finds heaps of stones piled up under solitary trees or under trees with a conspicuous shape. They are called *sŏnghwangdang* (literally: shrine or temple of the site of the fortress or city, [586]). Even today passersby often throw a new rock on the heap. Hung on the tree as votive offerings are scraps of paper, on which characters may be written, remnants of clothing, old rubber shoes and the like [587]. Sometimes there is in the immediate vicinity a *changgun* post or a tiny shrine, in front of which *mudang* occasionally dance or picnics are held.

One last expression of the Shamanist sphere in the landscape is the clinging rocks (Tr. Kor. Br. R. As. Soc., 5, I, 1914, p. 47). Where there are almost vertical granite rocks or boulders along paths, passing women lay harder slabs of stone into natural depressions, rubbing these depressions out until the stone slabs stick to them. There is a fertility charm involved.

The religious sphere that molded the Korea of the Chosŏn dynasty most highly is Confucianism with its doctrine of the five relations among men (parent – child, master – servant, man – woman, older – younger, friend – friend). Whereas Shamanist concepts dominate the broad masses, Confucius' moral philosophy was the concern of the king, the government, the officials and the other *yangban*. With its burial rites and ancestor worship, though, it has a strong hold among the simple people and here it joins forces with geomancy. The most important Confucian sanctuary in Korea is the "Hall of Great Accomplishment" (Taesŏngjŏn) with the Confucian Academy (Munmyo) in Seoul, reopened a decade ago by the Japanese. Semiannually great solemn ceremonies, with the highest officials participating, are held in this temple, which does not differ greatly in its outward structure from the main Buddhist halls. As mentioned above, Confucian temples (*hyanggyo*) are scattered over the cities of the entire country. Nowadays they generally serve as schools or police stations. There are famous schools and lecture halls in Kaesŏng, Oksan (Kyŏngju–gun, today Wolsŏng–gun) and Tosan (Andong–gun). In Seoul the octagonal Temple of Heaven was erected as late as 1896–98, after the ruler of Korea had proclaimed himself emperor. It and the other royal buildings have little, monkeylike clay or metal figurines on the edges of the roof, the so–called gargoyles (*chapsang*), a Shamanist influence on the Confucian sphere.

Such influence is most evident in the realm of ancestor worship. Round about the capitals Kyŏngju, Kaesŏng and Seoul lie, remote and peaceful, the royal tombs of the different dynasties, which are entirely Chinese in style. There are 98 of them. In tall pine forests where all woodcutting is strictly prohibited, the tomb grounds extend up a hill that is covered by a green meadow. At the entrance to the meadow stands the graceful wooden Red Arrow Gate (*hongsalmun*), painted red, which is not unlike the Japanese *torii* in construction, but has vertical arrows between the two thin lintels. It is a typically Korean piece of architecture that is unrelated to any of the Chinese *p'ailou* types. Then a shrine sometimes follows. Directly at the foot of the grave stands a massive stone ceremonial table, in front of it, as a sign of Buddhist influence, a stone lantern and to both sides stone military and civil officials with their horses. To the right of the shrine is a small pavilion that contains the tombstone. The tomb itself is a large, hemispherical grass – covered mound of earth with a low, horseshoe – shaped wall enclosing it from the rear toward the mountain [571, p. 46ff].

The graves of the *yangban* are similar in style to the royal tombs, only less extensive and naturally without the symbols of royalty. They too are surrounded on three sides by pine woods. Numerically of course the graves of the millions of ordinary people predominate. But they too are laid out according to geomantic principles at auspicious sites, where the deceased will find peace and his spirit will not disturb his descendents. Frequently many hundreds of them lie side by side like grassy hemispherical mounds, here and there with arrow–shaped tombstones. But one does find solitary graves, occasionally on mountains that reach a height of 1000 m [591] or ones where truly Alpine skill is required to climb them. Twice a year on set days in the second and eighth months of the Chinese lunar calendar the Koreans make a pilgrimage to the graves of their ancestors to offer them the prescribed sacrifices, and then one sees the white figures crouching everywhere on the gray or green slopes. In addition, the male ancestors back to the fourth generation are honored by their families in the same manner each year on the anniversary of their death. Funerals take place as processions from the house of mourning to the grave, often a great distance, accompanied by larger or smaller numbers of persons, depending on the circumstances of the deceased. Prayer flags always precede, and the heavy bier, covered with a white awning, is carried by a large number of men. To violate a grave is considered one of the most serious crimes. Today the extensive fields of graves limit the space available for agriculture and impede the construction of roads and railroads. The government–general has therefore ordered unified cemeteries to be set up, a measure that has not met with much sympathy among the population, which is still firmly rooted in traditions. Nonreligious national customs too, particularly ones that are linked to certain days and are observed by the whole population simultaneously, suddenly give the Korean landscape a special tone, such as Children's Day, the early summer swinging games or New Year's Day.

Buddhism, which entered Koguryŏ in 372 A.D. and profoundly influenced Korea's political and cultural life for 1000 years, fell into disrepute and degenerated during the Chosŏn dynasty. Since 1910 the government–general has endeavored to preserve its remnants and to improve the financial situation and the moral character of the monks, but they have not been very successful. According to official statistics, in 1939 there were only 7244 monks and nuns, along with 194,800 registered believers. But no other religious sphere has had such a lasting influence on the landscape as Buddhism.

Since the end of the 4th century temples have been founded in Korea. The oldest still extant ones, however, such as Pusŏksa (Yŏngju–gun, now

Yŏngp'ung–gun), date back only to the middle of the 14th century, i.e., to the period during which Buddhism was at the height of its secular power but also shortly before its fall (p. 29f). Most have burned down and been rebuilt at the same site in a similar form several times. Taehŭngsa (Haenam–gun) experienced this fate no fewer than eight times. The magnificent pagodas that grace the temple yards, the Buddhist symbols of withdrawal from the world, often date back to the founding period, however, for almost all of them consist of solid stone and are thus a kind of monument. Only a single temple, Pŏpchusa at Songnisan, has a wooden building designed like a pagoda with five tile roofs.

Of the immense number of Buddhist places of worship that once existed in and near the cities as well, the only ones that still remain are for the most part those that eke out a secluded existence in the mountains. There are 31 main temples and 1305 smaller ones. The bigger ones (Kor. –sa) lie in valleys at the foot of high mountains, with which they are closely connected. The magnificent album on the 31 main temples [596] includes the names of the mountains in the headings to the descriptions. Some of them are large–sized establishments with more than 100 monks and an even larger number of lay brothers. They attain the size of villages, but can be distinguished from these at the first glimpse by the regular arrangement and the dimensions of the buildings and the fact that all of the buildings have tile roofs. They include Taehŭngsa, Songgwangsa (Sunch'ŏn–gun, now Sŭngju–gun), Sŏnamsa, (same county), T'ongdosa (Yangsan–gun), Yujŏmsa (Kŭmgangsan) and Sŏgwangsa (near Wonsan). Entrusted to their care are the smaller temples distributed over the slopes and the heads of the valleys, including the hermitages (Kor. –am), which often consist of only one building and sometimes can be reached only via mountain paths.

The temples are distributed very unevenly across the country. There are very few of them in the northeast, since it did not become a part of Korea until the Chosŏn dynasty (p. 30), which did, however, allow a few large temples to be founded in the vicinity of Hamhŭng, the home of its founder. In the northwest there are only few temples today, presumably due to its particularly difficult historical fate. In central Korea only the coastal margin of the Sea of Japan and the environs of Seoul and Kaesŏng have an abundance of temples. Thus the majority are concentrated in southern Korea, with the exception of Chejudo (map in [583]).

The location deep in the mountains imparts to the temple environs an air of solemnity and seclusion from the outset. By planting dark firs (*Abies holophylla*) the monks have accentuated this mood even more. In contrast, avenues lined with cherry trees lead to the temples situated in regions of low hills, and in April they blossom pinkish white. Some temples were under royal protection, even during the Chosŏn dynasty. The road approaching them leads through a Red Arrow Gate. This is soon followed by the massive, tiled main gate, which is sometimes built into a wall (Clark [580, p. 47ff]). The entrance to the inner confines of the holy buildings is the Gate of the Four Heavenly Kings, two of which are portrayed by pictures or sculptures on the right and two on the left of the gateway [867]. In these inner confines there are a number of large, rectangular buildings with colossal, soaring, half–hipped roofs, resting on wooden posts painted red. On the front side the interstices between the timbers are filled with delicate latticework, on the other sides with the usual mud plastering, which, however, is always painted and sometimes has colorful pictures on it. The main beams and the bizarre bracketing system supporting the widely overhanging eaves are also painted in garish colors. "It is not easy to describe the paint on a Korean palace. If the reader will imagine that a rainbow has been dashed against the house and that fragments of it have adhered to every exposed piece of wood he will gain a faint idea of how it looks" (Hulbert [43, p. 250]). At the ends of the eaves hang small brass bells, all in tune with each other. Attached to each clapper is a little chain ending in a thin leaf of brass shaped like a fish. When the wind blows against the leaves, the bells ring out in delicate tones. The loftiest buildings surround a rectangular courtyard, the center of which is occupied by a tall stone pagoda. On the back side of the courtyard the sometimes double-roofed "Main Hall" stands elevated on stone steps. Its roof, supported by mighty timbers, covers a high, wide, windowless room, whose dim semidarkness is pervaded by the scent of glimmering incense sticks. This room contains at least three gilded Buddha statues. The walls are covered with gaudily colored pictures lacking in perspective that depict Buddha surrounded by monks, wise men and warriors. In a second hall there is a Bodhisattva in the company of the ten judges of the underworld, and the walls are hung with pictures of purgatory. A third one is dedicated to the worship of the "Seven Spirits of the Great Bear" constellation (Kor. *Ch'ilsŏng*), a fourth and always very small one, to that of the "Lonely Saint" (Kor. *Toksŏng*). A fifth one sometimes combines 500 small, almost identical, Buddha statues of bronze or porcelain (Kor. *nahan*). In large temples there are often even more main halls. The other buildings contain the monks' cells, the study hall, the guest rooms, the kitchen. In addition there is the open hall with the votive tablets of the pilgrims, who flock to the temple festivals in great numbers and often indulge in quite worldly pleasures there today, and there are a likewise open hall for relaxation and the small building in which the tremendous bronze temple bell hangs just above the ground. In the forest, in the immediate vicinity of the temple, lies the cemetery with huge, urnlike memorial stones, the so-called sari pagodas (Kor. *sarit'ap*, [580, p. 57]).

Generally the monks wear the universal white clothing, but their hair is shorn and around their necks hangs a rosary. The abbot is distinguished by the bluish gray color of his clothing. The mission of the monastery's inmates is not to care for spiritual welfare, preach or ameliorate earthly needs, but exclusively to worship Buddha. The main religious ceremony is held during the night or at dawn. Then the temple bell booms out deep and threateningly, struck by a horizontally suspended wooden ram. The little bells, giant drums and gongs join in, so that a prodigious din resounds far through the quiet wooded valley. The monks slip into the dark halls and crouch down along the walls, and from all sides sounds an unmelodious chant using the words "*namu amit'a bul*" ("I put my trust in Amitabha").

Here and there in the temple grounds are the gigantic images of Buddha known as *Mirŭk* that were mentioned earlier (p. 29). They are often whitewashed and are in bas relief or statues. Such "White Buddhas" are sometimes also found far from temples as the only vestiges of Buddhist style in secular areas. On one day in the year, though, Buddha's sun shines even on secular areas, namely on the eighth day of the fourth month, when in celebration of Buddha's birthday colorful paper lanterns are hung over the streets and paper fish billow in the wind in honor of those born during the past year.

Of the newer sects we need mention only Ch'ŏndogyo (the "Doctrine of the Heavenly Way"), which developed out of the Tonghak movement (p. 35) in 1906 (Clark, [580, p. 144ff]). It is a religion combining Buddhist, Taoist, Confucian and Christian elements. The movement, which had a large number of members for a while, maintains temples and meeting rooms in the cities.

Korea is the country in East Asia in which Christianity has gained the strongest foothold, relatively speaking. This development goes way back in the history of traditional Korea (p. 33), so that it is best treated at that point. In 1938 the Christian confessions numbered 500,300 adherents; 113,833 of them were Roman Catholics, 287,200 belonged to the American Presbyterian mission, 54,628 to the Anglo–American Methodists. The missions have founded churches, meeting halls, schools and hospitals in a large number of cities. Many of them, especially the churches with their tall steeples, are landmarks of the cities where they stand. In small settlements, such as To–dong on Ullŭngdo, the meeting room is simply a straw–roofed Korean house marked by the Christian cross. The styles of these Christian buildings are as varied in origin as are the mission societies. American, French, English and German influences form a motley hodgepodge. The German mission is represented exclusively by the Roman Catholic Benedictines, and the majority of the 40 Germans resident in Korea today belong to this order. Since 1927 their activities have been restricted to the two Hamgyŏng provinces (p. 205). They are run by the abbey of Tŏgwon near Wonsan (p. 308). Here, in Wonsan, Kowon, Yŏnghŭng, Hamhŭng, Ch'ŏngjin and Hoeryŏng, there are brick churches with German stylistic elements.

The buildings created by Buddhism and Confucianism in Korea are among the greatest achievements brought forth by Korean art. A. Eckardt is correct in seeing the specifically Korean element in these buildings as the expression of their ability to practice moderation with great tact and true delicacy. "Korea owes this classicism, in addition to its own talents, to the fortunate position of the country between China and Japan" [571, p. 72, 204]. Thus the geographical and physiognomical significance of such buildings is more than the sum total of their dimensions and numbers.

The last mentioned spheres are the only ones in which Korea's settlement departs from its broad base, farming. Otherwise the relationships between the two are so close that one cannot be fully understood without the other. Some have thus already been mentioned above in the section on settlements. Others will follow in the next section on agriculture [603, 605, 607, 609–612]. As we have said, more than three quarters of the Koreans live by farming the land. Toward the end of the Chosŏn dynasty this percentage was even higher. Thus no other economic activity influences the landscape more than farming. Compared with the rest of East Asia some branches of animal husbandry are relatively highly developed, but by European standards livestock raising plays a very minor role here as well. Above all, it is not organically linked to agriculture, so that apart from the draft power of the oxen, it does not play an integral role in farming.

Since Korea is a steep, mountainous country, the difficult relief and climate closely limit agricultural activities, but the pioneers of Korean agriculture, the fire–field farmers, have overcome these difficulties to the highest possible degree, as we shall show below (p. 226ff), using the Kaema upland as an example. The upper limit of cultivation is sustained by them alone. In central and southern Korea I encountered the uppermost fire fields at an elevation of 1150–1400 m, in the Kaema upland at 1500 to 1850 m. Permanent cultivation, which alone will be discussed here, stays within much narrower limits. In central and southern Korea it extends upward as far as 800–1000 m, in the Kaema upland to around 1300 m. But even at these elevations the average slope angle is quite steep and the soil accordingly rocky. Most of the fields are thus concentrated in the elevations between 0 and 400 m, and even these are never entirely occupied by fields. The sight of broad plains representing an uninterrupted sequence of cultivated fields is rare in Korea, and it is limited to the infrequent extensive alluvial plains along the lower reaches of the large rivers and to parts of the Nangnang plain, i.e., especially east and south of P'yŏng–yang. Otherwise everywhere one looks the view includes hills covered with shrubs and woods, which impart a lively and charming air to the scenery. Even today only 20.8% of the country is covered by fields, and during the dynastic period the cultivated area was even smaller.

The best soils are the loamy alluvial ones on the lower reaches of the rivers. Upstream they usually rapidly become sandy, even gravelly. Even then, in large parts of the country they are the relatively most favorable areas for farming, because the residual soils are very rocky, and the Koreans have not yet overcome this difficulty to any great degree. Naturally the number of rocks increases with the slope angle. As a result, the fields are highly concentrated in the valley bottoms and on the lower slopes. A map of the distribution of cultivated fields thus closely resembles the map of population distribution (Map 3, cf. editors' note on p. 7). The best residual soils are formed on limestone, which is particularly widely distributed in P'yŏng–annam–do. They are covered by a loamy terra rossa. These soils have the disadvantage, however, that the subsoil is permeable to water, so that irrigated paddy farming is difficult on them. The same is true of the brown forest soils on the basalt sheets that often fill the valleys of central and northern Korea. Much less favorable are the highly leached red soils on the crystalline rocks of the south and the fringe of the Sea of Japan. The poorest soils are the yellow ones resulting from the weathering of the so widely distributed granite (p. 79).

The most important climatic bases for cultivated plants are the same as for natural ones, and they were outlined at the beginning of the chapter on plant geography (p. 127). The growing season, calculated as the length of the period during which temperatures exceed 5° C, decreases from south to north, from the coasts to the interior and from lower to higher elevations. On the coast of the Sea of Japan it is somewhat longer than at the same latitude on the Yellow Sea coast. Other climatic data are also important, however, for the short, annual cultivated plants and for the preparatory work done on the fields. No publications on the climate near the ground in Korea have come to my attention. The figures on the earliest and latest date on which hoarfrost occurs and the length of the period between these two dates, which are included on p. 98, do, however, provide certain points of reference. They show that the period without frost is subject to the same great regional fluctuations as the growing season. In Cheju it is 278 days, in Chunggangjin, in contrast, only 132 days, etc. This period is much shorter than the growing season, which is determined from the mean daily temperatures at a height of 1.50 m. Naturally data on the length of time during which the ground is covered with snow would also be very important. The snowfall data (p. 98) provide a certain amount of information. As we pointed out there, the snow cover of course lasts for a much shorter time than the period between the first and last snowfall. Particularly important for planting and harvesting are the dates of the last and first night frosts. For these too the only observations available are ones made at the standard height. They are listed for selected places in Table 11 (after [620, p. 11]).

The average frost–free period almost doubles from north (Kapsan) to south (Pusan). It is shorter at stations lying at high elevations, such as Chŏrwon, than at lower elevations at the same latitude, such as Haeju. Near the coast (Ulchin) it is longer than inland (Ch'ŏngju), and on the coast of the Sea of Japan (Kosŏng, Pusan) longer than on the Yellow Sea coast (Haeju, Mokp'o). The midsummer heat in Korea is practically independent of the latitude and the distance from the coast, and even in highlands such as the Kaema upland it is only slightly diminished. There is a fundamental difference between the months at either end of the agricultural season, though, as the above remarks again showed. In the north the temperature curve has a very steep peak, in the south it is much flatter. Consequently, the agricultural climate of the different parts of the country varies greatly even in the thermal sphere. In southern and central Korea it is mild enough to allow some grains to be planted in autumn; in northern Korea, in contrast, the winter cold is too great. All cultivated plants there are spring crops, which closely limits the variety of crop rotations and the forms of intercropping.

	Station	Name of station	Altitude (m)	Average date of last frost	Average date of first frost	Average length of frost-free period (days)
Northern Korea	Inland	Kapsan	810	12 May.	9 Sep.	119
		Onsŏng	83	20 Apr.	23 Sep.	155
	Coastal	Hamhŭng	38	26 Apr.	15 Oct.	171
		Sŏngjin[a]	31	16 Apr.	18 Oct.	184
Central Korea	Inland	Ch'ŏrwon	220	6 Apr.	6 Sep.	152
		Suwon	37	24 Apr.	10 Oct.	168
		Taejŏn	50	24 Apr.	10 Oct.	168
	Coastal	Haeju[b]	50	25 Apr.	12 Oct.	169
		P'yŏng-yang[b]	6	23 Apr.	14 Oct.	173
		Kosŏng[a]	9	29 Mar.	19 Oct.	203
Southern Korea	Inland	Ch'ŏngju	44	21 Apr.	10 Oct.	171
		Chŏnju	51	18 Apr.	20 Oct.	184
		Taegu	51	11 Apr.	26 Oct.	197
		Kwangju	39	25 Apr.	18 Oct.	175
	Coastal	Ulchin[a]	10	7 Apr.	1 Nov.	207
		Mokp'o[b]	31	6 Apr.	12 Nov.	219
		Sunch'ŏn	18	29 Mar.	31 Oct.	215
		Pusan[a]	13	26 Mar.	9 Nov.	227

[a] On the coast of the Sea of Japan. [b] Near the coast of the Yellow Sea.

Table 11. Late and early frosts in Korea

Not only for wild plants, but also for cultivated ones, it is of utmost importance that the period of high temperatures is also the period of high precipitation, so that 80–100% of the latter falls during the growing season. The absolute amount of rain that falls during this period is sufficient on the average for all field crops cultivated in Korea, including rice; after all, upland rice is also grown in Korea. But, as we already pointed out on p. 93, in many years the distribution of precipitation does not correspond to the average annual cycle. The average variability in the annual figures is great. For cultivated plants it is more serious if the heavy rains fail to fall during the six summer months than if there is an excess, and in this connection we must again stress that it is not the summer monsoon itself that brings the real precipitation, but rather typhoons and cyclones (p. 102, 108). Summer months with a constantly developed monsoon are thus dangerous for agriculture. Since typhoons and cyclones bring the main precipitation, large amounts often fall in short periods of time. Such flood disasters cause more damage than they do good. We have already mentioned the flood damage in the valley bottoms on p. 119f. But the fields on the slopes also suffer, partly due to erosion or increasing sandiness of the soils, partly due to actual destruction of the plants, as caused by cloudbursts or hail. The typhoons, to be sure, being purely tropical cyclones, are without hail. But it is not uncommon for extratropical cyclones to bring hail, especially in the spring months, when the delicate plants are most easily damaged. According to the detailed calculations of the government-general, damage in 1931 was: 128,282 ha due to floods, 23,663 ha due to hail, 625,438 ha due to wind, 520 ha due to insects, i.e., a total of 764,538 ha, or 16.6% of the cultivated area [605, p. 116], and the estimated value of this damage was 7,7 million yen. Even months with a large amount of precipitation are therefore not always favorable for agriculture; it depends on how the precipitation is distributed during these months.

If the edaphic and climatic bases of agriculture already present some serious disadvantages, in addition to the great advantages, the picture becomes considerably darker in the realm of mental predispositions. I. Bishop, who was certainly an impartial critic, already mentioned [207, Vol. I, p 185] "As compared with the exquisite neatness of the Japanese and the diligent thriftiness of the Chinese, Korean agriculture is to some extent wasteful and untidy." Indeed Korea's entire economy still suffers today from the two and a half centuries of isolation following a period of frightful devastation. In accordance with the general mentality (p. 152f), only the most urgent agricultural work is done. Therefore the fields often look very untidy. Nowhere, with the exception of Chejudo, does one find the fields surrounded by walls of stones collected from them. Instead, the stones, often large ones, remain lying for years on end, and frequently the roots and stems of the last year's crop are not even completely removed. The weeds are often left standing. Heydrich, who wrote an excellent treatise on Korean farming implements, writes [558, p. 21]: "The Korean farmer's implements are wretched and often crudely and clumsily made." Certainly farming is inconceivable without hard work, and a day of rest on Sunday is unknown in East Asia. There is thus less idle leisure, referred to on p. 152f, in the villages than in the cities. But nowhere in traditional Korea was anything more done than what was absolutely necessary to assure a modest yield, and even today it is rare for a person to do more on his own initiative. The motivation for prosperity and for intelligent foresight in farming has been lost.

This attitude is all the more serious because the distribution of property was already unfavorable during the dynastic period [614]. Particularly in southern Korea a large proportion of the cultivated land was in the hands of landlords, who rented the fertile paddy land to poor tenant families under the oppressive conditions of sharecropping and with a lease that was renewable every year. The much less fertile ground of northern Korea belonged to the farmers themselves to a much greater extent. Very often there was a manager or intermediate tenant between the tenant and the landlord, who enjoyed a comfortable life far away in the city. The Korean Sang–uk Han says about this in 1927 [607, p. 59]: "Probably one of the greatest evils in our country is that the landowner can set the rent rather arbitrarily and unilaterally and usually levies it not as a predetermined amount of cash, as in Europe, but as a share in the crop, usually without paying any heed to the outcome of the harvest. In addition to this, only in the rarest of cases is there a written contract that can be referred to in the event of any controversies. Even if it is perhaps possible to get along with the owner himself, there is generally a manager between him and the tenant, who is given free rein and who can impose more difficult conditions of his own accord. Even though there has been a movement for some years now to check all of these abuses and to take remedial action everywhere, it must not be forgotten that our tenancy system has been in force for centuries, that the people know no other system and that it is of course very difficult to break with tradition."

The size of farms is small everywhere, by European standards; therefore it is not because of the farm size that Korean farming is often rather extensive today and was even more so during the dynastic period. An individual farm generally has from two to six fields. Paddies are smaller than dry fields. The average size therefore decreases from north to south. According to Hoon K. Lee's investigations [605, p. 101] the average size is 109 ares in Hamgyŏngbuk–do, but only 24 ares in Chŏllanam–do. The average for Korea is 40 ares. These farms were and still are today run by the families or clans under the direction of the head of the family or clan. Jobs that the family cannot handle by itself with the necessary speed are done cooperatively (Kor. *p'umashi*). In this system 4 to 12 neighbors join up and take turns working for each other for an equal amount of time.

In agriculture, too, there are close connections to China. This is particularly true of northwestern Korea, for understandable reasons, while in the south, particularly on Chejudo, ties extend to Japan. In the section on history we mentioned, in appropriate contexts, the origin of the crops cultivated in Korea (p. 20ff, 31, 47). These remarks showed that many crops were imported from China, while others may have been domesticated in Korea itself. The forms of many farmimg implements and their use and the methods of cultivation frequently reveal relationships with the Middle Kingdom.

The crops cultivated in Korea include such decidedly subtropical ones as rice, kaoliang, sweet potatoes, sesame, cotton, ramie, paper mulberry and bamboo. The perennials among these are limited to the south with its milder winters. For the rest they mirror the summer heat at lower elevations in the entire country. Korea's agriculture thus exhibits subtropical features much more strongly than do the woody wild plants, which must withstand the winter cold. At the boundary between the plants of the subtropical zone and those of the temperate zone we have foxtail millet, maize, soybeans, hemp, mulberry and melons. Barnyard millet and broomcorn millet, naked barley, wheat and most fruit trees belong in the southern part of the temperate zone. The least demanding plants are barley, oats, potatoes and buckwheat, which extend up to the upper limit of cultivation, where they are grown mainly by the fire–field farmers. The agricultural scene is thus quite varied. The Japanese statistics for the present time list 44 crops, almost all of which were already known in traditional Korea.

Since ancient times rice has been the most important cereal. By far the most of it is grown on irrigated paddies (paddy rice, Kor. *nonbyŏ*, Sino–Kor. *sudo*, [618, 620–623]). As we have already mentioned, upland rice (Kor. *patpyŏ*, Sino–Kor. *yukto*) is also cultivated in Korea. Artificial irrigation thus serves the purposes of securing and increasing the yield. Between irrigated rice paddies and upland rice fields there are all kinds of transitions. At the two ends of this continuum are the dry fields, on which upland rice is intercropped with other dry–field crops, and the constantly irrigated paddies, in which only rice grows, in bunches of several plants. In the latter the plants have been transplanted from a seedbed. There are also paddies, however, in which the rice is sown by hand, and in which it does not always stand in bunches, and dry fields, on which it is transplanted in bunches from a seedbed. And the majority of the paddies are flooded only by standing rain and groundwater and not irrigated by flowing water. The decisive factor in the distinction between irrigated paddies and dry fields is thus the preparation of the field to hold a layer of water, and this consists in grading the field and enclosing it with a low earthen dam to hold back the water. Fields that are dependent on rain or groundwater only were much more in the majority in traditional Korea than they are today. Toward the end of the Chosŏn dynasty only one–sixth of the rice fields were irrigated (p. 411f). In so mountainous a country with so much rain, where such irrigation works can usually be constructed with relatively modest means, this is a clear indication of a lack of energy in the agricultural sphere.

In traditional Korea very different varieties of rice were grown, eight of which are listed by Hulbert [557, p. 108]. They include glutinous varieties that strike the eye in the period before ripenening because of the blackish-brown color of the husks. The Koreans make rice cakes from the flour of these varieties by mixing it with water and spicy wild herbs and beating the cold mass for a long time with wooden hammers. Nonglutinous rice, in contrast, is eaten only cooked and forms the main item in the diet of the wealthier classes. The yields of the traditional Korean varieties are moderate. The old, like the new, varieties ripen at different speeds and are thus finely adapted to the different climatic conditions. At higher elevations, however, the summers become so cool and short that not even the fast ripening ones can be grown. Rice farming has an upper limit, therefore, that lies at approx. 750 m in southern and central Korea. The very rare rice paddies in the Kaema upland extend upward to 1000 m. The main focus of rice production in traditional Korea was already in the south, for climatic reasons. According to censuses in 1404 and 1769 about three–quarters of the total paddy area were in the six present–day southern provinces. Toward the north late and early frosts, to which an aquatic plant such as rice is of course particularly sensitive, severely limit the opportunities for cultivation.

The range of paddy rice farming has topographical and geological bases as well, though. The low intensity of traditional Korean agriculture is manifested, among other things, in the fact that tiers of irrigation terraces perched steeply above each other on slopes are very rare. In this respect the irrigated areas of Korea look very different from those of the Malayan island world. The paddies are restricted for the most part to areas with only slight natural slope angles that are easily graded. Foremost of these are the alluvial valley bottoms, which simultaneously have the best soil. Map 4 (cf. Editors' note on p. 7) shows clearly how the areas with paddies are lined up along many river systems and structural valleys and in the plains behind the sandbars along the coast of the Sea of Japan. Extensive alluvial plains are found only on the western side of the peninsula and in the Naktong basin. The paramount position of the western and southern side of the peninsula in irrigation agriculture is due to the basins and adjoining marshlands there. In areas with permeable strata, i.e., particularly basalt flows in structural valleys and limestone areas, there are of course generally few irrigation works.

The chief constituent of the Korean rice farming areas is the paddy (Kor. *non*), which is leveled and enclosed by earthen ridges approx. 25 cm high. Its size and shape varies, depending on the terrain. The average is probably approximately 20 ares. There are also tiny "fish eye rice paddies" (Kor. *tomi nunkkal non*), though. In northern Korea soybeans or sesame are often planted on the ridges. Paddies that lie at the summer flood level of the rivers are filled by groundwater from below and flooded for an unregulated period of time. The same is true of paddies flooded by rainwater. When the water stagnates in these paddies for a long time in the summer heat, it turns into an evil–smelling liquid, murky green from algae, in which the unaerated rice plants cannot thrive. For this reason many such paddies have at least a groundwater hole. The water is dipped out of it with a tremendous wooden scoop (Kor. *yongdure*) that hangs from a three–legged frame and holds up to 30 liters.

For paddies that are to be irrigated the water is diverted from a river by means of a canal. At the point where the river is tapped, a weir made of stones or of poles and twigs (Kor. *po*) is often built across the river. Where no such river is available, the water from creeks is stored in artificial ponds (Kor. *pangch'uk*). The canals, contained between banks, flow along the slopes with a very slight gradient, and the irrigation ditches branch off from the main canal to the fields. Occasionally the main canals are conducted on wooden trestles along rocky cliffs and across valleys. In side valleys and on old alluvial cones it is not uncommon for the entire stream to be used for irrigation, and an irrigation canal is then unnecessary. In these cases, though, care must be taken that the debris transported by the stream does not cover the fields and that surplus water can drain off. The drainage ditches built for this purpose usually run around the outside of the paddy complex. They raise their beds by depositing sand and debris and thus flow along as much as 5 m higher than the paddies. Poplars or acacias are planted on their dams.

In the Korean dynastic period the ends of the bays filled with tidal flats on the western and southern coast were already dammed off here and there with earthen dams. Floodgates permit the water from the land to flow off at low tide. The thus reclaimed silty marsh soil yields very good rice harvests, if irrigated sufficiently to prevent salt water from infiltrating into the root region of the rice plants.

Korean rice farming is so closely tailored to the summer temperature conditions that nowhere are two rice crops a year possible. There is no intercropping, as on the dry fields. The paddies are instead devoted exclusively to rice farming every summer year in and year out. In southern Korea, however, a winter dry–field crop is often grown on the drained paddies, especially barley, naked barley, rye or wheat, and in Kyŏngsangnam–do hemp as well. They are planted in October and harvested in June. Even before the harvest the rice seedbeds are started in some corner of the paddies, so that immediately after the winter crop is harvested the plowing and transplanting of the rice plants can begin. All other paddies in southern Korea and all paddies in the other provinces lie fallow all winter.

As soon as the weather permits in early spring, Korean farmers begin to prepare these fields and their seedbeds. Often they have been broken up the first time in autumn with hook plows (Kor. *chaenggi*) pulled by brown or black oxen. Now in spring, i.e., in March and April, the earthen ridges around the paddies and seedbeds are repaired. For this purpose the big three–man spade, a typical Korean implement (Kor. *karae*), is used. After one man has thrust the spade into the ground, the other two pull it up by ropes attached to the side of the blade. The ridges are often subsequently packed down with big wooden hammers. Paddy and seedbed are then plowed again. After the surface has been flooded, if possible, the mud is harrowed by means of a tilted board with wooden teeth (Kor. *ssŏrae*) pulled by men or oxen. In these jobs the seedbed takes precedence over the paddy. This work on the paddies continues on into May or June. Ashes, rice straw, leaves and old roofing from the huts are put into the seedbed and later into the paddy as fertilizer. Finally the drained seedbed is divided up by means of strings into rectangular sections so narrow that they can be reached by the human arm from the strips between, which are approx. 20 cm wide. The mud in these sections is leveled with the help of boards. Then the presprouted rice grains are sown densely by hand, and the water is immediately allowed to flow over the seedbed again, where it remains until the plants are transplanted.

Three events form the basis of the annual rhythm of work in the paddy: sowing in the seedbed, transplanting of the seedlings into the paddy and finally the harvest. Sowing must be done earlier in the north than in the south because of the shorter summer. It takes place in the first or second third of April, whereas in the south it can wait until the first half of May. Transplanting is done when the seedlings have reached a height of 25–30 cm, which is 35–40 days for the fast ripening varieties and 45–50 days for the slow ripening ones. Thus this stage is reached in the north even farther in advance of the south. In the north transplanting is done between the beginning and end of May; in the south it drags on from early June to early July. The harvest is generally between the beginning and end of September in the north, while in the south it is prolonged until early November.

After sowing the care of the seedbeds goes on. One month later a topdressing of pulverized chicken droppings and plant ashes is added. Birds are frightened away with bushes, rags or scarecrows stuck in the ground. At night open lamps are sometimes burned over the seedbeds to lure and catch the moths of *Chilo simplex* Butl., the most dangerous rice pest in Korea [623]. During the entire period in which the rice is growing in the seedbeds, they are constantly weeded.

Because the seedlings are very sensitive to dryness, transplantation from the glaringly green seedbeds into the paddy must be done very rapidly. For this reason it is a cooperative project. The plants are carefully pulled out of the mud of the seedbed by hand and gathered into bundles. On the very same day the bundles are spread around the field, which must under all conditions be flooded approximately hand deep, and planting begins immediately. Bunches of eight to ten seedlings are stuck by hand into the same hole. The bunches are lined up exactly by means of a string with red cotton threads knotted into it at equal intervals (about 20 cm). In rapid, rhythmical work the bunches are planted at the spots marked by the red threads.

The irrigable paddies are kept constantly under water from now on, as much as possible. The plants develop in them in luxuriant green, whereas those on unirrigable fields become frail and yellowish green in many summers. During dry periods the irrigable fields are guarded at night so the neighbor does not steal the water. Everywhere one then sees the pipes of the chatting watchmen glowing through the night like fireflies. About 1 month after transplanting the field is weeded for the first time with a short–handled hoe (Kor. *homi*). Industrious farmers weed again a few times. The unpretentious yellowish–green blossoms develop in Korea before the onset of the typhoon season. As the grains grow larger, the danger that the heavy heads will be knocked down by rain increases accordingly. The farmers then sometimes tie three or four bunches together, so that they will support each other. The heads turn brownish–yellow and finally stand almost as high as a man.

About 1 month before the harvest the irrigable fields are drained. At harvesttime the bundles are cut with a sicklelike tool (Kor. *sonnat*) and spread out on the ridges to dry. If soybeans, which are generally harvested later, are planted on the ridges, the bundles are stacked on drying racks set up on the paddies. Finally the bundles are gathered into sheaves and the sheaves are formed into cross–shaped stacks. Threshing is done in the farmyard by different methods, for instance by means of a hackle, but always without a flail, which is reserved for the dry–field crops.

In addition to rice there are other paddy crops, but they take up an extremely small space. Especially worth mentioning is barnyard millet, which is a dry–field crop for the most part (p. 178). There is also the lotus (*Nelumbo nucifera* Gaertn.), whose root, highly seasoned, is a popular supplement to the daily rice diet of the city–dweller, and whose large red or white blossoms adorn the ponds in July, and the rush *wanggol*, whose reedlike leaves are woven into mats, slippers and hats.

This very small number of paddy crops, among which rice overwhelmingly predominates, contrasts with a very large number of dry–field crops. Nagai, Nakagawa and Takasaki have devoted a detailed study to them [624] and Tanaka has summarized them [600]. In the rice– growing areas the irrigated paddies fill the lowest parts of the terrain, the valley bottoms, niches in the side valleys and the alluvial cones, while the dry fields extend up the slopes.

The most important dry–field food crops are the millet varieties. They, together with barley, take the place of rice among the poorer classes of the population. According to Hulbert [557, p. 205] six varieties were grown in old Korea: (1) *ch'ajo* (glutinous millet); (2) *mejo* (dry millet); (3) *kijang* (yellow glutinous millet); (4) *megijang* (yellow dry millet); (5) *ch'ŭngjŭngme* (green glutinous millet); (6) *p'i* (panicled millet). One and 2 are varieties of foxtail millet (*Setaria italica* Beauv.), 3 and 4 of broomcorn millet (*Panicum miliaceum*), 6 is barnyard millet (*Echinochloa crus galli* B.P., var. *frumentaceum* Roxb.). I did not become acquainted with variety 5. The two varieties of *cho* (*jo*) were very common in traditional Korea and still are today. They grow to a height of 1.50 to 2.00 m and have relatively wide, long leaves. The leaves of the one variety turn wine red in autumn, the other greenish yellow. Both have bristly, dense panicles not unlike a foxtail. The spikes of the first variety are about 15 cm long and brown, the second 25 cm long and accordingly thicker and yellow. The *cho* varieties are followed in frequency by barnyard millet, *p'i*. Werth [554, p. 57] reports that it is also grown by the Koreans in the Ussuri region, often as their most important food crop. The brownish gray spikes consist of widely spaced, dense spikelets. Least often grown is broomcorn millet. It has hanging, long–stemmed panicles with shiny yellow grains and yields the most costly millet, which brought almost the same price as rice in traditional Korea. The millets are less demanding than rice with respect to temperature, precipitation and soil. They ripen faster and thus predominate in the northern part of the country as food crops. Among other things, their cut up straw serves as fodder for cows and horses. In the north the *cho* varieties are sown between the last third of March and the last third of April and harvested between the last third of August and the second third of September. The growing season thus lasts 5 months. Barnyard millet has a somewhat shorter growing season, especially when it is irrigated. Broomcorn millet has the shortest growing season and is thus more widely distributed in the Kaema upland than the other two. On Chejudo, where there is another center of foxtail millet cultivation, sowing is not done until the second third of June and harvesting in the last third of October. Thus the growing season there is only 4.5 months. (In earlier publications in which I mentioned the names of Korean varieties of millet, some of the German and botanical names are mixed up, because at first I was certain only of the Korean names. It took a comparison of the panicles I had brought back with me with the data and pictures in the literature to clarify the European names. The Korean names were always correct in these earlier papers.)

Kaoliang (*Andropogon sorghum* Brot., Kor. *susu*) is often included among the millets, but it is quite distinct in its physiognomy and its requirements. In its initial stages, during which it closely resembles maize, it grows very slowly and needs plenty of precipitation, but later it develops very deep roots so that then it is not affected by long dry spells. It reaches a height of more than 3 m and has very large panicles. According to Hulbert [557, p. 204] three varieties were common in traditional Korea. One of them has loose panicles (var. *effusus* Körnicke, Kor. *nulsusu*) with purplish black hulls and a second one has dense panicles and brownish red hulls (var. *contractus* Körnicke, Kor. *mŏksusu*). In the latter the panicle has the shape of a club. Because of their long roots and thick stalks kaoliang and maize are often planted on alluvial plains where there is danger of flooding, since they withstand floods better than other plants. Kaoliang can survive on poor soil, but it needs greater warmth than the other Korean millets. As a result it is practically absent in the Kaema upland, is sown and harvested later and has a growing season as much as 3 weeks longer than the others. The seeds are used for food for humans and animals. The long, thick stalks are used in house construction.

Whereas all of the above mentioned crops are exclusively spring crops, barley, naked barley, wheat and rye are grown in southern and central Korea as winter crops. Wheat is even mainly a winter crop, sown in September and harvested around late June or early July. There is no winter wheat in northern Korea. The main area in which it is grown is on the good soils of the provinces Hwanghae–do and P'yŏng–annam–do, which get little precipitation in early summer. On the coast of the Sea of Japan, in contrast, wheat growing suffers due to the heavy early summer rain. In southern Korea it is grown partially on the paddies that are drained in winter. The same is true of naked barley, which is more sensitive to cold and whose paddy area lies outside of the −4° C January isotherm. It is used partly as animal fodder, partly as food, while the different varieties of common barley are used exclusively for flour. The latter is the grain with the most versatile forms of cultivation. In northern Korea it is grown exclusively as a spring crop, sown in March or April, depending on the elevation and harvested between the beginning of June and the middle of July. Hence the growing season lasts only 70 to 100 days. Dry weather in spring is unfavorable for its development. In central and southern Korea barley is a winter crop. The two regions overlap in the mountains of central Korea, in which spring and winter barley are grown. The farther south one goes, the later the autumn sowing is done in the winter barley regions. In the central Korean mountains it is done in early September, on Chejudo at the beginning of November. The rate of winter growth is much faster in the south, so that the harvest begins already in early May on Chejudo, but in central Korea not until the end of June or early July. In central Korea the intensive rainfall quite often makes the harvest more difficult. A very large portion of the winter barley is grown on paddies in southern Korea and even on the Yellow Sea fringe of central Korea.

Only recently B. Laufer called attention to the fact that rye is grown in East Asia and especially in China [625]. He considers it not unlikely that rye entered China during the Tang dynasty (618 to 906 A.D.) from the Pamirs. For my part, I was surprised to find rye in Korea. Its area of cultivation is even relatively large (p. 414). It lies partly in the southern provinces, particularly on the islands and peninsulas of the southern coastal fringe, partly in the coastal region of the P'yŏng–an provinces. In the former region it is grown as winter rye and is also partly grown on paddies. Hulbert [577] does not mention rye among the crops of traditional Korea. I do not know whether it was brought to Korea in olden times from China or recently from Japan or North America. The latter is more likely, because rye is called by the English name *rai* in both Japan and Korea.

Tobacco too reached Korea via Japan, and maize presumably likewise, though soon after the Hideyoshi invasion (p. 31). In traditional Korea almost every farmer grew his tobacco himself (628), for the love of smoking was and is great. The plants are planted in June onto fields where the winter grain has been harvested and the tobacco can already be harvested in September. Maize (Kor. *oksusu*) is attacked by a pest in southern Korea and consequently only very small amounts are grown there. In contrast, it is grown abundantly in the northwest and in the central eastern part of the country, in areas with a mean August temperature of 22–25° C. In the Kaema upland it extends up as far as 1200 m. Maize requires bountiful rainfall; therefore it is not sown until the middle of April in northern Korea. The harvest is between mid–August and mid–September;

hence the growing season is only 4 months in the hot summer. In traditional Korea very little was grown.

In Korea oats have nothing to do with the other Central European grains mentioned above. Instead they are part of a crop combination including buckwheat and potatoes. They are a mountain plant and require cool summers. The largest share is grown in the Kaema upland at temperatures below 22° C. They were already highly esteemed as a food grain in traditional Korea; they are cooked like rice or mixed with rice. From the straw an excellent yellow paper is made. Besides oats the most important fire–field grain is buckwheat (Kor. *memil*), with its modest demands on climate and soil. In the plains it is grown as a second crop after winter or even spring grain, as it has the shortest growing season of all Korean crops (50 to 70 days, depending on the elevation). It is sown in June and harvested in September. Buckwheat flour is used, among other things, to make a kind of macaroni known as *kuksu*. Whereas buckwheat, because it is used as a second crop, extends out of the Kaema upland and the other mountains far down into the lower elevations in all of Korea, the potato, which was not introduced until 1832 (p. 47), is concentrated mainly at higher elevations and in the north, similarly to oats. In the mountains wild boars often wreak havoc on the potato fields. Planting is done between the end of March and the middle of April, digging between the middle of June and the beginning of October, depending on when they are needed. The varieties with red tubers ripen earlier than the white ones.

Here and there at low elevations in the western and southern provinces, particularly in the vicinity of large cities, the normal potato is replaced by the sweet potato, which is very sensitive to cold. On some islands off the southern coast, such as Wando and Kŏjedo, and on Chejudo they are grown much more extensively. In provinces where the spring is cool the tubers are sprouted on the *ondol* heating first, because this subtropical plant does not have enough time to ripen here otherwise.

The role of pulses as crops second to the cereals is much more important in East Asia than in Central Europe. In Korea the most important by far is soybeans (Kor. *k'ong*), which make few demands on the soil [626]. Tanaka [600, p. 18] rightly emphasizes that Korea's climate is actually ideal for them. Because of the larger amount of precipitation in summer it is even better than Manchuria's. Soybeans are today an indispensable part of the various dry–field crop rotations. After the work on the paddies is finished, they are planted on the ridges too, where they grow exceedingly well without being fertilized. They are grown everywhere in Korea, but least in the Kaema upland because of the warmth they require. On dry fields they are sown between the end of March and the beginning of May in northern Korea. Toward the south sowing is delayed until the end of June. Harvesting is done in October, except in Chejudo, where they are not harvested until the beginning of November. Soybeans are the crop that is left standing longest, and they are not harvested until the leaves have turned brown and started to fall. They are used as a very nutritive food for humans and horses. Cooked soybeans are also used as fertilizer for the rice paddies. Above all, though, the hot brown soy sauce, which is never lacking on any Korean or Japanese table, is made of them, using salt and pepper and a process of fermentation. Bean curd (Kor. *tubu*, Jap. *tofu*) and soybean paste (Kor. *toenjang*, Jap. *miso*) are also made of them.

There are many varieties of other beans as well. Hulbert [557] names 12 kinds of round beans and nine kinds of long–seeded ones. The most important ones are the adzuki bean (*Phaseolus radiatus*, Kor. *p'at*), the mung bean (Kor. *noktu*), which is sown in Hwanghae–do, its principle area, as a second crop after winter wheat, the fast ripening peanut (Kor. *ttangk'ong*), the brown bean (Kor. *pamk'ong*) and the undemanding kidney bean, the latter of which is grown almost exclusively in northeastern Korea and in the Kaema upland and is also a fire–field crop there.

Cotton is by far the foremost of the fiber crops [627]. In traditional Korea only the so–called native varieties were grown, which probably entered Korea from China (p. 29). They differ from the American varieties introduced by the Japanese in their rapid growth, short, coarse fibers and great durability. In northern Korea they are sown in April and harvested from the end of August on, depending on when the bolls open. The harvesting of the individual fields continues on into October. It must be completed prior to the first frost. Cotton was already the most important textile crop in traditional Korea. Most of the clothing is made of it (p. 154).

The second most important fiber crop in Korea in ancient and in recent times is hemp (Kor. *sam*), which is grown everywhere, even at high elevations, but least in the central western provinces. Sowing begins at the end of March, except in the mountains, where it does not begin until May, because the young plants are sensitive to frost. Most is harvested to produce fibers for clothing. In northern Korea this occurs in July or August. While buckwheat is then grown on the main part of the fields, the hemp plants along the edges are left standing until October to let the seeds ripen. Their fibers are used in making ropes. The varieties grown in southern Korea are ready for use later. The hemp plants grow as tall as 3.70 m. The pale yellow stalks, from which the fibers have been removed, are used for making fences or as roofing material. The fibers are removed in pits in the ground by an antiquated process, in which steam is conducted through stacks of stems. White clouds of steam escape from the twigs and dirt loosely covering the pits. Then the fibers are stripped off and dried on lines, like clothing.

Another textile crop is ramie or China grass (*Boehmeria nivea* and *viridis*, Kor. *moshi*). This plant is very sensitive to cold and is restricted for the most part to the western provinces of southern Korea.

In the complete absence of butter, the cultivation of oil–seed crops is important. They are grown, however, only to a very limited extent. Most important is the very ancient sesame. The plants resemble our foxglove and stand out because of their white and pink bell–shaped blossoms. The oil made from wild sesame (Kor. *tŭlkkae*) is used only in lamps and for making oil paper. There are two varieties of cultivated sesame (Kor. *ch'amkkae*), white and black, and they are grown for cooking purposes. The seeds of perilla (*Perilla ocymoides*), a labiate with white blossoms, yield an oil that is used in making lacquer. It is grown all over Korea, intercropped with other crops, but the cultivated area is very small. Finally there is the castorbean, whose tremendous plants are scattered singly around the melon fields, where they simultaneously serve the purpose of driving away certain insect pests with their smell.

Finally, the melon fields themselves are a characteristic element of the dry–field areas. The melons are planted around the end of March or beginning of April. From the middle of June on the fruits are eaten. At the edge of the melon fields there are wooden watchtowers, which are kept occupied by members of the family as soon as the melons begin to ripen, to prevent theft.

Because the average size of the fields is so small, there is no fundamental difference between agriculture and horticulture. There are all kinds of transitional sizes, and many garden crops are also grown in fields. The only real difference is the fence, not the size. Every farmer has a garden close to his house, in which he grows his daily needs. Near the cities tiny unfenced lots with garden crops crowd together. Here, thanks particularly to the Chinese who immigrated during the last decades, vegetable farming has begun to develop. The most important garden crops are different varieties of turnips, cabbage, radishes, cucumbers, melons, squashes, tomatoes, lettuce, onions and red pepper. Many of them are used uncooked to make *kimch'i*, a kind of highly spiced sauerkraut, hardly edible to Europeans, which is never missing at a Korean meal.

A culture of a very special type, closely related to horticulture, is that of ginseng [630, 631]. This extremely highly esteemed root is not only gathered as wild ginseng (p. 142); it was already cultivated in traditional Korea. Even the earliest European travelers (e.g., 178) report on ginseng cultivation near Kaesŏng. Since ginseng is a shade lover and very sensitive to driving rain, heat and cold, it is grown under long, low protective shelters made of pine branches or straw, which are open to the north. The plantings are surrounded by withered brown fences of leafy branches to protect them from thieves. The shelters are divided into sections (*k'an*) of 1.50 x 0.75 m with 32 plants in each. The roots are dug between the middle of September and the end of October of the sixth year. The best specimens are steamed while still fresh, causing the formerly snow–white color to turn red. Red ginseng is exported to China. The roots that are not as good are dried, packaged by weight and sold in Korean pharmacies.

Because stock raising plays such an insignificant role, little manure is available for fertilizer, nor was green manuring practiced to any great extent in traditional Korea. Fertilizing of the paddies was already mentioned above. Compost prepared from household refuse, ash, weeds, peat, litter, sod and leaves is sometimes used on the dry fields. By no means all of the crops are fertilized, however. Fields of vegetables, melons and squash are fertilized with human excrement brought to the country from the cities. Far from the cities, public toilets, erected by the farmers along the roads, invite passersby to assist in procuring fertilizer.

Heelless hook plows similar to the types found in China and Japan are generally used. The heavy soil of the plains is worked with big yoke plows pulled by two oxen. In the mountains, lighter plows with one ox predominate. Plows are also used for turning the soil between standing rows of plants in fields that are intercropped. Occasionally these light plows are pulled by humans, and it is characteristic of the role of the Korean woman that she does this hard work while the man guides. The seeding of the dry–field crops is done by hand. On the whole Korean agriculture is still characterized today by laborious manual work; the use of machines is unknown. Because they are so small, millet seeds are spread on the soil with a simple gourd sower (Kor. *ch'ach'o t'ong*, or *sshiat t'ong*). The dry fields are weeded several times between sowing and harvest by diligent farmers with short–handled hoes. They are also cared for otherwise when necessary, though by no means to the same extent as the more valuable paddies. For instance, several times I observed women using brooms to sweep caterpillars off the leaves into gourds. Harvesting is done with the sickle mentioned previously. The bundles are spread out on the ground in parallel rows to dry and then stacked up into cross–shaped or conical stacks. In the mountains where there is an especially large amount of precipitation they are covered with a protective cap made of straw. The dry–field grain crops are threshed by treading out the grain with the feet, with flails or by means of water–powered mortars (p. 233). The chaff is separated from the grain by winnowing. The domestic portions of these tasks and the weeding are done by women.

Whereas the paddies remain completely level, even when dry–field crops are grown on them in winter, dry fields are often plowed so that troughs and ridges result (ridging). There are also level dry fields, however. In the north the troughs are made deeper and wider. In the south their dimensions are about twice those of our potato fields after they have been hilled up. The troughs and ridges provide different conditions. During the rainy months the troughs collect the moisture, and in winter they are filled with snow so that the winter crop in them is protected. The ridges are especially warm and relatively dry locations in summer. In level country the troughs are dug from east to west if possible, the plants standing in them being better protected from the north winds. On slopes the troughs are dug horizontally to reduce erosion. Both troughs and ridges can be used for growing crops, though never does the same crop grow in both. In moist mountainous areas it is fairly common, however, for the particularly wet troughs to be left unplanted.

The irrigated paddies are always planted with only a single kind of grain. Many of the dry fields in other parts of the country are also one crop fields. They include, e.g., most of the fire fields and the kaoliang and maize fields along the banks of the rivers. Mixed plantings occur but are relatively uncommon; occasionally foxtail and barnyard millet are found growing together, and beans climb up the stalks of kaoliang and maize.

Much more common, however, is intercropping, which involves sowing and harvesting different crops between and after one another on the same dry field. The different crops are planted in rows 0.5–1 m wide. Intercropping is often combined with ridging, in which case the troughs and ridges form the rows. Level fields are, however, also intercropped. Winter grain on dry fields is almost always intercropped, and, for the reasons just mentioned, the winter crop is planted in the troughs and the spring crop on the ridges. Intercropping is, however, also prevalent north of the zone where winter grain grows. Then two or more spring crops are planted after each other in rows and harvested. On fields that are ridged both troughs and ridges may be used, or only the ridges. The number of intercropping variations is extremely large. Among spring crops alone the following are common: (1) beans/millet, beans/maize, beans/kaoliang, sesame/cotton, sesame/buckwheat, maize/kaoliang, maize/potatoes, castorbean/melons (p. 181), soybeans/hemp, beans/potatoes, (2) foxtail millet/barnyard millet/maize, sesame/castorbeans/cotton, sesame/beans/broomcorn millet, sesame/beans/hemp, sesame/melons/maize, beans/melons/maize, (3) sesame/cotton/beans/melons. There can thus be as many as four different crops in the intercropping system. In winter cropping there are fewer variations, the winter crop consisting of wheat, barley or rye, the spring crop of soybeans, other beans or cotton. In southern Korea one occasionally even finds perennial woody plants, such as mulberry or bamboo between the rows of the winter crop. More than two crops are rare when one is a winter crop. I noticed barley/melons/maize.

A variety of intercropping that is especially common in northern Korea and is related to the climate consists in planting crops that require a lot of sun in strips 0.5–2 m wide around fields that are otherwise planted with only one crop. Such borders are planted particularly with sesame, but also with beans, castorbeans or buckwheat. They are thus predominantly low–growing plants. Borders are planted especially around fields where tall plants are growing, such as kaoliang, maize or foxtail millet. In this manner the plants in the borders ripen faster and better than they would between the rows of tall plants.

In their previously mentioned standard work [624, p. 11] Nagai, Nakagawa and Takasaki list four systems of crop rotation on dry fields. The first two consist only of spring crops; the other two also include winter crops. They are as follows: (1) A single crop of spring grain or beans followed by a 2– or 3–year crop rotation, such as: first year maize, foxtail millet or barnyard millet, second year barley or kaoliang or: first year barley, second year foxtail millet, third year kaoliang, beans or melons. (2) Intercropping of spring crops of the many variations mentioned above. (3) Winter and spring crops in a 2–year rotation with three harvests, e.g., winter wheat planted in the troughs in October and harvested in June, prior to it foxtail millet on the ridges and in the next spring beans (soy) on the ridges. (4) Winter and spring crops in a 2–year rotation with four harvests, e.g., millet planted on the ridges in May, beans or maize in between in June; both harvested in September, one after the other; immediately barley in the troughs, which is harvested the next June, shortly before that buckwheat on the ridges. The following 1–year system with winter crops is a fifth combination: in October barley planted in the troughs, harvested in June; beforehand kaoliang or cotton on the ridges, afterward soybeans in the troughs. With intensive fertilization lacking, longer or shorter fallow periods are left between the crop rotations even today.

All paddy and dry–field crops are annuals due to the cold winter. But the cultivated plants also include a number of shrubs and trees. They occupy a much smaller area than the field crops, but physiognomically they play an important role, especially in central and southern Korea. The first that should be mentioned is the white mulberry (*Morus alba* L.), which is grown as a shrub in East Asia [633]. A large number of varieties are cultivated. I found it growing up as high as 700 m several times. Only in the parts of northern Korea with extremely cold winters is it lacking, its frequency increasing southward on the peninsula. As described above, mulberry bushes occasionally stand in rows on fields that are intercropped; more often they are found by themselves in small plantations. In traditional Korea the leaves of the wild mulberry were also used. The leaves of the mulberry are used in Korea too for raising silkworms. In traditional Korea silk clothing was already very common among the *yangban* in summer (p. 154), and silk was among the kingdom's few articles of trade (p. 32f). There are two cocoon crops, one in spring right after leaves have appeared on the bushes and one in autumn, the spring crop being much larger. The spring breeding begins in early May in southern Korea, between the 10th and 15th of May in central Korea and between the 20th of May and the middle of June in northern Korea. The time of hatching is regulated by prior refrigeration of the eggs. The caterpillars begin to spin after about 25 days and pupate on pine branches. In P'yŏng–anbuk–do the raising of wild silkworms, which was previously more widespread, has been preserved. The farmers let the leaf–green caterpillars hatch in their houses and then release them outdoors onto oak trees (*Quercus acutissima* or *Quercus glandulifera*). Here they are watched day and night to scare off the voracious birds. Pupation takes place on the same trees. The thread produced by these silkworms is very thick and durable but without luster.

The Moraceae also include the paper mulberry (*Broussonetia kazinoki*, Kor. *tak namu*), which is quite sensitive to cold and is therefore cultivated primarily in southern Korea and in particular in Kyŏngsangbuk–do and Chŏllabuk–do. Its bast is used in making an excellent paper that does not rustle.

In the vicinity of the cities and farmsteads there are many fruit trees. With the exception of orange trees, which grow only on Chejudo, they are also all summergreen. Only rarely are they found in groves, such as the apple trees along the lower Taedonggang, particularly around Hwangju, and the pear trees that were already mentioned during the period of the Three Hans (p. 25) in the area around the mouth of the Naktonggang. Peach and apricot trees are also quite common. Southern Korea's most wonderful fruit tree is the tall persimmon tree (*Diospyros kaki* Thunb.), which bears delicious, juicy, orange–colored fruits in late autumn. All told about a dozen *Diospyros* varieties are cultivated in Korea. Plum trees are sometimes only ornamentals. The blossom of one variety was a national symbol in the Korea of the Chosŏn dynasty. The Paulownias (*Paulownia imperialis* S. and Z. and *Paulownia coreana* Uyeki) are also ornamentals with their large heart–shaped leaves and their big, fragrant, purple blossoms. They are limited to regions with relatively mild winters. The fruit trees also include two varieties of edible chestnut (*Castanea bungeana* and *Castanea crenata*), which do not only grow wild in the virgin forests but are also planted singly or in small groves in the vicinity of the villages. The same is true of the Manchurian walnut, a large tree resembling an ash with gigantic feather–shaped leaves. Both extend fairly high up into the mountain valleys.

One last physiognomically important cultivated woody plant is bamboo. Whereas the wild bamboo species generally grow as low shrubs and extend up to high elevations (p. 137), the cultivated species form poles with evergreen leaves that change colors in severe winters. They reach a height of 10 m. The species in question are *Phyllostachys bambusoides* S. and Z. and *Pseudosasa japonica* Makino. Both are significantly more sensitive to the climate than the wild species are. The former extends up only to 300 m, even on Chejudo, the latter up to 500 m and on Ullŭngdo to only 200 m. On the west coast I found timber bamboo as far as the T'aean peninsula, on the east coast as far as Kangnŭng. The fact that it extends 1° of latitude farther north on the east coast is explained by the milder winters there. Timber bamboo is restricted almost entirely to the lower elevations in southern Korea, where it forms thick groves at the foot of slopes and around the straw– roofed villages, which they conceal. As all over East Asia, the poles are used in house construction, for reinforcing sails and for implements.

Let us conclude our examination of Korean agriculture with a list of types of agriculture. This ensues naturally from what we have said above.

1. Areas without cultivation (high regions in the mountain ranges).

2. Fire–field farming, mostly including some permanent fields (buckwheat, oats, potatoes – spring barley, foxtail millet, broomcorn millet, maize, soybeans, kidney beans, hemp. No mulberry bushes).

3. Permanent dry–field farming without double cropping (foxtail millet, kaoliang, maize, buckwheat, soybeans, adzuki beans, potatoes, hemp, spring wheat).

4. Permanent dry–field farming with double cropping, low intensity (winter barley, winter wheat, foxtail millet, barnyard millet, kaoliang, rice, oats, maize, buckwheat, soybeans, adzuki beans, potatoes, hemp, cotton).

5. Permanent dry–field farming with double cropping, medium intensity (winter barley, foxtail millet, barnyard millet, kaoliang, broomcorn millet, rice, oats, soybeans, adzuki beans, kidney beans, peas, potatoes, hemp).

6. Permanent dry–field farming with double cropping, high intensity (winter wheat and rye, buckwheat, soybeans, adzuki beans, mung beans, sweet potatoes, native cotton, apples).

7. Permanent dry–field farming with two crops of grain (winter barley and wheat, spring barley and wheat, foxtail millet, kaoliang, oats, maize, rice, buckwheat, soybeans, adzuki beans, potatoes, native cotton, hemp).

8. Intensive paddy farming and dry–field cultivation without double cropping (rice, spring rye, spring barley, foxtail, barnyard and broomcorn millet, kaoliang, maize, soybeans, adzuki beans, native cotton, hemp).

9. Intensive paddy farming without winter crop on paddies (rice, winter barley, winter wheat, foxtail and barnyard millet, buckwheat, soy, adzuki and mung beans, potatoes, native cotton).

10. Intensive paddy farming with winter crop on paddies (rice, winter barley, winter naked barley, winter wheat and winter rye, upland rice, soy, adzuki and mung beans, sweet potatoes, perilla, American–type cotton, hemp, ramie, persimmon, paper mulberry).

11. Intensive paddy farming with winter crop on paddies and cultivation of subtropical woody plants (rice, winter barley, winter naked barley, winter wheat, winter rye, upland rice, soy, adzuki and mung beans, sweet potatoes, American–type cotton, hemp, ramie, paper mulberry, persimmon, bamboo).

12. Dry–field farming and cultivation of subtropical woody plants (winter barley and wheat, foxtail millet, upland rice, buckwheat, soybeans, sweet potatoes, American–type cotton, bamboo, oranges, tea – maize, potatoes).

The ranges of these 12 types are shown in Fig. 38. The boundaries of the ranges are shown according to their current state and thus include the development that took place during the last generation under Japanese guidance. All of the types probably already existed in traditional Korea, however.

Little need be said about animal husbandry because of its slight significance [635]. To be sure, the animals are not fed exclusively with foliage and grass, but also with refuse from agricultural produce, especially with rice husks. Horses are even given plenty of soybeans and cows some millet, particularly in winter. But nowhere does regard for the livestock determine the agricultural procedures. Apart from the chickens found in most farming households, cattle are foremost in number and significance. Since they are indispensable as draft animals, the

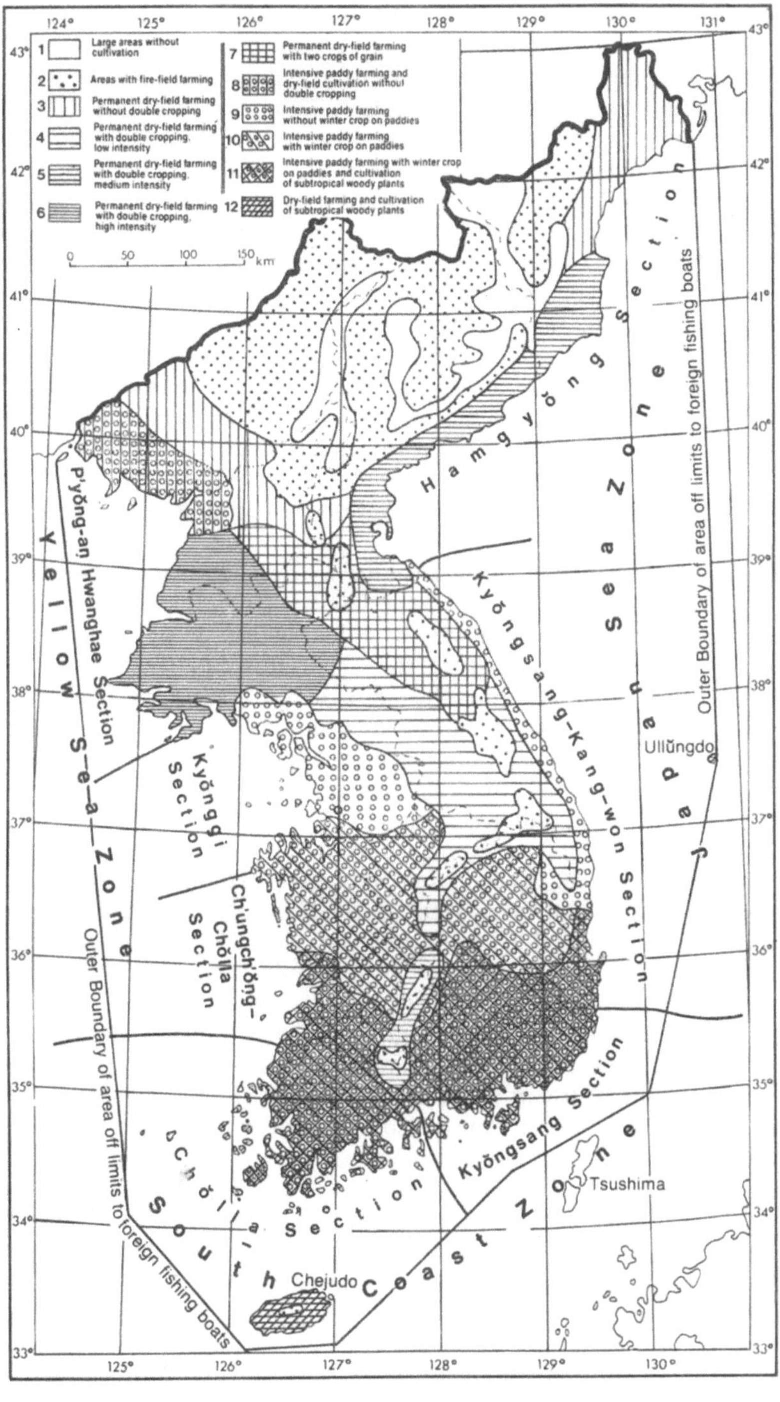

Fig. 38. Agricultural regions and fishery zones

majority of the farm families have at least one cow. The Korean cattle breeds are very powerfully built and strong, by East Asian standards. Cattle are used as pack and draft animals; in traditional Korea they did the heavy transportation almost exclusively. Keeping cows for their milk is virtually unknown. The horses are small, stubborn animals, but very powerful with a lot of endurance [575]. They are closely related to the Mongolian pony and were probably already brought to Korea in early Korean times. The Mongols greatly improved Korean pony breeding (p. 373). Even today pony breeding is particularly common on Chejudo. In recent times it has begun to develop in the northeasternmost corner, where there are modern horse and cattle farms. The ponies are riding and pack animals. Recently they can be seen pulling wagons on graded roads. Only on Chejudo do they also pull plows; on the mainland this would be considered ridiculous. Pigs are kept in all parts of Korea, even in the Kaema upland, fewest on the east coast of the peninsula. The intensity is also slight, though, by European standards, for the Korean pig is very small (maximum weight 37 kg) and grows slowly and is therefore uneconomical. The keeping of sheep was forbidden in the Korea of the Chosŏn dynasty. The few sheep needed by the king each year for sacrifices were imported from China. Goat raising is concentrated in southern Korea, especially on Kŏjedo.

Not only was the Kingdom of Korea as a whole self–sufficient, for want of any substantial foreign trade; each individual family also produced almost all of the items it needed itself. Thus there was but little room for skilled trades, and many of the crafts that flourished during the first century of the Chosŏn dynasty perished or fell into decay during the subsequent wars (p. 31). During the last generation the country was first flooded by imported Japanese manufactured goods and more recently by industrialization, so that there was little opportunity for trades to flourish again. Nevertheless a few handicrafts were left over from olden times, and here and there they are expressed in the landscape. The list of traditional Korea's tribute goods and exports already showed this (p. 32f).

In traditional Korea the existence of rich deposits of bituminous and anthracite coal was so unknown that even Gottsche, who traversed Korea for the express purpose of finding exploitable mineral deposits, claims in his travel account [188] that "coal is lacking." Consequently charcoal production played quite an important role. At the same time, though, it severely ravaged the remaining forests, particularly in the densely populated provinces. It is still flourishing today, although mainly for the Japanese population, which has not accepted the *ondol* heating system with its use of brushwood, except in the north where the winters are extremely cold. Therefore in the mountains of southern and central Korea, such as Chirisan, Kyeryongsan and Ch'ŏnmasan near Kaesŏng I quite often came upon smoking charcoal kilns. The Forest Experiment Station in Ch'ŏngnyang–ri near Seoul maintains an experimental and demonstration kiln.

Paper making likewise still flourishes today. Korean paper, particularly the oiled paper, is excellent. According to Lapeyrière [260, p. 243] there were 53 kinds of paper in traditional Korea, 14 of them oiled paper. It is made from the bast of the *Broussonetia* (p. 185), oat or barley straw, or wastepaper. The finished sheets are spread out to dry in the open air. Occasionally entire villages are occupied with paper making, and then the sheets laid out to dry on the sunny slopes shine forth for many kilometers. There are very many uses for paper even today:

coverings for doors, wallpaper, newspaper, book and writing paper, fans, rain protection for hats (p. 154), raincoats.

Salt was extracted from seawater in traditional Korea also. Along the coast of the Sea of Japan such old saltpans are still in operation today. Small canals have been dug through the narrow sandbars there to allow salt water to enter. They conduct it to level, firmly pounded surfaces about 1 are in size with a layer of loose dirt about 5 cm thick spread out by means of an ox–drawn harrow. In summer when the weather is good men sprinkle water from the canals over the ground with gourd dippers twice a day. After 5 days enough salt has accumulated by evaporation. The dirt containing the salt is piled up in heaps and put in earthen kettles with permeable wooden bottoms. Here fresh salt water is poured over it, leaching the salt out of the dirt, so that a concentrated solution collects in the receiving vessels. This is evaporated over a wood fire. The Korean salt obtained in this manner still contains rather large residues of bittern salts and is therefore of poor quality.

Many trades were distinctly itinerant. They included iron production and still include pottery [568]. The latter moves from place to place, using the clays resulting from the weathering of shale and gneiss. These clays are formed into pottery on the spot on a potter's wheel and then fired in a kiln temporarily erected nearby. The potter's wares also include the "male" and "female" tiles (p. 159). The male ones are made by cutting narrow pipes in half lengthwise, the female ones by cutting wider pipes into three pieces. The achievements of the potter's trade, gigantic pots as big as a man for storing soy sauce, *kimch'i*, water and grain, can still command respect today.

Traditional Korean mining was also more of a trade than a technical operation. It amounted to nothing more than superficial digging. In the old reports on the country gold is mentioned time and again, often in fantastic exaggeration (p. 39f), and gold dust was one of the tribute gifts and articles of trade (p. 32f). But even the old gold mines were hardly more than primitive placer mines, and a scattered few still exist today. At other places one can still find waste heaps that are remnants of the former digging activities [569]. Beginning in 1895 concessions were granted by the king to foreign companies, which started mining gold from primary deposits on a large scale even before the annexation of the country. The three biggest enterprises were those at Unsan (P'yŏng–anbuk–do, American), Suan (Hwanghae–do, English) and Chiksan (Ch'ungch'ŏngnam–do, Japanese). Among these concessions were also two German ones, Kŭmsŏng near Tanggogae in Kangwŏn–do and Sŏnch'ŏn in P'yŏng–anbuk–do, which, however, soon ceased operating.

The wares produced by the artisans are transported by peddlers as far as several days' journey and sold in the open in front of the shops. The device used by men for carrying all type of burdens is the *chige*, a primitive pack frame. Women carry burdens in baskets on their heads. Wares of this type are: clay pots, doors with paper panels, chairs, chests, wooden bowls, baskets and brassware, woven mats, fabrics, firewood, brushwood, salt, rice wine, rice cakes, sweet rice curd, bean curd and roasted chestnuts.

All of these goods are also sold at the periodic markets, which are very popular even today. In traditional Korea at least five markets had to be held each week in each county, alternating between the different villages and different types of goods. The Japanese have retained the main features of the market regulations. The white clad men and women walk to market in separate groups, laden with their burdens as described above. Heavy goods are transported on two–wheeled oxcarts or on the backs of oxen and ponies, which carry huge baskets or sacks on either side of a packsaddle. In the region of fire–field farming and its vicinity the sled is the usual means of transportation (p. 232). High–ranking Koreans still travel overland on horses or donkeys today. In traditional Korea officials and other *yangban* were carried in palanquins. Even today one occasionally sees this antiquated means of transportation. There were also 780 cattle markets in Korea, which were held six times a month. Particularly famous was and is that at Suwon. The area served by a market had a radius of 8– 12 km in traditional Korea, in the more spacious north even 20–24 km. Today it is even larger, since the roads have been improved.

The roads were in an incredibly poor condition in traditional Korea. In the last decades before the opening of the country only one single overland route was passable for carts in places, the Imperial highway from Seoul to Ŭiju. All old travel accounts are full of complaints about the condition of the roads, which posed great difficulties even for porters and pack animals. Connections that are of secondary importance are still in the same condition today in this mountainous country. The most important overland routes in traditional Korea can be seen on the map of the travel routes of explorers (Fig. 10), since they used chiefly the main roads. The routing naturally depends strongly on the terrain, and the roads often follow structural valleys (p. 72f) for long distances. The larger rivers are crossed by ferries. Smaller ones are crossed by frail wooden bridges that were often dismantled during the rainy season or destroyed by the floods. During floods the ferries were sometimes not usable for weeks either. The most important roads met in the capital. Along them there were supposed to be accomodations for humans and animals every 12 km, but these caravansaries were often not in a usable condition. There were also places along these roads where government officials could be lodged. The term for these places (Kor. Ch'aryŏn–gwan) is still preserved as a place name today. Quite a few settlements owe their development to their function as a market or a place of lodging [600].

The transportation facilities were improved somewhat by the fact that some of the river systems are navigable to a limited degree (p. 120). The riverboats are long, narrow freight vessels with shallow drafts and no deck. They have a cotton sail with transverse battens of bamboo, since when ascending the stream they make use of winds from the western or southern quadrant. The boats used on the different rivers are built somewhat differently, the details being adapted to the nature of the river in question (p. 121). The boats on the Han–gang are as much as 24 m long and have a majestically upswept prow. The crew consists of three men, two to pole, tow or row and the helmsman. The boats on the Taedonggang have a high bridge, or platform, from where the helmsman operates the enormous rudder, and are also manned by four rowers [576].

Finally there is a last communication facility of traditional Korea lasting into the present that should be described, the beacon fires (Kor. *ponghwa*) that were in use until the telegraph was introduced in 1894. During the Chosŏn dynasty there were five lines with their branches starting at the coastal and land boundaries and converging on Namsan near Seoul. On capes and high mountains low, sturdy towers were erected of stones fitted without mortar. The ruins of many of them are still left. The five lines consisted of 696 towers (map in [579]). The capes and mountains lie within sight of each other. Fire signals were passed on by night, and smoke signals by day. The messages consisted of one to five signals with a predetermined meaning; one signal meant peace, five meant war with an invading enemy.

In the preceding we have discussed only the anthropogeographical traits of traditional Korea that have continued to have a substantial influence into the present time. They reveal a very distinct individuality. It has been emphasized repeatedly that many of these traits came from or were at least influenced by China. Part Four will describe the radical changes that Korea's geographical character has undergone during the last generation as a result of Japanese influence. Thus in the anthropogeographical realm as well, the geographical present contains emanations from both of the other East Asian spheres, as could be expected from the outset from Korea's intermediate position and its small size. Above and beyond the old Chinese and the new Japanese spheres of influence, however, Korea has an immense number of very original features, as this chapter has clearly shown. In the realm of physical geography the transitional character from continental to maritime forms predominates. The influence of physical geography on the anthropogeographical sphere, however, is very often manifested only in quantitative and seasonal differences. Korea's anthropogeographical individuality is surprisingly similar in its basic features in all parts of the country. Thus this country may be a small one in the larger region of East Asia, but it still is a very strong geographical personality. It is only beyond all of the individual features of this personality depicted in the preceding chapter that the regional differences begin. These we shall investigate in Part Three.

PART THREE:

THE KOREAN REGIONS

The Principles Behind the Regionalization of Korea

As Braun and Hartnack (Pomm. Geogr. Ges. *Gedenkheft zum fünfzigjährigen Bestehen,* Greifswald 1932) emphasized, every land area upon whose countenance man has impressed distinct traces of his activity contains both "totalizing" and "centralizing" features. The totalizing features are present more or less everywhere in the land area in question and therefore give it the same areal character. Adjacent areas, however, have different totalizing features. Centralizing features, in contrast, proceed from a point, for example a city, or from a line, for instance a river valley, or a section of railroad. These points or lines may lie on the border of the previously considered areas, which stand out as uniform individualities because of their geographical totality. The centralizing features radiate from the points or lines into the various areas that resulted from the totalizing view.

If one wishes to divide Korea exclusively from the standpoint of centralizing regional features, one could hardly find a better division, in its principle characteristics, than the eight or twelve provinces (p. 12ff). The Korean provincial boundaries have stood the test of time for more than half a millennium and admirably suit the needs of the administration and the economy. The fact that here and there the modern development will make local changes in the provincial boundaries necessary in the near future, e.g., at the mouth of the Kŭmgang, is a matter of secondary importance that can be neglected here. He who wishes to stress only the centralizing features of the Korean landscape will view the provinces as geographical regions and treat them as such. The most comprehensive geography of Korea, which is found in Volumes 16 and 17 of the Japanese Imperial Geography *Nihon Chiri Fūzoku Taikei* [299], follows this course, and since its emphasis lies very heavily on the spheres of anthropogeography and folklore, this was quite appropriate in that context.

In the present book, however, the same significance is ascribed to the physical geography of Korea as to its anthropogeography, primarily on the strength of the author's travels and research. In sparsely populated areas the former often influences the physiognomy much more strongly than the anthropogeography. From this the necessity arises to devise a new regionalization in which the totalizing features are given their due weight.

If one wanted to consider only the totalizing features, the main division would be a bipartite one resulting in a southwestern and a northeastern Korea. Such a division is indicated in the same Japanese Imperial Geography (Vol. 16, p. 46). In it a dividing line running northwest–southeast from above Ŭiju on the Yalu to the southern tip of Kang-won-do separates an "exterior" Korea from an "interior" Korea. Exterior Korea is the region consisting primarily of basin and hill country along the Yellow Sea and the Korea Strait, interior Korea the area of high mountain ranges and plateaus in continental Korea and the part of the peninsula bordering on the Sea of Japan. As I have pointed out previously [231, p. 258f], a dividing line of this type also plays a decisive role in the sphere of man's most important landscape - forming activity: agriculture. It would not be difficult to devise a subdivision of exterior and interior Korea on the basis of such a fundamental dividing line.

A regionalization of Korea like this would, however, be based entirely on totalizing features and completely ignore the centralizing features. The primary dividing line would split the river basins of the Yalu, Taedonggang and Han - gang completely in two, and the system of land transportation by river, road and rail, which is extremely important for the economy, would be neglected. I therefore prefer a regionalization that attempts to give equal consideration to both the totalizing and the centralizing features.

In the natural sphere Korea has confronted us again and again as a transition zone that is interposed between the vast spaces of continental East Asia and the both horizontally and vertically extremely segmented areas along the periphery. This transitional character is most strongly apparent in the map of the annual range of temperature (Fig. 24). But it manifests itself in many other climatic features as well and thus in hydrography and natural vegetation, in agricultural crops and methods, in housing and in the seasonal changes in the landscape. Since this transition occurs from the center of northern East Asia toward its periphery, it has a trend from north– northeast to south–southwest. The axis of the peninsula follows the same direction, and it is for exactly this reason that the transition is so very obvious on it. The primary dividing lines must thus run approximately from east to west. Only by drawing the primary dividing lines in this manner do we give the centralizing features the weight that is their due. This way the regions that are closely linked to each other by land and river connections are left together, and the horizontal division that still plays an important role in the anthropogeographical sphere, considering traditional Korea's unfamiliarity with the sea, is given the consideration it deserves.

The separation of continental or northern Korea from peninsular and insular Korea deserves to be considered first when these east–west boundary lines are drawn. Peninsular Korea spreads out so long in the direction of the transition that it is divided into central and southern Korea by a further primary dividing line. The dividing line between northern and central Korea runs from Sŏhanman to Tonghanman. The details of how it is drawn are determined by my intention to avoid dissecting the Taedonggang basin and the area around Wonsan and Yŏnghŭngman, on the one hand, and the Nangnim Range, on the other hand. Therefore it twists back and forth between 39°30' and 40° northern latitude (Fig. 39). This line is the northern limit of wheat cultivation to date.

There can be varying opinions as to where the dividing line between central and southern Korea should be drawn. I start it in the east in such a manner that the main part of the T'aebaek Range is entirely in central Korea and the Naktonggang basin and the Sobaek Range just as entirely in southern Korea. In the west the horizontal indentations of the coast provide the best basis here, too. I draw the dividing line into the southeastern tip of Kyŏnggiman, the estuary of the Ansŏngch'ŏn near Asan, so that the large T'aean peninsula to the south is entirely in southern Korea. Thus drawn the boundary between central and southern Korea lies far enough south to give the centralizing features radiating out of the

Fig. 39. Geographical regions. The northern Korean regions are represented by *dots*; the central Korean ones by *crosses*; and the southern Korean ones by no special marks. The individual regions within these three main divisions are distinguished from each other by different types of *hatching*.

capital, Seoul, the necessary scope. The boundary between central and southern Korea therefore fluctuates back and forth around the 37th parallel (Fig. 39). Drawn in this manner this line simultaneously takes a number of important features from other regional spheres into account. We may mention, e.g., that this is the northern limit of cultivation of American varieties of cotton and of the use of paddy fields for winter wheat and the southern limit of oats and maize cultivation. Several varieties of paddy rice also have their northern or southern limit at this line.

This primary dividing principle, emphasizing the regional changes in the direction of the axis of the peninsula, is followed by a secondary principle based on the differences between the Yellow Sea, Interior and Sea of Japan regions. Each of the three main divisions is subdivided accordingly. The interior regions are naturally most extensive in continental northern Korea. They are represented there by the Kaema Upland and the Paektusan Area. Otherwise only one region each along the Sea of Japan and the Yellow Sea need be distinguished in northern Korea. Let us call the latter the P'yŏngbuk Region. Near its northern edge central Korea is so narrow and the Main Korean Range so low that maritime influences from both sides extend all the way into the interior. Thus it is not necessary to distinguish a separate interior region here, and central Korea's only interior region, the Upper Han - gang Region, begins where the peninsula starts to widen out again, south of the 39th parallel. Along central Korea's Sea of Japan coast two regions are distinguished, the T'aebaek Range and the Wonsan Area. On the much more highly dissected and densely populated Yellow Sea side there are three regions: the Taedong Basin, the Hwanghae Peninsula and the Kyŏnggi Region. In southern Korea the Sobaek Range is an interior region that stands out immediately by virtue of its location and relief. Adjoining on the Sea of Japan side is the Naktong Basin, and on the Yellow Sea side the Ch'ungch'ŏng–Chŏlla Region, and along the Korea Strait these three south Korean landscapes give way to the Southern Coast Borderland. Finally the islands of Chejudo and Ullŭngdo each deserve separate descriptions (Fig. 39).

The reasons for drawing the boundaries between these regions within the three main divisions, for choosing their names and for emphasizing their individuality will be the subject of the following 16 chapters. Finally it is worth mentioning that the degree of division of a country naturally depends on the scale of the presentation, and in the case of this book this was defined from the outset. Naturally a further subdivision is possible. For the most spacious part of the country, northern Korea, this has been done in passing in the text and in the map. A few selected subregions in central and southern Korea are also given special, detailed treatment.

On the basis of the above considerations we arrive at the following regionalization of Korea:

I. Northern Korea.

1. North Korean Japan Sea Borderland. 2. Kaema Upland. 3. Paektusan Area. 4. P'yŏngbuk Region.

II. Central Korea.

5. Taedong Basin. 6. Hwanghae Peninsula. 7. Kyŏnggi Region and the Capital, Seoul. 8. Upper Han–gang Region. 9. Wonsan Area. 10. T'aebaek Range including Kŭmgangsan. 11. Ullŭngdo.

III. Southern Korea.

12. Naktong Basin. 13. Sobaek Range including Chirisan. 14. Ch'ung-ch'ŏng–Cholla Region. 15. Southern Coast Borderland. 16. Chejudo.

1 The North Korean Japan Sea Borderland

As we stressed in the morphological overview in Part One, the slope of the Kaema upland down to the coast of the Sea of Japan forms a much broader fringe than does that of the T'aebaek Range. Consequently the former has enough regional autonomy to necessitate separate treatment, in contrast to the latter. It is very difficult, however, to draw a perfect regional boundary between the Kaema upland and the Japan Sea borderland. The individual features that are characteristic of the coastal fringe reach up to different elevations in the valleys and mountains of the southeastern slope of the Hamgyŏng Range and are gradually replaced by those of the Kaema upland. At an elevation of 1000 m this transition has generally been completed; therefore the 1000 m contour line can serve as a makeshift boundary. The entire Hamgyŏng Range, which rises high above this elevation on all sides, is thus included in the Kaema upland; it forms the southeastern edge of the same.

The interior boundary of the North Korean Japan Sea Borderland runs more or less parallel to the coast at an average distance of 40 km. In the northeast this increases to 80 km and more. Together with the Hamgyŏng Range and the coast, the Japan Sea fringe swings from the northeasterly direction that characterizes its southern half to a north–northeasterly one. The southern half belongs to the province of Hamgyŏngnam–do, the northern half to Hamgyŏngbuk–do. There are rather significant differences between these two halves in other respects as well. Toward the northeast the Japan Sea fringe, together with the country of Korea, extends up to the big bend of the Tumen, which, however, does not form a pronounced regional boundary except in some areas of anthropogeography. Apart from these there are very close relationships to the adjacent portions of Manchuria and the Ussuri region. Toward the southwest we shall have the North Korean Japan Sea borderland extend to the low transverse ridge that runs north of Yŏnghŭngman and reaches the coast at Paegan. It is thus 450 km long and has an area of roughly 17,000 km^2.

In accordance with its borderland character, by far the largest part of the region is dominated by simple consequent rivers that rush in their short courses from the Hamgyŏng Range to the Sea of Japan. Some straight stretches of valley appear to follow old tectonic shatter zones and thus represent structural valleys. They include some that are not perpendicular to the coast (p. 73). For the valley of the Chuŭlch'ŏn this origin is demonstrated by a line of hot springs [347], for that of the Sŏngch'ŏn–gang by collapsed Tertiary deposits [750]. Primarily consequent streams are the Sŏngch'ŏn–gang, the two Namdaech'ŏn ("Great Southern Stream"), one near Pukch'ŏng and one near Kilchu, the Puktaech'ŏn and the Chuŭlch'ŏn. The system of the Namdaech'ŏn near Tanch'ŏn has a more complicated structure with stretches of longitudinal valley. The Susŏngch'ŏn, which flows into the sea at Ch'ŏngjin, flows along the eastern foot of the northern spur of the Hamgyŏng Range in a north–south direction. Still an unsolved mystery is the course of the Tumen with its bend swinging far northward to 43°. This gives the river such a slight gradient that it meanders highly, thus differing from most of the other streams mentioned. Inside the big bend of the Tumen the Oryongch'ŏn develops a completely abnormal system. It rises at a distance of only 10 km from the coast and heads north, flowing into the Tumen not far from Shinasan, 55 km from the coast.

The extreme northeastern part of Korea encompassed by the Tumen, which we shall call the "Hoeryŏng–Unggi section," exhibits an independent hydrography deviating from its purely borderland character. This special position is also revealed in its morphology. There are several ridges parallel to the coast here that can probably be attributed to recent uplifting. On top of the southernmost of them, the Shinbong Ridge, lies the drainage divide between the Oryongch'ŏn and the coastal creeks. Similarly independent in its hydrography and orography is the Kilchu–Myŏngch'ŏn section, which bulges out eastward in the form of a broad peninsula.

In their lower reaches most of the rivers in the north Korean Japan Sea borderland have widened their valleys very highly via lateral erosion and created large alluvial plains in them. They transport a tremendous amount of debris, because they come partly from the crest region of the Hamgyŏng Range, which towers above the timberline, and otherwise flow through largely deforested regions in which weathering and mass wasting are also very intense. For most of the length of their courses they flow in disorderly beds of gravel and sand. Above the mouth they generally split up into several arms. The upper reaches form narrow V–shaped valleys. The spurs that the Hamgyŏng Range sends out toward the coast are for the most part erosion remnants. The Mach'ŏllyŏng Ridge (1031 m), which consists of highly metamorphic sediments and forms the boundary between the two Hamgyŏng provinces, has a special character; it probably follows a transverse axis of uplift (p. 6).

Japanese geologists have discovered a number of faults running parallel to the coast. Tateiwa [750] combines them in what he calls the Hamgyŏng system. They are most distinctly developed in the Kilchu–Myŏngch'ŏn section. Nowhere do they display direct morphological significance, however, though some later uplifts have developed in this direction. At other places structural valleys follow them. The structures running parallel to the coast that are created in this manner are of only secondary importance in the relief; for transportation, however, they are of primary importance (Fig. 16).

The forms of the North Korean Japan Sea coast were determined much more highly than in central Korea by the recent ingression, which presumably owes its origin to the postglacial eustatic rise in the sea level (p. 77). There are still quite a few large, deep bays, although deposition by the rivers has made up for far more than half of the sea's previous gains. The southwesterly current (p. 124) has already carried some of the sediments along the coast and created narrow bars, whose infertile sands are covered by thickets of red pines and masses of grassy grave mounds (p. 166). Here and there rocky islands have become reattached to the land in this manner. Occasionally the bars dam off areas, some of which are lagoonlike formations, others bays. Behind the bars there are often alluvial plains, while at other places mountain spurs jut out into the sea in the form of rocky, cliff–lined headlands. Cliffs also surround most of the small rocky islands, many of which have an abrasion platform on top. This type is represented particularly nicely by the "egg island" (Nando) near Unggi. An exception is the largest island, Mayangdo, with its jagged profile towering up to 179 m, which thus resembles the islands off the southern and western coast. Abrasion platforms are found up to an elevation of 180 m on the mainland as well. They continue up the rivers in the form of alluvial terraces. The succession of events was thus: sudden rise of the coast, interrupted by the formation of abrasion platforms, dissection of the platforms due to fluviatile erosion, marine ingression, straightening processes that are attuned to the present - day sea level. As A. Schultz' remarks show, the coast of the southern Ussuri region is of the same type (Das *Ussuriland,* Königsberg 1932, p. 11).

The Japan Sea borderland is the Korean region that is blocked off most strongly to the west and northwest by high mountains. Therefore its climate varies greatly from that of the rest of Korea. The most important of these peculiarities are: extremely little winter precipitation, summer precipitation likewise relatively slight and low summer temperatures. The small amount of summer precipitation is explained by the fact that rain is not generally brought by the sea monsoon, which has a frequency of 49.8% in July in Sŏngjin, but by cyclones approaching from the west, which have already lost their moisture when they arrive here in the lee. The sea monsoon lowers the summer temperatures, however, and occasionally brings the coast fog that can last for days. In winter this part of Korea is strictly governed by the continental high, so that lows bringing precipitation are forced to follow a more southerly track. The land monsoon then prevails with considerable wind velocities. In Sŏngjin it has a frequency of 66% in January. Presumably winds of the bora type also develop in winter in the valleys running down from the edge of the Kaema upland to the coast. Definite proof is lacking thus far, however.

Additionally there are rather large regional differences. Whereas the coast has milder temperatures than the Yellow Sea coast between the Yalu and the Taedonggang, continental influence increases rapidly inland in both summer and winter. Thus the annual range also increases rapidly in this direction. And because the North Korean Japan Sea borderland extends over 3°30' degrees of latitude, there are of course considerable differences between its two ends. Toward the southwest both temperature and precipitation increase in all seasons. Precipitation intensity also increases in this direction. This is demonstrated by the figures given in Table 12, representing one station each in the interior and on the coast and in the northeast and the southwest (for the principles behind the climatic tables, cf. p. 83f).

Along with high relative humidity (Sŏngjin 89%), heavy cloudiness (Unggi 8.1) and a low percentage of possible sunshine (Unggi 35%), the amount of precipitation that falls in August is between one-fifth and more than one-quarter of the annual total. It rains at times on more than half of the days in July. But the amount of summer precipitation and the precipitation intensity remain low, by Korean standards. Hoeryŏng has the lowest average annual precipitation on the Korean mainland (508 mm). The winters are relatively dry, fair and sunny (January in Unggi: 49.5% relative humidity, 3.3 average cloudiness, 69% possible sunshine). The weather on land then shows no relationship to the sea next to it. Due to the low winter temperatures the northeasternmost of the bays, particularly

Unggi, freeze over in some winters and even along the coast more than a quarter to a third of the days in January bring snowfall. Sŏngjin has an annual average of 43.4 days of snowfall, and here, as in Unggi, only the months from June to September are always without snowfall.

Near the coast, the widespread areas of granite hills rising above the alluvial lowlands weather to yellow soils. This leads to the formation of masses of sand and grus, dissected by gullies (p. 79). In the southwest, which is warmest and has the most precipitation, as we have pointed out, the residual soil is even orange or reddish in color. Dome - shaped and sugarloaf mountains do not appear above the zone of yellow soils, however, apart from first signs that they are being formed. In regions with crystalline limestones of the Mach'ŏllyŏng series we find typical rendzina soils. On the steeper slopes in higher regions rocky skeletal soils predominate.

At low elevations the alluvial bottomlands that are not filled with disorderly beds of gravel are almost all cultivated; at higher elevations they are quite often still covered by bottomland forests containing a large number of species in their

Station		Temperature						
		January			August		July	
	Altitude (m)	Mean temp.	Mean min. temp.	Absolute min. temp.	Mean temp.	Mean max. temp.	Absolute max. temp.	Annual range
Chongsŏng	49	−13.6	−19.2	−38.4	22.7[a]	28.0[a]	40.0	36.3
Unggi	64	−9.1	−13.0	−26.0	21.2	24.6	36.4	30.3
Hamhŭng	38	−6.0	−12.5	−28.7	23.8	28.0	38.2	29.8
Mayangdo	67	−3.9	–	–	22.1	–	–	26.0

[a] July.

Station	Precipitation								
	January				August				
	Average amount (mm)	per 1000	Average no. of days with precip.	Precipitation intensity	Average amount (mm)	per 1000	Average no. of days with precip.	Precipitation intensity	Annual amount (mm)
Chongsŏng	4.3	8	2	2.2	114	219	14[a]	10.6	521
Unggi	5.6	7	4	1.4	191	249	18[a]	11.9	769
Hamhŭng	14.7	16	4	3.7	242	271	15[a]	17.3	893
Mayangdo	14.2	23	7	2.0	131	211	13[a]	10.9	620

Table 12. Climatic data for the North Korean Japan Sea borderland

original composition. The deciduous trees of the adjacent slopes are joined here by trees that prefer moist habitats, such as willows and their relatives, the chosenias. Typical of these bottomland forests is the very luxuriant growth of lianas. In the valley of the Susŏngch'ŏn above Komusan I discovered the following composition at an elevation of 500 m: *Betula schmidtii*, *Betula chinensis*, *Populus maximowiczii*, *Populus suaveolens*, *Quercus mongolica*, *Ulmus davidiana* var. *japonica*, *Ulmus macrocarpa*, *Alnus hirsuta*, *Maackia amurensis*, *Salix rorida* Lacksch., *Salix integra* Thunb., *Chosenia bracteosa* Nakai, *Acer ginnala*. In this dense summergreen deciduous forest, in which the red pine is the only conifer, one finds lespedezas, beautiful spireas and the Daurian rose as undergrowth. Grapevines and *Clematis serratifolia* Rehd. wind their way up the tall trees.

The slopes have generally been artificially deforested up to high elevations and are covered by low shrub and grass vegetation. The shrubs are lespedezas, hazel, oak, rhododendron, ash, pine and larch. Here and there they were already replaced by splendid red pine forests during the Korean dynastic period, and the Japanese have energetically continued this development by artifically reforesting the barren granitic grus slopes with the same pine. On soils with more moisture larches have been planted quite successfully. From approx. 500 m upward deciduous trees and shrubs mingle with the pure pine forests, such as the oak *Quercus mongolica*, the maples *Acer pseudosieboldianum* and *Acer mono*, the buckthorn *Rhamnus davurica*, the Manchurian walnut (*Juglans manchurica*), the barberry *Berberis amurensis*, the common ash (*Fraxinus rhynchophylla*), the lindens *Tilia manchurica* and *Tilia amurensis*, the wild pear (*Pyrus ussuriensis*), the Korean crab apple (*Malus baccata* var. *manchurica*), the Amur corktree (*Phellodendron amurense*), the spindle tree *Euonymus macroptera*, etc. At an elevation of 700 m the red pines cease to grow and birches begin to appear in this deciduous forest, particularly *Betula schmidtii*, *Betula chinensis* and *Betula dahurica*. They are joined by alders, the wild apricot (*Prunus ansu*), poplars, maples and the mountain ash *Sorbus commixta*. This slope forest is also characterized by a lush, tangled growth of climbing grapevines and other lianas, such as *Pueraria hirsuta*. It belongs to the oak zone of our general classification (p. 138). At an elevation of around 1000 m, i.e., at the upper limit of the Japan Sea borderland, the oaks disappear, and in the subsequent maple zone more and more of the northern Korean conifers begin to appear (p. 131). In the southwest, in accordance with the change in climate in the longitudinal direction of the coastal fringe, the styrax zone inserts itself under the oak zone (Fig. 36). In the northeast both the bottomland and the slope forests are closely related to the corresponding formations in the Ussuri region (A. Schultz, *Das Ussuriland,* p. 36f). B.W. Skwortzoff includes the northeastern corner of Korea down to Ch'ŏngjin in his Manchurian floral region (Bibl. Sbornik 2, 1932).

The alluvial plains near the coast and, in decreasing intensity, the valley bottoms, upward to their upper ends at quite varying elevations, are used for rice cultivation. Its intensity is, however, nowhere near comparable to that of the Yellow Sea coast at the same latitude (Fig. 87). Along with light - hulled rice the glutinous, dark hulled rice is also grown here. In the northern part of the coastal fringe the fastest ripening varieties of rice, such as Odayo, are grown, and even in the southern part such fast ripening varieties as Soseitaia and Kamenoō are cultivated at best. Soybeans grow on the ridges between the paddies. Here and there in the northern half barnyard millet, which is likewise irrigated, takes the place of paddy rice. To some extent a crop rotation between millet and rice is also practiced. Another paddy crop that is occasionally grown in the Japan Sea borderland is *wanggol* (p. 177). Everywhere irrigation canals are diverted from the rivers.

Barnyard millet is simultaneously a dry-field crop. Others include the other varieties of millet, particularly foxtail millet, both varieties of which are grown here, the one that has red leaves in autumn and the one with green leaves, both varieties of kaoliang, barley, oats, beans, soybeans, peas, cabbage, sesame, hemp, red pepper, maize and potatoes. Buckwheat cultivation extends over from the Kaema upland into the higher parts of the valleys. Like there, the automatic grain mortar (p. 233) is a typical phenomenon along the creeks. These dry fields are found on the lower parts of the slopes; higher up shrubs and woods predominate. In some of the higher parts of the valleys farming has decreased in the past decades, because the government has refused to recognize Koreans who had preempted land here. The abandoned fields are covered by proliferous shrubbery. Because of the cold winters and the short summers, several crops that are typical for the rest of Korea, such as wheat, sweet potatoes and cotton, are lacking here, as in the Kaema upland. In contrast, the cultivation of mulberry bushes and various varieties of fruit has been developing quite nicely in recent years.

Cattle and horse raising are more advanced here than in many other parts of Korea. They had already begun during the Korean dynastic period and have been energetically developed by the Japanese. Particularly in the Hoeryŏng–Unggi section there are large farms with big fenced-in pastures on which hundreds of fine horses and oxen graze. The year-round use of the ox-drawn sled is very common in the Japan Sea borderland (p. 232).

Agriculture forms the basis of livelihood for the population all over the country, but along the coast fishing is an important subsidiary occupation, which is not done with the same intensity on any other stretch of coast (p. 443). The most important fishing harbors are Sŏsura, Unggi, Ch'ŏngjin, Ŏdaejin, Tokchin, Sŏngjin, Shinch'ang and Sŏhojin. Because of the changes between high and low water temperatures, the sea off this stretch of coast is particularly rich in both boreal and subtropical species. Most important is sardine fishing, which is carried on from June to October and has assumed an enormous volume during the last decade (p. 434ff). Second most important is the Alaska pollack, which is fished during the winter. In addition there are other boreal fish, such as herring, cod, gray mullet, flounder and chum salmon. The other subtropical fish, such as mackerel, yellowtail, sea bream, etc., are of comparatively lesser importance. Increasing amounts of herring and especially sardines are processed in the small fish oil pressing factories described on p. 442, which are distributed everywhere along the beaches where the fishing boats are drawn up on land. The fish oil is sent to Japan in oil drums from the many ports, whereas part of the fish cake is used in Korea itself as fertilizer. Recently cold storage depots for fish and factories for fish oil, fish meal and canned fish have been built in the port cities, particularly in Ch'ŏngjin, but the older methods of preserving fish, salting and drying, are still very common. In 1937 14,500 persons were employed in fishery on the coast of the North Korean Japan Sea borderland. They manned 5641 fishing vessels, 371 of which were equipped with engines. In 1936 the coast of Hamgyŏngbuk-do alone supplied 3.6 million barrels of fresh sardines (13.7 million yen), 3.2 million drums of fish oil (10.7 million yen), 710,470 sacks of coarse fish meal (6.0 million yen) and 225,000 sacks of fine fish meal (1.1 million yen).

Here and there mining is done in the Japan Sea borderland, particularly in the immediate vicinity of the coast, which facilitates transportation. Most important is the mining of Tertiary lignite. In addition, gold, iron, silver and nickel ores are mined, particularly in the counties of Tanch'ŏn–gun and Iwon–gun [343, 763]. In Sŏngjin–gun the graphite schist and the magnesite deposits of the Mach'ŏllyŏng series are exploited [762]. The area around Chuŭl has gained significance due to its hot springs on granite and granite gneiss (Chuŭl, up to 56.5° C; Chuŭlonch'ŏn, up to 60.8° C; Ch'ŏnp'yŏng, 44° C; Posang–dong, cf. [347]).

Particularly strong economic impulses have been generated for the past decade by the development of hydroelectric power. The tremendous differences in elevation between the Kaema upland and the coast have made this very profitable. The most important of these projects are the Pujŏn–gang project of the Chōsen Suiden Kabushiki Kaisha, which was completed in 1932, and the Changjin–gang project, which had just been started during my stay in Korea [748, 749]. The former has a capacity of 200,000 kW, the latter will have a capacity of 320,000 kW when completed. A third project of the same type with a capacity of 220,000 kW is planned for the headwater region of the Hwangsuwon–gang (p. 218). The first two lie only 30 km apart, north of Hamhŭng. Because of the steep gradient, the dams are not on the Japan Sea rivers, but on the other side of the divide in the wide valleys of the Kaema upland, which slope gently northward. The water is diverted over the escarpment to the Sea of Japan through penstocks and into intake conduits. The Pujŏn–gang reservoir covers an area of 24 km^2 and holds 670 million m^3. The penstock is 26.5 km long. The pressure gradient of this plant is no less than 999 m. While I was there, the dam of the Changjin–gang project was just being built at a narrow part of the valley near Kujin– ri. These two big projects have helped greatly to boost the economy in the valleys radiating out from Hamhŭng. The water is used for irrigating the paddies after its power has been used. High tension lines run from Hamhŭng to Ch'ŏngjin, P'yŏng–yang and Seoul. The electricity is used not only for supplying light and power for northern and central Korea, but also for the chemical industry. In Hŭngnam, just south of Hamhŭng, a tremendous nitrogen–fixation plant has been built, in addition to other factories. A second plant lies along the high tension line to Ch'ŏngjin near Kŭktong. Since these power plants are not sufficient to cover the Japan Sea borderland's demand for power, which has increased enormously in the past decade, construction of a further power plant with a capacity of 25,000 kW has begun in Puryŏng–gun. In contrast to the above mentioned plants, it uses water from a river flowing into the Sea of Japan, the Susŏngch'ŏn.

Compared with the Kaema upland the Japan Sea borderland is a zone of convenient transportation, though primarily only in the longitudinal direction. One of the main roads of dynastic Korea (p. 190) already ran in this direction, the one followed by Delotkevitch [237] and other Russian explorers. It alone linked the extreme northeastern part of the country with the capital. The Japanese did a good job of improving this road at first, with an eye to its military advantages. This highway, planned according to modern concepts, crosses the Mach'ŏllyŏng Ridge in a pass no less than 725 m high, where it has a truly Alpine character. It runs from Wonsan via Hamhŭng and Sŏngjin to Ch'ŏngjin. From there several good roads radiate out into the Hoeryŏng–Unggi section. Side roads go up through the consequent valleys across the passes of the Hamgyŏng Range and into the Kaema upland. Especially important are the roads from Hamhŭng through the valley of the Sŏngch'ŏn-gang across Hwangch'oryŏng into the Changjin-gang basin, from there across Kŭmp'aeryŏng into the Hŏch'ŏn-gang basin and from Pukch'ŏng across Huch'iryŏng into the same river basin. They are almost the only efficient road connections between the Kaema upland and the rest of Korea. On them rice and other basic necessities are transported by oxcart or truck up into the Kaema upland, and wood and wood products are transported down to the coast.

In many places the just mentioned main road is passable only with great difficulty today, because the northeast Korean rail line, which was completed in 1928, was built parallel to it, and road maintenance was neglected. From Hamhŭng to Sŏngjin the tracks stick close to the coast, for the main part. In the alluvial lowlands the only hindrances are the flood-prone rivers, and they are crossed on bridges that are several kilometers long. In the foothills, however, a number of viaducts and tunnels are necessary. Beyond Sŏngjin the Kilchu– Myŏngch'ŏn graben (p. 210ff) affords a shorter and technically easier passageway that is removed from the coast. From the mouth of the Ŏrangch'ŏn to Nanam, where our region again becomes a narrow fringe, the railroad must follow the coast once more. Ten kilometers beyond Nanam, in Susŏng, this new line connects with the older one coming from Ch'ŏngjin, leaves the coast for good and runs north via Hoeryŏng, crossing the Tumen into Manchuria on a 317 m long bridge not far from Namyang. This line is a link in the long distance connection Seoul – Changchun or Harbin, and is thus an exceptionally important knot in the bond linking the parts of the East Asian continent that are dominated by Japan. The section Ch'ŏngjin – Namyang is simultaneously a link in the shortest connection between Tōkyō and Harbin, which includes the ship route between Tsuruga or Niigata and Ch'ŏngjin. For this reason this section has been operated by the South Manchurian Railway since 1 October 1933. The same is true of the section Najin – Unggi – Namyang, which follows the lower reaches of the Tumen and likewise serves primarily as a connection between the Japanese islands and Manchuria.

Quite similar to the road system, branch lines run from the main line into the mountain valleys. Due to the technical difficulties of the ascent to the passes of the Hamgyŏng Range, some of these lines are narrow-gauge. The most important of them is the line Kilchu – Hyesan opened in 1937, with a further line branching off at Paegam toward Musan. It penetrates into the heart of the Kaema upland. The branch line Komusan – Musan was recently expanded to normal gauge, because it transports the ore from the meanwhile energetically exploited iron ore deposits of Musan (p. 218).

The many bays provide harbors, sufficently deep, protected from storms and rarely frozen over, which have been developed by the Japanese to modern ports in the past 2 decades. They are used for fishing, coastal shipping and for connections to Japan, particularly for exports of fish, fish products and industrial products. The northeasternmost of them, especially Unggi, Najin and Ch'ŏngjin and their cities, are developing, for Korea, in an unusually rapid tempo. They have become the export harbors for central Manchuria, particularly for soybeans. The

Japanese are also planning to route passenger services from Europe to Tōkyō partially via Ch'ŏngjin.

The traffic in these ports and on the connecting railroads has increased enormously during the last 10 years. Under the influence of the incorporation of Manchuria into the pan–Japanese economic sphere, the northeastern tip of Korea is experiencing an upswing unlike any other part of Korea, most particularly since the beginning of the war with China. Here a gateway opens up between the eastern end of the Kaema upland and the southwestern end of the Russian Ussuri region. Through this gateway traffic can flow from eastern Manchuria, without crossing any really difficult terrain, to a conveniently located coast and from thence through the bay of the Tumen and the Sea of Japan to home ports in Japan (Fig. 95).

At the same time Korea's northeastern tip has a definite military significance, for it represents the farthest advance of Japanese territory toward the Soviet Ussuri region on the northwestern shore of the Sea of Japan. Nanam is the headquarters of one of the three Japanese divisions stationed in Korea. The three already mentioned large ports can also prove to be important for the landing of troops. Najin, which lies at a distance of only 170 km from Vladivostok, was presumably developed not only as a commercial port but also as a military one. The deep bay protected by the high island of Taech'odo is excellently suited to this purpose.

Energies of all kinds are concentrated in the immediate vicinity of the coast. As Map 3 (cf. editors' note on p. 7) shows, the inhabitants crowd together here in a very confined space, whereas in the valleys leading up into the Hamgyŏng Range the population figures decrease rapidly. This dense settlement of the narrow strip of coast gives the counties in the Japan Sea borderland a much higher population density than the adjacent ones in the Kaema upland, although the former also generally extend up to the drainage divide of the Hamgyŏng Range and thus include a very sparsely settled hinterland (Fig. 85). Apart from the Hoeryŏng–Unggi section and the Kilchu–Myŏngch'ŏn section, all of the larger settlements lie on or in close proximity to the coast. Due to the mining, industrial, maritime and military significance of the coastal fringe the two Hamgyŏng provinces are among the four provinces with the largest percentage of Japanese in the total population (Fig. 83). They are concentrated in the new settlements that were founded under Japanese rule or developed from tiny Korean villages. In Unggi the proportion is 12, Hoeryŏng 13, Sŏngjin 15, Hamhŭng 17, Najin 19, Ch'ŏngjin 20, Hŭngnam 31 and Nanam 33% of the total population. The last mentioned figure represents the highest proportion of Japanese in any city in Korea.

In these cities the Japanese generally live in separate residential areas, which can be distinguished from the Korean districts at a glance because of the Japanese–style houses. In the Korean districts Japanese activity is also evident in the straightening and widening of the main streets. Along them lie the public buildings constructed of stone or wood: administration buildings, schools, banks, shops, hospitals, barracks, hotels, kisaeng (geisha) houses. The Korean houses in the cities are of the same type as in the villages; they are one-story houses with low roofs. Roofs with Korean tiles are restricted to the cities, but there are also still many huts with thatched roofs in the cities, and it is not even uncommon for the Korean houses to be covered with birch bark held down by a layer of stones. On these roofs dozens of specimens of the rosettelike plant *Cotyledon japonica* Max. sprout between the stones. In many cities Christian churches, which are always among the tallest buildings, bear witness to the German Catholic and the Anglo–American Protestant mission [747].

A detailed geographical description of the North Korean Japan Sea borderland must be based on the differences pointed out above between the northern and the southern Hamgyŏng fringes. The former consists of the two relatively wide flanks of Hoeryŏng–Unggi and Kilchu–Myŏngch'ŏn and the narrow central section of Chuŭl.

The northern half of the Hoeryŏng–Unggi section is composed mainly of the shales, conglomerates and calcareous sandstones of the P'yŏng-an system and the shaly sandstones, conglomerates and shales of the early Tertiary period [731]. These rocks form mighty mountains with long, undissected slopes and widely spaced valleys. It is a landscape with a grand scale. The summit regions often consist of basalt, which occurs to a much greater extent here than the geological map 1:1 million shows [730]. The highest peak, Chŭngsan (1041 m), forms a sharp ridge running west–eastward and is composed of several perlitic flood basalts. Between these high mountains Tertiary basins with low-lying floors are interposed, particularly those of Hoeryŏng, Haeng-yŏng and Kyŏng-won, the last of which extends eastward across the border into Manchuria, where its center, the city of Hunchun, lies. In it the Tumen splits into a number of arms. Rice is still grown in these basins. Dry fields with the crops mentioned on p. 201 extend up to unusually great elevations on the gentle slopes, yielding on top to very scrawny brush that is rarely taller than a man. The summit regions bear 6–8 m high secondary deciduous forests of approximately the composition mentioned earlier (p. 200).

The settlements stretch upward into the tributary valleys of the Tumen in the form of long linear villages. Towns, such as Chongsŏng, Onsŏng and Kyŏng-won, prefer locations near the river. The center of this section is, in contrast, sparsely populated. Hoeryŏng (22,821 inhabitants) lies at the point where the main northeast Korean railroad reaches the Tumen. Rafting on the river has led to wood-processing industries here. Nearby to the west and to the north are coal mines, the former in Tertiary and the latter in P'yŏng-an strata [731, 732].

The southern half of the Hoeryŏng–Unggi section consists of the above mentioned mountain ranges trending southwest–northeast and composed of granite and P'yŏng-an shale. The northwestern one culminates in T'ap'yangbong (826 m), the southwestern one in Songjinsan (1146 m). They are still partially covered by splendid secondary forest; partially they have been reforested with larches and pines. Both extend across the Tumen into the Manchurian–Ussuri boundary region. Between their northeastern ends lies the basin of Kyŏnghŭng, which for its part extends all the way through the narrow southeastern tip of Manchuria at least as far as Karantin in the Ussuri region.

The Tumen flows through the same in two mighty meanders. It breaks through both ranges in probably antecedent valleys, in which the floodplain completely disappears in places. Eight kilometers below the Korean border town of Kyŏnghŭng the mouth of the river begins. Prior to the eustatic rise of the sea level following the Würm glacial stage this was a very hilly area composed of granite and quartzite, but the rising sea split it into an archipelago. The clayey deposits of the debris-laden Tumen have cemented the islands together again. Many of them still have inactive cliffs on the side toward the sea. Some shallow basins that were left out of the process of aggradation form lagoons closed off by offshore bars. Offshore bars also connect the active cliffs of the outer islands with each other. One of them even blocks the Korean half of the mouth of the Tumen.

Due to the small amount of precipitation, for most of the year the Tumen has a very slight discharge for a river that is 516 km long. Since weathering and denudation are very intensive in the region, it is extremely heavily laden with debris and thus is aggrading very highly from its northernmost bend at Onsŏng on. Fifty kilometers above its mouth it is still only 200–300 m wide. During floods it deposits a steep submarine delta into the Sea of Japan. Thus it is navigable only up to the basin of Kyŏng-won–Hunchun, i.e., for a distance of 100 km, for Korean junks with very shallow drafts. Traffic is, however, singularly light, because for more than half a millenium the river has formed the boundary between two nations that have hermetically isolated themselves from one another. With the expanding Japanese influence in Manchuria the boundary character will probably become less crass, although the threat of bandits in the latter is still a severe obstacle thus far. In its last 25 km the river forms the boundary between the Japanese and the Russian spheres of influence, and crossing is prohibited here by both sides on pain of death. The only good-sized town in the Korean part of the region around the mouth is the fishing port Sŏsura, built on a protected site on the lee side of the last island that has become connected to the mainland.

In the basin of Kyŏnghŭng–Karantin a structural valley begins. It first diagonally crosses the range that is closer to the coast of the two mentioned above, but from Unggi on it runs directly parallel to the coastline. The shatter zone that it follows presumably originated in connection with the collapse of the Sea of Japan (p. 70). It characterizes the coastal region of the entire Hoeryŏng–Unggi section from Unggi to Susŏng. The deeper of the ingression bays extend into it, so that it consists of separate stretches of valley with passes up to 355 m in elevation between them. The route Kyŏnghŭng – Unggi – Najin – Ch'ŏngjin and the railroad Unggi – Najin follow it. For the land connections between the rapidly developing north Korean port cities it is therefore of great importance.

The city of Unggi, with 22,085 inhabitants, stretches out in a southeast–northwesterly direction at the foot of the granite mountains lining the bay. Its axis is formed by the Japanese business street Honmachi and its northwestward extension, which is lined with Korean houses. The harbor basin, protected by two mighty breakwaters, is 6 m deep, and on its land side it is surrounded by large depots and sawmills. The main import is rice, the most important exports are wood, fish oil and fish meal fertilizer. Recently it has developed into a transit station for soybeans, soybean oil and cakes of pressed soybeans from Manchuria. With a maritime trade volume of 21.7 million yen (1935), Unggi is ninth among Korea's ports.

More rapid still is the development of Najin [735–738]. When I visited the settlement in autumn 1935, where formerly a small Korean hamlet had stood, the feverish development of city and port was in full swing, directed by the South Manchurian Railway. This development is progressing in three stages and is expected to be finished in 1947. Everywhere in the innermost corner of the bay, houses with corrugated metal roofing were sprouting out of the ground. Large numbers of residential buildings, shops, department stores and restaurants were already finished. 3 km to the southwest a new brickyard with 22 kilns had started operating. The cumbersome two-wheeled Korean oxcarts brought tremendous tree trunks into the city. At the nearby bay where the village of Huch'ang-dong lies granite blocks were being quarried and transported to Najin in barges to build the quays and piers. Meanwhile three piers, each 300–500 m in length, are completed, and the port now already has a capacity of 3 million tons/year. Mighty dredges were in the process of removing the marshy land near the shore to increase the depth. In 1938 the settlement, which does not even appear in the census of 1931, already had 24,508 inhabitants. Trade had already reached a value of 11.8 million yen in 1935, and in 1938 1.73 million tons of goods were transshipped. Najin is one of the ports that are open to international traffic, so that shortly before the beginning of the present World War German and English ships loaded soybeans for Europe here. The new port is connected with Tsuruga by a direct passenger steamship line. The railroad connection to Najin caused great difficulties, however. The planned stretch Ch'ŏngjin – Najin could not be built yet because of the terrain, but its place is taken by a busy road built for a large traffic volume. Since 1935 Najin has been connected with Namyang via Unggi and the Tumen valley by the already mentioned railroad. On this stretch too, however, it was necessary to build a 3.8 km long tunnel between Najin and Unggi.

Even more impressive is the development of Ch'ŏngjin (Fig. 40). Prior to the Russo–Japanese War it was a small fishing village with about 100 inhabitants. In 1908, after the establishment of the Protectorate, the Japanese opened up a commercial port, which expanded rapidly following the annexation. In 1913 Ch'ŏngjin had 8,000 inhabitants. The upswing advanced at a more and more rapid pace after Ch'ŏngjin was connected with the Korean railroad network in 1928 and after the line to Changchun was opened in 1933. In 1928 the city had 23,407 inhabitants; in 1938 72,353, and this development is still continuing. The older part of the city lies in the eastern tip of the ingression bay, which is protected by a mountainous peninsula jutting out 2 km south–southeastward. Up to now an older breakwater has protected the fishing fleet. While I was there they were in the process of closing off the entire bay with a mighty straight breakwater and had reclaimed large expanses from the sea for storage facilities and for the railroad lines connecting to the port. The big steamships to Tsuruga or Niigata can now moor directly at the quay, where the express train Changchun – Ch'ŏngjin simultaneously stops. In 1934 the turnover of outgoing and ingoing goods amounted to 36.5 million yen, in 1938 to 93.3 million yen. Ch'ŏngjin is thus today the leading Korean port on the Sea of Japan. The harbor described here, which was nearing completion in 1933, was to be supplemented by two further ones abutting on it to the west. The last one has meanwhile been dredged out of the deposits of the Susŏngch'ŏn and services the 500 vessels of Ch'ŏngjin's fishing fleet, while the first one is reserved for ocean-going vessels.

As traffic increased, the city expanded accordingly. From its beginnings between the steep, sparsely wooded mountains composed of serpentine [739] and the coast, it has grown northwestward into the narrow neighboring valleys and onto the hills between them. Separated from the rest of the city by the mountain on which the radio station stands, a suburb has developed at the edge of the plain of the Susŏngch'ŏn, where the roads from Susŏng and Nanam cross.

Ch'ŏngjin is today "one of the largest landing places for fish in the East and in the world in general" [653]. In contrast to Najin and Unggi is has also been developed as an industrial city in the past 10 years. Ironworks process the ores from Musan (p. 218); the abundant wood in the hinterland led to the construction of a rayon mill; the Manchurian soybeans are processed to oil and soybean cakes; some plants are concerned with producing fish oil and fish meal, others with the hydrogenation of fish oil, still others with canning sardines and other methods of preserving the catch for human consumption. The fish oil and fish meal produced in Ch'ŏngjin alone had a value of 20.7 million yen in 1937. Ch'ŏngjin's maritime trade has a very different character from that of Najin; in Najin transit goods from Manchuria make up 76% of the trade volume, in Ch'ŏngjin only 21%.

The narrow central section of the northern Hamgyŏng part of the Japan Sea fringe begins at the triangular aggradational plain through which the many arms of the Susŏngch'ŏn flow and at the corners of which the cities of Ch'ŏngjin, Susŏng and Nanam lie. Nanam, the new capital of Hamgyŏngbuk-do, is a new Japanese administrative and military city (21,258 inhabitants). In the plan four

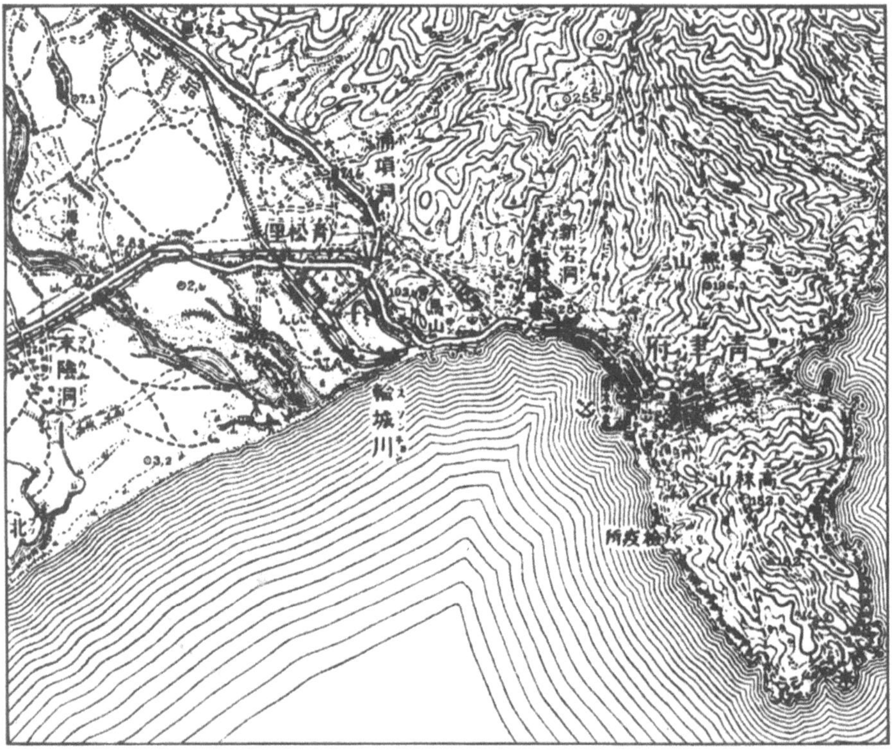

Fig. 40. The Bay of Ch'ŏngjin (unaltered section of the topographical map 1:50,000)

sections can be clearly distinguished (Fig. 41). In the southwest lies the Korean part of town with its village character. The likewise cramped Japanese shopping district with its one-story buildings stretches out along the Nanamch'ŏn. The main part of town north of it is spacious, laid out on a rectangular grid and very clean and consists of elegant Japanese and European style villas in large, beautiful gardens, new hotels, imposing school buildings and banks. The north and northwest, finally, are the site of the actual military quarters with likewise widely spaced barracks for infantry, cavalry and artillery and the many buildings of the hospital. Nanam is thus in its function, as in its town plan, a unique phenomenon in Korea.

The structural valley mentioned above is interrupted by the triangular plain of the Susŏngch'ŏn and begins again at Nanam, from where it continues on to the northern end of the peninsula of Kilchu–Myŏngch'ŏn, always following the outer coast at a distance of 8–10 km and repeating its seaward concave curvature (Fig. 16). It has a total length of 165 km. Just south of Nanam and north of Chuŭl it contains terrestrial Tertiary strata with several lignite coal mines [741] and has the character of an exhumed graben. The mountain wall on the side toward the sea towers up as high as 542 m. Tertiary deposits also extend upward in the straight valley of Chuŭlonch'ŏn, which cuts through the main structural valley at an angle. It is known for its hot springs and is presumably of the same type. At the point where these two valleys intersect, and simultaneously at the northern edge of the fertile aggradational plain of the Chuŭlch'ŏn, the good-sized street settlement of Chuŭl (22,448 inhabitants) spreads out. Halfway between Nanam and

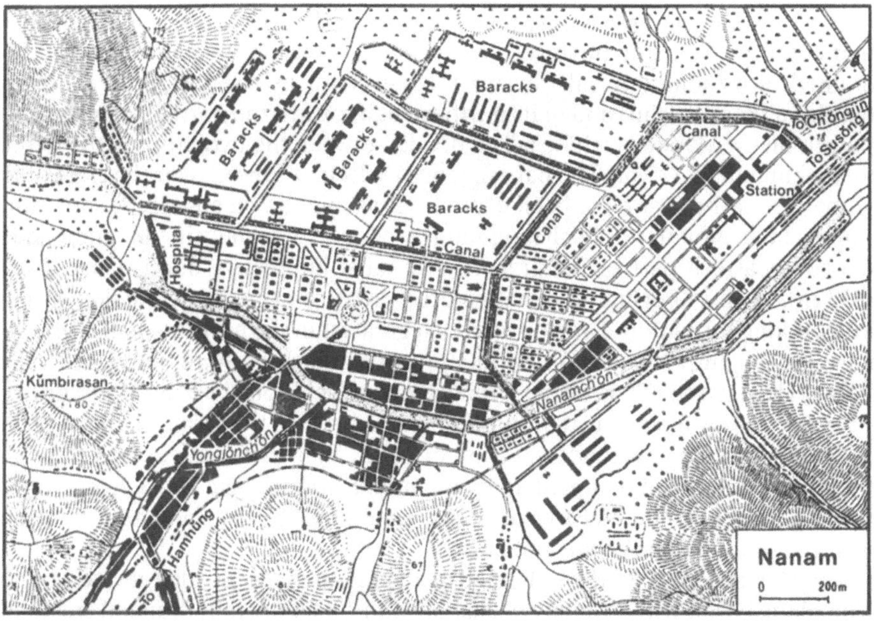

Fig. 41. Nanam, capital of Hamgyŏngbuk-do (after an official plan)

Chuŭl and also along the main north Korean railroad lies the old capital of Hamgyŏngbuk-do, Kyŏngsŏng, which has been outdistanced by Nanam today (Fig. 91).

The Kilchu–Myŏngch'ŏn section is tectonically and morphologically a separate area, which stands out from the rest of the Japan Sea borderland as a distinct identity of its own (Fig. 12). A thorough geological investigation has been done by I. Tateiwa [744], F. Yamanari [742, 743] and J. Makiyama [746]. Its mid-Tertiary deposits consist of Yongdong basalt and sedimentary formations ("Myŏngch'ŏn series": sandstones, shales, conglomerates 1800 m thick, Table 46, Figs. 42 and 43). The uppermost member of the latter is the Manhodong formation, whose lignite coal is mined not far from the train station of Myŏngch'ŏn. These mid-Tertiary deposits have collapsed into a graben approx. 20 km wide and still 75 km long, trending north–northeast to south–southwest, which opens in the northeast into the large bay of Kyŏngsŏng. This graben was formed around the time of the transition from the Middle to the Late Miocene. The fault along the eastern side separates the graben from the Ch'ilbosan horst, while in the west there are complicated step faults. During the Late Miocene planation occurred in the entire area (p. 68f), offsetting the direct morphological effects of the graben formation, if there ever were any such effects. During the Pliocene and Pleistocene, alkali basalt, alkali rhyolite, trachyte and granite porphyry flowed over the erosion plain, covering the former edges of the graben. In the south they were joined by the Lower Pleistocene, approx. 150 m thick, sedimentary Namdaech'ŏn formation.

Since then, i.e., since about the Middle Pleistocene, the Kilchu– Myŏngch'ŏn area has been strongly uplifted, and both the relatively soft Middle Tertiary strata filling the graben and the strata of the Namdaech'ŏn formation have been partially removed, so that the graben appears in the relief as a zone of low elevation (Fig. 43). Today the area is not a plain, however, but a very diverse hill and mountain country. Nor is it traversed by one single river in its length; instead it is drained in its southern half toward the southeast by the system of the Namdaech'ŏn (near Kilchu) and in its northern half toward the northeast by the Kŭktongch'ŏn (= Myŏngganch'ŏn). The divide between the two drops in the middle to a minimum of 425 m. The hard igneous rocks that flowed over the edge of the graben from the east have been carved out as mighty, steep-edged, sometimes sloping table mountains. Such mountains are found especially in the center and in the southeastern part of the graben. The most impressive one is Chaedŏksan (833 m). With 250 m high lava walls it drops away to the more gentle slopes of the Manhodong formation. Erosion has been greatest in the northern half. Here solitary basalt plugs shaped like towers jut up as much as 90 m above the valley floors. Here too the Pleistocene Ŏrangch'ŏn and Kŭktong basalts flowed into the graben, forming a plateau at around 200 m, into which the Ŏrangch'ŏn has again cut itself with many meanders. Another valley that has been cut into the same flood basalt has been dammed up by a sandbar, forming the lake Chang-yŏnho. Other such plateaus of Pleistocene basalt, called "mesas" by Yamanari, are found near the southern end of the graben.

The western edge of the exhumed graben is the jagged ridge formed by the Yongdong basalt with its many peaks, beyond which the tall, rounded crests and plateaus of the Hamgyŏng Range tower. Both trend, like the graben, south-

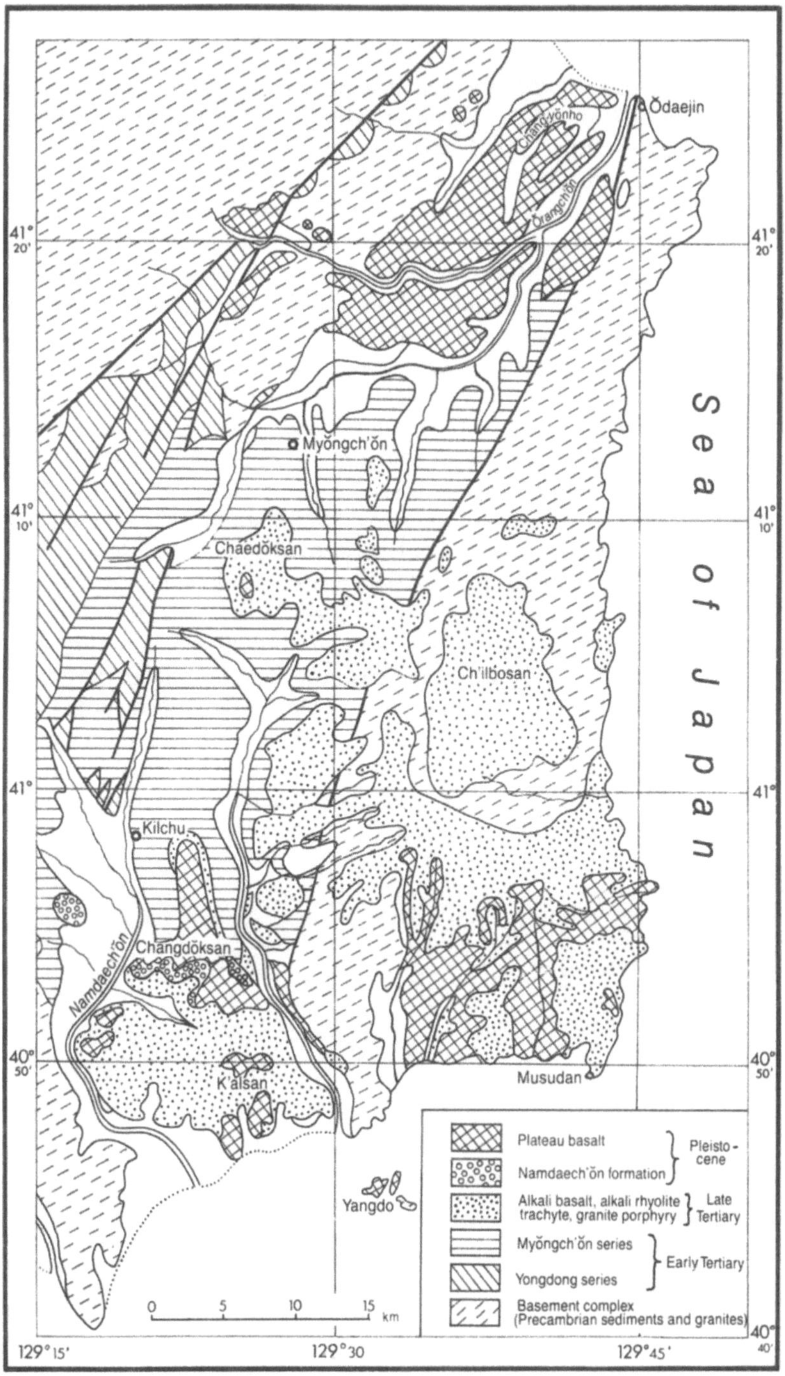

Fig. 42. Geological–morphological sketch of the Kilchu–Myŏngch'ŏn graben (after F. Yamanari, with additions by the author)

southwest. As a result of the exhumation of the graben and a renewed local Pleistocene uplift, the Ch'ilbosan horst today gives the appearance of a massive mountain range. It occupies the eastern third of the peninsula and drops off steeply to the sea. In Ch'ilbosan itself it towers up to 907 m, and in Sang-ŭngbong even up to 1103 m. The old granite, which was faulted by the formation of the graben, is covered by Late Tertiary granite porphyry and the sedimentary Saamch'ŭng formation. The latter, which consists of volcanic breccia, has been etched by erosion into bastions, towers and pillars, resulting in views that are reminiscent of the part of Germany known as the Elbsandsteingebirge. On the southern escarpment of the uplifted Ch'ilbosan massif there are Pleistocene basalt flows ending at the coast in cliffs lined with pillars. The most recent of these fill the bottoms of today's V–shaped valleys.

The entire Kilchu–Myŏngch'ŏn area leaves a rather barren impression. Only here and there on the sedimentary rocks of the graben does one find small stands of red pine or Wilson larch. The table mountains and mesas are covered primarily by grasslands, which feed large herds of oxen. On them one finds barns for sheltering the animals at night and sheds for storing hay. Thus livestock is driven up seasonally to Alpine pastures, as in the Alps. On the summit plateau of Chaedŏksan is an old fortress with four gates and a cistern blasted out of the solid rock for the garrison. The cultivation of the dry-field crops mentioned on p. 201 occupies only relatively small areas in the Kilchu-Myŏngch'ŏn section. Rice is practically nonexistent. Consequently the average population density is low (Fig. 85). The population clings to the bottomlands, which allow good farming in the eroded valleys, particularly around the two county seats (Map 3, cf. editors' note on p. 7).

The above mentioned importance of the Kilchu–Myŏngch'ŏn graben for transportation is revealed not only by the fortress that we have just described, but also by the presence of the two large markets in Kilchu and Myŏngch'ŏn. Kilchu lies at the spot where several rivers from the graben join the Namdaech'ŏn, in a

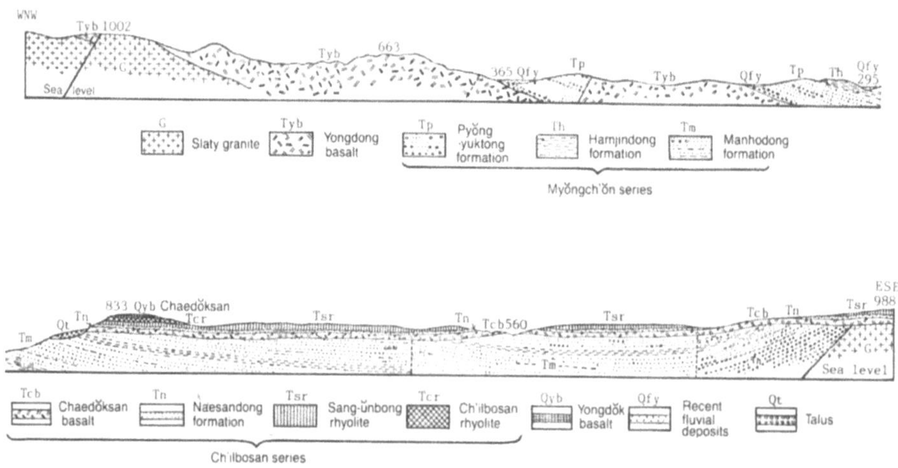

Fig. 43. Geological profile through the Kilchu–Myŏngch'ŏn graben (after I. Tateiwa). Scale approx. 1:75,000. The lower section of the profile is the extension of the upper one to the right. The legend refers to both parts.

basin that these rivers have formed in the easily eroded strata of the Namdaech'ŏn formation and the Tertiary sediments. It has been famous for its cattle market since olden times.

The southwestern half of the North Korean Japan Sea borderland can be dealt with much more briefly, for there are no sections in it that display an independent significance beyond their borderland character. Thus everything we said in the introductory part of this chapter is particularly true of the southwestern half, and the features described there are practically the sum total of its being.

As we have already pointed out, of the many spurs that advance toward the coast between the valleys of the Japan Sea fringe, the Mach'ŏllyŏng Ridge is worthy of special mention. Only the southern end of this mountain range, which extends north as far as the Paektusan area, will be considered here. It consists particularly of crystalline limestone shale and dolomite. The ridge runs north–south, developing in these rocks a steepness akin to the Alps and forming several sharp pyramidal peaks 800–1000 m in height. On the rendzina soils fire-field farming is still practiced today, despite their steepness. The meager fields bear buckwheat, sesame, potatoes and oats almost up to the peaks, and at lower elevations the millet varieties as well. Much larger areas are desolate fallow lands on which masses of *Artemisia messerschmidtiana* grow rampantly. Otherwise the slopes are covered by the brush that has grown up on previously fallow areas, and the steeper the slope is, the higher the brush generally is, because there it is least often ravaged. It consists of oaks, maples, poplars, lindens, willows, elms, birches, the red pine, hazel, the common ash, *Phellodendron amurense*, *Weigela praecox* and *Weigela subsessilis*, spireas, lespedezas and blackberries. At the top of the pass (Mach'ŏllyŏng = "Heaven Touching Pass") a few splendid trees from the primary forest have remained standing around a small temple. They include the common ash, the wild cherry and *Micromeles alnifolia*. The same species occur in the nearby brush, and the virgin forest probably consisted of these same species.

On the coast of the southwestern flank of the Japan Sea borderland there are several small ports that are used for coastal shipping and for transporting ores, fish oil and fish fertilizer to Japan. They preferably occupy the densely populated alluvial plains; however, they do not lie on the sandbars because of the danger of silting, but instead on the rocky stretches of coast nearby. Sŏhojin (13,915 inhabitants) is the port of Hamhŭng and Hŭngnam and Shinch'ang (19,399 inhabitants) that of Pukch'ŏng (20,829 inhabitants) and its hinterland, which extends across the pass of Huch'iryŏng into the Kaema region. The largest of these ports is that of Sŏngjin (23,496 inhabitants), northeast of the Mach'ŏllyŏng Ridge and still in Hamgyŏngbuk-do. Two breakwaters protect the small fishing harbor, while the steamships from Japan have to anchor outside in the roads. With 8.2 million yen trade value (1934) Sŏngjin is No. 11 among Korea's ports. Recently the town has gained industrial significance, especially because of a magnesite processing plant.

By far the largest settlement in the southwestern flank is Hamhŭng, capital of Hamgyŏngnam-do from time immemorial (Fig. 44). Together with the closely adjoining Hŭngnam it had 128,793 inhabitants in 1938. It lies on the eastern edge of the Sŏngch'ŏn–gang plain, the largest delta plain of the Japan Sea fringe, 45 km in length from north to south. The city spreads out directly above the river, which presses toward the eastern edge of the plain at this point. Protected against flooding, it covers the foot and slope of a mountain spur projecting from the north, Pallyongsan, which pushes its way between the delta plain and the broad side valley of the Horyŏnch'ŏn. The red residual soil on this spur is covered by a lovely red pine forest, through which the remains of the city wall run. At the geographical latitude of this mountain spur the delta plain narrows down, and the river does not begin to split up into its main arms until here. Seen from the sea this is thus the first convenient crossing point for road and railroad. The valley of the

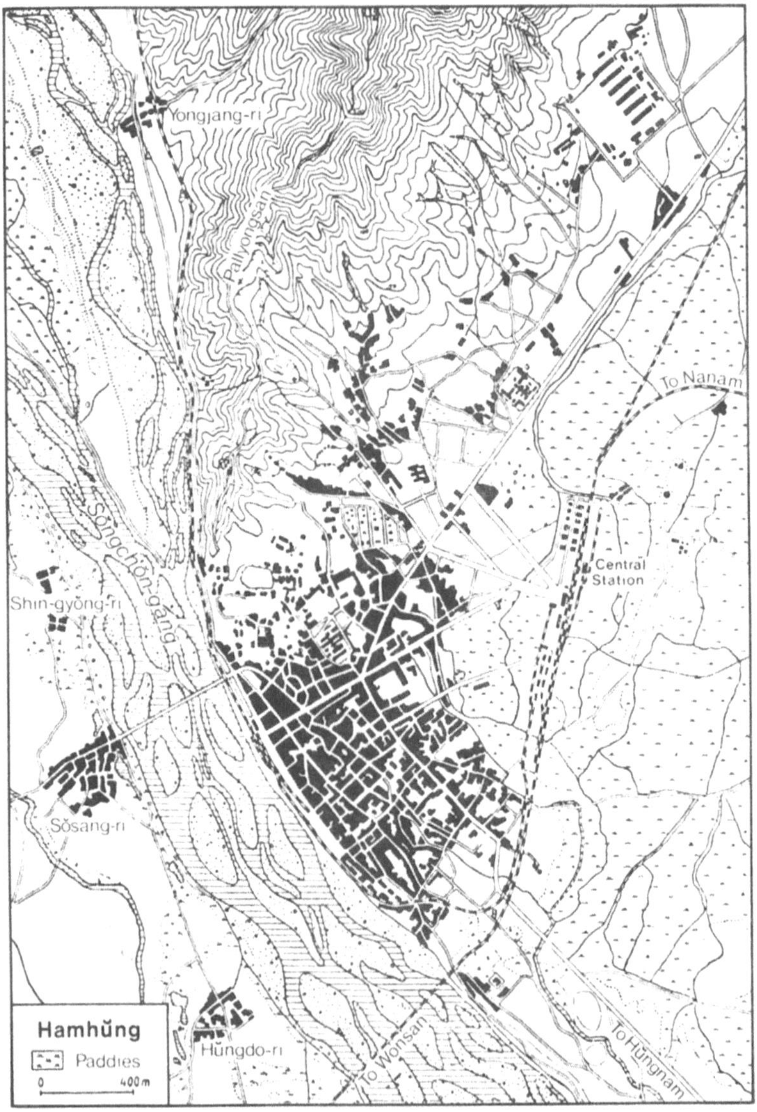

Fig. 44. Hamhŭng and environs (after an official plan)

Horyŏnch'ŏn and that of the Sŏngch'ŏn-gang above Oro-ri are parallel structural valleys [750], through which the hinterland of the city extends far northeastward into the eastern Kaema upland (Pujŏllyŏng, Kŭmp'aeryŏng). At the northern tip of the delta plain the valley of the Hŭngnimch'ŏn joins that of the Sŏngch'ŏn-gang, and the route across the pass of Hwangch'oryŏng into the western Kaema upland runs north through the former. In this manner Hamhŭng dominates the entire province. Historical sites, such as the grave of the parents of the first Chosŏn dynasty ruler (p. 30) and the Confucian school testify to the ancient significance of the city. The southern part of the city is purely Korean. Forming the center is the business district, which extends westward from the train station. The northeastern part adjoining Pallyongsan is the Japanese city with the buildings of the provincial administration and the military, the Japanese schools and part of the Canadian Presbyterian mission. More and more Japanese shops are gradually advancing into the Korean city.

The extreme southwestern part of the North Korean Japan Sea borderland is a hill country consisting of granite that weathers to red soils and metamorphic schists. It is covered by sparse red pine forest interspersed with bushes of McCormick oak and hazels. The southwesterly current has transported the sediments of the Sŏngch'ŏn-gang southwestward, forming a foreland several kilometers wide consisting of a number of beach ridges lying parallel to the coast. These sediments have bypassed a lake, Kwangp'o, that was part of the former ingression bay in the extreme southwest. Extensive, well-irrigated plains covered with paddies spread out around it. The center of this charming landscape, with the wall of the Main Korean Range already rising in the background, is the county seat of Chŏngp'yŏng.

Here the North Korean Japan Sea borderland merges with the area around Wonsan and Yŏnghŭngman without a sharp boundary. Wonsan, however, has such close ties to Seoul through the Ch'ugaryŏng furrow that it is better to include it and its environs in central Korea.

2 The Kaema Upland

Rising in the southeast with the Hamgyŏng Range, the Kaema upland extends north as far as the upper reaches of the Yalu and the Tumen (Fig. 39). Here it ends at the basaltic plateau of the Paektusan area. In the west it is cut off from the P'yŏngbuk Region by the Nangnim Range, which is still a part of this region. Thus it is the hinterland of the North Korean Japan Sea borderland, with which it is closely connected structurally, morphologically and climatically, but only loosely from the viewpoint of transportation. It too stretches out lengthwise from northeast to southwest, measuring almost 400 km in this direction. From northwest to southeast it extends for a maximum of 160 km. It has an area of approx. 30,000 km^2.

As we pointed out (p. 6f and 70), it forms the crest and the sloping surface of a large fault block that is tilted toward the northwest, or of a correspondingly formed asymmetrical geanticline. The Nangnim Range is the northern end of the crest of the main Korean tilt block. These two crests isolate the Kaema upland to such a degree that it is the most highly individualized part of Korea.

This isolation is revealed first of all in the hydrographical character. Only two wild, winding canyons connect its river system with the outside world, that of the Yalu between Shin-galp'ajin and Much'ang-dong and that of the Tumen between Musan and Songhak-dong. In the gorge of the Yalu at Songjŏn one looks up almost 1000 m to the steep peaks towering immediately south of the valley. Since the main direction of slope of the tilt block is toward the north, the hydrographical thalweg formed by the two main rivers lies at the northern edge, and the Kaema upland is drained in the same direction. Here the tributaries Changjin-gang, Pujŏn-gang, Hŏch'ŏn-gang and Sŏdusu flow into the main rivers. The color of these streams is as varied as that of the Danube and its tributaries. When I was there in late summer in 1933 the uppermost stretch of the Yalu itself had dark, clear water. The first big tributary, the Hŏch'ŏn-gang, which together with its tributary, the Hwangsuwon-gang, is much longer than the upper Yalu itself, had light, turbid water, so that below the confluence the river was brownish white and opaque. The second great left-bank tributary, the Changjin-gang, had dark brown, clear water, as did other smaller tributaries on the same side, evidently from the moors in the area, and the two colors remained in juxtaposition far below the confluence.

As a result of this hydrographical pattern, all parts of the Kaema upland are farther from the ultimate base level than any other Korean region. Therefore, the continuous uplift since the Miocene (p. 70) has not had an erosive effect except in the lower, northern portions, and even there the effect has been diminished. Consequently, the Kaema upland is by far the highest region in Korea. As Map 1 (cf. editors' note on p. 7) shows, only the northern valleys are lower than 1000 m, and for long stretches not only the two crests but also the interior mountain ridges rise above 1500 m, even above 2000 m.

The contrast between the slope on the Kaema side of the Hamgyŏng Range and that on the Japan Sea side is one of the most impressive features of the Korean landscape (Fig. 45). On the latter the slopes are very steep with an average of 50° and more and with straight profiles dissecting each other in very sharp crests, and there are many valleys close together. All of this is the result of the intensive erosion that went hand in hand with the asymmetrical uplift. The slopes are covered by coarse, angular talus and show evidence of massive landslides. The valley bottoms shine from afar because of the active boulder, debris and gravel beds of the rivers. The erosion is extremely intensive here, especially above the timberline and in deforested areas.

On the Kaema side, in contrast, there are extensive plateaus at very differing elevations. They extend from the north as far up as the summit of Kwanmobong (2541 m) and other peaks in the Hamgyŏng Range. Wide-bottomed valleys with gentle to moderately steep slopes and a very slight gradient have cut into them. The valley bottoms at an elevation of 900–1400 m are often swampy. The valleys lie far apart. At high elevations solifluction mantles move downslope, at lower elevations ravines and talus cones also. In the northern half of the Kaema upland, which is already affected by recent headward erosion, these broad-bottomed valleys are lacking, however. Here the Yalu has entrenched itself with many meanders. Its water level lies at 700 m in Hyesan and at only 500 m in Much'ang-dong. The same is true of the Tumen. The erosion valley of the Yalu runs just along the edge of the big Paektusan basaltic plateau (p. 235). The basalt

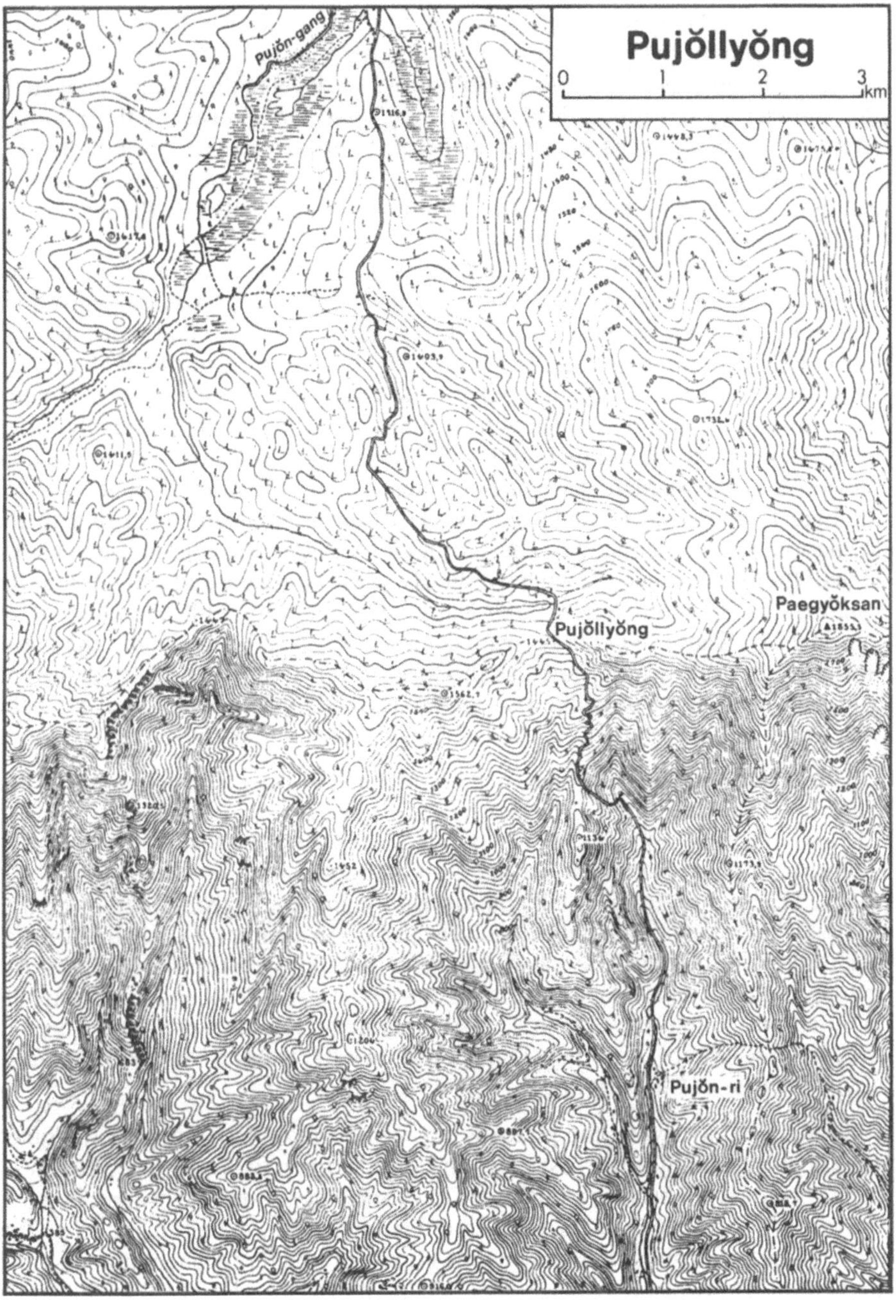

Fig. 45. The contrast in forms between the Kaema upland and the slope down to the Sea of Japan near the pass of Pujŏllyŏng (section of the topographical map 1:50,000)

flows have pushed the river to its southern edge. Their surface lies at 1000 m near Hyesan and at 800 m near Chunggangjin; the Yalu valley lies 300–500 m lower. The mountains of the Kaema upland, in contrast, rise to 1300–1600 m just south of the Yalu and thus tower considerably above the level of the basaltic plateau. In the Tumen region the basaltic plateau extends past the Sŏdusu and the adjoining stretch of the main river somewhat farther eastward. Here, too, the mountains of the Kaema upland rise steeply above the basaltic plateau. For us this morphological boundary is simultaneously the regional boundary between the Kaema upland and the Paektusan area (Fig. 49).

The western half of the Kaema upland consists entirely of granite gneiss and "highly metamorphic sediments" (gneiss, phyllite, schist, graywacke, crystalline limestone) trending in the Liaodong direction (p. 62). Here and there they are intruded by Kogurian granite. In the eastern half, in contrast, the old basement complex is covered in places not only by the Taedong and Kyŏngsang sediments but above all by basaltic and rhyolitic igneous rocks. These are extremely varied in age. Some of them can be found at an elevation of 2000 m and higher, where they form extensive plateaus, some of which are swampy. They probably date to the Late Tertiary (p. 74f) and were raised to these great heights since then by the formation of the geanticline. The recent flows follow the present-day valleys. One of them runs down the valley of the Hwangsuwon-gang and extends into the valley of the Hŏch'ŏn-gang as far as Kapsan [763]. Another fills the valley of the Samsuch'ŏn. These rivers have again cut as much as 120 m deep into the basalt flows, so that the surface now has the form of terraces. From the area around Turyusan three lava flows, likewise cut into terraces, extend southward down the valley of the Puktaech'ŏn to the Sea of Japan at different elevations [762].

The Hamgyŏng Range begins in the north in the area near the gorge of the Tumen at Songhak-dong. The northernmost section, which extends as far as Ch'ayuryŏng (914 m), does not yet have the above described asymmetry in its full intensity, because the difference in elevation between the two base levels is still slight. This section forms a hill country with many knolls, composed of granite gneiss, with no remaining old plateaus. Near Musan magnetite–hematite ore has been found in the quartzite schist, and with the improved transport conditions, mining of this ore has begun. The ore reserves are estimated at no less than 400 million tons, so that the Musan deposits are among the most productive ones in the Japanese empire [759, 760]. Transportation is by rail from Musan to Komusan across the pass of Ch'ayuryŏng.

South of Ch'ayuryŏng the plateaus begin to develop. In Tojŏngsan the range rises above 2000 m for the first time, and from here on to its southwestern end it displays the asymmetry in its full intensity. North–northwest of Kwanmobong, in the vicinity of Sŏllyŏng (2442 m), are the cirques that we mentioned on p. 75f. Whereas the plateaus in this area overlie granite or granite gneiss, a short distance to the south, between Kwesangbong (2333 m) and Mant'apsan (2205 m), there is an area in which the plateaus forming the divide consist of flood basalts. Southwest of Mant'apsan the valley of the Namdaech'ŏn (near Kilchu) extends through the Hamgyŏng Range far to the northwest, past Paegam. Otherwise, too, recent headward shifts in the divide and stream piracy on the part of the Japan Sea rivers are quite common.

Southwest of the trench of the Namdaech'ŏn as well, the plateaus forming the divide are usually composed of basalt. They are overtopped by the mighty Turyusan (2309 m), which has been carved out of a sheet of rhyolite. Due to headward erosion by the Puktaech'ŏn and the Namdaech'ŏn (near Tanch'ŏn), the flood basalt has been dissected into separate lobes here, under which the highly metamorphic sediments of the Mach'ŏllyŏng series are exposed to a great depth. Near Tongjŏm (Kapsan-gun) the latter contain rich copper ore deposits that are awaiting the development of transportation.

Toward the southwest the Hamgyŏng Range drops in elevation. North of the quartzite pyramid of Hŭisabong (1760 m) the flood basalt ends, and from there on the plateaus forming the divide are again composed of the old rocks. Beyond Sambong (1987 m) the range is so low that the passes through it are very good connections between the Japan Sea borderland and the Kaema upland. The most important are Huch'iryŏng (1335 m), Kŭmp'aeryŏng (1676 m), Pujŏllyŏng (1445 m, Fig. 45) and Hwangch'oryŏng (1200 m) (p. 203).

Between the tributaries of the Yalu and the Tumen there are side ridges trending northward that are often even higher than the points at which they branch off from the main range. Most important for the relief of the Kaema upland is the often mentioned Mach'ŏllyŏng Ridge (p. 7), across which both the divide between the Yalu and Tumen and the boundary between the Hamgyŏng provinces run. It begins in the north, deep in the Paektusan area, with Pukp'ot'aesan (2239 m), Namp'ot'aesan (2435 m) and Changgunbong (2108 m), three Alpine peaks that rise precipitously as high as 1000 m above the basaltic plateau of Paektusan. In Namp'ot'aesan, which is composed mainly of granite, the prebasaltic relief towers to a great height, surrounded by the recent extrusions. For the rest, the northern half of the Mach'ŏllyŏng Ridge is formed for a distance of 80 km by a rhyolite flow that has been highly uplifted and fractured, being brought into very different elevations in the process, and is enveloped by basalt. It often forms swampy plateaus. Its lowest point is at the pass of Ch'oegaryŏng (1572 m), which is crossed only by a pony trail thus far. Farther south follows the most important pass across the Mach'ŏllyŏng Ridge, Namsŏllyŏng (1790 m), which leads from the source of the Namdaech'ŏn (near Kilchu) into the Hŏch'ŏn-gang region and has been used by the railroad Kilchu – Hyesan since 1937. Beyond this pass the Mach'ŏllyŏng Ridge and the Hamgyŏng Range intersect, and the Mach'ŏllyŏng Ridge has a new culminating point in the area around Turyusan, which is composed primarily of flood basalts. In the vicinity of Anbandŏgi, with its abundance of gold and mica, the ridge rises above 1800 m for the last time [762]. The mica deposits, which are intercalated with gneiss of the Mach'ŏllyŏng series, are being mined. At Anbandŏgi the flood basalt ends, and the 60-km-long southern end of the ridge runs almost entirely in the highly metamorphic sediments and granites (p. 213).

The highest of the side ridges runs between the Hwangsuwon-gang–Hŏch'ŏn-gang and the Pujŏn-gang. It has more the character of a sprawling mountain country and is composed almost exclusively of granite gneiss. In an area extending 50 km from north to south rounded peaks rise above 2400 m. The highest of them are Ch'ailbong (2506 m) and Puksubaeksan (2522 m). I saw Puksubaek Ridge only from the west and at a considerable distance. It is not unlikely that cirques exist on the eastern side of the higher peaks, although they are not indicated on the topographical map 1:50,000.

The ridge running north between the Pujŏn-gang and the Changjin-gang culminates in Yŏnhwasan (2335 m).

Finally this series of side ridges includes the westernmost mountain range in the Kaema upland, the Nangnim Range. It has, however, a different position in the orography of Korea as a whole, as pointed out on p. 5. This explains a very important, exceptional feature that distinguishes it from the other side ridges of the Kaema upland described above. Its eastern slope still has the same character as the other side ridges: long, gently rising, hardly dissected slopes attaining great heights with wide, broad-bottomed valleys between them. The western slope, in contrast, is affected by the headward erosion of the left-bank tributaries of the Yalu, the Chasŏnggang and the Tongnogang, the Ch'ŏngch'ŏn-gang and the Taedonggang, for it lies much closer to the ultimate base level than the eastern slope does. Although it is part of the main dip slope it has much steeper slopes for this reason. Valley bottoms are lacking or are very narrow. The slopes dissect each other in serrated ridges and peaks. At a distance of only 25 km from the divide the valleys on the western side have an elevation of only 500–700 m, those on the eastern side, in contrast, 900–1300 m.

Beginning at the mouth of the Changjin-gang, the Nangnim Range initially trends southwestward, attaining a height of 2185 m in Hŭisaekpong. Between Karŭngnyŏng (1493 m) and Adŭngnyŏng (1479 m), passes connecting Changjin with Chasŏng and Kanggye, respectively, it turns southward and rises to its greatest height, the rounded Maengbusan (2214 m) and the Wagalbong group (2261 m), in the latter of which there are a sharp-pointed horn and a cirquelike ravine. The pass of Sŏrhallyŏng (1427 m), which follows, was famous in ancient Korea and was already used by the first European explorers. At the massif of Sobaeksan (2184 m) the Myohyang Ridge branches off toward the southwest. Its tectonic situation has already been mentioned several times (p. 7). It culminates in Nangnimsan (2014 m), and still reaches a height of 1909 m in Myohyangsan, more than 60 km from Sobaeksan, only to drop off rapidly after that. On the topographical map 1:50,000 it alone bears the name Nangnimsanmaek. Myohyangsan is a very steep-sided mountain with steep valleys, composed of granite with giant feldspars, rising directly beside the Ch'ŏngch'ŏn-gang, whose valley bottom lies at an absolute elevation of only 50 m. The lower portions of the slopes have developed into rounded rocky surfaces with waterfalls rushing over them here and there, the higher ones into jagged ridges that clearly show the bedding structure and the jointing of the rock. The main Nangnim Range runs farther southward from Sobaeksan and ends at Sasusan (1747 m). Here the Wonsan arc of the Main Range begins (p. 5).

These many side ridges give the Kaema upland considerable relief energy. Because valleys are few and far between, the individual slopes are so long that the difference in elevation between the peaks and the adjacent major valleys is as much as 1500 m despite the relatively slight slope angle.

This stronghold of northern Korea has a special climate with characteristics that do not recur anywhere else in the country. The climatic characteristics that we became acquainted with in connection with the Japan Sea borderland, which dominate the entire northeastern part of Korea, are joined by new ones caused by the upland nature of the region and by the fact that it is cut off from the sea. To date it is not possible to analyze all of the features of this climate, to be sure, for as yet there are only six meteorological stations here. Although they include Korea's highest station, P'ungsan (1150 m), as a rule the stations are in valleys, namely, in county seats and border towns. Settlements extend up to 1550 m, however, and farming as high as 1850 m, while the timberline lies at 2000–2150 m. Climatically the northeasternmost part of the P'yŏngbuk region also belongs to the Kaema upland. Therefore two (one) important stations from this region are also included beneath the line in Tables 13 and 14.

As we already pointed out (p. 115), nowhere does the Korean monsoon climate have as pronounced continental features as in the Kaema upland and the ad-

Station		Temperature						
		January			August		July	
	Altitude (m)	Mean temp.	Mean min. temp.	Absolute min. temp.	Mean temp.	Mean max. temp.	Absolute max. temp.	Annual range
Musan	470	−12.9	−19.2	−30.2	22.2	28.6	38.7	35.1
Hyesan	725	−19.6	−28.5	−42.0	21.3	28.8	38.0	40.9
P'ungsan	1150	−16.4	−23.9	−41.5	18.2	24.3	34.5[a]	34.6
Changjin	950	−19.8	−29.1	−43.0	19.1	25.4	34.8[a]	38.9
Huch'ang	585	−18.7	−27.0	−40.0	22.0	28.1	39.1[a]	40.7
Chunggangjin	313	−21.0	−28.9	−43.6	22.6	28.9	38.0	43.6

[a] August. [b] February. [c] December. [d] July.

Station	Precipitation								
	January				August				
	Average amount (mm)	per 1000	Average no. of days with precip.	Precipitation intensity	Average amount (mm)	per 1000	Average no. of days with precip.	Precipitation intensity	Annual amount (mm)
Musan	4.0	7	3	1.3	117	216	16	8.4	542
Hyesan	5.2[b]	9[b]	4[b]	1.3	135	241	14	11.3	559
P'ungsan	7.6[c]	12[c]	3	2.3[b]	174	272	15	13.4	647
Changjin	13.7[b]	17[b]	7	2.0[b]	189[d]	227[d]	17	12.3	829
Huch'ang	12.2[b]	13[b]	5	2.4[b]	228[d]	234[d]	17	14.3	971
Chunggangjin	9.6[b]	12[b]	7[b]	1.4[b]	183	225	18	12.2	812

Table 13. Climatic data for the Kaema upland

joining region to the northwest. The combination of being cut off from the sea to the west and to the southeast and being open toward the north has an astonishingly strong effect. The figures presented in Table 13 show that the winter temperatures are extremely low. This holds particularly for the basins in the Yalu valley, in which stagnating cold air pockets evidently form. In Chunggangjin, the station in Korea that is farthest from the sea, with a minimum distance of 220 km from the coast, this phenomenon is more pronounced than in the 400 m higher Hyesan. In general the winters in the Kaema upland become milder with increasing altitude. This is demonstrated especially by the station of P'ungsan, where the January mean temperature is 4.6° C higher than in the 837 m lower Chunggangjin. Changjin does not show this astonishingly intense temperature inversion as clearly, because, in contrast to P'ungsan, it lies in a high mountain basin. The mildest winters are found in Musan on the Tumen, only 70 km from the sea and connected with it by the low saddle of Ch'ayuryŏng. Nevertheless, even its winter temperatures are significantly lower than on the nearby coast of the Sea of Japan. From this temperature inversion we can conjecture that the Kaema upland has essentially an autochthonous radiation climate in winter. The fact that in Chunggangjin on the average 55% of the days in January are without wind confirms this conjecture. The most frequent wind directions in Chunggangjin in January are northeast and east–northeast. They demonstrate that there is no lack of influence from the land monsoon from Manchuria. It is, however, much less frequent and intense in the basin of Chunggangjin than on the coast of the Sea of Japan. On the actual upland both are presumably much greater. But wind observations are lacking for the stations there.

Table 13 simultaneously demonstrates that the winter precipitation in the Kaema upland is as abnormally low, even by Korean standards, as in the North Korean Japan Sea borderland. Accordingly the average cloudiness also has its minimum in January (Chunggangjin 3.8), and the number of hours with sunshine its maximum in February (63%). In Chunggangjin all of the winter precipitation already falls as snow. This station registers an annual average of 72 days with snowfall, i.e., almost twice as many days as Unggi. Despite the southerly latitude of less than 42°, snow lies in Chunggangjin with brief interruptions for a good half year. In P'ungsan only 3.5 months are without frost, on the average, as Table 14 shows, in the worst case only 2.5. The other stations listed in Table 14 are only slightly better off.

	First frost			Last frost			First ice formation		
	Earliest date	Latest date	Average date	Latest date	Earliest date	Average date	Earliest date	Latest date	Average date
Hyesan	22 Aug.	25 Sep.	12 Sep.	17 May	19 Apr.	29 Apr.	18 Sep.	15 Oct.	1 Oct.
P'ungsan	28 Aug.	18 Sep.	9 Sep.	15 June	22 Apr.	22 May	21 Sep.	12 Oct.	1 Oct.
Changjin	12 Sep.	28 Sep.	17 Sep.	21 June	4 May	25 May	12 Sep.	12 Oct.	23 Sep.
Chunggangjin	16 Sep.	5 Oct.	24 Sep.	2 June	3 May	17 May	14 Sep.	10 Oct.	25 Sep.

Table 14. Dates of first and last frost and first ice formation

Summers are warm in the Kaema upland, considering its elevation. Despite its altitude of 725 m, Hyesan has higher temperatures in July than Unggi, for in this season, too, the influence of the sea is slight there. In Chunggangjin, e.g., the southwest monsoon has a frequency of only 16.6% in June, the month in which it is most highly developed. The continental character of the summer is also expressed in the fact that, in contrast to almost all other parts of Korea, July is already the warmest month, on the average. Unlike midwinter, the summer temperatures decrease in the upland with increasing elevation. The annual temperature range is thus smaller at higher elevations than in the valley of the Yalu. Chunggangjin has the highest figure at 43.6° C. The data on the length of the period without snow and frost show, however, that the summers are short. The temperature curve has an extremely steep peak (Fig. 27), and usually only July and August are really warm. Although this continental temperature cycle exists even in summer, this season brings relatively frequent precipitation. About half of all of the days in July bring rain, and the precipitation intensity is just as high as on the coast. Here, too, it is mainly cyclones coming from the west that bring precipitation, particularly those on track b (p. 104f). For them the Nangnim Range is of course an obstacle. Consequently both summer and annual precipitation are much higher west of it, in the P'yŏngbuk region, than in the actual Kaema upland. Huch'ang has twice as much rain in July as Musan. Compared with western and southern Korean figures the amount of precipitation in the Kaema upland, as large as it is, is thus small everywhere. This too is an expression of its continental character. Associated with the midsummer precipitation maximum are maximum cloudiness (Chunggangjin, July 7.1) and a minimum amount of sunshine (August 44%), as is usually the case in Korea.

The very pronounced continentality of the climate, by Korean standards, shows that the Kaema upland is closely related to the adjoining southeastern part of Manchuria. The same relationship is seen in the vegetation. Nowhere else in Korea, with the exception of the Paektusan area, is the Manchurian–Daurian admixture in the vegetation as strong as here. A second trait is of course the altitudinal zonation of the vegetation, which corresponds to the altitudinal decrease in temperature in the summer half-year.

On the gentle slopes of the central Kaema upland, where farming has spread upward to the very highest elevations permitted by the climate, the virgin forest has been pretty much destroyed. Large, continuous stands remain only above the upper limit of cultivation. On the steep outer sides of the Nangnim and Hamgyŏng Ranges and on the steep slopes along the entrenched streams in the northern half of the upland it still covers extensive areas, however. Aside from the Paektusan area, the Kaema upland is thus the Korean region with the most virgin forest (Map 4, cf. editors' note on p. 7). The Japanese statistics list the productive forest area in the Korean part of the Yalu region as 16,100 km^2 and of the Tumen region as 4100 km^2 [299, vol. 17, p. 356], and by far the largest part of this area lies in the Kaema upland and the adjoining mountain ranges. In the Yalu region 29% of the area and in the Tumen region 24% is covered by predominantly deciduous forests. Thus approximately three-quarters are coniferous forests.

Let us describe the composition and the altitudinal changes on the basis of three examples. Climbing up the valley of the Namhach'ŏn (Editors' note: the Puk'ach'ŏn is evidently meant, since Lautensach's description refers to this northern tributary of the Poroch'ŏn) northwestward from Ch'ŏnp'yŏng toward Kwanmobong, up to an elevation of approx. 1000 m one is still in the region of the Japan Sea borderland, according to the definition given on p. 196, and thus of predominantly deciduous forests of the composition described on pp. 200. At 900 m the oaks disappear and the fir *Abies nephrolepis* begins to appear. Somewhat higher the spruce *Picea jezoensis* begins. Soon the larch *Larix dahurica* and the Korean nut-pine join in. Deciduous trees, such as the small-leaved linden, the birches *Betula schmidtii* and *Betula chinensis*, the common ash, the giant poplar *Populus maximowiczii*, the maples *Acer pseudosieboldianum* and *Acer tschonoskii* and *Fagara schiniifolia* extend up from lower elevations. We are still in the maple zone of the general classification. In addition there are rhododendrons (*Rhododendron schlippenbachii*, *Rhododendron mucronulatum*), Siebold's hazel and the lianas *Actinidia* and the wild grapevine as undergrowth. At 1050 m the white-trunked *Betula ermanii* appears, and it extends from here to the timberline. The festoons of the giant lichen *Usnea longosissima* become increasingly dense the higher one goes. Around 1600 m one climbs through almost pure coniferous forest in places, consisting of the already mentioned firs, spruces and larches with sporadic ashes and birches, in which the undergrowth has given way to a carpet of moss. At other places there are stands of a single species, occasionally quite sparse ones, of *Larix dahurica*, *Picea jezoensis*, or *Betula ermanii*. At an elevation of 1900 m the latter begins to predominate. Around the <Asa> lodge (for the significance of angle brackets cf. note in foreword) (2100 m) only *Larix dahurica* is still left along with it. The trees are still as much as 20 m high here. Mosses, reindeer moss and cranberries form a dense, multicolored ground cover. Not until 2200 m do the larches disappear and the beeches become stunted. On the north side of Kwanmobong the timberline often lies at 1900 m, however, and it is only along the edges of ravines that spruce forests (?) extend up to 2100 m. In the vicinity of the uppermost birch shrubs the dwarf pine *Pinus pumila* appears. The summit region is covered by a very low subalpine plant association, consisting of grasses, mosses, *Rhododendron parvifolium* and *Rhododendron chrysanthum*, white mountain avens (*Dryas octopetala* var. *asiatica*), the gentian *Gentiana algida* Pallas, the stonecrop *Sedum telephium*, cranberry and bog huckleberry, the cinquefoil *Potentilla fruticosa*, reindeer moss, etc.

Maengbusan is an example from the Nangnim Range. Whereas Kwanmobong consists of hard, coarse-grained granite containing very much quartz and little mica, the entire Maengbusan region is composed of gneiss. In contrast to the south side of Kwanmobong, on the east side of Maengbusan stands of primary forest are found only above 1300 m, for at its foot there are broad, settled valley bottoms at 1000–1200 m, and around them the virgin forest has been destroyed. The lower limit of the remaining virgin forest lies above the upper limit of oaks. As an unusual feature, *Thuja koraiensis* Nakai occurs here from 1500 m on as a shrub with a maximum height of 1 m. Of the deciduous trees, the ashes are the first to disappear. Instead *Betula ermanii* and *Picea jezoensis* become more frequent here too. The latter and *Abies nephrolepis* extend up to about 2050 m on the southeast side of the mountain. Above this stretches nothing but pure, stunted, open birch woodland, in which grasses as tall as 1.50 m grow rampantly. Above this birch zone, covering a vertical span of only about 100 m, grassy meadows continue upward, with cushions of *Rhododendron chrysanthum* scattered about in them. Just below the ridge (2200 m) stands of *Pinus pumila* begin, which hug the ground, getting to be 20 cm at the most and cover the different peaks of the Maengbusan group. Thus on the whole the plant cover on Maengbusan is composed of much fewer species than that on Kwangmobong, which can probably be attributed to the more continental character of its climate.

Myohyangsan, finally, which extends far toward the southwest, rises in a climatic region that is much milder in winter and has much more precipitation in summer. Here the primeval forest, unaltered or only little changed by man, extends down almost to the valley floor, i.e., 200 m, in the environs of the temple of Pohyŏnsa, which was founded in 968 (787). As we pointed out already on p. 132, for many centuries the temples had the power to prevent the forests from being destroyed by the nearby inhabitants. In contrast to the first two examples, however, Myohyangsan (1909 m) does not tower above the climatic timberline. The lack of forest in its summit region is instead due to the steepness of its rocky peaks. Still it offers the rare opportunity to study a natural north Korean vegetation profile with a vertical span of 1700 m. At all elevations the extraordinarily large number of species and the lushness of the forest are striking. At 200 m its character is determined mainly by the many species of oak (*Quercus mongolica*, *Quercus aliena* Bl., sometimes var. *acuti-dentata* Max., *Quercus serrata*). In addition there are in this "lower oak zone" the elm *Ulmus japonicus*, the small-leaved linden, the Korean mountain pear, the Manchurian cherry tree, Siebold's maple and *Acer mono*, the common ash, the Manchurian walnut and such trees that are sensitive to cold as the frequent *Castanea bungeana* Bl. and *Clerodendron trichotomum* Nakai. The only conifers in this lowest zone are the red pine and *Abies holophylla*. This zone extends up to approx. 600 m and ends with the two last mentioned conifers. Up to 1250 m one finds the common oak. Also characteristic of this second story ("upper oak zone") are the already mentioned maples and *Acer tegmentosum*, *Betula schmidtii*, the splendid hornbeam *Carpinus erosa*, the cornel cherry *Cornus controversa* Hemsley, Hercules' club (*Aralia chinensis* L.), the likewise thorny lacquer tree *Rhus vernicifera* and *Maackia amurensis*. Beautifully blooming shrubs and lianas form the undergrowth. The third zone extends up only to about 1400 m (maple zone). Above 1400 m conifers festooned with lichens already predominate, particularly the Korean nut-pine, *Picea jezoensis* and *Abies nephrolepis*. The undergrowth, consisting of *Rhododendron*, the lilac *Syringa palibiniana* Nakai, the honeysuckle *Lonicera sachalinensis* Nakai, the Korean snowball and the climber *Tripterygium regelii* is lush. In the shrub vegetation in the vicinity of the highest peaks the dwarf pine *Pinus pumila*, the Korean arbor vitae (*Thuja koraiensis*), the alder-leaved mountain ash, *Betula ermanii*, field senecio (*Senecio campestris*) and the Korean edelweiss (*Leontopodium coreanum*) occur.

Below the climatic limit of cultivation the actual Kaema upland affords plenty of opportunity to investigate the secondary and tertiary formations that have taken the place of virgin forests after they have been destroyed by fire once or repeatedly. They have developed as wood, shrub or weed formations. The more luxuriant of them consist basically of the same plant species as the primary forests, only in other growth forms and in other frequencies. With the exception of the very fire-resistant *Larix dahurica*, the shade-requiring conifers are not found in these second growths, however. Oaks, in contrast, which love warm soil, are very common in them, particularly *Quercus mongolica*, and they extend up to higher elevations than in the primary formations. In southern exposures they occur up to 1500 m. Likewise very numerous are birches, which sprout up most rapidly following fires. They are joined by great masses of certain shrubs that are rare in primary virgin forests but take advantage of the dry soil and the abundance of light to occupy ground in the second growths, especially *Lespedeza* species and *Rhododendron schlippenbachii*. The generally open secondary forests thus consist primarily of oak, birch and larch.

The shrub formations on good soils are often composed of a greater variety of species. On the summit of P. 1247.6 south of Hyesan I observed a knee-high plant association on shale consisting chiefly of the common oak, Siebold's hazel, the Daurian rose and *Lespedeza bicolor*. Associated plants are the poplar *Populus davidiana*, the small-leaved linden, red birch, the hawkweed *Hieracium umbellatum*, several species of aster, bracken, *Thalictrum thunbergii* D.C., *Cimicifuga heracleifolia* Kom. and *Artemisia messerschmidtiana*. In other such shrub formations the larch *Larix dahurica*, birches, poplars, ashes, maples, willows, the hawthorn *Crataegus maximowiczii* Schneid. and the honeysuckle *Lonicera sachalinensis* occur along with some of the just mentioned plants. Many grow on highly eroded and thus depleted soil, however, and the number of species and individual plants is very scant and their development very scrawny.

The poorest stands consist entirely of weeds, particularly the above mentioned *Artemisia messerschmidtiana* (Kor. *ssuk*). The inhabitants of the Kaema upland put dried, crushed pieces of this plant on the skull of children suffering from epilepsy and set this on fire. The terrible pain arising as it burns down is supposed to cure the epilepsy. In fact, I occasionally encountered people with a bald spot about the size of a silver dollar just above the hairline, particularly in the Kaema upland, but also in other parts of Korea. It is the result of this cure.

The second growths described here are in some cases only stages of succession. On former fields weeds are the first thing to grow. If no new intrusion occurs on the part of humans, after a number of years brush grows up, gradually becoming much taller than a man, and finally secondary forest can develop. The weaker and rarer human influence remains, the more species it will contain and the greater will be the similarity to the virgin forests. If one and the same area is frequently burned over and cultivated, the soil finally becomes so depleted due to leaching and erosion that without sound forest management it is incapable of supporting more than shrub or even weed formations. In this case these have become semi-spontaneous end stages.

The development of these second growths can only be fully understood if we observe the form of economy that makes the intrusions into the natural vegetative sequence. It is fire-field farming [559]. The type of farming involving the use of fire-fields (Kor. *hwajŏn*, Jap. *kaden*) is not restricted to the Kaema upland, to be sure; it exists at high elevations in the mountains of central and southern Korea as well (p. 161f). The former is, however, the region in which it is most widespread. In Fig. 46, showing the distribution of fire fields in northern Korea, the Kaema upland and the adjoining area to the west stand out clearly. It shows the situation in 1916, however, a year in which the intensity of fire-field farming in the counties west of the Nangnim Range still exceeded that in the actual Kaema upland. Meanwhile large numbers of fire-field farmers (Kor. *hwajŏnmin*, Jap. *kadenmin*) have also settled in the latter. The following example shows how enormously fire-field settlement has increased in the central part of the Kaema upland [561, p. 26]. In 1924 the village of Kaeunsŏng-ri (Samsu-gun) had 87 farmsteads, 114.9 ha of fire fields, 53.5 ha of permanent fields, thus a total cultivated area of 168.4 ha. Owing to flood damage in the coastal areas of the Japan Sea fringe, there was a large influx of population in the next few years. In 1928 the village already had 366 farmsteads, 1088.4 ha of fire fields and 600.3 ha of permanent fields, a total cultivated area of 1688.7 ha. The number of farmsteads had more than quadrupled during these 4 years, and the cultivated area had increased tenfold. Such examples are not unique, and so it is understandable that between 1916 and 1926 the percentage of fire fields in the total cultivated area increased from 9.2 to 16.1% in P'ungsan-gun, from 15.0 to 19.6% in Kapsan-gun, and in Changjin-gun even from 17.8 to 56.7%.

Fire-field farming has thus contributed greatly during the past decades to the rise in the population density of the Kaema upland, which is still low today. The abundance of forest for clearing, the vastness of the hitherto extremely sparsely populated region, the small average inclination of the slopes, compared with all other forested regions, all of this must have attracted those Koreans who had lost their means of livelihood in the lowlands and were determined to build a new life in a struggle with the forces of the mountains and in opposition to all laws. As we

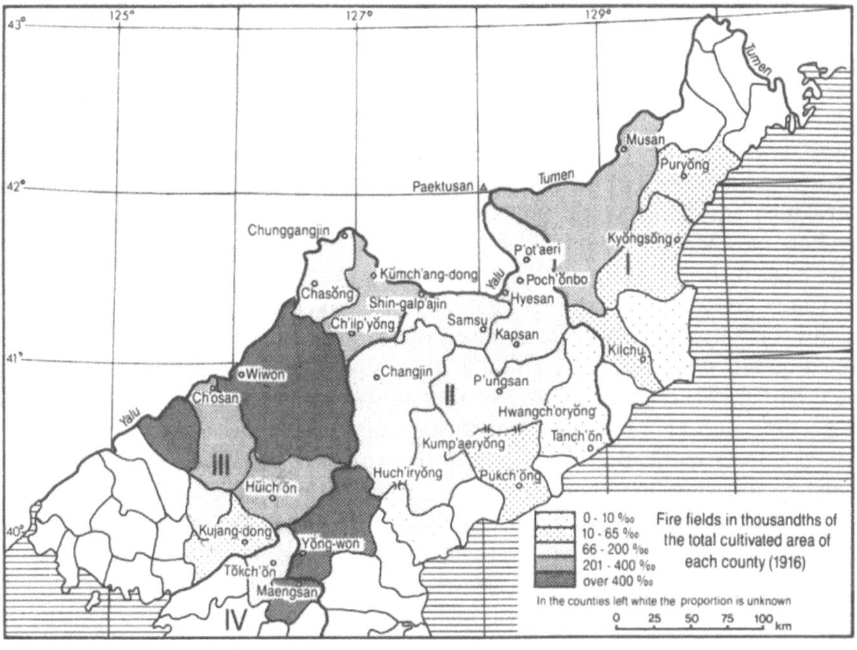

Fig. 46. Fire fields in northern Korea in per thousandths of the total cultivated area of each county. **I** Hamgyŏngbuk-do; **II** Hamgyŏngnam-do; **III** P'yŏng-anbuk-do; **IV** P'yŏng-annam-do

mentioned on p. 161, everywhere in Korea the fire-field farmers have for centuries been the social stratum that settled at the highest elevations, pushing the virgin forest back further and further. The old Korean maps show (p. 38f) that the favored locations with warm summers in the lower valley bottoms were settled at a relatively early date in the Kaema upland as well, perhaps also originally by fire-field farmers who have long since changed over to a sedentary form of existence and permanent farming. Today's fire-field farmers thus settle at high elevations here, too, and must endure a very harsh climate. This is even truer today than during the period of Korean independence, when Sieroszewski wrote [220, p. 76]: "It has become an indubitable truth that due to the sufferings that the people had to endure ... the general upper limit of cultivation ... has shifted much higher in rough, rocky Korea than in blessed Japan. The sluggish Korean, who is not able to overcome his own officials, has coped with the storms, the frost and the snow of his native mountains." A maximum settlement limit of 1550 m and a cultivation limit of 1850 m exploit every last opportunity afforded by the short summer, and even the slightest downward deviations from the average climatic data presented above can lead to complete crop failures.

In primary natural forest the geographically very important process of clearing the forest takes place as follows. The fire-field farmer recognizes the best soil on the basis of slight differences in the composition of the forest. Mixed or deciduous forests are preferred to coniferous forests. Nevertheless the uppermost fire fields extend far into the coniferous forest zone. In autumn, when the rains cease, the selected forest is burned. Prior to this they peel a chest-high strip of bark 60–150 cm wide off the medium-sized trunks or cut a deep notch into the really thick ones with an axe. When primary forest is fired, the trees never burn down to the roots. Stumps always remain standing, and the ground is covered with half charred remnants of branches. If rain extinguishes the fire prematurely, the black, branchless, bare stumps poke up as much as 10 m and more into the air. Nowhere is it customary to dig the stumps out. Naturally the only way to loosen the soil and mix it with the fertilizing wood ash is by means of hoes on these fields, from which the half-burned remnants of branches are not removed. The soil soon becomes depleted, since the fire-field farmers are too poor at the beginning of their operations to buy oxen and thus no manuring is done. Therefore, after 3–5 years the field is abandoned and becomes covered by second growths. Frequently the same fire-field farmer returns to the primary field, on which the stumps or tall dead trees are still standing, and clears the brush or secondary forest. Not until this process has been repeated several times does the field gradually become clearer. Meanwhile the farmer has perhaps been able to buy a yoke of oxen and then tills the soil with a hook plow. He plows around the stumps, which are still standing even then. The furrows are plowed horizontally to decrease erosion. Sieroszewski reports [220, p. 76]: "Some fields were so steep that even to till them by hand appeared to me the supreme expression of the most toilsome diligence; and yet I later met Koreans who worked with a yoke of oxen on such steep slopes that the hooves of the higher animal were above the back of the lower animal."

The fire fields are predominantly rectangular, almost square. Two sides of the rectangle run parallel to the contours, the other two follow the direction of the steepest gradient. Thus the rectangles are superbly fitted to the relief.

Everywhere in regions of advanced fire-field farming the solitary fields stand out on the shrub-covered slopes. In winter they are white against the gray background of the branches of the bushes, in spring brown against a bright green background, in summer and autumn the colors of the fields alternate, depending on the crop and the stage of development, between pale green, greenish-white (buckwheat blossoms), yellow, yellowish-brown and wine-red (autumn coloring of the leaves of one of the foxtail millet varieties).

The fields that were originally cleared one by one in the forests, or the second growths on them, come closer and closer together as the clearing proceeds, and the primary forest disappears completely from immeasurably large areas. Since fire-field farming generally progresses from the permanent fields at low elevations upward into the heads of the valleys and up the mountains, when hiking one observes these temporal changes in spatial succession going up the valleys or mountains. This kind of fire-field series consists of the following parts: (1) Valley bottom or basin with rice, maize, oat, millet, hemp, potato and soybean fields, in permanent cultivation, almost without fallow. (2) Gentle slopes with former fire fields that are now permanently cultivated with no shrub formations between them. The lavish use of wood from conifers in the houses and their palisade fences indicates an abundance of forests in the recent past. (3) Back slopes or upper parts of valleys with fire fields on which there are still low charred stumps with brush or secondary forest between them. (4) Plateaus or remote parts of valleys with fire fields in the secondary forest on which tall, charred stumps are standing and burnt branches are lying around. (5) Continuous secondary forests and partially destroyed remnants of primary forest with solitary primary fire-field clearings. (6) Untouched primeval forest.

The most important fire-field crops at lower elevations are foxtail and barnyard millet, barley, soybeans, kidney beans, hemp and sesame. With increasing elevation oats, potatoes and buckwheat become more and more prominent, and at the highest elevations they are the only crops. The yields decrease more and more with increasing elevation, and early and late frosts and wild boars are severe hazards. In zones 4 and 5 the fire-field farmer changes his homestead when he changes his fields, so that one often finds partially or completely ruined huts in them. These migratory fire-field farmers have often exhausted their meager harvests by spring and live all summer on berries, nuts, mushrooms, herbs and roots. Beekeeping is popular among all fire-field farmers and among the sedentary farmers as well.

As we have pointed out, fire-field farming has contributed greatly to the settlement of the Kaema upland. But up to the present day it has continued to push the virgin forest back and in the process has largely destroyed a national possession that is becoming more and more precious. Therefore, immediately after the annexation the Japanese administration took up the struggle against the fire-field farmers. But only since 1934 has this struggle been successful, at least to the extent that the clearing of fields by fire in primary forests has been practically stopped. The imprints that fire-field farming has made on the geographical countenance of the Kaema upland will never completely disappear.

The Japanese forestry commission has taken over the management of the remaining forests. In the natural forests wood is cut selectively and only as much as grows back. Conifers are cut preferentially, however, so that the composition of the forest is changing in this way as well. In winter the trees that have been felled are dragged by oxen to the logging roads or directly to the sawmills.

If the creek there is not navigable by raft, the trees, cut into logs approx. 7 m long, are transported downhill through canals made of boards and wickerwork until a navigable river is reached. The logs are fastened together into a wedge-shaped, flexible raft by means of oak switches pulled through holes hacked into the ends of the logs. Often small rafts are built at first, and they are converted into big ones when a larger river has been reached. On the banks of the Yalu and the Tumen one finds many such transshipment points, which are characterized by gigantic masses of wood, some of it swimming in the water. Next to them settlements have sprung up. The rafts are navigated by one or several courageous Japanese with great energy and skill, and this type of raft construction is Japanese in origin. The raftsman stands on the second section of the wedge and turns the first one from there by means of a steering beam. The raft is moved sideward by rowing with might and main. Hanging on the cliffs and sandbanks along the rivers there are not infrequent remnants of rafts, attesting to the danger of the trade. On small rivers rafting is done most intensively during the months in which they are swollen by melting snow or summer rain. Weirs hold back the water temporarily in the smallest streams so that they can be used. East of Maengbusan, for instance, wood is collected and cut at the forestry station of Sŏsang-ri in the valley of the Sŏdonggang. Each year rafting is done in April and May, and sometimes into June. Directly at the station the Sŏdonggang is dammed up, and here begins one of the wooden canals running parallel to the creek. In the just mentioned months 900–1000 logs daily are floated individually the 10 km to Changjin and made up into small rafts there. At some points Japanese logging companies have even built narrow-gauge railways to transport the logs.

The Kaema upland is not populated exclusively by fire-field farmers. The number of sedentary farmers living in the villages and hamlets in the valleys is much larger. Paddy rice cultivation is not completely nonexistent, but for climatic reasons, it is less significant here than anywhere else in Korea (Fig. 87). In Musan-gun paddies make up only 0.6% of the total cultivated area, in Kapsan-gun 2.4%, in Samsu-gun 1.7%, in P'ungsan-gun 0.1% and in Changjin-gun 0.3%. Rice cultivation reaches an elevation of almost 1000 m here, the highest in Korea. Since the growing season for the rice plants is only 5 months, the varieties grown here must mature very rapidly. These are, however, simultaneously the varieties with the lowest yield. To what extremes the opportunity to grow rice has been driven can be seen from the fact that the mean temperature of the 5 months from May to September is 14.8° C in Changjin, in the immediate vicinity of these highest rice paddies. In Asahigawa, in contrast, the point of northernmost rice cultivation on Hokkaidō, it is 16.1° C. Otherwise the crops on the permanent fields are generally the same as the ones mentioned above in connection with fire-field farming, just as there is in general a continuous gradation between these two types of agricultural landuse. For climatic reasons oats and potatoes are grown with more intensity than anywhere else in Korea (Fig. 88). Many of the crops, especially sesame, are preferentially planted in borders around fields planted to other crops (p. 184). Only rudimentary beginnings of intercropping have developed, but crop rotation from year to year is common. All crops are spring crops, because the climate makes winter crops completely impossible.

Many sedentary settlements have developed out of former fire-field farmsteads or hamlets into good-sized villages in the past decades. Usually they spread out along a path following the valley (Fig. 47). In regions that are prone to flooding the creek is fringed by remnants of the former bottomland forests (p. 199f). There are few meadows, for the oxen are fed mainly from the shrub formations. Pigs, which the fire-field farmers raise along with chickens, are fed with the refuse from the harvesting of field crops and the acorn harvest of the woods.

Since wood is more abundant in the Kaema upland than anywhere else in Korea, it has left its mark in all buildings, even in the larger towns. This wooden architecture was initiated by the fire-field farmers and is still used most by them today. Because even the larger villages developed out of fire-field settlements, it predominates in them as well and is retained even in new buildings, as long as there is still enough wood in the vicinity. The house of the "wild," migratory fire-field farmer is always a primitive log cabin made of notched logs with strongly protruding ends. Like all simple Korean houses it has no windows. The gently sloping gabled roof generally consists of shingles and is weighted down with stones in the Alpine manner. In the simplest cases the interior consists of a single room. The larger part of the floor is raised, and under this raised part, built of mud, run the flues of the floor heating system. They end at the gable side of the house in a chimney extending considerably beyond the roof, which is made out of a

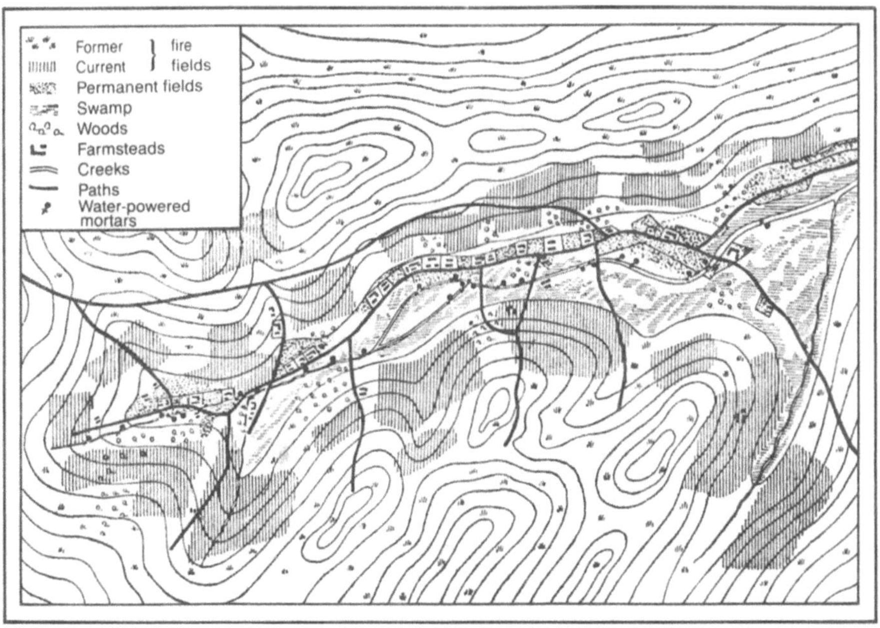

Fig. 47. Fire-field village Sŏhŭng-ri, Shinnam-myŏn, Kaema upland (after U. Oda). Scale approx. 1:25,000

hollow log and rises separately beside the house like the smokestack of a factory. Because of the cold winters the cracks between the logs of the cabin are filled with mud. All houses in the Kaema upland are derived from this primitive basic type. As the fire-field farmers become sedentary, and particularly as they switch over to permanent cultivation, the houses improve and increase in size, but the log construction is retained, as long as there are still forests in the vicinity. The house becomes longer and consists of several *kan*. In front of them runs the raised passageway that was brought up from the lowlands, overhung by the shingle roof borne by wooden pillars. The roof is often a hipped roof like the tile and thatched roofs in the lowlands. The house is joined by outbuildings, which gradually increase in number (shed, storeroom, pigsty, cow stall, doghouse, privy, well). They are also built roughly of logs (Fig. 48).

Both fire-field farmers and sedentary farmers finally have to give up constructing log cabins in areas that no longer have enough wood. The side walls are then made of wickerwork plastered over with mud, in the manner that is customary in the lowlands. The board roof is often covered with earth on which weeds proliferate. The chimney is quite commonly made of four boards nailed together at the corners.

The cultural geography of the Kaema upland is also characterized by other special traits that were probably originally items of fire-field culture but were later retained by the other cultural spheres or, like the shingle-roofed houses, have even gradually spread southeastward and westward into the lowlands. Transport is primarily by means of a two-runner sled pulled by oxen. It is also used in summer, when it scratches up the roads. The runners for this conveyance, which naturally wear out rapidly, are made at almost every farmstead. Thin, flexible trunks of suitable trees are soaked in water and given the desired shape by weighting them down with stones and allowing them to dry.

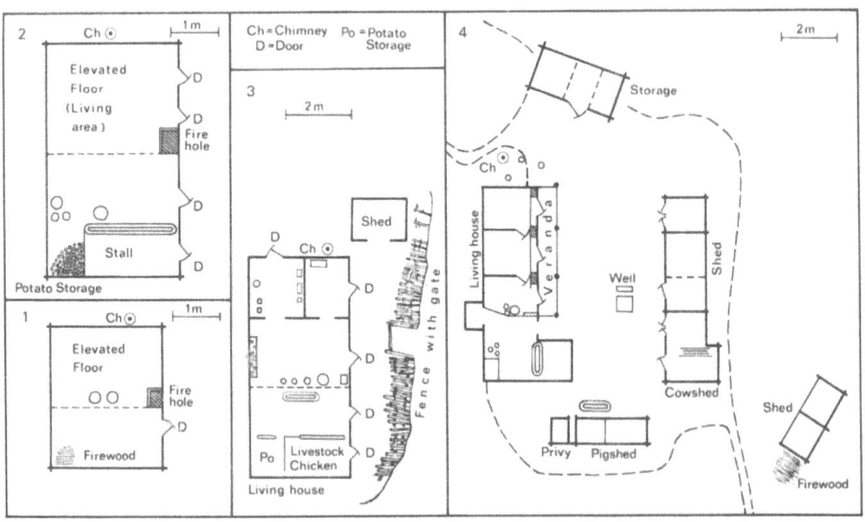

Fig. 48. Floor plans of log cabins of fire - field farmers in the Kaema upland (after U. Oda)

Another characteristic cultural element is the automatic grain mortar. It is a water-driven pestle (Kor. *mul pang-a*), which has been described frequently [203, p. 75, 558, p. 34]. A wooden beam approx. 5 m long is set up as an unequal lever. The short arm of the lever ends in a water box with a slanting back side so that it will empty completely and rapidly. The box is filled from a water pipe fastened above it, sinks, raising the longer counter arm in the process, and empties as soon as it has sunken low enough, causing the counter arm to come dashing down with great force. A wooden pestle attached to the end of the long arm transmits the blow to the grains lying in a receptacle embedded into the earth. The long arm of the lever, along with the grain receptacle, is usually protected against rain by a steep roof.

On the larger rivers in the upper stretches of valleys without rapids one also finds long, narrow dugouts. A last wooden landmark of the Kaema upland is finally the asymmetrical wooden bridges [203, p. 124]. They are supported by two rows of pillars. Because of ice drift the pillars on the upstream side are set so that they slant highly, at an angle of 30° from the horizontal at the most. The downstream pillars are nearly vertical and connected directly to the former by notches. The surface of the bridge lies above the upper part of the slanting pillars, brought into a horizontal position by wooden wedges shoved underneath it.

The farms of the Kaema upland, whether run by fire-field or sedentary farmers, are practically self-sufficient. Otherwise there are, apart from wood-processing industries, only a few copper and iron mines and the reservoirs of the power stations mentioned on p. 202. The population density is accordingly extremely low, considering harsh climate and the poor soil, but it is now increasing relatively rapidly. With 9.7 and 13 inhabitants/km^2, respectively, the counties of Musan-gun and Changjin-gun have the lowest population densities in Korea (Fig. 85). There are no cities of any caliber. Changjin, Samsu, Kapsan and P'ungsan have been county seats and market towns since the Korean dynastic period, and from that period some of them still have walls and gates today. Musan, Hyesan and Shin-galp'ajin, being border settlements, are simultaneously garrisons. Their character is thus fairly strongly influenced by the Japanese. Transportation to and from southeastern Manchuria via these towns has been insignicantly slight thus far, but should gradually increase, as soon as the danger from bandits has been remedied there. Within the Kaema upland transportation has not yet developed very highly, because of the primarily self-sufficient character of its economy.

Musan lies at the end of the railroad from Komusan on a basalt terrace 10–20 m above the Tumen. The town, with approx. 5000 inhabitants, forms a tightly closed quadrangle. Shin-galp'ajin, built on a grid, extends upward on a talus cone, surrounded to the north by a big bend of the Yalu. Southeast of the present-day town lie the remnants of the old Korean border fortress. The most important settlement today is Hyesan, which was only a tiny hamlet in 1890 [197, 203], but already had 11,795 inhabitants in 1936. It lies across from the Manchurian county seat of Changbai Xian in a basinlike wide spot in the upper Yalu valley that was eroded out of the thinly laminated gray shales and grayish-white tuffs of the Kyŏngsang system, just above the confluence of the Hŏch'ŏn-gang and the Yalu. Hyesan is an important junction point, and since 1937 it has been connected with Kilchu by rail via the pass of Namsŏllyŏng (1790 m) and Paegam

(p. 203). It nestles up against a small table mountain, whose surface, about 50 m high, is probably a remnant of an old terrace. In the center, which is built entirely in Japanese style, lie the wooden buildings of the constabulary and the police, the post office and some Japanese stores. Not far from the road to Kapsan rise the taller modern brick buildings of the baracks, the hospital and the head forestry station. The peripheral parts of town consist of Korean village-type houses.

3 The Paektusan Area

North of the Yalu and the Tumen the Korean–Manchurian border zone also consists primarily of Precambrian, intensely folded gneiss, phyllite, hornblende schist and quartzite, intruded by Kogurian granite, trending in the Laodong direction and overlain by slightly folded Paleozoic and Mesozoic sediments and by porphyritic flood basalts. This structural region northwest of the central and lower Yalu is the Taizi He region, referred to by the Japanese geologists as Taitzuhō (p. 62). The parts of Manchuria adjoining Korea's Hoeryŏng–Unggi area, i.e., the province of K'ientao, are also of the Taizu He type, with the difference that here and there northeast trending basalt flows conceal the basement. Orographically these two flanks of the Korean–Manchurian border zone form a hill country dissected into innumerable knolls by highly meandering rivers. None of them even remotely approaches an elevation of 2000 m. The Taizi He region is morphologically still a part of the main tilt block of Korea. Accordingly the height of its peaks increases toward the northeast, rising to 1600–1700 m at the level of Chunggangjin or the Manchurian town of Linjiang on the opposite bank. Similar heights are attained in the K'ientao region.

Lying between these two flanks the Paektusan area represents a completely different type [775, 776]. Although only its southeastern sector is still Korean territory, the entire area will be dealt with here, since in its physical geography it is an unusually uniform region, and the anthropogeographical contrasts that otherwise characterize the northern Korean boundary are quite insignificant, due to the almost complete lack of sedentary population. Only the southeastern sector has been well studied scientifically, though, and I, too, traveled only in it, so that the description will refer mainly to it. I have dealt with the Paektusan area in more detail elsewhere than is possible here [778].

Between the hill countries of Taizi He and K'ientao there is a tremendous sprawling basalt plateau with a nearly elliptical contour. Its larger axis trends west–east and is about 190 km long. The smaller north–south axis measures approx. 140 km. Thus the entire Paektusan area has an area of about 22,000 km^2. As we have already mentioned, it is bounded to the south by the big bend of the Yalu between Hyesan and Linjiang, while to the east it extends somewhat beyond the Sŏdusu. Beginning at the point where the Sŏdusu flows into the Tumen, the border in Manchuria follows more or less the line Samsang-dong – Antu – Liangjiangkou – Taipingchuan – Jingyu – Linjiang (Fig. 49). Roughly in the eastern focus of the ellipse lies the central mountain ring of the Paektusan volcano, which soars up to 2744 m, forming the highest point in Korea, and encircles the subsidence caldera with the lake of Ch'ŏnji (p. 7f).

From its foot, at an elevation of 1800–1900 m, the basaltic plateau slopes down periclinally in all directions. Everywhere in the east and south its outer edge lies much lower than the peaks of the Kaema upland and the P'yŏngbuk region. The Paektusan area forms the lowest part of the northward sloping Kaema tilt block. It is a depression in which the old rocks have been completely covered by enormous masses of basalt 50–200 m thick (Fig. 50). The bulk of the basalt flows probably occurred during the Early Pleistocene epoch. Since then the depression has risen again, so that the Yalu and the Tumen and their tributaries have sawed through the basalt flows and entrenched themselves deep in the old basement. Former uplifts of the latter are revealed here and there along the valleys even within the basaltic plateau in the form of windows exposed by erosion. All along these valleys the basaltic plateau breaks off with a sharply defined drop-off exposing magnificent basalt columns. Southeast of Linjiang the Yalu has separated a tongue from the main flood basalt, and it forms the plateau of Osudŏksan (800 m, Fig. 51) on the Korean side. The highest parts of the plateau in the area surrounding the central mountain ring are covered by a layer of pumice on top of the basalt.

The columnar structure of the basalt makes it permeable. Large portions of the plateau are therefore without surface runoff. Other parts are, in contrast, completely impermeable, perhaps because the columnar structure is lacking in the uppermost layer of basalt, or because the basalt has already weathered to clay on the surface. Because the inclination of the basalt surface is so gentle, water stagnates here, forming extensive swamps, which are the breeding grounds of billions of mosquitoes in summer. Here and there the plateau has small closed basins, which probably formed directly during the solidification of the lava. In impermeable

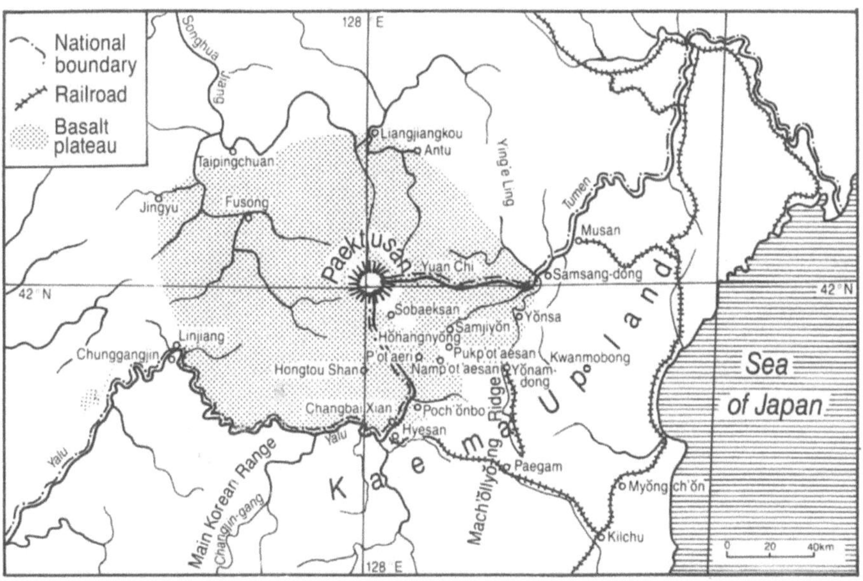

Fig. 49. Generalized map of the Paektusan area

areas they contain lakes. The largest of these in the Korea sector is Samjiyŏn. They also include Yuan Chi (Manchur. Bōlhuri), which lies just north of the upper reaches of the Tumen along the political boundary and, according to Manchurian traditions, is the place of origin of the Manchu dynasty.

The valley system of the basalt plateau has a radial structure. The most important of these radial furrows are the valleys of the upper reaches of the Yalu, Tumen and Sungari (Songhua Jiang). The latter flows out of Ch'ŏnji and has forced its way northward through the mountain ring with a deeply entrenched barranco. These radial streamlets also include the right-bank tributaries of the big bend of the Yalu, which the Chinese name by numbering them from Linjiang upstream. The last one is the "Twenty-third River" (Chin. Ershisandaogou). In the Korean sector, particularly in the vicinity of the central mountain ring, most of these valleys are dry valleys, in which one can only expect to find flowing water during the summer rainy season. Judging from James' description [190, p. 255, 260], the creeks in the northwestern sector have a greater and more continuous flow.

Here and there the regular slope of the plateau is interrupted by eminences that represent local eruption centers. North and west of the central mountain ring they are all low. But in the southwest one of these old volcanic skeletons, Hongtou

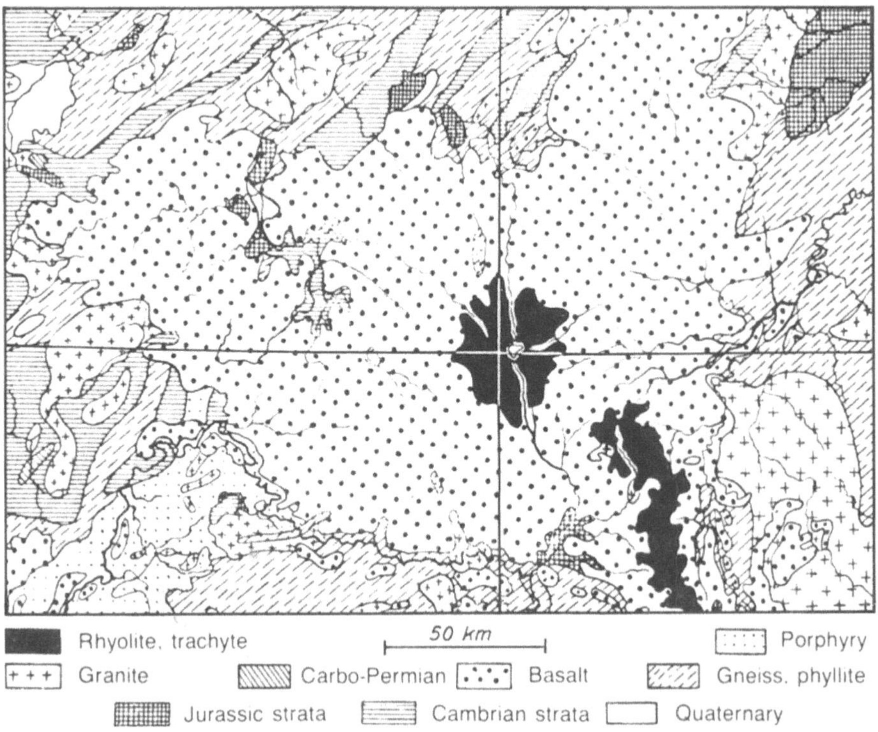

Fig. 50. Geological map of the Paektusan area (after Geological Atlas of Eastern Asia)

Shan, rises to a lonesome height of approx. 1900 m. An almost uninterrupted chain of elevations runs southeastward. It begins in the immediate vicinity of the central mountain ring with Taeyŏnjibong (2360 m) and the mountain group of Sobaeksan ("Little White Mountain", 2174 m), which are still composed of the same volcanic rocks as the central mountain ring. There then follows a shallow saddle, Hŏhangnyŏng (1401 m), beyond which the Mach'ŏllŏng Ridge, described on p. 219, begins with the mighty P'ot'aesan peaks.

It is difficult to evaluate the climatic conditions in the Paektusan area, since I have no other data than the means of the stations in the depths of the Yalu valley, Hyesan and Chunggangjin (p. 221), the observations of a Japanese winter expedition in 1934/35 [776] and my own notes on the weather. According to this material, the climate of the Paektusan area appears to be very similar to that of the Kaema upland. Temperatures are evidently very low in winter. The winter expedition measured temperatures as low as −41.0° C on the basaltic plateau. Nevertheless, here too there appears to be a temperature inversion in winter in relation to the depths of the Yalu valley. The winter northwest monsoon evidently blows much more constantly on the unprotected plateau than in

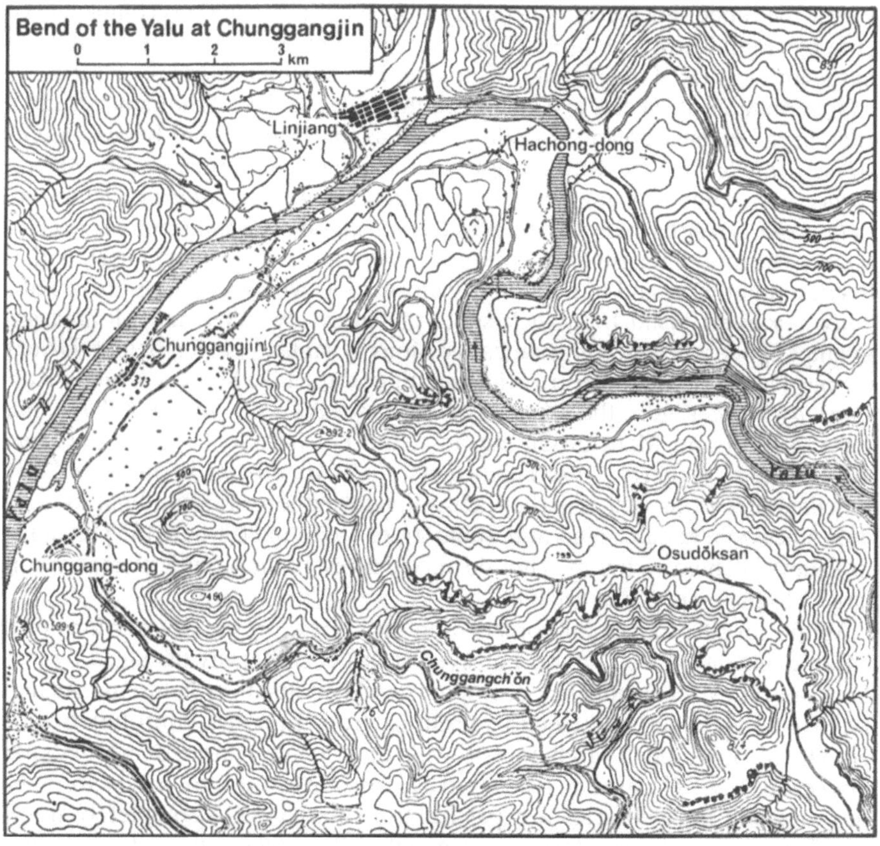

Fig. 51. Bend of the Yalu at Chunggangjin

Chunggangjin (p. 222). According to the observations of the winter expedition, the snow cover in the southeastern sector near the mountain ring is rather light. On its southeastern outer side it is often completely blown off the ridges, so that there are intermittent dark stripes in the white blanket. On the windward side there is presumably more snow. Kawasaki [768] reports that the last snow in the mountain ring generally does not melt until the second third of August, and new snow begins in the first third of October. The various European travelers found Ch'ŏnji frozen over from the end of October until May, and even in July there is ice floating on it. Midsummer, in contrast, most likely brings quite considerable temperatures to the basaltic plateau. On 30 August (1933) I measured 21.5° C still at 4 P.M. in P'ot'aeri (1100 m), in a valley entrenched into the basaltic plateau. Summer is of course the rainy season here too. S. Bergman [229, p. 78] had problems with tremendous amounts of rainfall on the plateau at the end of July 1935. The central massif is sometimes hidden by clouds for weeks on end.

All travelers who describe the view from the rim of the crater write of the deep impression that the tremendous expanse of forests on the basaltic plateau made on them. The upper timberline lies at 1900–1950 m on the southeastern side of the central massif, in other words, lower than in the Kaema upland, which can probably be attributed to the permeability and dryness of the layer of pumice. Below the timberline the whole country in its natural condition is completely covered by virgin forests. Still today they grow most luxuriantly in the moister spots in the lower parts of the plateau. Here one finds summergreen mixed forest with a large number of species in a composition similar to that of the same altitudinal zone in the Kaema upland. Conifers, such as the Daurian and Korean larches, *Abies nephrolepis*, the Korean nut-pine, the flat-leaf spruce and the Korean spruce, are joined by maples (*Acer tegmentosum*, *Acer ukurunduense*, *Acer tschonoskii*, *Acer barbinerve*), birches (*Betula japonica*, *Betula costata*, *Betula ermanii*), the mountain ash *Sorbus amurensis* Koehne, which adorns itself with red berries in autumn, and in particularly moist spots willows (*Salix sibirica* Pallas, *Salix starkeana* Willd.). In such spots the forest dwindles and Molinia grass swamps appear, in which my ponies occasionally sank up to their bellies. Drier locations have a dense undergrowth consisting of the lilac *Syringa velutina* Kom., *Rosa acicularis* Lindl., the raspberry *Rubus idaeus* L. var. *microphyllus* Turcz., the Ussurian and Manchurian currants, the honysuckle *Lonicera chrysantha* Turcz., the bird-cherry *Prunus padus*, etc. Oaks are missing in this mixed forest for climatic reasons, just as in the Kaema upland in the same altitudinal zone. The only places where oaks, elms and walnut trees are found is where the plateau and its valleys fall below 1100 m. Thus the mixed forest of the basaltic plateau belongs largely to the maple zone (p. 138).

Virgin forest of a similar composition once grew on the dry portions of the basaltic plateau as well. Here, however, it has been heavily damaged by forest fires caused by the negligence of trappers and bandits or by lightning. As in the Kaema upland, natural selection resulting from fires has contributed to the growth of pure, open larch forests with only a few scattered white-barked birches (p. 139). The ground is covered by a layer of low shrubs, in which cranberries and bog huckleberries are particularly common. On completely treeless clearings masses of grass, ferns, the cinquefoil *Potentilla fruticosa* L. and our native fireweed thrive.

At an elevation of about 1500 m most of the deciduous trees have disappeared from these natural forests and above this predominantly pure coniferous forests begin. For edaphic reasons *Betula ermanii* is lacking, while in the Kaema upland it occurs up to the timberline. These forests consist almost exclusively of the above mentioned conifers, with the exception of the Korean nut-pine, which requires a milder climate. This mixed coniferous forest extends up to roughly 1700 m. It is highly impassable, and this is increased by the intense growth of the lichen *Usnea longosissima* fostered by the high relative humidity even in winter.

The upper forest zone, finally, consists of pure, open larch forests, which extend up to the timberline, in contrast to the Kaema upland. Thus for edaphic reasons the Paektusan area lacks both the birch forest zone and the dwarf pine zone. These larch forests are of natural origin. The wide spaces between the trees are covered by a thick, undulating carpet of low plants. Reindeer moss imparts a snow-white color to large areas. In between bright green mosses, cranberries and huckleberries, the dark-green leaved *Rhododendron chrysanthum*, dwarf juniper and the crowberry proliferate. This carpet is particularly colorful in autumn, when the leaves of some of these low shrubs turn pale green, dark red, pink and purple. In the higher regions the Paektusan forests are interspersed with meadows here and there that glow in early summer with a luxuriance of blossoms enthusiastically extolled by F. Younghusband, the famous portrayer of Himalayan vegetation.

The fauna is adapted to the primeval nature of the basaltic plateau. Physiognomically it does not stand out very much, although it is not as scarce as it has been in the Kaema upland for several decades. Large mammals are rare, and the small mammals and birds are primarily species that live in the treetops or on the ground and are easily hidden. Of the former I saw vigorous red and roe deer, gigantic wild boars and a blackish-brown bear. Actual predators, particularly the Manchurian tiger and the leopard, are rare today. Only the lynx is frequently seen, as are packs of wolves. Occasionally one hears the pecking of the great black woodpecker reverberating through the quiet forests. The black grouse feeds on young larch shoots and small larch branches, the various tits on the seeds of the conifers. There is a small insect eating bird of prey with gorgeous bluish-green plumage called *Eurystomus orientalis calonyx* [229], which is rare in the rest of Korea. The owls live on mice, the eagles on wild rabbits and other ground-living animals. Other animals that live on the ground are the Siberian partridge (*Perdix barbata*), the hazel grouse and the black woodcock. The various magpie species tend today to remain in swarms in the proximity of settlements and thus at the edge of the basaltic plateau, because they are fond of grain, but also penetrate into the forests. The magpie is the sacred bird of the Manchu. In the legend about the origin of their imperial family it twice plays a decisive role [774].

When the state of Pohai flourished for several centuries during the second half of the first millennium A.D. (p. 28), the basaltic plateau of Paektusan and its valley system were probably more densely populated than they are today. This can be conjectured from Matveeff's description [77]. That a denser population was and is possible is demonstrated by the fact that in the Kaema upland the fire fields extend up to 1850 m. The report of an expedition by Umuna, an envoy of the Chinese Kangxi emperor, in 1677 shows that the forests on the northern side were completely deserted at that time [129]. The first Europeans to climb Paektusan, however, the Englishmen James, Younghusband and Fulford [190], encountered Chinese farmers, ginseng hunters, sable hunters, woodcutters and gold hunters there in 1886, who lived highly dispersed in log cabins and had joined together in guilds. This settlement occurred despite the ban on the part of the Chinese government, who wished to maintain the area around the sacred mountain of the Manchu dynasty as uninhabited territory and imperial hunting preserve. As the travel accounts by Campbell [197] and Cavendish and Goold–Adams [203] show, on the south side settlements did not extend beyond the geographical latitude of Hyesan at that time and had thus not yet crossed the Yalu. In the next two decades, however, i.e., between 1890 and 1910, good-sized Korean hamlets and villages pushed northward into the valleys and even began to climb up onto the plateau. In 1911 an uppermost hamlet by the name of <Tenshudō> (for the significance of angle brackets cf. note in foreword) was founded only 24 km south of the crater rim [768]. As in the Kaema upland, this expansion of Korean settled area is the result of fire-field farming. In the valleys the settlements that have come into being in this manner extend up to 40 km north of Hyesan today. Fire-field farming has led here as well to the familiar destruction of natural forests and the emergence of diverse second growths on the slopes of the valleys and the peripheral parts of the plateau, and here the same fire-field series has also developed (p. 229). Here, too, the fire-field villages in the valley bottoms have become permanent settlements, and the same crops are grown. In recent times one also finds the same lumbering and rafting supervised by the head forestry stations in Musan and Hyesan [770, 771].

In the past 2 decades, most particularly since the death of the warlord of Manchuria, the Chinese general Zhang Zuolin (1928), the population of the plateau has decreased greatly, because large numbers of the former soldiers of Zhang Zuolin have become bandits. Banditry, which has become a severe political problem in all of Manchuria, has reached its prime in the forests of Paektusan and has encroached into the Korean sector. The hamlets inhabited by Chinese and by Koreans are raided, the harvest robbed; not infrequently the houses are set on fire and the inhabitants killed. The bandits' log cabins, roofed with moss or sod, are scattered throughout the forests. On Sobaeksan there are supposedly whole "villages" of them. Because of the danger from bandits the fire-field farmers have left the plateau and retreated to the larger settlements in the valleys. In the Korean sector these are protected by Japanese police stations, which have been converted into small fortresses. On the Manchurian side the protection is less complete thus far, and the Chinese farmers there suffer severely. It will be possible to get rid of the bandits only when the plateau has been opened up to traffic. So far the natural forests are crisscrossed by narrow hunting paths only, and the road from Hyesan to Musan via Hŏhangnyŏng, which has been planned for a long time and which does not run into any basic difficulties with the relief, has not been built yet.

At an elevation of 1800–1900 m the plateau character ceases. Around the central mountain ring the relief begins to become bolder. Dome-shaped elevations composed of solidified pumiceous sands or ash begin to appear (Fig. 52). Tremendous volcanic bombs are often scattered around on them. Additionally, on their slopes striped ground has developed as distinctly as if this were a model. The slopes are dissected by erosion rills and partially leveled by denudation. In areas where such recent denudation has occurred the pumice is gleaming white. The white color of the snow in winter and the pumice in summer led to the name of Paektusan (= White Head Mountain). Wherever the slopes are covered by sparse vegetation, the white color immediately stops, giving way to a light brown and dull green. The vegetation is lusher in the steep, but not deeply dissected ravines, in which grows an Alpine plant association with variegated blossoms in early summer: Alpine burnet (*Sanguisorba alpina* Bunge), the stonecrop *Sedum telephium*, the campion *Silene repens* Patr., the pinks *Dianthus morii* Nakai and *Dianthus sinensis* L., the hawkweed *Hieracium umbellatum* L. and the daisy *Chrysanthemum sibiricum* Fisch. var. *alpinum* Nakai.

The periphery of the crater rim has a maximum diameter of 4.5 km, and the diameter of the central massif including the outer slopes is at least twice that much (Fig. 53). The 900 m by which it exceeds the plateau are thus insignificant compared with its breadth. The view from outside does not reveal anything of the subsidence caldera in its center, and one thus has the impression of a broad, rounded massif with slopes dissected by small ravines and several separate summits on top. These slopes are composed primarily of acidic lava. Those on the southeastern side are almost completely barren. Only in the hollows and ravines are there still grasses, cushions of *Rhododendron chrysanthum*, blooming golden yellow in early summer, *Rhodiola augusta*, which is red in autumn, and the above mentioned stonecrop.

The rim of the crater has extremely varied elevations. The differences are partially due to variations in the thickness and bedding of the lava from a more recent period of major activity and partially to the ravines forming in the outer slopes. The lowest point (around 2450 m) is in the southwest. In addition to the highest peak, Taejŏngbong, a number of other peaks also exceed the 2700 m limit.

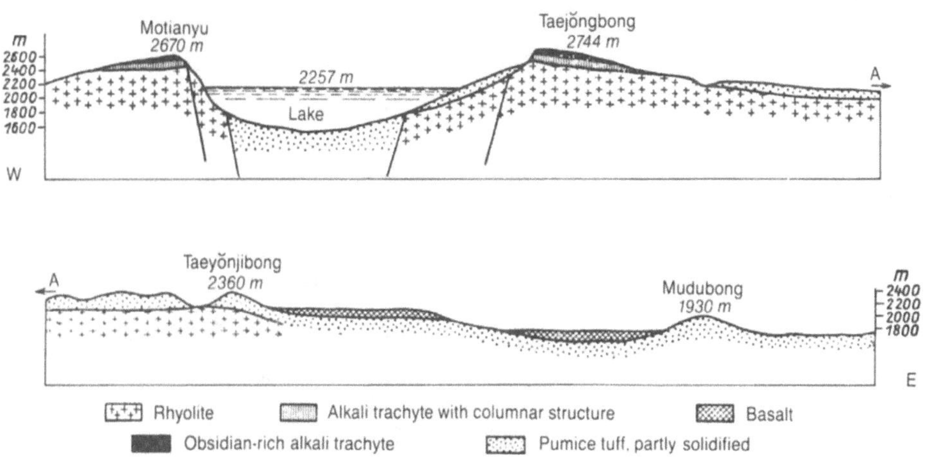

Fig. 52. Geological profile through Paektusan (after Nihon Chiri Chizu). The lower half of the profile forms the right–hand continuation of the upper one.

The extremely steep, up to 500 m high inner walls of the caldera clearly show the structure of the central mountain ring. Rocks from an earlier period of major activity can be distinguished from those of a more recent one. The former are rhyolites high in quartz that take on a bright coloring as a result of weathering. Their former surface was altered during a period of erosion between the two periods of activity. Above the erosion surface more recent lavas follow; of these the layers near the peaks are rich in obsidian and very hard. In some places the latter are capped, finally, by the solidified pumiceous sands. When these different periods of activity occurred is uncertain. According to Chinese records, the volcano reportedly erupted for the last time in 1702 (p. 75), but the center of the eruption is unknown, and Umuna described the caldera in 1677 exactly the way it looks today [129].

There can be no doubt that the wide basin in the center of the mountain ring is a subsidence caldera. The "Heavenly Lake" (Kor. Ch'ŏnji, Chin. Tianchi) filling it lies at 2257 m. Its depth is unknown to date, but it must exceed 442 m. It is definitely not unlikely that Ch'ŏnji will join the nine lakes on earth known to exceed a depth of 500 m as number ten. The hot sulfur-iron acidic springs lined up along the northern lakeshore like a string of beads probably rise in the fissures that developed during subsidence. From Taejŏngbong a jagged ridge juts radially into the caldera, crowned by a serrated rock that is fantastically steep. Probably the lava of which it is composed represents a dike. It projects far out into the lake, interrupting the circular form. Evidently the rock was so highly reinforced by this dike that it was unaffected by the subsidence.

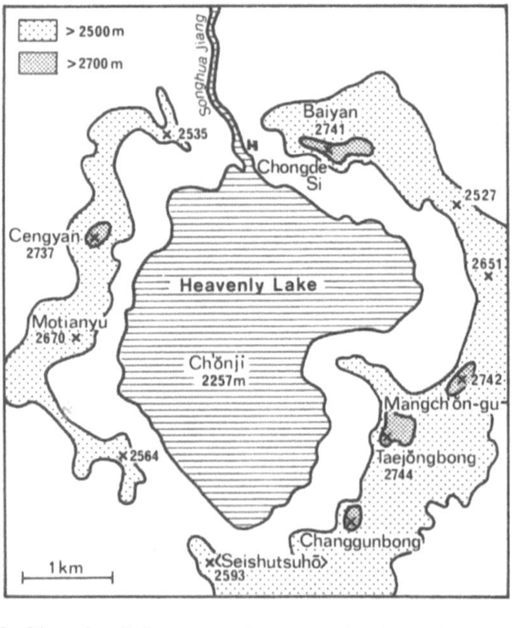

Fig. 53. Sketch of the central mountain ring of Paektusan

Beneath the vertical precipices of the hard more recent lava tremendous talus slopes extend down to the water level. Most of them are inactive today. Between the boulders low vegetation forms a continuous cover. Ash-gray lichens and light-green mosses are its most important elements. Occasionally this cover is interrupted by large patches of *Rhododendron redowskianum*, which is 5 cm high at the most. In between are very low specimens of the bog huckleberry and stonecrop. This dwarf association is similar to that on the peak of Kwanmobong (p. 224), but with much fewer species. Down the old, inactive talus slopes run active ravines lined by fresh, multiply branching levees. The forms remind one highly of the gullies formed by subpolar mud flows. The old talus slopes are most highly developed in the armchair-shaped niches that interrupt the even curvature of the caldera walls at three points. As we noted on p. 77, T. Kano [391] interpreted them as cirques, which appears doubtful to me [778].

On a 20-m-high terrace to the right of the outlet of the Songhua Jiang out of the lake a small Buddhist temple named Chongde Si has been standing since 1906. It was established during the last years of the Manchu dynasty, on what was Chinese soil at the time, in a region that is sacred according to the tradition of the neighboring peoples. No monks live in it, however, and only from time to time do solitary Chinese Buddhists or Taoists, and occasionally Koreans, make the arduous pilgrimage up to the Heavenly Lake.

Although Paektusan lies at the latitude of Rome, many elements of the landscape show a close relationship to much more northerly regions, because of the high elevation and the cold winters of its monsoon climate. It forms a transitional region between the Kaema upland and northern Manchuria, indeed even Transbaikalia, and together with the former it contrasts quite highly with the rest of Korea. Overlooking the vast expanses around the central mountain ring in the region where birches form the timberline, one no longer has the impression that he is in central East Asia. And many of the physiognomically important plants mentioned above have a range that extends far northward and they themselves, or very close relatives, occur in northern Europe; as relics of the Ice Ages they even exist in the higher of our subalpine mountains. Within Korea these plant associations form an alien high altitude zone that is otherwise found only in a much more restricted range on the highest peaks in the Kaema upland. Finally, striped ground and mud-flow levees form a link to regions closer to the pole.

The observer standing on the peaks of the mountain ring of Paektusan sees a view of monumental grandeur, whether he looks outward or inward. Gazing outward, he scans over the slopes with their white patches and downward to the sheer unending plateau with its immense forests. Gazing inward his eye looks down 500 m over steep precipices to the broad surface of the lake, which appears to be motionless, even when storms are raging overhead. In good weather it is a radiant dark blue, and the forms and colors of the caldera walls are reflected in perfect clarity. The red of the lower lavas, the gray and black of the higher ones, the gleaming yellowish white of the pumiceous sand appear double their actual size in the reflection and are all the more impressive. All observers agree that the contrasts that this twofold view unites make the scene viewed from Paektusan one of the most enthralling sights on earth.

4 The Yalu and the P'yŏngbuk Region

The extreme northwestern part of Korea, occupied by the province of P'yŏnganbuk-do, will be called the P'yŏngbuk Region after this province. This region does in fact coincide for the most part with the province and has, like it, an area of approx. 28,000 km^2. In the extreme eastern part, however, we have included the entire Nangnim Range in the Kaema upland, whereas the provincial boundary follows its crest. In the south where the provincial boundary follows the lower Ch'ŏngch'ŏn-gang, I have the region extend up to the divide between it and the Taedonggang, which runs southeastward in the extension of the Myohyang Ridge parallel to the Ch'ŏngch'ŏn-gang at a distance of 10–20 km. Thus I avoid dividing the lower Ch'ŏngch'ŏn-gang valley, which forms a unit from the standpoint of economics and transportation. The P'yŏngbuk region is then the northwestern Korean frontier zone between the Nangnim–Myohyang chain and the Yalu. From southwest–northeast it is 310 km long. The greatest width, not far from the Yellow Sea coast, is 125 km. Toward the northeast it narrows down to 50 km (Fig. 39).

The P'yŏngbuk region, the Kaema upland and the Paektusan area are linked by Korea's most important river, the Yalu [779]. It encircles the P'yŏngbuk region on two sides. Almost its entire course is characterized by intense meandering. Because of this and because of the giant bend along the southern edge of the Paektusan plateau, the total length is drawn out to 813 km, although the distance from the source to the mouth is only 400 km as the crow flies. With a clarity befitting a model, the amplitude of the meanders increases downstream with the discharge. The broadest one is the last one, that at Ch'angsŏng, 80 km above the mouth. Many of the tributaries flowing out of the P'yŏngbuk region, such as the Chasŏnggang, the Tongnogang, the Ch'ungman-gang and the Sapkyoch'ŏn, also meander highly, and it is always very impressive to see the small diameter of the meanders of the tributaries in the immediate vicinity of the giant meanders of the main river (Fig. 54). Meander necks can be observed in all stages of development, and cutoff spurs are common. The nicest example is the region around Wiwon (Fig. 55). The meander there has developed in almost vertical, highly metamorphic schists and encircles a meander core with a width of 2.5 km, a length of 4.5 km and a relative height of about 300 m, while the meander neck is only 250 m wide. Only a little bit farther downstream lies the magnificent cutoff spur of P. 326. At the time when the lower terrace was formed, the river flowed in a meander 3.5 km long around a vertical dike of porphyry, shaping it into a wall. Later the cutoff occurred in the middle of the wall at a narrow spot in the dike. Now the river squeezes its way through a vertical chasm only 150 m wide and has rapids that are a danger to navigation.

Rocky islands in the river are restricted to the upper stretch, as far downstream as Yŏnp'o-dong. In contrast, bars created by aggradation begin just below Shin-galp'ajin and continue up to the mouth. The grain size decreases in this direction from mighty imbricated boulders to gravel, sand and finally clay. There are two kinds of bars. Along straight stretches they are shaped like a narrow wedge. The tip pointing upstream emerges gently out of the water; the broad downstream side forms a dropoff as much as 20 m (!) high. In stretches with meanders the very many bars are crescent-shaped. They lie between the outer

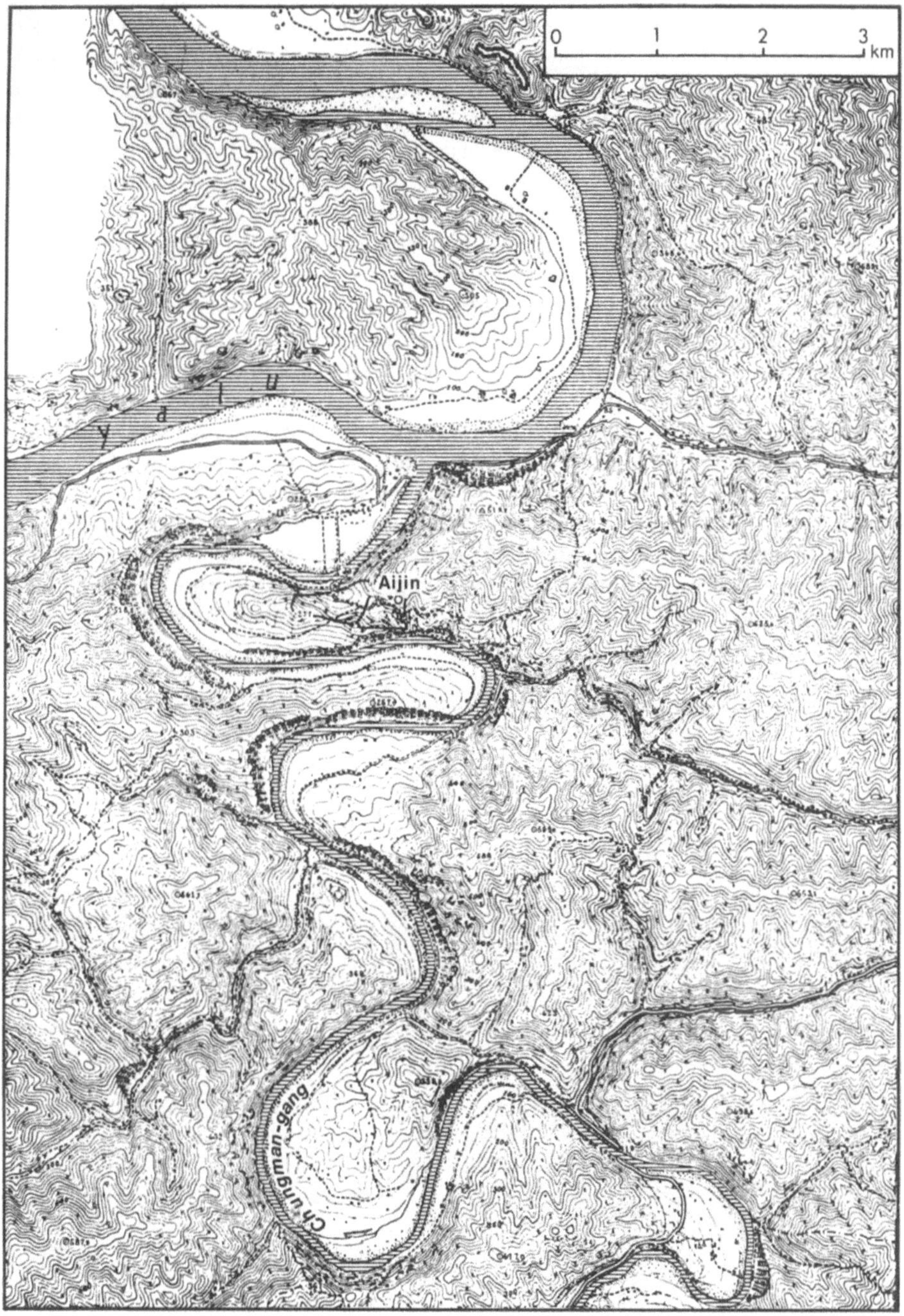

Fig. 54. Meanders of the Yalu and its tributary, the Ch'ungman-gang (section of the topographical map 1:50,000)

curve of the river, which follows the slipoff slope, and a secondary inner channel line following the chord. This kind is also very common in other meandering Korean rivers. On the stretch downstream from Chunggangjin low terraces begin to appear, and from the meander at Ch'angsŏng downstream they continue without interruption. The large bottomlands and islands well covered with fields of kaoliang and other crops along the last 65-km-long stretch of the river consist of such terraces. Above them are rocky terraces, some covered with gravel, at various heights above the river. The most prevalent is the 100-m level. A systematic investigation of the terraces will probably lead to very important findings on the geological history of the Yalu valley.

From a morphological viewpoint the course of the river is distinctly divided into three segments. From its source to Hyesan the Yalu flows through one of the radial furrows in the volcanic area of Paektusan (p. 236). From there to Linjiang–Chunggangjin it flows around the periphery of the basaltic plateau and simultaneously follows the much higher northern edge of the Kaema upland and the P'yŏngbuk region (Fig. 49). From Linjiang to its mouth it is one of the rivers flowing southwest on the dip slope of the main Korean tilt block. Along this segment, too, it is deeply entrenched below the level of the peaks in the Korean–Manchurian border zone. On the Korean side they still rise to a height of more than 1000 m (Ch'ŏnmasan = Heaven Touching Mountain 1169 m) only 60 m from the coast. Thus the Yalu is a mountain stream all the way to its mouth.

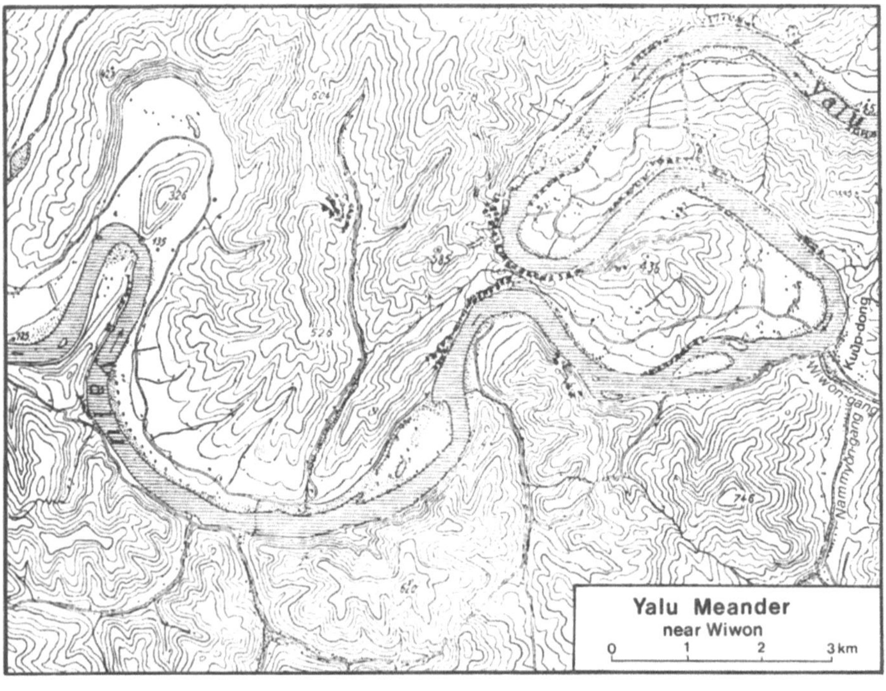

Fig. 55. Meanders of the Yalu near Wiwon and to the west (simplified section of the topographical map 1:50,000)

Naturally the height of the peaks decreases toward the channel due to erosion by the main river and its tributaries. Thus one cannot, as Kotō does [352], see this fact as proof of a tectonic inflection of the dip slope that determined its course downstream from Chunggangjin.

Along the two upper segments there are still immense natural forests today, and they often extend down to the water level (p. 254). Lovely coniferous and deciduous forests follow the banks of the river in some places even in the southwestern stretch, and this is more frequently the case on the Korean side than on the Manchurian side, where the Chinese have carried the destruction of the forests much farther than the Koreans on their side. The rectangular fields of contemporary or former fire-field farmers extend on both sides of the river as far as the region below Pyŏktong and can be observed from the river on the slopes up to an elevation of 900 m and more.

The abundant forests along its upper reaches have made the Yalu Korea's foremost river for rafting. The conflict over the right to exploit these resources was one of the superficial reasons for the Russo–Japanese War (p. 35). In addition to the Japanese-style rafts manned by Japanese, which often bear the banner of the rising sun, there are large numbers of Chinese rafts. These are made on the Manchurian side and anchor only on it at night during the trip. The Chinese rafts are much more cumbersome than the Japanese ones; they consist of an inflexible succession of sections, rectangular in shape, with a log cabin on them for the family of the raftsman. The trip from Hyesan to the mouth takes 3 weeks. There the Japanese rafts are taken apart in Shinŭiju, and the Chinese rafts in Dandong on the opposite bank.

Thus in rafting the boundary character of the river is already apparent, and it is expressed in many other anthropogeographical spheres as well. The fact that for its entire length the Yalu forms the boundary between two countries that had a continuous tendency toward complete isolation until the Japanese occupation of Manchuria in 1931 is particularly underlined by the fact that until 1938 it was not crossed by a single bridge on the entire 700 km of the main stretch from Hyesan to Shinŭiju. Since then it is crossed at Manp'o by the railroad Shinanju – Kanggye – Tonghua, which is now under construction (p. 258). The villages and hamlets on the Korean side have the Korean house types described above for the Kaema upland. On the Manchurian side, however, there live not only Koreans but many Chinese. The Chinese houses represent a different style; their houses have gabled roofs and windows and are never half-hipped. The roof is steeper than the shingle roof of the Koreans and is thatched with straw that is carefully clipped at the edges.

The nature of the winding valley makes it difficult to use it for overland transportation, and the boundary character has only increased the difficulty. Nowhere on the Manchurian side does an improved road follow the river to date. On the Korean side, on the whole, only short stretches of the valley are used for roads connecting the settlements along the river to the traffic network. Such settlements linking the routes to the valley are in particular Hyesan, Shin-galp'ajin, Kŭmch'ang-dong, Chunggangjin, Manp'o, Wiwon, Ch'osan, Pyŏktong, Ch'angsŏng and Sakchu. Only on the last 40 km above Shinŭiju does the road continuously follow the broad, straight valley. The just mentioned settlements are connected by roads running parallel to the river, to be sure, but these run high up through the back country across passes rising as high as 1198 m between Hyesan and Shin-galp'ajin, more than 800 m (Samjŏnghyŏn 825 m) between Chunggangjin and Chasŏng and still way above 700 m (T'allyŏng = Coal Pass 797 m) between Ch'osan and Pyŏktong.

By far the majority of these Korean settlements developed out of former border fortresses (e.g., Chunggangjin = Middle River Fortress), and some still have the old stone walls and covered gates. They also contain a number of other buildings from the Korean dynastic period, such as lovely pavilions in scenic locations above the river (e.g., Segŏmjŏng near Manp'o). The Japanese have added barracks and hospitals, the American missionaries churches and meeting halls. Many of these border towns do not lie directly along the Yalu, but in the mountains in its vicinity for better protection. None of the traditional Korean counties bordering on the Yalu, with the exception of Ŭiju-gun, has its seat directly on the banks of the river. The larger of the Manchurian towns in the Yalu valley still have well-preserved stone walls in rectangular form with whitewashed fortified towers.

Near Shin-galp'jin the discharge of the Yalu is so highly increased by the Changjin-gang that navigation begins. It is always very much endangered, however, by the rocky shoals and gravel and sandbars we have described, even in the lower reaches. Additionally, the 150 km long stretch between Shin-galp'ajin and Chunggangjin is very rushing, because in it the river drops from 560 to 320 m, and for several months navigation is stopped by freezeup (p. 120). In their shipping too the nations differ. Chinese junks can be distinguished from afar by the horizontal bamboo poles reinforcing the cotton sails, whereas on Korean boats the sails are held by vertical ropes.

These vessels are moved upstream by laborious towing. The towlines are secured to the mast or to a special support in the bow. One to six men tow the vessels upstream on the banks of the slipoff slope, where the current is not as strong. Thus the towpath switches from one side of the river to the other and is very primitive, despite the brisk traffic, so that the crews quite often have to climb over rocks while towing. That these difficulties have not been overcome even today, although many thousands use the towpath annually is likewise a consequence of the boundary character. In the inflection points of the meanders the ships change sides by rowing and with sails set, losing a lot of headway in the process. In places where the current is particularly rushing, such as at the rapids known as Namban (= Egg Rock) upstream from Yŏnp'o-dong, the junks gather, and the towing is done in teamwork.

For the last decade the Japanese have used so-called propeller boats on the Yalu. These are boats with a shallow draft approx. 2 m wide and 16 m long at the maximum, which are driven by an aircraft propeller instead of a ship's propeller. They too move upstream along the foot of the slipoff slope. The upstream passage from Shinŭiju to Shin-galp'ajin takes 4 days, the downstream passage 2. They land only on the Korean side. Propeller-driven boats, junks and rafts make the Yalu a very busy river, despite its adverse natural conditions and its boundary character.

When the development of Manchuria under Japanese guidance progresses, the boundary character of the Yalu will become accordingly blurred. An important beginning of this development is the Sup'ung power plant, which is to be one of the largest in the world. At the village of Sup'ung-dong 80 km above Shinŭiju a tremendous dam has been built across a narrow part of the Yalu valley. When the main floodgate is closed, which was planned for June 1941, 230 km^2 will be flooded, two-thirds of which lie on the Korean side. Beforehand 80,000 people will have to be resettled. The plant will have a capacity of 640,000 kW. It is the first of a total of seven Yalu dams with a planned total capacity of 1,585,000 kW The others are to be constructed at Huch'ang, Chasŏng, Manp'ojin, Wiwon, Ŭiju and Hunjiangkou. Since 1937 the administration and electrification of the Yalu are supervised by a technical commission that is composed equally of representatives of Manchukuo and Chōsen.

The Yalu rises on the southeastern slope of the central mountain ring of Paektusan and in its first segment it flows through a canyon that is initially slightly entrenched in the pumice layer and then highly entrenched in the basalt and its basement. Settlements and transportation already avoid it here and follow the eastern tributaries. Rafting, however, soon begins on the main river as well. Downstream from the basin of Hyesan, at the confluence of the Hŏch'ŏn-gang and the Yalu there is a dam across the river constructed of fascines with rock packings. In the middle it has an opening through which the larger rafts made up here can pass.

In the first half of the peripheral meander segment Hyesan – Linjiang the valley usually has a floor which is covered with paddies. The dry fields and the hamlets of the farmers extend upward on the slipoff slopes, whereas rafting settlements lie directly along the river. Shin-galp'ajin is the entranceway to the northeastern Kaema upland. Just downstream from this endpoint of navigation the Yalu grazes the northern end of the Nangnim Range. Here the river, which was already as much as 300 m wide, narrows down to 40 m several times. East of Kŭmch'ang-dong steep undercut slopes have developed in the thick-layered limestones of the Chosŏn system, out of which a wild maze of vertical pillars and slabs has been sculpted by erosion. Between them grows lush green forest, truly Korean mountain scenery.

At Linjiang–Chunggangjin the conglomerates and shales of the Taedong and Kyŏngsang systems have again formed an elongated wide stretch that is advantageous to settlement and transportation (Fig. 51). In the lower segment, which begins here, straight, relatively wide sections of valley alternate with rugged, narrow, meandering gorges. In the splintery eruptive rocks, which are incorrectly all shown as prophyry in the geological map 1:1 million, the slopes are extremely steep and often rocky. Downstream from Ch'angsŏng the Yalu valley begins to widen out for good. The slopes become more gentle and the high mountains gradually recede. The river splits into several arms, and downstream from Sup'ung-dong the tides begin to make themselves felt. At a narrow part of the valley flat, on the high, pine-clad Korean bank, lies Ŭiju, the famous old entranceway to Korea, where the Imperial highway from Beijing to Seoul (p. 32) crossed the Yalu. Today Ŭiju is a modest county seat and market town, which, however, still preserves some nice remnants of buildings from olden times. The population (10,392 inhabitants) is decreasing, since Shinŭiju has taken over Ŭiju's role as crossing point and capital of P'yŏng-anbuk-do.

Fifteen kilometers downstream from Ŭiju the various arms of the Yalu join again to form a single riverbed that is 700 m wide at low water. Here a 5-km-wide floodplain protrudes out undissected toward the northwest from the Korean edge of the valley. Below this point islands again begin to form and the stream channel, through which powerful tides flow, begins to widen out into an estuary with broad expanses of shining mud flats at low tide. At this location the main Korean railroad, built by the Japanese soon after the war of 1904/05, crosses the river on a bridge 947 m long that is also open to automobile traffic. The railroad runs to the old Manchurian city of Dandong (1939: 219,777 inhabitants), where it connects up with the simultaneously constructed line Dandong – Shenyang (Mukden). Here they built the city of Shinŭiju (= New Ŭiju) (Fig. 56). The streets run in a monotonous grid partly parallel to the riverbank and partly perpendicular to it.

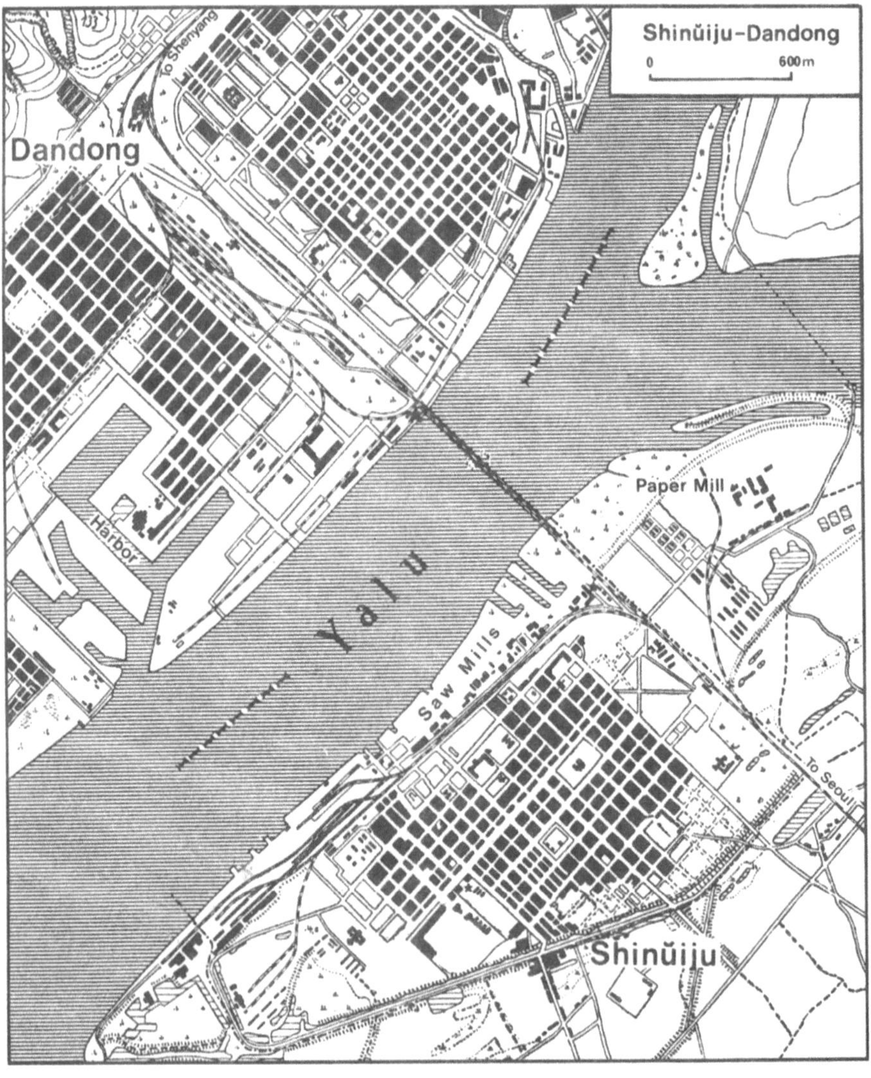

Fig. 56. Shinŭiju–Dandong (after an official plan)

Toward the south the city is protected by a levee. Lined up along the Yalu are the big wood-processing plants, in front of which the rafts are taken apart. The Ōji Company's large paper mill is also located there. The constantly growing city had a population of 52,384 in 1938, 16% of which were Japanese and 12% Chinese.

With a total trade volume of 73.1 million yen Shinŭiju was No. 5 among Korea's ports in 1935. It is, however, much more a river port than a seaport, since the Yalu's narrow flood channel poses great difficulties to maritime shipping and can be negotiated only by ships with a maximum draft of 4 m. On the Korean side of this flood channel lies the county seat of Yong-amp'o (13,556 inhabitants, 6% of them Chinese). It is Shinŭiju's outer harbor, though thus far very incomplete, and an important fishing port. A few one-story stone houses with bow windows date back to the period when the Russians had a foothold here and Yong-amp'o was geopolitically a sensitive spot of the highest order (p. 35). In the vicinity expanses of marsh, covered with rustling reeds (*Fragmites communis* Tr.), spread out in front of the crystalline, hilly old red landmass. Large numbers of crystalline islands jut up out of the mud flats off the mouth as testimony to the recent ingression.

In accordance with the dip-slope character, the western slope of the Nangnim Range lacks a pronounced foot. The transition from the steep, sharply dissected mountains there to the mountain country in the northeastern P'yŏngbuk region filling the big bend of the Yalu at Chunggangjin occurs without a distinct dividing line. This mountain country lacks dominant lines. Its peaks rise to more than 1500 m (Paeksambong 1556 m), in its southeastern part even higher (Paeksan 1875 m). Southwest of the Tongnogang begins the Chŏgyuryŏng Ridge (Sungjŏksan 1994 m), which runs parallel to the Yalu and forms the divide between it and the Ch'ŏngch'ŏn-gang. This ridge falls off very gradually toward the south to an extensive hill country with wide areas not exceeding 100 m in elevation, composed of granite gneiss, granite and metamorphic schist. It fills the obtuse angle between the Yellow Sea coast and the Ch'ŏngch'ŏn-gang, and the Taeryŏnggang and the Kuryŏnggang meander through it in the direction of slope. This hill country, which has weathered to a red soil, is probably a recently dissected remnant of the Nangnang plain (p. 69). Along the rivers that were mentioned and their tributaries very low alluvial land extends far northward between the chains of hills. Here and there in the hill country dome-shaped granitic mountains with a sharp knick jut up incredibly steeply to great heights. An example is Samgaksan (937 m) near T'aech'ŏn, which has rugged ravines and almost undissected, rounded rocky slopes 500 m high. The phenomenon of dome-shaped mountains (p. 79f) has reached perfection here.

The most striking feature in the relief of the P'yŏngbuk region is a thalweg that runs from the mouth of the Ch'ŏngch'ŏn-gang to Kŭmch'ang-dong on the Yalu and in places acquires the character of an alignment of valleys. It follows the Myohyang and Nangnim Range at some distance and crosses the entire length of the P'yŏngbuk region. Its first segment, the lower Ch'ŏngch'ŏn-gang valley, is strikingly straight up to the junction of valleys at Hŭich'ŏn. It runs along the boundary between the already mentioned granite gneiss and Paleozoic sediments that were affected by the Taebo orogeny. On its left flank Chosŏn limestone is folded into granite gneiss and Mesozoic rocks into the Chosŏn limestone. Probably this old structure is the cause of the straightness of the lower Ch'ŏngch'ŏn-gang valley. Upstream from Hŭich'ŏn the thalweg follows the meandering side valley of the Myŏngmunch'ŏn upward and crosses over two valley passes with a maximum elevation of 815 m to the central section of the Tongnogang, which, in contrast to the upper and lower reaches, has the north–northeasterly direction of our alignment of valleys. Thus the junction of valleys at Kanggye is reached, which has the same significance for transportation in the northern P'yŏngbuk region as that at Hŭich'ŏn has for the southern part of the region. From there it swerves eastward across Chingnyŏng (1057 m) into the central Chasŏnggang valley and finally across Kasallyŏng (925 m) to Huch'ang and Kŭmch'ang-dong. Whether the northern part of this thalweg has a structural basis as well is uncertain. The particles of sediment folded into the granite gneiss here and there with the same trend leave this possibility open. Considering the difficult terrain in the P'yŏngbuk region in its longitudinal direction, the thalweg is of great significance for transportation, although there are five passes in it.

On the left slope of the lower Ch'ŏngch'ŏn-gang valley near Kujang-dong there is a good-sized cavern in the limestones of the Chosŏn system called Tongnyonggul. The total accessible length to date is 1463 m [788]. It consists of a series of passages following crevices and chambers as much as 70 m high and 150 m wide. During the rainy season a creek flows through the lowest parts and lakes fill the floors of the lower chambers. The passages, and particularly the chambers, are embellished by an abundance of stalactites and stalagmites. Kawasaki estimates that it took 60,000 years to form one of them. The large masses of cave earth contained bones of bears and wild boars that possibly date back to the last Ice Age (p. 75ff). Most interesting of all are the formations on the floors of the lowest chambers: streaks and sinter terraces formed by dripstone. The latter have raised edges so that water remains standing in them. They are therefore appropriately called "rice paddy terraces."

The coast of the very shallow Sŏhanman rounding off the P'yŏngbuk region is a highly indented ingression coast. The hill country often extends directly to the sea. At the coast and not far from the confluence of the Taeryŏnggang and the Ch'ŏngch'ŏn-gang, Ch'iraksan rises to 366 m. It is composed of hard granite gneiss and surrounded by block streams. The granitic peninsula of Ch'ŏlsan sticks far out toward the south. Off it the Shinmi archipelago rises out of the mud and sand flats with their many tidal inlets. Marshes fringe the hill country and extend far northward along the rivers and creeks flowing into the bays. They are most extensive around the bay formed jointly by the mouths of the Ch'ŏngch'ŏn-gang and the Taeryŏnggang and most particularly on the peninsula of Ipsŏk-ri bordering it to the south.

In the chapter on the Kaema upland the climate of the northeastern P'yŏngbuk region was described as the most highly continental climate in all of Korea. The climatic data for the stations Huch'ang and Chunggangjin demonstrated this fact (p. 221f). The ice cave of Namshin near Tongjŏm, which is not even 900 m high, is impressive evidence of the cold winters. Approaching the Yellow Sea all climatic elements become decreasingly continental. But even on the Yellow Sea coast the summers are hotter and the winters colder than at the same latitude on the coast of the Sea of Japan. In the Japan Sea fringe Hamhŭng has an annual temperature range of 29.8° C. At the same distance from the Yellow Sea coast, Anju's range is 34.7° C. Owing to the influence from Manchuria, the stations on the Yalu are more continental than the respective stations on the Ch'ŏngch'ŏn-gang. All of this is demonstrated in Table 15.

Whereas in the north (Chasŏng, Ch'osan) July is still the warmest month on the average, as in the Kaema upland, approaching the coast this role passes on to the month of August. July, however, is the month with the most precipitation almost everywhere, in contrast to the Japan Sea borderland and most of the Kaema upland. Near the coast it is simultaneously the month with the highest relative humidity (Yong-amp'o 87.9%) and the least sunshine (Shinŭiju 39%). Close to the coast the summer monsoon is much stronger and more constant than in the

Station		Temperature						
		January			August		July	
	Altitude (m)	Mean temp.	Mean min. temp.	Absolute min. temp.	Mean temp.	Mean max. temp.	Absolute max. temp.	Annual range
Chasŏng	330	−17.5	−25.1	−38.7	23.6[a]	29.4[a]	40.0[a]	41.1
Ch'osan	190	−15.8	−22.8	−36.1	24.5[a]	29,8[a]	37,6[a]	40.3
Hŭich'ŏn	140	−11.5	−19.2	−36.0	24.1	29.4	37.0[a]	35.6
Shinŭiju	6	−9.5	−14.5	−27.3	25.0	29.2	39.1	34.5
Kusŏng	108	−9.7	−15.5	−31.0	24.3	28.9	36.5[a]	34.0
Taehwado	102	−6.5	–	–	24.1	–	–	30.6

[a] July. [b] January. [c] December. [a] August.

Station		Precipitation							
		January				August			
	Average amount (mm)	per 1000	Average no. of days with precip.	Precipitation intensity	Average amount (mm)	per 1000	Average no. of days with precip.	Precipitation intensity	Annual amount (mm)
Chasŏng	11.4	12	5	2.2[b]	228	246	17	15.2	926
Ch'osan	11.3[c]	12[c]	5	1.9[c]	224	231	15	16.9[d]	958
Hŭich'ŏn	14.1	11	5	2.8	390	303	17	23.0	1290
Shinŭiju	10.6	10	3[b]	2.7	265	258	16	16.8[d]	1028
Kusŏng	15.0[c]	13[c]	5	3.5	405	296	16	25.3	1366
Taehwado	9.6[b]	14[b]	4[b]	2.0	158[d]	231[d]	13	13.1	686

Table 15. Climatic data for the P'yŏngbuk region

northeast. In Shinŭiju 67.5% of the July observations are of wind directions between southeast and south–southwest. In Ch'osan, in contrast, 55.9% of the August observations are windless. Summer precipitation increases toward the coast, and near it, it increases toward the south. Kusŏng has for northern Korea the record July precipitation of 405 mm. In Kanggye Mills counted 24 days of rain each in June and July 1911 [781, p. 11]. As everywhere in Korea, the annual amount of precipitation conforms in its distribution to the summer precipitation. In the vicinity of the coast it is far more than 1000 mm everywhere. Only the islands off it have smaller amounts here as well (Taehwado).

Winter precipitation is, however, very slight everywhere. In the interior windless days predominate. Ch'osan holds the Korean record with 61.5% windless days in January. Near the coast, though, the land monsoon is very constant. In Shinŭiju 83.6% of the December observations are of the wind directions northeast to north–northwest. Relative humidity has its minimum then (Yong-amp'o 58.5%), and sunshine its maximum (72%). The number of days with snowfall decreases toward the coast and is only 27 per year there, i.e., barely a third of that in Chunggangjin.

The practically uncleared virgin forests covering the western slope of the Nangnim Range, whose composition was described on p. 224f in connection with Maengbusan and Myohyangsan, continue on into the northeastern hill country and on the Chŏgyuryŏng Ridge, but are interrupted here along the valley furrows by fire-field clearings and their second growths. Here and there even summit plateaus have been cleared, and they are covered today by light-green grassy meadows and shrubs, a sight reminiscent of the glades in our subalpine mountains.

To describe the changes in the composition of the natural forest with increasing altitude we shall take Paeksambong as an example. It is composed of hard biotite gneiss and rises from the floor of the valley of the Chasŏnggang at 750 up to 1556 m. The first zone extends up to 1150 m and is formed by deciduous forest in which oaks, particularly *Quercus manchurica*, are the leading element (oak zone). They are joined by the maples *Acer mono* and *Acer barbinerve*, the Manchurian walnut, the wild cherries *Prunus maximowiczii* and *Prunus maackii*, the birch *Betula costata*, *Euonymus pauciflora* and, as infrequently appearing conifers, *Abies holophylla* and *Taxus cuspidata*. Between 1150 and 1300 m conifers become more frequent. The Korean nut-pine, which occurs here and there at lower elevations, has its widest range here. The duo *Abies nephrolepis* and *Picea jezoensis* begins to appear here and gains the upper hand more and more toward the summit. The small-leaved linden extends up to an elevation of 1400 m, where the maple zone ends; the birch *Betula ermanii*, the mountain ash *Sorbus amurensis*, the maples *Acer ukurunduense* and *Acer tegmentosum*, the lianas *Actinidia kolomikta* and *Tripterygium regelii* and bushes of *Spirea prunifolia* var. *simpliciflora* and *Syringa robusta* extend up to the summit region. The flat summit, which was cleared for surveying purposes, is covered by a jungle of giant hemlock and the shrub *Thalictrum tuberiform* Max. The virgin forest around Kasallyŏng (925 m) is composed of even more species, and conifers are already quite common despite the low elevation. Like the forests on Myohyangsan, these forests differ from those of the Kaema upland in the large number of species of which they are composed [782, 783]. Regarding the plant associations following the river courses, I should like to refer the reader to the comprehensive ecological study of the valley of the Tongnogang by Mills [781].

About the second growths in the P'yŏngbuk region we can say essentially the same as for the Kaema upland. Toward the southwest, however, with the increasingly favorable climate, they too contain more different species. In the shrub formations at T'allyŏng (725 m) the masses of *Lespedeza* species predominate. Second to them are the oaks, lindens and alders, and third the red birch, the elm *Ulmus japonica*, the hazel *Corylus heterophylla* Fisch., the Siebold maple, the aspen, the Schlippenbach rhododendron, ferns and grasses. Lianas make large stretches of this jungle impenetrable. Approaching the Yellow Sea fringe less hardy species begin to appear in these shrub formations, such as *Palura paniculata* Nakai var. *pilosa* Nakai, *Philadelphus tenuifolius* Rupr. and Max., which is related to our mock orange, the snowball *Viburnum pubinerve* Bl. var. *puberulum* Nakai, the feathery-leaved nut-gall tree, *Rhus javanica* L., etc. Lower elevations, especially the hills in the southern part of the P'yŏngbuk region, are not infrequently covered by pure red pine forests. More frequent, however, are scrawny, dry scrub formations consisting of red pines, the juniper *Juniperus rigida*, oaks and chestnuts, which are ruthlessly exploited by the densely settled population.

As Table 32 shows, P'yŏng-anbuk-do is the province in which fire-field farming is most widespread. When one enters the province from P'yŏng-annam-do on the road from Tŏkch'ŏn to Kujang-dong, this fact and its effect on the landscape immediately strike one very strongly. This form of economy extends down from the north as far as the northern boundary of the southern hill country and is still fully operating today even in the steep granite dome-shaped massifs there, such as Samgaksan. Thus many elements of the landscape are the same as in the Kaema upland. With the increasing destruction of the primary forest toward the southwest these features recede, however, at least those that are based on an abundance of wood. Only in the northeastern hill country are they as frequent and pure as in the Kaema upland.

The difficulties inherent to fire-field farming are even greater here than there. In addition to the other obstacles, the slopes are often extremely steep. I observed fire fields on slopes with an angle of up to 60°. On ridges and peaks forest or brush is left standing everywhere to diminish erosion, which often digs deep gullies into the slopes with fire fields nevertheless and devastates the valley bottoms, turning them into swamps. Along with the generally lower absolute altitude of the surface, fire-field farming also shifts down to a lower elevation. I encountered the highest fire fields at 1150 m. Toward the southwest, with the increasingly milder climate, a larger variety of crops is grown on the fire fields. On the 477.5 ha of fire fields belonging to the village of <Emmedō> (Hŭich'ŏn-gun, for the significance of angle brackets cf. note in foreword) the following crops were grown in a year that is not mentioned [561, p. 8]: foxtail millet 50.0%, adzuki beans 11.5%, potatoes 11.4%, maize 6.9%, buckwheat 6.9%, kaoliang 6.5%, soybeans 4.3%, radish 2.2%, hemp 0.5%, etc.

Permanent cultivation increases from year to year, as the fire fields gradually become permanent fields. Intercropping and the planting of certain crops as borders is widespread on them. Joining the fire-field crops mentioned for <Emmedō> on the permanent fields are: upland rice, mung beans, melons, vegetables, even the Korean varieties of cotton and some sweet potatoes. The attempt to grow American varieties of cotton around Hŭich'ŏn was, however, a failure. Ŭiju-gun is the Korean center of maize growing, but it is so common in the rest of the P'yŏngbuk region that P'yŏng-anbuk-do is the number one producer of maize in Korea (p. 414).

Paddies are found in all valleys up into the northeast and as high as 750 m. With the increasingly favorable climate and soil they increase in extent toward the southwest, and near the coast rice cultivation completely dominates the valley bottoms and the marshy areas (Map 4, cf. editors' note on p. 7). In this and many other aspects of the cultural landscape the southwestern section near the coast is related to the central Korean Yellow Sea regions, so that it can be separated from the rest of the P'yŏngbuk region as a discrete section (Fig. 39). In Chŏngju-gun rice is cultivated on 19.3% of the total area, in Yong-amp'o-gun on 28.0%, and almost half of the fields are paddies. In addition to rice, *wanggol* is grown in paddies here. The coastal fringe is an important rice surplus area that ships rice to Japan.

The intensified rice cultivation in the marshes is primarily the work of irrigation associations run by the Japanese. The Indō and Taishō Companies have carried out large reclamation and irrigation projects in the area around the mouth of the Ch'ŏngch'ŏn-gang; the latter and the Fuji Company have created one of the largest projects of this type, the rice polders southeast of the mouth of the Yalu in Yong-amp'o-gun. These polders were laid out on former mud flats that were interspersed with a few crystalline islands. Straight dikes with an outer surface of stones connect an outer chain of islands. Thus two separate, rectangular polders were formed, the larger one closer to the mouth of the Yalu being finished in 1916, the other in 1926. Together the two encompass an area of 11,200 ha. On the 4950 ha belonging to the Fuji Company 1700 Korean rice-farming families live. The water for irrigation is drawn from the Sapkyoch'ŏn by means of a canal that tunnels under the divide between the stream and the polders. A reservoir has been constructed in the upper reaches of the Sapkyoch'ŏn. When there is a water shortage the Kojin-gang, which has a reservoir with a water capacity for 30 days, is also tapped. A second canal services the extension project. By cleverly taking advantage of the natural slope of the former tidal flats, systems of irrigation canals crossing each other at right angles were dug through the two polders. Five floodgates in the dikes regulate drainage. To be near a drinking water supply the hamlets are preferably built at the foot of the islands. One man with agricultural training is in charge of each 30 families. The administrative office lies roughly in the middle of the older project, surrounded by experimental fields and seedbeds. The farmers turn over the surplus production beyond what they themselves need to the office, and it is sold at a time when it will command a good price.

The energies of the P'yŏngbuk region are also increased by a second feature of economic life, mining. In addition to copper mining in Tongjŏm (p. 66) and iron and coal mining, e.g., near the county seat of Kaech'ŏn, there is above all gold mining. In gold production too P'yŏng-anbuk-do heads all Korean provinces (p. 447). There are gold mines near Hagwang-dong (Ŭiju–gun), near Chasŏng, near Kilsang and <Chokeidō> (Sŏnch'ŏn-gun, for the significance of angle brackets cf. note in foreword) and Kyo-dong and Samsŏng (Kusŏng-gun). By far the oldest and most productive are in the northern part of Unsan-gun, however, after which they are usually named, and in the adjoining southeastern part of Ch'angsŏng-gun near Taeyu-dong (Fig. 57). The center of the Unsan mines, run by the American Oriental Consolidated Mining Company since 1895, is near Pukchin (786). Near Pukchin the remarkably straight valley of the Kuryonggang follows a shatter zone cutting through granite and granite gneiss, which extends northwestward across Ŏjaryŏng (474 m) into the T'aech'ŏn area for a total length of 35 km. These rocks are additionally permeated by quartz veins running in varying directions that are impregnated with auriferous chalcopyrite. The company has two separate shaft systems near Pukchin. One is that of the Taebau mine, the other that of the Kyodong and Chindong mines. Each one has a stamp mill. The two stamp mills, connected by a narrow-gauge railroad, lie 2 km apart on either side of the side valley that converges from the north at Pukchin. Crushing is done in two phases. The auriferous quartz sand resulting from the second crushing process is floated to inclined, fluted shaker tables, where the gold is concentrated by segregation. Following further concentration procedures the concentrate is leached in a mercury reduction plant, and more recently in a cyanide reduction plant. Old Pukchin lies in the angle between the two valleys. New Pukchin spreads out in the side valley opposite the lower mine, separated from Old Pukchin by the enormous mountain of tailings. The population (19,619 inhabitants) is still growing rapidly. Between 1895 and 1938 the Unsan mines produced no less than 80,000 kg gold. The Taeyudong mine is a subsidiary of the Unsan mines, which has been in operation since 1910 and later changed over into the hands of another foreign company. It lies 20 km farther to the northwest. In 1937 its gold

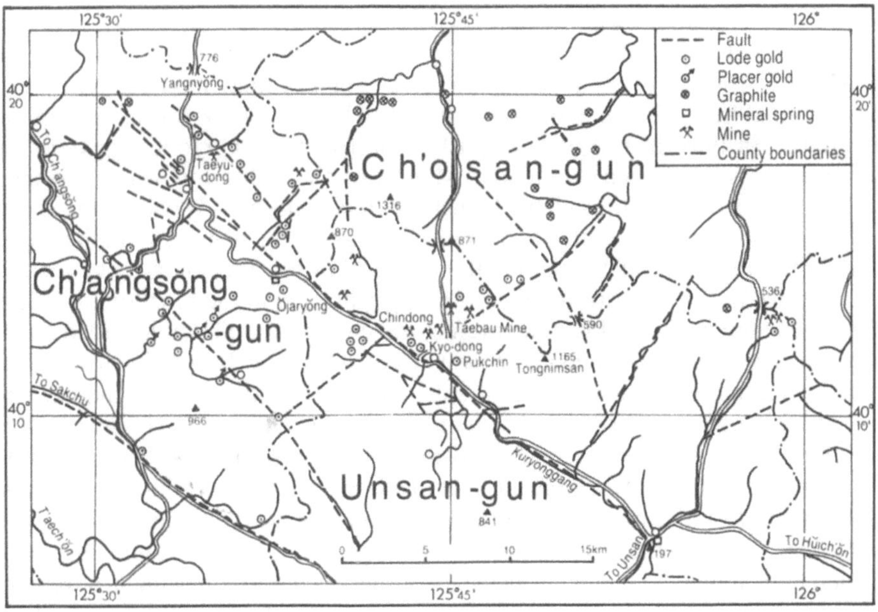

Fig. 57. The Pukchin and Taeyu-dong gold mines (Unsan-gun and Ch'angsŏng-gun, after Y. Kinosaki)

output surpassed that of the Unsan mine for the first time, and it is now Korea's most productive gold mine. Both enterprises have been in Japanese hands since 1939.

On the Ch'ŏngch'ŏn-gang and the Taeryŏnggang rafting is done. On the latter the rafts consist of crooked pine logs as thick as a thigh at the most, since this river rises in the red pine region. On this river mainly Korean rafts are seen. These have the form of a small, narrow rectangle, and the logs are fastened together in a rather tangled and disorderly fashion. There are rudders fore and aft. The Ch'ŏngch'ŏn-gang, in contrast, transports mainly Japanese-style rafts made of straight fir and spruce logs. The wood is processed in Pakch'ŏn, Kujang-dong, Anju and Shinanju. The Ch'ŏngch'ŏn-gang is navigable to boats as far as Hŭich'ŏn.

The fish catches on the coast of the P'yŏngbuk region are the most important of the entire Korean Yellow Sea haul (p. 443). In spring flounder and corvenia are caught, in summer croaker, Spanish mackerel, hairtail and red sea bream, in spring and autumn shrimps; eels in autumn only and in autumn and winter mullet.

Apart from the Yalu there are two major transportation routes in the P'yŏngbuk region. The more important one runs through the coastal fringe. This was the route of the old Korean Imperial highway, which crossed the Ch'ŏngch'ŏn-gang at the attractive old county seat of Anju. Today it has been replaced by the main railroad, the Korean "trunk line" which runs closer to the coast and crosses the Ch'ŏngch'ŏn-gang at Shinanju. The relationship between Shinanju and Anju is the same as that between Shinŭiju and Ŭiju. The vigorously growing Shinanju (13,029 inhabitants) will gradually outdistance Anju (18,284 inhabitants). On the main railroad line there are also the county seats of Chŏngju (12,502 inhabitants) and Sŏnch'ŏn (19,453 inhabitants).

The second route follows the thalweg described above, which represents the overland traffic axis of the P'yŏngbuk region. Till now there has been only a secondary road here with an enormous amount of bus and truck traffic, by Korean standards. In 1934 the railroad was pushed on from Shinanju to Kanggye. In the past few years it has been extended to Manp'o on the Yalu, from where it is soon to be continued through southern Manchuria to Changchun. It will be the shortest connection from Seoul to Changchun. Kanggye is the main town in the upper P'yŏngbuk region (17,898 inhabitants). It huddles together on a rocky terrace 20 m above the northeastern bend of the Tongnogang in the junction of valleys described above. The old Korean prefectural city had a wall that included Namsan, which was once fortified and now is covered by a lovely red pine forest. Impressive remnants of the wall and the splendid North Gate are still standing. The one-story Korean houses along an irregular network of streets have shingle roofs weighted down with stones for the most part here, too. In the otherwise highly Japanized town there are a large number of administrative and military buildings. Kanggye is after all a main base of Japanese power in northern Korea.

Out of the way of the modern traffic routes the county seat of Yŏngbyŏn is concealed high above the meandering valley of the lower Kuryonggang in a tremendous, well-preserved old Korean fortress sprawling out across the surrounding, up to 490-m-high, dome-shaped granite mountains ([789], Fig. 58). Yŏngbyŏn-gun is known for its silk production.

From what we have said it is easily understandable that the population density of the P'yŏngbuk region increases greatly from northeast to southwest (Fig. 85). The most sparsely populated counties lie in the northeast and in the middle (Huch'ang-gun 17 inhabitants/km^2, Kanggye-gun 27, Hŭich'ŏn-gun 26), the most densely settled counties on the Yellow Sea coast (Chŏngju-gun 161 inhabitants/km^2, Yong-amp'o-gun 221). On the average the P'yŏngbuk region still belongs to the sparsely settled Korean regions (p. 14).

5 The Taedong Basin

The region that we are calling the Taedong Basin coincides almost exactly with the drainage basin of the Taedonggang. Its boundaries thus run along the drainage divides of this northernmost river of the actual peninsula. Only the river system of the Chaeryŏnggang will be considered as part of the Hwanghae peninsula. It does not join the Taedonggang until the southern tip of the big loop of the latter, which is affected by the tides, and can just as well be considered an independent hydrographical area. In the east this region extends just to the boundary of the province of P'yŏng-annam-do, which also ends at the divide almost everywhere. In the west it borders, like the province, on the Yellow Sea between the mouths of the Ch'ŏngch'ŏn-gang and the Taedonggang. In the northwest it does not reach the provincial boundary (p. 244), but in the southeast it extends considerably beyond the boundary into the province of Hwanghae-do. In northwest–southeasterly direction its maximum length is 130 km, in northeast–southwesterly direction 220 km, and its area is approx. 16,000 km^2 (Fig. 39).

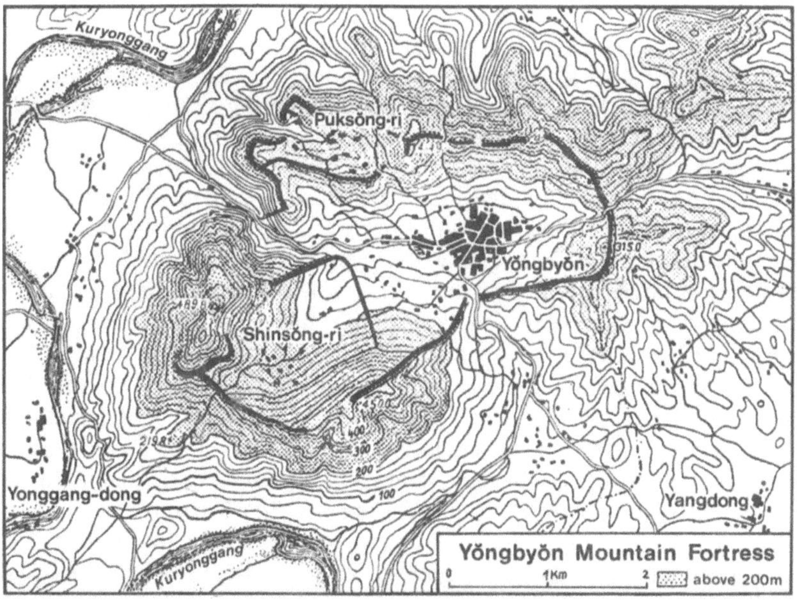

Fig. 58. The fortress of Yŏngbyŏn in P'yŏng-anbuk-do (after Mukoyama)

The 431-km-long Taedonggang rises on Sobaeksan in the fork of the Nangnim Range. In contrast to its close neighbor, the Ch'ŏngch'ŏn-gang, which consistently retains its southwesterly direction, the Taedonggang changes four times from a north–southerly to an east–westerly direction before it reaches P'yŏng-yang. In the process it remains in the northwestern half of its drainage basin. Therefore the larger tributaries come from the east and rise in the Wonsan arc of the Main Korean Range. They are the Mat'an-gang, the Piryugang, the Namgang and the Hwangjuch'ŏn. They, too, especially the two middle ones, consist mainly of segments trending south–southwest and west–northwest, i.e., perpendicular to each other. This very striking feature of the hydrographical character appears to be related to the direction of the Taebo orogeny (p. 64) and thus to have a structural origin. All of these rivers have a strong tendency to meander in the rocks of the Sang-won and Chosŏn systems. In granite gneiss and in the P'yŏng-an system, in contrast, the predominant tendency is toward straight segments. We have already pointed out this fact for the Piryugang and its tributary, the P'yŏng-wonch'ŏn, on p. 74. The most magnificent Piryugang meanders have developed near the county seat of Sŏngch'ŏn in the limestone shale of the formation of the Sang-won system named after this town (Table 46). The neck of the southernmost of these meanders, with which the Piryugang changes from a southerly to a westerly direction, is only 8 m high and 50 m wide at the lowest spot. For almost 1 km it is 100 m wide at the most. Its crest is jagged. The Koreans distinguish 12 "mountains" in the base, neck and lobe of the meander. In view of them on the other side of the river at the edge of town lies the magnificent pavilion Kangsŏllu constructed 640 years ago, and in each of its wooden pillars the name of one of these mountains is carved in Chinese characters.

Just below P'yŏng-yang, which is still almost 100 km from the mouth, the till then generally narrow floodplain of the Taedonggang widens out, and the river forms branches. As soon as it enters the steep-hilled terrain of the Kyŏngsang system, however, which again rises above 200 m, the river course narrows down to 500 m anew, and the floodplain ceases completely. In this configuration the river forms its giant loop with its southward pointing vertex, into which the Chaeryŏnggang flows (p. 74). Directly below this loop the 45-km-long estuary begins. Small peninsulas composed of solid rock project toward it from both north and south. The shallow bays between them are filled with mud flats and tidal inlets. The massive growth of the goosefoot *Suaeda japonica* Makino, which corresponds ecologically to our glasswort, gives the mud flats a reddish color. The outer estuary is lined for miles on the northern side by a strip of sand flats. With the ebb current the water of the Taedonggang, which is almost always overloaded with suspended sediments, is transported far out into the Yellow Sea.

The effect of the recent ingression is not limited to the estuary. It extends up to P'yŏng-yang. Whereas upstream from P'yŏng-yang the Taedonggang has a depth of only 1–2 m, it increases below the city to 8 m; in the above mentioned defile where it enters the Kyŏngsang strata it increases to 18 m, and in the loop it is 11–14.5 m below mean low tide. The strong tidal current reaches up past P'yŏng-yang. At Songnim the mean spring range is still more than 4 m, at P'yŏng-yang 1.5 m. Navigation by 3000 t, ocean–going vessels is thus possible without artificial assistance far upstream, to Posan, and could be extended as far as P'yŏng-yang. But upstream from Chinnamp'o the Taedonggang freezes over each year for several months (p. 120), so that even then it would be necessary to maintain an outer harbor. A good 240 km of the river course is navigable by small junks. On the main river this shipping extends up to Ŭnsal-ri, on the Piryugang to Sŏngch'ŏn and on the Namgang to north of Koksan. Rafting begins even farther up, but because coniferous forests are much less abundant, it is not as important as on the Yalu. It was presumably already done during the Lolang period, however (p. 23).

The geological structure of the Taedong basin differs completely from that of the P'yŏngbuk region. All along its western coast run the highly metamorphic sediments of the P'yŏng-won massif, their maximum width reaching 40 km (p. 62). Otherwise it is dominated by the P'yŏngnam depression, in which the strata of the Sang-won, Chosŏn, P'yŏng-an and Taedong systems were intensely folded by the Taebo orogeny ([791–807] and p. 62, 64). The axes of the folds trend primarily in an easterly direction. As the Table of Geological Formations (Table 46) shows, the first two systems consist primarily of thick limestone shale and dolomite. The P'yŏng-an strata, in contrast, are composed particularly of sandstone, shale and seams of anthracite. They are restricted to two separate areas. The southern one overlies the metamorphic sediments of the P'yŏng-won massif west of P'yŏng-yang in a pronounced angular unconformity and extends 56 km eastward from there. It ends near Samdŭng on the lower Namgang. The northern one lies on the central section of the Taedonggang [792, 793] and culminates in Chang-ansan (1248 m). Near P'yŏng-yang the Taedong strata are preserved only to a limited degree.

Here and there islands of granite gneiss protrude up through this enormous group of folded strata ranging in origin from the Proterozoic era to the Dogger epoch. Above all, however, it is interpenetrated by intrusions of "young" granite. The largest of these lies in the headwater region of the Namgang and the Piryugang and culminates near its southern edge in Haramsan (1486 m). In the contact zones of these intrusions mineral hot springs occasionally issue. One of these is, for instance, in Yongt'aek on the southern foot of granitic Sunghwasan (685 m), northwest of Sŏngch'ŏn. This spring, known as Sŏngch'ŏnonch'ŏn, is a sodium chloride acidic spring with a water temperature of 48° C and is tapped by a cement well 6 m deep in which the water rises by its own pressure, filling the six basins of the bathhouse erected over the spring.

Downstream from P'yŏng-yang on the Taedonggang the folded strata of the above mentioned systems are overlain by a band of the Kyŏngsang system consisting of porphyry, tuff, sandstone and conglomerates. Because these are predominantly hard strata it forms the above mentioned steep-hilled area around the big loop of the Taedonggang, rising up as high as 332 m close to the river in the isolated Haeapsan.

Like the P'yŏngbuk region, the Taedong basin forms a strip in the dip slope of the main Korean tilt block, from the crest in the northeast to the Yellow Sea coast in the southwest. Consequently the elevations decrease in this direction, on the whole. The Taedong basin is, however, far from being a simple, tilted and dissected tabletop-like peneplain. In general it is a mountain country with quite complicated, not easily comprehensible relief, and its individual features have yet to be interpreted. If we have nevertheless chosen the name Taedong basin for this region, it is the character of its core area, the region around P'yŏng-yang, that justifies the choice. Here the Nangnang plain spreads out (p. 69), and particularly south of P'yŏng-yang it is quite extensive. It indiscriminately truncates the wildly agitated structure of the subsurface, is covered by terra rossa and slightly dissected by the rivers and creeks that are tributary to the Taedonggang, and it extends from the south bank of the river, where it has an elevation of about 30 m and where the field of tumuli from the Lolang period lies (p. 22ff), 50 km to the south and southeast via Chunghwa and Hwangju. In the south its boundary is the mountain range of Chŏngbangsan, which runs east–west and rises to almost 700 m, and along which I draw the boundary with the Hwanghae peninsula. In the southwest the border is formed by the steep-hilled area of Kyŏngsang strata breached by the Taedonggang, in the northwest by the mountain and hill country of the P'yŏng-won massif, which rises to more than 500 m despite its closeness to the coast. This setting gives the Nangnang plain the character of the floor of a basin. Offshoots of it extend up the Taedonggang, Piryugang and Namgang and are occasionally covered by extensive masses of gravel. The permeable nature of most of the rocks underlying this peneplain have contributed greatly to its preservation.

Framing the entire Taedong region at a greater distance are the Myohyang Ridge with its southwestern extension and the southern end of the Nangnim Range, along with the mountains of the Wonsan arc. The latter rise to 1724 m in the northern Paeksan, 1095 m in Ch'ŏrongsan, 1452 m in the southern Paeksan and 1324 m in Turyusan, but repeatedly sink below 900 m between them. The lowest parts of the drainage divide lie northwest and east of Yangdŏk. They are Kŏch'aryŏng (557 m) and Ahobiryŏng. Via them the railroad and road from P'yŏng-yang to Wonsan leave the Taedong basin. The mountains of the northeastern Taedong basin, which rise within the frame set by these deeply dissected, occasionally gorgelike valleys, are generally lower than the just mentioned peaks of the drainage divide. Exceptions are particularly Chang-ansan and Haramsan.

The relief features in the limestone region are rugged. One frequently finds vertical cliffs 100 m and more high. The mountains have often been carved out of inclined strata and their infaces and backslopes are then asymmetrical. Frequently the terrain is dissected into many individual, sharply jagged peaks and steep needles. In particularly hard strata of the Sang-won and Chosŏn systems on top of weaker ones, hogbacks have not infrequently developed. The riverbeds in the limestone region are often empty until well into the rainy season in midsummer, so that the disorderly white masses of debris shine from afar. The peaks composed of granite and granite gneiss, in contrast, form round domes that are recognizable by their form from a great distance. At lower elevations they are covered by sandy yellow soil and a very scrawny plant cover consisting of red pine and oak shrubs. On moderate and gentle slopes the limestone series, in contrast, is covered by clayey red soil that becomes saturated with water and is thus almost completely green with secondary forest, shrubs or grass.

In permeable limestone older features are retained for a long time, and the valleys cut into them with a sharp edge. The main drainage divide in the east runs primarily across limestone, and along it such ancient features have been preserved at great heights. Such a case is, e.g., the high-lying valley of Sansŏng-ri, which is tributary to the Yonghŭnggang, and thus the Sea of Japan. The divide runs around its northern, western and southern sides. This is the explanation for the 15-km-long, tubelike extension of the province Hamgyŏngnam-do toward P'yŏng-annam-do, which appears in many of the included maps. The floor of the Tongmyŏn-gang valley, which belongs to the Taedonggang system, lies here at 250 m, that of the Sansŏng-ri valley at 550 to 650 m. The creeks flowing into the Tongmyŏn-gang advance victoriously toward the Sansŏng-ri area along a broad front. In places the divide is only the upper edge of a steep precipice, no longer a ridge. At the head of this interesting high-lying valley, the western end of the tube, Ch'ŏrongsan, has, however, remained standing, and it drops off precipitously toward the west, toward Maengsan.

Where gently sloping surfaces in the limestone region lie high enough above the groundwater table, karst phenomena have developed. They are particularly typical in the limestone shale of the "Great Limestone Series" (Table 46). Large caves have repeatedly developed in them [808]. Koryŏngsan (543 m) forms a high zone pitted by dolines. On both sides of the Taedonggang just below the mouth of the Piryugang lies the karst region of Hadan-ri. It forms a hilly plateau belonging to the Nangnang plain with a relative elevation of 75 m above the river. From the Taedonggang several bluntly shaped valleys reach into this plateau. They are separated from it by steep, often rocky slopes and head in rocky cirques. Gushing karst springs rise at the foot of the slopes. The valley floor is deeply filled with red soil and splintery weathered fragments. The plateau is dissected into a system of karst basins as much as 40 m deep. The bottom of these basins is covered by red soil with masses of gray blocks of limestone shale sticking out at the edges.

The differences between the climate of the Taedong basin and that of the P'yŏngbuk region are mainly gradual, but not fundamental ones. Therefore we can deal with the climate briefly on the basis of Table 16.

The first three stations lie in the eastern half of the Taedong basin, Yangdŏk and Koksan simultaneously closer to the coast of the Sea of Japan than to the Yellow Sea coast, Chinnamp'o not far from the latter and P'yŏng-yang in the center. The thermal continentality decreases from northeast to southwest. The fact that it is weaker in Yangdŏk and Koksan than in Tŏkch'ŏn is a result of the influence of the Sea of Japan. When equal distances from the Yellow Sea coast are compared, it is weaker in the Taedong basin, which is part of the peninsula, than in the P'yŏngbuk region, which borders on Manchuria. A comparison, however, of Chinnamp'o and Suwondan at the same latitude on the Sea of Japan coast shows that here too the Yellow Sea coast is more continental than that of the Sea of Japan. Suwondan has a mean January temperature of −2.1° C and a mean August temperature of 24.0° C, i.e., an annual range of only 26.1° C, compared with 31.6° C in Chinnamp'o. How hot the lower Taedong basin can get in midsummer I discovered on 6 July 1933, when I measured 42.5° C at 1 P.M. in the big loop of the Taedonggang. In contrast to the P'yŏngbuk region, the amount of precipitation is much larger in the mountainous eastern portion of the Taedong basin than in the west. Thus the zone of diminished precipitation near the coast, which is restricted to the islands off the coast in the P'yŏngbuk region, extends far onto the mainland here. July receives an even larger share of the precipitation here than there. In Koksan it increases to one-third of the annual total, and the precipitation intensity of this month is correspondingly high.

In the eastern and particularly in the northeastern parts of the Taedong basin, i.e., in the region of the Main Korean Range, there are still lovely virgin forests.

Since it is lower here than in northern Korea, however, in many places second growths extend up beyond them. On Haramsan the boundary between predominantly deciduous forest and predominantly coniferous forest lies as low as 1100 m in places and not until 1400 m in other places. Because of the large amount of summer precipitation the deciduous forest is unusually luxuriant and varied. At an elevation of 900 m (oak zone) it is composed here of oaks, maples (*Acer pseudosieboldianum* and *Acer mono*), aspens, willows, alders, the Korean mountain pear, wild cherries (*Prunus maackii* and *Prunus serrulata*), the Manchurian walnut, the common ash, the red birch, the cornel cherry *Cornus controversa* and the corktree *Phellodendron amurense*. In the undergrowth Hercules' club occurs as a small tree along with the lianas *Actinidia* and the wild grapevine.

Station		Temperature						
		January			August		July	
	Altitude (m)	Mean temp.	Mean min. temp.	Absolute min. temp.	Mean temp.	Mean max. temp.	Absolute max. temp.	Annual range
Tŏkch'ŏn	135	−12.5	−20.1	−36.1	24.0	29.3[a]	38.0	36.5
Yangdŏk	263	−10.9	−17.8	−34.0	23.4	27.9	36.6[a]	34.3
Koksan	157	−8.7	−15.5	−32.0	24.6	29.7	37.6[a]	33.3
P'yŏng-yang	51	−8.1	−12.6	−28.5	24.5	29.4	36.4	32.6
Chinnamp'o	10	−6.6	−10.9	−23.0	25.0	29.1	35.5[c]	31.6

[a] July. [b] December. [c] June.

Station	Precipitation								
	January				August				
	Average amount (mm)	per 1000	Average no. of days with precip.	Precipitation intensity	Average amount (mm)	per 1000	Average no. of days with precip.	Precipitation intensity	Annual amount (mm)
Tŏkch'ŏn	13.1	11	4	3.3	355	301	15	23.7	1183
Yangdŏk	17.1	17	5	3.4	280	268	15	18.7	1048
Koksan	19.8	16	5	4.0	420	334	15	28.6	1258
P'yŏng-yang	13.5	14	5	2.5[b]	258	277	15	17.2	937
Chinnamp'o	12.6	18	5	2.5	193	273	14	13.8	705

Table 16. Climatic data for the Taedong basin

In the zone of predominantly coniferous forest the vegetation on Haramsan is very similar to that of Myohyangsan (p. 225).

At the edge of a clearing in this forest there stand two small wooden temples roofed with slabs of limestone shale and an altar. On the 15th day of the 6th month according to the Chinese lunar calendar the inhabitants of the nearest village perform a sacrificial ceremony here. The left-hand temple contains an appealing painting of a Buddhist monk. In it a vegetarian sacrificial offering is made according to Buddhist rites. In front of the right-hand temple, which is dedicated to the Mountain Spirit of Haramsan and contains pictures of the male and female mountain spirits, a sacrificial animal bought with money collected from the inhabitants is slaughtered. After the ceremony the meat is divided up among those present. From late summer on many ginseng hunters worship here, to beseech the Mountain Spirit for luck in their search.

The summit plateau of Haramsan at an elevation of more than 1440 m is covered by a very strange second growth. It is a practically impenetrable jungle of trees and shrubs, most of which are not higher than a man. The weight of the snow in winter has caused these plants to grow in a sprawling shape. This is most clearly seen in the moss-covered yews, whose trunks, as much as 20 cm thick, bend over horizontally at a height of about 1 m. After all, the yew, being the only conifer in this formation, suffers most under the weight of the snow. Second in frequency is the common oak, which thus occurs at unusually high elevations here. Third is *Acer tschonoskii* var. *rubipes*, fourth the mountain ash *Sorbus commixta*. In the middle of this pathless wilderness stands a monument bearing the inscription Shimadae (Horse Testing Ground) in Chinese characters. According to Korean traditions the first king of the Chosŏn dynasty supposedly rode horses up here. Today the jungle is so terrible that it took us 1.5 h to cover the 500 m from the summit to the monument. Perhaps we are witnessing Korea's oldest second growth here, which remains unchanged in this stunted form because of the altitude and the abundance of snow.

Here, too, the valley floors are covered by nature with bottomland forests. I studied their composition near Tŏkch'ŏn on the banks of the Taedonggang. In midsummer the air in this very luxuriant forest is like a stifling hot greenhouse. The upper tier consists of elms and willows, the middle one of ashes, the lower one of elderberry bushes of the species *Sambucus latipinna* and *Sambucus pinnatifolia* and of *Princepia sinensis* Oliver, *Zelkova serrata* Makino, "Virginia creeper" *Ampelopsis brevipedunculata* Koehne, the birthwort *Aristolochia contorta* Bunge, the strangler *Celastrus orbiculatus* Thunb., the buckthorn *Rhamnus schneideri* Lev. and Vnt. and the lilac *Syringa amurensis* Rupr. var. *genuina* Max.

In the dense second growths in the limestone mountains calcium-loving shrubs, such as the lilac *Syringa dilatata* Nakai and *Belamcanda punctata* Moench are very common. On the slopes of the peak of Ch'ŏrongsan, which is composed of limestone, I studied a varied secondary shrub forest at an elevation of 900–1095 m. In addition to some of the plants encountered at the same elevation on crystalline Haramsan I collected the Chinese birch, the lilacs *Syringa amurensis* var. *genuina* and *Syringa dilatata*, the maple *Acer triflorum*, the Korean box tree, the mock orange *Philadelphus lasiogynus* Nakai, *Deutzia parviflora* Bunge, *Weigela subsessilis* Nakai and spireas. The *Lespedeza* and *Rhododendron* species that are so common in such formations on crystalline rocks are missing. On all rocks in the Taedong basin open shrub formations and secondary forests have developed at lower elevations. Red pines predominate in them or even stand alone.

At lower elevations the deforestation of large areas is presumably thousands of years old, although from the analysis of the wood in the coffins of Lolang we can infer that at that time forests of the trees mentioned on p. 23 still existed around P'yŏng-yang. The red soil of the Taedong basin is particularly good for dry-field farming because of its capacity to absorb water and because of its fertility. The finds from the "coastal culture" (p. 17), the wealth of the Lolang graves and the role that P'yŏng-yang has played time and again in Korea's history since the days of Kija (p. 20ff) are indications of an early dense population in the environs of P'yŏng-yang. The variety of crops and the refined agricultural methods still reveal Chinese influence today.

Irrigation is relatively little developed in the Taedong basin, however. The proportion of paddies in the total area, at 16.9%, is even much lower in P'yŏng-annam-do than in P'yŏng-anbuk-do (p. 414). The reason for this is the permeability of extensive areas whose climate and gradient would very easily permit paddy cultivation. The increase in the intensity of paddy farming toward the coast resembles the situation in the P'yŏngbuk region (Fig. 87). In the valleys of the P'yŏng-won massif there are many well irrigated fields. In many other parts of the Taedong basin, however, one finds only rainwater paddies in which a murky liquid is stagnating in midsummer. On the dams of the paddies soybeans or sesame are planted here too, to make use of even the tiniest available areas. To compensate for the small amount of paddy rice, relatively much upland rice is grown in P'yŏng-annam-do. It accounts for 11.4% of the total area of rice cultivation in the province. Usually it is sown in bunches right away, so that a level upland rice field differs from a rainwater paddy only in that the ridges are lacking. The main area lies north of P'yŏng-yang.

Since 26% of the total area of P'yŏng-annam-do is covered by fields, the province belongs to the well cultivated ones. This is due to the relatively favorable climate, relief and soil. For the first time we encounter double cropping on dry fields, namely barley, naked barley and wheat. With the exception of the northwestern and the northeastern edges, the Taedong basin belongs to the main area of wheat cultivation. Foxtail millet and kaoliang are also grown in large quantities. On the Nangnang plain around P'yŏng-yang the Japanese attempted to grow sugar beets. Because the results were too poor, this attempt was abandoned in 1932. Other crops that are frequently grown are maize, which like the potato and hemp is grown mainly in the east, where there is more precipitation, and occurs here as high as 900 m. Still others are the Korean varieties of cotton, which have their main center in the southwest, barnyard millet, buckwheat, soybeans, adzuki beans, green beans, peanuts, castorbeans, sweet potatoes, melons, radishes and vegetables. The east is also one of the Korean tobacco-growing regions. Mulberry bushes are also cultivated. And finally the cultivation of fruit trees is highly developed. Apples from Hwangju are famous all over Korea. They are partly exported via Chinnamp'o to Japan, China and Manchuria. In the vicinity of the settlements there are often plantations of peach, apricot and chestnut trees. In the melon fields and the fruit plantations there are watchtowers that are kept occupied by day and by night during the period just before the fruits ripen.

Intercropping of spring crops is very common. There are extraordinarily many variations (p. 183). Beans, and particularly peanuts, can be found in mixed cultures with maize or kaoliang, and even with upland rice. Especially in the mountain valleys that receive little sunshine in summer it is very common to plant borders around the fields (p. 184). Crops planted in such borders are castorbeans, beans, wild and cultivated sesame and even buckwheat.

Fire fields still exist today, but are nowhere near as frequent as in the P'yŏngbuk region (p. 255). The consequences of former fire-field farming for the landscape are still very apparent though in the eastern mountains here as well. The second growths described above are probably all the results of fire-field farming. On Ch'ŏrongsan I encountered the highest fire fields at 900 and 1000 m, where potatoes and sesame were being grown, respectively. The Japanese forestry service is trying very hard today to stop fire-field farming, not only by decrees and surveillance but also through information and by taking advantage of Korean superstitions. On a pass south of Tŏkch'ŏn I observed a fir tree that had been felled except for a stump about 1.50 m high. It had a piece of bark on top as a roof to protect it from decay. Attached to the stump was a sign with official regulations on forest conservation. Underneath it, on a spot where the bark had been removed, was a picture of a tree spirit in the figure of a traditional Korean mandarin in black on white with the inscription: "*Shōwa* 7th year, 9th month (i.e., October 1932). Tree–Bush Sacrifice Great Supervisor."

As in the P'yŏngbuk region, the economic situation in the Taedong basin is characterized not only by the paramount agriculture, but also by mining. There are a great many gold deposits in the contact between the granite intrusions and limestone (p. 66). They are mined, for example, near Sŏgang and Sunan. The oldest gold mine is that of Suan–Holgol south of the central section of the Namgang. Kotō devoted a study to it, on the basis of research by two other Japanese [312b, cf. also 310, 806, 807]. Here the granite towers up to 1120 m again in Ŏnjinsan, which is visible for a great distance. On the northern slope, where the surface intersects the contact, the mine lies at an elevation of 400 m in a still very mountainous, gullied, sparsely wooded area. It is connected with the Namgang by road. Following earlier attempts at mining that were later abandoned, it has been operating continuously since 1896 under various owners. The mining process is similar to that at the Unsan mines (p. 256f).

Much more important for the economy of the Taedong basin is the mining around the lower reaches of the Taedonggang. On the eastern side of the giant loop of the river, and thus capable of being reached by sea-going ships up to 5000 tons, the town of Songnim (20,051 inhabitants, 12% of them Japanese) lies in the middle of the highly sculptured hill country. Only a few kilometers to the east are four iron mines in limestone belonging to the Chosŏn system [817]. Narrow-gauge railroads connect them with Songnim, which is also connected with the Korean network at Hwangju by a standard-gauge railroad. On the banks of the river an ironworks stands with three blast furnaces founded in 1919 and beside it the Mitsubishi Company's steel and cast iron plant with its tremendous smokestacks that are visible from afar. Songnim lives exclusively from the smelting, processing and shipping of iron. It is the most important center of this type in Korea. The Japanese name of the city, Kenjiho, honors the Japanese officer Kenji

Watanabe, who landed here with troops and equipment during the Sino–Japanese War in 1894.

The most important mines in the Taedong basin lie in the immediate vicinity of P'yŏng-yang on both sides of the river. Here there are anthracite deposits in various formations of the P'yŏng-an system. The total reserves are estimated at 600 million tons [796–804]. Anthracite from the Hongjŏm formation is mined most actively. Due to the intense folding and overthrusting, which caused the anthracitization, the bedding is complicated, and mining is difficult. It is done in underground mines. The most productive mines lie across from P'yŏng-yang on the Nangnang plain near Sadong, a few kilometers to the east near Sŭngho on the banks of the Namgang and 15 km northwest of P'yŏng-yang, with which all are connected by railroad. Since the anthracite is quite friable, a large proportion of the production is made into briquettes in the vicinity. Some of it is shipped downstream in junks.

The start of coal mining led to the development of an important industrial complex outside the city gates on the Nangnang plain (Fig. 59). In Sŏn-gyo-ri stand the buildings of the Western Chosōn Electric Power Company and the Nippon Sugar Company, the latter of which formerly processed the beets from the area, and in Sŭngho the Onoda cement plant. Other industrial establishments in P'yŏng-yang are engaged in rice polishing, producing rice wine, beer brewing, tobacco processing, spinning, manufacturing rubber shoes and stockings, pottery, brickmaking and producing corn products and starch, the latter of which was in American hands till recently. In addition, P'yŏng-yang has a large railroad repair shop.

P'yŏng-yang's geographical location is the most advantageous one in the northern half of Korea [809–813]. It lies on a river that could be navigated up to the city by the sea-going ships of ancient times and on which southern Chinese wood was probably already shipped in during the Lolang period (p. 23). The rivers of the basin converge on this area. Here the ancient Imperial highway from Seoul to Beijing crossed the river. Low passes make it possible to cross the Main Range to Wonsan. Intensive agriculture is possible in its environs. P'yŏng-yang is rightfully named the "city of tradition." It has the longest history of all Korean settlements, and time and again it was the capital of states in the northern half of the country. In our historical chapter we pointed this out for the time of Kija, the Lolang period and the Koguryŏ period. Before the national boundary lay at the Yalu, P'yŏng-yang also played the role of northwestern gateway. Possession of the city has a military significance that has been revealed in many wars, particularly during the Hideyoshi invasion and the Sino–Japanese War of 1894/95. Even today bullet marks on the pavilions in the northern part of the city bear witness to the battles that raged here in 1894.

The city lies in a sheltered location on a peninsula projecting southward between the Taedonggang and its right-bank tributary, the Pot'onggang, at the foot of the hills to the north, which are composed of massive beds of graywacke that drop off toward the Taedonggang with a rocky, undercut slope. The channel line stays close to the west bank near the city, while on the Nangnang plain side wide sand flats appear at low tide. The level terrace on which P'yŏng-yang lies does not rise enough above the water level in the old part of the city, however, to rule out floods. The last time that it was endangered by flooding was in 1925 [452]. From

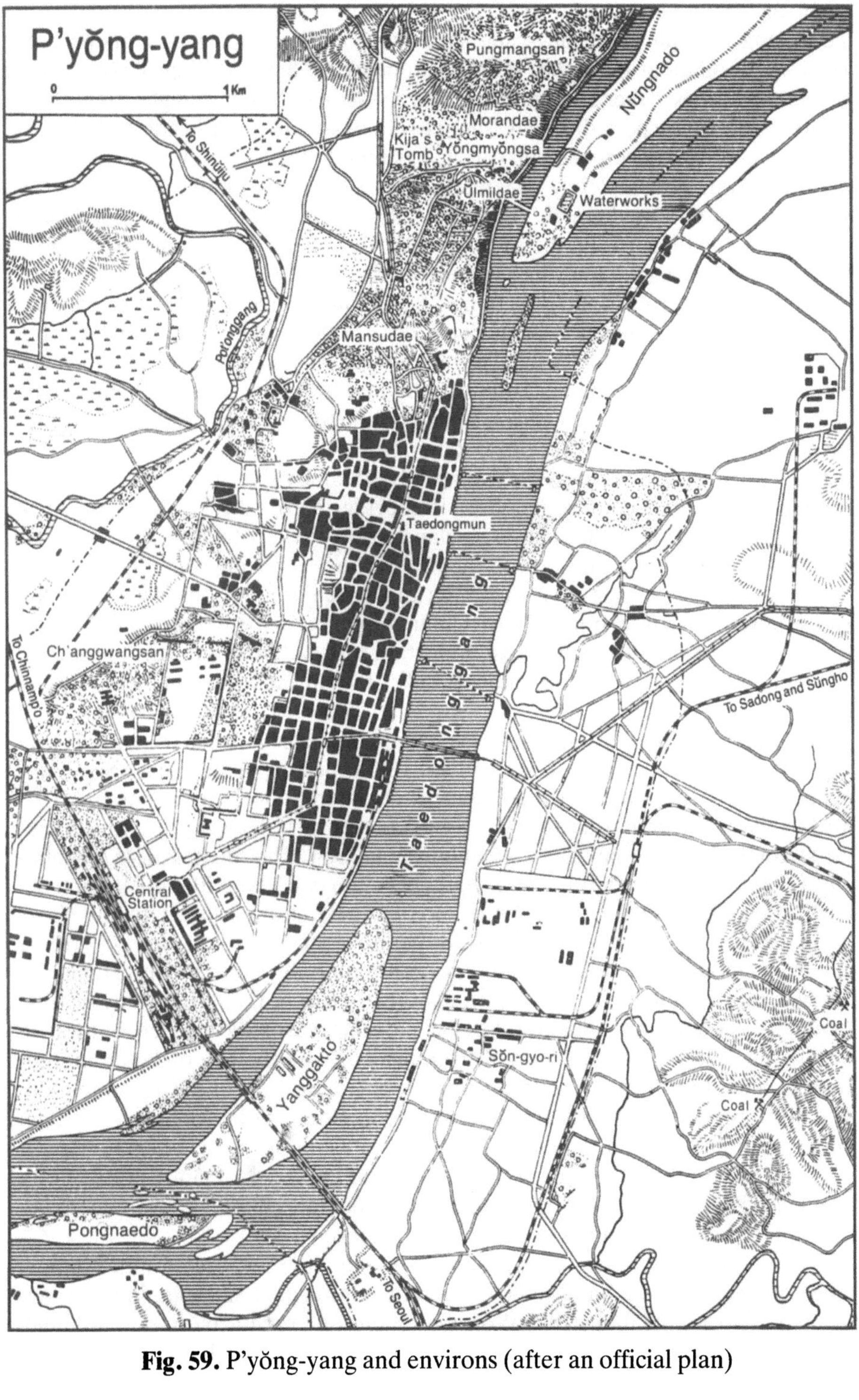

Fig. 59. P'yŏng-yang and environs (after an official plan)

the foot of the hills the old part of the city spreads out southward for 3 km on the bank of the river. An approximately 300-year-old city map hanging in the Lolang museum in P'yŏng-yang shows that the built-up area extended only to the South Gate at that time. It also demonstrates, however, that P'yŏng-yang, in contrast to most other Korean cities, had a system of broad, straight streets long before the Japanese period. Today this network of streets, some heading parallel to the river and others perpendicular to it, has merely been improved, fortunately, without becoming completely systematic. Inside this network of main streets is a maze of little traditional Korean alleys. P'yŏng-yang has a large number of buildings from the pre-Japanese period as well, such as the double-roofed Taedong Gate, which stands at the spot where formerly the main ferry crossed the river in the direction of Seoul, and beside it a tall stone pagoda. Beautiful old Korean pavilions with lovely views of the city, the river and the fertile plain on the opposite side crown the northern hills. The old part of the city still ends today at the Great South Gate. Southwest of it lies the new, entirely Japanese part of the city with barracks, the division headquarters, the provincial capital of P'yŏng-annam-do, the hotels, and the central station (Fig. 59).

P'yŏng-yang is the center of a good road system, by Korean standards, whose two halves are connected by a modern iron bridge across the Taedonggang. The Korean trunk line uses the first large island below the city to cross the river on a bridge 1200 m long. In addition to the branch lines servicing coal mines and industry, the lines to Kanggye via Sunch'ŏn and Kaech'ŏn, to Wonsan via Sunch'ŏn and P'aŭp and to the port city of Chinnamp'o branch off here. This makes P'yŏng-yang Korea's second most important railroad junction. It is the second largest city (1938: 234,726 inhabitants, 13% of them Japanese) and is still growing.

The seaport of the Taedong basin, Chinnamp'o was only a small fishing village at the time of the Sino–Japanese War [814–816]. Its development began when it was opened up by the Korean government as an international treaty port in 1897. Meanwhile Chinnamp'o has grown to be Korea's third largest port, which handled a total trade volume of 114.9 million yen in 1934. The city has 61,457 inhabitants (1938), 12% of them Japanese. It lies between two bays with tidal flats on the north side of the estuary, at the blunt end of a rocky promontory with hills rising up to 100 m. The harbor basins afford medium-sized steamers direct berthing and have modern stevedoring facilities with railroad sidings. Unfortunately, every winter Chinnamp'o, and especially the approach, is obstructed by drifting ice on the Taedonggang for several weeks. For this reason the Chōsen Anthracite Mining Co. intends to ship its coal from the ports on the Sea of Japan in midwinter, once the diagonal railroad line P'yŏng-yang – Wonsan is completed. The most important exports are coal, iron ore, cotton, iron products, soybeans, apples and paper. Simultaneously Chinnamp'o is important as a fishing port. It also has industrial facilities for polishing rice, milling and the manufacture of iron products. Twelve kilometers northwest of Chinnamp'o, in Kwangnyangman, are the most important salt pans in the northern half of Korea.

Compared with the two cities described above all others are of minor importance. None has more than 14,000 inhabitants. Some have silk spinning or weaving and other textile industries. Worthy of mention are Sunch'ŏn, railroad junction and county seat in the northwest, and particularly Sŏngch'ŏn, which played an important role during the Koryŏ dynasty, but is bypassed by the long-distance routes today. A flourishing little county seat nearby, P'aŭp, lies in a junction of valleys in the eastern mountains that plays an important role for transportation. On the main railroad and road to Seoul lie the local markets and bridge locations Chunghwa and Hwangju. The houses in all of these rural towns are of village types here, too. The very common thatched roofs have the straw braid along the ridge line, its ends often weighted down with stones. In summer they are half overgrown by climbing gourds and cucumbers. There are, however, also hipped, sweeping roofs covered not only with Korean tiles, but also with slabs of slate and limestone shale.

In its average population density the Taedong basin corresponds almost exactly to the average for all of Korea (p. 14). In the counties that extend up to the drainage divide in the Main Range, the density is naturally low (Yŏngwon-gun 22, Yangdŏk-gun 31). Toward the coast it increases quite regularly and reaches its highest values in the southwest (Yonggang-gun 151, Map 3 [cf. editors' note on p. 7] and Fig. 85).

6 The Hwanghae Peninsula

The southwestern part of the province Hwanghae-do occupies the largest peninsula on the Korean Yellow Sea coast. Due to its location far to the west it has so much individuality that it deserves to be treated as a separate region, although there are close relationships, both natural and cultural, to the neighboring regions, particularly the Taedong basin. This individuality extends only to the actual peninsula, however, which is separated by the estuaries of the Taedonggang and the Han-gang. Toward the north and the east it is difficult to define the boundary. We drew the southern boundary of the Taedong basin basically according to morphological and hydrographical criteria (p. 259, 262). With this we also defined the northern boundary of the Hwanghae peninsula. In the east the drainage basin of the Yesŏnggang causes problems, because it lies in the transitional area between the peninsula, on the one side, and the Kyŏnggi and upper Han-gang regions, on the other side. I include it in the former, since for 550 years most of it has belonged to the province of Hwanghae-do, and thus in the realm of cultural geography it has developed more connections to the Hwanghae peninsula than to the adjoining regions to the east, although the drainage basin of the Yesŏnggang can scarcely be considered part of the Hwanghae peninsula itself. The Yesŏnggang region also includes the extreme northwestern part of the province of Kyŏnggi-do, in which the former Korean capital Kaesŏng lies. The more or less rectangular peninsula thus has an average length of around 160 km and a width of 90 km. It has an area of about 13,000 km^2 (Fig. 39).

As the remarks on p. 261 showed, the western third of the Taedong basin already cannot be considered simply a part of a dip slope that tilts southwestward. The Hwanghae peninsula lies completely outside this system and develops pronounced morphological traits of its own (Fig. 60). The structural trends of the relief are oriented partially east–west and partially north–south. North–south runs the narrow, 100-km-long drainage basin of the Yesŏnggang, south–north the 60-km-long basin of the Chaeryŏnggang, which rises only 18 km from the south coast, but flows into the big loop of the Taedonggang. The morphological nature of these two drainage basins varies greatly. The Yesŏnggang has entrenched itself in a region of mountains and hills, in which isolated eminences exceed a height of 700 m even near the coast. The Chaeryŏnggang, in contrast, flows through a broad plain, and only in the east is it followed by a pronounced mountainous rim.

Outside of these two regions the Hwanghae peninsula is dominated by east–west trending mountain ranges, which though not very high in absolute ele-

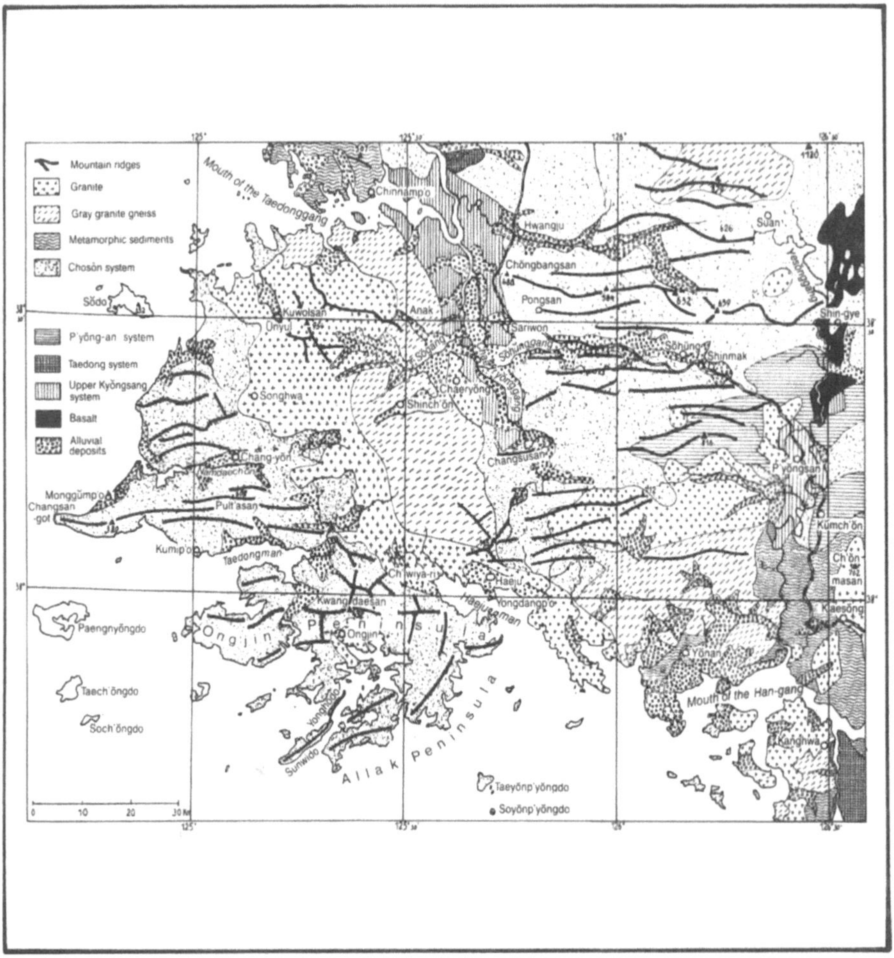

Fig. 60. Geological–morphological map of the Hwanghae Peninsula

vation are occasionally almost Alpine in their steepness, with partly narrow meandering valleys, partly wide floodplains running between them in the same direction. This structural principle is most clearly visible in the northeast and the southwest. In the northwest, in contrast, broad, dome-shaped elevations and gently undulating hills with less clearly defined trends predominate. In the southeast the latter prevail with alluvial marshlands connecting them. The route of the Korean trunk line shows the two main trends of the relief of the Hwanghae peninsula very clearly; between Hwangju and Kaesŏng it is composed exclusively of segments running west–east or north–south.

The fact that the northeast and the southwest consist primarily of east–west ranges is a consequence of the geological structure. As in the Taedong basin, in these two parts of the peninsula the strata of the Chosŏn system are widely distributed. Here quartzite and shale are more prominent than there, however. The folds have the same west–east trend, for the most part, and those jagged west–east ridges are primarily steeply upfolded, thick-bedded quartzite ridges, which have been selectively sculpted out of their surroundings. The epigenetic creeks and rivers repeatedly break through them. It is possible that an older peneplain that was later uplifted was the starting point from which this relief developed. Far below the quartzite ridges lie more recent, also dissected erosion surfaces, which probably correspond to the Nangnang plain. In the northwest and southeast, however, granite and "gray granite gneisses" predominate, the latter of which consist of alternating layers of very coarse-grained gneiss, thinly laminated fine-grained gneiss and quartzite with varying strike directions. Consequently the mountain ranges generally do not trend in a uniform direction. Where a uniform direction does exist here and there, e.g., north of Haeju, it is probably a result of horst formation. The phenomenon of sugarloaf and dome-shaped mountains takes on splendid forms in the towering, solitary granite massifs.

Naturally, the farther one proceeds to the southwest, the more maritime traits the climate has. The annual temperature range decreases rapidly in this direction. The smallest range (25.7° C) is found on the lighthouse island of Soch'ŏng-do off the southwestern coast. As Fig. 24 shows, the annual range is smaller on the Hwanghae peninsula than in the Taedong basin. Despite its location, protruding far out into the sea, the annual range is still larger than in corresponding locations on the south Korean coastal fringe, however. Here too precipitation decreases toward the west with the diminishing height of the mountain ranges. The actual peninsula between the two estuaries and especially the islands off the coast have the lowest figures for all of Korea (Map 2, cf. editors' note on p. 7). All of this is demonstrated by Table 17. For the other details the reader is referred to the appropriate comments on the Taedong basin.

During the Koryŏ dynasty Kaesŏng, under the name of Songdo (p. 29), was the center of the state for almost half a millennium. The anthropogeographical reshaping of the natural landscape, which is already demonstrable for prehistoric times (p. 16ff), made great progress on the Hwanghae peninsula during these 500 years. Even today one encounters a large number of temples, hidden away in the mountain ranges, that were founded during this golden age of Buddhism. In their environs there are still remnants of virgin forest that have not been completely destroyed, but even they have not remained exempt from human intrusions. Nowhere on the Hwanghae peninsula did I encounter unspoiled virgin forest. Generally the plant formations on the mountain ranges are highly degraded, because everywhere there are densely settled valleys and basins in the vicinity with a large demand for wood (Map 3, cf. editors' note on p. 7). Often one sees only very low shrub formations consisting of the various oaks, lespedezas and grasses, with extensive, barren talus slopes between them. The oaks of which these formations are composed belong to many different species. Whereas in northern Korea the common oak predominates, on the Hwanghae peninsula,

Station		Temperature						
		January			August		July	
	Altitude (m)	Mean temp.	Mean min. temp.	Absolute min. temp.	Mean temp.	Mean max. temp.	Absolute max. temp.	Annual range
Shin-gye	60	–7.3	–13.1	–36.4	24.8	30.1	40.1	32.1
Chaeryŏng	30	–6.7	–12.0	–26.0	25.3	30.3	41.0	32.0
Kaesŏng	60	–6.1	11.6	–24.5	25.1	29.8	38.0	31.2
Ŭnyul	28	–6.2	–11.5	–25.5	24.8	29.4	36.5	31.0
Sŏdo	94	–3.4	–	–	23.4	–	–	26.8

[a] December. [b] August. [c] July

Station	Precipitation								
	January				August				
	Average amount (mm)	per 1000	Average no. of days with precip.	Precipitation intensity	Average amount (mm)	per 1000	Average no. of days with precip.	Precipitation intensity	Annual amount (mm)
Shin-gye	15.4	14	4	3.3[a]	302	274	15	21.6[b]	1105
Chaeryŏng	10.2	12	3	2.9[a]	212	246	13	17.4[b]	858
Kaesŏng	19.2[a]	16[a]	4	2.7[a]	330	281	15	22.0[c]	1185
Ŭnyul	11.9	16	3	3.4[a]	196	269	10	19.6	728
Sŏdo	6.3	13	4	1.9[a]	107	223	11	11.5[b]	480

Table 17. Climatic data for the Hwanghae Peninsula

where the upper limit of the oak zone is never reached and where extremely varied habitats exist at lower elevations, a large number of other species are also quite prominent. They are the species that prefer dry southern slopes, *Quercus acutissima,* the broad-leaf oak *Quercus dentata*, *Quercus aliena*, *Quercus mongolica* Fisch. var. *manchurica* Nakai and *Quercus serrata*. In inaccessible ravines in the quartzite ridges and granite mountains there are still good and varied second growths, which already contain a number of southern Korean species. From the dominating trees and shrubs that we have mentioned here we can infer that the Hwanghae peninsula still belongs to the styrax zone of our general classification (p. 137f).

As mentioned, the severe destruction of the original vegetation on the Hwanghae peninsula dates back to an earlier period. Nowhere did I encounter active fire–field farming. The only reason that the statistics for the province of Hwanghae–do show a small number of fire fields (p. 405) is that the province extends northeastward far beyond the boundaries of the peninsula into the high mountains.

Hwanghae-do is the province with the highest proportion of cultivated land in the total area (34.5%). It owes this rank exclusively to its main portion, the peninsula, with which we are dealing here. Outside of the peninsula there is little agriculture in the province. The large county of Koksan has only 14.3% cultivated area; the counties of Shinch'ŏn and Yŏnbaek on the peninsula in comparison, have 46.5 and 57.6%, respectively (Fig. 86). As in the P'yŏng-an provinces, the proportion of paddy area increases toward the coast. But it is not much larger than in the two of them. Dry fields predominate. Only 8.3% of the total area of the province is devoted to paddies, but 26.2% to dry fields. In individual counties the proportion of paddies is naturally much higher, such as in the low-lying areas north of the mouth of the Han-gang (Yŏnbaek-gun 34.3%). The predominance of dry fields is explained here, as in the Taedong basin, by the wide distribution of permeable rocks. In addition to limestone there is also basalt in the northeastern part of the Hwanghae peninsula.

The varieties of dry-field crops and the methods of growing them are just as diverse as in the Taedong basin, and in the sphere of agricultural geography the Hwanghae peninsula strongly resembles the southwestern part of the Taedong basin. Far ahead of all others is wheat, which is grown with more intensity than anywhere else in Korea (Fig. 88). In addition foxtail millet, kaoliang, barnyard millet, upland rice, barley, buckwheat, soy and other beans, sweet potatoes, cotton, melons, squash and vegetables are grown. More extensive cultivation of cotton is being heavily propagated here by the Japanese, particularly by the agricultural research station in Sariwon. As in the Taedong basin, oats are almost completely lacking. Maize, hemp and potatoes are only sparsely represented. The variations in intercropping are just as diverse as in the Taedong basin. Even more barley and wheat are grown as winter crops, because the winters are milder. Like there they are planted in the troughs, and the spring crop (cotton, beans or millet) is planted on the ridges a few weeks before the winter crop is harvested. In the vicinity of the settlements flourishing groves of fruit trees and chestnuts and mighty specimens of *Abelicea serrata* grow. The chimneys beside the farmhouses are often made of unfired clay and have a covering of straw as a protection against rain.

From the coasts with their many bays fishing is carried out, the catch varying greatly from year to year. The year that I was there was a very good one. The net proceeds of a single village on the Allak peninsula, Pup'o-ri, were 100,000 yen, of another village on the island of Taeyŏnp'yŏngdo, 600,000 yen. Important, but thus far insufficiently developed fishing ports are Monggŭmp'o on the west coast, Kumip'o, Pup'o-ri and Yongdangp'o on the south coast. During the fishing season (p. 442f) large fleets of junks gather in them, as on Taeyŏnp'yŏngdo. The once famous pearl fishing done actively off the coasts of the Hwanghae peninsula (p. 24f) has waned because the banks have been depleted. Winter is the calm season during which only a few whales are caught. In spring cod, sharks, sea bream and corvenia are fished, in summer mackerel, sardines, shrimps and eels, in autumn yellowtail and hairtail, while croaker is caught during the entire summer half-year and Spanish mackerel, sea cucumbers, flounder and gray mullet are fished in both spring and autumn.

Apart from village workshops, the industrial development is only in its beginnings thus far. Mining, in contrast, plays a remarkable role, as in the Taedong basin. Gold mining is highly developed in Ongjin-gun. Otherwise gold is secondary in importance to iron ores. The iron mines here also exploit primarily sedimentary ores from the Chosŏn system, but also ones from the Kyŏngsang system. Many of them have railroad connections, such as those at Hasŏng and Naet'o on the upper Chaeryŏnggang and Ch'ŏlsan near Chaeryŏng. The latter produced a good 1 million tons of 43–62% ore between 1910 and 1927 [817]. Others lie near Ŭnyul, Tongch'ang-ri and Kŭmgang-ri (Anak-gun) and near Songhwa [819]. All of these are open pit mines, but some of them have dug into the depths in five to six levels. Some of the mines we have mentioned do not work continuously, however. Of future importance, finally, are the mineral hot springs that rise in gneiss and granite, such as those of Ongjin, Paekch'ŏn, Songhwa, Anak and Shinch'ŏn [346–348].

The Hwanghae peninsula, too, has an average population density that is almost equivalent to the average for all of Korea (p. 14). The span between the northeast and the southwest is not as great as in the P'yŏngbuk region and the Taedong basin, however, because the peninsula does not extend into the inhospitable regions of the Main Range. Of the counties that belong entirely to this region, Shin-gye-gun on the upper reaches of the Yesŏnggang has the lowest density, with 59 inhabitants/km^2, Yŏnbaek-gun on the mouth of the Han-gang, filled with rice-growing lowlands, has the highest, with 175 inhabitants/km^2 (Fig. 85).

The Hwanghae peninsula begins in the extreme northeast with a section whose physical geography we are acquainted with from the Kaema upland. Between Koksan and Shin-gye a basaltic mass consisting of several stratified lava flows was deposited into a north–south trending valley, which was eroded during a previous period and across which the divide between the Namgang and the Yesŏnggang ran. It was probably the structural extension of the upper Namgang valley. South of Koksan this basaltic mass begins with a 100-m-high escarpment. Above this escarpment its surface lies at 260–280 m above sea level, and it dips toward the south. An extension that continues beyond Shin-gye into the central Yesŏnggang valley finally has a height of only 30 m. The total length of the flood basalt is 52 km; the maximum width, 17 km. The masses of basalt have pressed the upper Yesŏnggang upward on the right-hand flank of its former valley, and it has cut a diversion channel here since then. Since the basalt is permeable, the surface remains almost undissected. The divide between the Namgang and the Yesŏnggang is very difficult to determine. Only the Chisŏkch'ŏn, a tributary of the Yesŏnggang, has entrenched itself about 60 m deep, without sinking through the basalt, and exposes two lava flows with magnificent columns. The prebasaltic surface, which consists of schist and limestone shale of the Chosŏn system, juts up through the basalt flows in the form of a large number of small, rounded hills. The often tabletop-like surface of the basalt is covered by a fertile, well-cultivated layer of red soil, which becomes saturated during the summer rains. On the old mountains, in contrast, only thickets of oaks and lespedezas grow.

The Yesŏnggang then cuts through the beds of red sandstone and conglomerate of a collapsed block of the Kyŏngsang system and finally flows into the estuary of the Han-gang west of Kaesŏng, with a mouth widened by the tides.

Of the larger-sized Korean cities, Kaesŏng is the one that has preserved its Korean character most strongly. Only 3% of its 56,595 inhabitants are Japanese (Fig. 61). The city lies at an elevation of 40 m in a basin that developed in coarse-grained biotite granite and is framed by a bright-green setting of the same rock rising to 488 m (Song-aksan) in the north. Large stretches of the wall

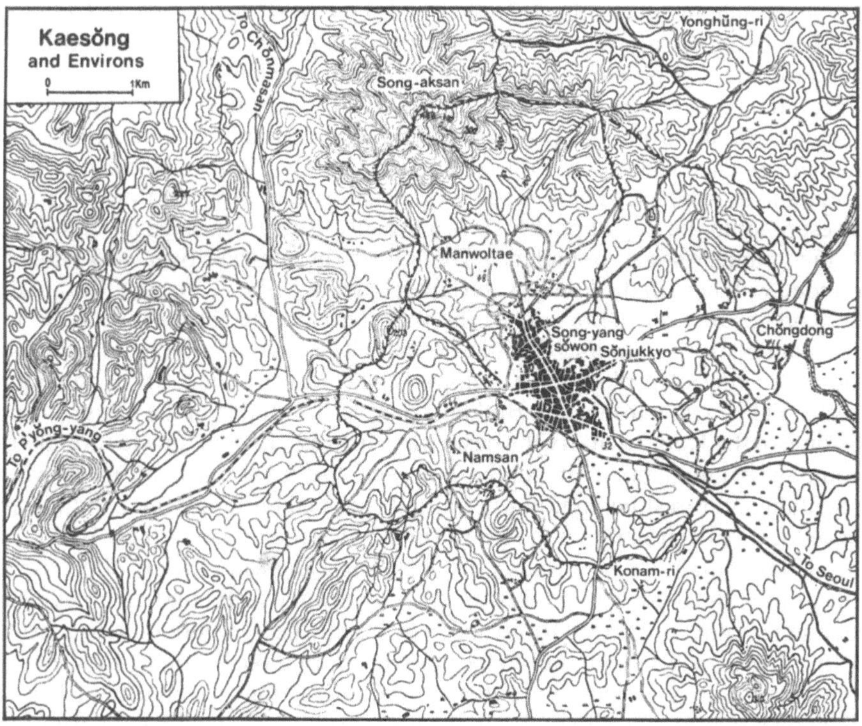

Fig. 61. Kaesŏng and environs (simplified section of the topographical map 1:50,000)

of the capital of the Koryŏ dynasty are still in good condition, and in the north it runs along this framework and up across the highest point. The area surrounded by the wall was never completely filled with houses. Toward the south, however, the city grew far out into the plain after the construction of the wall. Here the old wall runs east–west, dividing the city precincts in two, and today the handsome "Great South Gate" lies in the center, at the intersection of the two main streets. A more recent fortification, of which only portions of an earthen wall still remain, encircles the entire basin to the east and south. Kaesŏng's historically important sites embellish the older, northern half of the city and its charming environs, such as two attractive archery grounds, the school Song-yang sŏwon, built in honor of the esteemed last minister of the Koryŏ dynasty, the stone bridge Sŏnjukkyo, on which the latter was murdered by the followers of General Yi, the large memorial hall Chŏng Mongju ch'ungbigak close by, the Confucian school Sŏnggyun-gwan, the grave of the founder of the Koryŏ dynasty and the walls of the old royal palace on the hill Manwoltae. Kaesŏng has a government ginseng office. The foremost center of Korean ginseng cultivation is in the environs of Kaesŏng, especially southeast of the city in the hill country interspersed with paddies that extends to the Imjin-gang (p. 422). From the Koryŏ period porcelain manufacture has been preserved in the environs, the seat of trade likewise being Kaesŏng [822].

The same granite intrusion that forms the basin of Kaesŏng and its northern boundary towers up farther north in the dome-shaped massif of Ch'ŏnmasan (= Heaven Touching Mountain, 762 m). Toward the west, north and east Ch'ŏnmasan has been dissected extremely sharply out of its cover of blackish-gray metamorphic schists. The latter form a hill country with moderately steep slopes everywhere covered by vegetation. Ch'ŏnmasan rises out of this with massive, rounded, often barren, steep slopes around 300–500 m high. Its highest elevations are arranged in a circle with a diameter of barely 3 km, and the center of the mountains is occupied by a basin, out of which the water flows northward over the lowest point in the ring of peaks. Here a broadly protruding ledge has developed at the granite–schist boundary. The Ojoch'ŏn plunges over a 50-m-high, rounded wall here in the lovely waterfall of Pagyŏn p'okp'o. The admiration that the Koreans feel for the beauty of the waterfall and the basin is expressed in a large number of inscriptions in Chinese characters carved into the domed walls. Along the ring of peaks runs a well-preserved fortification with its main gate next to Pagyŏn, and inside two Buddhist temples lie hidden in the forest. During the time of the Koryŏ dynasty the basin was a fortress. Except for a few monks the mountain is uninhabited. Recently, however, the royal house of Yi, to whom mountain and forest belong, has permitted charcoal burning again, and this has caused severe destruction to the forests in the higher elevations of the basin.

In Ch'ŏnmasan the gray rounded slopes, sugarloafs and towers often rise directly out of the dense green forest. At the top the massive rounded slopes disappear under a cover of brown forest soil, which again bears woods or dense shrubs. Below it water runs down the slopes in sheets during the summer. Between the light-colored areas washed by water are darker sections covered by lichens or green ones overgrown with moss. In the lush shrubbery in the higher parts of the basin the predominant species are *Lespedezas*, *Rhododendrons* and the oaks mentioned on p. 275, the maples *Acer pseudosieboldianum*, *Acer ginnala* and

Acer mono, the red pine and the mountain ash *Sorbus commixta*. Less common are shrub forms of the Kaempfer larch, the common ash, the ox-horn hazel nut (*Corylus manchurica* Rupr. and Max.), the Korean blueberry, of *Castanea crenata* S. and Z. var. *dulcis* Nakai, *Micromeles alnifolia* Koehne var. *typica* Schneid., *Polygonatum officinale*, *Benzoin obtusilobum* Kuntze, *Stephanandra incisa*, *Styrax obassia*, *Rhus trichocarpa* Miquel and the liana *Akebia quinnata* Decne. In the forests around the two temples many of these shrubs occur as trees. They are joined by the Korean wild pear, the hornbeam *Carpinus erosa*, the hawthorn *Crataegus pinnatifida* Bunge, the oleaster *Eleagnus crispa* Thunb. var. *typica* Nakai, *Fagara shiniifolia*, *Palura paniculata* Nakai var. *pilosa* Nakai, *Securinega fluggeoides* Müller and *Callicarpa japonica* Thunb. Many lianas climb up the trees, such as *Pueraria hirsuta*, *Smilax sieboldii*, *Tripterygium regelii* and the wild grapevine. This list of flora shows that the magnificent Ch'ŏnmasan forest still belongs to the styrax zone of our general classification. A large number of representatives of the oak zone are, however, already found in it.

West of the mouth of the Yesŏnggang up to Ch'wiya-ri at the base of the Ongjin peninsula follows a former archipelago composed of granite gneiss, its parts now connected to the mainland by the formation of coastal marshes. The material for this intensive alluvial formation was provided by the debris transported by the Han-gang. North of the low-lying marshes the hills unite to form a gently dissected hill country covered by red soil. Rice paddies spread out on the coastal marshes for the most part, some of them admirably irrigated. Several Japanese irrigation associations, such as Seiyōsha, have rendered outstanding services here. In the hill country behind there are several large reservoirs. In Yŏnbaek-gun, whose main town is Yŏnan (15,884 inhabitants), excellent black peat is dug at the bottom of former rice paddies.

Near the western end of this section, away from today's main transportation routes, lies the capital of Hwanghae-do, Haeju (48,649 inhabitants, 9% of them Japanese), founded around 1400 and moved to its present location in 1467. For several years it has been connected with Sariwon by a private railroad following the upper reaches of the Chaeryŏnggang. The extension of this railroad connects the city with the nearby port of Yongdangp'o. Haeju lies 2.5 km from the coastal mud flats in a valley parallel to the coast, nestled up against the foot of a 945-m-high granite mountain ridge that already runs east–west. In the city plan the strict rectangular boundaries of the traditional Korean city are still quite evident [821]. Haeju has outgrown them in all directions. On the main east–west street stands the old Chinese-style amusement hall Puyongdang, built into a lotus pond. As the magnificent view from the Haeju shinsa shows, the northern half of the city is dominated by Japanese tile roofs, the southern half by Korean straw ones. The fortress that was indispensable in olden times lies hidden in the mountains to the north. It was laid out already between 907 and 960 A.D. [131, p. 95]. At the foot of the mountains, which are adorned by lovely old red pine forests, three tasteful pavilions standing just above the rocky bed of the river bear witness to the tranquil life of city dwellers in traditional Korea.

North of this coastal region the mountain ranges composed mainly of Chosŏn and P'yŏng-an strata trend east–west with the direction of the Taebo orogeny. The northernmost one ends in the west with the precipitous summit of Chŏngbangsan (480 m), which is crowned by a fortress, and many of its peaks come close to 700 m. The southern foot of the Chŏngbangsan range is followed by a valley that is used by the road and the railroad Kaesŏng – P'yŏng-yang. In it lie the market and farming towns of Shinmak and Sŏhŭng. The Sŏhŭnggang meanders through this valley, except for a 10-km-long central segment, in which it breaches the next range to the south in a rocky defile. Near the western end of this valley lies the Tertiary coalfield of Pongsan, which is being intensively exploited today [818]. The valley finally widens out like a funnel and merges with the basin of the Chaeryŏnggang, which is rounded off to the south by the rugged quartzite wall of Changsusan (747 m). In the angle between the Sŏhŭnggang valley and the Chaeryŏnggang basin, in the middle of paddies, dry fields and orchards and on the main road and railroad, lies the busy market town of Sariwon (30,389 inhabitants), in which milling has also been developing in recent times.

These ranges composed of Chosŏn rocks end in the west at a straight, north–south trending fault scarp. Along it the block composed of porphyry, porphyry conglomerate, tuff, schist, limestone and sandstone of the Kyŏngsang system that is familiar to us from the big loop of the Taedonggang has been downfaulted (817). These rocks compose the hilly pediment along which the railroad and road from Sariwon run, past the foot of Chŏngbangsan northward toward Hwangju. The western edge of the Kyŏngsang fault block is not evident in the landscape, in contrast to the eastern one. The hill country spreading out there extends right across the main fault out of the Kyŏngsang strata and into granite gneiss. The centers of this well-cultivated red hill country, above which the installations of many of the above mentioned iron mines rise, are the market towns of Chaeryŏng (17,566 inhabitants), Anak (19,171 inhabitants) and Shinch'ŏn (17,239 inhabitants). Between the hill country and the pediment of Sariwon–Hwangju lies the most deeply depressed part of the Kyŏngsang fault block. The postglacial rise of the sea level has caused the Chaeryŏnggang and its western tributary, the Sŏgang, to aggrade highly. The tides affect both rivers and their many branches far upstream. Levees have had to be built along them so the rich floodlands could be used for rice paddies. At high tide the water level in the rivers lies high above the floodlands. The rice farmers live in dispersed farmsteads scattered across the floodlands. The parts where no levees have been built yet are still uninhabited reed jungles today.

In the extreme northwestern part of the Hwanghae peninsula one of the big granite intrusions of central Korea is exposed. Its dry soil and barrenness contrast strikingly with the bright green colors in the region of the Chosŏn system, where the destruction of the natural forest has had less disastrous effects. The low granite mountains and hills are covered by the usual scrawny red pine scrub on otherwise eroded, bare yellow soil. The plain of Shinch'ŏn, extending far westward into the granite region, has, however, been turned into a rice-growing area by a Japanese company, which has built a reservoir at its southern end.

The mountains bordering the plain and the yellow hills consist of the same granite. They include the highest peak on the Hwanghae peninsula, Kuwolsan (954 m), one of the five sacred mountains of ancient Korea. In the central part of this mountain the granite is very fine-grained and highly jointed. Consequently the rounded forms make way to very steep dropoffs without any vegetation. Like Ch'ŏnmasan, near the summit there is a high-lying basin, and the stream draining it plunges a total of 300 m into the depths in a three-tiered waterfall. The

Kuwolsan basin also once served as a fortress, and the walls still run along the surrounding ridges. The bottom of the basin is farmed by a hamlet that still bears the name Sansŏng (p. 163). Due to the elevation finger millet takes the place of foxtail millet. The lovely forest in the valley of the waterfall consists primarily of red pines and Korean nut-pines. The deciduous trees of the former natural forest are represented only in the form of shrubs and young timber. The forest belongs to one of the main temples, P'aeyŏpsa, which lies in a state of disrepair at the western foot of the small mountain.

The southwestern sector of the Hwanghae peninsula is again dominated by the Chosŏn folds, and with them begin the ranges that also characterize it. They, like the longitudinal valleys separating them, trend east–west. The center of this section is the county seat of Chang-yŏn (13,463 inhabitants) on the Namdaech'ŏn. The ranges are most superbly developed between the Namdaech'ŏn, the main river, which flows through such a longitudinal valley, and the southern coast, i.e., the bay Taedongman. Directly above the basin of Chang-yŏn, in which a peneplain with a relative elevation of 50–70 m truncates steeply inclined, laminated schist, the quartzite mountain of Pult'asan rises like a wall. Some of its peaks tower above 600 m on the scarp slope with mighty precipices, while the average elevation remains at 400 m. The parallel range to the south has no outstanding peaks. It extends much farther westward, though, and forms the 20-km-long peninsula stretching out to the west from Hwanghae-do, ending in the cliff-girded cape of Changsan-got, which is dangerous because of its tidal currents. Between these two ranges there is a longitudinal valley following a thick interfolded layer of shale. South of them, around Kumip'o, bars have begun to form, and north of the Changsan peninsula the process of coastal straightening has progressed to a degree that is rare on the Korean Yellow Sea coasts. Already around the turn of the century the Europeans in Korea built a summer resort with single-family houses near Kumip'o that is known as "Sorai Beach" [823]. Wherever barren quartzite cliffs do not predominate, Pult'asan is covered by a shrub formation about twice as tall as a human. It is composed of the familiar oaks and lespedezas. Among the latter *Quercus acutissima* is very commonly found in dry, sunny locations on the south side. Between these shrubs many grasses and flowers grow, including a gorgeous fiery red lily (*Lilium amabile* Palibin). In the ravines on the north side blackish brown, slippery humus has formed between the quartzite blocks, and in these moist, shady habitats a magnificent and varied deciduous forest has been preserved. It is composed primarily of *Benzoin obtusilobum* Kuntze, here a massive tree with oil-bearing fruits. Other good-sized trees are *Carpinus erosa*, *Acer mono* var. *savatieri* and *Sambucus latipinna* Nakai. The familiar lianas make this deciduous forest well-nigh impenetrable.

Between the bay Taedongman in the west and that of Haeju in the east the land connection narrows down to 25 km. Here a secondary peninsula shaped like a bunch of grapes begins, which widens out to 77 km and projects 45 km to the south. It is called the Ongjin peninsula. Here the crenulation of the central Korean Yellow Sea coast resulting from the recent ingression has reached perfection. Tertiary peninsulas hang onto the relatively small main body of the peninsula in all three possible directions. The largest one is the southeasternmost one, which we shall call the Allak peninsula after the main village.

The Ongjin peninsula is also composed mainly of Chosŏn strata. In the center, however, is another granite intrusion. Along the contact of the granite and the Chosŏn strata the peninsula rises to its highest elevation (Kwangdaesan, 432 m). Korea's third largest gold mine operates in it. The core of the granite, however, forms a basin in which the salty hot springs of Ongjin (30–40° C) rise [348]. The approximately 30-km-long Allak peninsula, which trends in the strike direction of the folds (southwest–northeast), has near its center an isthmus that is only 1 km wide. The peninsula is composed of a saccharoidal white sandstone belonging to the Chosŏn system with embedded quartzite deposits and consists of dozens of peaks, which, however, rise only to 217 m. The intruding tidal bays continue on as alluvial plains, welding the peaks together. The northwestern inner coast is an unchanged ingression coast, with the elongated islands of Sunwido and Yonghodo, a former mountain range, offshore. On the southeastern outer side coastal straightening has begun in the form of cliffs and bars. The practically uninhabited outer end of the Allak peninsula is covered almost entirely by open, low red pine forest, with a lush undergrowth consisting of the familiar oaks in association with *Sorbus commixta*, *Rhododendron mucronulatum* and lespedas. Toward the northeast hamlets, dry fields and paddies increase as alluvial plains and low hills with red soil become more common.

Between the cape of Changsan-got and the southwestern tip of the Allak peninsula there is a group of islands off the coast of Hwanghae-do that was formerly known to Europeans as the Sir James Hall Group (p. 11). It is called Paengnyŏng Group today, after the largest island. The southeasternmost of these islands is the just mentioned Sunwido. These too are densely populated rocky islands, rising to more than 200 m, with cliffs and ingression bays. Paengnyŏngdo and the two neighboring islands to the south, Taech'ŏngdo and Soch'ŏngdo, had a total of 7459 inhabitants in 1936, who lived by farming and fishing. The postglacial age of the ingression that separated these islands from their extensions, the arms of the Ongjin peninsula, is proven here by the tooth of an elephant found on the bottom of the sea near Paengnyŏngdo [820].

7 The Kyŏnggi Region and the Capital, Seoul

Between the Hwanghae peninsula in the north and the T'aean peninsula in the south the coast of the mainland recedes almost 100 km toward the northeast, and here Kyŏnggiman spreads out, with the islands of the Tŏkchŏk archipelago scattered around in its southern half. The coast of the mainland is dissected by a number of good-sized funnel seas, the largest of which lie at the two ends of this stretch of coast. They are the estuaries of the Han-gang and the Ansŏngch'ŏn, the latter known as Asanman. The hinterland is drained almost exclusively by the Han-gang. The upper reaches of many of the branches of the Han-gang emphasize the dip slope character of central Korea in their southwesterly direction. From Ch'ungju to its mouth, however, the main river flows northwestward, i.e., perpendicular to the direction of dip, for 200 km. In its middle reaches, the largest tributary, the Imjin-gang, takes the opposite direction. Both flow through hill country and plains above which isolated peaks rise more or less steeply. The mountains that rise above 1000 m do not begin until 80 km from the coast, i.e., northeast of the central portions of the Han-gang and the Imjin-gang. This region between the higher mountains and the coast of Kyŏnggiman has a central position in Korea's economy, transportation network and politics. We shall call it the Kyŏnggi Region, since it coincides for the most part with the province of the capital, Kyŏnggi-do. On its three landward sides the Kyŏnggi region does not extend quite as far outward as the province. The boundary with the Hwanghae peninsula was elaborated on p. 271. In the southeast the narrow strip along the railroad Ch'ŏnan – Changhowon, which is oriented toward Ch'ŏnan and thus the Ch'ungch'ŏng provinces, is included in the Ch'ungch'ŏng–Chŏlla region. In the northeast I have the Kyŏnggi region extend only to the foot of the higher mountains, thus drawing the regional boundary just northeast of the central portions of the above mentioned rivers, while the provincial boundary ascends to an elevation of 1468 m in Hwaaksan. With these boundaries the Kyŏnggi region is 130 km long and has an average width of 85 km and an area of around 11,000 km^2. The Tŏkchŏk islands, which are scattered as far as 95 km from the coast toward the southwest, are included (Fig. 39).

If we include the drainage basin of the Imjin-gang, which flows into the funnel-shaped estuary of the lower Han-gang 50 km above its mouth, the Han-gang system drains the western slope of the Main Korean Range for no less than 320 km (Fig. 62). The longest branch of the river (482 km) is that of the Namhan-gang (South Han River). It rises in the Main Range on Odaesan (1563 m) and describes an enormous curve toward the south as far as Tanyang, beginning the northwest course mentioned previously at Ch'ungju. All other large members of the Han-gang system enter from the right. Near Yŏng-wol (Kang-won-do) the P'yŏngch'anggang flows into the Namhan-gang, followed downstream from Kangch'ŏn-ri by the Hoengsŏnggang and finally 40 km upstream from Seoul by the Puk'an-gang (North Han River). The latter rises in Kŭmgangsan and sticks primarily to a south–southwesterly direction. Near Ch'unch'ŏn the Soyanggang flows in from the left, as does the Hongch'ŏn-gang 42 km downstream. The Imjin-gang, finally, flows more or less parallel to the Puk'an-gang and to the west of it, and the Komit'anch'ŏn, the P'yŏng-anch'ŏn and the Hant'anch'ŏn are its left-bank tributaries.

All of the members of this highly branching river system have in common that they meander strongly in granite gneiss, Chosŏn strata and metamorphic sediments. The meandering stretches are connected by remarkably straight stretches of river, the nature of which will be discussed further on. For most of their length these rivers are entrenched into the dip slope of the tilt block in the form of V–shaped valleys and gorges with rock terraces running along above them at various levels. Generally the riverbed completely fills the bottom of the valley, so that floodplains are lacking. The Han-gang still flows in such a canyon as far as Ch'ang-u-ri, 30 km above Seoul, and even close to the capital the riverbed fills the entire width of the valley floor. Right between these narrow stretches there are, however, scattered spots where the valley has widened to intensively farmed valley flats. The most extensive ones are those of Ch'ungju, Ch'unch'ŏn and Ttuksŏm downstream from Ch'ang-u-ri.

Only below Seoul does an uninterrupted floodplain follow the river for 35 km. Then it merges into tidal flats, and at high tide the river again washes against rocky banks on both sides. Just below the mouth of the Imjin-gang this tidal bed

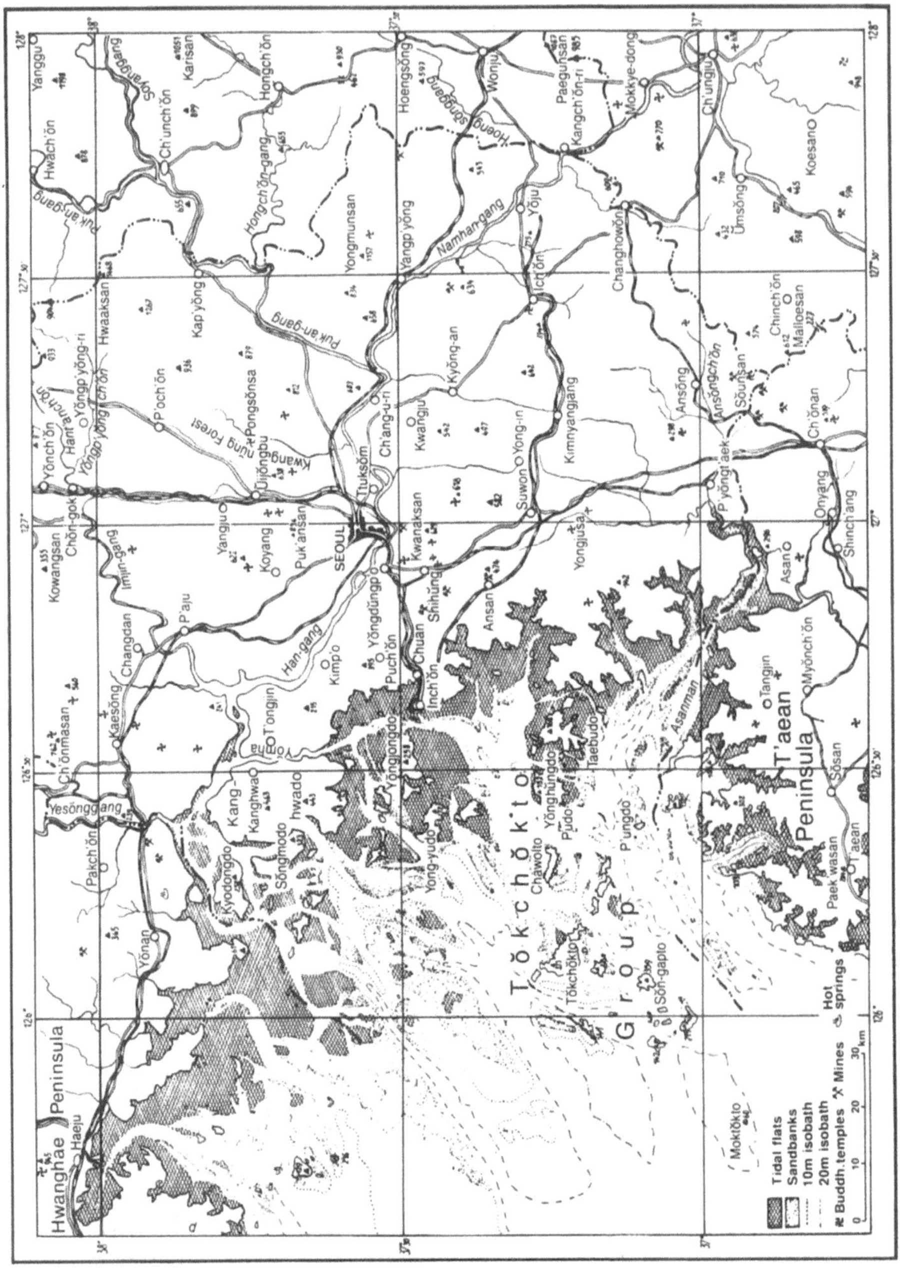

Fig. 62. Sketch of the Kyŏnggi region and Kyŏnggiman

is already 3 km wide. North of the island of Kanghwa it has increased to 9 km. As the European expeditions during the 1860s discovered, much to their dismay, it is generally very shallow, and the sand and mud bars change their positions at every flood. These summer floods still inflicted severe damage on the banks in 1925 and 1936 [452, 824]. Finally the Han-gang divides into several arms between rocky islands, because the recent ingression has inundated several deep zones in the estuary. A narrow arm still resembling a river, named Yŏmha, separates the large (280 km^2) island of Kanghwa from the peninsula of Kimp'o. Rapid tidal currents (8 knots) rush through it. Northwest of Kanghwa lies the actual estuary, which again is divided into several branches by the islands of Kyodongdo and Sŏngmodo.

The effects of the tide are felt upstream as far as about 20 km above Seoul (Fig. 32). In winter the river is still frozen over at the capital for 2.5 months. Navigation is very limited; only as far as Seoul–Yongsan is the river navigated by small river steamers. The seaport of the capital is Inch'ŏn, which is connected with it by a 39 km long railroad and a modern road built in 1936. Riverboats of the type described on p. 190 travel as far as Yŏngch'un on the Namhan-gang and Hwach'ŏn on the Puk'an-gang. In this fashion the Han-gang is navigable for a distance of 300 km.

The Kyŏnggi region belongs to the structural region of the same name, which forms a massif, in contrast to the adjoining areas to the northwest (Fig. 12). It consists of a series of south–southwest to north–northeast trending zones. Between three zones composed predominantly of granite, strips of gneiss and metamorphic schist trending in the same direction are exposed. The latter belong to the Yŏnch'ŏn group (Table 46). They consist of mica schist, phyllite, shale and quartzite, that is, of rocks of extremely varied morphological properties.

Faults and shatter zones, primarily with the same trend, run through the massiv (Fig. 66). The most tightly packed bundle begins at Asanman and runs across to the bay at Wonsan, Yŏnghŭngman. Long stretches of these shatter zones have been sculpted into straight valleys that are often remarkably wide, compared with the creeks and rivers flowing through them. It is not uncommon for the streamlets in them to flow toward each other, so that some of the sections of the valleys are inclined in a direction opposite to that of the main dip slope. This is true, for instance, of the section of valley between Yangju and Chŏn-gok. It is part of one of the longest alignments of valleys running from Ttuksŏm near Seoul via Ŭijŏngbu to Taegwang-ri and then further north–northeast to Yŏnghŭngman. This is the often mentioned Ch'ugaryŏng furrow (p. 5, 73f, 75). Thirteen to 20 km east of it is another valley 100 km long running from east of Ŭijŏngbu via P'och'ŏn and along the upper Hant'anch'ŏn to southwest of Hoeyang and containing four drainage divides. The road from Seoul to Wonsan follows it to just short of Kŭmhwa. The granitic Kwanaksan (629 m), south of Seoul, is split in two by a straight valley like this. In gneiss and schist the rivers following these structural lines occasionally oscillate around a straight line, and above these meandering stretches the straight valleys can be traced in well-developed terrace levels. In granite, however, even the rivers in these valleys are fairly straight. The above mentioned main river branches use the structural valleys only for relatively short stretches. In the Kyŏnggi region, the absolutely straight lowest section of the Puk'an-gang that is entrenched into Yŏnch'ŏn quartzite is the best example.

The hill country, which is quite extensive along the coast and extends eastward to the Namhan-gang in the southern half of the Kyŏnggi region, is seldom higher than 200 m. It probably developed as a result of dissection of the Nangnang plain, which was traced by Kobayashi in the vicinity of Yŏju, Ich'ŏn and Changhowon. Near the coast alluvial plains lie between the ranges of hills, the largest one being in the extreme south, with the county seat of P'yŏngt'aek as its center.

The fuel requirements of the capital have repeatedly deforested this hill country for great distances, so that erosion has left terrible scars. Where they are composed of granite, the deforested slopes have been washed bare. The deeply weathered rock and the grus-covered slopes composed of the eroded material are bright yellow. Only the Japanese reforestation has begun to control this dreadful devastation, which fills the rivers with sand to a dangerous degree and destroys the fields on the valley floors. Gneiss and schist weather primarily to red soils.

The same granite that forms the hill country and the basins after having weathered to yellow soil towers up above it here and there to grayish-black, dome-shaped mountains of Alpine steepness. Only in places where recent spalding has occurred due to surface-parallel exfoliation from below, does the rock have a pale yellow color. Where the granite is full of steep joints in the summit regions, as in Kwanaksan, it forms sharp ridges and towers. The summit forms in Puk'ansan (836 m) north of Seoul are quite different; it is characterized by barren domes and sugarloafs with slopes of 60°. As on Ch'ŏnmasan (p. 278) and Kuwolsan (p. 280 f), in Puk'ansan they surround a high-lying basin resembling a natural fortress.

The hill country extends into the western peninsulas. But even in the granite of the island of Kanghwa there are still mountains as high as 468 m. Only in the innermost tips of the bays have small areas with coastal marshes developed. Some of these were already reclaimed during the Korean dynastic period by means of man-made dams; otherwise tidal flats with their tidal inlets completely fill the bays. Nowhere on the Yellow Sea coast is the tidal range greater, and thus the tidal flats wider, than here. Tremendous areas of continuous mud and sand flats extend out to the islands closer to the mainland, such as Sŏmando, Taemuŭido and Yŏnghŭngdo. Sandbanks extend from there on out to the outer edge of Kyŏnggiman, and even the isobaths representing 20 m under the mean low water at spring tide reveal the influence of sedimentation from the mainland (Fig. 62). Through the entire bay run deeper channels, which begin at the mouth of the Han-gang and open up like funnels toward the southwest. Despite these unfavorable conditions, at the wish of the urban population, several beaches have been developed along the coast. The Japanese prefer the island of Wolmido across from Inch'ŏn (Fig. 39). Except for Inch'ŏn the coast has no harbor of any significance. In the bays filled with tidal flats there are several modern salt pans. The largest is near Chuan in the immediate vicinity of Inch'ŏn. Due to the unfavorable oceanographical conditions, the fishing is the poorest in all Korea (p. 443).

The Tŏkchŏk islands are of exactly the same type as the Paengnyŏng group (p. 282). They, too, are surrounded by tidal flats. Sŏn-gapto rises up to an elevation of 359 m. The outermost island is the lighthouse island of Moktŏkto. These islands, populated by fishermen and farmers, had no fewer than 28,164 inhabitants in 1936, but only 10 of them were Japanese.

The above described forms and colors of weathering and erosion, which influence the character of the Kyŏnggi region more highly than any other part of Korea, are dependent on three factors, the large areas of coarse-grained granite with appropriate joints, the immense human interference with the natural vegetation, dating back thousands of years, and the specific features of the climate, particularly the still severely cold winters and the extremely high precipitation intensity in summer. These and the other important climatic traits are demonstrated in Table 18, as far as they have been measured.

A comparison with the climatic data for the Hwanghae peninsula shows that the Kyŏnggi region has still milder winters, due to its more southern location. The severity of the winter decreases here too from the interior toward the coast and is

		Temperature						
Station		January			August	July		
	Altitude (m)	Mean temp.	Mean min. temp.	Absolute min. temp.	Mean temp.	Mean max. temp.	Absolute max. temp.	Annual range
Kap'yŏng	62	–6.1	–12.8	–28.0	25.2	30.1	38.0[a]	31.3
Seoul	86	–4.7	–9.6	–23.1[c]	25.8	30.5	37.5	30.5
Inch'ŏn	69	–4.1	–5.6	–21.0	25.1	29.2	36.9	29.2
Moktŏkto	66	–1.3	–	–	23.7	–	–	25.0

[a] July. [b] January. [c] December. [d] August.

	Precipitation								
Station	January				August				
	Average amount (mm)	per 1000	Average no. of days with precip.	Precipitation intensity	Average amount (mm)	per 1000	Average no. of days with precip.	Precipitation intensity	Annual amount (mm)
Kap'yŏng	19.2[b]	14[b]	4	4.8	434	324	14	31.0	1337
Seoul	23.9	18	6	4.0	395	295	16	24.7	1340
Inch'ŏn	21.1	19	5	3.2[c]	308	275	15	20.5	1106
Moktŏkto	15.3[b]	21[b]	5	2.1[c]	171[d]	232[d]	11	17.1	732

Table 18. Climatic data for the Kyŏnggi region

least on the islands. The cold season is not quite as lacking in precipitation, however, as there. Part of the winter precipitation falls as snow. Seoul has an annual average of 32.5 days with snowfall, and Inch'ŏn 30. Here, too, only the months from May to September are always without snow. Snow falls on an average of 8–9 days in December and January. In April 1933 I still found patches of snow as low as 100 m, though after an unusually cold winter. Thus in the Kyŏnggi region as well winter is inconceivable without snow-covered mountain slopes, pines and straw roofs. This winter landscape often glows in a bright light. The relative percentage of possible sunshine in February is 65% in Inch'ŏn. The Kyŏnggi region is then also under the influence of the land monsoon. Of the January observations in Inch'ŏn 54.3% are of wind directions between northwest and north–northeast. Due to the orographical influence on the wind direction this land monsoon character is not as clear in the valleys and basins of the interior. The wind velocities are also much lower there. But on the coast, e.g., on the island of Kanghwa, the "*hanŭi param*" is greatly feared, and the Korean villages hide from it in the folds of the terrain [829, p. 9].

Midsummer is only slightly hotter than on the Hwanghae peninsula. Consequently the annual temperature range is smaller in the Kyŏnggi region than there. The amount, and with it the intensity, of the summer precipitation is much greater everywhere, however, and can only be compared with the figures for the Taedong basin. At the station of P'och'ŏn, 40 km northeast of Seoul, the mean July precipitation is 468 mm. That is the highest figure reached by any month in Korea. Here, too, it is evident that the coasts and the islands receive less summer and annual precipitation than the mountainous interior parts of the country. The relative humidity rises in summer to figures that are unpleasant for Europeans, considering the high temperatures (Inch'ŏn: July 85.9%). These high degrees of humidity are spread across the land by the sea monsoon. In Inch'ŏn winds from south–southeast to southwest account for only 44.7% of the observations in July, however, and the fact that wind directions from west–southwest to northwest are represented with 25.3% even then indicates that the massive summer precipitation is brought predominantly by cyclones.

The Kyŏnggi region is a very ancient region of high culture and dense population. Consequently the vegetation is scrawnier than in any other Korean region. The barrenness of the hills and dome-shaped mountains strikes every traveler who enters Kyŏnggi-do by land. It was stressed already time and again by early European visitors (e.g., [176, 179]), whereas the report of a Chinese mission in 1487 mentions "myriads of pine trees" in the environs of Seoul [836, p. 35]. I did not encounter untouched virgin forests anywhere, and even tall secondary forests are rare today.

Relatively most frequent are good stands of red pine. They embellish, for instance, the sacred Namsan, at the southern edge of the city precincts of the capital, a few hills near Suwon, and Hwasan (Flower Mountain) near the temple of Yongjusa, not far from the road from Suwon to P'yŏngt'aek. The last mentioned forest even contains the Korean nut-pine and *Quercus acutissima*. The moss carpet on the ground is covered by magnificent bushes of *Rhododendron mucronulatum* with purplish red blossoms. Near the city of Suwon, the Potsdam of the Chosŏn dynasty, 35 km south of Seoul, the red pine forests date back to King Sejo (1455–68), who had 5.5 million pine trees planted here around his father's grave. Much closer to the natural condition is the wonderful forest of Kwangnŭng northeast of Ŭijŏngbu. It contains the temple of Pongsŏnsa and the graves of the just mentioned king and his wife and had therefore been preserved for 460 years, until the Forest Experiment Station of the Government–General recently took over its protection and care (Kwangnŭng = radiant tomb). The most common trees are the red pine, *Abies holophylla*, the Korean nut-pine, the oaks *Quercus mongolica*, *Quercus neoglandulifera*, *Quercus dentata* and *Quercus aliena*, the Korean wild cherry (*Prunus serrulata* var. *intermedia*), the Manchurian cherry and several species of hornbeam. The detailed list of flora published by the Forest Experiment Station in Kwangnŭng [826] contains 13 oaks, 6 maples, 5 wild cherries, 4 willows and 3 hornbeams. Twenty-two percent of the species also occur spontaneously in the Japanese islands. The forest belongs to the styrax zone of the general classification (p. 137).

The scrub that predominates in the rest of the Kyŏnggi region is so scrawny that the yellow and red color of the weathered soil can be seen through it everywhere. Here, too, the common oak, the Manchurian oak and the red pine predominate. In April the rhododendrons bloom pink, red and mauve. The yellow slopes recently terraced and reforested by the Japanese forest authorities are covered primarily with red pines, the Siberian alder, *Quercus acutissima*, the hazel *Corylus heterophylla* Fisch. and the legume *Indigofera kirilowi* Max.

Below the scrub-covered or completely barren slopes of the hill country the dry fields spread out. The rice paddies cover not only the bottoms of the main and side valleys like a graceful system of arteries, but also niches in the slopes and gentle old talus cones. Paddy farming, which in the more northern regions of the Yellow Sea fringe is limited primarily to the moist lower parts of the main valleys, becomes quite extensive in the Kyŏnggi region. Paddies cover one-sixth of the total area of Kyŏnggi-do and more than half of the cultivated area (p. 405, 414). Even in mountainous counties such as Yangp'yŏng-gun, in which only 16.4% of the total area is cultivated, paddies make up 48.8% of the cultivated area, and in the southwesternmost county, Chinwi-gun (now P'yŏngt'aek-gun), paddies cover 34.3% of the total area and no less than 67.5% of the cultivated area (Fig. 87). The larger rice growing regions follow the rivers and cluster near the coast; in fact, on the map showing the distribution of paddies (Map 4, cf. editors' note on p. 7), several straight structural valleys even stand out. Most of the rice production here supplies the domestic market. In the Kyŏnggi region, too, particularly on Kanghwado, the fiber plant *wanggol* is grown quite commonly in paddies.

Dry-field farming is done with just as much care and versatility as on the Hwanghae peninsula. Because of the relatively mild winters more winter grain can be grown than in the northwestern regions. Barley is grown here almost exclusively as a winter crop, indeed it already occurs in this form here and there on rice paddies, a type of cultivation that is so characteristic of southern Korea. Wheat too, of which large amounts are grown particularly in Kap'yŏng-gun, is primarily a winter crop. For the rest, soybeans, adzuki beans, green beans, foxtail millet, upland rice, rye, kaoliang and buckwheat dominate the dry fields. American varieties of cotton already occur in the extreme south of the Kyŏnggi region. Oats and barnyard millet are grown practically only in the north. This region is definitely a transitional region between northern and southern Korean crops and agricultural methods. Well thought out systems of intercropping are developed here too, including triple cropping, e.g., simultaneous cultivation of barley, maize and melons. Truck farming is very important (radishes, red peppers, lettuce, cabbage, leeks, cucumbers, squash, watermelons, lotus roots), especially for supplying the urban population. Approaching the capital the vegetable-growing areas increase noticeably. The excellent market afforded by the cities makes the painstaking cultivation of early vegetables worthwhile. In April I saw cucumber fields on which, as a protection against the still existent danger of night frosts, the farmers had covered each little plant with a cone of oil paper, weighting the edges down with earth to keep the wind from blowing it over. Finally, plantations of fruit trees and mulberry bushes are common on the lower slopes here as well. In the Kyŏnggi region, in the picturesque old county seat of Suwon (25,252 inhabitants), lies Korea's first agricultural experiment and training station. From it scientific, carefully tested farming methods are disseminated among the Korean population [610].

Fire–field farming has died out in the Kyŏnggi region. The statistics for Kyŏnggi-do do show slight traces (p. 405), but they are found in the high eastern mountains, which we are not including in the Kyŏnggi region.

In accordance with this high state of economic development is the population. With more than 200 inhabitants/km^2, the Kyŏnggi region is the most densely populated region in Korea. Here also the eastern counties are the ones that are relatively sparsely settled (P'och'ŏn-gun 81 inhabitants/km^2, Yŏju-gun 100), the western ones the most densely populated ones (Kanghwa-gun 195 inhabitants/km^2, Chinwi-gun [P'yŏngt'aek-gun] 199). None of these rural counties has the average density mentioned above. That figure is due alone to the weight of the cities Seoul and Inch'ŏn; according to the census of 1938 more than 37% of the population of the Kyŏnggi region lived in these two cities.

Even today Korea's capital is better known in Europe and America under the name of Seoul than Keijō [833–843]. The former is the name that is still preferred by Koreans and simply means "capital." The city has changed its name frequently in the course of its history, or several names have been used simultaneously, for instance, the name preferred by the Chinese and used by Carl Ritter [133], Kong ki tao, and the Korean Hanyang. Since the Japanese occupation two Chinese characters have come into official use that are pronounced Keijō in Japanese (= walled capital). The Korean pronunciation, which would be Kyŏngsŏng, is not customary usage.

When King T'aejo moved the seat of government from Songdo (Kaesŏng) to Seoul in 1392, he demonstrated outstanding political and geographical foresight (Fig. 62). Seoul lies in the center of the capital province and its hill country, in the center of the western half of Korea, which is highly favored in its surface forms and summer climate, not far from the largest river of the peninsula, at the focal point of the natural overland transportation routes, which converge here from the north–northwest and the south parallel to the Yellow Sea coast, from the north–northeast across the Ch'ugaryŏng furrow, from the east across Taegwallyŏng and other low passes in the T'aebaek Range, from the southeast across Ch'up'ungnyŏng and Ihwaryŏng in the Sobaek Range. The fact that it lies 30 km from the coast was of little significance formerly, considering the inhospitableness of the Yellow Sea and the traditional Korean unfamiliarity with the sea.

Just as advantageous is the local situation. Because of the Han-gang's dangerous floods and for better military protection, Seoul was not built on the river, but in a basin a few kilometers to the north. This basin was formed primarily in coarse-grained granite; its floor has an elevation of 20–30 m, and it measures 6.5 km from west to east and 5 km from north to south. It is open only to the east, and in that direction it is drained by the Ch'ŏnggyech'ŏn toward the basin of Ttuksŏm. Otherwise it is surrounded by imposing mountains. In the south Namsan rises to a height of 265 m, in the northwest Inwangsan to 338 m, in the north Pugaksan to 342 m. The more northern of these mountains have precipitous, grayish-black rounded slopes and peaks above moderately steep yellow slopes covered with pine scrub, and despite the low absolute elevation, the view from Namsan across the basin to these mountains has something of a wild Alpine character, especially because the domes and sugarloafs of tall Puk'ansan can be seen in the background. In addition to the opening to the east the mountainous framing surrounding the basin has a few gaps that play an important role in transportation. Through the Peking Pass in the northwest (74 m) enters the main route along the coast, the route that was used by the annual Chinese mission (p. 32). Most important is the wide gap in the southwest, only 37 m high, that connects the city with the Han-gang, 4 km from here, with Inch'ŏn and with the southwestern provinces.

Long before the time of T'aejo this basin and its surroundings were already inhabited. The prehistoric finds at Ŭngbong are evidence of an Eneolithic settlement on a hill in the southeastern part of the mountainous framing of the basin (p. 16f). The historical tradition, collected and translated by Gale [836], begins shortly thereafter. The first recorded mention of the area around Seoul dates back to 18 B.C. 12 years later Onjo, a son of the king of Koguryŏ, built the capital of his newly founded state, which was to become the kingdom of Paekche, in the high-lying basin of Puk'ansan. This fact is proclaimed by an inscription in a Buddhist temple at that location. From 371 to 475 Seoul was the capital of Paekche under the name of Hansŏng, and in the following centuries the city played an important role in the struggles between the Three Kingdoms (p. 25). From 1101 on it was rebuilt on the basis of geomantic considerations. At that time it was not the capital, however, but belonged to Shilla until 935 and then to Koryŏ.

The authenticated beginnings of the present-day city development commence with the year of 1392. Since then Seoul has continuously been the capital of the politically united peninsula. At that time the city wall was built anew as it stands today. The location of the wall was determined according to geomantic instructions given by a monk named Muhak. In the southwestern gap of the basin a heavy snowfall that ended at a sharp line supposedly indicated the course of the wall. To build the wall, which was begun in the first month of 1396, 119,000 laborers were summoned from the northern and western provinces. After they had been dismissed, 79,000 laborers from the southern provinces worked on it, and in the ninth month of the same year the work was completed. Around the same time and also according to geomantic principles the palace known to foreigners as the Summer Palace (Kor. Kyŏngbokkung, i.e., Palace of Shining Happiness, city plan, Fig. 63, C/D, 1/2) and several others were built. They are all still standing in the original form or in reconstructions carried out according to the old plans. In

Fig. 63. City plan of Seoul (after an official plan)

1468 the giant city bell was hung in a low pavilion named Poshin-gak (D 4) at the intersection of Chongno (Bell Street), the main east–west street, and the north–south street that begins at the main gate of Kyŏngbok Palace. Thus the essential features of the cityscape are more than 500 years old, a very long time for the wooden East Asian cities. Many of them are influenced by Chinese geomancy and their names are based on Chinese philosophy, which influenced Korean intellectual life highly (p. 30), particularly at the time of T'aejo.

Always closely connected with the fortified settlement in the basin was the fortress on Puk'ansan. Several times it served as a refuge for the kings residing in Seoul. The structures that are still standing today were not built until 1711, under King Sukchong, however [837]. The high-lying basin there is surrounded by a wall 10 km long with double gates. Inside this wall lie the ruins of palace and fortress, monuments, the hamlet Puk'an-ri with its fields, woods and a couple of run-down Buddhist temples. From the time of the Manchu invasion on, i.e., during the entire isolation period, the safety of Seoul was also guaranteed by the "five fortresses," which lay far out in the country, one on each of the main roads. They were Kanghwa, Kaesŏng, Ch'unch'ŏn, Kwangju and Suwon. In most of them the old walls and gates are still a characteristic feature today. This is particularly true of the island fortress of Kanghwa, which for its part has repeatedly been the refuge of the kings since the Mongol invasion [829].

In the 1880s, immediately after the end of the isolation period, Seoul presented a view whose basic features were hundreds of years old. T'aejo's wall still surrounded the basin, and still surrounds it today, with a total length of 23 km. It climbs up above 300 m several times, and its height varies from 7.5 to 12 m. Of the eight city gates the Great East Gate (Tongdaemun, I 3) and the Great South Gate (Namdaemun, C/D 6) are still standing. The latter stands in the southwestern gap; isolated and covered with ivy, it now embellishes a traffic rotary. The maze of crowded houses had already spilled out past the South Gate toward the Han-gang in 1880, but even then in the higher parts of the mountains surrounding the basin it never extended up to the city wall, which often runs along high above rounded rocky slopes. Around 1880 Seoul had only two straight, wide streets, one of which was narrowed down by several rows of market stalls. Most of the houses then were straw-roofed, village-type houses. As late as 1903 a census showed 42,821 houses, 71% of which were covered with straw, 19% with tiles and 10% with a combination of both (Kor. Rev. 1903, p. 361). Isabella Bishop draws a vivid picture of this maze of houses on the basis of her impressions in 1894/97 [207, Vol. I, p. 36]: "Etiquette forbids the erection of two-storied houses, consequently an estimated quarter of a million people are living 'on the ground,' chiefly in labyrinthine alleys, many of them not wide enough for two loaded bulls to pass, indeed barely wide enough for one man to pass a loaded bull, and further narrowed by a series of vile holes or green, slimy ditches, which receive the solid and liquid refuse of the houses, their foul and fetid margins being the favorite resort of half-naked children, begrimed with dirt, and of big, mangy, blear-eyed dogs, which wallow in the slime or blink in the sun."

In 1880 only the buildings of the royal palaces rose above this maze of houses. At dawn the city bell was struck 33 times for the 33 heavens of Buddhist mythology, in the evening its deep, resounding, threatening tone rang out over the basin of Seoul 28 times for the 28 principal stars. With the evening bell everyone but the blind and officials were required to disappear from the streets, which shrouded themselves in impenetrable darkness. During the day the alleys and streets, especially around Chongno, were and are full of white-clad men adorned with black top hats. Few women showed themselves on the streets in 1880, and they covered their heads with a green silk coat. Peddlers carried goods of all kinds piled up high on their backs and hawked them or displayed them in open shops along the edge of the street: pipes and tobacco pouches, hats, fans, shoes, paper, clothing fabrics, small dining tables, salt, grain, beancurd, brass bowls, clay jars, baskets, sliding doors covered with paper, wooden cabinets and chests with brass furnishings. Bulls with high loads of fuel ponderously pushed their way through the crowds of chatting, smoking, idle people. "Gray in gray" is how E. von Hesse-Wartegg [204, p. 49] portrays the city in 1894, "hardly distinguishable from the ground, the straw roofs of the tens of thousands of low shacks line up close together, like gray mounds in a graveyard, without streets, without outstanding buildings, without temples... without trees and gardens – an unutterably sad but still strange sight, and only the vast expanse of the city and the heights rising in terrible wildness lend it something of an air of grandeur."

That was Seoul's appearance and its life shortly before the end of the isolation policy. After Japan and the large Western powers had forced trade treaties on Korea, the first changes were the legation buildings that were constructed in the northwestern part of the city. They were the first stone buildings in European styles with more than one story. Most of these buildings still exist as consulates-general today. The Japanese legation went up in flames during an uprising in 1882, however, and a new one was built in the southeast, on the north slope of Namsan, a move that was to turn out to be extremely important for the entire city development since then. With the official representatives of the Christian countries missions also gained a firm foothold in Seoul and began to influence the appearance of the city. In 1893 the French cathedral was built in simple Gothic style on a hill in the southern half of the basin. It dominates the city from afar (E 5). Outwardly the American Protestant mission has been even more successful than the Catholic mission in Seoul, too. Its most important buildings are the massive Romanesque style church (D 4) and the big Severance Hospital east of the main station (C 6).

In the wake of the nominal declaration of independence of 1895 the old reception gate on the road to Peking Pass, where the king used to welcome the envoy of the Chinese emperor, was torn down except for the two main pillars. In its place the mighty Independence Arch was built in a majestic Renaissance style. Today it is on the brink of collapse. In one of the imperial gardens the marble "stone-made hall" (Sŏkchojŏn), supported by Ionic columns, was built (C 4), which now serves as a museum of East Asian art. Around the turn of the century Seoul was the only city in East Asia that had telegraph and telephone, a public water supply, an electric streetcar and electric lighting simultaneously. After the annexation in 1910 a completely new development began; indeed, if one wishes to exaggerate the development of the city slightly, one could claim that under Japanese leadership a completely new city has developed. In one generation the city has experienced the metamorphosis from old Seoul to modern Keijō.

The Japanese-led city administration has ambitiously continued the transformation of the street network begun during the last epoch of the dynastic period.

As the city plan shows, two straight, wide east–west streets cross the sea of houses. They are intersected by four wide north–south streets, of which the two western ones converge on the South Gate. From here the main route out of the city to the south runs through the suburb of Yongsan to the Han-gang and crosses it on a 450 m long iron bridge. In the extension of the northern east–west street, the "Street of the Great West Gate" (Sŏdaemunjŏng, now Shinmunno) brings the major traffic out of the walled inner city into the western suburbs and across Peking Pass into the northwestern provinces. Similar arterial roads lead out from most of the other gates. All of these main inner and outer streets are excellently paved today. Along them run the lines of the electric streetcar and the many masts with power lines, so characteristic of East Asian cities, towering far above the low masses of houses.

As Korea's overall development has progressed under Japanese leadership, the functions of the capital have accordingly multiplied and become complicated. These functions are divided up among the various city precincts, thus gradually resulting in more and more individualization of the different city districts. The northwestern part of the city center south of Kyŏngbok Palace is the seat of the government (C/D, 2/3). It is dominated by the enormous granite building of the Government–General. This was built into the southern part of the yard of Kyŏngbok Palace in an Americanized Renaissance style crowned by a dome 55 m high and conflicts flagrantly with the architecture of the other buildings belonging to the former royal residence. South of it lies the district with the consulates–general, the houses of the Europeans and some of the Protestant mission buildings (C/D, 4/5). In the section between the triangular City Hall Plaza (D 5) and the bend in the "Great South Gate Street" (Namdaemunno) the stylistic contrasts of the Kyŏngbok Palace district are multiplied (D/E 5). Here stand abruptly beside one another one-story old Korean shops, two-story Japanese stores and multi-storied Americanized or Europeanized buildings: banks, department stores, the Central Post Office, City Hall, Citizen's Hall with its tall tower, which was not completed until 1935, and the Chosŏn Hotel. West and east of this district the streets often take on a decidedly Chinese look. In the south, where the Japanese legation was constructed in 1885, the Japanese section spreads out on the slopes and at the foot of Namsan (E/F, 5/6). Here stand one and two-story pure Japanese wooden houses with gray tile roofs in little ornamental gardens. Here, too, rises the residence of the Governor–General. The narrow Honmachi (= Main Street, now Myŏngdong), which is closed to automobile traffic, winds its way through the middle of this section in an east–west direction. Along it multi-storied Japanese shops and restaurants are lined up side by side. Life and activity proceed exactly the same as on a similar street in Japan. On the western slope of Namsan with its nice views the Japanese have built a shrine to the Meiji emperor. It is called Chōsen Jingū and is the most important Shinto temple in Korea (D/E 7). The northeastern part of the city center is the district of learning (F/G/H, 1/2/3). There are the zoo and the botanical garden, the university hospital, the Imperial University with its buildings, institutes and laboratories, and the trade college. What is left of the city center, that is the east, the center and the central northern part, is still filled by the maze of Korean houses, and the scenes in these sections still resemble those in the old Seoul of 1880, although the Japanese love of tidiness cleans the alleys to some extent.

To the east, west and southwest, i.e., everywhere where the terrain allows, suburbs extend out from the walled inner city. Due to their proximity to the Han-gang and the railroad, the western and southwestern suburbs have come to be more conveniently located. Here lie the factories and the larger workshops (rice mills, factories for manufacturing wheat flour, soap, rubber shoes and paper, silk mills, breweries, ice cutting from the Han-gang, machine and automobile repair shops, power plant, government print shop). Recently this industrial development has even expanded across the Han-gang to Yŏngdŭngp'o, where there are a gigantic pottery and tile factory, railroad workshops, cotton mills, breweries and iron mills; in fact it has begun to spread out along the main Korean railroad toward Anyang and Suwon. Yongsan, the southwestern suburb, is simultaneously the seat of the head office of the Korean government railways and the largest railroad repair shop in the country. Above all, however, it is Korea's military center with European-style barracks for infantry and artillery and the headquarters of the Japanese army in Korea. The suburbs on the Han-gang, finally, are home primarily to Korean fishermen and boatsmen.

The first real population census, taken in 1911, showed a population of 277,711, in which 86% of the inhabitants were Koreans, 13% Japanese and 0.2% Chinese. There were only 45 Westerners. In 1933 Seoul had a population of 394,572, 71% of them Koreans, 28% Japanese, 1.5% Chinese and 0.1% others. The absolute figures have increased considerably for all groups. Relatively, the number of Koreans has decreased and that of Japanese and particularly Chinese strongly increased. In April 1936 a large number of neighboring towns inhabited primarily by Koreans were incorporated, so that the city of Seoul had no fewer than 737,214 inhabitants at the end of 1938. The number of Japanese sank relatively in the process and amounts to only 19% now. The characteristic composition of the population in the city center and the actual suburbs is disguised by this. Already during the Korean dynastic period Seoul was the generally admired center of gravitation for the entire country, where many landowners consumed their rent. Today Seoul is additionally the center of industry and commerce and the seat of the civilian and military administration of the Japanese possession, Korea.

Seoul can play this role only by simultaneously being the center of the modern Korean transportation system. When we deal with the overland transportation routes Seoul's central position will be stressed (p. 467). The excellent natural location of the city within the country made it easier to improve the traffic routes, and in the age of railroads it develops its capacity to the full. In Yŏngdŭngp'o the southern branch of the Korean trunk line, coming from Pusan, meets the line from Inch'ŏn. Both then cross the Han-gang on a 650 m long double bridge, and in the railroad station in Yongsan they join the line from northeastern Korea, which utilizes the Ch'ugaryŏng furrow and follows the northern bank of the Han-gang on the last part of its route. Seoul's central station lies very advantageously in the mountain gap at South Gate. Originally it was a terminal station, and the northern branch of the main railroad went via Yongsan. During World War I it was made into a through station for the trunk line by relocating the line toward Shinŭiju, a technically difficult job.

As the center of the Korean railroad network, Seoul's long-range influence extends beyond the country's borders in both south and north. From a regional point of view the short-range influence of the city on its immediate environs is greater still. The barrenness of the surrounding mountains has already been mentioned. Within a radius of 15 km one finds old graveyards of the inhabitants of the city. The tombs of the kings extend even farther out, such as that of Sejo at Kwangnŭng and that of his father in Suwon. In quite an area around Seoul one still finds traces of the original faith of its Korean inhabitants today: spirit trees with heaps of stones around them, *changsŭng* posts, which simultaneously serve as milestones (p. 164), dancing grounds of shamanist priestesses. Contrasting most decidedly with these are the modern Japanese research stations in the vicinity of Seoul, such as the Forest Experiment Station in Ch'ŏngnyang-ri and the Agricultural Experiment Station in Suwon.

In 1902 S. Genthe [217, p. 201] called Seoul "the strangest of all Asian capitals in its mixture of unadulterated barbarism and modern veneer." Seoul has lost none of the diversity of its cultural scene since then. As the above description has shown, it has, on the contrary, become much more varied. In its architectural characteristics it is most likely the most contradictory city in East Asia.

Whereas Seoul can look back on an ancient, great tradition, until the end of the Korean isolation period there were only insignificant villages at the location of its seaport, Inch'ŏn. One of the hilly peninsulas extends westward as far as Inch'ŏn, and there are tidal inlets running between the islands off the coast and into the immediate vicinity of the city (Fig. 64). Until the 1880s the Korean fishing village of Chemulp'o lay here between the last, 70-m-high hill of gneiss and the next tidal

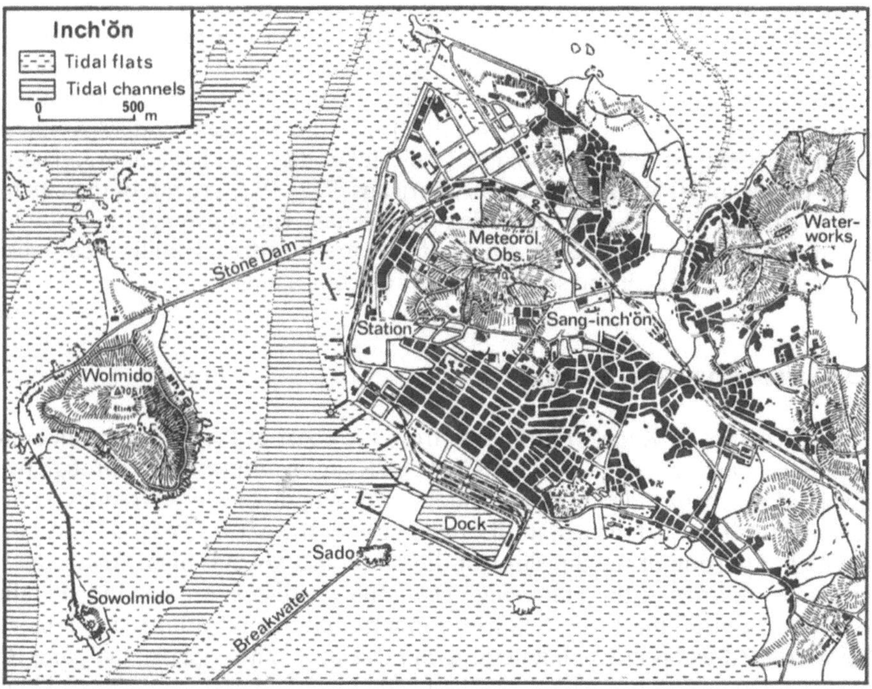

Fig. 64. Inch'ŏn and environs (after an official plan)

inlet, while the farming village of Inch'ŏn spread out on the eastern backside of the same hill. The Europeans still call Korea's fifth largest city after the fishing village that has since disappeared, whereas in Japanese (and Korean) circles only the name of the farming village is customarily used. It has retained its old character.

The development of Inch'ŏn began in 1883, in the year in which the place was conceded to the Japanese and Europeans as a free port. At the location of the fishing village a city district with a network of streets crossing at right angles spreads out today. It is the constantly bustling business and port district and consists primarily of one and two-story Japanese or Chinese houses with a few tall Europeanized stone buildings interspersed between them. To the northwest of this modern city center lies the Chinese section of town. On the seaward slope of the range of hills and on its summits lie the villas of the Europeans, where they have a panoramic view of the tidal flats and the high islands in them. Many of these are now in Japanese hands. Gone are the days in which European trade with Korea via Inch'ŏn flourished, and even the tiny group of Germans who have worked here valiantly since the conclusion of the trade treaty are in dire straits.

Near Inch'ŏn the tidal range increases to 9.60 m (p. 122). Consequently, unloading at sea is fraught with difficulties, and for this reason the Japanese have built a dock with railroad sidings. It is not accessible at all hours of the day, because the tidal channel approaching it has a minimum depth of only 4 m despite constant dredging. A second dock is under construction. The principal exports are rice, soybeans, animals for slaughter, hides and charcoal. The chief imports are wheat, sugar, machines, oil, textiles, coal and cement. Inch'ŏn's maritime trade amounted to a total of 160.3 million yen in 1934, making it the country's second most important port. Since then the 200 million yen mark has been surpassed, owing to the events in China. The port has regular passenger and freight ship connections with Lüda (Dairen), Qingdao (Tsingtao), Shanghai and Taipei (Taiwan). Of its 108,774 inhabitants (1938) 13% are Japanese and 3% Chinese. In 1900, in contrast, Inch'ŏn had only 14,073 inhabitants, 31% of them Japanese, 8% Chinese and 0.5% Europeans–Americans [217]. The growth of the city has thus resulted primarily from the influx of Koreans. The convenient location has led in recent times to the development of industries, especially mills, rice-polishing plants, sake, match and soap factories and cotton mills.

8 The Upper Han-gang Region

The province of Kang-won-do rises from the coast of the Sea of Japan along a wide front and extends across the T'aebaek Range southwestward through two-thirds of the peninsula. It thus includes two geographically extremely varied regions, the T'aebaek Range itself, together with the narrow coastal fringe, and the higher portion of the dip slope of the tilt block, which is highly dissected by the upper reaches of the Han-gang. The latter possesses so much individuality and is so far from the Japan Sea fringe in both nature and culture that it merits being treated as an independent geographical unit. This Upper Han-gang Region is a good-sized area with a length of 240 km and an average width of 90 km, extending from the sources of the Imjin-gang to the big southern bend of the

Namhan-gang and covering an area of approx. 20,000 km^2. The boundary with the T'aebaek Range is drawn such that its highest peaks, including the uppermost source valleys of the Han-gang system, are left outside. The upper Han-gang region thus covers two thirds of Kang-won-do, peripheral portions of Kyŏnggi-do and Ch'ungch'ŏngbuk-do and even the southwestern tip of Hamgyŏngnam-do, which extends into the Imjin-gang area (Fig. 39).

It is the only region in central Korea that does not border on the sea and thus corresponds to the Kaema upland in northern Korea. Its geomorphological situation on the higher portions of a dissected tilt block also resembles that of the Kaema upland. But the ultimate base level is much closer here than there. The branches of the Han-gang are shorter than those of the Yalu, and as a result the recent dissection is much deeper in the upper Han-gang region, although its summits have nowhere near the absolute heights of the Hamgyŏng Range and its neighboring ranges. Not a single one of the county seats lies at a higher elevation than 350 m, but the highest peaks rise to 1300–1400 m. In fact, in Kariwangsan, between the Namhan-gang and the P'yŏngch'anggang, the upper Han-gang region rises to 1561 m, and in the Imjin-gang area Maebongsan is even 1665 m high. When we described the Han-gang system (p. 283) as a whole, we discussed the hydrography of the upper Han-gang region. The deep erosion along its branches has dissected the dip slope into a number of mountain ranges branching off southwest and south–southwestward. They are crowned by many high peaks, so that for long distances the relief is completely void of large unifying features.

As the northeastern continuation of the Kyŏnggi region, the upper Han-gang region belongs for the most part to the same structural region, the Kyŏnggi massif (Fig. 12). The same granite and gneiss–schist zones run through it as through the Kyŏnggi region. The granitic intrusion of Ch'unch'ŏn is enveloped by quartzites of the Yŏnch'ŏn group that were indurated in the contact. Puk'an-gang and Soyanggang converge in the granitic region. Erosion, starting from these rivers, has removed the granite, so that the area of the intrusion forms a basin 10 km wide. The edges of the basin, which consist of quartzites, rise to 655 m in the west (Samaksan) and 899 m in the east (Taeryongsan). The river breaches them in the southwest below the confluence in a narrow canyon. The basin has two floors. The lower one lies at an elevation of 65–80 m and is formed by the floodplain, which is as much as 8 km wide. Above it, at an elevation of 125–200 m lies the second one, which is composed of granite. It is dissected into a hill country with yellow soils (cf. p. 79).

In the southeast, however, the upper Han-gang region extends far into the region of the Okch'ŏn depression. Here the Chosŏn, P'yŏng-an and Taedong strata are compressed into the imbricate folds of Yŏng-wol (p. 64f). This complex extends from the extreme southwest near Ch'ungju northeastward across the T'aebaek Range and to the Sea of Japan. Between Tanyang and Yŏng-wol the Chosŏn strata consist of shales, quartzite schists, dolomites, some of which are saccharoidal and weather to a purplish–red color, dolomitic limestones, limestone conglomerates and marls [844–847]. Northwest of Tanyang the granitic intrusion of Kŭmsusan (1116 m) is sculpted in the shape of a cockscomb, its southern end breached by the Namhan-gang in a wild gorge full of rapids. Another region of folded Chosŏn strata lies in the north between the Komit'anch'ŏn and the Puk'an-gang. In these regions with Chosŏn strata meander formation is again particularly intensive. In them, too, karst phenomena have developed, as in the limestones of the P'yŏng-an strata, with dolines, caves and underground watercourses [848, 849].

The structural valleys in the upper Han-gang region are of virtually decisive importance for most of the region. The shatter zones that we encountered in the Kyŏnggi region run through the upper Han-gang region as well and even penetrate into parts of the T'aebaek Range. The upper Imjin-gang follows the northwesternmost one. The shatter zone then extends north across Muhŭllaeryŏng into the valley of the Saryŏulgang, which rushes toward Yŏnghŭngman. The easy passage afforded by this pass is the reason that the province of Hamgyŏngnam-do extends 37 km beyond the divide into the straight valley of the Imjin-gang.

A bit farther to the southeast the province has a second, even longer southward bulge. It follows the Ch'ugaryŏng furrow, which we have briefly mentioned several times (p. 5, 73f, 75). In particular we stressed that it does not represent a tectonic graben, in the sense meant by Kotō, but rather a structural valley, most of which is filled by basalt, giving it a level floor. It begins at the southern end of Yŏnghŭngman and runs in a completely straight line 120 km south– southwestward to Taegwang-ri. From here it continues due south for another 80 km in the Kyŏnggi region (p. 285). The mountain ranges that follow the main part of the Ch'ugaryŏng furrow rise above 1000 m several times and are composed primarily of granite and secondarily of Chosŏn strata and Yŏnch'ŏn schists. The northern end of the furrow is occupied by the wide delta plain of the Namdaech'ŏn (near Anbyŏn), which flows through it from the divide on. Nine km from the coast a flood basalt appears from underneath the layers of the delta. It extends 33 km up into the furrow, its surface rising to 280 m along the way. The Namdaech'ŏn has cut through it in a canyon, whose steep edges are formed by basalt columns. The slopes of the furrow dip under the basalt flow at a sharp angle. This flood basalt is quite frequently so permeable that the creeks in the side valleys are absorbed by it. Large stretches of it are completely without surface drainage and are crossed only by shallow, dry valleys. Today water from the side valleys is brought into these dry valleys in concrete-lined canals, so that the red soil washed together in them can be used for rice cultivation. The railroad and the road from Wonsan to Seoul follow the basalt surface without obstacles. South of this flood basalt follows an 11 km long narrow stretch of furrow without basalt, which clearly shows the erosive character of the entire system. The Namdaech'ŏn, which meanders in basalt, flows through this section in a completely straight line. The narrow floor of the furrow ascends steeply along this stretch. The railroad follows it here as well, whereas the road takes the eastern detour via the pass of Ch'ŏllyŏng (685 m) and through the source region of the Puk'an-gang.

In the Ch'ugaryŏng furrow a new basalt flow then begins at an elevation of 580 m and it increases toward the south to a tremendous expanse. The Namdaech'ŏn has entrenched meanders in it as well, and the railroad follows them along a stretch with many tunnels. The recent erosion has not yet reached the sources of the river, however, which lie in a swampy part of the lava surface. Here, at an elevation of just under 600 m, lies the highest point of the furrow and of the railroad line. Due to the permeability of the lava there is no definite divide. From here to P'yŏnggang the lava surface sinks to 400 m and from there to the area north of Ch'ŏrwon to 250 m. Along the last mentioned stretch islands of the underlying gneiss and granite repeatedly rise up through the basalt, which is evidently less thick here and as much as 13 km wide. On the less permeable parts of this section Yamadanōjō and other irrigation associations have built large reservoirs, into which the creeks from the edges of the furrow and the islands are diverted through concrete-lined canals. Similar reservoirs are also situated in the side valleys where there is no basalt. Thus the area around P'yŏnggang and Ch'ŏrwon has become a first-rate center of paddy cultivation (Fig. 87).

West of this part of the Ch'ugaryŏng furrow the system of the Komit'anch'ŏn and the P'yŏng-anch'ŏn is much more deeply entrenched than the surface of the basalt-filled furrow. After the outpouring of lava the side valleys of these rivers broke through the western edge of the furrow, thus creating a relief that could not have existed at the time of the lava flow. The framework that once determined the extent of the basalt flow is now missing here. The upper P'yŏng-anch'ŏn has even gnawed its way straight across the basalt flow and rises at its eastern edge. North of Ch'ŏrwon such breaches had already occurred in several places prior to the outpouring of basalt. Here tongues of basalt extend through the edge of the furrow far toward the west. The lava flowed like a waterfall over the step formed by such a previous stream capture, and this is still expressed today in a strong increase in its surface gradient. A similar lava fall runs from the same area into the valley of the Hant'anch'ŏn, breaching the eastern edge. The lava flowed out through these two exits only. In the furrow itself it did not flow further southward. For the next 18 km the furrow is again entrenched in Yŏnch'ŏn schist. The Ch'at'anch'ŏn and the railroad follow its narrow floor, which lies at an elevation of 120–180 m and on which the county seat of Yŏnch'ŏn is located.

Just east of this southern third of the actual Ch'ugaryŏng furrow the Hant'anch'ŏn, which itself also rises in a structural valley, flows at a much lower level (p. 285). Its valley is also filled by a basalt flow that follows it for 60 km until it joins the Imjin-gang and then continues down along the latter until 32 km above its mouth. Both rivers have dissected the lava flow into terraces, which are excellently irrigated here as well. This basalt flow crosses the Ch'ugaryŏng furrow south of Yŏnch'ŏn and shoves a tongue upward into it, forcing the Ch'at'anch'ŏn to aggrade. With this the basalt that fills the Ch'ugaryŏng furrow ends. In contrast to it, the basalt flow of the Hant'anch'ŏn fills a valley upon which long side valleys converge. The lava has also penetrated a certain distance into each of them, blocking the mouths. Above these blocked mouths lie broad floodplains of fluvial origin. They are most extensive around the road junction of Kŭmhwa and on the Yŏngp'yŏngch'ŏn around Yŏngp'yŏng-ri. They, too, are artificially irrigated. The irrigation canals are diverted from the above mentioned rivers at weirs. Because red pine logs are rafted down the rivers, special bypasses are built into the weirs.

Further to the east and southeast follow other structural valleys. Only that of Ch'angdo-ri, which crosses the upper Puk'an-gang, is again filled with basalt for a distance of 17 km. Of particular importance for transportation is another one that runs from the central part of the Soyanggang, just where it turns west and begins to form big meanders, southwestward to the central part of the Hongch'ŏn-gang. It connects the county seats of Inje (12,215 inhabitants) and

Hongch'ŏn and stands out clearly on the map of rice cultivation (Map 4, cf. editors' note on p. 7).

The upper Han-gang region fills the center of the actual peninsula and thus its climate has stronger continental features than the adjacent regions to the east and west. This is particularly true of its southwestern half, whereas in the northeastern half the proximity of the Sea of Japan is clearly felt in spite of the T'aebaek Range in between. Thus in the climatic sphere, too, there is a certain similarity to the Kaema upland, although due to the more southerly latitude and the narrowness of the peninsula the continental features are expressed to a lessened degree only. This is shown in Table 19.

Station		Temperature						
		January			August		July	
	Altitude (m)	Mean temp.	Mean min. temp.	Absolute min. temp.	Mean temp.	Mean max. temp.	Absolute max. temp.	Annual range
Sep'o-ri	540	−12.1	−18.4	−28.1	20.9	24.9	32.0	33.0
Inje	200	−5.9	−11.3	−30.0[a]	24.4	29.1	37.4[b]	30.3
Chŏngsŏn	307	−5.4	−10.8	−27.1	24.5	29.1	37.0	29.9
Ich'ŏn	115	−7.7	−14.5	−34.0[a]	24.4	29.3	37.4[d]	32.1
Ch'unch'ŏn	154	−6.9	−13.9	−36.0	25.5	30.9	37.9	32.4
Wonju	128	−6.0	−12.2	−29.9	25.9	31.0	38.1	31.9

[a] December. [b] July. [c] January. [d] June.

Station	Precipitation								
	January				August				
	Average amount (mm)	per 1000	Average no. of days with precip.	Precipitation intensity	Average amount (mm)	per 1000	Average no. of days with precip.	Precipitation intensity	Annual amount (mm)
Sep'o-ri	10.7[a]	10[a]	4	1.5[a]	395	354	16	24.7	1115
Inje	18.5[a]	17[a]	6	3.1[a]	335	313	15	22.4	1076
Chŏngsŏn	22.8[c]	22[c]	5[c]	3.8[c]	287	274	14	19.1	1047
Ich'ŏn	18.0	15	4	4.5	384	319	15	25.6	1203
Ch'unch'ŏn	22.8	19	6	3.6[a]	374	311	15	25.1	1205
Wonju	23.2	20	5	4.6	395	335	16	24.7	1186

Table 19. Climatic data for the Upper Han-gang region

The last three stations belong to a line running approximately along the central axis of the peninsula, the first three to one farther to the northeast, i.e., closer to the Sea of Japan. Of these, Sep'o-ri, which lies at the highest point of the Ch'ugaryŏng furrow, is definitely a high altitude station, at which the relationship to the Kaema upland is especially evident. In both series the winters become milder toward the southeast and the summers warmer. The annual temperature range decreases on the whole in this direction, but is higher than in the two adjoining maritime regions. The annual amounts of precipitation are generally somewhat lower than in the Kyŏnggi region, although it has lower elevations. The cyclonic winds from the southwestern and northwestern quadrant evidently lose a considerable amount of their humidity along the way. Still the upper Han-gang region receives much more precipitation than the Kaema upland, despite the much lower elevation, because it lies on the dip slope of the main Korean tilt block. Precipitation is highly concentrated in midsummer. Almost everywhere around a third of the annual amount falls in July.

Considering the similarity between the physical conditions in the upper Han-gang region and the Kaema, upland it is understandable that a phenomenon that was particularly characteristic and had especially serious consequences there has its second main area of distribution here. It is fire-field farming (p. 226ff). The province of Kang-won-do is number four in Korea in the intensity of fire-field farming (p. 405), and the vast majority of Kang-won-do's fire fields are found in the upper Han-gang region. Here too fire-field farming is restricted to the uppermost zone of agricultural activity. Along the road from Wonju to Kangnŭng I saw active fire fields extending up as high as the summit of Korup'ogisan (1238 m). The general impression given by these regions influenced by fire-field farming is the same as in the Kaema upland. The most important fire-field crops are potatoes, beans, buckwheat and oats here as well. The latter have their second most important area of distribution in the upper Han-gang region (Fig. 88). Since fire fields can extend down to lower elevations here than in the Kaema upland, and since the climate at equal elevations is much milder, the number of frequently grown crops is greater than there. As in the P'yŏngbuk region, maize, barley, soybeans, foxtail millet, sesame, castorbeans, hemp, etc. are also grown, and rye, which is otherwise rare, is often seen here as a fire-field crop.

Fire-field farming is older in the upper Han-gang region than in the Kaema upland. The changes it has caused in the natural vegetation are accordingly more intensive. Remnants of natural forest are rare, and I did not investigate them. They are most likely quite similar to the magnificent forests that are preserved in the T'aebaek Range (p. 316f). Most of the upper Han-gang region is covered by secondary shrub formations similar in composition to those of the Kyŏnggi region.

In the higher parts of the region permanent farming is almost as meager as fire-field farming. The stones have not been removed from the permanent fields either, and plowing is done so superficially that the roots and stalks of the previous year's maize still lie around on the fields. A more pleasing sight to see is the permanent farming in the transitional area toward the Kyŏnggi region, for instance, around Ich'ŏn, Yŏngp'yŏng-ri, Ch'unch'ŏn, Hongch'ŏn, Hoengsŏng and Wonju and in the lower valleys in the hinterland. The same crops are grown as in fire-field farming, only with more emphasis on those with more exacting requirements. Wheat, barnyard millet, kaoliang, tobacco, cotton and vegetables are additionally grown. Here too fruit trees and mulberry bushes have a certain degree of significance in the valleys. Around Tanyang there is a local center of ginseng cultivation.

As the above descriptions, Map 4 (cf. editors' note on p. 7) and Fig. 87 show, paddy rice is grown to a remarkable extent and with modern methods in some valleys and basins. The basalt areas of P'yŏnggang and Ch'ŏrwon, the alluvial floor of the basin of Ch'unch'ŏn, the structural valley northeast of Hongch'ŏn, the valley of the Hongch'ŏn-gang and the environs of Hoengsŏng and Wonju deserve special mention. All of these regions lie in the low southwestern part. The higher northeastern part has only few paddies, although they extend up to 750 m in the valleys here. In Ch'ŏrwon-gun rice paddies make up 15% of the total area and 40.2% of the cultivated area; in Chŏngsŏn-gun, in contrast, only 0.6% of the total area and 5.1% of the cultivated area.

All other branches of the economy are of negligible importance. Only the coal mining that has recently been developed in P'yŏng-an strata in Yŏng-wol-gun is worthy of mention. Since the autumn of 1937 a tremendous thermal power plant has been in operation near this county seat; it supplies southern Korea's large cities with power and has a capacity of 100,000 kW.

Thus far the upper Han-gang region has been severely handicapped by its disadvantageous natural location. Navigation is extremely limited on its rivers, due to their mountainous character (p. 120). There are as yet very few roads and even they are as yet quite unimproved, with few exceptions, e.g., the road from Seoul to Ch'unch'ŏn. The most important long-distance route by far is the road from Seoul via Hoengsŏng, Kach'ŏn-ri, Taehwa, Hajinbu and the pass of Taegwallyŏng to Kangnŭng, the center of the east coast of the peninsula. The only direct railroad line in the upper Han-gang region follows the Ch'ugaryŏng furrow, and the details of its routing have been described above. It has brought increased economic activity only to the towns of Yŏnch'ŏn, Ch'ŏrwon, Pyŏnggang and Pokkye-ri. The largest center among these is Ch'ŏrwon (20,744 inhabitants), an important market for rice, soybeans, cattle, wood and charcoal. Here the electrified private railroad to Kŭmgangsan via Kŭmhwa branches off from the main line. It was built for touristic purposes, but also serves economic ones. For these reasons Ch'ŏrwon-gun is by far the most densely populated county (103 inhabitants/km^2). Otherwise the population density of the upper Han-gang region is low, due to the mountainous nature and the inconvenient location, although nowhere near the extremely low figures for the Kaema upland. The average density is around 50 inhabitants/km^2.

Ch'unch'ŏn is Korea's only provincial capital that has no railroad connection as yet. It is a backward little town with only 16,959 inhabitants (13% of them Japanese) and a location that scarcely equips it to be the capital of Kang-won-do. It still has this function from the Korean dynastic period, when the coast of the Sea of Japan was the neglected backside of the non-seafaring kingdom. At that time a location that placed the provincial administration as close to Seoul as possible did have a lot to speak for it. Today, however, the coast of Kang-won-do is slowly beginning to awaken, and as soon as the railroad along the coast of the Sea of Japan is completed, the administration will probably be moved to Kangnŭng. Ch'unch'ŏn lies in direct proximity to the floodplain of the Soyanggang in a small basin inside the basin described before. The center fills the alluvial valley floor; the peripheral parts extend up onto the surrounding granite hills. The city is an inorganic mixture of Korean straw shacks, Japanese houses, stores and administrative buildings with roofs of tile or corrugated iron and a few multistoried European-style buildings. In the peripheral parts of town the network of narrow alleyways is a veritable maze.

9 The Wonsan Area

If we disregard the islands of Ullŭngdo and Chejudo, the Wonsan area is the smallest of the regional units distinguished here. It stretches out about 150 km in a southeast to northwesterly direction in the background of Tonghanman and extends from the coast to the main divide, which follows the Wonsan arc of the Main Korean Range (p. 5). Thus it has an average width of 45 km, so that its area is a good 6500 km^2 (Fig. 39). The bay at Wonsan, Yŏnghŭngman, is a unique feature of the Korean coast; 20 km deep and 40 km across, it forms the landward extension of Tonghanman and thus represents the westernmost tip of the Sea of Japan. Located at the neck of the Korean peninsula, never impeding transportation because of ice, subject to the low tidal range of the Sea of Japan and with convenient connections to the capital via the Ch'ugaryŏng furrow and with P'yŏngyang via Ahobiryŏng, it is of outstanding importance from the viewpoint of political geography. This became suddenly apparent when the Russian frigate "Pallas" sailed along the Korean east coast in 1854 and discovered along Yŏnghŭngman "Port Lazarev," which appeared very convenient for permanent occupation (p. 34). In 1879 Wonsan became the second of the ports that the Korean government was forced to open to Japanese trade. With this year the development of the city began. As late as 1885–87 the Russians Shestakov and Ostolopov did oceanographical research here. Since the annexation of Korea, the Japanese have made the environs of Yŏnghŭngman one of their strategic zones. Consequently topographical and goegraphical sources are scarce.

The outer side of the bay Yŏnghŭngman has a very gentle seaward concave curvature (Fig. 66). From the north a long tombolo with narrow lagoons extends southward, connecting a gneiss island, which rises as high as 268 m, with the mainland. This was the origin of the 16 km long Hodo peninsula. In the southeast two short bars protruding out at right angles toward a former small gneiss island form the cape of Kunsan-ri. Between it and the southern tip of Hodo the entrance to the bay is only 24 km wide. In its middle rises the schist island of Yŏdo. The course of the 10-m isobath emphasizes the gentle concave curvature of the outer side, but north of Yŏdo it makes a bulge far into the bay, so that the harbor of Wonsan is accessible to ships with a draft of up to 7 m. This smooth outer side of the bay probably resulted from the flexurelike downwarping of the Korean Japan Sea fringe. The bay itself owes its existence to a eustatic ingression of the Sea of Japan into a local depression. In the middle of it a large number of gneiss and schist hills have become islets due to the ingression. Many views definitely remind one of the Korean Yellow Sea coast. But the semicircular bay as a whole is a phenomenon that is alien to Korea's western and southern sides. Many former islets have been connected to the mainland by deltas and bars, just like Hodo. Of this type is the Kalma peninsula, which projects 5 km into the bay from the south and protects the harbor of Wonsan from waves from the east. In what was formerly the innermost part of the bay recent riverine alluviation has tied schist islands to the mainland. Behind them the alluvial valley of Tŏgwon stretches out today. Immediately after the postglacial ingression the bay was twice as large as today. From the south the Namdaech'ŏn (near Anbyŏn), and from the northwest the Yonghŭnggang and the Saryŏulgang have deposited their deltas into the bay. Along the stretches of the inner part of the bay where there are no river mouths, the primary ingression forms have remained unchanged.

A letter from Wonsan to Petermanns Mitteilungen from the year 1880 (Pet. Mitt. 1880, p. 317, 369) describes the condition of the town and of trade during the period just after the port was opened by the Japanese. Wonsan consisted at that time only of the shabby fishing and farming village with its crooked alleys that, as Wonsan-ri, forms the southeastern part of the city today and spreads out along the bay for almost 2 km. The main export articles at that time were cotton and cotton fabrics from the western and southern parts of the country, salt, beans, sea grass, and dried fish from Hamgyŏng-do, gold dust and tiger and leopard skins from the Kaema upland, hemp and hempen clothing, tobacco, pipes, pears, chestnuts and persimmons from the environs. The natives lived primarily on millet, wheat and beans at that time. Rice, which today is the most important crop in the vicinity, was rare. Today (1938) Wonsan is a city with 67,363 inhabitants, 19% of whom are Japanese and 2% other non-Koreans. It consists of two only loosely connected parts, the former village of Wonsan-ri, which meanwhile is divided straight through by the road to Seoul, and the modern port and administrative city of Wonsanp'o (Fig. 65). Connecting them is a narrow strip with urban buildings stretching out along the mountain slope, in front of which stands the railroad station for the line running from Seoul to northeastern Korea. This strip contains the German church and school of the Benedictine mission. Wonsanp'o lies at the eastern foot of a land-tied mountainous gneiss island, Ch'angdŏksan. It is urban and in part European in its buildings. The harbor is equipped with quays and warehouses and is protected by a long, stone breakwater against waves from the north. It serves as a port both for short-range coastal shipping and for overseas passenger and freight traffic, particularly to Japan. In addition Wonsan is an important fish landing place. Since the northeastern ports have begun to flourish (p. 206ff), Wonsan is no longer the foremost Korean port on the Sea of Japan coast, although its maritime trade is still important (1934: 24.5 million yen). The most important exports are dried fish, fish oil, fish cake, graphite, gold, livestock, beans and wood.

The Wonsan area consists of four clearly distinct rock and relief zones. In the west rise the sharp peaks and ridges of the mountains composed of limestone shale belonging to the Chosŏn system that form the divide between the rivers flowing into the Sea of Japan and those flowing into the Yellow Sea. They reproduce the arc of the depression that forms the bay. East of the divide as well many peaks exceed the 1000-m limit. In its immediate vicinity a structural valley runs 70 km to the south, and in this the two headwaters of the Yonghŭnggang meander toward each other. In the counties of Munch'ŏn and Kowon shales and sandstones of the P'yŏng-an system containing anthracite and bituminous coal deposits are folded into the Chosŏn limestones [851, 852]. Mining of them has been begun. In front of this backdrop, which looks like a wall of high mountains from the bay, granite foothills spread out with hilltops that are rounded but steep-sloped and covered by pine forests or pine-oak shrub formations. Toward the sea a zone of low hills follows. They are composed of metamorphic schists and drop off sharply toward the alluvial plains. Finally there is a zone of hills with red soils mainly on gneiss that has for the most part been dissected by the ingression into the islands and peninsulas described above.

The strip of metamorphic schists and the adjoining granite hills to the west, along with the alluvial plains and deltas running through them, are the zone of highest economic intensity. Most of the latter have been put to use for rice farming (Map 4, cf. editors' note on p. 7). Even in the map showing the proportion of artificially irrigated paddies by counties (Fig. 87), the intensity of the paddy farming in these alluvial plains stands out, despite the size of the counties, which extend into the mountains up to the main divide. The adjoining hills are attentively cultivated with dry-field crops, especially kaoliang and millet. The population crowds together in this zone (Map 3, cf. editors' note on p. 7) and Fig. 85).

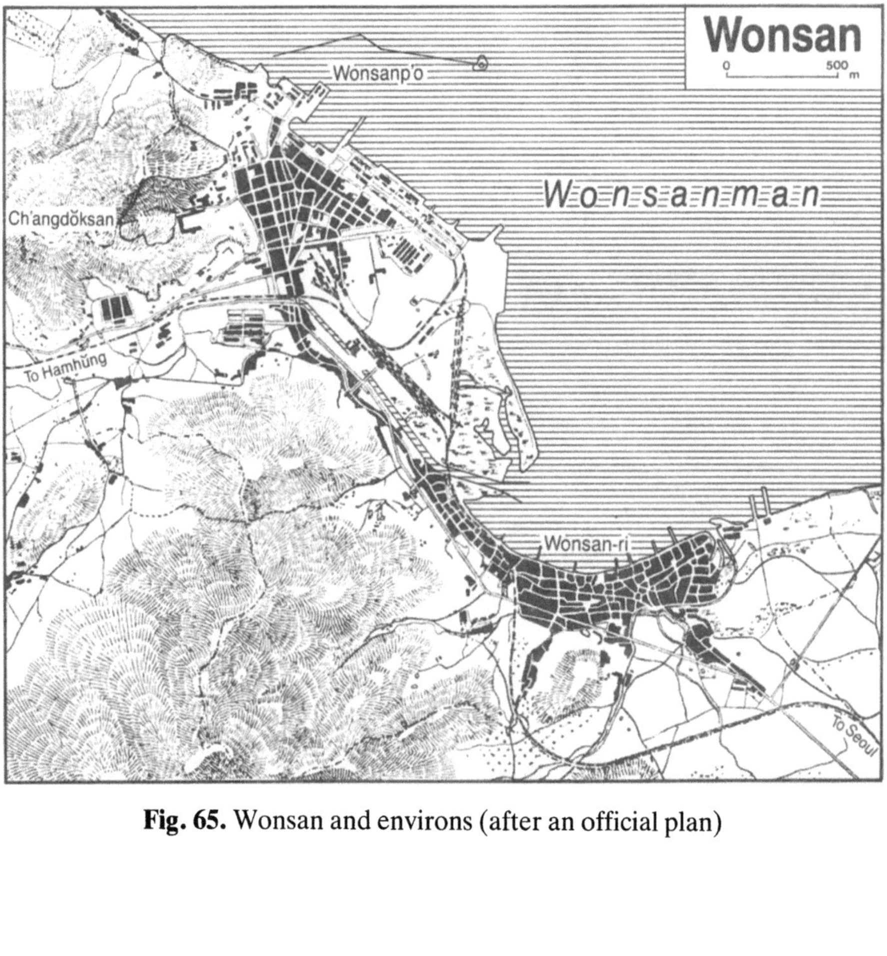

Fig. 65. Wonsan and environs (after an official plan)

In it there also lie the county seats and market towns of Anbyŏn, Munch'ŏn, Kowon and Yŏnghŭng, in several of which Christian churches tower above the low sea of houses. Since 1927 the center of the German Benedictine mission in northeastern Korea has been in Tŏgwon, where the abbey stands, a brick building with a church in pure Lower Saxon Romanesque style. All of the just mentioned towns are lined up along the main railroad line from Seoul to northeastern Korea via Hamhŭng and Wonsan, from which the line to P'yŏng-yang branches off at Kowon. A testimony to the significance of the Wonsan area in ancient Korea is the huge Buddhist temple of Sŏgwangsa, lying a short distance from the old road from Wonsan to Seoul at the beginning of the Ch'ugaryŏng furrow and at the foot of its western mountain edge, which already rises here to almost 1300 m. This temple still contains relics of the first ruler of the Chosŏn dynasty.

To fully understand the advantageous location of Wonsan's harbor and the prime agricultural conditions in its environs we must finally evaluate the climate

		Temperature						
Station		January			August		July	
	Altitude (m)	Mean temp.	Mean min. temp.	Absolute min. temp.	Mean temp.	Mean max. temp.	Absolute max. temp.	Annual range
Yŏnghŭng	35	-4.5	-9.2	-21.9[a]	24.3	28.6	38.5[b]	28.8
Wonsan	35	-4.2	-7.9	-21.9	23.6	27.4	38.2[b]	27.8
Yŏdo	67	-2.9	-	-	23.6	-	-	26.5

[a] February. [b] July. [c] September.

	Prexipitation								
Station	January				August				
	Average amount (mm)	per 1000	Average no. of days with precip.	Precipitation intensity	Average amount (mm)	per 1000	Average no. of days with precip.	Precipitation intensity	Annual amount (mm)
Yŏnghŭng	10.9	10	4	2.7	292[b]	273[b]	14[b]	20.9[b]	1070
Wonsan	29.3[c]	22[c]	6	4.9[c]	301	227	17[b]	18.8	1334
Yŏdo	18.7	19	6	3.1	210	214	15[b]	15.0	978

Table 20. Climatic data for the Wonsan area

of the Wonsan area. A. Defant drew the attention of European meteorologists to it already in 1908, by reproducing the results of a meteorological study by Okada [853]. It is distinguished by frequent warm, dry foehn winds that descend from the passes along the main divide to the coast of the Sea of Japan when a high lies over northern China and a low somewhere over the Sea of Japan. On 27 April 1905, under the influence of foehn conditions, the temperature in Wonsan rose from 4.6° to 22.4° C between 6 A.M. and 2 P.M., while the relative humidity dropped from 96% to 24% and the wind turned from southwest to west, increasing in force from 1.4 to 6.5. Another characteristic case was on 26 June 1904, when the temperature even rose to 38.2° C under similar conditions. Foehn winds occur most frequently in Wonsan in spring and autumn. They are of great importance for the early development and the ripening of crops.

To characterize the average climatic conditions, the data from the station in the interior, Yŏnghŭng, the coastal station at Wonsan and the island station on Yŏdo are listed in Table 20.

The greater thermal continentality at the station in the interior, compared with the island station, and the relatively small amount of precipitation on the island are traits that are repeated everywhere in Korea and have been pointed out frequently. The winters in the Wonsan area are very much milder and the summers somewhat cooler than on the coasts of the Taedong basin, which lie at the same latitude. In all seasons the winds coming from the large, deep Sea of Japan have a moderating effect, whereas this is not true to the same degree of the winds from the shallow, narrow Yellow Sea.

Compared with the Taedong basin, spring and autumn are also warm in the Wonsan area. The mean maximum November temperature in Yŏnghŭng is 2.2° C higher than in the correspondingly located Yŏng-yu, but the mean maximum August temperature is 0.4° C lower. This is obviously an influence of foehn winds, which are lacking in midsummer. Another characteristic of the Wonsan area is the relatively high percentage of the total precipitation that falls as winter rain. This is probably due to the fact that the winter monsoon can also bring orographical precipitation here, particularly when it comes from northeast. The winter monsoon is not frequently demonstrable in Wonsan, however; in winter too the prevailing winds are from southwest, west-southwest and west. In January such winds have a frequency of no less than 69.4% (p. 85). During this season they are always associated with the lowest temperatures. The winter snow is never brought by these winds, however, but rather by northeasterly winds.

10 The T'aebaek Range, Including Kŭmgangsan

The Japanese geographers and geologists, in particular B. Kotō [352], have designated the southern half of the Main Korean Range by the name T'aebaek-sanmaek (Big White Mountain Range) (p. 5). It begins just south of Yŏnghŭngman and follows the east coast, repeating its gentle curvature. In this manner it extends to the wide structural valley of P'ohang–Kyŏngju–Taegu, and it has a length of 380 km in this direction. With its 120-km-long, fairly low southern end it extends far into southern Korea, forming a part of the backdrop of the Naktong basin. This section will therefore be treated in connection with the latter, so that here only the northern and central sections will be examined. The central section only just includes the tangle of mountains that is T'aebaeksan itself and extends to the source of the Naktonggang. In contrast to the northern Korean fringe of the Sea of Japan, the central Korean fringe is so narrow everywhere that it is not capable of forming a region of its own. Geographically it can only be considered as the eastern slope of the T'aebaek Range and will therefore be treated as part of it. Thus delimited the T'aebaek Range has a length of 260 km, an average width of 40 km and an area of around 10,000 km^2. It is restricted for the most part to the northeastern third of the province of Kang-won-do, and only its southern end extends a short distance into Kyŏngsangbuk-do (Fig. 39).

The T'aebaek Range contains the crest, the steep eastern slope and the highest parts of the more gentle western slope of the main Korean tilt block. This observation includes fundamental traits of its relief features, to be sure, but only very few of them. Kotō therefore divided it up into three parallel ridges [352]. But even this concept is much too undifferentiated. It is even more erroneous to conceive of such ridges as long, narrow fault wedges. For the time being it is absolutely impossible to interpret the individual features. Often we shall have to be content with a morphological description of the main features.

The contrast between the steep eastern slope and the much more gradual western slope is the basic feature of the relief of the T'aebaek Range, which stands out almost everywhere and which many of the sheets of the topographical map 1:50,000 show excellently. This contrast is also firmly embedded in the consciousness of the people. Since traditional times the eastern slope of the province of Kang-won-do has been known as Yŏngdong (i.e., east of the mountain passes) and the western slope as Yŏngsŏ (i.e., west of the mountain passes). Yŏngdong consists of the six long, narrow counties of T'ongch'ŏn, Kosŏng, Yang-yang, Kangnŭng, Samch'ŏk and Ulchin. Taegwallyŏng (865 m), the pass across the drainage divide on the road from Kangnŭng to Seoul, is only 18 km from the Sea of Japan. West of the pass in Yŏngsŏ the road stays above an elevation of 560 m for a stretch that is just as long, although it constantly utilizes the valley floors. At this elevation it reaches the uppermost part of the Namhan-gang. Whereas in Yŏngdong it winds its way up extraordinarily steep slopes, in Yŏngsŏ it leads through a fairly gently sloping old landscape with remnants of still older erosion surfaces at different elevations on its summits. The divide separates these gentle forms sharply from the recent steep slopes, sharp ridges and talus-filled gorges of the uppermost region of Yŏngdong. Where the divide runs through the limestone shales of the Chosŏn system, which is particularly the case in the south, in the region of the Okch'ŏn depression, the old surface of the land near it is pitted by dolines.

As a result of the great differences in elevation of the base lines, the rivers of Yŏngdong are rapidly shifting the divide westward. The nameless pass across the main drainage divide at an elevation of only 530 m that is used by the road from Kansŏng to Inje is formed by creek debris, and the gentle, straight valley leading up to it from Yŏngsŏ continues on the other side of the divide in terraces that lie far above the deeply entrenched gorge of the Pukch'ŏn (near Kansŏng), which flows into the Sea of Japan. Due to the rapid progress of the headward erosion parts of the old level landscape are often almost completely surrounded by the steep escarpments of the eastern slope, so that they form erosion remnants. Indeed, east of the divide one encounters completely detached remnants of surfaces forming monadnocks. One of these is Yukpaeksan and its surroundings (1220–1267 m), whose old surface lies higher than the drainage divide 10 km farther to the west.

As explained on p. 69, Kobayashi studied the area around Yukpaeksan in particular detail and named the uniform erosion surface, which he tries to trace through the entire peninsula and which is supposedly warped up as an asymmetrical geanticline, after it [380]. In reality the situation is very much more complicated. The remnants of the old erosion surfaces lie at very different levels and do not belong to a single geomorphological cycle. Thus the summit plateau of Hwangbyŏngsan, northwest of Taegwallyŏng, lies at 1400 m and the just mentioned fields of dolines southeast of Noch'usan at 800–900 m. East of pass P. 530, which we described above, there is a remnant of an erosion surface at 630 m that is almost as flat as a tabletop and spreads out for a length of 4 km. It is unlikely that all of these unconnected surfaces with such varying elevations belong to a single erosion level, although obviously we have to expect that the intensity of uplift varied greatly from region to region in the T'aebaek crest area.

Just as in the upper Han-gang region, running through the T'aebaek Range in a north–south direction are a large number of straight structural valleys, which often cross the main drainage divide. Around them the divide naturally drops to particularly low heights. Such a case is pass P. 530, which is the lowest pass across the entire range and forms a wide passageway. North of it, in the gorge of the Pukch'ŏn, enormous landslides have developed on the highly jointed rock walls. When I was there the road running along them from Kansŏng to Inje could be passed only at the risk of one's life. The Sea of Japan rivers that use these south–north structural valleys naturally flow at a very acute angle into the sea, since the coast runs in a north– northwesterly direction. There are very many such cases. The Namgang of Kŭmgangsan first flows 20 km toward the south in a structural valley, breaks through to the next structural valley east of there and then follows the latter for 25 km to the north. Near the northern end of these two valleys the difference in elevation between the valley floors is no less than 630 m, at a horizontal distance of 10 km. In the headwater region of the Namgang the above described contrasts in the relief between Yŏngdong and Yŏngsŏ are consequently lacking. These Yŏngdong valleys approaching the sea at acute angles from the south, which often begin at particularly low parts of the main drainage divide, contrast with the pure consequent valleys directed toward the northeast (Figs. 19, 66).

The T'aebaek Range starts out in the north right away with the massive Hwangnyongsan (= Yellow Dragon Mountain, 1268 m). After a slightly lower section in the headwater region of the Puk'an-gang there follows Kŭmgangsan (Diamond Mountains), which rises to 1638 m in Pirobong. The divide drops to an extremely low elevation in the vicinity of the passes P. 530 and 641, 40 km farther on. Nearby mountains tower up to 1271 m in Yŏngsŏ, however. A bit farther to the south the main drainage divide itself rises to the highest peak in the entire range, Sŏraksan (1708 m), whose granitic forms resemble those of Kŭmgangsan and which, like Kŭmgangsan, has a special atmosphere imparted to it by a large number of Buddhist temples [857]. In the headwater region of the Namhan-gang, west of Taegwallyŏng, the highest peaks in Yŏngsŏ are again

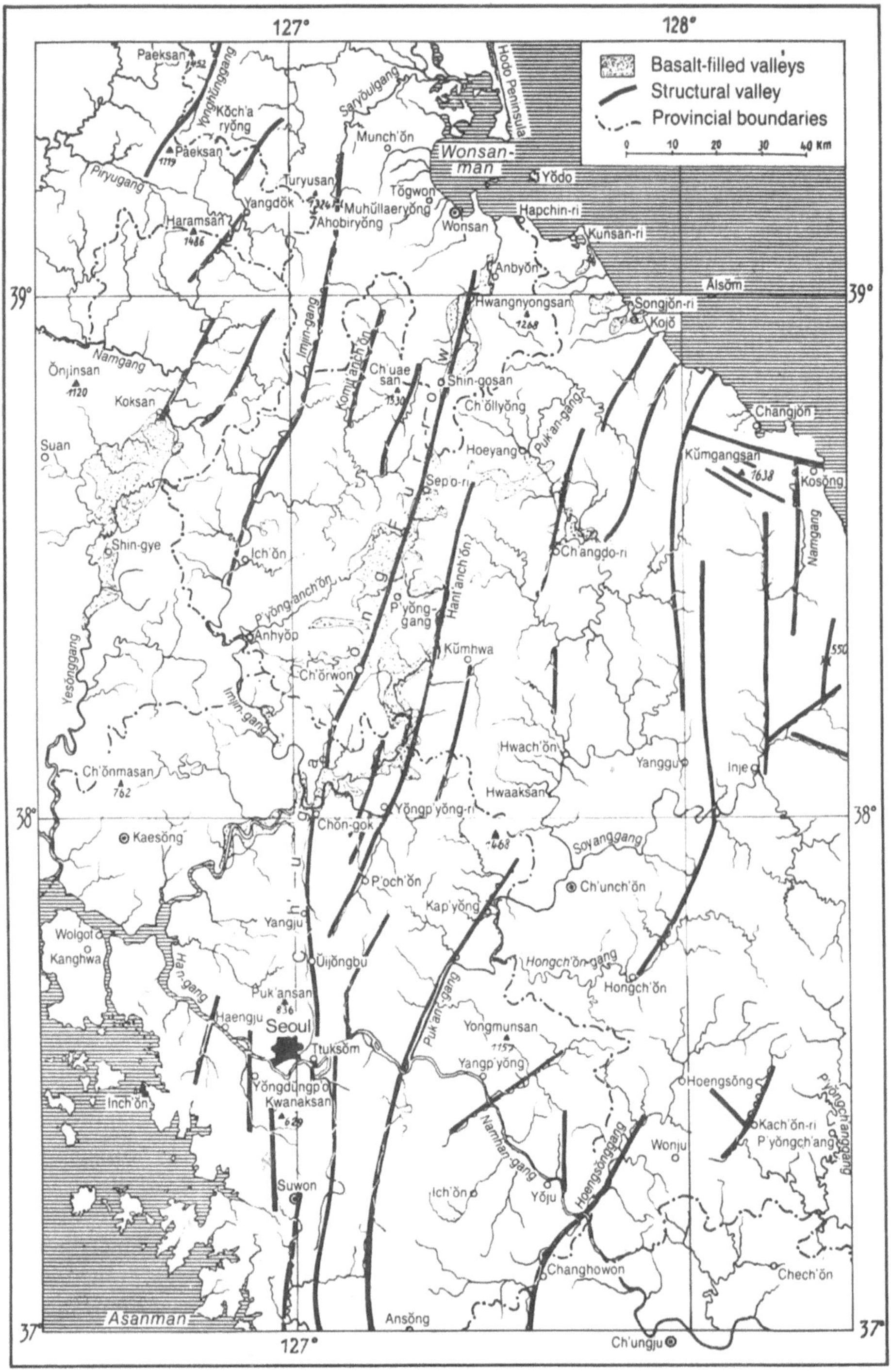

Fig. 66. Structural features and basalt-filled valleys in central Korea

more impressive than those along the divide. Here the marvellous Odaesan rises to 1563 m, Kyebangsan to 1577 m, and Parwangsan, which is composed of P'yŏng-an strata, to 1458 m. Farther southeast in a straight granitic structural valley, the divide drops again to 651 m. The last high peak along the main drainage divide is Myŏnsan (1245 m). Higher are the mountains to the west crowning the divide between the Han-gang and the Naktonggang, Hambaeksan (1573 m) and T'aebaeksan (1561 m). In these summit massifs the Sobaek Range branches off from the T'aebaek Range in a southwesterly direction. The latter drops to lower elevations rapidly after that. Only very few of the peaks in the southern third, in Kyŏngsang-do, exceed the 1000-m limit.

In Yŏngdong side ridges often extend out into the immediate vicinity of the coast. Two such ridges, together with the main ridge, enclose the granitic basin of Kangnŭng, which is perhaps tectonic in origin. The brownish–yellow, weathered hills on its floor are separated by alluvial plains that become broader and broader toward the coast. A bar of coarse sand closes the bay off from the sea.

In many places in Yŏngdong Kobayashi's Yŏdo plain is exposed below the side ridges at an average elevation of 200 m (p. 69). It is dissected into a large number of rounded hills with steep slopes that are gullied by trenches and ravines. The remnants of this surface are most common in the south, where they reach a maximum width of 10 km near the cape of Yongch'ugap. Due to the frightful deforestation in the lower regions of Yŏngdong there is tremendous erosion, and the valleys are filled with jumbled beds of gravel that carry water only during periods of heavy precipitation.

Between Ulchin at the southeastern end of Yŏngdong and Kangnŭng the coast is almost exclusively a cliff-lined coast dissected by shallow corrasion embayments. Even good-sized valleys approach the coast with rather narrow alluvial bottoms. These alluvial plains end toward the sea with short bars, which occasionally leave only a very narrow opening for the paltry rivers. Above the cliffs there are elevated marine terraces with marine deposits at various elevations up to 150 m. The abrasion terrace of Kŭmjin–Chŏngdongjin, which truncates the folded P'yŏng-an strata, is up to 1 km wide and 60–80 m high. The higher and older of these terraces are dissected into hills, just like the Yŏdo plain. In the rearward continuation of the beach platforms, different levels of river terraces, particularly at 20, 70 and 100 m relative elevation, extend up the valleys of Yŏngdong.

North of Kangnŭng evidence of the recent coastal uplift is rarer. All the more frequent are the traces of the eustatic postglacial ingression and its consequences. The alluvial plains filling former bays at the mouths of the rivers become broader and broader, the bars longer and longer. Behind the latter lagoons likewise increase in number toward the northwest. Between the bars rocky capes ending in cliffs jut out. Good-sized islands have become connected to the mainland by the plains, others still lie off the coast. Alsŏm, a rocky isle truncated by an abrasion surface, still rises up to 108 m 17 km from the mainland. Large ingression bays, such as that of Changjŏn near Kŭmgangsan, take the place of delta plains in places where no big river flows into the sea. The most recent of these ingression forms have been preserved off Kŭmgangsan. The extreme northwest, which extends up to the southern end of Yŏnghŭngman, has a straight stretch of coast where the process of straightening has been completed.

In the Yŏngsŏ part of the T'aebaek Range there are no meteorological stations. The climate there is presumably similar to that of the northeastern upper Hangang region, particularly the high altitude station of Sep'o-ri. Winter is probably still colder and summer still cooler, though. The monks at the temple of Wolchŏngsa below Odaesan told me that snow generally lies there at an elevation of 650 m from the second third of November until the first third of April and that at times it gets to be waist-deep. There are also winters, however, in which the snow cover increases to a depth of 3 m. Nakai reports similar conditions in the high region of Kŭmgangsan [860]. It is completely frost-free only in July and August, and I myself measured −2° C on hoarfrost-covered Pirobong (1638 m) on 14 October 1933 at 7 A.M. Baelz still encountered "numerous snow patches, some of them many feet deep" on 21 May 1903 from 900 m on [218, p. 230]. Evidently the relatively large amount of winter precipitation in Yŏngdong extends across into the adjoining parts of Yŏngsŏ. Midsummer presumably brings colossal amounts of precipitation to the high regions. While climbing Odaesan on 13 June 1933, after the day before had been hot and sultry, I experienced the passage of a storm front during which torrential rains prevailed for 3.5 h and hail the size of pigeon eggs fell at Wolchŏngsa.

In Yŏngdong there are many meteorological stations, but all of them lie at low elevations and directly along the sea, so that we only know about the climate of the narrow maritime fringe. The stations listed in Table 21 will serve as examples.

Here, too, winters are extraordinarily much milder in the coastal region and summers are slightly cooler than at the same latitude on the west coast. The January mean temperature is 3.5° C higher in Kangnŭng than in Seoul, that of Chukpyŏnman is 3.9° C higher than in Tangjin on the T'aean peninsula. Greater still is the difference between the mean minimum temperatures of the same month. Kangnŭng had 88 days with frost and only 4 days with ice in the winter of 1931/32; Seoul, in contrast, had 115 and 14 days, respectively. The August temperatures at the two Yŏngdong stations are, however, 1.2° and 1.8° C lower than at the Yellow Sea stations. The annual temperature range is thus much smaller in Yŏngdong than on the Yellow Sea. Due to the mildness of the winter, in the extreme southern part of Yŏngdong January mean temperatures above 0° C occur; on the west coast this is not the case until 190 km farther south. This maritime character of Yŏngdong is also revealed by the relatively high proportion of winter rain in the total precipitation. The ratio of the amount of rain in the summer month with the most precipitation to the winter month with the least precipitation is 16.5:1 in Seoul and 4.0:1 in Kangnŭng. This winter precipitation can be attributed partly to Lc_2 cyclones and partly to Mwb weather types. On the annual average the mountains evidently receive extremely large amounts of precipitation. The abundance of rain at Onjŏng-ri at the entrance to Kŭmgangsan is an indication of this. It is the station with the most precipitation in all of Korea. Protruding capes, such as Chukpyŏnman, in contrast, receive relatively little precipitation. Characteristic of the Yŏngdong summer is also thick fog that often disperses on the other side of the main drainage divide [857, 460].

We have pointed out the importance of foehn for Yŏnghŭngman in spring and autumn. No similar isolated observations have been published for Yŏngdong. Its orographical situation is, however, the same as that of Yŏnghŭngman, and there are a number of climatographical facts indicating that foehn winds also play an important role here during the winter half-year. The only station in Yŏngdong that publishes wind observations is Kangnŭng. There, as in Wonsan, it is not the northern monsoon that has the largest proportion of the wind directions in winter, but rather winds from southwest to westerly directions (p. 85). They amount to 63.9% of all wind directions observed in November, 72.5% in December, 70.5% in January and 62.6% in February. The mean maximum temperature in these months is around 1.5° to 3.7° C higher than in Seoul. The average relative humidity is lower in Kangnŭng than in Seoul from October to April, despite the northern monsoon. In January it amounts to 66.8% in Seoul, but only 54.9% in Kangnŭng. All of these facts indicate that foehn winds play an important role in the winter climate of Yŏngdong.

As a result of the mild winter climate, many physiognomically important plants extend much farther northward in Yŏngdong than on the Yellow Sea coast (Fig.

Station		Temperature						
		January			August		July	
	Altitude (m)	Mean temp.	Mean min. temp.	Absolute min. temp.	Mean temp.	Mean max. temp.	Absolute max. temp.	Annual range
T'ongch'ŏn	23	−3.8	−8.2	−28.5[a]	24.0	27.9	38.4	27.8
Onjŏng-ri	104	−3.0	−8.7	−21.0	23.9	29.6	35.0[c]	26.9
Kangnŭng	15	−1.2	−5.3	−20.2	24.6	28.8	37.4	25.8
Chukpyŏnman	49	+0.7	–	–	23.8	–	–	23.1

[a] February. [b] August. [c] July. [d] January. [e] March.

Station	Prexipitation								
				Januarx			August		
	Average amount (mm)	per 1000	Average no. of days with precip.	Precipitation intensity	Average amount (mm)	per 1000	Average no. of days with precip.	Precipitation intensity	Annual amount (mm)
T'ongch'ŏn	24.4[a]	19[a]	4[a]	6.1[a]	227	219	13[b]	23.1	1262
Onjŏng-ri	38.7	22	5[d]	2.8	348[b]	199[b]	15[b]	13.3[b]	1755
Kangnŭng	59.2[d]	46[d]	6	6.8[c]	236	182	15	15.7	1297
Chukpyŏnman	41.5	43	6[d]	5.0[c]	164	170	13	12.6	965

Table 21. Climatic data for the Yŏngdong fringe of the T'aebaek Range

36). They include several wild bamboo species, which can be found as far up as Kŭmgangsan. Near the settlements one also finds small stands of tall cultivated bamboo here and there. In the foundation legend of the temple of Naksansa, which stands on an abrasion platform on top of a cliff near Yang-yang, bamboo plays an important role [857, p. 12], and several of the rocky cliffs off the coast of Yŏngdong bear the name of Chukto (Bamboo Island). Yŏngdong, at least its southern half, belongs to the "bamboo zone" of the general classification (p. 137).

The natural vegetation of the coastal region was thus probably quite luxuriant and varied. Today very little of it remains. Here and there one finds tall, old specimens of the red pine and the juniper *Juniperus sargentii* Takeda. Deforestation has proceeded to an enormous degree. The hills near the coast are for the most part covered only by small red pine shrubs, between which the eroded, barren soil can be seen everywhere. In the northern half of the coastal region there are also oaks, especially the broad-leaf oak, *Quercus dentata*, and rhododendrons. Reforestation by the Japanese forestry authorities has already produced gratifying results, but there are tremendous tasks ahead.

Toward higher elevations the vegetation in the T'aebaek Range becomes denser. With the wide distribution that fire-field farming had here, too, in former times, it consists primarily of shrubs or secondary forest at the most. Let us describe the composition of a secondary forest, using Nŭnggyŏngbong (1123 m), just south of Taegwallyŏng, as an example. Splendid specimens of the common oak, the small-leaved linden, the Siebold maple, the red-barked *Acer tschonoskii*, the hornbeam *Carpinus erosa* and the maplelike member of the Araliacea, *Kalopanax pictum*, their crowns densely merging, form the main elements. As for conifers, there are only a few specimens of the red pine. The undergrowth consists of some exquisite smelling bushes that bloom white or violet, additionally Hercules' club, the Siebold hazel, the member of the Spireacea, *Stephanandra incisa*, and, finally, various rhododendrons, lespedezas and ferns. Everything is luxuriantly entwined by wild grapevines. This secondary plant association differs from the primeval forests mainly in that the number of conifers is greatly diminished. It is related to the oak zone of the primary vegetation.

On the steep slopes of the T'aebaek Range near the main drainage divide there are still remnants of old virgin forests here and there. Mrs. Bishop [207, Vol. I, p. 152) describes such a forest in the area west of Kŭmgangsan at an elevation of 400–500 m. The mountain side there is "covered with luxuriant and odorous vegetation, specially oak, chestnut, hawthorn, varieties of maple, pale pink azalea, and yellow clematis, interspersed with a few distorted pines." The most nearly natural forests are those in the areas that have belonged to temples for many centuries, as on Sŏraksan, Kŭmgangsan and Odaesan.

The forests around the latter have belonged at least since the middle of the 15th century to the temple of Wolchŏngsa, which maintains a large number of hermitages in them. In a magnificent location chosen with excellent taste the tomb of the "Sakyamuni Buddha" lies solitary in the forest at an elevation of 1100 m. Since this temple was founded already in 645 A.D., it presumably has had forest rights since that time. Even today settlement, hunting and, naturally above all, the felling of trees are prohibited in the forests of Wolchŏngsa, which surround the main headwater valley of the Namhan-gang, and even gatherers of herbs and mushrooms need a permit. In this entire forest region splendid, tall deciduous trees predominate, with an undergrowth of conspicuously blooming and often pungent smelling shrubs. In the summit region of Odaesan (1563–1450 m) the wild apple *Malus baccata* var. *manchurica*, the yew *Taxus cuspidata*, the fir *Abies nephrolepis* and *Rhododendron schlippenbachii*, which blooms here in a gorgeous pink in June, predominate. Wild grapevines grow particularly lushly in the summit region. This summit region belongs to the zone of predominantly coniferous forests (p. 138f). Below 1450 m there predominate the Manchurian oak, the small-leaved linden and the maples *Acer pseudosieboldianum* and *Acer barbinerve* and under them shrubs of Amur lilac, the small-leaved magnolia and the Siebold hazel. The only conifers of any importance are *Abies holophylla*, which form wonderful trees as much as 40 m tall, and the Korean nut-pine. Many of the plants that dominate in the summit region extend down across the lower parts of the slopes to the temple. The entire belt from 1450 m down to the temple belongs to the oak zone. As in southern Korea, there is no special maple zone. In these forests, which are for the most part pathless and difficult to penetrate, live the black bear, wild boars, very many wild doves and graceful chipmunks. Once I saw one of these "*taramjwi*" taking rice grains from an altar of the hermitage of Sajaam in a clearing surrounded by tremendous lindens and oaks, in front of the backdrop of the steep, forest-covered mountains, and it reminded me of paintings by Ludwig Richter.

The settlements are crowded closely together in the coastal region, particularly in the alluvial areas in the northern half (Map 3, cf. editors' note on p. 7 and Fig. 85). Whereas the higher parts of the T'aebaek Range are very sparsely settled, here there are large villages. The alluvial plains are used foremost for rice cultivation (Map 4), the rushing streams often being diverted around the irrigated fields, because of the amount of debris they transport. By expanding artificial irrigation, the yield could be increased considerably, but for this reforestation is a prerequisite. Many rice paddies are dependent on ground or rainwater. In the northern part of the coastal region one therefore finds many dry fields in the plains as well. The most important crops grown on them are barley, wheat, foxtail millet, buckwheat, soybeans, adzuki beans, peas, potatoes, maize, cabbage and onions. Maize is often planted as a border. Intercropping is not nearly as highly developed as on the Yellow Sea coast. Agriculture in general is rather backward. The isolated location has hindered stimulating impulses from outside thus far.

In the coastal region the gathering of marine products equals agriculture in importance. Masses of edible seaweed are torn loose from the bottom of the sea by means of whisks or small scythes. This is done partially from primitive rafts. Women divers, who come from Chejudo, gather the seaweed that is swimming around in the water, which is then dried on the sandbars. A large portion of the sardine catch is made into fish oil and fish cake in boilers constructed on the bars. Large numbers of metal cans and straw sacks for transporting these products to Japan are landed in the small harbors. In winter boreal fish, which come with the current from the north, are caught, particularly herring, shrimps and eels, while cod and Alaska pollock are fished from autumn until spring. In summer subtropical fish brought from the south by the Tsushima current are caught, such as sardines, mackerel, Spanish mackerel, yellowtail, sea bream and abalone. Gathering of sea mustard and shark and sea cucumber fishing are restricted to spring.

The coast of Yŏngdong is, however, still very backward in this as in other economic spheres.

As far as overland transportation is concerned, the situation is by nature very unfavorable, and energetic attempts to overcome these difficulties have been made only in the past decade. The skillfully routed coastal road from Wonsan to P'ohang, which utilizes the Yŏdo terrace and the beach platforms, is thus far, however, in a rather unfinished condition. More important are the roads crossing the mountains, some of which are very busy, such as those from Seoul – Hajinbu – Kangnŭng via Taegwallyŏng, from Inje – Kansŏng via P. 530 and from Kŭmhwa – T'ongch'ŏn. The very important coastal railroad line had been pushed by the Japanese from Anbyŏn to Yang-yang by 1938. Its continuation to P'ohang is a question of only a few years. Until it is completed short-range coastal shipping plays the major role in passenger services. For freight traffic with Wonsan, Pusan and Japan it will always retain its importance. The most important ports of call are from northwest to southeast: Kojŏ, Changjŏn, Kŏjil-ri, Taep'odong, Chumunjin, Samch'ŏk and Chukpyŏn. Many of these harbors lie in bays or are protected by small peninsulas jutting out immediately to the north of them. This smooth downwarped coast is completely devoid of large natural harbors. Ullŭngdo is reached by an 8–h boat trip from Chukpyŏn three times a week. The harbors are quite often inconveniently located in relation to the major settlements, so that between harbor and county seat a fairly long overland trip is necessary. This is true of Kosŏng (11,906 inhabitants), Kangnŭng (18,431 inhabitants) and Ulchin (14,476 inhabitants), whereas Samch'ŏk (12,203 inhabitants) has its port in the immediate vicinity.

No part of Korea, with the exception of Paektusan, has been described so often in the Korean, Japanese and Western literature as Kŭmgangsan (the Diamond Mountains). Gale [864] has translated passages from Korean travel accounts dating back to 1489 A.D. that show impressively how vividly the Koreans have been able to grasp the special character of these mountains. A Korean author writes [864, p. 1]: "From ancient times kings have wondered over it; priests of the Buddha have extolled its beauties; great scholars have sung its praises; artists have painted its views, but none have done it justice. It was the wonder of East Asia in the past, now it is gradually becoming a wonder of the world." Its name is found already in a Chinese document from the Tang period (617–935 A.D.). When he was sent as an envoy to the Chinese court in 1396, the Korean scholar and statesman Kwon Kŭn was invited by the emperor to write a poem about Kŭmgangsan, which evidently already enjoyed a reputation in China at that time. Of the Japanese literature on Kŭmgangsan, T. Nakai's monograph on plant geography [860] is excellent. The first European travelers to visit Kŭmgangsan were C.W. Campbell, E. Brass, Lord Curzon, Isabella Bishop and S. Genthe [197, 201, 206, 207, 217, 865]. Later the German abbot Norbert Weber wrote a comprehensive travel book on Kŭmgangsan [867].

This abundance of literature has a twofold basis: the special significance that these sacred mountains have for Korean Buddhism and the outstanding scenic beauty, which surpasses everything else that the country has to offer in the way of aesthetic appeal, and that is truly not little. Where today the touristic–aesthetic viewpoint is the motive for a large number of trips and writings, in previous times the religious viewpoint predominated. But the monks also developed the innate

Korean appreciation of aesthetic scenic values to perfection, and a host of geographical names for the mountains had their origins in their feeling for beauty and truth as well as in Buddhist mythology. Examples are: Valley of Ten Thousand Different Figures, Valley of Ten Thousand Waterfalls, Blue Crane Peak, Flying Phoenix Falls, Nine Dragon Falls, Fire Dragon Pool, Dragon King's Palace, Hell Mirror Rock, Pool of the Yellow Shades (hell), Three Buddha Rock, Nirvana Hill. Indeed, even the name of the mountains themselves probably goes back to a classical Buddhist scripture, the Diamond Sutra (Kor. Kŭmgang Kyŏng), "which, cutting sharp as a diamond, proclaims that all earthly things are void" (O. Nachod, Pet. Mitt. 1921, p. 317). At any rate it has nothing to do with actual precious stones, of which the primarily granitic mountains are devoid.

The following legend tells of the advent of Buddhism to Kŭmgangsan [858]. In 4 A.D. 53 Buddhas, who came from Punjab in India, landed on the coast of the Sea of Japan near Kosŏng and immediately moved into the mountains. At the spot in the Namgang valley where the temple of Yujŏmsa now stands, they sat down on the branches of a gigantic elm tree and struck their bells. Thereupon nine dragons appeared from a nearby pool and demanded that the monks should leave the valley, for valley and mountain had belonged to them since time immemorial. The monks refused and explained that the great Buddha had sent them from India to establish their headquarters and live according to his law. At this the dragons threatened to use force. Immediately a terrible storm arose, the heavens darkened, lightning flashed, and the mountains shook to their very foundations. When it turned light again, the elm had been blown away, but the 53 monks still sat calmly at the same spot and struck their bells. One of them wrote a couple of magical words on a piece of paper and threw it into the pool. The pool then began to seethe and boil, and with loud cries of woe the dragons escaped into the far depths of the mountains, where they settled down at the Nine Dragon Pool. The mayor of Kosŏng, who had followed the 53 Buddhas, built a temple, with the permission of the king, on the spot where the elm tree had stood. It was named Yujŏmsa (= Elm-tree Rest-house). One of the halls of this temple contains a very old, still anxiously guarded carving of the Buddhas on an artificial tree. The Korean scholar Yi Won encountered it already in its present form in 1489 A.D. [864, p. 46]. Only, of the 53 Buddha figures, 20 have meanwhile disappeared. This is the Korean Buddhist version of the story of the victory of faith and knowledge over the brute forces of nature.

Since Buddhism did not come to Korea until 372 A.D. (p. 25), the legend anticipates the events. At any rate, Yujŏmsa is presumably the oldest temple in Kŭmgangsan. Of the three other large temples, Shin-gyesa was also founded in the 4th century, Chang-ansa (Temple of Eternal Peace), which later rose to glorious ascendency due to royal protection, in the 5th century, and P'yohunsa in 677 A.D. During Buddhism's prime under the Koryŏ dynasty, as many as 180 temples are supposed to have stood in the mountains. At that time thousands of pilgrims traveled to Kŭmgangsan annually, because in the Korean Buddhist world there was a widely held belief that he who visited its sacred precincts would be spared the trip to hell, and even from China the faithful undertook the arduous journey. In the immediate vicinity of the highest peak lies the tomb of a crown prince of the Shilla dynasty, and high up into the interior of the mountains the cliffs are covered with the chiseled names and poems of early visitors, some of them dating back to the 13th century [863]. All of this indicates that very early, occasionally even in the first millennium A.D., the mountains were already permeated into their innermost corners by monastic inhabitants and pilgrims.

The main drainage divide of the T'aebaek Range runs in a north–south direction right through the middle of Kŭmgansan (Fig. 67). Its eastern half thus belongs to Yŏngdong and has extraordinarily steep slopes. It is known as Outer Kŭmgang (Kor. Oegŭṃgang). The more gently sloping western half belongs to

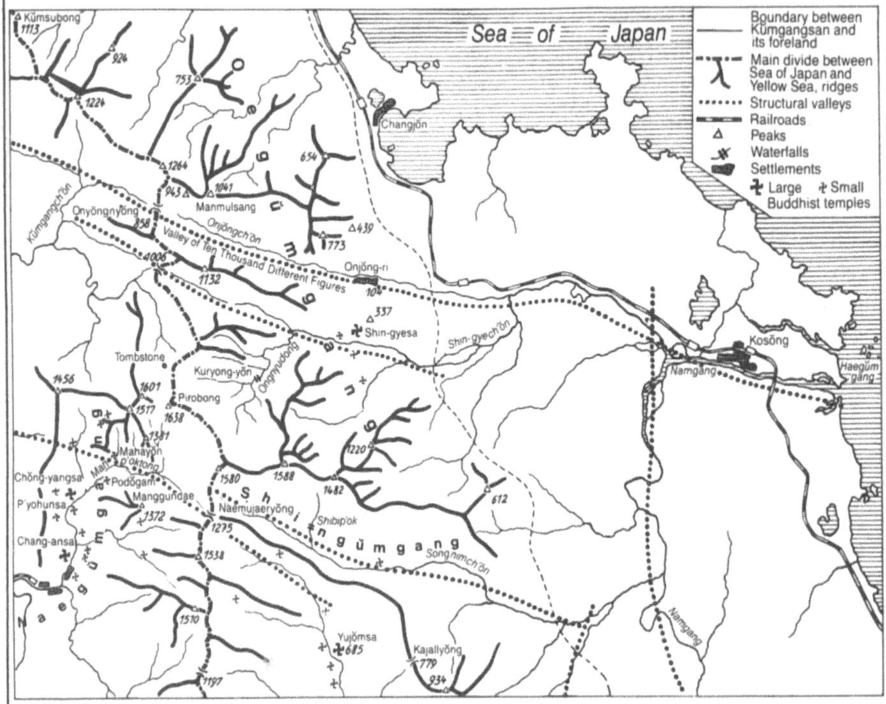

Fig. 67. Generalized map of Kŭmgangsan

Yŏngsŏ. It is drained by the western and eastern Kŭmgangch'ŏn, which flow into the Puk'an-gang. The southwestern sector is called Inner Kŭmgang (Kor. Naegŭmgang). It consists exclusively of the valley system of the upper eastern Kŭmgangch'ŏn. The mountain range is only 25 km long from north to south and from east to west.

Two facts form the basis of the unique world of forms that is Kŭmgangsan. It is composed primarily of a generally coarse-grained biotite granite with intense jointing and a large number of thick quartz veins, and it is a region of intensive recent uplifting at a very short distance from the coast, even for the T'aebaek Range. That is the reason why the main drainage divide lies at a distance of only 10 km from the sea and rises to more than 1600 m. The Outer Kŭmgang is thus the scene of tremendously strong downcutting. Only in the south is the distance from the ultimate base level unusually great, due to the giant bend of the Namgang (p. 311). Yujŏmsa lies here already at an elevation of 450 m. The Inner Kŭmgang is, however, also subject to downcutting, though it is not as extensive as in the main part of the Outer Kŭmgang, since the base level in the valley of the eastern Kŭmgangch'ŏn at the edge of the mountains lies at 450 m. The northwestern sector is composed primarily of gneiss and differs little in its wealth of forms from the normal conditions in Yŏngsŏ.

The valleys in the Outer Kŭmgang are probably predominantly structural valleys. In the case of the valley of the Onjŏngch'ŏn this is certain, for on its floor near Onjŏng-ri six radium springs issue with a temperature of 40–45° C, yielding an average of 12,000 hl per day [347]. This straight structural valley begins at the coast east of Kosŏng and runs west–northwest into Kŭmgangsan, forming here the enormously deeply dissected "Valley of Ten Thousand Different Figures." In its course the main drainage divide drops to 858 m in the pass of Onjŏngnyŏng, the lowest elevation within the Diamond Mountains. Finally this structural valley continues on into Yŏngsŏ for another 8 km. One can assume that a number of other valleys in the Outer Kŭmgang have the same origin, because they are so straight and so deeply entrenched and because of the zones of intense jointing in them. They include the valley of the Shin-gyech'ŏn, which runs parallel and 2 km to the south of that of the Onjŏngch'ŏn, that of the Songnimch'ŏn, whose environs were only recently discovered aesthetically and touristically and are therefore known as New Kŭmgang (Kor. Shin-gŭmgang), and finally that of the upper Namgang above Yujŏmsa. Beyond the pass of Naemujaeryŏng (1275 m) the latter continues on in the Inner Kŭmgang in the valley of the eastern Kŭmgangch'ŏn.

Following these structural lines and stimulated by the recent uplift of the granitic mass, erosion has cut extraordinarily deeply into Kŭmgangsan. The slopes of these valleys are Alpine in steepness; their little side valleys often enter the main valley with good-sized rock steps, and in the side ravines there are rock slides that can be seen from afar with their yellowish white color. Starting from these deep cuts, the forces of weathering and denudation have created the immense wealth of rock forms that give Kŭmgangsan its unique personality in East Asia and, indeed, in the entire world.

As on the rest of the peninsula one can distinguish three different altitudinal regions of weathering and erosion forms (p. 79ff). The region of predominantly chemical weathering to yellow soils exists only at the granitic northeastern foot and can also be found in the gneiss hills on the seaward side of the mountains. Above it lies the zone of rounded, blackish–gray slopes, giant domes and sugarloafs. At the top, finally, is the region of splintery, razorlike ridges, ribs and precipitous slopes with their rugged pillars, towers, pinnacles and needles. These are often clad with a covering of lichens, a thin but physiognomically very effective grayish white garb. Where the covering of lichens is missing, the weathered granite has a purplish brown color in this region.

These three altitudinal zones certainly have a climatic origin. The elevation of the boundaries between them also depends, however, on the grain size and jointing of the granite and on the major relief features. Few joints and coarse grains encourage the formation of rounded forms, and where these conditions exist, the black rounded slopes and domes can extend down to sea level and up into very high elevations. The northernmost mountain of Kŭmgangsan, Kŭmsubong, forms one single gigantic dome, despite its height of 1113 m. Small grains and many joints foster the development of pillar shapes. In the Valley of Ten Thousand Different Figures such shapes thus extend down to 500 m in the rocky labyrinth that is known as Kumanmulsang. There small grain size is combined with relatively large spacing between joints, so that wide, tall, sharp-edged towers have developed. The nearby Shinmanmulsang (943 m), in which large grain size is combined with closely spaced joints, is a confused maze of pillars and needles that not only top sharp ridges, but also extend down a long way on the precipitous slopes. This maze is interrupted by ravines in which boulders tumble into the depths, enveloping the slopes below with talus. In Ŏngmanmulsang (1041 m) grains are generally coarse and the spacing between joints medium-sized. There stacks of rectangular blocks shaped like pillows resembling the German klippes predominate. The famous Manmulsang north of the upper Onjŏngch'ŏn valley is really the area in which Kŭmgangsan's unique character is intensified to greatest abundance and highest perfection. Manmulsang means "View of Ten Thousand Different Figures." Ten thousand is in this case the mystical number of Buddhism and is naturally not to be taken literally. It is better translated with "myriads." On the average the boundary between the splintery forms and the rounded ones probably lies at approximately 800 m.

In the valleys the rounded forms extend higher up than on the ridges. Whereas on the slopes they have a convex curvature, in a cross section of the valleys they have concave curvatures. These have developed most clearly in the high water beds of the rivers. In the structural valleys rounded forms do not occur in the riverbeds, however; they are restricted to the cases in which the direction of the valley has no relationship to the joints. These valley sections are very steep and consist of a series of individual steps down which the rivers plunge in wild falls, swirling out deep pools at their feet, in which the water appears dark green. The grinders at the bottoms of these potholes are moved only by floods. The nicest examples of such valley sections are the valley of the upper Shin-gyech'ŏn, known by the name of Ongnyudong, in the Outer Kŭmgang and the "Valley of Ten Thousand Waterfalls" (Kor. Manp'oktong) in the Inner Kŭmgang. The latter is the section of the eastern Kŭmgangch'ŏn valley that lies above P'yohunsa. The wild valley of Ongnyudong contains the Nine Dragon Fall, which drops over a rock ledge more than 50 m high. The forms of these stepped valleys remind one amazingly of glacial forms, but have nothing to do with the same.

At the highest spots on the domes and pillars in coarse-grained granite there are quite frequently small pits that were formed by solution. The "Angels' Rouge Pots" in Shinmanmulsang are of this type. In coarse-grained rocks solution also rounds the edges of the pillars and forms them into all manners of human and animal shapes, after which the Valley of Ten Thousand Different Figures is named.

The tendency to form all of the aforementioned rock formations is greatest, finally, where the recent erosion is strongest and the slope gradient is thus steepest. This makes the profound contrast between the Outer and the Inner Kŭmgang comprehensible. In the Outer Kŭmgang one pillared ridge follows another, one dome the next one. Continuous forest stands cover only the talus slopes and extend up the ravines. In the Inner Kŭmgang, in contrast, medium-steep, talus covered slopes with beautiful forests predominate, out of which only occasionally solitary pillars, walls or rock battlements protrude, primarily on ridges and peaks.

Next to the relief features, the second landscape element that is of prime importance for Kŭmgangsan is its forests. Due to the power of the temples and the almost complete lack of farming settlements, they have been preserved in their natural condition for the most part, or at least in tall, continuous stands. Deforested by farmers and only covered with shrubs of red pine, broad-leaf oak, lespedeza and rhododendron are the eastern gneiss foreland, the northern end of the granite mountains, where there are no temples, and the northwest. To give an impression of the overwhelming wealth of tree species of which the otherwise predominating forests are composed, let us mention that Nakai's monograph [860] contains the names of three species of pine, three elms, four lindens, five birches, six wild cherries, eight oaks, nine maples and ten willows. In the stands of trees in the foothills, the red pine predominates. Even in historical times this low forest zone was much more extensive. In places where today variegated shrub formations prevail, in 1489 Yi Won still found dense primeval forests, so that for miles he could hardly see the sky [864, p. 44].

Otherwise tall mixed forests dominate Kŭmgangsan. In the Outer Kŭmgang they are often sparse because the slopes are so extremely precipitous, and on slopes with a southern exposure pines predominate, on those with a northern exposure, deciduous trees. In the Inner Kŭmgang and the uppermost Namgang region the forests are denser and lusher, and deciduous trees, in the just mentioned abundance of species, are prevalent. Only in a few spots in the Inner Kŭmgang, for instance in the area surrounding Naemujaeryŏng, conifers predominate for some unknown reason, such as 30 m high yews that are covered all over with red berries in autumn, gigantic Korean nut-pines, Chinese junipers and flat-leaf spruce trees. In the vicinity of the temples there are also gorgeous specimens of conifers, among them *Abies holophylla*. Temple records indicate that this is the result of the monks' ancient predilection for the grave coniferous trees. The summit of Pirobong (1638 m) and its environs are covered by a very dense, low formation on only medium-steep slopes that is the consequence of early human interference, for in the immediate proximity stands the monument to the crown prince of the Shilla dynasty. It is difficult to fit Kŭmgangsan into the general altitudinal classification of summergreen mixed forests (p. 137f). In the depths of the valleys and in the foreland the styrax zone is clearly recognizable, to be sure, but the oak, maple and coniferous forest zones do not contrast sharply with one another.

As Nakai determined [869, p. 9], there are certain differences in the composition of the vegetation of the Outer and Inner Kŭmgang, probably for climatic reasons. The more highly oceanic Outer Kŭmgang reveals certain relationships with Japan, the more continental Inner Kŭmgang with Manchuria.

Tall, vigorous forests, emerald green pools, foaming white waterfalls and rivers, black domes and whitish gray or purplish brown, serrate ridges, these are the elements of which the Kŭmgangsan landscape is composed. Under the serenity of a quiet autumn sky, in the first half of October, the coloring of the foliage is an additional, very special element. It is extraordinarily variegated, and in this respect it surpasses everything that our country has to offer. It is especially for its sake that the scenery in Kŭmgangsan is so famous. The pink of the enormous specimens of the common oak and the Manchurian oak, the dazzling yellow of the lindens and birches, the blazing red of the ashes and particularly of the Siebold maple, together with the dark bluish green of the Korean nut-pine and the grayish–white of the rock pillars, create a kaleidoscopic effect that attracts many hundreds of visitors annually.

The third sphere that shapes the landscape of Kŭmgangsan is Buddhism, even today. In addition to the four main temples there are 21 small ones in Kŭmgangsan, and in all of them together there still live 210 monks and 30 nuns. The temples own paddies and dry fields of their own and are able to maintain themselves by leasing them. The Japanese government has confirmed these ownership rights. This has not checked the decrease in the number of monks, however. Of the four major temples two lie at the edge of the Outer Kŭmgang, the other two in the Inner Kŭmgang. The small ones, in contrast, are not that evenly distributed across the mountains; rather, most of them are in the Inner Kŭmgang, where the Buddhist sphere is therefore much more prominent. The four main temples lie on the floors of the main valleys, whereas the small ones are scattered around in the side valleys and on the slopes. In the latter case, the artificial clearings surrounding them afford beautiful views of forests and rocky peaks. Particularly famous is the "View of the Twelve Thousand Peaks" from Chŏng-yangsa, which has lain on the right slope of the valley of the eastern Kŭmgangch'ŏn since the 7th century. This temple is known for the vividly colored paintings on the outer walls of a hexagonal temple depicting tigers, herons, elephants and other animals. The hermitages, finally, which mostly consist of only a few small buildings, sometimes lie in such precarious locations that they can be reached only by mountaineering paths. Since the 8th century the small temple of Podŏgam has been hanging over an almost vertical precipice above the Valley of Ten Thousand Waterfalls on a tiny rock platform, a bronze pillar supporting the overhanging part.

Even far from the temples the visitor to Kŭmgangsan encounters Buddhist influences in the landscape. The names and poems chiseled into the rocks in Chinese characters often date back to pious pilgrims, who bequeathed donations to the temples for this purpose. They fit tastefully into their environs of rock, moss and forests. In the Inner Kŭmgang there is also a group of three standing Buddhas carved into the vertical side of a giant block and a relief of a seated Buddha about 20 m high chiseled out of a rock wall. The Korean Yi Chŏng-kwi, who passed through these innermost defiles of the Oegŭmgang in 1603, discovered the latter exactly the way it is today [864].

The Korean farming villages and fields generally end at the edge of the Outer and Inner Kŭmgang. Only near Yujŏmsa have approximately 20 potato farmers settled down, who collect the cones of the Korean nut-pine high up in the mountains in autumn. Onjŏng-ri was formerly a farming village in character, but today it lives mainly from visitors to the mineral springs and from the tourist trade. It is the center of tourism for the Outer Kŭmgang, which is now easy to reach with the coastal railroad line. The center of the Inner Kŭmgang is Chang-ansa. Here the electrified Kŭmgangsan railroad from Ch'ŏrwon ends. Chang-ansa has been the entranceway to the mountains on the landward side since ancient times. Between the temple and railroad station hotels and houses for summer vacationists have sprung up, facilities that are otherwise still unknown in Korea.

From these two centers a network of tourist paths provide access to the mountains. They are as yet rather primitive, however, so that trips to Kŭmgangsan are still quite arduous. In the interior there are lodges that are designated with the German word "Hütte" on the Japanese signposts and maps. The first one stands on top of Pirobong and was named after the German engineer, Dr. Kumme. Along the main tourist paths one finds teahouses as resting places here and there. Otherwise the inner part of the mountains is not used very much for economic purposes. Once in a while one encounters woodcutters, and occasionally one finds a scheelite quarry high up in the mountains.

On the coast off Kŭmgangsan near Kosŏng coarse-grained granite crops out again, forming domes and rounded slopes. They are particularly nicely formed north of the mouth of the Namgang on islands that have become connected to the mainland and on others that have remained islands. Some of them bear picturesque parasol-shaped pines and have steep cliffs shaped by the salt water into solution forms. On them the granite even forms pillars and towers again. This area, which is a miniature Outer Kŭmgang transplanted into the sea, is known as "Sea Kŭmgang" (Kor. Haegŭmgang).

Fifty kilometers farther to the northwest near Kojŏ a 30-m-high, narrow basalt promontory juts out into the sea, forming a vertical cliff, out of which the surf has carved gigantic perpendicular bundles of basalt pillars. These "standing" pillars are joined by others that are known in East Asian linguistic usage as "sitting" and "lying" pillars. The lying ones consist of the columns of basalt dikes that later pierced the main surface perpendicularly or at a slant. The columns of these dikes, which for their part are perpendicular to the solidification surface, have a horizontal or slanting position. The sitting columns are the stumps of perpendicular ones that have been truncated by the surf. This cliff, with the parasol-shaped pines and the attractive pavilion of Ch'ongsŏkchŏng on the basalt platform above it, forms an outpost of the natural beauties of the Diamond Mountains.

11 Ullŭngdo

From the mountains of the T'aebaek Range near Ulchin one can see the peaks of the island of Ullŭngdo far to the east on a clear day (p. 2). It rises out of the depths of the sea, which reach at least 2150 m all around it. The steepness of the surrounding sea bottom continues above ground, and with boldly soaring ribs the pentagonal island, whose diagonals do not exceed 12 km, rises to 984 m (Fig. 68). Administratively it has belonged to the province of Kyŏngsangbuk-do since 1914, because it was settled by people from there. Geographically it is more closely related to the T'aebaek Range. In its coastal region, however, especially on the southern side, there are features that remind one of the extreme southern part of Korea, for the island climate is much milder than that of Yŏngdong at the same latitude (37°30'). Otherwise it also has an extremely large number of highly pronounced insular traits that make it a small world of its own and justify separate treatment [877–884].

As the *Samguk sagi* [102] reports, in the centuries of the Shilla dynasty the island was initially an independent political unit by the name of Usan, until it was forced by a trick to submit itself to the kings of Shilla in 512 A.D. [878]. Since that time it has borne the name it does today, Ullŭngdo. In 1160 Kim Ju-Rip, the governor of Kang-won-do undertook a veritable expedition to Ullŭngdo, during which he discovered traces of former settlements on the uninhabited island. Around 1500, at the latest, the Japanese reached the island and began to use the magnificent giant pines, after which they named the island Matsushima at the time. La Pérouse, the European explorer (p. 45), found the island uninhabited again in 1787, but he found a few Korean boat yards on the rare sandy beaches in the inner parts of the small bays. Dense primeval forest covered the steep slopes from the highest peak to the upper edge of the vertical cliffs that almost completely surround the island [160]. Not until the second half of the last century did the Korean government settle the people from Kyŏngsang-do whose descendants still inhabit the island today. No other part of central Korea is thus as recently settled as Ullŭngdo. The only parallels are in the fire-field areas of northern Korea, and accordingly there are similarities in cultural geography, owing to the abundance of wood still available here today, as there, and some of these similarities belong to the sphere of fire-field farming.

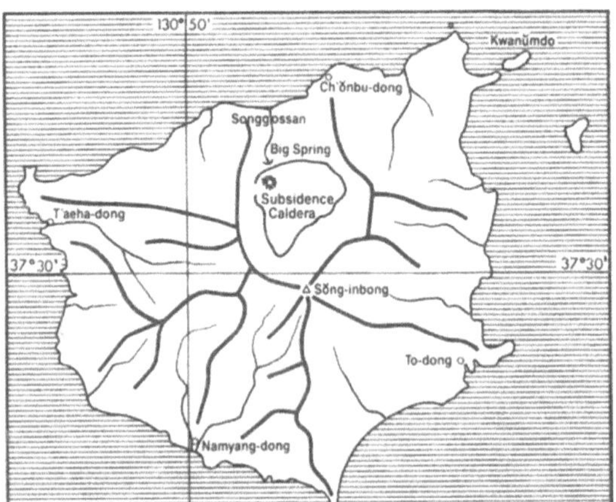

Fig. 68. Topographical sketch of Ullŭngdo

Ullŭngdo is a volcano, which rises 3150 m above the bottom of the Sea of Japan, but only approximately one-third, with an area of only 73 km^2, projects above the sea level (Fig. 69). Nothing is known for certain about its age thus far. The volcanic events still influence some of the forms so highly today, however, that the last one cannot under any circumstances be pre-Pleistocene in age. The morphological and structural center of Ullŭngdo is a subsidence caldera, which occupies the center of the northern half. In it the most recent volcanic formation on the island, a trachyte cone surrounded by crescent-shaped walls of tuff, rises to 611 m. The mountain ring surrounding the caldera often rises to more than 900 m in the peaks in its southern half. They include the highest elevation, Sŏng-inbong, which lies in the center of the island. The thick-bedded trachyte and phonolite lava and tuff flows forming the mountain ring are bedded periclinally to the caldera. They have been dissected by deep headward erosion from outside and play varying morphological roles, depending on how resistant they are. The multitudinous resulting forms are increased even more by perpendicular, radially arranged lava dikes, which form colossal walls.

The second basis of Ullŭngdo's special character lies in its equable island climate. Especially the winter is extremely mild for this latitude. To-dong, the main town in the southeast, has a mean January temperature of +1.7° C, almost 2° C more than Samch'ŏk, which lies at the same latitude on the coast of Yŏngdong. The evenness of the island climate is also expressed in the annual precipitation cycle. Whereas in mainland Korea winter has a pronounced precipitation minimum, on Ullŭngdo the absolute maximum falls during this season (Fig. 27). The winter monsoon takes on moisture on its way across the Sea of Japan, and cyclones on tracks a–c (p. 103ff) bring abundant precipitation. The island's climate is thus quite similar to that of the northwestern coast of Honshū. But summer also brings abundant, mainly cyclonic rain. In To-dong the first snow falls in the second third of December, on the average, the last in the last third of February. At higher elevations good-sized amounts accumulate, similar to Honshū.

Since clearing began only very recently on Ullŭngdo, seven-tenths of the island are covered by forests. According to Nakai [881], there are 372 vascular plants, 9% of which are endemic. Of the 89 species of woody plants, 16% are even indigenous, and they include such frequent and characteristic plants as *Fagus multinerve* Nakai, Korea's only beech, which grows luxuriantly from 200 m up into the summit region, the small-leaved maple *Acer okamotoanum* Nakai, the majestic, firlike *Tsuga sieboldii* Carr. and *Tilia insularis* Nakai, with its pale gray bark. The likewise indigenous island pine *Pinus parviflora* S. and Z. has become rare, however, because of the above mentioned exploitation. This high degree of

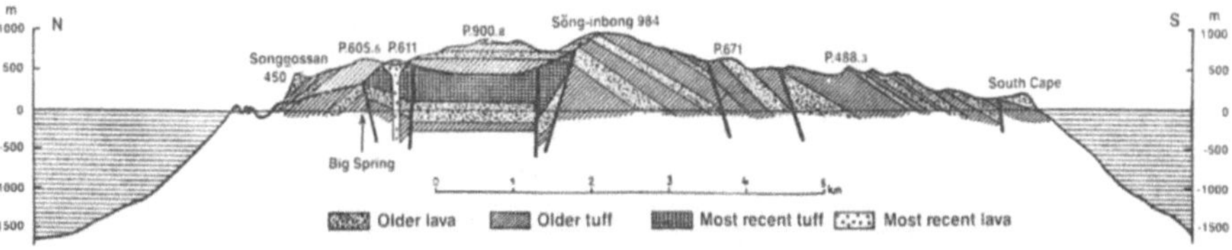

Fig. 69. Structural profile through Ullŭngdo (no exaggeration)

endemism indicates that Ullŭngdo has been isolated for a long period of time (p. 71), and this is also confirmed by the fact that there are neither tigers, foxes and hares nor frogs and snakes. On the whole Ullŭngdo's vegetation is more closely related to that of the Japanese islands than that of the Korean mainland.

The constant abundance of precipitation, the warm summers and the mild winters encourage the growth of vigorous summergreen deciduous and mixed forests. The deciduous trees that particularly stand out include, in addition to those already mentioned: *Acer mono* var. *savatieri*, *Kalopanax pictum*, *Styraz obassia* S. and Z. and *Sorbus commixta* Hedl. The last two are covered in midsummer by large clusters of wonderfully fragrant blossoms. This summergreen forest can be classified as belonging to the styrax zone and the bamboo zone, although there are certain irregularities in its altitudinal structure, as Wilson already pointed out [496]. Under it on the coasts, except in the north, there is a plant association with shiny, leathery evergreen leaves that resembles the corresponding formation in the southern coast borderland (p. 369f, Fig. 70). This is a very striking result of the mild winters in the bottom zone on the sides that do not face the land monsoon. Since this bottom zone was particularly subject to the deforestation of the past 100 years, evergreen plant formations now exist almost exclusively on very steep slopes.

Eighty percent of the 12,030 inhabitants of Ullŭngdo earn their livelihood by farming. How intensive this farming must be is demonstrated by the fact that the population density in relation to the cultivated area is no less than 547 inhabitants/km^2. Regional and economic contrasts often collide quite unexpectedly on this recently settled island. A rice paddy, the expression of the most intensive farmwork in East Asia, can lie directly beside practically untouched virgin forest. Although all creeks constantly carry a good amount of water, irrigation plays only a very slight role, due to the immense steepness of most of the slopes. Irrigated paddies constitute only 1.7% of the total cultivated area. Dry fields are distributed in approximately the following manner: maize 26, potatoes 25, soybeans 13, barley 11, cotton 6, mulberry bushes 6, wheat 3, hemp and ramie 3% [892]. By Korean stan-

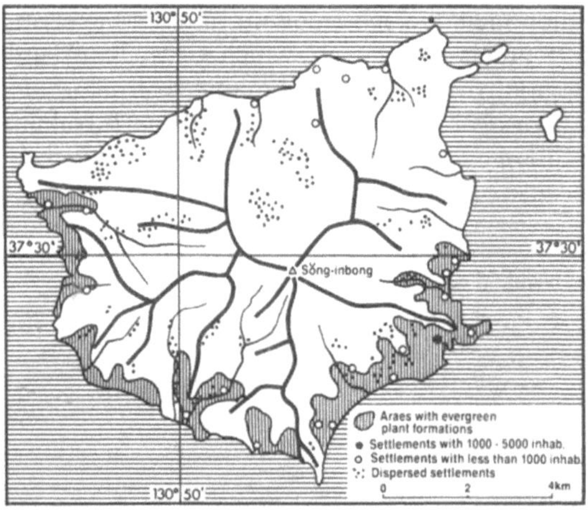

Fig. 70. Vegetation and settlements on Ullŭngdo

dards this is quite a modern agricultural situation, and with the wide distribution of crop rotations and intercropping methods it reveals a delicate adaptation to the local conditions.

Next to agriculture fishing is an important subsidiary occupation, 10% of the island population being employed in it. Natural harbors, however, are extremely scarce. In the few landing places one sees women repairing the green fishing lines everywhere. The lines are then laid out ready for use on woven plates. To keep the lines from getting tangled, the hooks are stuck into the straw edge of the plate. The most important commercial fish is the mackerel. Squid and abalone also bring good catches. Further, the gathering of plants in the natural forests as a dietary supplement is still very common. Woodcutting is, however, strictly supervised by the Japanese authorities.

The huts of the Korean farmers and fishermen are often still covered with shingles. Nor are the wooden houses familiar to us from the Kaema upland lacking. Most of the nearly 500 Japanese inhabitants live in To-dong, the administrative and commercial center of the island (Fig. 70).

When one climbs through the almost impenetrable undergrowth, through lianas and bamboo jungle to the highest point on the edge of the caldera in midsummer, one looks down all around at the rounded treetops of the virgin forest, their uniform green pleasantly interrupted by white clusters of blossoms. Suddenly the billowing jungle ceases, and on the other side of a yawning abyss the bright blue Sea of Japan sparkles up from unfathomable depths and expanses. When in winter ones gaze roams from the coast upward, in contrast, the forms of the individual trees, bare of foliage, stand out against the snow–white ground, and the evergreen patches of vegetation near the coast contrast distinctly with their barren winter surroundings. This seasonal contrast, the volcanic nature and the recent settlement determine the basic features of Ullŭngdo's special character.

12 The Naktong Basin

In Part Two the southeastern basin of the peninsula showed up time and again as a region with a unique character (e.g., p. 6, 63, 93). It is cut off from the rest of Korea by the Sobaek Range. The eastern foot of the Sobaek Range is generally quite distinct, so that it is not difficult to draw the boundary between it and the Naktong basin. This boundary runs approximately along the line P'unggi – Hamch'ang – Kimch'ŏn – Chirye – Ch'angch'ŏn-dong – Koryŏng – Sanch'ŏng – Hadong. The northern half of the Naktong basin is sealed off toward the east by the southern third of the T'aebaek Range. On the other side of the thalweg Taegu–P'ohang follow T'ohamsan (745 m), Kajisan (1240 m) and Paegyŏksan (942 m), blocks that are part of the echelon structure. Toward the south, toward the Korea Strait finally, the basin is also blocked off by mountain ranges and peaks that are often as much as 700– 800 m high, although there are passages between them that are lower than 200 m leading to the bays of the south coast. These coastal mountain ranges in the south, together with the adjoining peninsulas, bays and islands, are of the same type as that predominating farther toward the west and up to the southwestern tip of Korea, and together with this western continuation they form the region Southern Coast Borderland. In my regionalization, therefore, the Naktong basin extends only to Chinhaeman (Fig. 39). With these boundaries it measures 230 km north–south and a maximum of 160 km east–west and has an area of about 25,000 km^2.

In the interior the elevations are remarkably low along the river courses even for quite a distance to the north. Near Andong, which lies on the Naktonggang about one-quarter of its length from the source, the water gauge already has an elevation of less than 100 m. It is likely that these low elevations did not develop entirely by erosion, but were the result of subsidence postdating the basin-forming processes of the Cretaceous period (p. 63). Probably it consisted of a series of local movements that simultaneously influenced the main features of the river system. These have continued on into the present, as is shown here and there where an older erosion relief has been buried. e.g., on the Naktonggang near Indong.

The main river draining the basin is the 526-km-long Naktonggang. It alternates between a southerly and a westerly or easterly direction, making right-angle bends four times. From the mouth of the Namgang on it flows parallel to the south coast for 65 km. This curious feature of the hydrographical character is intensified by the Namgang itself, which follows the coast for the 95 km of its lower reaches, although at the beginning of this stretch it is only 8 km from the same. Other good-sized right-bank tributaries rising in the Sobaek Range are the Hwanggang, the Kamch'ŏn and the Naesŏngch'ŏn, whereas the Panbyŏnch'ŏn and the Kŭmhogang, the left-bank tributaries, rise in the T'aebaek Range. Finally, south of Yangsan, the Naktonggang makes an almost right-angle bend toward south–southwest into a structural valley, divides up into several arms and deposits a delta into a former bay that is flanked to the east by the above named Paegyangsan, to the west by Pulmosan (801 m, p. 18). Because of the very slight gradient, the Naktonggang is navigable for 340 km, i.e., up to above the mouth of the Naesŏngch'ŏn, by Korean riverboats. Since the mean spring range at Pusan is only about 2 m, the river tide advances only a few dozen kilometers upstream.

The interior of the basin has a surface that does anything but slope evenly toward the thalwegs of the main river branches. Rather, it is just as mountainous as the interior of the Bohemian basin, and like there, the basin character is actually produced by the higher mountains framing it, to which the river system has adapted itself. This vigorous relief in the interior is the result of both the just mentioned differential tectonic movements and selective erosion, which allows the tremendous variations in the hardness of the rocks to come into full force. The Naktong region belongs for the most part to the area of Cretaceous sedimentation in the Tsushima basin (Fig. 12), in which the exceedingly thick rock sequences of the Kyŏngsang system were deposited. As the Table of Geological Formations (Table 46) shows, the Kyŏngsang system of the Tsushima basin is divided into the Naktong, Shilla and Pulguksa series. The first of these consists primarily of shales, bituminous clays, sandstones, graywackes, quartzites and conglomerates and forms a 10 to 20-km-wide strip in the western part of the basin. The center and the east are composed mainly of rocks belonging to the Shilla series, which consist of sediments of the same petrographical character and of enormous porphyry and tuff flows. Both the sedimentary rocks and the effusive rocks of this series are immensely thick. Thus the porphyry of the Chusasan formation is 1000 m thick, the shale of the Taegu formation, 2000 m thick. The Pulguksa series consists almost entirely of volcanic rocks. The effusive felsophyre and granite porphyry flows are found particularly in the eastern half of the basin, whereas the "young granite" intrudes the two older series of the Kyŏngsang system everywhere in the Naktong basin. Finally, in a strip along the coast of the Sea of Japan, which is 30 km wide at the most, Tertiary rocks are widespread, namely, both the conglomerates and shales of the Yŏnil series and more recent basalt, andesite and trachyte flows along with their associated tuffs.

The volcanic rocks are generally very hard and form high ridges. The "young" granite intrusions, together with their contact zones, often have been sculpted out in the form of broadly arching hills. In the porphyry flows the valleys are entrenched with steep edges. The shales, in contrast, have allowed the development of local basins. Low erosion surfaces truncate their regions, and under them broad alluvial plains extend along the rivers.

The southeastern part of the Naktong basin is crossed by two systems of straight structural lines, one in a southeasterly direction and the other in a south–southwesterly direction. Along these shatter zones straight structural valleys have developed, and they run through the various rocks without a bend. In regions with sedimentary rocks and in granite they are often wide and have alluvial bottoms, or intersecting talus cones. The hot spring of Tongnae rises in one of these valleys 15 km north of Pusan [345]. The straight east coast from the cape of Changgi to east of Pusan has the same direction as one of the systems of structural lines and is probably the result of recent faulting. An indication of this is another hot spring, that of Haeundae, directly on the coast northeast of Pusan [346].

In the northwest and the north the Naktong basin protrudes into the Yŏngnam massif (Fig. 12). Here it is composed of granite gneiss and granitic intrusions, many of which are probably Precambrian in age. They form primarily hilly country and rolling erosion surfaces, which are quite extensive on the Naesŏngch'ŏn and on the adjoining section of the Naktonggang.

The Naktong basin was the heart of the State of Shilla. As such it has been densely populated at least since the time of Christ, although the assertion that the capital of Shilla, Kyŏngju, had a population of 790,000 for a time is likely to be highly exaggerated. This dense population has led to intensive destruction of forest and soil, as in the Kyŏnggi region. Especially in the northern half the slopes are covered by extremely scrawny second growths today. They often consist only of very pitiful red pine shrubs without even grass between them. At higher elevations gullies cut deeply into the rocks, giving the mountains sharp, craggy forms, and block streams extend downward in ravines. At lower elevations in granitic regions, the deforested soil weathers intensely to yellow soils here as well, whereas the other rocks tend to weather to red soils. The torrential rains of summer wash masses of weathered material into the rivers, which thus have broad, shallow beds. With great effort the side creeks are diverted around the paddies. By aggradation they raise their beds to 2–5 m above the paddy areas and are dammed by levees of sand and gravel planted with poplars and acacias. Gullies dig deeply into the deforested, weathered slopes. Heavy rains lead to severe flood disasters. During my stay the Naktonggang near Waegwan rose 5 m between 20 and 21 May

1933. The water had a reddish brown color and was covered with innumerable patches of whitish brown foam as much as 1 m in diameter and 10 cm high, floating like ice floes on the raging stream. Since the flood disaster caused by a typhoon in 1936, the government-general has begun to construct massive levees. Simultaneously the reforestation of the most heavily devastated parts of the basin has been continued energetically. The main species used for reforestation are the red pine, the alders *Alnus tinctoria* Sagent. var. *glabra* Call., *Alnus firma* S. and. Z. var. *multinervis* Regel and *Alnus sibirica* Frisch., and *Lespedeza cyrtobotrya* Miquel.

Compared with the same latitudes in the Ch'ungch'ŏng–Chŏlla region, the climate of the Naktong basin is relatively mild in winter, while the summer is just as hot as there. That is why the annual temperature range is smaller here. The Naktong basin thus also shows up as part of the maritime outer side of the peninsula. Naturally the annual range decreases within the basin as one approaches the sea, chiefly because the winter temperature increases noticeably in this direction. All the same, the winter is cold enough everywhere to produce a snow cover even at low elevations. On 25 March 1933, after an abnormally cold winter, to be sure, I found snow patches in ravines down to 100 m, and the mountains over 1000 m in elevation were still completely white with snow.

The northern part of the basin is characterized by what are remarkably small amounts of precipitation for the peninsula. This is because it lies in the rain shadow of the Ld_2 and Le_2 cyclones, which come from the west. If the summer monsoon itself were the bringer of precipitation, particularly this part surrounded by a semicircle of mountains in the north should receive large amounts of orographical precipitation. The southern part, in contrast, gets its rain especially from strongly developed Lf_2 cyclones and Ld_1 typhoons, so that the amount of precipitation increases greatly toward the south. All of this is demonstrated in Table 22. Of the stations listed, Ch'ŏngsong is a county seat in the northern part of the basin. Yŏngdŏk lies only 28 km farther to the east, but separated from Ch'ŏngsong by the T'aebaek Range, and not far from the coast of the Sea of Japan. Kŏch'ang lies in the Hwanggang valley directly at the foot of the Sobaek Range. Ulgi is a cape southeast of Ulsan; Chinju a county seat on the Namgang.

As we have already pointed out, only very little is left of the original deciduous and mixed forest. The best clues as to its composition are again the environs of the great number of large temples, especially those that have still continued to exist into the present. The forest of the temple of T'ongdosa in Yangsan-gun, founded in 645, contains, according to my investigations, at an elevation of 100–200 m: the oaks *Quercus dentata*, *Quercus aliena*, *Quercus mongolica* and *Quercus variabilis*, the hornbeams *Carpinus laxiflora* and *Carpinus tschonoskii*, the member of the elm family *Zelkova serrata*, the hackberry *Celtis japonica*, the oleaster *Eleagnus crispa* Thunb. var. *typica* Nakai, the edible chestnut *Castanea crenata* S. and. Z. var. *dulcis* Nakai, the Siebold ash and *Micromeles alnifolia* Koehne var. *typica* Schneid. All of these trees are summergreen and most can also be found in central Japan. The only conifers are *Abies holophylla* and the red pine. The latter also forms artificial stands of only one species in this forest. Underneath grow deciduous shrubs, such as *Ligustrum amurense* Koehne, *Rhododendron schlippenbachii*, *Benzoin obtusilobum*, *Weigela subsessilis*, *Weigela florida* and *Lespedeza maximowiczii* Schneid. var. *tomentella* Nakai.

Clematis patens and other lianas climb from tree to tree; on the ground bamboo grasses luxuriate. Many of the listed species are typical representatives of both the bamboo and the styrax zones.

The Cock Forest (Kor. Kyerim) near Kyŏngju is a small copse associated with a legend about the origin of the last Shilla dynasty, which is recorded in the *Samguk sagi* [102]. Occasionally one finds its name applied in a mystical, allegorical sense to the state of Shilla, or even to all of Korea. According to this tradition it has existed at least since 65 A.D. Obviously it has been highly altered and depleted

Station		Temperature						
		January			August		July	
	Altitude (m)	Mean temp.	Mean min. temp.	Absolute min. temp.	Mean temp.	Mean max. temp.	Absolute max. temp.	Annual range
Ch'ŏngsong	195	−3.4	−9.3	−25.5	25.1	30.2	39.0	28.5
Yŏngdŏk	20	+0.3	−4.2	−17.0	25.8	30.0	38.5	25.5
Kŏch'ang	199	−2.5	−8.2	−19.6	25.8	30.8	37.4[a]	28.3
Kyŏngju	39	−0.2	−5.2	−16.2	26.6	31.8	42.0[a]	26.8
Ulgi	44	+0.4	–	–	22.9	–	–	22.5
Chinju	25	0.0	−4.9	−16.4[b]	26.5	30.5	36.9	26.5
Pusan	13	+1.7	−2.1	−14.0	25.8	29.6	35.3	24.1

[a] July. [b] December. [c] August.

Station	Precipitation								
	January				August				
	Average amount (mm)	per 1000	Average no. of days with precip.	Precipitation intensity	Average amount (mm)	per 1000	Average no. of days with precip.	Precipitation intensity	Annual amount (mm)
Ch'ŏngsong	23.2	22	4	5.8	196	230	11	17.8	852
Yŏngdŏk	25.5	25	4	6.4	182	198	12	15.2	1026
Kŏch'ang	28.0	24	3[b]	9.3	248	216	13	19.1	1113
Kyŏngju	25.7[b]	25[b]	5[b]	5.0	192[c]	189	12	17.4[c]	1014
Ulgi	43.7	35	6	6.5[b]	219	176	13	16.1	1245
Chinju	29.6[b]	22[b]	5	5.4[b]	253	187	15	17.7[c]	1353
Pusan	31.5[b]	22[b]	6	5.2[b]	277	195	13	21.3	1418

Table 22. Climatic data for the Naktong basin

by man, but it contains only trees that also exist in the forest of T'ongdosa. The same is true of the much more varied forest of the temple of Pulguksa, also near Kyŏngju, which was destroyed in 1592. Particularly remarkable is the fact that despite the southerly latitude these forests are completely lacking in broadleaf evergreen trees. The winter temperatures are still much too low for them. Apart from the conifers, only the low bamboo grasses in the undergrowth form an evergreen element. The first evergreen plant formations do not begin to appear until the southern edge of the Naktong basin toward the southern coast borderland.

The second growths are not all as scrappy as those described above. On porphyry, tuff and schist there are more or less continuous shrub formations with a greater variety of species. Basically they contain the same species as the forest of T'ongdosa.

Because the Naktong basin is so mountainous and because land that is actually arable is frequently devastated by flooding, the proportion of cultivated land in the total area is smaller than in the western regions of central Korea, despite the favorable climate. It is 21.4% in Kyŏngsangbuk-do and 23.7% in Kyŏngsangnam-do, compared with 34.5% in Hwanghae-do and 32.3% in Kyŏnggi-do. But the farming done on these areas is definitely southern Korean. The proportion of paddies in the cultivated area is very high. In Kyŏngsangbuk-do it is 48.8%, and in Kyŏngsangnam-do even 61.5%. The paddies cover especially the alluvial plains in the basins and structural valleys and the terraces to the sides of them. In the direct proximity of the larger rivers they are often lacking because of the danger of flooding. The relative intensity of rice cultivation increases on the whole from north to south. The traditional Korean rice varieties have been almost completely replaced by higher yield Japanese varieties. Of these, the medium fast to slow-ripening varieties predominate, as everywhere in southern Korea, such as Kokuryodō, Soshinreki, Miyako and Chushinreki. Nowhere else in Korea is as large a percentage of the paddies planted to other crops in winter, particularly to barley, naked barley and wheat. The proportion of double-cropped paddies in the total paddy area is 46.4% in Kyŏngsangbuk-do and 48.4% in Kyŏngsangnam-do. Everywhere in late spring one sees the garish green rice seedbeds in a good state of development, while directly beside them the winter crop begins to ripen and turn yellow. In many counties more than half of the barley, naked barley and wheat that is harvested is grown by double cropping on paddies.

The other crops are restricted to the dry fields here too, however, since they require the warmth of summer. They include especially soybeans, adzuki beans, mung beans, sweet potatoes, perilla, hemp, tobacco, cotton and paper mulberry. In the northern half of the basin there are local centers of ginseng cultivation. Cultivation of mulberry bushes and silkworm raising are highly developed. Cottage industries and factories produce raw silk, rice wine, straw sacks, textiles, but also brassware and furniture. Foxtail millet extends just barely into the Naktong basin from the north. Only small amounts of kaoliang, broomcorn millet, buckwheat and potatoes are grown; northern Korean crops, such as barnyard millet, oats, maize and kidney beans are completely lacking. The villages are often surrounded by groves of apple and other fruit trees, bamboo thickets, squash, melon and vegetable patches.

On the entire coast from Ulchin to Masan, particularly around Pusan, fishing with set nets or by trawling is highly developed. Of particular importance, as elsewhere on the coast of the Sea of Japan, is the production of fish oil and fish fertilizer, but fishing for food fish is also done intensively. Many subtropical fish, such as yellowtail and shark, are caught from the coast of Kyŏngsangnam-do even in winter, because the surface temperatures are high even then. Sardines, mackerel, horse mackerel and sea bream are fished primarily in spring and autumn. Fishing for Spanish mackerel and hairtail is done mainly in autumn, whereas boreal fish such as herring, cod, gray mullet, flounder and whales are caught almost entirely in winter. In Changsaengp'o on Ulsanman a Japanese whaling station has its seat. The most important fishing centers lie in the passenger and freight shipping ports of P'ohang, Pusan and Masan, but fishing is carried out from all of the small corrasion embayments lined up along the coast of the Sea of Japan as well. Ports of this type used exclusively for fishing are Kuryongp'o and Pang-ŏjin.

The area around the mouth of the Naktonggang is known to be the maritime gateway to the land bridge of Korea. It lies closer to the western end of Honshū and the northern end of Kyūshū than any other section of the Korean coast. Simultaneously the bays of Ulsan, Pusan and Masan provide good opportunities for landing and for the construction of ports. Far up into their tips they have depths of 12–18 m, and because of the low tidal range, they are almost free of tidal flats. Farther to the northeast both tides and mud flats are completely lacking, to be sure, but the faulted and downwarped coasts there have almost no deep ingression bays. Farther to the southwest bays increase in number, but due to the large tidal range their inner parts are covered by extensive mud flats. Besides, in this central section only small creeks flow into the bays, so that there is little deposition. The entire hinterland is dominated by the Naktonggang system, and its silting does not affect these bays. The current significance of the Naktonggang gateway for Japan's military position on the Korea Strait can be seen from the fact that the entire region forms a strategic zone with the naval port of Chinhae in the center.

The Naktong basin is the direct hinterland of this gateway and was therefore a passageway of the first order during all periods of conflict (p. 27). Its high culture has fallen into ruin again and again in the process. But in times of peace as well, important overland routes ran through the Naktong basin from the mouth of the Naktonggang northwestward across the Sobaek Range and toward the capital. In the Korean dynastic period there were three main ones: via Ulsan, Kyŏngju and Andong and across the pass of Chungnyŏng to Ch'ungju, via Taegu, Sangju and the pass of Ihwaryŏng likewise to Ch'ungju and via Ch'angnyŏng, Kimch'ŏn and the pass of Ch'up'ungnyŏng to Ch'ŏngju. They were also used by the first European travelers (Fig. 10).

Today these old pack trails have been replaced by the railroad. The Korean trunk line, which connects with the steamship line from Shimonoseki in Pusan, first follows the Naktong valley to Samnangjin, then turns northward and crosses the divide to the Kŭmhogang in a long tunnel. Then it continues via Taegu, Waegwan, where it crosses the Naktonggang, and Kimch'ŏn to Ch'up'ungnyŏng, the lowest pass by far in the Sobaek Range (254 m). Standard and narrow-gauge lines branch off from this main line to Chinhae, Masan, Chinju, Kyŏngju,

P'ohang, Ulsan and Andong. A fairly dense network of roads serves as feeders for these railroad lines.

Considering the very great importance of rice growing for the economy of the Naktong basin as a whole, it is not surprising that the population density corresponds fairly closely to the proportion of paddies in the total area. The counties with the smallest paddy area are also the ones that are least densely populated. They are Yŏng-yang-gun (58 inhabitants/km^2) and Ch'ŏngsong-gun (65 inhabitants/km^2), both in the north. The counties with the most intensive irrigation agriculture are among the most densely populated and lie in the south (Miryang-gun 264 inhabitants/km^2, Ch'ang-won-gun 217, Kimhae-gun 191). The high density of the counties of Tongnae-gun near Pusan (174 inhabitants/km^2) and Talsŏng-gun near Taegu (223 inhabitants/km^2) demonstrates, however, that industry and transportation are also responsible for the, on the average, quite high population density, which amounts to around 180 inhabitants/km^2.

The northwestern part of the basin extends along the foot of the Sobaek Range from Kimch'ŏn via Sŏnsan, Sangju and Yech'ŏn to Andong and Yŏngju. It consists of hill country with scrub vegetation, part of which has recently been reforested with red pines. The hills often do not exceed 300 m, while the valleys have cut down only 30–150 m. The area is composed of granite gneiss and biotite granite, which has weathered and been eroded in the manner described above. All of the resultant phenomena have developed here to perfection. Many valley bottoms are filled for the best part of the year by waterless beds of sand. The rice–barley fields are scattered around the hill country like little dots in the side valleys and on terraces. Extensive, continuous paddy areas are found only here and there, particularly in the lowlands followed by the railroad line from Kimch'ŏn to Andong. All of the larger settlements along this line are county seats and busy market towns. Kimch'ŏn (18,892 inhabitants) is additionally an important rail and road junction on the left-hand river bluff of the Kamch'ŏn. Sangju (29,915 inhabitants) and Andong (21,686 inhabitants) are larger to be sure, but more remote and thus less highly urbanized and Japanized. They are large villages with one-story straw-roofed houses.

The 120-km-long southern end of the T'aebaek Range has the characteristic features of the main part, but to a lesser degree. In particular the contrast between the steeply sloping eastern side and the gently sloping western side exists here as well, and at high elevations there are remnants of strongly uplifted old erosion surfaces. The peaks diminish more and more in height toward the south. East of the uppermost part of the Naktonggang the range runs through granite gneiss and the metamorphic sediments of the Yŏngnam massif. At Irwolsan, which is again 1219 m high, it enters the Kyŏngsang strata, however. In the area north of Yŏngdŏk, which has been well investigated geologically [888], there are several step faults parallel to the direction of the range on the steep eastern side. They have dislocated the sandstones and conglomerates of the Shilla series in relation to the intrusive "young" granites. This has not influenced the surface forms, however, with the exception of a few stretches of structural valley that were carved out subsequently. In one of them the hot springs of Paegamonch'ŏn issue. South of the important road from Ch'ŏngsong to Yŏngdŏk, which crosses the main drainage divide at an elevation of only 300 m, no peak of the T'aebaek Range reaches a height of 1000 m.

The coastal region off the southern third of the T'aebaek Range also resembles the northern section in all of its main features, so that we can refer the reader to the description on p. 309ff. The larger settlements, Ulchin, P'yŏnghae, Yŏnghae, Yŏngdŏk and Ŭich'ang, lie at the edge of small rice-farming plains near the coast. None of them reaches a population of 15,000.

In the central part of the basin between the drainage basins of the Panbyŏngch'ŏn and the Kŭmhogang there are good-sized elevations, some of them higher than the T'aebaek Range, with which they fuse in the east. The highest ones are lined up along the divide between the two just mentioned tributaries of the Naktonggang. They include Pohyŏnsan (1124 m), composed of Shilla strata, and the enormous, jagged granite peak of P'algongsan (1192 m) northeast of Taegu. The Naktonggang cuts through this ridge near Waegwan. West of the water gap follows the extremely steep walled Kŭmosan (977 m), a mountain topped by a porphyry flow. It dominates the central of the above mentioned old Korean overland routes, and for this reason it was built into a mighty fortress. Some of its walls, gates, wells and cisterns still exist. Close to the highest point, inside the fortified summit area, lies a temple that is visited by many pilgrims. It also encloses a hamlet at an elevation of 800 m whose inhabitants grow meager amounts of potatoes and millet. They have given up the rice farming that was formerly attempted here.

South of this massive range the basin of the Kŭmhogang and Hyŏngsan-gang system spreads out for 90 km from west to east from Sŏngju past Taegu and Yŏngch'ŏn to An-gang and Kyŏngju. It lies for the most part in the thick, often thickly bedded black shales of the Taegu formation. Granite and diorite intrusions and remnants of porphyry flows rise above the Taegu hills. In this scrub-covered hill country there are many small reservoirs for irrigation. Japanese irrigation associations, such as the Ankōri Company have also built larger reservoirs. Between the county seat and market town of Yŏngch'ŏn (17,559 inhabitants) and Taegu there are extensive areas with rice paddies in the wide floodplain of the Kŭmhogang. Near the city the floodplain is also covered by well laid out orchards and fields of vegetables.

Taegu is the center of transportation for the entire Naktong basin. As such and as the capital of the province of Kyŏngsang-do it played an important role already during the Korean dynastic period in both war and peace. From 1931 to 1938 its population increased, partly due to the incorporation of surrounding areas, from 93,319 to no fewer than 172,040 inhabitants. The fact that one-quarter of these inhabitants are Japanese demonstrates Taegu's significance as the capital of Kyŏngsangbuk-do, as a center of transportation and commerce and as a garrison. The city lies for the most part on a level alluvial terrace on the southern edge of the Kŭmhogang valley, into which ranges of hills extend here and there from the south and the west. In the city center one-story Japanese houses with bluish gray tile roofs or sparkling corrugated iron ones predominate, whereas in the suburbs the Korean straw roofed houses prevail. The main streets have been straightened and widened by the Japanese. Along them stand two to four story stone buildings, which have been set into the muddle of low houses in a disorganized fashion. In the southeast the two towers of a Protestant mission church soar above the straw roofs. Nothing is left of the rectangular wall that is frequently described in the old travel accounts. The earthworks of the old Korean fortification in the western part of the city do, however, still exist and are now covered by an attractive park with a Shintō temple. In the west and south of the old city center are squares on which the famous large markets have been held at the beginning of September and in January for ages. Taegu is the most important center of silk reeling and weaving in Korea and has one of the most important cigarette factories in the country.

In contrast to Taegu, Kyŏngju has gradually languished since the fall of the Shilla dynasty in 935, after having been the residence of 56 kings for a period of 992 years. Kyŏngju (21,878 inhabitants, Fig. 71) lies in the center of a system of valleys that cross each other, providing convenient connections with P'ohang, Taegu, Pusan and Ulsan, in a rice paddy plain surrounded by deforested mountain slopes. A square wall measuring 600 m on each side encloses part of the present-day city, which has grown far beyond the wall toward the east and the south. The ancient royal city appears to have covered the entire plain toward the south as far as the Namchŏn and toward the east as far as the Pukch'ŏn. Remainders of the highly Sinified ancient Shilla culture are still standing as ruins in the landscape as far as 20 km out and continue to have a firm influence on the character of the landscape. In the southern part of the present-day city and its environs tremendous hemispherical earthern tumuli covered with grass or woods rise as much as 25 m above the plain. On the northern bank of the Namch'ŏn lie the ruins of the ancient royal palace in the form of a broad crescent-shaped embankment (Panwolsŏng = Half Moon Fortress). Northwest of it stands the astronomical observatory, built in 647 A.D. and thus the oldest observatory in East Asia, a bottle-shaped stone tower more than 9 m tall (Ch'ŏmsŏngdae). In the environs one encounters shrines and graves of kings and ministers, the marvellous turtle tablet honoring King Mu-yŏl, who unified the peninsula, and remnants of former temples in the form of fundaments, stone pagodas, stupas and artistically very important stone Buddha figures. The finest of these temples is Pulguksa, which was founded more than 1500 years ago, destroyed in 1592 and partially restored a few decades ago. It lies 6 km southeast of the city on the slope of T'ohamsan. On the seaward side of the same mountain a round stone grotto was carved into the weathered rock, in the middle of which a gigantic granite statue of Buddha sits (Sŏkkuram). The temple was obviously fashioned according to Chinese models and is one of the most brillant achievements of Shilla art.

The 15-km-long, triangular Changgi peninsula is a unique feature on Korea's Japan Sea coast. Its origin is probably connected with the echelon structure of the southeastern part of the Naktong basin. It forms the northern continuation, ending with the cape of Changgi, of the Pulguksa fault block, which consists here of Tertiary andesites, basalts and trachytes and their garish green tuffs. The present-day surface forms are, however, the result of more recent events. The eastern fringe of the peninsula is occupied by an abrasion platform up to 850 m wide, rising inland and ending at an old cliff that has been raised as much as 60 m. The old mountain, which forms the backbone of the peninsula, rises behind it to 215 m. It drops off precipitously toward the northwest to an unraised active cliff. Thus the peninsula has recently been warped up asymmetrically. The abrasion platform is covered by barley fields in winter and bean and millet fields in sum-

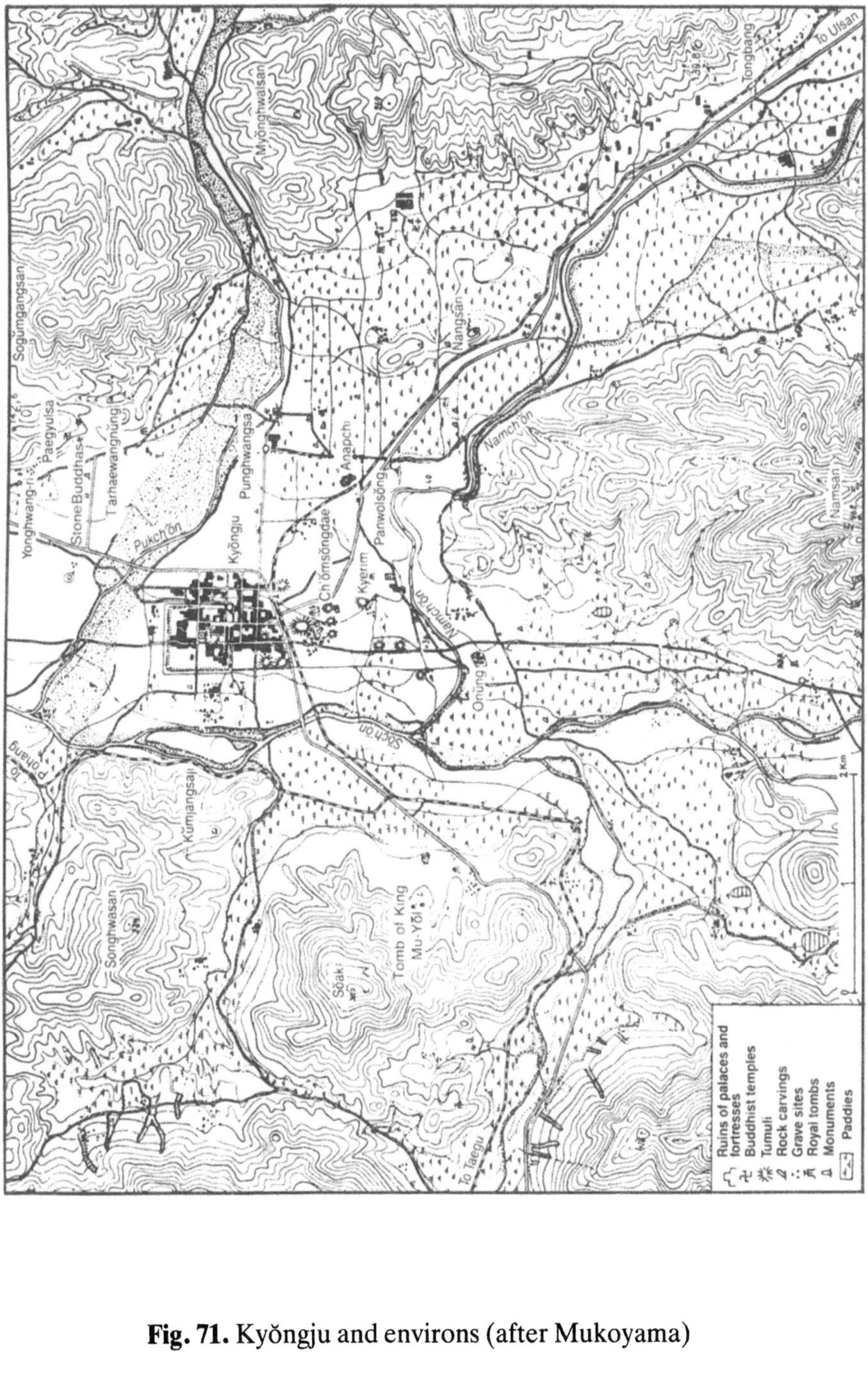

Fig. 71. Kyŏngju and environs (after Mukoyama)

mer. There are fishing villages all around the coast, and from them fish are caught for oil in spring, particularly sardines. In winter herring is fished, and all year round seaweed is zealously gathered. The dirty little Korean fishing villages on the northwestern coast consist of farmsteads enclosed by palisades of bamboo, boards and twigs, on which the fish are hung up to dry.

Closed off by the Changgi peninsula is the bay of Yŏng-ilman, the northern half of which is more than 10 m deep. Landward it ends at a 9-km-long bar overgrown with roses and bindweed, behind which the Hyŏngsan-gang has deposited a triangular delta. On the northwestern end of the bar, at the end of the higher terrain, which drops off steeply toward the delta plain, lies the fishing and maritime trading town of P'ohang (15,250 inhabitants, 17% of them Japanese). This location gives it a striking resemblance to Danzig.

South of the basin of Taegu the southern interior of the basin becomes mountainous again. The higher elevations are again within the granite intrusions or the porphyry flows. Thus the highest peak, Kajisan, rises just at the boundary between granite and porphyry. The valleys in porphyry, augitophyre and felsophyre are really extraordinarily steep-sided. A prime example is the Naktonggang gorge between Samnangjin and Yangsan. In areas with sedimentary strata belonging to the Naktong and Shilla series the terrain is generally lower. Along the river small basins have been eroded out here and there. A broad floodplain follows the Namgang within the Naktong strata. Above it, safe from floods, lies Chinju (35,380 inhabitants), which was the capital of Kyŏngsangnam-do until 1925. The largest alluvial plain in this section is the 15-km-long and up to 17-km-wide delta of the Naktonggang (Fig. 4), on the northern edge of which the county seat and market town of Kimhae (22,512 inhabitants) lies. The surface forms of the eastern half of this area are strongly influenced by the above described structural valleys. In one of them lies Ulsan (15,587 inhabitants), whose particular significance lies in the fact that it has an airport for civilian traffic. The air route Seoul – Tōkyō, avoiding the militarized zones of Pusan–Kimhae, on the one hand, and Shimonoseki, on the other hand, runs via Ulsan and Fukuoka. Ulsan is visible from afar with the 60 m high towers of its radio station.

With the bay of Ulsan, the ingression coast of the Korea Strait begins. As pointed out above, its first section reaches to Chinhaeman. This section alone affords really good conditions for traffic to Japan. The bay of Ulsan no longer plays an important role, however. Only the harbors of Chinhae, Masan and Pusan are of importance. Chinhae is primarily a naval port. The settlement is constructed around a multiple street junction (18,988 inhabitants, 24% of them Japanese). Masan's time of grandeur is past. It was during the first years of this century (p. 35), and the former Russian consulate, an unwieldy stone building with bow windows, dates back to that period. In 1907 Masan's maritime trade turnover was one-quarter that of Pusan. Meanwhile, however, the port of Masan is no longer open to international trade, since its access passes too close to Chinhae. Despite its location on a deep bay almost devoid of mud flats it has primarily local significance today as a port of exportation for the rice, soybeans and tobacco produced in the vicinity (32,411 inhabitants, 18% of them Japanese). Old Masan is a Korean farming village. Two km farther out along the bay, at the end of the railroad line, on the breakwaters and quays, New Masan spreads out. It is constructed exclusively in Japanese style.

Almost the entire passenger traffic between Japan and Korea and a considerable portion of the freight traffic is concentrated in Pusan today. Large steamships with connections to the fast trains from Tōkyō and Seoul cover the 224 km in 9.5 h twice a day in either direction. A ferry connection would not bring any advantages for the time being, because the Korean and Japanese rail networks have different gauges. There is also busy cargo shipping between Pusan and the Japanese ports opposite it. In addition, Pusan is the center of Korean coastal shipping. It is therefore Korea's foremost passenger and cargo port by far. Finally, it is Korea's most important fish market and also supplies nearby Japan with an abundance of fresh fish [653, p. 300f].

Already in 1443 King Sejong concluded a trade treaty with the daimyo of Tsushima pertaining to three southeastern Korean ports, including Pusan [900]. This soon developed into a permanent Japanese settlement close to Pusan. During Hideyoshi's war against Korea the small town became Japan's most important military base on the mainland. From then on, with brief interruptions, Japanese fishermen and traders have lived near Pusan. In 1678 the Korean government assigned them a special concession 9 km south of the Korean town of Pusan. The Japanese settlement has not left this place on the mainland across from the island of Yŏngdo since then. When the Korean government opened the country to Japanese trade in 1876, the Japanese settlement of New Pusan became the island empire's first large base in Korea. That year marks the beginning of the impressive development made by city and port due to Japanese diligence and flair for organization. In 1910, in the year of annexation, Pusan had 38,217 inhabitants. In 1931 it was 146,098 and in 1938 213,744, among them no fewer than 59,231 Japanese (28%).

The city lies on the central part of a bay that is enclosed by steep, barren mountains and dominated by the 7-km-long island of Yŏngdo (Fig. 72). The latter has recently been connected to the mainland by a 400-m-long iron bridge, since the settlement had spread out onto its northwestern end. Old Pusan (Pusanjin) nestles in the innermost corner of the bay, overtopped by a fortified mountain. It has remained a small Korean town. The modern residential city spreads out in a rectangle 2000 by 500 m opposite Yŏngdo and at the southern foot of the barren Kubongsan. It consists of various blocks with straight streets crossing at right angles, between which the hill of Yongdusan with its decorative park rises. Most of the tall buildings stand in the crowded district around it: government palace, hospital, fish market hall, chamber of commerce, police department, the administration buildings of the Japanese shipping lines, banks, city hall, hotels and theaters. The port, railroad and industrial facilities stretch out for 7 km between New and Old Pusan, between Kubongsan and the bay, on ground that has partially been reclaimed from the sea with great effort. The four quays at which large steamships can dock directly, because the water next to them has a depth of 11 m, have a total length of 1368 m. Passenger services already transported 806,980 persons in 1928, 64% of them Japanese and 35% Koreans. In 1929 9955 steamships with a tonnage of 1.34 million and 15,247 sailing ships with 0.14 million tons entered the harbor of Pusan. The total trade amounted to 269.4 million yen in 1934. The primary exports are rice, raw silk, beans, marine products and livestock, the main imports are Japanese industrial goods, in particular machines, pharmaceutical products, fertilizer, cotton yarns, ginned cotton and petroleum. Korea's modern industrialization has spread to Pusan as well, where cotton, hemp, linen, porcelain and enamelware factories have been established in the past years. Nearby Tongnae (18,985 inhabitants) is where Pusan's residents go for holidays and recreation.

13 The Sobaek Range, Including Chirisan

For a distance of 280 km the Sobaek Range (Kor. Sobaeksanmaek = Small White Mountain Range) runs diagonally through southern Korea. It maintains now a southwesterly, now a more purely southerly direction, and for the entire distance the boundary between the Ch'ungch'ŏng and Chŏlla provinces, on the one side, and the Kyŏngsang provinces, on the other side runs along the top of it. In general the mountains drop off more steeply toward the Naktong basin than toward the Ch'ungch'ŏng-Chŏlla region, in whose eastern fringe there is also mountainous country crossed by the Mihoch'ŏn, the Kŭmgang, the Sŏmjin-gang

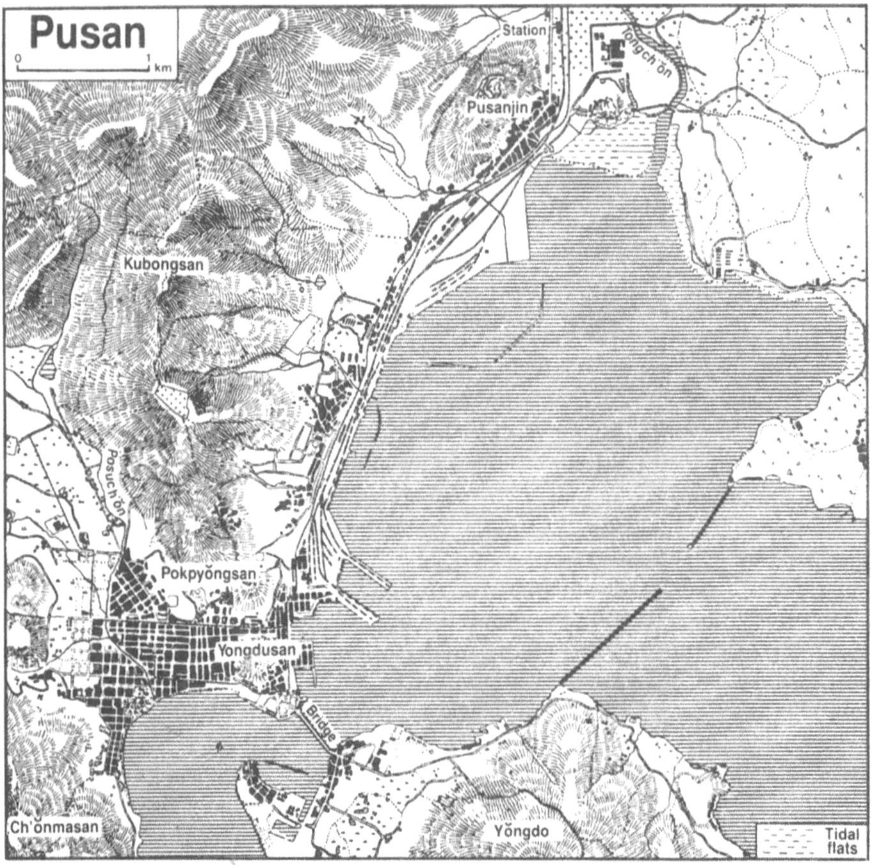

Fig. 72. Pusan and environs (after an official plan)

and the Posŏnggang. To a diminished degree, therefore, it has the same asymmetry as the T'aebaek Range. Whereas the latter trends in a markedly discordant direction to the geological structure, however, the former basically follows the Yŏngnam massif, and only in the middle of its northern half does it extend a short distance into the Okch'ŏn fold belt (Fig. 39). Thus it has basically a concordant trend. Accordingly it is composed primarily of granite gneiss and granite, and only in the just mentioned place does it have a share in the Shilla and Pulguksa strata and the metamorphic sediments. It can probably be interpreted as a posthumous up-arching of the Yŏngnam massif, for in its summit regions there are also remnants of old, gently sloping erosion surfaces. The valleys are generally deeply entrenched and steep-sided, the ridges often jagged and serrated.

The first section of the range, which culminates in Kungmangbong (1421 m), extends to Chungnyŏng (= Bamboo Pass, 689 m). The latter lies in the course of a structural valley utilized by the road from Tanyang to Yŏngju. It is followed by a granitic part of the range trending almost due westward as far as Hanŭlchae (Hanwŏllyŏng), or Ihwaryŏng. Here the range is again crossed by structural valleys, in which the divide drops down to 519 and 548 m, respectively. The most important connection between Seoul and Pusan in ancient times and that often described in the accounts of the first travelers leads across Ihwaryŏng. At its eastern foot in a junction of valleys lies the county seat of Mun-gyŏng, in the vicinity of which anthracite deposits belonging to the P'yŏng-an system are mined.

The next section trends almost due south and remains much lower in elevation. The highest peak is Songnisan (1057 m), whose porphyry flows form magnificent rock ledges and columns, jutting up out of marvellous mixed forest. The mountain belongs to the temple of Pŏpchusa and is considered sacred. Then comes, exactly in the middle of the whole range, the extremely deeply incised pass of Ch'up'ungnyŏng [254], which is utilized by the Korean trunk line and is therefore the only really important pass for overland traffic in the Sobaek Range today. It is also crossed by the road from Taejŏn to Taegu, connecting the capitals of Chŏllabuk-do and Kyŏngsangbuk-do. Directly beyond this pass the range rises again to 1111 m in Hwanghaksan. This more than 100-km-long section of the Sobaek Range reaches a height of 1290 m in the andesite peak of Taedŏksan and even 1594 m in the mighty granite gneiss ridge of Tŏgyusan. Toward the Naktong basin it sends out a 30-km-long offshoot, which culminates at its eastern end in the imposing Kayasan (1430 m), and ends about at the road Namwon – Unbong – Hamyang. Then the range rises to its highest elevation (1915 m) in the gigantic massif of Chirisan, only to drop immediately thereafter to below 30 m in the gorge of the Sŏmjin-gang. Just south of the latter it again rises to 1218 m in Paegunsan, ending shortly thereafter near Kwang-yang. The last two sections extend only 30 and 20 km, respectively, in the longitudinal direction of the range, but the widest part of the Sobaek Range is at Chirisan.

Observations on the climate of the higher altitudinal zone are lacking, since the few meteorological stations are located in the larger towns in the valley bottoms. Two of the typical stations have the average figures shown in Table 23.

Winter is quite cold, considering the southerly latitude, so that the slight precipitation that does fall falls as snow. The higher mountains have a snow cover for 5–6 months. While climbing the summit of Chirisan I still found patches of the snow cover on 8 May 1933, and in the nights before and after, which I spent at

Chirisan, the temperature still dropped below 0° C at 1400 m. In the Namgang valley in the northern part of Chirisan, however, at 250 m, snow falls only between the end of November and the beginning of March, generally melts within a week at the most and rarely reaches a depth of 30 cm. Summer is hot. It brings enormous amounts of rain, and the peaks are hidden by clouds and fog for weeks on end.

As in the T'aebaek Range, fire-field farming is still generally widespread at the upper limit of settlements in the Sobaek Range. In the spring of 1933 I hardly climbed a peak of the Sobaek Range from which the smoke of burning forests or scrub areas could not be seen somewhere. The significance of fire-field farming for the landscape varies greatly, however. Evidently the authorities in Kyŏngsangbuk-do have been much more successful at preventing fire-field farming than in Ch'ungch'ŏngbuk-do. Consequently the provincial boundary around Chungnyŏng forms a striking dividing line. On the Ch'ungch'ŏng side one

Station		Temperature						
		January			August		July	
	Altitude (m)	Mean temp.	Mean min. temp.	Absolute min. temp.	Mean temp.	Mean max. temp.	Absolute max. temp.	Annual range
Mun-gyŏng	175	–4.0	–8.0	–20.2	25.5	30.0	37.2	29.5
Hamyang	169	–1.5	–6.9	–18.5	26.0	30.8	38.6[b]	27.5

[a] February. [b] July.

Station	Precipitation								
	January				August				
	Average amount (mm)	per 1000	Average no. of days with precip.	Precipitation intensity	Average amount (mm)	per 1000	Average no. of days with precip.	Precipitation intensity	Annual amount (mm)
Mun-gyŏng	26.0[a]	21[a]	6[a]	4.3[a]	329	266	15	21.9	1237
Hamyang	33.4	28	5	6.7	261	219	14	18.6	1194

Table 23. Climatic data for the Sobaek Range

sees only steep brown fields, brush, scrub and destroyed woodlands, while the Kyŏngsang side is covered by tall oak and pine forests. This fire-field region strongly resembles those in the T'aebaek Range and in the Kaema upland. I encountered the highest fire fields at an elevation of 1150–1400 m, i.e., just below most of the peaks. Only a bit lower the hamlets and dispersed settlements of the fire-field farmers begin.

Thus the living space of the fire-field farmers is extraordinarily limited here. Permanent cultivation extends far upward out of the deeply incised valleys with their large villages. The warmth of summer allows an abundance and variety of crops to be grown. The no longer active talus cones are divided into rice terraces, and above the high-water bed the valley floors are also covered by rice paddies. Some of the irrigated paddies are used here, too, for growing barley in winter. For the types of dry-field crops the reader is referred to the discussion on the Naktong basin. Worthy of mention is the intensive tobacco cultivation, particularly in the valleys on the northwestern side, ginseng cultivation near Muju, the rows of mulberry bushes, sometimes growing in the same fields with rows of barley and spring crops, the groves of timber bamboo and the apple, pear, peach and persimmon trees around the settlements.

This rich economic situation, for such a mountainous area, is supplemented by occasional mines, of which the gold mines of Naesŏng near Ponghwa in the extreme northeast [890], Kŭmsan above Muju and Ch'onam near Kwang-yang deserve special mention. The Muju–Kŭmsan mine is in a quartz vein with an average width of 3 m containing auriferous chalcopyrite in the contact of granite and old schists.

When examining the map of population density or population distribution of Korea (Fig. 85 and Map 3, cf. editors' note on p. 7), one discovers that there is a strip with low values running through southern Korea along the Sobaek Range. Because of the varied economy we have just observed, however, settlement is much more intense than in any part of the T'aebaek Range. Changsu-gun (93 inhabitants/km^2) should provide a reference point for evaluating the average density.

As a result of the many centuries of fire-field farming, large parts of the mountains have been divested of their natural forests, in some cases up to the high peaks, and are covered by more or less scrawny second growths. As an example let us describe the vegetation on the summit of Taedŏksan and on its eastern slope. The highly depleted plant association on the summit consists of grasses and shrubs of *Lespedeza bicolor* Turcz., *Rhododendron schlippenbachii*, *Kalopanax pictum*, the honeysuckle *Lonicera sachalinensis*, the common oak, the small-leaved linden, the common ash and the Korean blueberry. Much more varied and lush is the shrub vegetation on the massive talus slopes consisting of andesite blocks covering the medium-steep eastern slope right next to those old fire fields, because fire-field farming is impossible on them. Here I collected Hercules' club, a wild rose, the wild grapevine, the just mentioned *Lespedeza* species, the wild pear *Pyrus ussuriensis*, the Siebold hazel, the hawthorn-leaved blackberry, *Fagara schiniifolia*, *Stephenandra incisa* and several lianas. What is remarkable is that the low grasses extending up to the summit of Taedŏksan include some bamboo species. This is an exclusively southern Korean characteristic of former fire fields. The species I encountered most frequently in the Sobaek

Range and on Chejudo are *Sasamorpha purpurascens* Nakai and *Sasa albomarginata* Makino and Shib. These species have nothing to do with the tall-stemmed bamboo groves cultivated in the low parts of the valleys around the villages and isolated farmsteads.

In the environs of the many temples, on steep rocky slopes and in particularly inaccessible, remote mountain areas, good-sized stands of primary forest have been preserved. Around Tusolbong (1314 m) just south of Chungnyŏng it has the following composition. Magnificent oaks and maples predominate, less frequent are the Siebold ash and the common ash, the hornbeam *Carpinus erosa*, the small-leaved linden, *Weigela florida* S. and Z. and *Palura paniculata* Nakai var. *vilosa* Nakai. All in all this is a pure deciduous forest of the oak zone, which already contains a number of southern Korean elements. On a neighboring peak with almost exactly the same elevation there are, however, also firs.

As the widest and by far highest part of the Sobaek Range Chirisan deserves special attention (Fig. 73). It does after all have a large number of individual features that could not be mentioned in the previous paragraphs. Its backbone forms a west–east trending ridge only 25 km long, along which the drainage divide between the Namgang and the Sŏmjin-gang runs. It culminates in the west in Nogodan (1507 m); in the middle the pyramid of Panyabong (1751 m) lies just off it to the north, and in the east it ends in P. 1682. A transverse ridge branching off from the latter toward the north contains the very steep summit pyramid of Ch'ŏnwangbong (1915 m), splintery, jagged, dissected by steep ravines, and towering far above everything else. Other transverse ridges run both northward to the Namgang valley and southward to the Sŏmjin-gang valley. The mountains are composed of hard black and white gneiss, diorites, which weather to a grayish blue color at the edge of the high-water beds, and biotite granite. Ch'ŏnwangbong is composed of the latter. Rounded slopes are generally lacking, but there are enormous streams of blocks as high as a house.

North of the main ridge and its side branches, near Unbong in the headwater region of the Namgang, there are valleys with broad floors and moderate slopes with the, for southern Korea, very unusual elevation of 450–500 m. This is an old area of moderate relief, which was raised to its present location by the recent uplift, but has not yet been reached by the erosion that was rejuvenated by the uplift. Proceeding upstream along the Namgang, the latter has come only to just south of the junction of valleys at the temple of Shilsangsa. The system of surface forms of Chirisan can only be understood at all as the result of extensive recent uplift and antecedence. Downstream from Shilsangsa the Namgang cuts through the Ch'ŏnwangbong side ridge in a rugged gorge with a right angle bend, separating Sambongsan (1187 m) from it, although just to the north of the latter it could use a convenient, straight structural valley with its drainage divide at an elevation of only 513 m, through which the road from Namwon to Hamyang runs. A similar antecedent gorge is the deeply entrenched V–shaped valley of the lower Sŏmjin-gang. It begins west of the boundary of Chirisan in the Chŏlla hill country, and its central part skirts two elongated structural depressions, those of Namwon and Kurye, which have mighty, no longer active talus cones coming down from the high mountains. Between these two basins the likewise highly entrenched Posŏnggang flows into the Sŏmjin-gang. Paegunsan is merely a part of the same area of recent uplift that has been separated off by erosion. Remnants of a much older relief than that of Unbong can be found at an elevation of 1200–1600 m in various places in the main range, particularly northwest of Nogodan. On the granitic edges of the basin of Namwon we find weathered yellow soils and grus and gully formation. Above this region a zone of dark gray, rounded slopes follows here, too.

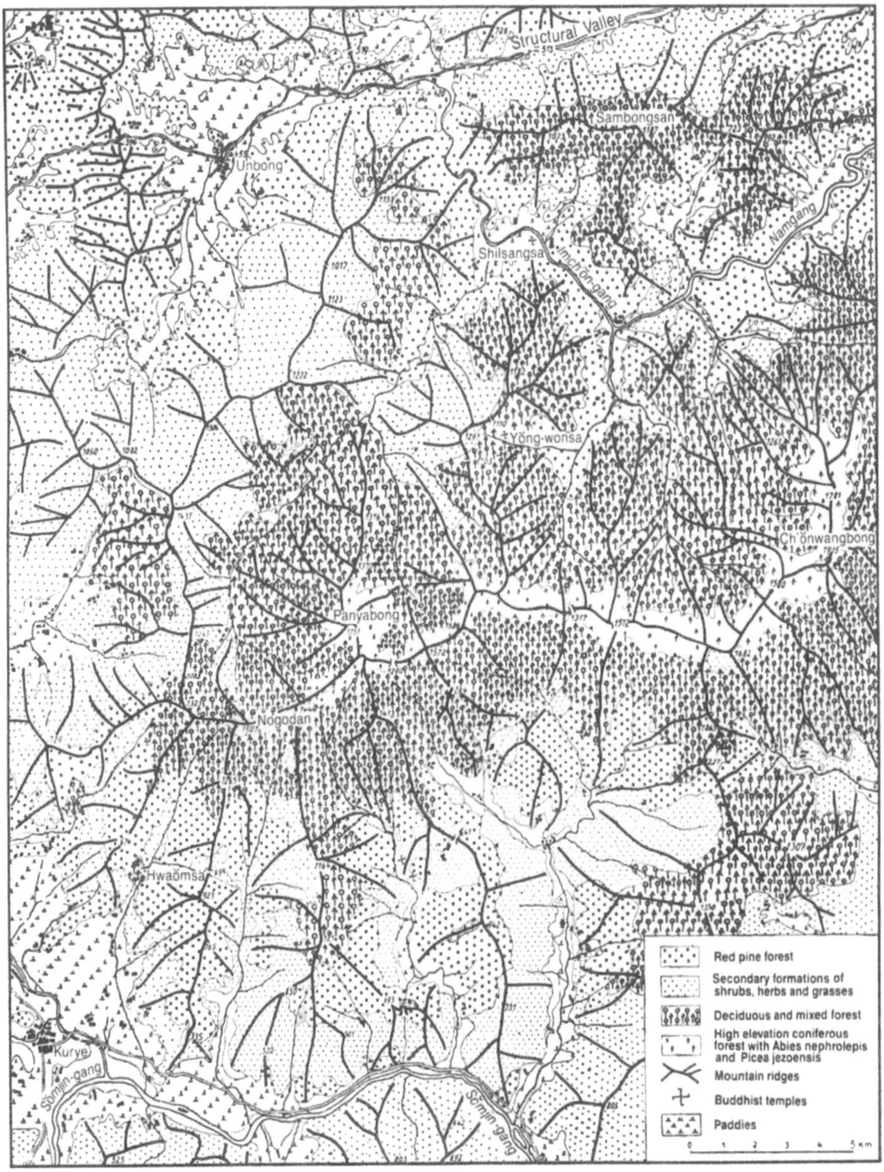

Fig. 73. Orography and vegetation in Chirisan (designed by the author on the basis of a map by Nakai)

Chirisan also has a large number of temples. It is one of the country's five sacred mountains. Their presence, the colossal steepness of the slopes and the remoteness of the central parts have promoted the preservation of primary forests to an extent that is not otherwise encountered in southern Korea. Nakai devoted the first of his monographs on Korean plants to the vegetation of Chirisan [902]. On the basis of his research and my own collections, the natural altitudinal zonation of the Chirisan forests is, broadly speaking, as follows. Of the most common forest trees, the red pine occurs from the foot of the mountain up to 800–1150 m, *Quercus glandulifera* and the broad-leaf oak *Quercus dentata* from the same to 500–600 m, *Quercus serrata* from 500 to 1000 m, the common oak from 900 to 1400 m, the birch *Betula ermanii* from 1350 m to the summit of Ch'ŏnwangbong, the Korean nut-pine from 300 m to the same, the Korean fir from 1000 m to the same. On the basis of the altitudinal distribution of these eight most important trees it is possible to establish three altitudinal zones that could be named for the red pine (up to 900 m), the common oak (900 to 1400 m) and the Korean fir (above 1400 m). The upper limits of these three zones naturally vary with the exposure and other factors.

The red pine zone corresponds to the combined bamboo and styrax zones of the general classification (p. 137f). The only places where it is really dominated by this conifer are where artificial reforestation has occurred, however. In the natural forest it occurs only sparsely. A wealth of deciduous trees is particularly characteristic there. On the south side of the mountain, e.g., in the forest of the temple Hwaŏmsa near Kurye, which was founded in 550 A.D. at the latest, the Japanese camellia occurs up to 500 m as the only evergreen tree in this zone. Among the deciduous trees, the following predominate: the above mentioned oaks, except for the common oak, further *Abelicea serrata*, *Castanea bungeana*, several *Celtis* species, *Prunus persica*, the pinnate-leaved nut-gall tree *Rhus javanica*, *Salix hallaisanensis*, *Malus toringo*, the wild mulberry and the Chinese walnut. Many of them blossom magnificently in white or pink in spring. The conifers occurring in this lower zone are the Korean nut-pine and *Cephalotaxus drupacea* S. and Z. It is thus primarily a summergreen deciduous forest with species that are already very different from the northern Korean ones. In the zone of the common oak the following occur in addition to it: several *Carpinus* species with their orange or yellow colored catkins, various wild cherries, maples and birches, *Cornus controversa*, the corktree *Phellodendron amurense*, *Symplocos crataegoides*, etc., and the conifers Korean nut-pine, Korean fir and *Abies holophylla*. Between the varied shrubs in the undergrowth the different bamboo species, up to waist high, luxuriate from the foot of the mountains to the upper limit of the oak zone. The uppermost zone contains much fewer species. In it the conifers predominate. The Korean fir, the Korean nut-pine, the flat-leaf spruce and the birches *Betula ermanii* and *Betula schmidtii* are the leading species. This forest extends high up on the steep slopes of the Ch'ŏnwangbong pyramid, reaching its uppermost limit just below the summit. The latter bears only rhododendron and birch shrubs. It probably lies just above the natural timberline, which is lowered somewhat by the local climate.

Nakai's list of plants on Chirisan contains 464 species. Of them 349 also occur on the Japanese islands, whereas only 57 are also found in northern China and Manchuria. Of the plants that simultaneously occur in Japan 34 have their northern limit in Chirisan, according to Nakai. These figures clearly show how closely the flora of Chirisan is related to that of Japan.

The extensive mountain forests still support a wealth of fauna. On the main ridge I encountered traces of leopards and saw an enormous tiger trap. The fire-field farmers are severely bothered by the large number of wild boars. Therefore the crops that the wild boars particularly like are grown only on the fields that are farthest away from the mountain forests, if possible. In the forests I found the nests built of bamboo stalks by the wild boars, in which the sows give birth to their young. The calls of the wild doves resound over great distances in the peaceful forests of the high altitudes. In the depths of the valleys one hears the cries of pheasants almost constantly during the day in summer.

The basins of Kurye and Namwon (14,814 inhabitants) and the broad, high-lying valleys of the Unbong zone are densely covered with villages and hamlets and bear rice–barley fields on talus cones and valley bottoms. The proportion of paddies in the total cultivated area is 72.4% in Kurye-gun and 76.4% in Namwon-gun (Fig. 87). The cultivated area is understandably only a small percentage of the total area, however, in all of the counties that have a share in Chirisan. In Sanch'ŏng-gun it is 14.7% and in Kurye-gun 14.8%. In the gorge of the Sŏmjin-gang there are quite a number of prosperous villages, some of them hidden in groves of bamboo. Beside the paddies, which are planted to alfalfa in winter, dry fields with barley and wheat spread out. On the blue, shimmering river there are many bustling riverboats with large cotton sails. From the depths of the Sŏmjin-gang valley the paddies extend up to an elevation of 750 m in Chirisan. In the interior of the mountains one sees the, for Korea, unusual sight of steep slopes parceled up into high-rimmed, irrigated terraces. The hamlets of the sedentary farmers follow the valley cuts upward to about 700 m, their fields extend up to 1000 m. Because the slopes are tremendously steep, each single row of crops (potatoes, wheat) is set on a separate small horizontal terrace. The uppermost fire-field farms lie on the old remnant of an erosion surface northwest of Nogodan at an elevation of about 1200 m, the uppermost fire fields at about 1400 m. The highest temple is probably Yŏng-wonsa (920 m). The influence of Buddhism on the landscape is much weaker in Chirisan than in Kŭmgangsan, however. In the remotest corners of the valleys lie the Japanese forestry stations, and in summer hunters, woodcutters, charcoal burners and gatherers of mushrooms, berries, herbs, roots and resin penetrate up to the highest ridges on narrow, difficult trails. Northwest of Nogodan at an elevation of 1300–1400 m the American missionaries in southern Korea have built themselves a summer resort with single-family houses.

From the depths of the valleys up to the highest elevations there are regular changes in the landscape that can be seen as a "cultural geographical series" here, too. When climbing up from the basin of Namwon southeastward into the high mountains, one finds the following elements of this series: (1) Rice–barley paddy plain; (2) yellow, weathered slopes and blackish gray rounded slopes, some completely barren, some with scrawny red pine secondary forest; (3) rectangular fire fields between secondary scrub; (4) secondary or primary mixed forest; (5) almost unaltered primary coniferous forest.

14 The Ch'ungch'ŏng–Chŏlla Region

Between the western edge of the Sobaek Range and the Yellow Sea coast a region spreads out that is the counterpart to the Naktong basin in southern Korea, on the one hand, but has close connections to the adjoining Yellow Sea regions of central Korea to the north, on the other hand. It begins in the southernmost fringe of Kyŏnggi-do and embraces the western half of Ch'ungch'ŏngbuk-do, all of Ch'ungch'ŏngnam-do, most of Chŏllabuk-do and the northern quarter of Chŏllanam-do. It includes all four capitals of the Ch'ungch'ŏng and Chŏlla provinces and will therefore be called the Ch'ungch'ŏng–Chŏlla region. Since, as we have pointed out above, the Sobaek Range does not have a pronounced western foot, one could debate as to the most expedient way to draw the boundary between it and the Ch'ungch'ŏng – Chŏlla region. The same is true of the southern boundary. I choose the lines Yŏnp'ung – Poŭn – Yŏngdong – Muju – Changsu – Sŏkkong-ri and Sŏkkong-ri – Hwasun – Naju – Hamp'yŏng. With these boundaries the Ch'ungch'ŏng–Chŏlla region has a maximum length of 210 km, a maximum width of 160 km and an area of about 21,000 km^2 (Fig. 39).

As part of the main Korean tilt block, the Ch'ungch'ŏng–Chŏlla region slopes toward the southwest. The majority of the large rivers flow in this direction: the Ansŏng-ch'ŏn, the lower reaches of the Kŭmgang below Pugang and its headward extension, the Mihoch'ŏn, the Man-gyŏnggang and finally, way in the south, the Yŏngsan-gang. The upper and central portions of the two longest rivers, the Kŭmgang (402 km) and the Sŏmjin-gang (212 km), do not fit into this pattern. Both run parallel to the Sobaek Range on its western side, one toward the north and the other toward the south. In these segments both, but particularly the Kŭmgang, have closely crowded, widely oscillating meanders like no other river in the peninsula. In contrast to the Sŏmjin-gang, the Kŭmgang flows only in the Ch'ungch'ŏng–Chŏlla region. From its mouth near Kunsan the Kŭmgang is navigable by riverboats for 121 km upstream to Pugang, just to the beginning of the meandering central segment. The 50 km long lowest segment from Kanggyŏng (17,009 inhabitants) on is widened by the tides, so that it can be navigated by small sea-going vessels. The tides reach 20 km further upstream, to Puyŏ.

The eastern part of the Ch'ungch'ŏng–Chŏlla region consists of sprawling mountain country with no pronounced longitudinal direction or ridges. It attains heights of 1187 m in the gently arching Mudŭngsan, southeast of Kwangju, the capital of Chŏllanam-do, and 1126 m in Unjangsan, northeast of Chŏnju. Roughly at the last named peak lies the beginning of the Noryŏng Ridge, which extends southwestward to Hamp'yŏng, not far from the Yellow Sea coast. It culminates at 763 m in Naejangsan and again at 620 m in Munjusan. The Ch'aryŏng Ridge runs parallel to it through the northern part of the Ch'ungch'ŏng–Chŏlla region. As an extension of the similarly trending ridges of the upper Han-gang region, it crosses the Namhan-gang downstream from Ch'ungju and extends out to the Yellow Sea coast near Taech'ŏn. These two ridges are named after the passes in which they are crossed by the traditional Korean overland route from Seoul to Mokp'o.

The mountain country and ridges form a backdrop for hill country that generally does not exceed an elevation of 200 m. Toward the south the hills are increasingly interspersed with alluvial plains, merging into marshy areas along the lower courses of the rivers. The fact that these plains are quite wide even along rather insignificant rivers is probably a result of backing up due to the eustatic rise of the sea level. The hilly terrain rises steeply above the alluvial plains.

The southwestern orientation of the river courses and ridges is not only a consequence of the corresponding direction of dip of the tilt block, but also of the geological structure, which is concordant to the main elements of the relief here (Fig. 12). In the extreme northwest the granite, granite gneiss and metamorphic schists of the Kyŏnggi massif predominate, in the extreme southeast the corresponding rocks of the Yŏngnam massif. The broad main part lies in the region of the Okch'ŏn depression, however. It consists of rocks of extremely varied character: granite, gneiss, metamorphic sediments, porphyry flows and strata belonging to the Taedong system and the Naktong and Shilla series of the Kyŏngsang system. They all have in common that they are arranged in stripes stretching out southwest–northeast. The Ch'aryŏng Ridge runs along the northwestern fringe, the Noryŏng Ridge primarily along the southeastern fringe of the Okch'ŏn depression. It is possible that they owe their origin to posthumous epeirogenic uplift of these fringes. This would mean that that there is a core of truth to Kotō's old concept (p. 67) in this point after all.

At low elevations red-colored soils predominate on the sparsely overgrown, deeply gullied slopes. Weathered granite does not have as strong a tendency to form red soils, and only here and there is it dominated by such, for example between Chŏngju and Chŏng-ŭp and in the basin of Tamyang–Naju with its especially mild winters. At elevations above 200 m yellow soils predominate everywhere on granite with their characteristic forms, and above this zone there is also a higher one dominated by jagged peaks and ridges or rounded slopes and sugarloafs, depending on the joint structure. Rounded forms also occur in other rocks, as long as they are unstratified, particularly in porphyry, conglomerates and quartzite. Peculiar to the granitic regions of the Ch'ungch'ŏng–Chŏlla region are broad, low-lying basins along quite small rivers closed off downstream by a low barrier of the same rock. Basins of this type are those of Ch'ungju, Choch'iwon, Taejŏn and Tamyang–Naju. It is possible that elongated, straight ones are highly widened structural valleys.

In the northwestern part of the Ch'ungch'ŏng–Chŏlla region the delicate crenulation of the ingression coast surpasses everything we have become acquainted with along the Korean Yellow Sea coast thus far. Kyŏnggiman and in particular its southeastern tip, the estuary of the Ansŏngch'ŏn near Asan, separate off the secondary peninsula that is named T'aean peninsula, after a town on it. South of it the coastal forms become somewhat coarser and the process of straightening has begun, in the form of occasional bars. Between the funnel seas mountainous peninsulas stretch out toward the west. The most extensive one is the one on which Puan lies (Pyŏnsan peninsula). The peninsula on which Kunsan lies, at the mouth of the sediment-rich Kŭmgang, is welded together of rocky islands and coastal marshes. All of these bays are filled with mud flats and narrow tidal channels.

Up to 50 km off the coast lie isolated mountainous islands. Only in the Kogunsan islands do they flock together in a group. With steep cliffs they rise above the shallow sea or tidal flats. They consist of the same gneiss, porphyry and metamorphic sediments as the adjacent parts of the mainland and are inhabited by fishermen and farmers. Wido rises up to an elevation of 255 m.

Since the Ch'ungch'ŏng–Chŏlla region occupies the side of southern Korea that faces the continent, winter is much colder than in the Naktong basin. The annual temperature range is correspondingly greater. The January temperatures increase toward the south and toward the coast, but only in the extreme southwest do the January temperatures lie above 0° C. Large amounts of precipitation fall, on the annual average, and since they are primarily brought by Ld_2 and Le_2 cyclones and in summer by Lb_1 and Lc_1 typhoons, they exceed the amounts in the Naktong basin. The Ch'aryŏng Ridge and the mountains around the basin of Tamyang–Naju are local areas with more than 1300 mm annual precipitation. Otherwise this value is surpassed only in the mountainous country in the southeast. As in the Naktong basin, the annual amount of precipitation increases on the whole toward the south. But the Yellow Sea coast and particularly the islands off it receive much less precipitation here too. The summer rainy season is distributed over a larger number of months. Table 24 contains data from two mainland stations each in the north, the center and the south, one of them in the interior and one near the coast. Finally, a northern and a southern island station are also included.

Like the Naktong basin, the Ch'ungch'ŏng–Chŏlla region was also the heartland of one of the "Three Kingdoms," namely, that of Paekche. Consequently the population density was relatively high at an early period here, too (p. 25). Today the Chŏlla provinces and Ch'ungch'ŏngnam-do are among the most densely populated provinces in Korea (p. 14), indeed if we subtract the urban population, they lead the whole country. The primarily rural county of Kimje-gun has a density of 321, the neighboring Iksan-gun 458. As a result the destruction of the primary forest has advanced to a high degree here, too. If no part of the Ch'ungch'ŏng–Chŏlla region makes such a wretched impression as many stretches of the northern half of the Naktong basin, it is primarily because reforestation has made much greater progress here than there. It had already begun during the Korean dynastic period. Along with the red pine that is used everywhere in Korea, one also encounters the black pine *Pinus thunbergii* in the reforested stands of this region with its milder winters. This tree has the advantage that is not as prone to infestation by the pine bug. But most of the hills and mountains are more or less skimpily covered by scrub vegetation here too and often appear completely barren from a distance. South of Namp'o on the coast, on the summit P. 417 I found relatively varied, dense shrub vegetation on conglomerate and sandstone strata. In it the red pine, *Quercus acutissima*, *Fagara schiniifolia*, lespedezas and wild roses predominated.

Remnants of primary forests have again been preserved primarily in the environs of the many temples. The lovely forest of Naejangsa on the Noryŏng Ridge southeast of Chŏng-ŭp can serve as an example. It is a mixed forest, and its broadleaf trees are bare in winter. Nevertheless it has a pronounced southern Korean character and belongs to the bamboo zone of the general classification. Among the oaks, the common oak is missing, and primarily southern Korean species, such as *Quercus variabilis*, but also *Quercus acutissima*, predominate. In addition there are several hornbeams, such as *Carpinus laxiflora*, the wild peach (*Prunus persica*), the maple *Acer palmatum* var. *coreanum*, the relative of the elm

Zelkova serrata and *Staphylea bumalda*. As far as conifers are concerned, the red pine is not lacking, it is true, but *Pinus thunbergii*, *Abies holophylla* and the member of the yew family *Torreya nucifera* are more common. The bamboo grasses, the hazel *Corylus heterophylla* and rhododendron are found as undergrowths. In the second half of April this forest blossoms out in gorgeous colors. The conifers are a rich green, the still leafless trees a dull grayish-brown, while the foliage of others is just breaking out in a delicate pale green. The oaks, which come into leaf more slowly, then have as a whole a greenish-gray color.

Station		Temperature						
		January			August		July	
	Altitude (m)	Mean temp.	Mean min. temp.	Absolute min. temp.	Mean temp.	Mean max. temp.	Absolute max. temp.	Annual range
Ch'ungju	190	–4.7	–10.8	–27.2	26.4	31.5	39.6	31.1
Sŏsan	36	–2.4	–6.4	–17.8[c]	25.6	29.8	38.9	28.0
Yŏngdong	111	–3.5	–8.6	–20.2	26.0	30.8	37.1	29.5
Kongju	20	–3.4	–8.1	–24.6	26.0	30.3	37.0	29.4
Kwangju	39	–1.0	–6.7	–19.5	26.5	31.1	38.3[c]	27.5
Yŏnggwang	35	0.0	–4.2	–17.7	26.5	30.6	36.5	26.5
Kaŭido	75	–1.4	–	–	23.1	–	–	24.5
Ŏch'ongdo	61	+0.4	–	–	25.3	–	–	24.9

[a] February. [b] August. [c] December. [d] October. [e] July. [f] March.

Station	Precipitation								
	January				August				
	Average amount (mm)	per 1000	Average no. of days with precip.	Precipitation intensity	Average amount (mm)	per 1000	Average no. of days with precip.	Precipitation intensity	Annual amount (mm)
Ch'ungju	23.5[a]	22[a]	6[d]	3.2[a]	324	297	14	23.1	1093
Sŏsan	21.0	18	4	5.2	303	257	13	23.3	1178
Yŏngdong	23.0	23	4[d]	3.8	223[b]	215	14	20.3[b]	1041
Konju	30.0[a]	22[a]	6[d]	3.4[c]	346	258	15	23.0	1340
Kwangju	35.5[a]	29[a]	7[d]	4.4[a]	238	194	16	14.9	1225
Yŏnggwang	30.8[f]	26[f]	6[f]	5.1[f]	218	183	14	15.6	1192
Kaŭido	17.7	22	6[f]	2.2	199	234[e]	13	15.3	819
Ŏch'ongdo	21.8	24	6[d]	1.8[c]	195	211	14[c]	16.2	925

Table 24. Climatic data for the Ch'ungch'ŏng–Chŏlla region

Orange-brown to greenish-yellow are the hornbeams, covered all over with catkins, deep red the wild peach, the shrubs of *Benzoin obtusilobum* glow in a bright yellow, the rhododendrons in pink or mauve.

The fairly wide alluvial plains, through which good-sized rivers flow from the mountains in the east, where there are large amounts of precipitation, and the warmth of the long summer have led to more intensive paddy farming than is found anywhere else in Korea. In Ch'ungch'ŏngnam-do 20.9% of the total area is used for rice cultivation, in Chŏllabuk-do 20.7%. In Iksan-gun this proportion rises to 62.9%. Rice paddies make up one half to two thirds of the total cultivated area. The two just mentioned provinces are again at the top with 65.8% and 71.5%. In Okku-gun, south of the mouth of the Kŭmgang, this percentage increases to the highest figure in Korea (86.6%). On the whole the intensity of rice cultivation increases from northeast to southwest. In Ch'ungch'ŏngbuk-do it is least developed, for morphological and climatic reasons. In this province, in Koesan-gun, paddies make up only 7.3% of the total area and 40.5% of the cultivated area. The paddies are most extensive along the lower reaches of the rivers between Nonsan and Chŏng-ŭp, particularly on the broad floodplains of the Kŭmgang and the Man-gyŏnggang. That was already the case in traditional Korea. Hulbert reports on this period [557, p. 52]: "The finest piece of rice-land in Korea is a board plain situated in the two districts of Keum-gu and Man-gyŭng in Chul-la Province. The two districts were named from the plain which is called Keum-gu Man-gyŭng Plain and means 'The Golden Valley a Boundless Sea of Waving Grain'." This area of traditional Korean rice surplus lies southwest of the Man-gyŏnggang around Kimje. The water for irrigation is provided by small reservoirs, which lie scattered about in the hill country and are still intact today. Meanwhile Japanese irrigation associations have increased the rice production in the area north of the lower reaches of the Man-gyŏnggang and the bay of the Man-gyŏnggang and along the lower reaches of the Kŭmgang around Yong-an, Kanggyŏng and Nonsan (Naep'o Plain) to the same intensity (Fig. 74). Dams in the mountains and hills beyond, such as the Taea-ri dam in Chŏnju-gun and the Hwangdŭng-ri dam in Iksan-gun, hold back the water and give the rivers the necessary discharge during periods of abnormal summer dryness. The water flows from the reservoirs and rivers to the fields through a large number of canals. Here, too, medium-fast to late-ripening, high-yield varieties, such as Kokuryodō, Tamakin, Soshinreki, Sekisanso and Chushinreki are preferred. Locally the cultivation of the lotus plant in paddies has a certain importance, such as near Naju.

What we said about dry-field crops and farming methods in the Naktong basin holds here, too. The Ch'ungch'ŏng–Chŏlla region is also outstanding in the production of soybeans, cotton and mulberry leaves. In addition, the cultivation of ginger and ramie deserves notice as a specialty. Barley, naked barley and wheat also occur here as winter crops on paddies, in Yŏnggwang-gun rye is grown on a large scale as well. The proportion of paddies used for dry crops in winter is smaller here than in the Naktong basin, however, presumably for climatic reasons. In Chŏllabuk-do it is still 33.2%, to be sure, and in Chŏllanam-do even 38.8%, but in Ch'ungch'ŏngbuk-do it is only 12.9%. Nor does intercropping have the same significance here as in the western part of central Korea. Of importance is that perennial shrubs are also included among the crops that are intercropped in

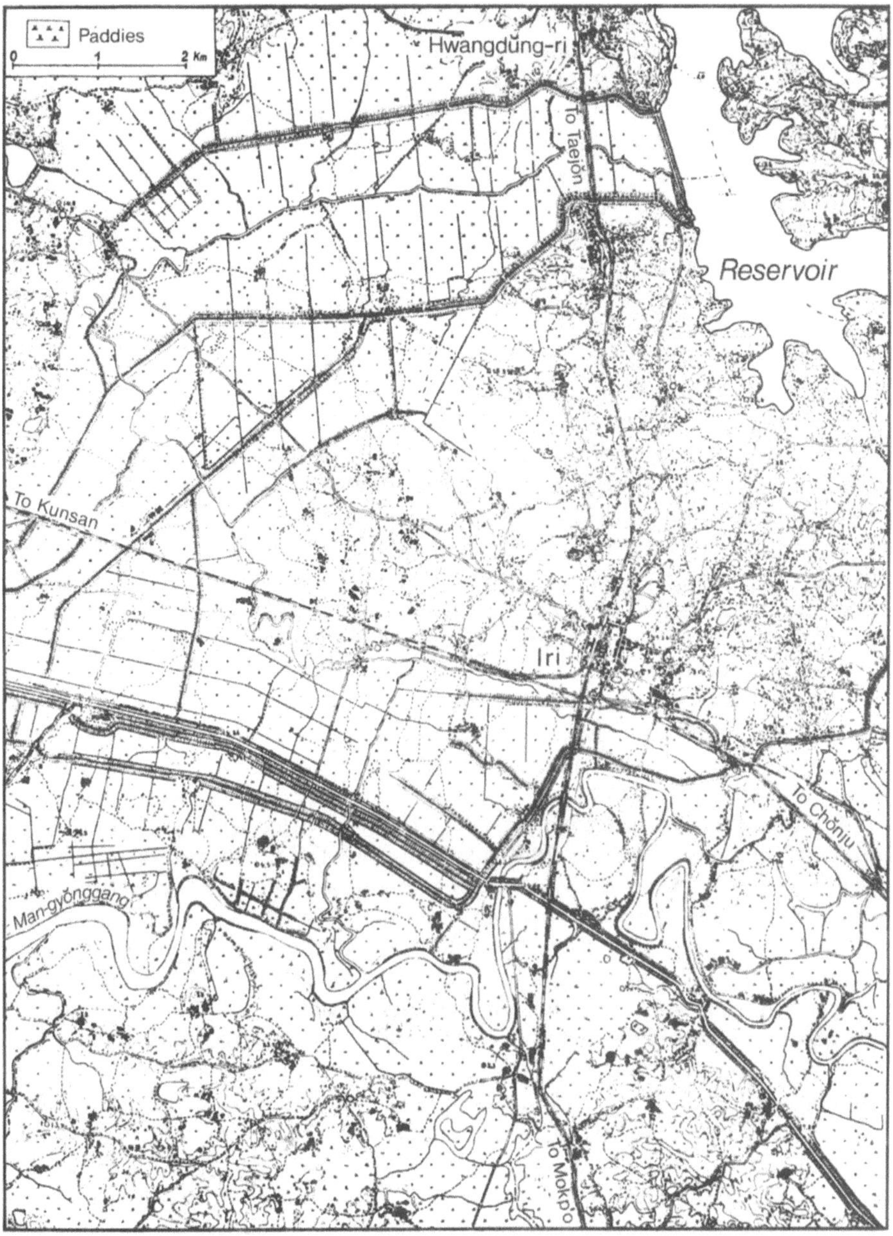

Fig. 74. Irrigated rice paddy plain around Iri (Chŏllabuk-do) (section of the topographical map 1:50,000)

rows here. Thus I encountered mulberry bushes in the same field as wheat and timber bamboo with peas. Groves of timber bamboo are found in decreasing frequency up to the central northern part of the Ch'ungch'ŏng–Chŏlla region. In the vicinity of the villages fields of red peppers, squash and melons and orchards of pear, apple, peach, chestnut and persimmon trees appear.

Here, too, fishing is important, Korean fishing is done from the small harbors lying at the tips of the mud flats on tidal channels. Examples of these are Kwangch'ŏn, Taech'ŏn, Namp'o, Sŏch'ŏn, Man-gyŏng, Chulp'o and Pŏpsŏngp'o. At low tide the sailboats generally lie aground on the mud in the tidal channels. Japanese fishery, which is done with small steamships, is concentrated in Kunsan at the mouth of the Kŭmgang. The main catches are made in spring and fall and consist of sea bream, yellow corvenia, hairtail, sea cucumber, but also shrimps, whereas in summer sardines, croaker and abalone are fished. Winter fishing for cod and mullet is meager. On the whole the catches are in accordance with the low figures for the entire Yellow Sea coast.

The relatively flourishing and varied industry is concentrated in the many large and small towns, which often still have the rectangle of the old Korean wall or a fortress towering above them. These branches of industry are based almost exclusively on agricultural production. They are devoted to making straw bags for transporting rice, rice polishing, cotton ginning and producing cottonseed oil, silk and cotton spinning and manufacturing bamboo basketwork, rice wine, beer, leather, ropes, paper, soy sauce and soybean paste. In the countryside there are also some mines, particularly gold mines, and clay pits, whose material is used on the spot for making tiles or pottery. The largest mine is the gold mine of Kimje.

Agricultural production, particularly of rice, is to a large extent for export to the Japanese islands. The most important port of exportation by far is Kunsan. Overland transportation has followed two groups of road since olden times. One comes from the south coast and runs parallel to the west coast to Seoul (p. 291). On one of these roads Hamel and his companions were brought from Chejudo to Seoul. Where it crosses the Noryŏng Ridge there is still a tremendous old Korean fortress of the same type as the Puk'ansan fortress. The other group begins at the gateway at the mouth of the Naktonggang and likewise ends in the capital. Both directions are followed today not only by roads but also by railroads. On the stretch Seoul – Taejŏn the Korean trunk line serves both directions together. In Taejŏn the line to Pusan and that to the southwestern port of Mokp'o part. Several branch lines fork off from the latter to the ports of the west and south coast.

As in the Naktong basin there is a certain correspondence between the population density and the intensity of rice cultivation. Iksan-gun, south of the lower Kŭmgamg, leads all Korean counties in both respects. The seaward adjoining counties of Sŏch'ŏn-gun (285 inhabitants/km^2), Okku-gun (270 inhabitants/km^2) and Kimje-gun (321 inhabitants/km^2) follow. With the intensity of rice cultivation the population density decreases on the whole from southwest toward northeast. The average density of the Ch'ungch'ŏng–Chŏlla region probably approximates that of the province of Ch'ungch'ŏngnam-do (186 inhabitants/km^2) and is thus only slightly less than that of the Kyŏnggi region.

The extreme northeastern part of the Ch'ungch'ŏng–Chŏlla region, which belongs to Kyŏngi-do and Ch'ungch'ŏngbuk-do, is the least favored part, climatically, morphologically and geologically. In Koesan-gun only 18.0% of the total area is used for agriculture. Sweet potatoes and timber bamboo are lacking in this section, but tobacco farming has been developed to an intensity otherwise unknown in Korea. In part a yellow-leaved American variety is grown. The yield for 1930 was 14,000 tons. Ginseng cultivation is also important near Ch'ungju. The county seat of Ch'ungju (26,210 inhabitants) lies in a basin in granite, through the northern part of which the navigable Namhan-gang flows. In the east high quartzite mountains tower above it. In the metamorphic sediments to which this quartzite belongs, 5 km west of the town, there is an iron mine. Ch'ungju is an important road junction. Nearby stands the pagoda of Chung-angtap (= Center Pagoda), which supposedly marks the center of the Korean peninsula. The railroad line connecting Ch'ungju with the Korean trunk line near Choch'iwon skirts Ch'ŏngju. This capital of the smallest Korean province is simultaneously the third smallest of the country's capitals (30,143 inhabitants, 10% of them Japanese). Surrounded by yellow, weathered, scrub-covered granite hills, the town lies in a rice paddy plain. Today it has a checkered pattern of mostly one-story houses and lives by administration, trade and a little industry.

The extreme northwestern part is the T'aean peninsula. From the north, the west and the south delicately crenulated ingression bays cut into it. Hanging onto its main body like a bunch of grapes is the tertiary peninsula of Anmyŏndo, which is interrupted in two places by narrow tidal channels. With an average width of only 5 km, it is 45 km long. A host of bays filled with mud flats carve it up into little quaternary peninsulas. It consists totally of metamorphic sediments that form low hills covered by pines. The actual T'acan peninsula also consists of shrub-covered and forested, very hilly red country on granite and metamorphic sediments (quartzite, mica schist). Its western half is dominated by the granitic dome-shaped mountain of Paek'wasan (284 m), whose summit is occupied by an old fortress. Here and there the tips of the bays have been dammed off and converted into rice paddy plains. The extreme southwestern tip of the T'aean peninsula, for its part, projects out another 10 km peninsulalike. On it the small fishing harbor of Anhŭng lies on a tidal channel, a large-scale old Korean fortress towering above it. In fact, this promontory does have quite some strategie importance.

The T'aean peninsula ends in the east at the rice paddy plain of the Sapkyoch'ŏn. East of the latter a north–south trending mountain scarp rises like a wall to more than 450 m in several places. We shall name it after the county seat lying in it, Yesan. The Yesan scarp in front of the Sapkyoch'ŏn plain has a striking resemblance to the Chŏngbangsan scarp in front of the Chaeryŏnggang plain (p. 279f) and is probably, like it, a fault scarp. North of the Yesan fault block the valley of the Kobundarich'ŏn stretches out straight from east to west; in it rises the medicinal spring of Onyang (51° C), much visited since the traditional Korean period [346, 349]. In the headwater region of the Kobundarich'ŏn the important junction of Ch'ŏnan (17,977 inhabitants) lies on the trunk line.

The Ch'aryŏng Ridge, which is crossed by the trunk line just south of Ch'ŏnan, is composed in its northeastern part mainly of gneiss, which rises as high as 700 m. In the contact of it and granite there is a profitable gold mine near Sŏunsan (Chiksan mine). In the southwest, near the coast, however, the Taedong strata, in the order listed in Table 46, become very extensive. They were laid into regular folds striking in the direction of the mountains by the Taebo orogeny. The detailed geological study of this area [905] clearly shows how the relief depends on the structure (Fig. 75). The most important ridge formers are the Kaehwa-ri conglomerate and the Ongmasan formation (sandstone and conglomerate with shale intercalated), the latter towering up to 680 m in the northwesternmost of the ridges. The lower Kŭmgang downstream from Kongju is followed by a downwarped syncline trending in the same direction composed of Shilla strata, its core, the Aengbongsan formation (brownish tufaceous shale and sandstone) forming a ridge that follows the river and rises to 308 m.

Southeast of the central part of the Ch'aryŏng Ridge rises an isolated small mountain, Kyeryongsan (828 m). It consists primarily of fine-grained granite. Nowhere did I encounter the large expanses of porphyry shown by the otherwise excellent geological map of Korea 1:1 million. The straight western boundary is formed by a deeply incised structural valley, which extends from east of Kongju to east of Nonsan. From it and from the Kŭmgang to the north V–shaped valleys, terraced and embellished by waterfalls, have cut extraordinarily steep walls into the granitic body. The ridges separating the side valleys are razor-edged and jagged. Gigantic, steep block streams fall from the heads of the valleys. Since there are several temples scattered across the mountain, it is still covered by lovely forests containing representatives of the styrax zone. Being a Shamanist place of worship, on the highest peak a bare, cloth-covered tree has been erected, along with a stone heap with a lantern hanging over it.

Northwest of Kyeryongsan lies the old capital of Ch'ungch'ŏngnam-do, east of it the new one. The former capital, Kongju, nestles into a shallow basin in the immediate vicinity of the poplar-lined Kŭmgang, but protected against flooding. Its topographical location is similar to that of Seoul. In recent times Kongju has fallen behind (12,054 inhabitants), but exactly because of that it has preserved much of its traditional Korean character, and it is famous for the magnificently blooming cherry trees lining its streets.

The new capital, Taejŏn, did not begin to flourish until Korea's third most important railroad junction was created here. When it was given city rights in 1904, it had 188 Japanese residents. In 1938 it had 9278, i.e., 23% of the total of 40,001 inhabitants. It has gone through a very fast development and is still growing rapidly. Since it is not linked to any Korean settlement, its architecture has an unusually strong Japanese element. It stretches out long and narrow with straight streets along the banks of the insignificant Taejŏnch'ŏn. Railroad repair shops and settlements for railroaders give it a special touch.

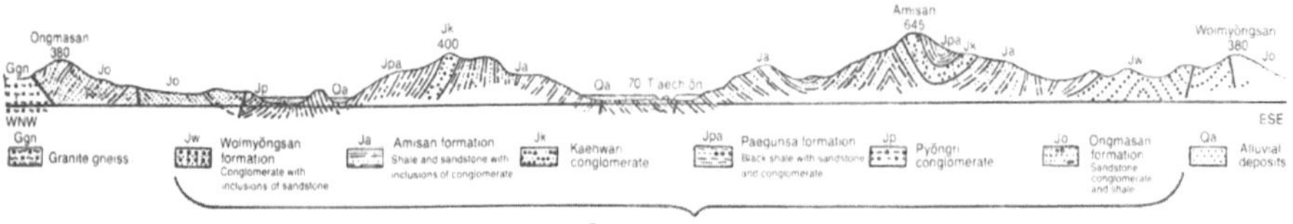

Fig. 75. Geological profile through the zone of folded Taedong strata north of the mouth of the Kŭmgang (after S. Shimamura). Scale approx. 1:60,000

Puyŏ on the lower Kŭmgang (13,049 inhabitants) was for centuries the capital of Paekche, but today it is a small county seat. Temples, pagodas, royal tombs, fortifications and fortress date back to that period. Twenty km farther southeast, near Nonsan, Korea's largest *Mirŭk* Buddha statue stands on a gentle, sparsely forested granite slope behind the temple of Kwanch'oksa. The figure, which is crowned by a giant, four-cornered pagoda-shaped hat decorated with bells, is 26 m high.

Kunsan lies on the southern bank of the funnel-shaped estuary of the Kŭmgang, i.e., on the side on which the brown riverwater flows and the estuary has its greatest depth. The town nestles against a red shale hill that was once an island and is now covered by a park. The very well-kept and tidy town is a completely new one with a rectangular pattern, which was created after the opening of the treaty port in 1899, and makes a predominantly Japanese impression. Two old Korean farming villages are connected to it purely superficially. The town, which is still growing rapidly today, had 44,284 inhabitants in 1938, 24% of them Japanese. The difficult problem of supplying drinking water was solved by pumping part of the water in the Fuji canal southwest of the town up 50 m and then running it through a filter plant before use. Kunsan's industry has the above described character, but manufactures metal and rubber products in addition. The harbor is a very modern, spacious installation. Three tremendous warehouses stand along the quays. Since the mean spring range rises as high as 7.5 m, large pontoons lie in front of the quays. They are connected to the land by loading equipment and a pair of hinged bridges each. In this way they can get along without a dock. Laborers are continuously busy loading cotton bales and especially rice sacks from the warehouses into the lighters mooring at the pontoons. In 1929 rice exports already amounted to 34 million yen. In addition to rice and cotton, soybeans, leather and straw sacks are important export articles. Among the much less important imports, textiles, coal, fertilizer, wheat flour and agricultural implements are foremost. With 74.4 million yen maritime trade volume (1934) Kunsan is the country's fourth most important port. On the opposite side of the Kŭmgang, which is 1200 m wide and is still the provincial boundary for the time being, the harbor of Changhang is beginning to develop, since the railroad through the western part of Ch'ungch'ŏngnam-do was extended this far in 1930.

On the foreside of the peninsula where Kunsan lies the rice polders of the Fuji Company have spread out since 1923. Basically they are set up the same as the Yong-amp'o polders (p. 256). Previously there were silty mud flats here as well, with a few pine-covered islands sticking out, some of them connected by bars [913, 914]. By means of mighty dikes, completely covered with stones outside and therefore very steep, which run straight for miles and use the islands as supports, two large polders have been wrested from the tidal flats. They are separated from each other by the largest island and a reservoir (Fig. 76). Three floodgates, one of which was blasted into the solid rock at the base of an old gneiss island, allow the mainland water to drain off. Each of the two polders contains 1000 *chŏngbo* (992 ha) of paddy land, and an additional total of 500 *chŏngbo* are occupied by buildings, roads and dry land. The paddy area of each polder is divided into rectangular fields of 5 *tanbo* (= 0.5 ha). From the direction of these rectangles one can recognize that in the north polder there are three blocks arranged differently to fit the gradient of the former tidal flats and thus the direction of the irrigation canals.

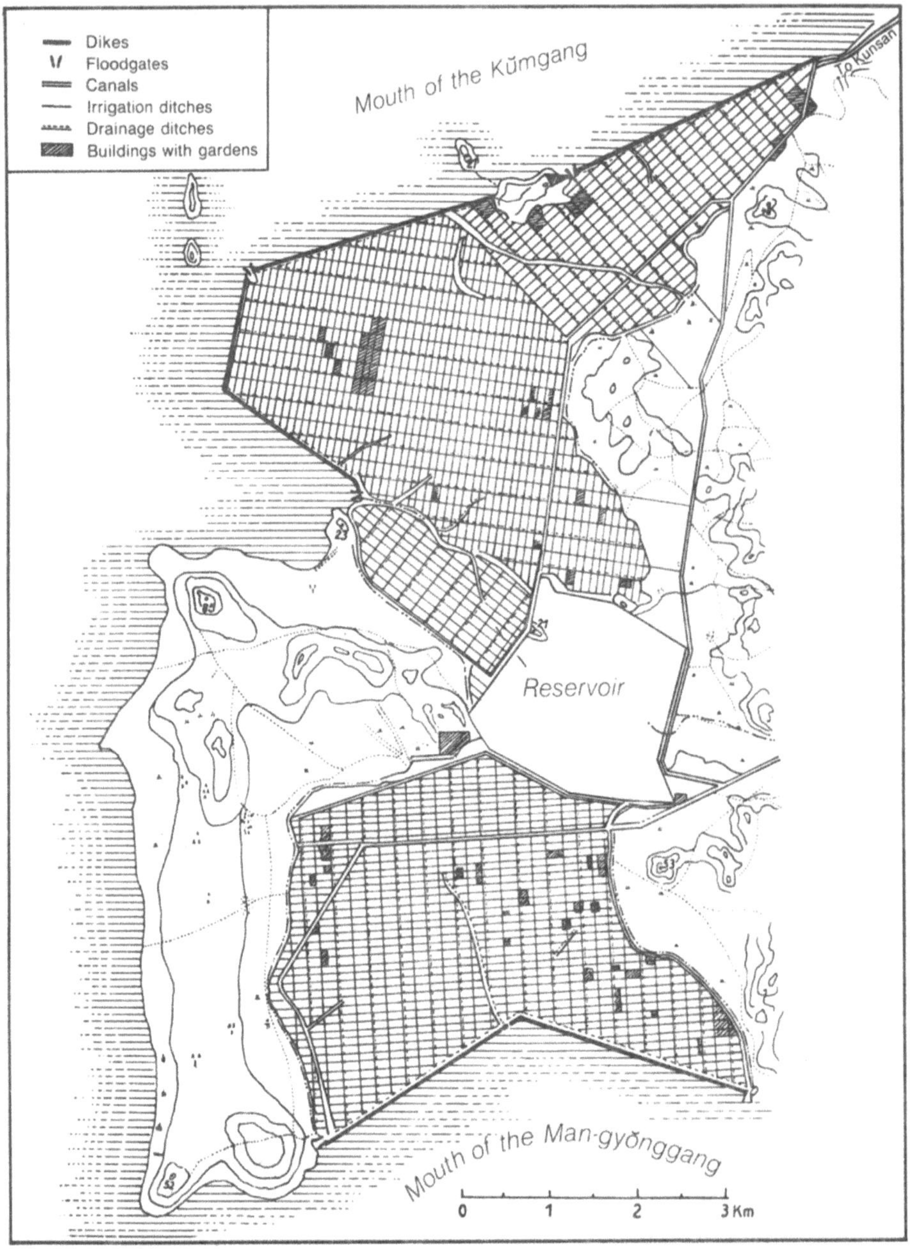

Fig. 76. The rice polders of the Fuji Company near Kunsan (after several maps belonging to the company)

The fields of the south polder form one single block. The latter are leased to Korean farmers, those of the north polder to Japanese farmers. In 1937 1708 Japanese were settled in the north polder, including family members. Each farmer has six fields, that is 3 ha, in lease. The soil is so fertile that rice yields as high as 6 *sŏk/tanbo* (109 hl/ha) are attained. In the Japanese polder there were 27 hamlets à 10 family homesteads distributed across the entire area in 1933. Each three families own a draft ox together. The rice must be irrigated very intensively, because the ground under the polders is salty. In places where nothing is growing on the ridges, particularly near the dikes, one can see the salt crystallizing out of the silt in the form of a white film. One of the tremendous canals flowing through the adjoining plain of Iri east of here leads up to the polders. From autumn to spring its water is pumped into the reservoir constructed between the polders, which has very high dams and an accordingly large volume. The surface measures 319 ha. From June on its water is used for irrigating the fields. A system of main and side canals crossing at right angles with irrigation ditches again branching off at right angles brings the water to the fields. The drainage ditches flow from the fields underneath the irrigation canals and meet at the floodgates.

The center of the paddy region around the lower Man-gyŏnggang is the railroad junction Iri (21,335 inhabitants, 20% Japanese, Fig. 74). Where this river leaves the Noryŏng Ridge, on a fertile plain with yellow, weathered granite hills rising above it lies Chŏnju (42,520 inhabitants, 14% Japanese), the capital of Chŏllabuk-do since time immemorial. From the Korean dynastic period there are several monuments still extant, above all the wonderful, double-roofed South Gate. The more recent development has been inorganic here as well. From the standpoint of transportation Chŏnju has an excellent location in the center of its province. It is the natural focus of the roads that cross the Noryŏng Ridge following antecedent river gorges.

The Noryŏng Ridge consists here of an 18-km-wide group of Shilla strata trending northeast–southwest. At the southeastern edge an at least 500 m thick, almost unstratified mass of conglomerates, 10 km long and 5 km wide, overlies this group. It consists of cobbles up to the size of a head, varying from ones with rounded edges only to ones that are completely rounded, of quartzite, gneiss, hornblende granite, biotite granite, phyllite and porphyry. Small to tiny debris down to the size of sand grains forms the cementing material. The block of conglomerates projects northeastward at a right angle, bounded by faults, into the gneiss area of the county seat of Chinan. The entire region in which the conglomerates are found is an area of pronounced sugarloaf formation. This is most spectacular at the northeastern end of the block in the two "Horse Ear Mountains" (Maisan, 673 and 667 m), which rise almost 400 m above Chinan. Their peaks are only 500 m apart, although the saddle connecting them lies 200 m lower. The rounded walls have slopes of up to 90°. Up to 70° they are sparsely covered by lush shrubs. Otherwise algae and lichens give them a grayish black color. On the south side, probably because of groundwater discharge, hemispherical cavities have been broken out, and their formation is progressing rapidly. Below the two Horse Ear Mountains, which are revered in Korea as a natural beauty of the first order, lie temples. On the summit of the western mountain there is an altar with five sacrificial bowls arranged crosswise in the cardinal directions and a prayer lantern.

At the Horse Ear Mountains the Sŏmjin-gang rises. At first it follows the direction of the Noryŏng Ridge southwestward. In a meandering, moderately incised course it flows, probably antecedently, through the Shilla sediments and Pulguksa porphyry of the above mentioned group of strata. In the loose talus deposit of Oegung-ri, which extends down to the valley floor, the cold air of winter is preserved the whole year, so that the cold air cellars established there can be used for the storage of silkworm eggs until the desired hatching date ([398], Fig. 77).

The Pyŏnsan peninsula is filled with a group of quartz porphyry flows. The harder ones form vertical walls on which the columnar jointing of the porphyry is visible. With enormous steepness and lush woodlands the mountains rise up to 424 m directly above the bay of Chulp'o.

The extreme southern part is dominated by the provincial capital of Kwangju (57,461 inhabitants, 14% Japanese). The extensive city lies in a southeastern bulge of the large basin of Tamyang–Naju, above which the mighty Mudŭngsan towers. The center is occupied by the pleasant Japanese town with the administration buildings. On the same main street that forms the axis of the town densely built Korean districts follow to the northwest and the southeast. The city makes its livelihood by cultivating the large paddy areas on the floor of the basin and the fertile dry fields in the granite backdrop, by silkworm raising, administration, textile industry and trade.

15 The Southern Coast Borderland

On every topographical map of Korea the part of the coastal fringe lying way in the south between Hamp'yŏngman on the Yellow Sea and Chinhaeman on the Korea Strait immediately stands out (Fig. 78). Certainly other sections of the Korean Yellow Sea coast have islands and peninsulas, but they generally do not play as great a role as here in either size or number. Only the Ongjin peninsula and the T'aean peninsula (p. 281, 357) have coastal forms similar to those of the extreme south. Nowhere else is the horizontal crenulation as deep and pronounced as here. The Haenam peninsula is 40 km long; the Kohŭng peninsula and the Yŏsu

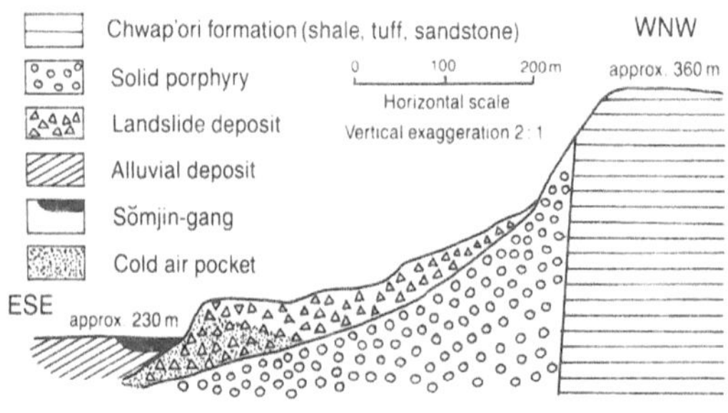

Fig. 77. Profile through the talus and the cold air reservoir of Oegung-ri (Chŏllabuk-do)

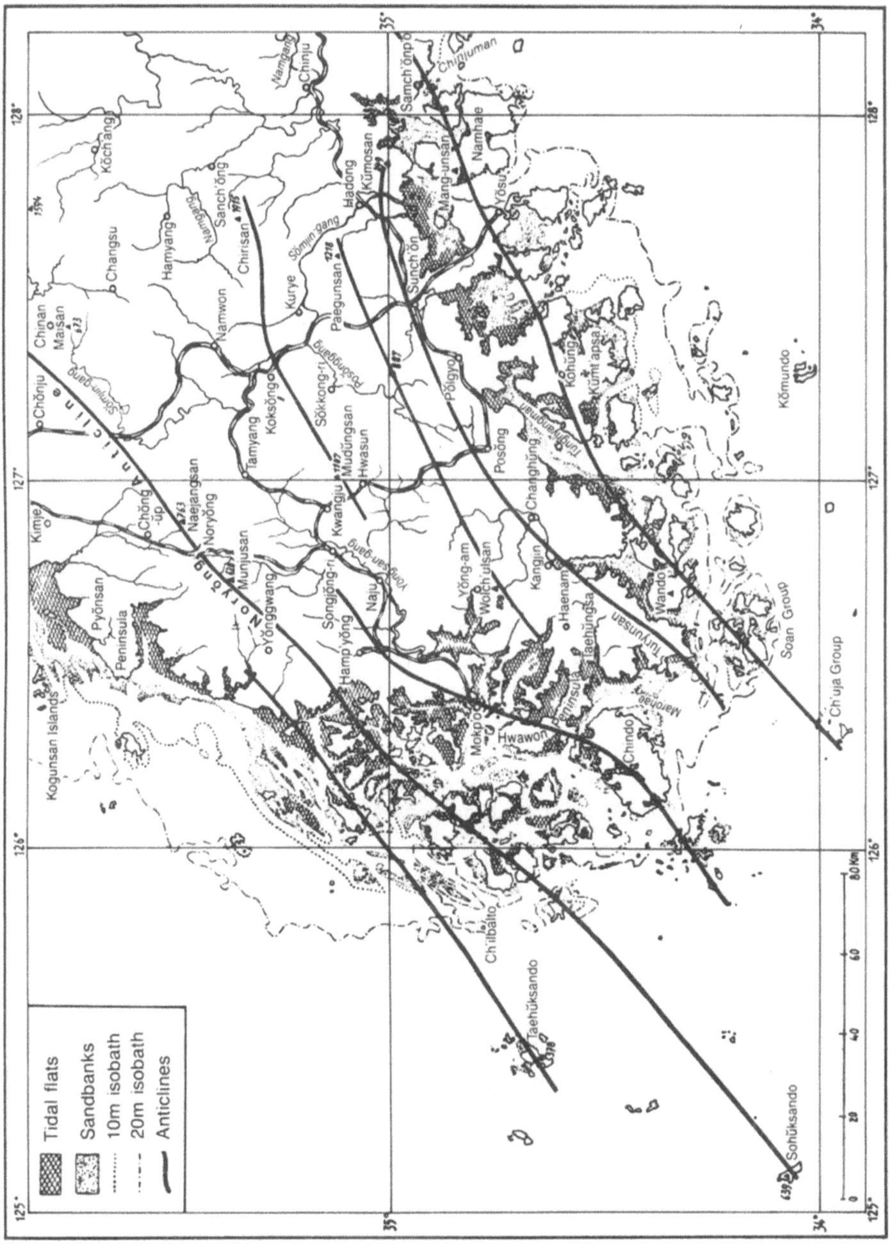

Fig. 78. Topographical–morphological sketch of the southern coastal fringe

peninsula only slightly shorter. Accordingly large are the bays separating them. The most important are Chinhaeman, Chinjuman, Kwang-yangman, Sunch'ŏnman, Tŭngnyangman (Posŏngman), Toamman and Marohae. Often large islands, separated from the mainland only by narrow sounds, take the place of the peninsulas. Kŏjedo (385 km^2), Chindo (315 km^2) and Namhaedo (305 km^2) are the largest islands that Korea has, apart from Chejudo, which we are not considering here. Since many peninsulas, such as those of Hwawon and Kosŏng, are welded to the coast only by very recent alluvial formations and many islands are connected to the mainland at low tide, there is no fundamental difference between the large islands and the peninsulas. The detailed location of the coast of the mainland is thus of no basic morphological significance. Peninsulas, large islands and bays follow each other closely, so that there is no room anywhere for a straight stretch of coast between them.

Partly off the large islands and peninsulas and partly in the outer expanses of the bays smaller islands follow. They form a second ring with a width of up to 35 km. The islands in it still lie quite close together and often flock together in groups. These include Tolsando, Wando and the Soan and Ui groups. Separated by expanses of sea without islands about 20 km in width, there finally follows an outer ring, consisting of sporadic small islands. The outermost of the latter, Sohŭksando, lies far out in the Yellow Sea, 126 km from the closest point on the mainland. A small group of three islands, which together enclose an excellent deep central bay, also belongs to the same ring. It is called Kŏmundo or Samdo (= Three Islands). This group with its good supply of drinking water was occupied by the English in 1885/86 during a severe crisis in British–Russian relations. They called it Port Hamilton [922]. The outermost ring is separated from Chejudo, which has a completely different character, by a strip of sea without islands 80 km wide. A total of 1029 islands with a circumference of more than 500 m belong to Chŏllanam-do. The provinces Chŏllanam-do and Kyŏngsangnam-do together have 2244 islands of all sizes [299, Vol. 17, p. 60]. Schröter [354] calculates that the island coasts account for 69% of the length of the Korean south coast.

The credit for having discovered this island world for the Western cultural sphere is due to the captains Murray Maxwell and Basil Hall (p. 45). They gave it the name Korea archipelago [162–164]. From one of the islands in the second ring the members of the expedition could "accurately count 135 islets at one glance" (C. Ritter [133, p. 622]). From a later period comes the vivid portrayal by S. Genthe, who crossed the Korea archipelago on the way back to the mainland from Chejudo on a trip full of adventures. He writes about the characteristics of the same zone [217, p. 340]: "To all sides our field of vision was now framed by islands of all sizes without interruption. Even the smallest of them appeared to be inhabited. With field glasses one could discover densely cultivated areas everywhere in the folds of the terrain and see the straw-covered houses of the natives sticking out of them, often betrayed by rising smoke. Everywhere I found rice, cotton, buckwheat, beans, millet and pepper fastidiously cultivated, and on the fields shone the white clothing of the farmers planting their winter crops and the colorful clothing of the women and children in the cotton plantations, harvesting the spotless white cotton wads bulging out of the opened bolls. The entire island kingdom gives the impression of great affluence and happy, unspoiled naturalness."

Neither in their cultural nor in their physical geography is there a fundamental difference between the island realm and the directly adjacent part of the mainland. We consider the latter as extending northward approximately to the straight line between Hamp'yŏngman and Chinhaeman (Fig. 39). The so delimited "Southern Coast Borderland" is at the same time the only part of Korea, with the exception of the volcanic islands of Ullŭngdo and Chejudo, in which evergreen vegetational elements are conspicuous in the landscape (Fig. 36). This is simultaneously an indication of a special climate. The Southern Coast Borderland is thus in various respects a well-individualized region. It fills the southern three-quarters of Chŏllanam-do and the extreme south of Kyŏngsangnam-do and has an area of about 10,000 km^2.

The difference between islands and mainland plays a small role in the horizontal structure, and it is equally unimportant for the vertical structure and the geology. The Yŏngnam massif with its gneiss, mica schist and quartzite, narrows down greatly here (Fig. 12). It ends at the southwestern tip of the mainland. The part of the southern coast borderland facing the Yellow Sea consists of rocks of the Okch'ŏn zone; the part facing the Korea Strait belongs primarily to the sedimentation area of the Tsushima basin, which extends far to the south here. The shale, conglomerate, sandstone, quartz mica schist, porphyry and tuff of the Naktong and particularly the Shilla series dominate wide areas in the two last mentioned zones on the mainland and on the islands. They are joined by the volcanic rocks of the Pulguksa series, which have developed an unusual petrographical variety here and appear as quartz porphyry, porphyry with feldspar, granophyre, syenite, diorite, alunite and granite [917, 918].

All of these rocks have been eroded according to their morphological properties. The Hwawon peninsula, which is for the most part very low, consists of Shilla tuffs. The nearby ridge of Turyunsan, in contrast, which rises at the southwestern cape of the mainland and extends northeastward for 40 km, jagged and precipitous and towering up to 703 m, is composed of coarse-grained Precambrian quartzite. The second highest mountain of the mainland part, Wolch'ulsan (809 m) near Yŏng-am, consists of granite containing very little mica and displays the granitic erosion forms marvellously, chimney rocks, sugarloafs and ridges studded with pinnacles. On the peninsulas and islands quartzite, granite and porphyry mountains attain similar heights. Thus Ch'ŏngdŭngsan (555 m) on the Kohŭng peninsula consists of a grayish black, very fine-grained quartzite of the Shilla series, and the outermost island of Sohŭksando, which rises up to 639 m, is composed of porphyry. Particularly in the eastern half of the coastal fringe mighty laccoliths of granite have been sculpted out of their covering of sedimentary Shilla strata. One of these attains a height of 786 m on Namhaedo (Mang-unsan), 555 m on Kŏjedo (Ongnyŏbong) and 849 m on the mainland (Kŭmosan). Both islands and mainland are very predominantly mountainous. Alluvial plains cover only very small areas, particularly on the former, which have only very short creeks.

During the last Ice Age the entire southern coast borderland was still mainland with valleys widened by the primarily lateral erosion. The height of the ridges separating side valleys was not very great in the peripheral parts, but it increased toward the present-day mainland and varied according to the hardness of the rocks of which these ridges were composed. The present-day appearance of the surface forms can be explained easily, if one imagines that a mainland mass of that sort was subjected to an ingression lying within the possible scale of a glacial eustatic rise of the sea level (p. 1). The peninsulas and large islands in the first ring correspond to the back parts of the old ridges separating side valleys and the area surrounding them, and the islands of the second and third rings to local monadnocks on the ridges of the periphery. As far out as the islets of the outermost ring the depth of both the Yellow Sea and the Korea Strait does not exceed 70 m, and only around Sohŭksando does it increase to almost 100 m. The bays correspond to the old wide valleys. They are very shallow, and their depth is out of all proportion to the heights attained by the peninsulas and islands enclosing them. The 20-m isobath does not curve into any but the easternmost of them. Occasional localized and isolated spots of greater depth in the bays generally lie in the narrow channels and can be attributed to the downward erosion of tidal currents. Standing on Yudalsan, a dome-shaped mountain composed of breccia very close to Mokp'o, and looking eastward, one can clearly imagine how the adjacent mainland would be divided into exactly the same maze of peninsulas and islands as the view toward the west and south shows today, were the ingression to continue.

Only the islands in the outermost ring are girded all around by high cliffs. Occasionally the last outposts are vertical rock needles. As a rule, the islands in the central ring have cliffs only on their outer sides, and they are 10 m high at the most. The coasts of the inner ring are completely intact, rocky ingression coasts. That is a result of both the recentness of the ingression and the protective effect of the outlying islands. The few alluvial formations on the islands are adapted to the present sea level. On the mainland, which has such long rivers as the Sŏmjingang and the Yŏngsan-gang, the alluvial areas are more extensive. They consist almost exclusively of fluvial aggradations and floodlands. Salt marshes have hardly developed as yet.

The mean spring range increases the farther west one goes in the Korea Strait and the farther north one goes in the Yellow Sea. It is 2.1 m at Pusan, 3.4 m on Kŏmundo and 5.4 m at Mokp'o [457]. These high figures and the great differences in the lunitidal interval for different points of the coast lead to tidal currents of a speed seldom experienced on earth. Particularly notorious are those in the strait of Mokp'o (10 knots) and in the sound between Chindo and the Hwawon peninsula, which has a minimum width of only 1200 m (9-11 knots). Being ignorant of the latter, the Japanese fleet suffered a decisive defeat at the hands of the Korean fleet during the Hideyoshi invasion in 1592 (p. 31). These tidal currents, the frequent cyclonic storms and typhoons with their waves as high as houses, the abundant fog in late spring and summer and the maze of islands and cliffs make the stretch of water in the Korea archipelago a region feared by seafarers of all nations.

The result of the increasing tidal range from east to west to north is that tidal flats increase in expanse in the same direction. Namhaedo is already connected to the mainland at low water of spring tide. On the north side of Kwang-yangman the mud flats are already quite extensive. The stretch of water west of Hamp'yŏngman around the Chido group consists almost entirely of tidal flats crisscrossed by wide tidal channels.

Even the major forms of the southern coast borderland are not completely explained by the above comments. Above all the question arises: Why is it that the recent positive shift in the shoreline has created these ingression forms of highest perfection only here, but not farther north on the Yellow Sea coast, where it flooded an old land area, exactly as in the south? It appears that the answer to this question can be found by contemplating the mainland fringe of our region. While dealing with the Naktong basin we already pointed out the peculiar fact that the Namgang, although it is only 8 km from the coast at Chinju, still turns east–northeast and becomes a tributary of the Naktonggang. From Chinju to the mouth of the Naktonggang its water covers an additional distance of 187 km. Exactly the same case is repeated farther to the west. The Posŏnggang rises only 5 km from the coast. It immediately flows through a wide floodplain with an absolute elevation of 200 m, crossing an old area of gentle relief that is absolutely foreign to the sea. In narrow meandering gorges it turns northeastward and flows toward the Sŏmjin-gang, which it reaches just in that river's antecedent gorge through the southern end of the Sobaek Range. Its water has to flow 170 km from its source to its mouth, departing as much as 35 km from the coast. The only explanation for this twofold case can be a recent uplift, whose axis follows the coast from east–northeast to west–southwest (Fig. 78). In fact, in this zone I encountered uplifted old erosion surfaces, for instance, between Sach'ŏn and Masan at a level of about 250 m, and another at a level of about 500 m. Kotō already pointed out that the islands and peninsulas of the south side have a hammer-shaped widened part in the middle, and he has one of the fault lines of his Hansan dislocation system running across them. To be sure, there is not the slightest ground for assuming the existence of a fault line, but this does seem to be one of the cases in which Kotō's basic idea can be endorsed. These two axes of recent uplift presumably have others running approximately parallel to them. On the basis of my observations from Yudalsan and Wolch'ulsan, I suspect that there are five others further to the northwest (Fig. 78). The next to the last one forms the continuation of the Noryŏng Ridge. All of these axes of uplift trend out toward the southwest into the boundary region between the Korea Strait, the Yellow Sea and the East China Sea and probably caused the advance of the Korean mainland and the island ranges in the same direction and thus the special type of coast of this region.

We already indicated above that the climate of the southern coast borderland also displays special characteristics. Table 25 provides information on this. It includes one mainland station each from the western and eastern halves, one island station in the west and one in the south and finally, way off to the southwest, Sohŭksando.

The southern coast borderland is, with the exception of Chejudo, the only region in Korea in which the mean temperatures of the coldest month lie above 0° C everywhere at low elevations. As a comparison of Mokp'o and Hadong and of Ch'ilbalto and Kŏmundo shows, here, too, the east is milder than the west in January on the average. Like everywhere else, the islands have higher mean temperatures in winter than the nearby mainland stations. Important for plant growth is further the fact that the average minimum January temperatures seldom fall below −3° C, and the absolute minimum temperatures nowhere reach −15° C. Summer is hot, as befits the southern location. The annual range exceeds the

25° C limit only in the north and lies around 22° C in the outermost island ring. These are extremely low figures, by Korean standards (Fig. 24).

The annual precipitation cycle for the southern coast borderland displays a certain equability even more clearly than that for the Ch'ungch'ŏng–Chŏlla region. The winter precipitation mimina, which do not always fall in January, are fairly high, both absolutely and relatively. The summer maxima are, relatively speaking, the lowest in Korea, with the exception of Chejudo and Ullŭngdo. The importance of the winter precipitation is expressed in the fact that on the islands the maximum number of days with precipitation in the annual cycle is often during a month in this season. The annual amount of precipitation is very large, especially in the south, because Ld_2 and Le_2 and especially the strongly developed Lf_2 cy-

Station		Temperature							
		January				August			
	Altitude (m)	Mean temp.	Mean min. temp.	Absolute min. temp.	Mean temp.	Mean max. temp.	Absolute max. temp.	Annual range	
Mokp'o	31	+0.9	–2.4	–14.2	26.2	30.5	37.0	25.3	
Hadong	20	+1.2	–3.9	–14.6	26.5	30.6	37.9[c]	25.3	
Ch'ilbalto	103	+1.8	–	–	25.9	–	–	24.1	
Kŏmundo	69	+3.4	–	–	25.5	–	–	22.1	
Sohŭksando	84	+3.7	–	–	25.5	–	–	21.8	

[a] October. [b] August. [c] July. [d] December. [e] February. [f] December and January

Station	Precipitation								
	January				August				
	Average amount (mm)	per 1000	Average no. of days with precip.	Precipitation intensity	Average amount (mm)	per 1000	Average no. of days with precip.	Precipitation intensity	Annual amount (mm)
Mokp'o	34.9	32	8[a]	2.7	200	181	15	13.6[b]	1103
Hadong	33.9[d]	22	4	6.8[d]	315	202	15	21.3[b]	1564
Ch'ilbalto	27.5[c]	37[c]	7[a]	2.0	138	184	14[f]	11.5	750
Kŏmundo	32.3[d]	28[d]	5[a]	5.4[d]	174	153	11	15.8	1135
Sohŭksando	40.0	36	7[a]	5.0	186	165	14[d]	12.4	1122

Table 25. Climatic data for the southern coast borderland

clones bring rain for most of the year. In midsummer and early autumn La_l, Lb_l and Lc_l typhoons join them. The west coast, and particularly the outlying islands, however, display the diminished precipitation that we have seen everywhere in the Yellow Sea fringe in all seasons, particularly in summer. Ch'ilbalto receives even less than half as much rain as Hadong.

Mokp'o, the only first-order station in the region, is completely dominated by the northern monsoon in winter. It comes across the Yellow Sea here, however, and is relatively moist. This fact combined with occasional passing cyclones is the reason for the high relative humidity (Jan. 70.4%), the small amount of sunshine (Dez. 44%) and the large cloud amount (Jan. 6.9) in winter. As we have already said, late spring and early summer are very foggy in the coastal and island region. Midsummer brings the monsoon to full force. July has the maximum relative humidity (85.3%) and cloudiness (7.7) and almost as little sunshine as midwinter (46%). The nicest season is the clear autumn, which has the minimum number of days with precipitation and cloudiness (Oct. 5.1%) and the maximum number of hours with sunshine (66%). This calm weather is occasionally interrupted by typhoons, however.

With the exception of Chejudo and Ullŭngdo, the southern coast borderland is the only Korean region in which the evergreen element gains importance in the vegetation. The reader must not imagine, though, that this means that at the northern edge of this region all summergreen plants are replaced by evergreen ones all at once at a sharp boundary. The evergreen element is restricted to the lowest altitudinal zone, which has a relatively small expanse in this mountainous country. The zone of intensive cultivation, which is devoted only to annual field crops and summergreen deciduous trees, except for bamboo, lies in the same altitudinal zone. Above all the following fact is important. The destruction of the natural forests has been just as extensive here as elsewhere in southern and central Korea outside of the high mountains. In the low semispontaneous second growths that have replaced the virgin forests, the evergreen vegetation elements, which are sensitive to cold, develop poorly at the northern limit of their range. The deforested soil cools off more in winter than when it is covered by tall forests. Therefore we can presume that very much primeval forest with a strong evergreen element has been destroyed and replaced by shrub formations that are primarily summergreen. Secondary forests are quite widely distributed, to be sure, and have developed magnificently on the islands in the central ring and even on Sohŭksando. But here, too, they are composed almost exclusively of pines.

Remnants of natural forests with a strong evergreen element can therefore be found almost exclusively in the deep, widened valleys that are temple property. I encountered such temple forests on the southern foot of Wolch'ulsan and most particularly on the northern slope of Turyunsan in the basin of Taehŭngsa. The slopes of this basin consist of mica schist with giant blocks of quartzite scattered over them from higher up. The temple, which was built during the Shilla period, owns lovely tall forests, which illustrate the zonation of the natural forest of the southern coast borderland excellently. The lowest zone consists of mixed forest with summergreen and evergreen species. The evergreen trees have fewer species, but as many individuals; locally they even predominate. The Japanese camellia is especially frequent and is covered all over by fiery red blossoms at the end of April. The evergreen oak *Quercus acuta* Thunb. grows taller still (about

15 m). Other evergreen trees are: *Machilus japonica* S. and Z. with broadly lanceolate, pinnately veined, thick leaves, *Eurya japonica* Thunb. var. *montana* Bl. with medium-sized, oval, slightly serrated leaves, *Neolitsea glauca* Koidzuma with large, leathery, lanceolate, pinnately veined leaves, *Textoria morbifera* Nakai with large palmate leaves that are palmately veined, *Eleagnus macrophylla* Thunb. with leathery, almost circular, pinnately veined leaves, *Daphniphyllum macropodum* Miquel with large, very thick, leathery, elliptical leaves. From a physiognomical standpoint, the member of the yew family, *Cephalotaxus drupacea* S. and Z. var. *pygmaea* Nakai deserves to be included, since its linearly arranged needles are unusually wide and therefore have the appearance of pinnate leaves. There are also a few other conifers, especially *Abies holophylla*, *Juniperus rigida* and *Torreya nucifera*. The evergreen character is intensified by lianas, such as the ivy *Hedera tobleri* Nakai, *Trachelospermum asiaticum* S. and Z. var. *intermedium* Nakai and by shrubs of the Korean box. The conifers and the summergreen trees and shrubs are almost exactly the same as the ones in the forest of Naejangsa (p. 352f), and the view in early spring is thus quite similar to there. Southern Korea's evergreen plants differ from the Mediterranean ones in their large leaves and the lack of aroma of their blossoms, which is related ecologically to the peculiarities of the summer moist subtropical monsoon climate. The camellia and several other evergreen trees extend up to 450 m in the forest of Taehŭngsa. Even before that the evergreen element decreases noticeably. The second zone following the lowest one, the so-called bamboo zone (p. 137), is almost exclusively summergreen and is characterized by many summergreen woody plants of the bottom zone. Around the hermitages in this zone one finds gigantic gingko trees, which were perhaps artificially planted. At 600 m the forest of Taehŭngsa becomes poorer, evidently due to human interference. Pine and rhododendron shrubs and bamboo grasses (*Sasamorpha purpurascens* Nakai) extend up to the peaks of Turyunsan.

In the secondary shrubs and forests of the temple Kŭmt'apsa on the Kohŭng peninsula I found very many of the above mentioned summergreen elements and conifers, but of the above evergreen woody plants only *Eleagnus macrophylla* and *Eurya japonica*.

The cultural geography of the southern coast borderland does not differ fundamentally from the adjoining regions to the north as highly as does its physical geography. Naturally fishing and shipping are more important here than there due to the immense interpenetration of land and sea. But the keystone of the economy remains agriculture here as well. As the above quotation by Genthe shows, it is performed industriously on the islands of the central ring as well. Indeed, even on Sohŭksando there are three hamlets that pursue a little farming in addition to a lot of fishing. Other islands in the Hŭksan group are, however, inhabited only by fishermen and are covered only by grass and brush. Only the small rock islands in the second and third ring are uninhabited.

The most important cereal here, too, is rice, which alternates with the same winter crops as in the neighboring regions. On Namhaedo almost all of the barley is even grown on paddies, and the same is true here and in Yŏsu-gun of rye. As Figure 87 shows, rice is cultivated less intensely on the islands and peninsulas than in the core area on the mainland. This is due to the lack of alluvial plains and irrigation water. Dry-field crops are cultivated all the more intensively. One quarter to one third of the total area, occasionally more than in the adjoining mainland counties, is used for farming in the island and peninsular counties (Fig. 86). Relatively intensive is the cultivation of plants that require warmth, such as the sweet potato, which is grown particularly on Wando and Kŏjedo and on the Yŏsu peninsula, and American varieties of cotton and ramie. Muan-gun near Mokp'o is famous for peas; the environs of Tŏkchin for lotus roots. Because of its mild winter, more timber bamboo is cultivated here than anywhere else in Korea. Seventy percent of Korea's timber bamboo area is in Chŏllanam-do, and by far the greatest part of this area is in the southern coast borderland.

From all inhabited islands and peninsulas fishing is done intensively by the Koreans in small, about 7-m-long, open boats. In these so dangerous parts of the sea it requires the greatest of courage, skill and perseverance on the part of the fishermen. All of the fish listed for the southern seas on p. 443 are caught. In winter the gathering of algae (laver, Kor. *kim*, *Porphyra tenera*) in the tidal flats plays an important role. It is done particularly on the coasts of Wando and in Kwang-yangman. Frames hundreds of meters in length are set up on the mud flats. Often a dozen such frames are lined up in parallel rows. As the tides change, floating algae get caught on the arched rods tied to these racks. On land the algae that have been gathered are then spread out on square woven trays, rinsed with fresh water and dried on racks, which are turned toward the sun and look like greenhouses. They form a valuable iodine-rich supplement to the Korean and Japanese meals. The abundance of fish in these waters simultaneously feeds great flocks of sea birds, particularly on the outlying islets and rocks. Collecting eggs and hunting birds are important side occupations for the Korean fishermen. The Japanese sea fishing is done primarily with special small steamships here as well and is therefore based in the larger ports, foremost in Ch'ungmu.

In contrast to Pusan and Masan, these ports do not lie in the interior parts of the bays, which run dry at low tide, but in their central sections, such as Samch'ŏnp'o (20,177 inhabitants) and Haech'ang-ri, or even at the tips of peninsulas, such as Ch'ungmu (23,826 inhabitants), Yŏsu (31,259 inhabitants) and Mokp'o. From Yŏsu and Mokp'o there are regular freight and passenger ship connections with the Japanese islands. Because of these close connections with the Japanese coasts, the proportion of Japanese is considerable (Yŏsu 11%, Ch'ungmu 13%, Mokp'o 15%). Kŏjedo will perhaps have a special importance later on. According to reports on hand, in which the island is called by the old Korean name of Cargo (e.g., Geogr. Zeitschr. 1941, p. 323), a tunnel planned from Kyūshū is to reach Korean soil there. Yŏsu and Mokp'o are connected to the Korean railroad network, Mokp'o by the main line to Taejŏn, Yŏsu by two branch lines that separate in the county seat of Sunch'ŏn (23,660 inhabitants) at the neck of the peninsula.

By far the largest of the ports we have mentioned is Mokp'o (1938: 65,572 inhabitants). It lies quite near the tip of a small, mountainous peninsula jutting out from the north, with another peninsula extending out from the southeast to within 1300 m of it ([459], Fig. 79). Both together enclose the large bay of Mokp'o with its many branches. The enormously strong tidal currents that circulate through this strait into the bay keep the depth of the former at more than 10 m. Mokp'o stretches out on the sheltered inner side of its peninsula at the foot of the above mentioned Yudalsan (228 m). Its large-scale development began after it was opened as a treaty port in 1897. The Japanese part of town consists of rectangular blocks of houses, between which barren cliffs of Yudalsan breccia suddenly jut up in several places, and has a number of multistoried public buildings. Exports had a total value of 22.6 million yen in 1934, imports half as much. The former consist primarily of cotton. Mokp'o is Korea's foremost port of exportation for cotton, as Kunsan is for rice. But in Mokp'o, too, rice is busily shipped. As in Kunsan, trading is done almost exclusively with Japan.

In Mokp'o and the other port cities a modest industry has developed on the basis of agriculture. Of mining operations particularly the alunite mines on the

Fig. 79. Mokp'o and environs (after an official plan)

Hwawon peninsula deserve mention. They yield a semitranslucent, soft, white rock that is partially speckled with gray and red. It is named Haenam jade after the neighboring county seat and is used partly like alabaster for artistic sculptures and partly as a raw material for the chemical and ceramics industries [917].

This varied and intensive economy allows a very high population density. On the islands and peninsulas it is even higher in places than on the mainland. In Namhae-gun it is 362 inhabitants/km^2, in T'ong-yŏng-gun 264 inhabitants/km^2, and even in the counties of Muan-gun and Chindo-gun, which include all the small islands of the extreme southwest, it is still 187 and 147 inhabitants/km^2, respectively. According to the census of 1936 the islands off the west side of the southern coast borderland alone had 171,102 inhabitants. On the more mountainous mainland Sunch'ŏn-gun manages only 144 inhabitants/km^2 and Hwasun-gun even only 127 (Fig. 85). The average density is probably hardly less than that of the Ch'ungch'ŏng–Chŏlla region.

16 Chejudo

In its size, elevation and plump outline Chejudo differs from the islands of the southern coast borderland at the first glance (Fig. 80). It has the form of an ellipse, whose axes are 76 and 32 km long, and thus an area of 1859 km^2. Out of a shallow sea, which seldom reaches a depth of 130 m, it rises to 1950 m in Hallasan.

Since Chejudo, like Ullŭngdo, is visible from the mainland, close connections have existed since early historical times. As the *Samguk sagi* reports [925], Chejudo was an independent state named Tamna until 1147 A.D., although it paid tribute to the various states on the mainland from 514 on. At that time it became a prefecture and was given the name it still holds today, Chejudo. In 1231 the island fell into the hands of the Mongols for the first time, who developed horse breeding on a large scale. They also exploited the abundance of timber to build ships for their two invasions of Japan. The Mongol rule lasted, with interruptions, until 1368. Since the sailing route Formosa – Nagasaki passes in sight of the

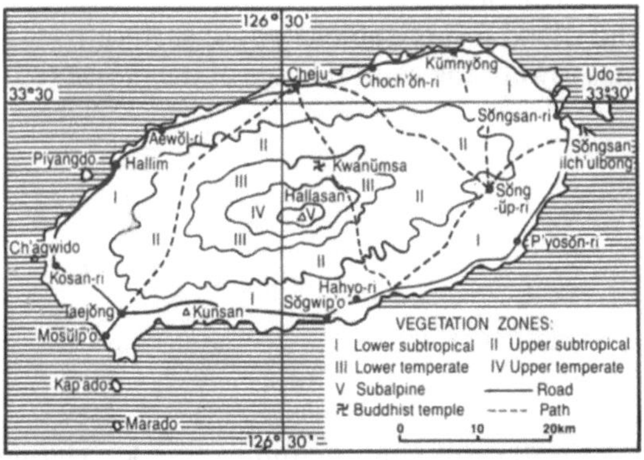

Fig. 80. Topographical sketch of Chejudo with the vegetation zones

mighty island, it was known to the Portuguese in the 16th century, who called it Ilha dos Ladrones (Isle of Thieves). B. Hoetink has demonstrated astutely that the island was probably sighted again in 1642 by the Dutch "quelpart" *de Brack*. The Dutch word quelpart referred to a type of sailing ship at the time, and the Dutch named the island after it [172]. In 1648 the name appeared for the first time in this sense, and 20 years later it was popularized all over the Western world by Hamel's book [171]. We owe the first graphic description of the view of the island from the sea to La Pérouse [160], the first coastal survey to Belcher [168]. The first non-Asian to travel on Chejudo as a free man appears to have been the American consul-general C. Chaillé-Long [195, 196]. S. Genthe's stay was more fruitful [217]. In 1926 R. Burnett Hall presented an anthropogeographical outline based on an abundance of material [927]. I myself have written a comparative regional geography of Chejudo and Ullŭngdo on the basis of observations made during my travels [883]. Of the Japanese scholars, above all T. Nakai [935] and K. Haraguchi [933] have written excellent monographs that can be put to good use in a geography.

Chejudo consists almost entirely of recent volcanic flows and its surface forms are caused primarily by them to a great extent (Fig. 81). Seen from Korea it thus forms a peripheral phenomenon, similar to Ullŭngdo, leading over to Kyūshū. With respect to climate and vegetation it also has a transitional character. According to Haraguchi's investigations, the results of which do not, however, appear to be totally verified, the lowest lava series consists of various trachytes. The oldest of these trachyte lavas forms the highest rim of the summit crater, Hallasan. A trachyandesite lava series follows, which already developed partially in the form of thick flows. The surface forms are, however, probably formed for the most part by Late Pleistocene basalt lavas, which represent the third eruption series. The Hallasan basalt covers almost half of the island. The summit crater out of which it poured lies very close to the center of the ellipse. One of the most recent eruption phases consisted in the formation of around 360 basaltic parasite volcanoes that sit on top of the lava flows. Appearing in groups, they determine the physiognomy of sections of the islands that are more remote from the central volcano. They are joined by a few cinder cones built up periclinally, like in a model, with outer slopes that drop off very precipitously, such as Sŏngsan ilch'ulbong in the extreme east. The most recent phases of volcanic activity evidently extended into the historic period [932].

On the whole the island forms a shield whose surface rises from all sides with concave profile lines toward the central bulge of the massif that forms the core of

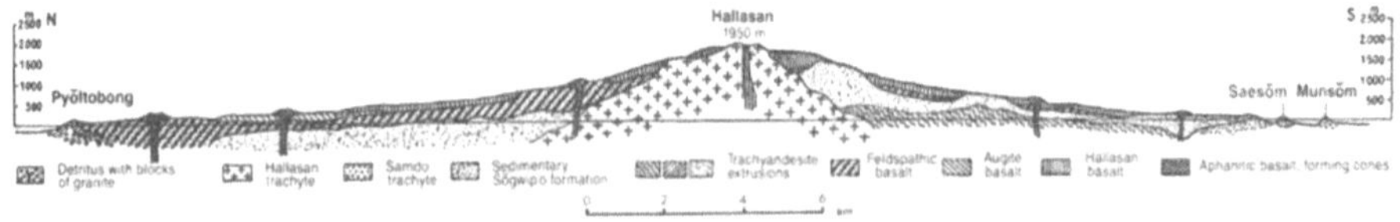

Fig. 81. Structural profile through Chejudo after K. Haraguchi (no vertical exaggeration)

the island. The lavas almost all have a columnar structure and are therefore very permeable. Consequently their surface is very dry. The jumbled gravel beds of the creeks, filled with giant boulders, are generally empty. The northwest storms have blown masses of shelly sand through wind gates on the north and northwest coast up to 4 km inland and formed them into dunes. They are practically the only nonvolcanic formations.

Since Chejudo is an outpost of Korea reaching far southward into the warm sea, it has the least continental climate of the entire country (Fig. 24). This is shown by Table 26, which draws on the only two stations: Cheju, the county seat in the center of the north coast, and the southern islet of Marado.

The winter land monsoon has the opportunity to warm up over the Korea Strait, through which a branch of the Kuroshio flows (p. 124). The mean January temperatures are thus the highest in all of Korea. Particularly well protected from the northern cold is of course the south coast, which lies in the lee of Hallasan.

Station		Temperature						
		January			August		July	
	Altitude (m)	Mean temp.	Mean min. temp.	Absolute min. temp.	Mean temp.	Mean max. temp.	Absolute max. temp.	Annual range
Cheju	17	+4.5	+2.1	−5.7	26.2	29.6	37.0	21.7
Marado	44	+5.8	–	–	26.4	–	–	20.6

[a] June. [b] August. [c] January.

Station	Precipitation								
	January				August				
	Average amount (mm)	per 1000	Average no. of days with precip.	Precipitation intensity	Average amount (mm)	per 1000	Average no. of days with precip.	Precipitation intensity	Annual amount (mm)
Cheju	59.1[c]	43[c]	17	3.5	220	160	13[b]	18.3	1380
Marado	44.4	38	12	4.5	190[a]	163[a]	11[a]	17.3	1166

Table 26. Climatic data for Chejudo

There the winter is even warmer than in Cheju. Chejudo is simultaneously the only part of Korea in which the average minimum temperatures of the winter months do not drop below the freezing point. Cheju has an annual average of only 18 days each with frost and snowfall. On the south coast snowfall is even one of the greatest rarities. With increasing elevation winter naturally becomes colder and more severe on Chejudo too. As Pieters [924] reports, snow lies in the crater until June. The northern monsoon is able to collect moisture while crossing the Korea Strait, and in addition, Chejudo is frequently struck by cyclones in winter as well, some of which follow the warm course of the Kuroshio (p. 103f). Consequently winter, especially on the north coast, is relatively moist and brings a fair amount of precipitation. For almost all of the winter Hallasan hides behind a blanket of clouds, whose lower edge is several hundred meters lower on the north side than on the south side.

The summer sea monsoon brings the island high temperatures and great humidity. On the south side of Hallasan the rising moisture condenses, forming a persistent blanket of clouds. The south coast is consequently cooler than the north coast in summer, and the annual temperature range is much smaller still in Marado than in Cheju. On the average it is only half as large in Chejudo as in the big northern bend of the Yalu (Fig. 24). The weather under the prevailing sea monsoon is, however, otherwise more or less clear. No other station in Korea has as high a percentage of possible sunshine in August as Cheju (61%). Nevertheless, summer is the main period of precipitation here as well. These summer rains are brought less by the sea monsoon itself than by Ld_2, Le_2 and Lf_2 cyclones, and toward autumn by typhoons, which often pass by in the immediate proximity of the island. Thus on the whole Chejudo has a climate that is very unfavorable for sailors, because of the frequent very high wind velocities and the heavy fog, particularly in spring. For the farmer, however, it is quite pleasant, by Korean standards.

With a population figure of 192,638 souls Chejudo has a density of 104. It is also supported here for the most part by agriculture. Fields compose 48.8% of the total area, the other half is divided approximately evenly between pastures and forests. The steep-sloped central massif, which covers approx. one-fifth of the total area, is still densely forested today, whereas meadows and fields dominate the flat peripheral parts. These have preserved remnants of their original vegetation only along the steep river cuts and on the cliff–girded islets. The island, which has been inhabited for thousands of years, has evidently already had this contrast between the center and the periphery in its cultural geography for a very long time. The boundary between the two lies at approximately 700 m. The central forest region is owned by the state today and is separated by a wall or a fire break from the peripheral private holdings.

Chejudo is characterized by an extremely impressive altitudinal zonation of its vegetation. I combine Nakai's ten altitudinal regions [935] into three main altitudinal zones, a subtropical one, a temperate one and a subalpine one (Fig. 80). These are separated from each other approximately by the 700 and 1600-m contour lines. The region belonging to the subtropical zone is the site of intensive agriculture and grazing. Thus when reconstructing the natural conditions, one encounters certain difficulties. Nevertheless, two subdivisions can be distinguished. The lower one is characterized by a predominance of evergreen plants. Because of the more favorable winter climate on the south side, it extends higher up there, to approximately 400 m, and the evergreen plants develop into sturdy trees as much as 20 m tall with dense foliage. Physiognomically the spreading shiino tree (*Lithocarpus cuspidata* Nakai) dominates. This is the only place in Korea where the camphor tree occurs. On the north side the same species often form only thin-stemmed little trees. Others are completely missing there. The climate appears to be least favorable for evergreen plants in the northwest, since this part of the coast is most strongly exposed to the winter monsoon winds. In the upper subtropical zone (bamboo zone) evergreen species, such as the camellia, occur everywhere only as shrubs and small trees, and summergreen woody plants predominate. They include several oaks and the Korean willow. The only frequent conifer in this and the following subzone is the black pine *Pinus thunbergii*.

The temperate zone, which is preserved in its full expanse as more or less virgin forest, consists of a varied summergreen deciduous forest, similar in composition to that on Kyūshū or the southern Korean mainland. Sporadic evergreen shrubs, such as *Daphniphyllum macropodum*, very common even on the north side, penetrate into the lower temperate zone (styrax zone), which extends to about 1100 m. Otherwise this forest is composed exclusively of often very good-sized summergreen deciduous trees, with lianas entwined about them and covered with epiphytes. Formerly, lovely stands of the hornbeams *Carpinus laxiflora* and *Carpinus tschonoskii* belonged to this zone. Most have been felled by the Koreans to cultivate the edible mushroom *shiitake* (Kor. *p'yogo*) on the rotting trunks. In the upper temperate zone (oak zone) evergreen elements are completely lacking. The summergreen deciduous trees of the previous subzone, such as the Siebold maple, extend more or less far up into it. Others begin in it as definite high altitude species, such as the oak *Quercus crispula* Bl. and the birch *Betula ermanii*.

At 1600 m on the north side the conifers begin to take over in a rapid transition. There are only three of them: the Korean fir, the yew *Taxus cuspidata* and the juniper *Juniperus sargentii* Takeda. As undergrowths *Rhododendron mucronulatum* Turcz., the black crowberry, *Empetrum nigrum*, and the bog huckleberry, *Vaccinium uliginosum* L., appear between mosses and grasses. This plant association extends up about to the rim of the crater, gradually becoming stunted.

The pastures that characterize the upper subtropical zone extend indiscriminately across the lava plains and the parasitic cones. No other part of Korea is as rich in pasture land as Chejudo, and animal husbandry is accordingly nowhere more important. Particularly in winter one sees here large herds of ponies, cattle and goats, a completely unaccustomed sight on the peninsula. The shaggy, stocky Cheju ponies climb up to the crater lake in summer. Chejudo is by far the foremost horse raising area in Korea, with 70% of the total stock.

In the lower subtropical zone, which is covered by fields for the most part, the porous lavas are covered by a brown forest soil in which there are still unweathered blocks everywhere. These have been gathered in centuries of tedious labor and piled up into fences 1–2 m high at the edges of the often tiny little fields. Such fences also line the paths. They keep the freely grazing animals from entering the fields and protect soil and crop from the stormy winds. Near the coast the fields are fertilized with seaweed and recently with fish cake from the Japanese fish oil factories as well. Fallow fields are thus rare in the lower subtropical zone.

In a most marked contrast to the Ch'ungch'ŏng–Chŏlla region, paddies make up only 0.9% of the total cultivated area of Chejudo (Fig. 87). Because the lava soils are so porous, Chejudo provides very few possibilities for irrigation agriculture. Thus the island has an agricultural feature that relates it to northern Korea. The composition of the dry-field crops also displays such features. Of the unirrigated area, 34% was planted to foxtail millet, 26% to barley, 8% to soybeans, 5.5% to buckwheat, 4.4% to upland rice and 4.4% to sweet potatoes in 1930. Despite the mild winters, only barley is planted as a winter crop. The strong preference for foxtail millet is a northern Korean characteristic. The prominent role of buckwheat, the almost complete lack of potatoes and the slight significance of wheat are ancient traits. Chejudo participated only in the older phases of development of the Korean culture, then entering a self-sufficient, peaceful insular seclusion, and therefore preserves old characteristics that have disappeared on the mainland or have been retained only in northern Korea. In 1937 the Japanese administration drew up an ambitious plan for stimulating the economy of Chejudo. The parts of it that are concerned with livestock are based on suggestions by the German Carl Raymon.

The very favorable climate is thus only very incompletely expressed in the agricultural character to date. Only the usually small gardens in the settlements in the lower subtropical zone, which are decorated with cultivated evergreen woody plants, reveal a bit of it. Nowhere else in Korea does one encounter orange and tangerine trees. In 1921 there were already 21,000 of them on the island.

In relation to the cultivated area, Chejudo's population density is 217. To support such a density with the primitive state of agriculture it must be supplemented by subsidiary economies. Very important is the collection of foodstuffs from the sea, which has an almost Japanese versatility. Its yield is to be increased five to sevenfold in 15 years, according to the just mentioned plan. Not only boats are used, but also primitive sledlike rafts which, according to Hulbert [925], come from Formosa. The fish caught by the people of Chejudo with lines and nets include sea bream, the three types of mackerel, sardines, yellowtail, tuna, skipjack, hairtail, abalone, shark and thus are understandably the most southern in character in the entire country. Boreal fauna are almost completely missing. On the coasts diving by women for marine foodstuffs is widespread. Once pearl oysters were the most highly prized object of this arduous activity. Around 1895 the Japanese began to collect oysters with modern diving equipment, and today oyster fishing on Chejudo is exclusively in their hands. The most important catch of the women divers today is tasty lobsters, which they sell to the newly established Japanese canning factories. They also dive here for edible seaweed. Many Chejudo women hire themselves out for part of the year on the coasts of the mainland.

Whereas up to 1937 Chejudo's agriculture had been influenced very little by the Japanese from the standpoint of their imperial economy, some large-scale Japanese fishing companies have existed for a long time. They include the whaling station at Sŏgwip'o, which is in operation from the beginning of September to the end of November and brings in about 30 animals a year, the largest of which are said to be up to 21 m long.

Racially the population of Chejudo shows much closer relationships to southern Japan than to the Korean mainland. A widespread legend about their origin also points toward Japan. For an unknown period of time, however, they have spoken the Korean language. For centuries the kingdom of Korea used Chejudo as a place for banishing criminals. Perhaps that is the foundation of the passionate, brutal, sly nature of the island population that emerges from Hamel's descriptions and those of the first Europeans who traveled freely. Today the strict Japanese adminstration allows such character traits little opportunity to come to the fore.

In its clothing, too, Chejudo differs from the mainland. Here the white cotton clothing of the mainland is dyed with native vegetable dyes. On the north side a dull reddish brown predominates, on the south side a bright blue. In this manner a lot of the labor of washing is saved, and the women participate in the major occupations in a manner that is otherwise unknown in Korea, not only in gathering food from the sea, but also in farming. For that they take off their skirts and wear wide pants in the colors described above.

The walls of the Chejudo houses are always built of dark basalt. Near the coast often only three of them consist of windowless, somber lava blocks. The front side, protected by the protruding side walls and the overhanging roof, is built like on the mainland and peeks out of this frame like a snail out of its house. The greatest problem for the settlements is the drinking water supply. Therefore the population flocks together in only about 150 villages, nine of which have the rather considerable size of 1000–5000 inhabitants, and in four towns with more than 5000 inhabitants (Fig. 82). The latter are still partially surrounded by high basalt walls today, and in front of their gates clumsy guardian figures hewn out of the bluish black basalt still stand here and there. The distribution of settlements clearly shows a heavy concentration at and near the coasts.

The main external traffic is still directed toward the Korean mainland today. Accordingly, on the north coast the capital Cheju has flourished (36,260 inhabitants = 18% of the island's population). In addition, the southern port of Sŏgwip'o has also been developing recently. It is the point of departure for the many people from Chejudo who leave the, for its current standard of civilization, overpopulated island to earn their living in Japan's industrial cities. It is said that 50,000 people from Chejudo live in Ōsaka, which is connected with Sŏgwip'o by a direct steamship line. The population of the island decreased by 5.5% between 1931 and 1936. The realization of the above mentioned plan for stimulating the economy by the Japanese authorities will certainly put an end to this unhealthy development.

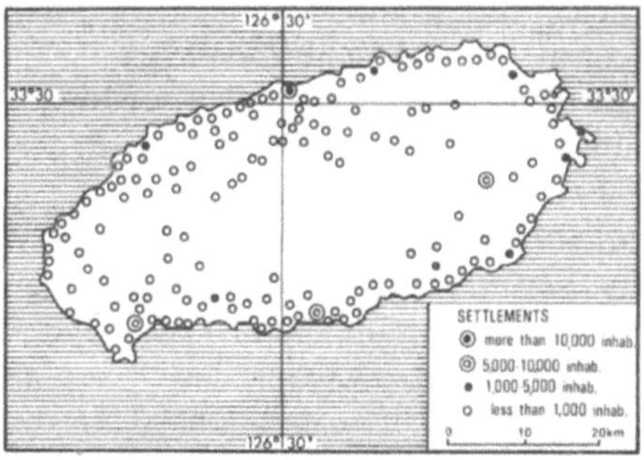

Fig. 82. Distribution of settlements on Chejudo (after R. Burnett Hall)

PART FOUR:

KOREA AS A JAPANESE FOREIGN POSSESSION

1 Background and History of Japanese Influence

Almost nowhere in Korea can one walk for even an hour without running into the geographical effects of Japanese activities. Although the broad fundament of the anthropogeography is still Korean, in the course of the last generation a cultural geographical superstructure has developed with astonishing rapidity that is either entirely of Japanese origin or has been the result of at least indirect Japanese influence. It is expressed in all spheres: in reforms in agriculture and the use of forests, in the development of cities and ports, in fishery and fishing methods, in the creation of mines and industrial plants, in the establishment of waterworks, roads, railroads and airlines. In the regional section this was pointed out time and again in all important details. Here we shall attempt to sum up the geographical effects of this recent development and to outline its results.

In the course of the last 2 millennia Japan undertook three large-scale attempts to incorporate the Korean land bridge, which forms the natural crossing point from the island empire to the continent, into its dominion, in the 4th–6th century (p. 27), at the end of the 16th century (p. 31) and around the turn of the last century (p. 34ff). This third attempt, which was compelled by the political constellation, having succeeded, Japan was faced with a situation that was difficult to master and to stabilize. These three attempts, particularly Hideyoshi's invasion, had left deep feelings of rejection and even hatred in the soul of the Korean people. Certainly the great majority of the Koreans had perceived their own government's maladministration in the last centuries as extremely oppressive. But "however unhappy their country might be, it was still their homeland. They had apparently concerned themselves very little with it as long as they had it for themselves, but when a foreign conqueror appeared, even the faintly smoldering patriotic spirit suddenly flamed up mightily." "One can definitely understand why the injustice of their own officials appeared less annoying to them than the strict justice of the foreign conquerors" (A.J. Brown, [705, p. 117, 122]). Certainly, China too had severely threatened Korea's independence time and again during the course of the centuries, and Mongolian and Manchurian conquerors who sat on China's imperial throne or were soon to gain it had undertaken devastating invasions into the unfortunate country. But China came simultaneously with the superiority of its high culture, which had the effect of attracting and winning the people. The Chinese classics had become for the cultivated Korean the only gateway to the world of higher education and religious views. The feelings of sympathy and solidarity they had created were all the stronger, because no other more highly civilized power had ever appeared on the Korean horizon. And for a good 1000 years the Chinese emperor's sovereignty was almost always exercised exceedingly mildly. During the Japanese invasions, China was always on Korea's side. Toward the Japanese, in contrast, the Koreans have never had the feeling of cultural inferiority. They still remember today that they transmitted high cultural assets to Japan in the 4th to 6th century (p. 27f) and forget only too easily that they themselves had originally received the same from China.

Outwardly their country was hardly in a more orderly condition when the Japanese began to develop it than a part of Africa subjugated by a European colonial power. And therefore the methods with which Japan began the development had to be basically similar to those of the European colonizing peoples of our age. In all official comments the Japanese diplomacy completely avoids the term "colony" for Korea's dependent relationship with Japan, to be sure, and indeed in 1942 complete political equality with the homeland was pronounced. In practice, however, there was no other way than to develop Korea like a colony. But in contrast to a Negro tribe, the Koreans have the spiritual prop of the awareness of a several 1000-year-old history rich in glory and culture. How strongly they feel this and how little their national consciousness has suffered under the terrible conditions of the last centuries showed up several times during the first contacts with the European powers in a downright grotesque manner. Even today the conviction that Korea cannot avoid a development toward modern civilization and that the country is not capable of this on its own is not widely held.

The Koreans' consciousness of their tradition is a difficult obstacle to understanding what Japan has achieved in Korea. In 1927 the Korean Sang-uk Han wrote on his fellow countrymen's attitude toward the agrarian reforms [607, p. 15]: "It must be stressed that the country people at home are still very backward and accordingly have a generally hostile attitude toward innovations; ... organizations of all types are also difficult to carry out, and thus the people are usually not at agreement with each other." It was inevitable that not only the reform of agricultural conditions, but also the construction of railroads and roads, the enactment of sanitary measures, the limitation of fire-field farming, with its destructive effects on forests, and many other things came into conflict with the attitudes of the Koreans, based as they were primarily on tradition and superstition. With the frugality instilled by centuries of need every urge toward better living conditions had been lost. Where such improvements did occur as a result of innovations, they were taken for granted without acknowledgment.

Add to this the fact that in Korea, as in every newly colonized land, many a representative of the colonizing country did not have the character or abilities to deal with the very subtle tasks at hand. Governor–General Ugaki himself has recognized this frankly. In a very informative special issue of the Ōsaka Mainichi [305] he writes in 1936: "The Japanese in Korea often treated their peninsular countrymen with contempt stemming from a superficial superiority complex." In another place he says [302, p. 7]: "The unwise and presumptuous manner with which some Japanese residents of Korea treated the Koreans contributed to the already existing aversion and the prejudice of Koreans against the local authorities."

A further reason for the Koreans' negative attitude toward Japanese measures lay in the Japanese authorities' love of regulations and the undifferentiated manner in which they were carried out by subordinate agencies. Governor–General Ugaki has also commented on this: "The Japanese were responsible for the Korean hostility. They applied too many regulations to the Koreans, who had previously hardly been bound by more than old orders. The publication of one modern regulation after another in rapid succession under the Japanese authorities caused animosity until they had finally become proficient in quietly accepting the laws."

Even worse was that Japanese financiers bought up Korean land and made tenants of the Koreans on it. Certainly the proportion of the cultivated area that is in Japanese hands is relatively small even today. Still the dissatisfaction with the previously described (p. 173) dismal land ownership conditions in traditional Korea acquired a nationalistic bias because of this. The Koreans' poor economic foresight and their lack of tenacity were also really partly to blame, for the Japanese authorities did not exert any direct pressure on them to sell their land. But the result, the further impoverishment of the farmers, was blamed on the Japanese alone. It is exactly in this point that the most passionate accusations have been made against the Japanese masters, accusations that are in their core not unwarranted, although often exaggerated (p. 400f).

Finally there were the consequences of the fact that the Koreans persisted in their economic passivity, despite their intelligence, and only a small percentage resolved to join in on the magnificent upswing that Korea was taking in all areas of modern economic life. Thus the economic leadership in private enterprises remained almost exclusively the domain of Japanese companies and entrepreneurs, and they alone derived the big profits. Certainly in the past few years quite a few Koreans have been educated in America, and a much larger number has gone through American mission schools in the country itself. Unfortunately most of them have gained only a superficial education, dominated by illusions, and consistently reject cooperation with Japan. Although they have come into close contact with the Western world, they have not yet gained the ability to become decisive economic leaders.

From 1905 to 1931 Korea was Japan's only large foreign possession. Taiwan and Karafuto (southern Sakhalin) have only the area of a large Prussian province, and their population is only one-fourth and one-sixtieth, respectively, of that of Korea. The Japanese thus concentrated most of their surplus national energy on the nearby peninsula. Again it was Governor-General Ugaki who frankly called Korea the criterion for measuring the future progress and prosperity of the Japanese empire [302, p. 6]. As everywhere, the Japanese applied their abilities fully and completely: diligence and skill, ambition and patriotism, perseverance and discipline, public spirit and talent for organization. The closeness of the homeland, the similarity of the physical milieu, the racial and linguistic affinity allowed them to use much of the experience gathered in the homeland in this, their first attempt at colonization on a large scale. They themselves had just learned Western methods of capitalistic economy and modern administration, though, and had thus had little time to observe the long-term effects of this development in their own country. It was therefore inevitable that many measures taken in Korea were too hasty and short-sighted. Above all the capitalistic development assumed blatant forms in Korea.

Korea was won to keep this peninsula, which ends only a good 200 km from Japan's coasts, from falling into the hands of the largest continental power on earth. In the period from 1895 to 1905 it was evaluated by Japan almost exclusively from such strategic vantage points, and the splendid topographical map 1:50,000 was later created for military purposes. Korea has not completely finished playing a strategic role today. This is demonstrated by the fact that in addition to an extensive police force and constabulary, which is quite sufficient for upholding internal order, three divisions are stationed in Korea (Seoul, P'yŏng-yang, Nanam). It is also demonstrated by the fortified zones around Pusan–Chinhae, Wonsan and Najin and by the fact that the topographical maps of the extreme northeast, the region north of the mouth of the Han-gang and part of the southwestern islands is published only in a version with shading but without contour lines and elevations. The construction of the first railroads followed strategic points of view, particularly the lines Pusan – Seoul and Seoul – Shinŭiju, as did that of the first improved overland roads. This viewpoint has gradually faded in importance, however, especially since the separation of Manchuria from China, which turned the Yalu – Tumen boundary for all practical purposes into an internal boundary within the Japanese-dominated part of the mainland. Only recently has the extreme northeastern part again become important as a sensitive boundary area with the Soviet Ussuri region. Near Kyŏnghŭng the Russian–Japanese border skirmishes of the second half of the last decade intensified in 1938 to an encounter lasting several days involving tanks and aircraft. The peninsula itself, though, has temporarily finished playing a role as a land bridge in the strategic sense. Troops were transported to Manchuria in 1931/33 primarily through the Yellow Sea via Lüda (Dairen).

The strategic viewpoint was immediately followed by the economic one, and the great majority of the changes in the sphere of cultural geography wrought by the last generation have had an economic origin. Korea was developed economically according to the needs of Greater Japan. First the country was made into an important producer of rice and soybeans. Soon cotton and silk production were added. After that mining was intensified, particularly of gold, coal and copper. As the third branch of industry, fishing was increased, for the purpose of supplying the empire with animal foodstuffs and raw materials. And finally, in connection with the China conflict, the closing of the Anglo - American markets and the entrance into the present war, it was time for Korea's industrialization. This economic development oriented toward the needs of the empire naturally went hand in hand with the reorganization of the settlement pattern of the towns, the construction of hydroelectric power plants and irrigation works, of high tension lines, telephone lines and air routes and finally the expansion of ports, railroads and roads.

The obvious, indeed astonishing success that Japan has had in this sphere, which must be rated all the more highly, because it took place for the most part without the sympathetic participation of the broad masses of the Korean people, is also acknowledged by those authors who primarily stress the drawbacks of the Japanese activities in Korea and come to a negative overall judgment [824, 725]. It was achieved on the basis of the most painstaking scientific research, which was oriented from the outset toward regional planning. As pointed out on p. 59ff, this research dealt with every field that came into question in any manner: the geological structure, mineral deposits, the meteorological conditions, the discharge of the rivers, the oceanographical and biological conditions of the surrounding seas, the customs and traditions of the people, the natural vegetation and its use by the Koreans, the traditional crops and the possibility of introducing new varieties, livestock raising and ways to increase it, sanitary conditions in town and country. The results of this research form important elements of the present book, but above all, they have guided the developments in the sphere of cultural geography during the last generation.

The regional planning based on this research is carried out partially by the government–general itself or by subordinate agencies; this is true of road and railroad construction, forestry and, recently, town improvement. Otherwise the government–general has entrusted this task to Japanese-run companies, some of which were founded by the government especially for this purpose, and some of which had already existed in the homeland and later expanded their operations to include the colonized country. Companies of the first type can be found particularly in the agricultural sector and in that of settlement, those of the second type in the industrial sector. The initiative of private individuals plays only a minor role, however. The companies of the first type include the Chōsen Takushoku Kaisha (Company for Developing Korea), the Sen–Man Takushoku Kaisha (Company for Korean–Manchurian Development), the Man–Sen Kungsu (Manchurian–Korean Promotion Society), the Chōsen Menka (Company for Korean Cotton) and the Fuji Nōson (Fuji Agricultural Company). The largest of these by far is the Tōyo Takushoku Kabushiki Kaisha (The Oriental Development Co. Ltd.), which even publishes a semiannual English language journal with summaries of its activities [XXVI]. The company was founded in 1908 with the express purpose of participating in Korea's development by improving agriculture and by opening up other economic sources. In 1917 the company was authorized to expand its activities beyond the Korean boundaries, but the old responsibilities remained the main focus. In 1933 two thirds of the total capital of 133.9 million yen was invested in Korean enterprises, and in 1938 the company's Korean capital investments amounted to 102.1 million yen. Of this there was invested: in agricultural holdings 33.5%, waterworks 20.4%, urban buildings 7.9%, mining 3.9%, land improvement 1.7%, brewing 0.6%, pottery making 0.4%, transportation and commerce 0.4%, sericulture 0.3%, spinning 0.3%, forestry 0.3%, orchards 0.3%, rice mills 0.3%, cultural activities 0.3%, oil refineries 0.2%, paper industry 0.2%, fertilizer industry 0.1%, electricity and gas 0.1%. In 1932 the company owned 38,243 ha rice paddies, 26,444 ha dry fields and 140,549 ha reforested areas. Most of the success that the Japanese have had in the economic sphere is the work of these companies, both their direct ventures and their indirect influence on the Korean farmers. That organizational mistakes occasionally occurred was inevitable.

The third standpoint from which the Japanese see their influence is the national one, and in this sphere the Eastern power is faced with very different tasks than the Western powers in their colonial regions. The Koreans are racially and linguistically related to the Japanese. The ultimate goal is thus assimilation, i.e., to win the 23 million Korean people for the Japanese national polity and eventually their Japanization. Such a high long-term goal can be achieved only through education. This begins, initially with short-range economic goals, in the many schools for training Koreans in agriculture. There is scarcely an important branch of agriculture for which there are no vocational schools today. Industrial and trade schools endeavor to prepare suitable Koreans for participation in the higher branches of industry. The general school system, which was developed by the Japanese in an exemplary manner, was set up entirely from the viewpoint of education toward assimilation [728]. The Japanese governors–general have fully recognized the significance of primary schools for training Koreans in diligence, perseverance, economic foresight, knowledge and a positive attitude toward the Japanese national polity and support them with all their means. But in 1934 only one fifth of the school-age children went to school. At an early point marriages were allowed between the Japanese and the Korean population elements, and the equal status of Japanese and Korean officials was proclaimed. Gradually the Korean element is to attain not only advisory, but also decision-making functions in the administration of communities and towns (p. 386). When I was in Korea six of the thirteen provincial governors were already Koreans, with Japanese vice governors as advisors. All of these steps are aimed at the goal of a self-government that is loyal to Japan and at economic prosperity. With the attitude of the Koreans such as we have described it, this goal still lies in a rather distant future, however. Recently the attempt is being made to get rid of the old clan system and to replace the monosyllabic Korean pronunciation of surnames with the polysyllabic Japanese pronunciation. Behind these efforts too lies the intention of assimilating the Korean nation.

When Korea was occupied militarily in 1904 at the beginning of the war with Russia it was at first subject to a purely military administration. It is a testimony to his insight into the size of the task ahead that the Tennō appointed the empire's most important statesman, Prince Ito, as resident–general. During his 3-year term of office, Ito established a civilian administration, set up unbiased courts, brought order into the Korean state revenue, initiated the gold standard and thus brought the Korean currency into line with the Japanese homeland, suppressed banditry, improved the first railroads and roads, which had been built in a great rush at the time of the Russian war, and took very strong action against those Japanese who were guilty of oppressing Koreans [712].

After the annexation in 1910, Count Terauchi became the first governor–general. In the 6 years of his office the first stage in the construction of harbors, roads, railroads and telegraph lines was completed. Water supplies, sewerage systems and hospitals were installed in the cities, and basic sanitary measures were carried out. The first agricultural schools were founded. Count Terauchi appointed a large number of specialists to positions in the provincial administration, for the purpose of instructing the Koreans in all important agricultural questions. He is also the founder of "Arbor Day," which is still celebrated today. Every year on 3 April, the anniversary of the death of the first Japanese emperor, everyone, from the governor–general himself and all officials down to the teachers with their schoolchildren, plants trees on the barren slopes. This is to help spread the idea of reforestation and of caring for trees and shrubs among the people. Terauchi also began the assimilation policy, by allowing marriages between members of the two peoples.

Despite these promising beginnings, the year 1919 brought the most difficult endurance test to which the Japanese presence in Korea was ever subjected. Under the impression made by President Wilson's promises regarding the independence of small nations, Korean patriots took advantage of the funeral of the Korean emperor, who had abdicated in 1907, to proclaim Korea's independence. The Japanese were taken unawares and suppressed the movement with a severity that again left deep traces in the soul of the Korean people.

The second period in the history of the government–general, which begins with the same year, did, however, gradually take a smoother course. This was due to the wise and conciliatory attitude of the new governor–general, Baron Saito. Under his leadership (1919–27) the resistance that had flared up during the events of 1919 lost a great deal of its scope and its ferocity. From these events Saito drew the conclusion that everything possible had to be done to win the soul of the people for the Japanese national polity. He is the actual founder of the assimilation policy. Already in 1919 he began to reorganize the administration of provinces, cities and townships (*men*, Kor. *myŏn*) in accordance with the idea of self-government. He created self-government councils with an advisory capacity, two thirds of which are chosen by election. And when he was governor–general for the second time, from 1929–31, he began to transform the town councils into ones with decision–making powers [721]. The towns that have been given this type of administration are called *yu* (Kor. *ŭp*), the others *ho-men.* By 1936 43 towns had achieved a *yu* status. The same decision-making sel-government was granted to the cities (*fu*, Kor. *pu*, now *shi*). In 1936 there were 18 settlements with this type of civic rights. They are primarily ones with a large proportion of Japanese in their population, including such newly founded ones as Najin. There are still four provincial capitals today (Haeju, Ch'unch'ŏn, Ch'ŏngju, Nanam) which do not have *fu* (city) status; they do belong to the *yu* (towns), though. The *fu* status of a settlement is thus not linked to its significance as a seat of government. It depends just as little on the size of the population; Najin had 23,695 inhabitants in 1936, Haeju, in contrast, 41,140.

It was also Saito who started the idea of education toward assimilation in the Korean primary schools. Unfortunately, during his period of office the Korean agricultural crisis increased to alarming proportions due to falling prices for agricultural products. The same was true in the Japanese homeland, but the Koreans blamed the government alone, and in 1926 on the occasion of the funeral of the last Korean emperor the independence movement again manifested itself in public.

The last developmental period of the Japanese foreign possession begins in 1931. Until 1936 General Ugaki was at its head. To diminish the severe poverty and indebtedness of the Korean farmers, he founded the "Rural Revival Movement." The revival movement consists in thorough individual counseling of every single farmer and tenant, with the goal of increasing his production, and in particular of controlling the food shortage that regularly occurs in spring, and further of organizing cooperatives, cottage industries and protective measures against usury. Above all, however, the individualized attention is expected to stimulate the farmer's willingness to work and his confidence in the future. In a 10-year plan all families that are in need are eventually to be included in the Rural Revival Movement. The success of this system was benefited by the rise in agricultural prices and the general stimulation of the economy that followed the incorporation of Manchuria into the pan-Japanese economic sphere. A similar revival movement was initiated for the Korean fishing villages. General Ugaki also organized the resettlement of Korean farmers from the extremely overpopulated rice-growing areas of the south to northern Korea and the adjoining parts of southeastern Manchuria and promoted cotton production in the south and gold mining. Under his guidance sheep raising was also begun in Hamgyŏngbuk-do.

He outlined the major goals of his leadership in several impressive talks [301, 302, 304, 306].

With the outbreak of the Chinese conflict the development of mining and industry took a tempestuous upswing. In the same years there was a succession of excellent harvests, so that the period from 1936–38 is virtually linked with the idea of a "Korean boom." Industry and mining absorbed a constantly increasing number of workers, who had previously earned an exceedingly meager living from agriculture, and as a consequence, the tremendous pressure that had weighed upon the agricultural sphere slowly decreased. Particularly Governor–General Minami, who has been in office since 1936, has been energetically pushing Korea's industrialization ahead. In 1941 the expansion of railroads and harbors was temporarily slowed down in favor of industry. The closing of the world markets to Japanese trade, which became total as Japan entered the present war, is now leading to the rapid development of all of Korea's economic capacity and welds the country more and more tightly to the pan–Japanese ecoomy. With this a certain amount of prosperity is beginning to enter even the farmers' huts, and accordingly the attitude of the majority of the people toward the Japanese leadership has improved, for the time being.

2 The Japanese Population Element

Korea does not have the vast continental space and the broad plains that have recently become the Japanese sphere of activity in mainland East Asia. It is a mountainous country with a charmingly diverse landscape, and almost everywhere one is close to the same seas as in the Japanese homeland. The summer is almost exactly as tropically hot and moist as there, rice and bamboo are characteristic plants, at least in the south, and only winter with its keen, dry frost brings an important difference into the environment. Almost everywhere at low elevations one encounters the characteristic shapes of the red pine; rhododendrons decorate the slopes in spring with their colorful blossoms, in autumn the foliage is as splendid as in Japan, and in spring the cherries blossom. Nevertheless, despite the pressure of overpopulation at home, the Japanese have by no means poured into the neighboring country in streams during the last generation. Rather, the number of Japanese in Korea has developed as shown in Table 27 (primarily after [VI]).

The number of Japanese was thus very small in 1890, and they were limited to the three treaty ports Pusan, Wonsan and Inch'ŏn and to the capital. There they had their own, exactly delimited districts, which were under the administration of the Japanese consulates. Between 1890 and 1900 the Sino–Japanese War was fought, mainly in Korea, and in its wake large numbers of Japanese poured into the country. As Hamilton's very realistic portrayal [219, p. 131ff] shows, it was exactly these elements who put such a strain on the relationship between the two peoples in the early years of Japanese domination. When the Protectorate was established, a large number of Japanese officials, entrepreneurs and merchants naturally entered the country all at once, so that the number of Japanese increased rapidly between 1900 and 1910. By 1920 it had doubled again. Since then the Japanese population has been increasing constantly, to be sure, but slowly. Its

	Number	Percent of total population		Number	Percent of total Population
1890	9,204	0.01	1925	424,740	2,2
1900	19,200	0.2	1930	501,867	2.5
1910	171,543	1.3	1935	583,439	2.6
1920	347,850	2.1	1940	633,288	2.6

Table 27. The number of Japanese in Korea

proportion of the total population therefore always remains small. Still the number of Japanese increased relatively more rapidly than that of Koreans until 1937, so that the percentages became larger. In 1937 the maximum was reached at 2.9%. The past years show a slight decrease in percentage, which is related to the strong Japanese ties to China. The garrisons are not included in any of these figures.

In view of this surprisingly slow development of the Japanese population element, despite the favorable nature of the environment, the question arises as to the reasons. They can be deduced from an examination of the employment structure of the Japanese population in Korea, as shown in Table 28 [after VI and XLIII].

From this it is evident that the Japanese participate very unequally in the different occupations held in Korea. Their role in agriculture and forestry is insignificant. Unfortunately these two branches are not separated in the statistics. Forestry, as an economic form run mainly by the state, employs quite a number of Japanese (p. 428f), so that the proportion of Japanese left for agriculture is accordingly smaller. It can in no case be more than 6500 families, which would correspond to a Japanese farming population of 30,500. The Japan Year Book 1940/41 lists the total number of Japanese family members supported by agriculture and forestry as 33,638. The Japanese fishing activity corresponds to the average

	Japanese employees		
	Number	Percent of all employees in a given occupation	Percent of all Japanese employees
Agriculture and forestry	7,222	0.24	4.5
Fishing and salt production	2,273	3.5	1.4
Mining	4,661	7.7	2.9
Industry	26,252	18.0	16.5
Commerce	31,992	9.8	20.2
Communications	9,089	18.0	5.7
Public services	64,365	32.2	40.6
Various occupations	5,240	1.5	3.3
Unknown	7,749	7.2	4.9
Total	158,843	3.7	100.0

Table 28. Employment structure of the Japanese population in Korea, end of 1938

of all Japanese employees residing in Korea. In mining and commerce the Japanese participation is already very considerable, and the leading posts are held predominantly by Japanese. In industry and communications almost one-fifth of the jobs are already in Japanese hands, and in public services (administration, constabulary, jurisprudence, education, research, etc.) almost one-third. Under the heading "various occupations" medical doctors and restaurant proprietors are heavily represented.

The distribution of Japanese employees in Korea according to occupation presents a rather different picture, since the total number of persons employed in the different occupations in Korea of course varies greatly. But by and large the occupational groups in which the Japanese are particularly prominent in Korea as a whole are the ones for which the Japanese ethnic group shows a strong preference. More than two-fifths of the Japanese are employed in public services, another fifth in commerce, one-sixth in industry, but only one-seventeenth in communications and one twenty-second in agriculture and forestry.

The attempt to settle a large number of Japanese as farmers in Korea was a failure, and the high hopes held around 1910 not only by official circles in Tōkyō, but also by critical Japan experts such as E. Grünfeld, proved to be unfounded. The Oriental Development Co., which was to organize this settlement of Japanese farmers especially, was able to settle only 3971 families between 1910 and 1929. M. Schwind recently evaluated the reasons for this failure (696). Korea itself is an extremely overpopulated country (p. 398), particularly in the southern half, which is the part that would primarily come into question for Japanese farming settlements, because of the milder winters. In addition, a single Japanese farmer is very reluctant to move into a Korean village, where he does not understand the language. Therefore the only possibility would have been compulsory resettlement of entire Korean villages. Since 1919 the government–general has never again resolved to do so, and even before then such cases were very rare. In addition, outside of his native country the Japanese prefers occupations that allow him to act like a lord. Most of the Japanese farmers thus settle on reclaimed land, particularly in the rice paddy polders along the west coast (p. 359ff), in exclusively Japanese villages or hamlets. Such land reclamation has not yet been carried out to a very great extent, though. The small number of Japanese in Korea is therefore a result of the fact that Korea has not been made into Japanese farming country to any noticeable degree. The other occupations still play a secondary role to agriculture, and only to the extent that they increase in importance will the number of Japanese be able to expand.

The three professions preferred by the Japanese, public services, commerce and industry, which together employ more than three quarters of the Japanese, are primarily urban occupations. The distribution of Japanese across the country is thus very uneven. This is shown by Fig. 83 (cf. also M. Schwind in 695). It shows the proportion of the Japanese population element in the total population by provinces and also contains pie diagrams showing the proportion of Japanese in the 18 cities and 43 towns (p. 386). The provinces with a large number of such towns, and particularly large ones, also have a relatively large proportion of Japanese. The others, in which this is not the case, such as Kang-won-do and Ch'ung-ch'ŏngbuk-do, have a much smaller proportion. These cities together contain no less than 69.4% of all the Japanese in Korea, and almost half live in the eight

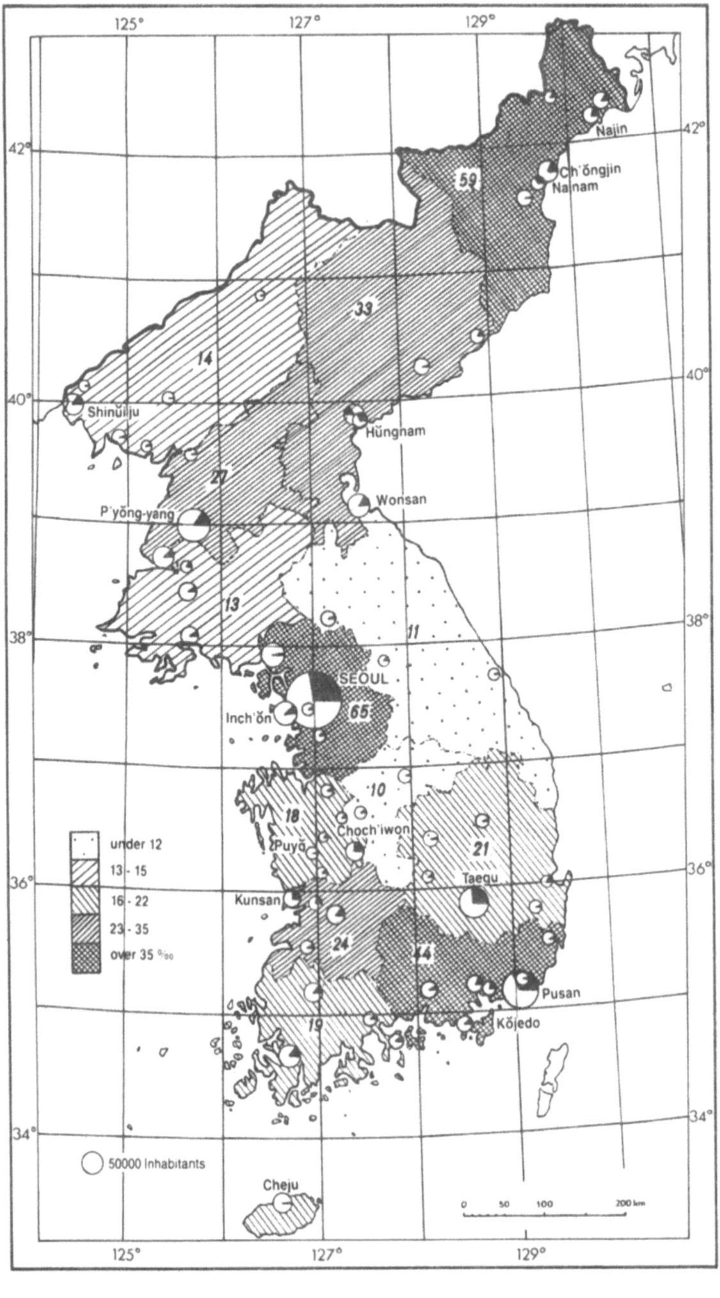

Fig. 83. Proportion of Japanese in the total population of the provinces and cities (end of 1938). The cities are represented by *circles* with an area proportional to the total population. The *black sectors* show the proportion of Japanese.

cities of Seoul, Inch'ŏn, P'yŏng-yang, Wonsan, Ch'ŏngjin, Kunsan, Taegu and Pusan, each of which has more than 10,000 Japanese residents. The proportion of Japanese in the cities and towns lies between 2% (Puyŏ, Cheju) and 33.1% (Nanam). Since southern Korea and Kyŏnggi-do have such cities in remarkably large numbers or sizes, the proportion of Japanese decreases on the whole from south to north. An exception to this rule is the two Hamgyŏng provinces, most particularly the northern one. This is a result of the most recent developments, for the North Korean Japan Sea borderland has had an impressive economic development in the past decade (p. 204). In Hamgyŏngnam-do the number of Japanese increased by 53% between 1930 and 1937 and in Hamgyŏngbuk-do even by 85%. The proportion of Japanese who live outside of the cities and towns is thus only 30.6%, i.e., for every 1 km^2 of the countryside there is one Japanese on the average. There is, however, not a single seat of a *myŏn* administration without at least one Japanese constable and one restaurant proprietor with their families, and in the coastal and boundary regions this is true of almost every large village.

In the cities the Japanese population element lives predominantly in separate residential areas with straight streets. These are recognizable from a distance because of their Japanese-style houses. They are charming wooden houses with sliding doors and bluish-gray tile roofs, often surrounded by the miniature gardens so familiar from their homeland. Despite the coldness of the Korean winter most of the Japanese here stick to their rather inefficient and not so harmless heating system with charcoal braziers. Only in the extremely cold northern interior has the Korean *ondol* heating system been partially accepted. In these parts of town the Japanese urban garb predominates. The men also prefer it in their private lives, whereas at work European clothing or uniforms prevail. One encounters rickshas pulled by coolies and other familiar phenomena of daily life in Japan. The Boys' Festival and the Girls' Festival are celebrated by raising flags shaped like fish, which billow up in the wind like balloons, their number and size corresponding to the number and age of the sons and daughters.

Often there is a Shintō shrine (Jap. *jinja*) in the vicinity of a place inhabited by a good number of Japanese. A straight road lined by cherry trees, often flanked by stone lanterns leads upward to the wooden or stone entrance gate (Jap. *torii*). The shrine stands behind it, half surrounded by trees. It is a one-story wooden construction with a gabled roof of bluish-gray tiles. Protruding rafters cross each other above the roofline. One looks straight through the empty center of the interior. Above the entrance hangs a straw rope. Often, particularly on national holidays, Japanese pray and sacrifice in front of the shrine. For civil servants the trip to the shrine is a precisely regulated duty. Additionally, a special holiday is devoted to each shrine. The State Shintō, to whom these shrines are devoted, consists in reverence for the Japanese national polity, the Imperial House, its divine ancestors and great national heroes, and has nothing to do with a specific creed. The most important shrine of this type in Korea is on Namsan near Seoul. It is the Chōsen Jingū, which is dedicated to the Sun Goddess (Amaterasu Omikami) and the Meiji Emperor. There are other *jingū* in the provincial capitals and other large cities. At present there are 58 of them. A total of 322 smaller *jinja* (Jap. *jinshi*) are distributed across the country [727]. In addition to State Shintō there are a number of Shintō sects with 301 prayer halls and 613 preachers. The number of believers is listed as 95,900 in 1939. The Shintō practiced by these sects developed out of the Japanese nature religion and must therefore be considered a kind of religion. Japanese Buddhism is also represented by nine sects and several hundred houses of prayer and temples.

3 Development and Distribution of Population

The Korean people have responded to the abolishment of their national independence with a biological answer that must be considered surprising after what we have said. Whereas there were 10 million Koreans, at the most, in 1876, the year in which Japanese influence began, and this had been preceded by centuries of stagnation and even decline (p. 151), at the end of 1940 23,653,039 Koreans lived in the country. The total number of Koreans is considerably larger, however, for in the interval of 60 years there has been a good deal of emigration (p. 394 f). In the country itself the Korean people have thus multiplied by 2.4, and the total growth is greater still. This impressive increase can be followed with exact statistics beginning with the year of the first population census, which Japan performed immediately after annexation. All older figures deserve only little confidence. According to Japanese calculations, the Korean people increased by 80.2% between 1910 and 1940. This corresponds to an average annual increase of 350,000 souls. By far the majority of the growth during the past 60 years thus occurred during the second half of this period, which is in a more limited sense the Japanese half. The birth rate was always high during this period. In 1923 it reached its peak with 40.2 per 1000, dropped to 28.5 per 1000 by 1936 and rose again to 33.6 per 1000 during the years of the "Korean boom." Due to the Japanese sanitary measures, the mortality rate decreased from 23.2 (1916) to 16.4 per 1000 (1938), so that there was usually a good-sized surplus of births. The population growth lay between 4.2 per 1000 (1920) and 19.7 per 1000 (1930), and in 1938 it rose again to 17.2 per 1000. The birth and growth rates of the Koreans are even much higher than those of the Japanese in Korea, whose numbers have increased more due to immigration than to natural increase. The results of this population development are shown in Table 29.

The figures in this table show the impressive growth of the Korean people. The growth per year did decrease gradually until 1938, partly as a result of emigration, but shot up to maximum amounts during the years of prosperity. Until 1938 the

	Absolute numbers				Percentage			Increase (+) or Decrease (–) in per 1000 per year			
	Koreans	Japanese	Foreigners	Total	Kor.	Jap.	For.	Kor.	Jap.	For.	Total
1910	13,128,780	171,543	12,694	13,313,017	98,6	1.3	0.10	+28.8	+103	+97	+29.9
1920	16,916,078	347,850	25,031	17,288,959	97.8	2.1	0.14	+19.2	+44	+179	+20.0
1925	18,543,326	424,740	47,460	19,015,526	97.5	2.2	0.25	+12.3	+36	+91	+13.1
1930	19,685,587	501,867	69,109	20,256,563	97.2	2.5	0.34	+15.9	+33	–30	+18.1
1935	21,248,864	583,439	58,888	21,891,191	97.1	2.6	0.27	+11.0	+29	–51	+13.1
1938	21,950,616	633,320	49,815	22,633,751	97.0	2.8	0.22	+38.8	–0.05	–99	+37.4
1940	23,653,039	633,288	40,000	24,326,327	97.2	2.6	0.16				

Table 29. Population development in Korea

number of Japanese increased relatively more rapidly, so that the percentage of Koreans in the total population decreased slightly. Between 1938 and 1940, however, it increased again. This development reveals a biological vigor that contrasts strikingly with the mental lassitude and raises the hope that this too can gradually be overcome.

Table 29 also includes the number of foreigners in Korea. Absolutely and relatively it has always remained small, in comparison with the number of Koreans and Japanese. It increased greatly until 1930, and then dropped sharply. By far the majority of these foreigners are Chinese [560, 698]. They live as very skillful, industrious and thrifty workers in transportation and mining, as merchants, truck farmers, restaurant proprietors and artisans in Seoul, P'yŏng-yang, Pukchin and Taegu and in the port towns of Shinŭiju, Inch'ŏn, Ch'ŏngjin, Wonsan, Chinnamp'o, Pusan, Kunsan and Mokp'o. In these cities they generally live together in special quarters, which have a northern Chinese milieu in a small, circumscribed area. In the western half of the country one can also find individual Chinese in many smaller towns. With the incorporation of Manchuria into the pan-Japanese economic sphere, a considerable number of them have returned to their own homeland, and this is the main explanation for the decreasing number of foreigners.

But the number of Europeans/Americans is also dropping sharply. There were still 1500 in 1938. Of these approx. 1100 are the families of the Anglo-American missionaries, who represent the various Protestant denominations, and about 200 are Roman Catholic priests and lay brothers. As pointed out on p. 169, the latter include about 30 Germans. The remaining Europeans/Americans are in commerce and mining, or they belong to the consulates—general and consulates in Seoul. The time of private European and American economic enterprises in Korea, which began promisingly after the conclusion of the trade treaties, is past. For the most part the European firms cannot keep up with the Japanese competition, and recently all mines were taken over by the Japanese (p. 445). There is only a single German firm still left, E. Meyer in Inch'ŏn.

The impressive growth of the Korean people did not occur everywhere to the same degree, for extensive internal migration took place simultaneously. During the period from 1910 to 1938 Kyŏngsangnam-do recorded the smallest amount of growth (56.2%), Kang-won-do the largest (99.8%). There was an above average population increase in six provinces: Hwanghae-do, Kyŏnggi-do, Ch'ungch'ŏngnam-do, Kang-won-do, Hamgyŏngnam-do and Hamgyŏngbuk-do. Hwanghae-do has by far the lowest population density in relation to the cultivated area and is thus still capable of absorbing population today, especially since its soil is excellent. In Kyŏnggi-do the two large cities of Seoul and Inch'ŏn have attracted population, and Kang-won-do, and the Hamgyŏng provinces are mountainous regions that have had a very much below average population density thus far. In the Hamgyŏng provinces, the rapid development of mining, industry and fishing has also highly promoted migration. Growth has been slightest in the four southern provinces, which have the highest densities in proportion to both total area and cultivated area. We can thus identify two types of internal migration. One is from the overpopulated southern provinces into the northeastern provinces, which are still capable of absorbing population; the other from the country into the cities. The latter is connected with a change in occupation, the former is not. The northward migration has an equalizing effect on the previous great contrasts in population density between south and north, the urban migration creates greater and greater differences between urban and rural districts. The former is a continuation of the migratory movement beginning during the Korean dynastic period (p. 151). Even today the north is a pioneer zone.

Most of the Korean counties, especially in the south, are highly overpopulated from an agricultural standpoint. Korea as a whole had a density of 493 persons per km^2 of cultivated area at the end of 1938. In Hwanghae-do the figure was 294, in Kyŏngsangnam-do, in contrast, 763. Despite the unpretentious nature of the Koreans, this large population increase has therefore led to emigration. Particularly large numbers of families emigrated after having lost their land due to excessive debts. The major country to which they emigrate is Manchuria, particularly the southeastern part bordering on Korea [699– 703]. Thus the northern Korean pioneer zone does not end at the national boundary, but continues on into the neighboring country. The Yalu and the Tumen probably never formed a sharp boundary between peoples. It is likely that a larger or smaller number of Koreans has lived beyond them for many centuries. Southeastern Manchuria is after all one of the core areas from which the Koreans came (p. 20). In the past centuries a remigration has commenced, and after the Russo-Japanese War of 1904/05 it was quite tumultuous at times. Northern Koreans moved into the regions just beyond the national boundary. Southern Koreans followed the railroad lines that were built in southeastern Manchuria. Many existed at a starvation level for years without a permanent residence, in the hands of Chinese usurers. Since the incorporation of Manchuria into the Japanese economic sphere, resettlement has been sponsored by the government–general with the aid of three companies founded for this purpose. Almost all of the Korean migrants to Manchuria remain farmers. They have the advantage over northern Chinese settlers that they have experience in growing rice, for which the climate of southeastern Manchuria is still favorable. In Manchuria they virtually have a monopoly on it. At present there are 484,000 Koreans residing in the province of K'ientao (corresponding approximately to the present-day Yanbian Korean Autonomous Prefecture, Jilin Province) on the other side of the Tumen. They form 75% of the total population there, and the entire cultural landscape of the province, with the exception of the cities, is Korean in character. According to popular tradition there was already a large number of Korean villages there at the time of King Sukchong (1674–1720). Until 1909 the boundary between Korea and China was not clearly defined here (p. 32). In 1890 there was renewed heavy migration following a famine in northeastern Korea. A further sudden increase in the Korean population of K'ientao occurred in 1907 after an office of the resident–general was opened in Yanji. There are 453,000 Koreans living on the other side of the Yalu, in the province of Dandong (now part of the Chinese province of Liaoning). The statistics for Manchukuo give the total number of Koreans as 1,152,127 in 1939. Like all rural inhabitants of Manchuria they have suffered heavily at the hands of bandits during the past decade.

The second area on the continent to which Koreans have emigrated is the Russian Ussuri region, particularly the southwestern part bordering directly on the province of K'ientao. I. Bishop visited and described the Korean settlers there as early as 1893 [207, Vol. II, p. 1–20]. They have settled particularly densely around Lake Khanka, where they grow rice and barnyard millet. Their number increased from 87,000 in 1936 to 180,400 in 1939. These Koreans have almost completely lost contact with their homeland, because crossing the border is prohibited by both the Soviet Russians and the Japanese on pain of death. Japan fears an influx of Communist ideas into its territory.

Finally, Koreans, who are after all also citizens of Japan, have streamed in gradually increasing numbers into the Japanese islands. In 1930 around 400,000 of them lived there; in 1937, in contrast, already 800,000. Most of them come from the southern part of the peninsula and Chejudo (p. 379). Thus there are already more Koreans in the Japanese islands today than there are Japanese in Korea. They work primarily at low-level jobs, such as in transportation and industry in Ōsaka, Kōbe, Nagoya, Yokohama and Tōkyō.

If one estimates the remaining Koreans living abroad, mainly in China, and otherwise in the USA, Mexico and Europe, at 60,000 at the maximum, one arrives at the sum of 2.19 million Koreans living abroad. The total number of Koreans in 1940 probably amounted to 25.8 million.

The present population distribution, which has emerged as the result of biological growth and migration within and out of the country, was already described in broad outline in the introductory remarks on provincial boundaries (p. 14ff). Although it has been directly influenced by the government–general only in a few rare cases, indirectly it is highly dependent on the Japanese measures. This will be evident from the remarks in the next chapters, which deal with the economic geography of Korea under the influence of the modern development of the country, and without which it is impossible to understand the individual features of the population distribution. In the regional section of this book these relationships were evaluated each time from the standpoint of the individual regions. Figure 2, which accompanies the introductory remarks, shows the population density by provinces. It increases from northeast to southwest. In the mountainous provinces of the northern and central eastern part it is below 60, in the rice-growing provinces of the southwest over 175. The average density for the government–general had increased to 110.2 in 1940, a very high figure for this mountainous country (Switzerland 1941: 103).

The individual features of the population distribution are shown in Map 3 (cf. editors' note on p. 7) according to the absolute method and in Fig. 85 according to the relative method. I drew up these plates on the basis of the official statistics for 1936 [691], which contain the population figures for the cities (*fu*, Kor. *pu*) and towns (*yu*, Kor. *ŭp*) and the 2322 townships (*men*, Kor. *myŏn*). On Map 3 one dot represents 1000 inhabitants. For settlements with 1500–19,500 inhabitants these dots are combined to form simple geometrical figures. Settlements with more than 19,500 inhabitants are drawn as circles with an area proportional to the population. The boundaries of counties are shown by red lines. Figure 85 shows the population density of the rural counties. The cities had to be left out because of their small size.

The first impression that strikes one when contemplating the dot map is the contrast between the dense population of the basins, valleys and coastal fringes and the sparse population, indeed, the deserted state of the mountainous heights. The major mountain ranges stand out clearly. The Paektusan plateau, the Mach'ŏllyŏng Range and the northeastern part of the Hamgyŏng Range unite to form an almost completely deserted area of about 10,000 km^2, within which only the trough of the Sŏdusu has 6000 inhabitants. Accordingly, Musan-gunis the least densely populated county in Korea, with 9.7 inhabitants/km^2, despite the settlements along the Tumen and its side valleys. The low population densities in the Kaema upland (Changjin-gun13, P'ungsan-gun18) are explained for their part by the fact that the northern side ridges of the Hamgyŏng Range are almost deserted. On the dot map the Main Korean Range also stands out clearly. This is least true of the highly dissected Wonsan arc and most true of the T'aebaek Range, which contrasts sharply with the densely populated Yŏngdong fringe of the Sea of Japan. Very much the same can be said of the Sobaek Range, the southwestern side ridges of the T'aebaek Range and the interior of Chejudo. On the whole, however, it is apparent that the influence of the relief is considerably greater in the north than in the south. The reason for this is the inclement climate of the north, which we have evaluated in detail a number of times. One must keep in mind, though, that the north is a pioneer zone that has only gradually been being filled with people during the past decades. This viewpoint holds especially for the extreme northeastern tip, whose relief presents relatively few difficulties and which is still quite sparsely populated to date (Chongsŏng-gun 30). The northeastern part of P'yŏng-nbuk-do also belongs to the pioneer zone (Huch'ang-gun 36, Chasŏng-gun 61, Kanggye-gun 63). It was not until the beginning of this century that the wave of fire-field occupation passed over it, followed by permanent settlement (p. 255f). And the same is true of the western slope of the T'aebaek Range (Inje-gun 32, Chŏngsŏn-gun 34, Hoeyang-gun 39).

Contrasting with these very sparsely populated mountain ranges are the densely populated coasts, valleys and plains. From Ch'ŏngjin to P'ohang a narrow but continuous band of settlements follows the coast of the Sea of Japan, and in the alluvial plains between Hamhŭng and Wonsan it even widens out considerably (Hamju-gun 140, Kowon-gun 92). Much wider still, however, is the densely populated fringe along the northern and central Korean Yellow Sea coast. From the Yalu to beyond the Ch'ŏngch'ŏn-gang the population gets more and more dense toward the sea (Yong-amp'o-gun 221, Chŏngju-gun 161). The same is true of the coastal region between Haeju and the mouth of the Han-gang (Yŏnbaek-gun 175). In P'yŏng-annam-do and Hwanghae-do, in contrast, the region with the maximum population does not lie directly on the coast. Instead, the densest settlement is crowded into the very fertile basin on the lower reaches of the Taedonggang and the Chaeryŏnggang between P'yŏng-yang and Chaeryŏng (Taedong-gun 165, Pongsan-gun 139, Shinch'ŏn-gun 145). The same picture is repeated between the mouths of the Han-gang and the Ansŏngch'ŏn. Here the zone of densest settlement follows the lowest part of the Han-gang and then extends past Seoul and Suwon to Chŏnan (Koyang-gun 167, Suwon-gun 181, Chinwi-gun 199).

On the coast of P'yŏng-anbuk-do and in the strip extending southeastward from Haeju it is above all the intensity of rice cultivation that is decisive for the dense settlement, whereas in the basin between P'yŏng-yang and Chaeryŏng dry-field and paddy farming are combined. In southern Korea rice cultivation is by far the most important factor supporting population. As Map 4 (cf. editors' note on p. 7) and Fig. 87 show, it is intensive here not only in the basins of the west coast, but also in the Naktong basin. Consequently the density rises to very high figures everywhere in the valleys. Regions that stand out as having very high concentrations near the Yellow Sea coast are: the basin of the Sapkyoch'ŏn (Yesan-gun 203), that of the Yŏngsan-gang (Naju-gun 241) and above all the entire region from Kongju on the Kŭmgang past Kunsan, Iri and Chŏnju to Chŏng-ŭp (Nonsan-gun 246, Iksan-gun 458, Kimje-gun 321, Okku-gun 270, Chŏng-ŭp-gun 246). Iksan-gun has the highest density in Korea. In the Naktong basin the most densely populated areas are the base of the mountains from Yŏngju past Sangju to Kimch'ŏn (Yech'ŏn-gun 205), the basin around Taegu-Yŏngch'ŏn (Talsŏng-gun 223), the lowest part of the Naktong valley (Miryang-gun 264, Ch'ang-won-gun 217) and the structural valleys near the east coast (Ulsan-gun 143).

On the whole, Korea's settlement is peripheral. A narrow fringe of high population density in the east and a wide one in the west contrast with a much more sparsely settled interior. This contrast is strongest in northern Korea. Toward the south it diminishes, until in southern Korea it is only weakly manifest in the narrow Sobaek Range. In the coastal regions there is a combination of low elevations, favorable climatic conditions, often good soils and a variety of other possibilities for economic development. These varied factors supporting settlement also contribute to surprisingly high densities on the peninsulas and islands of the south coast, despite their mountainous nature (T'ong-yŏng-gun 264, Namhae-gun 362, Yŏsu-gun 245). Even the small islands in the extreme southwest still support a good-sized population (p. 373).

Since more than three-quarters of the Koreans make their living by farming, the population density is naturally closely related to the ratio of cultivated land to the total area. This has already been indicated several times. A comparison of Fig. 86, which shows this ratio by counties, and the map of population density is thus quite informative. The counties with the lowest proportion of fields, Changjin-gun (32 per 1000) and Musan-gun (37 per 1000) in the Kaema upland, are also the counties with the lowest densities (13 and 9.7, respectively). The just mentioned county of Iksan-gun, which has the highest population density in all of Korea, also has the largest proportion of cultivated area (781 per 1000). There are, however, a number of departures from this basic rule. The fields have quite varying productive capacities. Paddies, the expression of the most intensive agriculture in East Asia, support more people than dry fields. Chongsŏng-gun (Hamgyŏngbuk-do), which has almost exclusively dry fields, has a density of only 30 with its cultivated area of 125 per 1000; because of its abundance of paddies, Yŏngdŏk-gun in northeastern Kyŏngsangbuk-do has, in contrast, a density of 112 with a cultivated area of 124 per 1000. Where the population is partially supported by other occupations, the density varies, despite equal agricultural conditions. Thus Unsan-gun, which has large gold mines, has a density of 69 with a cultivated area of 191 per 1000, more than the neighboring Kusŏng-gun, which only manages to have a density of 56 with a cultivated area of 208 per 1000. Nonagricultral occupations are particularly common in the counties around large cities. For this reason Koyang-gun near Seoul has a density of 167 with a cultivated area of 368 per 1000; Ich'ŏn, a bit farther upstream on the Han-gang, has only 140 with a cultivated area of 379 per 1000.

If the quality of the land and the demands of nutrition were the same everywhere, and if other occupations could be completely ignored, there would be no regional differences in the relationship between population density and cultivated area. Thus the differences in density that do in fact exist reveal the differences we have just touched on in soil composition, climate and landuse and cases of supplementary employment. On p. 394 we already pointed out that the absolute figures on population density in relation to the cultivated area are very high in Korea. Considering the slight significance of other occupations, there is thus a strong agrarian overpopulation pressure in Korea. Nevertheless the regional differences are still enormous.

The Kaema upland has again the lowest density, in relation to the cultivated area. With 165 inhabitants/km^2, P'ungsan-gun has the lowest figure in all of Korea. This is due to unfavorable climatic and soil conditions and the lack of other economic activities apart from forestry. But the big bend of the Tumen also has quite low figures, although the elevation is much lower (Chongsŏng-gun). Here there are still possibilities for settlement. The figures are already higher on the Hwanghae peninsula (282–374 inhabitants/km^2), where the cultivated area is relatively large but still consists primarily of dry fields. Everywhere the rice-growing areas are distinguished by high agricultural densities. Southern Korea has densities between 550 and 800. Only Chejudo is in its agricultural density, as in so many other connections, still at the level of northern Korea for the time being (217 inhabitants/km^2). In general the coastal fringes are distinguished by a higher agricultural density than the interior. Here low elevations, relatively fertile soils and supplementary occupations combine to increase the productivity per unit of area. For the urban areas this is self-evident a priori, but it is also true of the rural counties. Thus in northeastern Korea the coastal county of Pukch'ŏng-gun with 546 inhabitants/km^2 borders directly on the just mentioned county of P'ungsan-gun (165); in the southern Yŏngdong area the county of Samch'ŏk-gun (614) borders on Chŏngsŏn-gun (281). The entire east and south coast of the peninsula are distinguished by an almost continuous ribbon of peak values. In Yŏngdong let us point out Kosŏng-gun (740) and Yŏngdŏk-gun (903). On the Korea Strait Yŏsu-gun attains 996, Namhae-gun 1131 and T'ong-yŏng-gun even 1354. The last figure is the highest reached by a rural county in Korea. Fishing, fish farming, mining, transportation and commerce combine with agriculture here to attain these prodigious sums. On the whole these figures show with forceful clarity that under the influence of the Japanese order Korea has for the most part become a highly overpopulated country. It shares this development with many European colonies in Monsoon Asia.

4 The Fields

From the beginning the Japanese administration devoted special attention to the development of agriculture. This was done, on the one hand, to feed the increasing millions and lessen the dissatisfaction arising from poverty. On the other hand, the intention was to guide Korea's production according to the needs of the Japanese homeland and to gradually make the foreign possession a stable member of a pan-Japanese economic zone. Until 10 years ago this incorporation was sought almost entirely in the agricultural sector. Only since then have the other sectors of the economy, in particular mining, fishing and industry, been assigned corresponding roles. Until then Korea's part in the pan-Japanese economy was seen principally as providing agricultural foodstuffs and raw materials needed by Japan, especially rice, soybeans, cotton, raw silk, cattle and hides, and absorbing Japanese industrial goods in return.

Since more than three-quarters of the Korean people are farmers, their attitude toward the Japanese domination depends to a large degree on the development of the living conditions of just these farmers. The conditions were already exceedingly wretched in traditional Korean times (p. 173), but in the course of the last 2 decades of Japanese administration they have become even worse. This led to the uprising of 1919 and to other severe strains on Korean–Japanese relations. The distribution of property, already unfortunate during the Korean dynastic period, has become even more critical. More and more farmers have been forced to sell their property. In 1935 of 3,006,489 Korean farm families only 17.9% still owned their land: in contrast, 51.9% were tenant farmers and 27.7% partially owned and partially rented their land. Four-fifths of the farmers are thus subject entirely or in part to the terms of leases that have become even more oppressive since the occupation. Tenant farms are most common in the rice-growing areas of the south, where the soil is especially valuable. Since the farms are smallest there, the percentage of the area that is farmed by tenants in Korea is smaller than the proportion of tenant farms. Nevertheless it amounts to 57.5% of the total cultivated area (1937); specifically, 67.9% of the paddies and 50.9% of the dry fields. The conditions are least favorable in the foremost rice-growing province, Chŏllabuk-do, where rented land makes up 76.2% of the cultivated area. They are most favorable in Hamgyŏngbuk-do, where the paddy area is small (20.6%). The attitude of the owners toward the tenants has not gained in understanding since the end of the dynastic period. According to Hoon K. Lee it has become even more impersonal and thus harsher [605, p. 173]. The Japanese Y. Yagi, who wrote an excellent socio-economic study on rice farming in Korea [688], quotes a passage from the report *Tenant Customs in Korea* [614, p. 230], published in 1930 by the government–general, stating that "The attitude of the majority of the landowners and their managers toward the tenant farmers is downright arrogant. It is aimed at extorting fees. At any rate it is easy to imagine the great hardships under which the tenant farmers suffer due to the oppressive burdens."

These extraordinarily pressing rental conditions, together with the agrarian crisis, which affected Japan just as severely, led to a completely untenable farm household balance. As trustworthy a source as *The Japan Year Book* still provides the information shown in Table 30 on the "Conditions of Typical Korean Farmers" in its most recent volume [XLI]. It gives three examples, which are obviously averages from a fairly large number of observations.

Since 1924 expenditures exceed income for more than half of the farmers. Indebtedness has thus gradually increased, and consequently more and more farmers have sunk from the class of petty farm owners to that of tenants. It goes without saying that with such a balance expenditures for modern tools, fertilizer, etc. are impossible. Simultaneously one can see from the table that the farmers' labor potential is nowhere near to being fully utilized during the course of the year. The climatic conditions, together with the often mentioned life-style of the Koreans, reduce the number of days on which work is really done, despite their misery, to between 75 and 125. The Japanese farmer, living in a climate with much milder winters and being more industrious, works 200–250 days per year [XLI].

Hand in hand with the constantly increasing impoverishment of the broad masses went a decline in the standard of living. Rice is the most highly valued staple food. The increasingly unfavorable economic balance is forcing the Korean farm households to a greater and greater degree to sell their rice and consume millet, some of which is grown in their own country and some imported from Manchuria, and the barley that they themselves are growing more and more. The Japanese Y. Yagi and the Korean Hoon K. Lee demonstrate this development with almost exactly the same figures, as follows [688, 605]. In 1912 the average per capita rice consumption in Korea was 124.0 liters; in 1929 it was only 77.2 liters. Millet consumption, in contrast, rose from 42.1 to 65.3 liters during the same period. Yagi states further that the annual per capita rice consumption in Japan is 198 liters. If it were just as high in Korea, even the peak harvest of 1937 (p. 413) would not have been sufficient to meet the demand. Korea today is one of the foremost rice-exporting countries in the world, not because there is a surplus that the country does not need, but because of the described decline in the standard of living. The blame for this development has been placed on the Japanese nation, indeed on the government–general. This is justified, inasmuch as no effective remedial action has been taken.

The situation of the Korean farmers is not only a social question. It has simultaneously become one of nationalities, in that Japanese capital has found its way into rural property. The landowners are thus not only Koreans, but also Japanese, and the latter consist primarily of the irrigation associations (p. 401, 409f). The extent to which Japanese capital has entered, however, and this is not usually taken into consideration by biased critics, is not very great. According to such a reliable source as Lee [605, p. 147], in 1927 92.7% of the cultivated land was owned by Koreans, 0.5% by foreigners (mainly Chinese) and 6.8% by Japanese. No more recent figures are available. Since then the Japanese share has certainly increased somewhat. But even when one bears in mind that the Japanese primarily own valuable paddy land, the invasion of Japanese capital is not overwhelmingly great. Nevertheless it has provoked intense indignation among the Koreans and strained the relations between the two peoples, specifically for two reasons. First, in the local peasant revolts that have broken out time and again during the past 2 decades because of their plight, the Japanese police,

Number of family members	6	5	5
Family workers	2.9	2.1	2.4
Ownership	Partial owner	Tenant	Tenant
Size of farmed land	4.5 acres (1.82 ha)	3 acres (1.22 ha)	2 acres (0.81 ha)
Cash income per year	98.50 yen	52.70 yen	48.40 yen
Expenditures	121.70 yen	69.60 yen	59.20 yen
Deficit	23.20 yen	16.90 yen	10.80 yen
Debts	130.00 yen	49.00 yen	52.00 yen
Number of possible additional working days	290	260	240

Table 30. Conditions of typical Korean farmers

as guardian of law and order, have had to oppose the rebellious tenants. This has given the Korean farmers the impression that the Japanese authorities always support the landowners against the farmers, and thus the plight of the tenants and petty farm owners has generally been blamed on the Japanese domination. Second, the invasion of Japanese capital has increasingly been taking place in the sphere of large holdings, so that its social significance has increased greatly. This is shown by Table 31, which was also calculated according to the data given by Lee and Yagi.

Table 31 reveals the following facts: (1) Miniature and small holdings outnumber medium and large-sized holdings by far. (2) The fragmentation of property continued to increase between 1921 and 1928, with the exception of large holdings of more than 100 *chŏngbo* (= 99 ha). (3) Among the total number of owners the Japanese play a very minor role. (4) Their number is, however, relatively much larger in the bigger size classes than in the smaller ones. In the largest size class they already had the majority in 1921. (5) In all size classes the proportion of Japanese owners increased between 1921 and 1928. (6) This increase becomes relatively larger with increasing property size. (7) The number of Koreans owners in the size classes above 50 *chŏngbo* has decreased, not only relatively, but absolutely. A comparison between the nationalities turns out about the same when one considers the area and not the number of holdings. In the size class over 100 *chŏngbo*, for instance, the Japanese dominated in 1921 by 57.1% and in 1928 by 66.9%. By area, to be sure, this size class occupies only a very small part of the total cultivated land.

A very large portion of the Japanese holdings is in the hands of irrigation associations. There were 189 of them at the end of 1937 with an irrigated area of 215,345 ha and with a capital of 138 million yen invested in these works. Where they exist Korean small farmers have special difficulties holding their farms, because the use of the irrigation works naturally entails considerable expense. Therefore the majority of the Koreans' complaints are aimed at the irrigation associations, in particular the largest of them, the Oriental Development Co. But, on the other hand, it was they who gave the alluvial plains in the vicinity of the

	< 1 *Chŏngbo*[a]			1-5 *Chŏngbo*			5-20 *Chŏngbo*			20-50 *Chŏngbo*		
	Number	(%) K	J	Number	(%) K	J	Number	(%) K	J	Number	(%) K	J
1921	2,319,254	98.9	1.1	979,667	98.8	1.2	145,073	97.2	2.8	15,858	91.1	8.9
1928	2,646,543	98.6	1.4	1,043,538	98.2	1.8	157,044	95.6	4.4	17,681	86.8	13.2

	50-100 *Chŏngbo*			> 100 *Chŏngbo*			Total		
	Number	(%) K	J	Number	(%) K	J	Number	(%) K	J
1921	2,169	76.4	23.6	916	46.5	53.5	3,462,937	98.7	1.3
1928	2,300	70.3	29.7	888	37.7	62.3	3,867,994	98.3	1.7

Table 31. Number of landowners by size class and nationality

western river mouths their present thriving appearance and who have greatly intensified man's use of the gifts of nature. The fact that Korea's rice production has as much as tripled since 1910 is to a large extent due to their efforts.

Up until 1934 the government–general took few effectual measures to alleviate the need of the farmers, although it was not lacking in good will. It based its position too strongly on technical rights and placed too much emphasis on increasing production, that is on filling the needs of the Japanese homeland. From 1934 on a new decidedly pro-farmer policy began. Governor–General Ugaki became the founder of the "Rural Revival Movement" (p. 386), which has made gratifying progress meanwhile. The extremely favorable economic situation since the incorporation of Manchuria into the Japanese economic sphere has helped the movement to attain quick results. The need has been substantially allayed, although the vast majority of the farmers are by no means in a position to weather a crisis. As a part of the revival movement aid program, by August 1938 68 large grain elevators had already been constructed in the production centers and 11 export warehouses in the principal ports with capacities of 2.34 and 4.14 million hl, respectively. The farmers can store part of their harvest in them until the most favorable season for selling, and on the market serious surpluses and sudden drops in price can be avoided.

Despite the bitter poverty of the farmers, Korea has managed not only to more than double its population within a generation, but to feed most of this increase itself and at the same time to produce gigantic agricultural surpluses for export. It is obvious that this prodigious accomplishment had to involve a basic reformation of agriculture. From the first day of its activities the Japanese administration devoted itself with unceasing energy and indisputable success to increasing production. It always went about this discreetly. To begin with, extensive inventories were taken in every branch of agriculture. These began with the great land survey from 1910 to 1918 [553] and the work on the topographical map 1:50,000. In 1906 the large Agricultural Experiment Station in Suwon was founded. Today it has seven branch stations with extensive experimental fields in Poch'ŏnbo, Ch'aryŏngwan, Yonggang, Sariwon, Iri, Kimje and Mokp'o. A voluminous two volume work was published in 1931 on the occasion of the 25th anniversary of the Suwon station. It presents the methods and results of this experiment station impressively in great detail [610]. The projects include first of all the investigation of all Korean crops. Experiments are done with various foreign varieties, and, if necessary, new ones are cultivated. Thus the above report contains, for example, the information that in the province of Ch'ungch'ŏngbuk-do no fewer than 492 different varieties of rice were tested. The best methods of cultivation are determined. The suitability of individual crops for Korea's different climates and soils is determined. The insect pests affecting each plant are identified and methods for their control are worked out. Likewise the diseases to which the crops are subject and methods for curing or preventing them are studied. All of this research extends not only to each annual plant, but also to the mulberry tree, the bamboo grasses and fruit trees. In like manner the Korean farming implements are studied and improvements are devised. The soils are investigated, and the effect of chemical fertilizers on each individual plant is determined. Finally, livestock breeding and silkworm raising are studied correspondingly. The responsibilities of the seven branch institutes are divided by regions; Mokp'o is concerned especially with cotton growing, Iri with rice, Sariwon with the dry-field grains and fruit growing, Poch'ŏnbo with fire-field farming. The agronomical results attained in this manner must lead not only to model crops on experimental fields, but must also affect the agricultural practices of the individual farmer. The schools affiliated with the stations serve this purpose. These schools offer courses for the Korean farmers and send extension teachers into the villages. The sericulture schools train Korean women and girls especially, since on the farms they perform the major work in this branch of agriculture. The government–general further supports the formation of agricultural cooperatives in all towns and counties and subsidizes them financially. Such cooperatives exist for every single branch of agriculture, e.g., livestock breeding, sericulture, fruit growing, irrigation, etc. The umbrella organization is composed of dozens or hundreds of individual cooperatives, as the case may be. Their responsibility is to distribute seeds and make better tools available, to instruct by means of lectures and exhibitions, to award prizes for special products, etc. In contrast, the state does not support the traditional Korean mutual aid societies (Kor. *kye*), which put on lotteries to advance their members by means of the proceeds, because their system with its speculative basis does not accord with the government–general's strictly rational way of thinking [607].

The results of this generation of systematic work are enormous, in spite of the ponderous nature of the Korean farmer. The cultivated area increased from around 3.5 million ha in 1910 to 4.9 million ha in 1938 and thus now amounts to 22.2% of the national area. It is therefore about one-third greater than in 1910. The major part of this development appears to have taken place in the first years after the occupation. In 1918 the cultivated area already amounted to 4.5 million ha. The reason for this may be that in the first years of the new initiative a large number of previously fallow fields were farmed again. Probably only the development after 1918 is based on an increase in the actual cultivated area. According to the government– general [VI] there are now only 0.3 million ha of uncultivated land that could still be cultivated. Of this 0.1 million ha are to be used for settling fire-field farmers (p. 432). The remainder is to be rented to other farmers, who will be given the land free of charge if they are successful at clearing it. The value of the agricultural production increased from 241.7 million yen in 1910 to 1134.1 million yen in 1937, i.e., almost fivefold. Figure 84 shows the development of five of the most important agricultural products. The yields per hectare have thus risen considerably. As the surveys of the *Annuaire International de Statistique Agricole* show, they generally lie close to the world values today, indeed in the case of rice, hemp and a few other crops they even lie considerably above them. Only the Korean high altitude crops, such as oats, potatoes and soybeans produce substantially lower yields per hectare. It must be kept in mind, however, that a large portion of the dry-field crops, and especially soybeans, are intercropped, which naturally lowers the yield per hectare of each individual crop quite a bit, since after all the whole area is not available for this one crop. This explains, for example, the fact that the wheat grown between 1934 and 1939, most of which was intercropped, brought yields of only 6.7 to 9.8 q/ha, whereas the average world values for the same period lie at 9.0 to 10.9 q/ha.

Table 32 is a summary of landuse in the individual provinces. For the fields it is based on the data for 1930, because no more recent data were available for the provinces. For the forests it is based on the figures for 1937. Since the figures for forest area decreased between 1930 and 1937, the remainder, consisting of graves (still a total of 47,300 ha! cf. p. 165f), bodies of water, roads, etc. is in reality somewhat smaller than calculated in Table 31.

Here we are interested only in the part referring to the fields, while the part concerning forests will be discussed later (p. 427). It shows that the percentage of fields in the total area is highest in the Yellow Sea provinces of central and southern Korea, which can easily be explained on the basis of the relief and the remarks on p. 186. Hwanghae-do leads with 34.5%. In the provinces of the Kaema upland and the T'aebaek Range the proportion of fields is as little as one-third of this peak value. Ch'ungch'ŏngbuk-do and the two Kyŏngsang provinces hold the middle.

Figure 86 gives the proportion of cultivated land in the total area by county. It shows clearly that the upland counties of the Kaema upland and the ranges bordering it, the T'aebaek Range and even the Sobaek Range are short on cultivated land. The lowest level consists of the counties of Musan-gun, Huch'ang-gun, Changjin-gun, Hŭich'ŏn-gun and Yŏng-won-gun in northern Korea, where the difficult relief is combined with an inclement climate (p. 220ff, 252ff). The counties near the Yellow Sea coasts, in contrast, have maximum values in south-

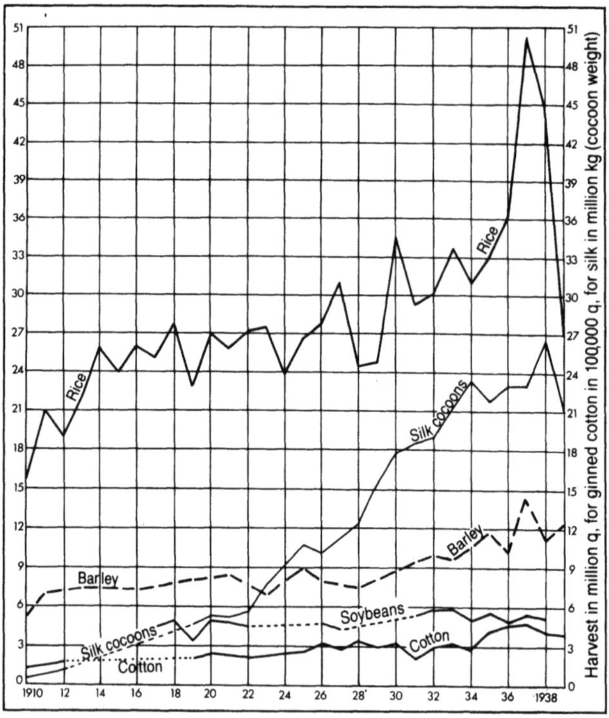

Fig. 84. Production of rice, barley, soybeans, silk cocoons and cotton, development from 1910 to 1939. (q = quintal = 100 kg)

ern, central and northern Korea. All of the counties in which more than half of the area is cultivated lie in this region (Yong-amp'o-gun, P'yŏng-won-gun, Taedong-gun, Yŏnbaek-gun, Chinwi-gun, Iksan-gun, Kimje-gun). Even in Chejudo almost 50% is cultivated land. The eastern fringe of the peninsula, in contrast, appears only here and there as an especially favored strip, since the counties in question all extend up to the drainage divide of the T'aebaek Range. In the wider North Korean Japan Sea borderland, however, there are a number of counties at a higher level than the neighboring counties in the Kaema upland. In southern Korea the contrasts between coastal fringe and interior and between west and east are much less pronounced, for here the eastern part includes the broad, fertile Naktong basin. Toward the end of the last chapter we already pointed out the connections between this map and the population density (p. 396f).

It is likewise unnecessary to go into the details of the physical basis of agriculture and the qualitative features of cultivation and farming. We have done justice to all of these viewpoints in the chapter on traditional Korea (p. 169–186). Here our purpose is only to outline the changes that Korean agriculture has undergone under the influence of Japanese development. Particular stress is to be laid on the quantitative conditions at the present time.

For statistical purposes Korean fields are divided into three types: paddies, permanent dry fields and fire fields. Table 32 contains the proportion of each type in the area of the province. The fire fields are a relic of traditional Korea. Their role in modern times is discussed in connection with the forests (p. 226, 431). Paddies are used almost exclusively for rice growing, as already pointed out on p. 177.

	Total area (km²)	Fields (%)[a]				Forests (%)				Settlements (%)	Others (%)
		Total	Paddies	Permanent dry fields	Fire fields	Total	Forests and brushland	Areas with saplings	Still barren		
Hamgyŏngbuk-do	20,347	10.8	0.6	9.6	0.6	82.8	61.7	9.6	4.4	0.2	6.2
Hamgyŏngnam-do	31,979	12.3	1.5	8.9	1.9	83.5	54.8	12.1	8.4	0.3	3.9
P'yŏng-anbuk-do	28,445	15.1	3.0	9.8	2.3	80.3	52.1	12.4	5.6	0.3	4.3
P'yŏng-annam-do	14,925	28.0	4.9	21.5	1.6	65.7	52.3	2.1	2.7	0.6	5.7
Hwanghae-do	16,732	34.5	8.3	25.9	0.3	58.7	43.8	8.7	2.5	0.8	6.0
Kang-won-do	26,263	13.2	3.1	8.9	1.2	82.5	47.5	17.3	10.7	0.3	4.0
Kyŏnggi-do	12,814	32.2	16.8	15.4	0.03	59.6	43.8	5.4	2.7	1.3	6.9
Ch'ungch'ŏngnam-do	8,106	31.8	20.9	10.9	0.02	60.8	41.8	11.9	4.4	1.5	5.9
Ch'ungch'ŏngbuk-do	7,418	22.2	9.9	12.1	0.15	71.9	54.0	10.1	3.4	0.9	5.0
Kyŏngsangbuk-do	18,989	21.4	10.7	10.6	0.05	70.0	59.4	3.5	3.8	0.3	8.3
Kyŏngsangnam-do	12,305	23.7	14.6	9.1	0.01	68.1	48.4	8.8	2.9	1.0	7.2
Chŏllabuk-do	8,531	29.0	20.7	8.3	0.04	63.9	51.7	5.2	1.9	1.2	5.9
Chŏllanam-do	13,887	30.9	15.1	15.8	0.02	62.9	48.9	6.0	3.3	1.2	5.0
Korea total	220,741	20.8	7.5	12.4	0.9	73.2	52.1	9.6	5.2	0.6	5.4

[a] All percentages refer to the total area listed in the second column.

Table 32. Landuse in the 13 Korean provinces in 1930 and 1937[a]

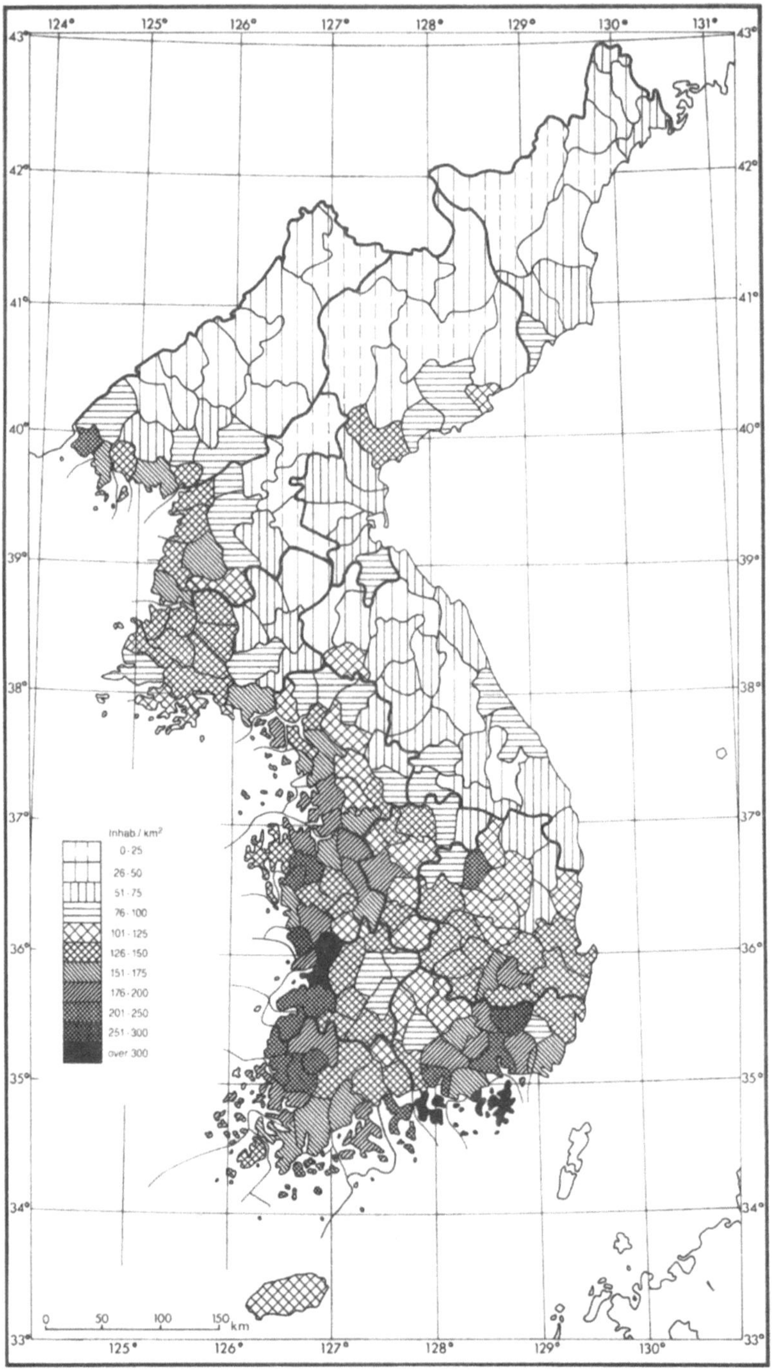

Fig. 85. Population density of the counties (1936)

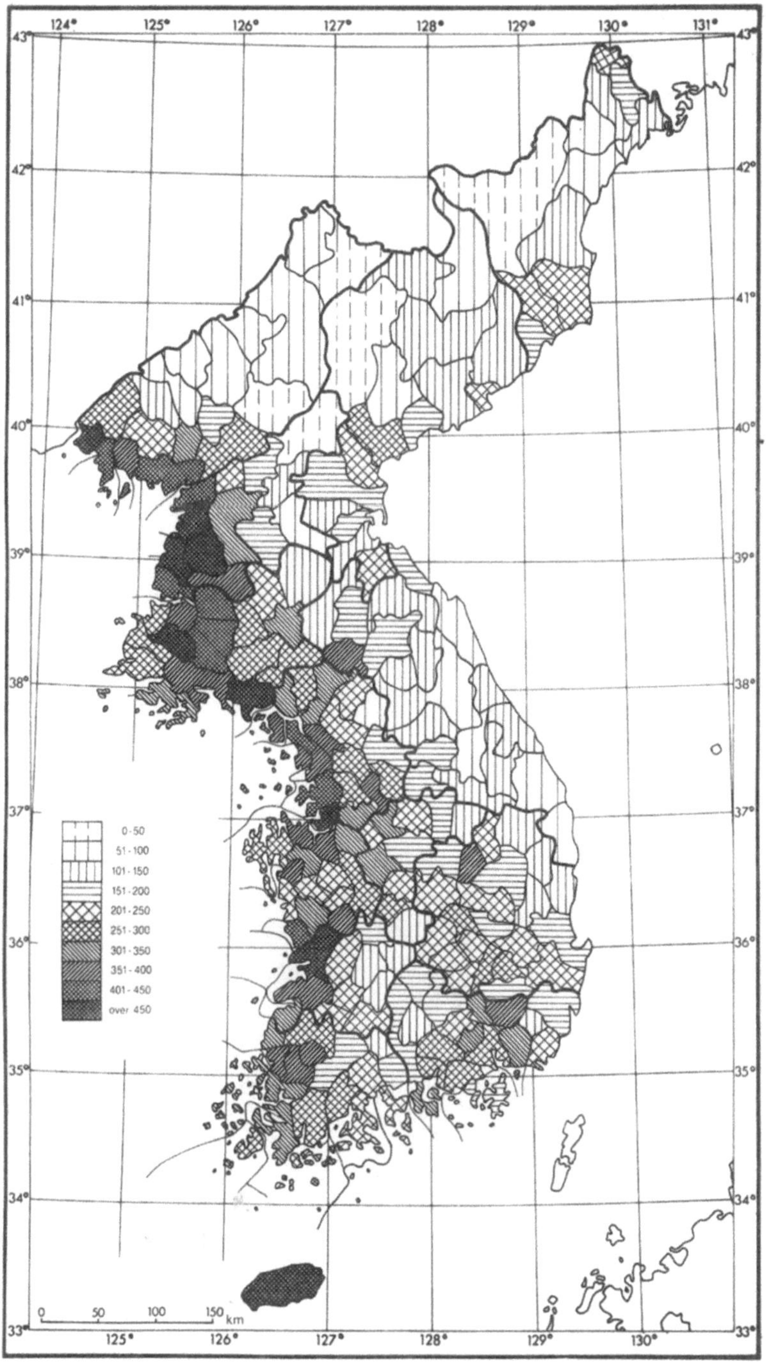

Fig. 86. Proportion of cultivated land in the total area of the counties in per thousandths (1930)

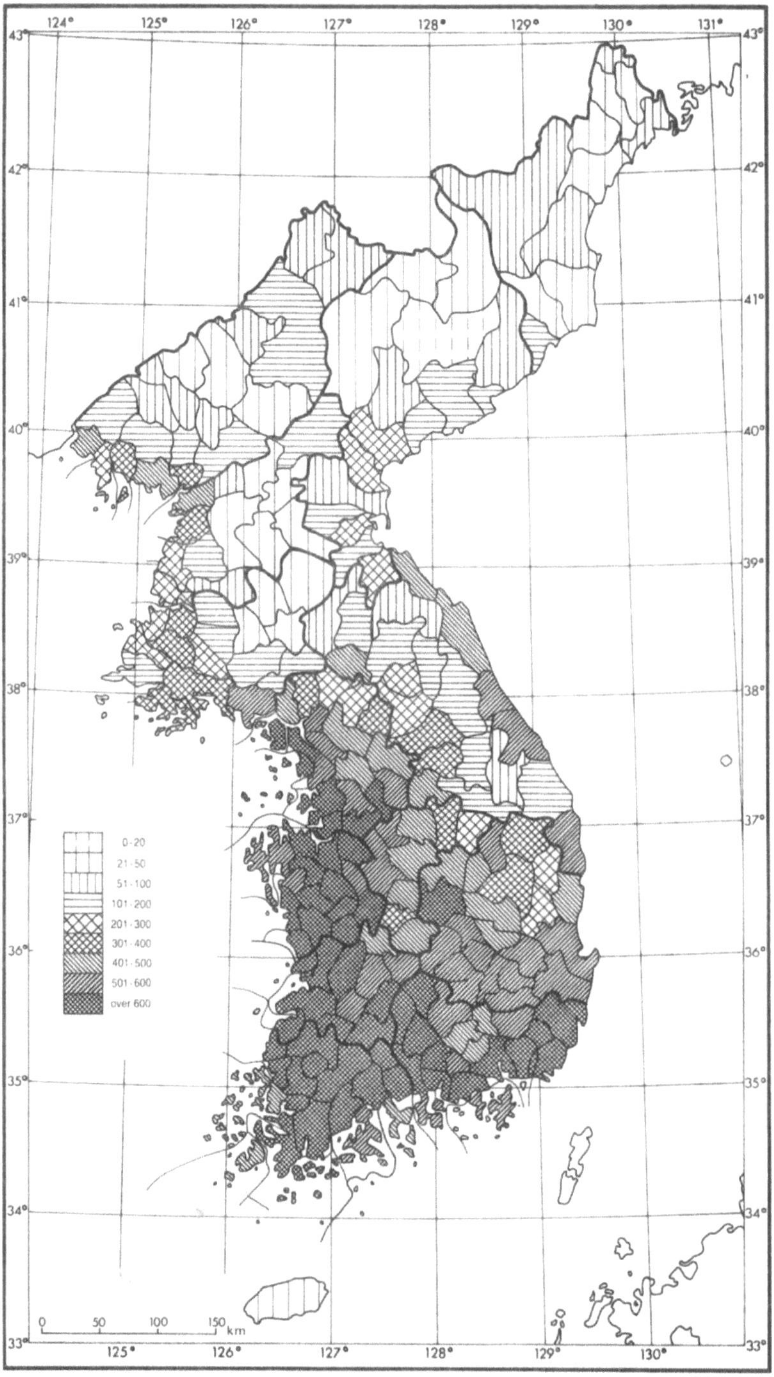

Fig. 87. Proportion of paddies in the total cultivated area by counties in per thousandths (1930)

Increasing Korea's rice production was one of the main economic goals of the government–general for 2 decades. The admirable results were achieved by three measures: selection of varieties, expansion of irrigation works and enlargement of the paddy area [618].

The climatic conditions, particularly in central and northern Korea, make a very careful selection of varieties necessary, although Korea does not yet lie at the northern limit of rice cultivation. Because of the warmth of summer, this boundary runs through Manchuria and the Ussuri region. The traditional Korean varieties have the disadvantage in the south that their yields are small and in the north that they are poorly adapted to the barely adequate temperature conditions. At the same time they are severely affected by insect pests and plant diseases. The Japanese agricultural experiment stations have made intensive efforts to select from the total number of known rice varieties the ones that are most advantageous for the individual Korean provinces, i.e., the ones that have the highest yields and are most resistant to damage by weather, disease and insects. These are being distributed to the farmers as seed to an increasing degree. In many provinces, especially in the south, the process of substituting these selected varieties for the native varieties has pretty much been completed. Here the Korean varieties have been driven back up into the valleys, where they still manage to scrape out a living for a short time in a state of vestigial endemism. In Kyŏngsangnam-do they were limited to 5% of the paddy area in 1930. In P'yŏng-annam-do, in contrast, they still occupied 82%. The national average was only 22% (maps in [304] and [620]). The exceedingly informative report by Nagai and Nakagawa on the most important varieties of paddy rice in Korea and their distribution (620) determines that in 1930 no fewer than 66 varieties of paddy rice were grown, each of which occupied an area of more than 1000 *chŏngbo* (992 ha), and contains maps of distribution for 51 of these varieties. The total number of varieties is very much greater still. In Table 33 we shall mention 13 of the most important of them, each with a ratio of more than 2% of the paddy area in 1930. Together the 13 varieties covered 77.2% of this area.

The medium-fast or late-maturing varieties predominate in the central or southern parts of the country and maintain a dominant position in relation to cultivated area. The varieties belonging to this group, 1, 2, 3 and 5, cover half of the paddy area. Following at a great distance are the rapidly maturing varieties (4, 6, 9), which are grown particularly in the northwestern coastal fringe and around Yŏnghŭngman. The varieties that mature latest (7, 8, 10) dominate the south coast. The fastest ripening varieties, which are simultaneously the ones with the lowest yields and are grown only in the north, cover such small areas that none of them is included in Table 33. The most characteristic of this group (Odayo) is limited to Hamgyŏngbuk-do and occupies only 0.6% of the paddy area.

The expansion of irrigation works is the work of irrigation associations, which were founded with strong support from the government–general. The leading association in this, as in many other economic spheres, is the Tōyo Takushoku Kabushiki Kaisha (p. 384). It has published a special report on the irrigation associations [619]. Immediately after it was founded, i.e., still during the time of the resident–general, the Rineki Company was also founded. The great majority did not follow until the period 1919–1931, after the government–general had started the "more rice plan". A law promulgated in 1917 regulates the organization of the Korean irrigation cooperatives, which have existed since olden times and were set up according to the principles of the *kye* (p. 403). In 1919 a further law proclaimed a government subsidy of 20% of the cost of improving existing irrigation works, 25% of the cost of converting dry fields into paddies and 30% of the cost of creating new paddies by clearing land in the interior or reclaiming land along the coast. Under the influence of these subsidies, e.g., 937 ha paddy land were obtained by clearing and 11,479 by reclamation in 1930 [605, p. 119 and 121]. Most of this work was done by those associations, which enjoy extensive rights in their areas. In 1926 there were 80 of them with an irrigated area of 116,292 ha, in 1932 191 with 219,929 ha, in 1937 189 with 215,345 ha. The most important of them, besides the ones already mentioned, are Fuji Nōson, Taishō, Ekiyoku and Chu-o.

Since small associations with an area of less than 10 *chŏngbo* (ha) are excluded from government support by the above mentioned law, only the wealthy and efficient associations, which are managed predominantly by Japanese, benefited from it. The large associations have been engaged firstly in constructing polders on the west coast with its large tidal range, two of which were described in the regional section of this book (p. 256, 359ff), and secondly in large-scale works in the interior. Figure 74 gives an impression of how strongly the latter have reshaped the natural landscape. In the mountains or hills modern dams with a large capacity have been built, some of which simultaneously provide electricity. Deep, wide

Variety of rice	Cultivated area (km²)	(%)	Characteristics	Main provinces of cultivation[a]	Trends in cultivation
1. Kokuryudō	3685	25.2	Ripens medium fast to late	VII, VIII, X, XI XII, XIII	Increasing strongly
2. Tamakin	1526	10.4	Ripens late, needs lots of water	SW of VII, VIII, XII, XIII	Almost constant
3. Soshinreki	1312	9.0	Ripens medium fast	S of VII and X, VIII	Decreasing very strongly
4. Kamenō	717	4.9	Ripens fast	S of II and III	Increasing strongly
5. Kumi	575	4.0	Ripens medium fast	IX, SW of VI	Increasing slightly
6. Hakuso	572	3.9	Ripens fast, needs little water	W and S of V	
7. Miyako	562	3.8	Ripens latest, very high yield	S of XI	Almost constant
8. Omachi	553	3.8	Ripens latest, very high yield	S of XIII	Increasing strongly
9. Hinode	448	3.1	Ripens fast to medium fast	S of IV, N of V and VI	Decreasing strongly
10. Chujukushin-reki	362	2.5	Ripens latest, very high yield	XIII	
11. Sekisanso	349	2.4	Ripens late, needs lots of water	W of XII	Increasing moderately
12. Benkei	302	2.1	Ripens late	N of XIII	
13. Chushinreki	300	2.1	Ripens late	S of VIII, XI	Increasing extremely strongly

[a] The Roman numerals indicate the 13 provinces in the standard order used in this book: I Hamgyŏngbuk-do; II Hamgyŏngnam-do; III P'yŏng-anbuk-do; IV P'yŏng-annam-do; V Hwanghae-do; VI Kang-won-do; VII Kyŏnggi-do; VIII Ch'ungch'ŏngnam-do; IX Ch'ungch'ŏngbuk-do; X Kyŏngsangbuk-do; XI Kyŏngsangnam-do; XIII Chŏllabuk-do; XIII Chŏllanam-do.

Table 33. The most important varieties of paddy rice in 1930 canals run in a straight line through the plains for long distances. In the rivers concrete weirs with iron floodgates regulate the diversion of water into the canals. The rivers are restrained by high levees to prevent the paddies from flooding. Instead of the wooden scoops used by the Koreans (p. 175), the much better and more efficient Japanese treadmills are used. They consist of a wheel with a ring of wooden boxes that dip under and scoop up water and are raised by the weight of the person treading on the periphery. Where large amounts of water need to be raised, steam or electric pumps are installed.

With these works the paddy area has not only been enlarged locally, but above all a good part of the paddies that were formerly dependent on groundwater or rain are now provided with irrigation water. This is shown by Table 34, which summarizes the irrigated areas by river basins (after [453]).

We said already on p. 175 that at the end of the Korean dynastic period only around one-sixth of the rice paddies were irrigated by flowing water and could therefore be covered with water for any desired period of time. The table lists this proportion as 17.1%. It shows that the most important irrigation areas were lined up along the smaller Yellow Sea rivers. In addition to the Yesŏnggang, the Kŭmgang, the Yŏngsan-gang and the Sŏmjin-gang there is the Man-gyŏnggang,

River system		Irrigated fields		Irrigation systems							
	Drainage area (km^2)	% of drainage area	km^2	Number	Area (% of irrigated fields)	Number	Area (% of irrigated fields)	Number	Area (% of irrigated fields)	Number	Area (% of irrigated fields)
Taeryŏnggang	3635	4.8	176.0	54	3.1	3	44.7	1	1.8	58	49.6
Ch'ŏngch'ŏn-gang	9778	2.4	230.7	200	7.2	–	–	1	2.7	201	9.9
Taedonggang	16,465	2.7	449.7	284	6.3	6	19.1	6	4.1	296	29.5
Chaeryŏnggang	3671	10.2	372.2	112	9.9	2	33.6	2	0.9	125	44.4
Yesŏnggang	4049	5.6	226.1	349	22.7	1	1.0	4	2.3	254	26.0
Imjin-gang	8118	6.2	507.2	653	12.0	9	25.6	12	10.4	674	48.0
Han-gang	26,355	6.1	1603.4	2777	10.8	11	4.9	11	1.5	2799	17.1
Kŏmgang	9886	15.2	1507.4	1564	15.1	11	3.7	6	0.3	1581	19.0
Man-gyŏnggang	1602	27.5	440.6	?	?	5	33.4	4	2.2	9	35.6
Yŏngsan-gang	2798	22.4	626.5	394	27.6	2	0.8	4	1.0	400	29.4
Sŏmjin-gang	4897	12.8	628.8	1797	19.8	4	2.3	–	–	1801	22.1
Naktonggang	23,860	11.0	2633.3	3718	16.9	21	6.4	16	1.2	3755	24.5
Yonghŭnggang	3397	1.5	49.6	2	10.9	–	–	2	3.0	4	13.0
Sŏngch'ŏn-gang	2338	7.1	165.7	75	34.1	–	–	1	0.2	76	34.3
Others	99,892	6.1	6094.7	14,962	18.2	42	9.9	143	4.6	15,147	32.7
Korea total	220,741	7.1	15,711.8	26,850	17.1	117	9.5	213	2.9	27,180	29.5

Table 34. Irrigated fields in 1928, by river basin

for which there are no figures for the traditional period. But even then the finest rice paddy plain lay in its basin (p. 354). On the whole there was a large number of very small associations. Each had an average of only 10 ha. The new associations in the table have an average size of 1273.8 ha, in comparison. They have substantially expanded the irrigated area above all in the river basins of the Taeryŏnggang, the Chaeryŏnggang, the Imjin-gang and the Man-gyŏnggang. The works set up by individuals since the occupation have an average size of 210.9 ha. Therefore, although there are 213 of them, they have expanded the irrigated area by only 2.9%. They lie particularly in the basins of the Imjin-gang, the Naktonggang and the Taedonggang. All in all 29.5% of the paddies were irrigated in 1928. This proportion has certainly increased since then and probably amounts to about one third. The official report of the government–general [VI] records 68%, but this cannot be correct unless all fields served by the primitive Korean groundwater scoops are included. This type of irrigation does not work, however, during long periods of drought. The areas that are actually irrigated have thus doubled under the influence of the country's development. That is a tremendous accomplishment. The curve representing the annual yields (Fig. 84) shows, however, how strongly the rice crop still depends today on the character of the weather of an individual year, that is, what weight the two thirds of the fields that are not irrigated, i.e., are only occasionally flooded by rain or groundwater, still have.

By building embankments along rivers that are prone to flooding, the government–general itself has done a great deal to increase the yields of the alluvial plains, which were previously endangered time and again due to the senseless deforestation of the country (p. 119f). In an 11-year program beginning in 1925, six large Korean rivers were controlled. The measurements and plans in the three volume publication on Korean rivers [453] form the basis for this and other projects. Smaller rivers have been controlled by the provincial governments with financial support from the government–general. When in the summer of 1936 the Naktonggang and other southern and central Korean rivers wreaked renewed havoc following excessive rainfall, radical large-scale measures were decided upon, and the Japanese Diet approved considerable amounts of money to execute them. Altogether no less than 300 million yen have been spent by the Japanese to get rid of this evil. This corresponds to almost one half of the entire national budget for 1939/40.

All of these Japanese measures permitted a very significant expansion of the rice-growing area. Above all, northern Korea was now incorporated into the Korean rice-growing area, whereas previously the emphasis had been on southern Korea alone. According to the old Korean censuses of 1404 and 1769 76% of the rice-growing area was in the six present-day southern provinces, in 1930, in contrast, it was only 61%. The various figures on the size of the paddy area for all of Korea in the individual years after the occupation vary quite considerably up till the present. In 1910 it probably amounted to 1.34 million ha. According to The Japan Year Book it had increased to 1.73 million ha in 1938, i.e., 7.8% of the national territory or 35.3% of the total cultivated area. Thus the increase compared with 1910 is 29%.

Rice cultivation has simultaneously been intensified. This was done by selecting better varieties, expanding irrigation and improving farming methods, in particular by more effective use of fertilizers. In southern Korea green manuring with alfalfa is now very common on paddies not planted to a dry-field crop in winter. The red blossoms of this nitrogen-fixing plant have become a characteristic feature of southern Korean paddies in winter. Simultaneously, the use of commercial fertilizers has been introduced, particularly ammonium sulfate, which is now produced in large amounts in the country itself (p. 451). The yields per hectare have increased substantially as a result. They amounted to 11.6 q/ha in 1910, 18.3 in the poor year of 1934 and 30.8 in the record year of 1937. The yield has thus increased considerably more than the harvested area. It was 15.6 million q in 1910, 31.0 million in 1934 and 50.1 million in 1937. The curve representing rice production in Fig. 84 shows this increase. Because the expansion of irrigation works is far from finished, this increase is interrupted by drastic reverses in dry years. In the average of the 6 years from 1934–39 the rice harvest was no less than 137% above that of 1910. Since under the pressure of the agrarian crisis the Koreans limited their rice consumption more and more, despite this impressive increase in the harvest (p. 400), a tremendous amount became available for export, particularly to Japan. The surplus for export averaged 11.9 million q during the period 1934–38, in 1910, in contrast, only 1.02 million q. Thus it increased almost twelvefold. As the figures in the Japan Manchoukuo Year Book 1940 show, exports were even higher than domestic consumption in 1934 and 1935.

Table 32 shows the proportion of paddies, and thus practically of all rice fields, in the provincial area. Foremost are the provinces of the northwestern part of southern Korean, Ch'ungch'ŏngnam-do and Chŏllabuk-do, in which paddies cover more than one fifth of the area. Hamgyŏngbuk-do has the smallest ratio. The short, cool summers, combined with the relief, restrict rice growing primarily to small, locally favored areas near the coast. Map 4 (cf. editors' note on p. 7) illustrates the absolute distribution of paddies at the present time.

Table 35 gives the proportion of paddies in the total cultivated area of the provinces. It shows that paddy rice is Korea's most important crop by far. Only in the two Hamgyŏng provinces is the paddy area much less than one-fifth of the total cultivated area. In four provinces of the central western part and the extreme southeast, namely Kyŏnggi-do, Ch'ungch'ŏngnam-do, Chŏllabuk-do and Kyŏngsangnam-do, more than one-half of the cultivated area is occupied by paddy rice. Chŏllabuk-do has the peak amount with more than seven-tenths of the cultivated area devoted to paddies. Chŏllabuk-do is simultaneously, however, the province with the greatest misery among the farmers and the province of frequent unrest.

Figure 87 shows the distribution of paddies across the country more exactly than these figures do. It shows the proportion of paddies in the total cultivated area by county. The counties in the Kaema upland and the ranges surrounding it have the fewest paddies. Toward the coast of the Sea of Japan and above all toward the Yellow Sea coast the intensity increases. In some northern Korean coastal areas along the Yellow Sea almost half of the cultivated area is already occupied by paddies. The low degrees of intensity of the Kaema upland continue on into the mountainous interior of central Korea. In the coastal fringes here the intensity is, however, incomparably greater. This is most particularly true of the Yellow Sea fringe. Absolutely, the paddy areas in P'yŏng-annam-do and Hwanghae-do are already quite large; but the for Korea unique extent of dry

1930	Farmed land (km^2)	Paddy rice	Upland rice	Barley	Naked barley	Wheat	Rye	Foxtail millet	Barnyard millet	Broomcorn millet	Kaoliang
I. Hamgyŏngbuk-do	2378	5.4	0.25	19.0	.. [a]	0.1	–	29.0	4.48	0.94	1.7
II. Hamgyŏngnam-do	4673	12.2	0.63	7.6	..	1.2	–	18.1	7.38	0.96	3.2
III. P'yŏng-anbuk-do	4715	19.9	0.48	1.3	–	0.05	2.2	25.1	3.27	0.63	4.0
IV. P'yŏng-annam-do	4172	16.9	2.01	4.1	1.08	9.3	1.9	29.5	1.12	0.45	5.4
V. Hwanghae-do	5575	24.2	0.73	2.6	0.34	23.4	0.6	31.2	0.85	0.31	2.3
VI. Kang-won-do	3754	24.6	0.21	9.8	0.13	8.4	0.2	20.9	2.20	0.18	0.9
VII. Kyŏnggi-do	3877	53.5	0.89	22.6	0.28	6.2	0.4	8.0	0.95	0.09	2.0
VIII. Ch'ungch'ŏngnam-do	2440	65.8	1.09	25.1	1.36	5.0	1.8	0.9	0.09	0.06	0.7
IX. Ch'ungch'ŏngbuk-do	1610	43.9	0.57	39.9	0.13	10.4	-	9.9	0.05	0.37	1.6
X. Kyŏngsangbuk-do	3855	48.8	0.38	50.7	3.00	10.5	5.0	1.3	0.03	0.13	0.3
XI. Kyŏngsangnam-do	2605	61.5	1.53	55.7	5.31	6.0	8.8	0.9	0.06	0.10	0.2
XII. Chŏllabuk-do	2384	71.5	0.38	21.1	5.90	4.0	10.0	1.0	0.13	0.09	0.3
XIII. Chŏllanam-do	4516	49.5	1.27	28.3	5.82	3.8	9.6	8.5	0.50	0.06	0.6
XIV. Korea total	46,555	36.2	0.82	19.0	1.66	7.4	3.0	16.8	1.83	0.35	2.0
XV. Cultivated area ranked for Korea as a whole		1	17	2	15	5	8	4	14	19	12

[a] - Crop not cultivated; .. Cultivated area less than 0.01 %.

1930	Maize	Oats	Buckwheat	Potatoes	Sweet potatoes	Soybeans	Adzuki beans	Green beans	Kidney beans	Peas	Cotton	Hemp	Ramie	Paper mulberry	Perilla
I.	1.5	6.2	0.5	4.45	..	26.5	14.8	0.21	1.92	0.91	–	0.72	–	–	0.35
II.	1.3	16.9	2.8	8.77	..	12.1	5.0	0.13	0.31	0.12	..	0.68	–	..	0.28
III.	11.5	0.5	3.6	1.48	0.07	15.5	0.7	0.61	0.08	0.01	2.0	1.15	–	0.01	0.30
IV.	4.9	0.2	3.8	1.33	0.20	12.8	10.8	0.88	0.01	..	7.3	0.25	–	0.05	0.11
V.	0.9	0.1	2.7	0.34	0.26	16.8	11.0	3.21	0.01	..	2.3	0.13	–	0.03	0.14
VI.	3.9	2.5	5.9	3.47	0.02	19.0	6.4	0.45	–	–	1.0	1.42	..	0.09	0.55
VII.	0.1	0.5	1.4	0.65	0.16	21.8	3.4	1.07	0.09	0.02	0.9	0.24	..	0.05	0.50
VIII.	0.2	0.2	0.9	0.37	0.25	18.5	2.7	0.85	0.10	0.01	4.9	0.40	0.18	0.01	0.42
IX.	0.3	0.1	1.7	0.79	0.06	20.3	4.2	0.99	0.01	..	8.5	0.23	–	0.28	0.68
X.	0.3	..	0.8	0.61	0.12	25.3	0.9	0.30	..	..	5.8	0.64	..	0.86	0.16
XI.	0.1	–	0.6	0.46	0.78	17.6	1.7	0.50	0.02	0.03	8.6	1.38	0.03	0.37	0.23
XII.	0.1	–	0.7	0.62	0.55	13.2	2.5	0.46	0.01	0.02	3.3	0.50	0.55	0.66	0.24
XIII.	0.1	..	1.5	0.07	1.57	7.4	1.4	0.55	..	0.34	11.7	0.41	0.10	0.05	0.13
XIV.	2.3	2.3	2.3	1.91	.0.31	16.9	5.1	0.89	0.15	0.10	4.1	0.61	0.03	0.15	0.28
XV.	11	9	10	13	20	3	6	16	22	24	7	18	25	23	21

Table 35. Acreages of most crops, by province, in percentage of the total cultivated area of the provinces

fields (Table 32) lowers the percentages here. At the mouth of the Han-gang the very high intensities of the southern Yellow Sea fringe begin. The southwest and the south stand out massively as the region of highest intensity (more than 60%). The county with the highest figure is Okku-gun near Kunsan (86.6%). The peninsulas and islands in the extreme south again have lower degrees of intensity. They are very mountainous and without large rivers and plains that are suitable for irrigation works. Because its basalt flows are so porous, the island of Chejudo stands way out with only 0.9%.

The figures on rice production given above include upland rice. The crop of 1935, however, amounted to only one 65th of the total rice harvest. The yields per hectare are much lower than the yields for paddy rice. Nevertheless, it is still grown in all Korean provinces today, most strongly in P'yŏng-annam-do, where the permeable soils on limestone highly limit artificial irrigation. Here 11.4% of the total rice growing area is even devoted to upland rice. The average for the country is only 2.5%. Still the area planted to upland rice increased relatively more strongly than the paddy area during the second and third decades of this century, because the rent paid for dry fields is much lower than that paid for irrigable land, so that upland rice can be produced more cheaply. In communities in which the topographical and geological conditions make it impossible to grow paddy rice, cultivation of upland rice is also the only possible way of obtaining the coveted foodstuff. Add to this the fact that Japanese research has provided the Koreans with an excellent variety of upland rice, Oiran. In 1935 the value of the total rice production had reached no less than 71% of the value of the agricultural production.

Although the efforts of the government–general have always been aimed primarily at rice growing, the great transformation is very clear in the permanent dry fields as well. In the country's agriculture as a whole they still lead by area. In 1910 their area probably amounted to about 2.0 million ha. In 1938 the 1.73 million ha of paddies mentioned above contrasted with 2.74 million ha of dry fields. The increase thus amounted to one third. The distribution of the permanent dry-field area by province can be seen again from Table 32. The following four types can be identified: (1) provinces with a small dry-field area due to their mountainous nature (Hamgyŏng-do, P'yŏng-anbuk-do, Kang-won- do); (2) those with a small dry–field area because paddies predominate (Kyŏngsangnam-do, Ch'ungch'ŏngnam-do, Chŏllabuk-do); (3) provinces with a very large dry-field area (P'yŏng-annam-do, Hwanghae-do); (4) provinces with an equal area of paddies and dry fields (Kyŏnggi-do, Ch'ungch'ŏngbuk-do, Kyŏngsangbuk-do, Chŏllanam-do).

As in traditional Korean times, many paddies and dry fields bear two crops a year today, too. Indeed, the frequency of double crops has increased strongly under the influence of Japanese management and in connection with the intensification of agriculture. On dry fields it is intercropping (p. 183f) that leads to two crops; on paddies it is double cropping of our native grains in winter (p. 176). A measure of the frequency of double crops can be obtained by dividing the sum of the harvested area of all individual crops by the actual cultivated area. By multiplying the quotient by 100 and subtracting 100 from the result, one obtains the percentage of the total cultivated area that is double cropped. Using the figures reported by Hoon K. Lee for 1930 [605, p. 111 and 150] we arrive at the percentages shown in Table 36.

Double usage of a field during the course of a year is thus much greater on dry fields than on paddies, where intercropping is impossible. It is more widespread in the south with its milder winters than in the north with its short summers. It is most frequent on both paddies and dry fields in Kyŏngsangnam-do.

It would lead us too far afield to deal with each of the dry-field crops here again. The reader is referred to the remarks regarding traditional Korea on pp. 177–182. For the rest I shall limit myself here to the cases in which the Japanese influence is particularly strong and exhibits special geographical expression. In Fig. 88 the distribution of the cultivated areas of the nine most characteristic dry-field crops is represented by dots, and Table 35 gives the proportion of each of the 25 most important crops in the cultivated area of each province in percent. This table shows the situation in 1930, since no more recent data were available for the provinces. Because of the double crops the sum of the percentages exceeds 100 everywhere, in some places quite considerably. In Kyŏngsangnam-do, for instance, the sum of paddy rice and barley alone amounts to 117.2. Finally, Table 37 shows the total yield and yield per hectare for the entire country for 1935 (p. 420, after [XLVI]).

By acreage and harvested amount barley is second of all crops. Both are only about half as large, however, as the area and amount of paddy and upland rice together. Figure 84 shows that the harvested amount has almost tripled since 1910. Barley is today, along with rice and the millet varieties, Korea's most important food grain (p. 400). The Japanese agricultural experiment stations have tested the suitability of 33 native and 28 Japanese varieties for cultivation in the different Korean regions [624]. The majority of the Korean varieties have proven to be more resistant to the winter cold than the Japanese ones; they are therefore preferentially grown as a winter crop. Of the total cultivated area 24.7% is on paddies that are drained in winter. In Kyŏngsangnam-do this percentage rises to 44.4% (1930). The main areas where it is grown are in the southern half of the country, as Table 35 and Fig. 88 show. In Kyŏngsangbuk-do the area planted to barley even exceeds the paddy area. Here and in Kyŏngsangnam-do it comprises more than half of the total cultivated area.

	Paddies	Dry fields		Paddies	Dry fields
Hamgyŏngbuk-do	0	19	Ch'ungch'ŏngnam-do	12.9	78
Hamgyŏngnam-do	0.1	13	Ch'ungch'ŏngbuk-do	21.5	83
P'yŏng-anbuk-do	0	21	Kyŏngsangbuk-do	46.4	83
P'yŏng-annam-do	0	24	Kyŏngsangnam-do	48.4	84
Hwanghae-do	0.6	31	Chŏllabuk-do	33.2	63
Kang-won-do	1.3	32	Chŏllanam-do	38.8	46
Kyŏnggi-do	1.5	5			
			Korea total	21.2	39

Table 36. Frequency of double crops in percentage of the paddy and dry–field area

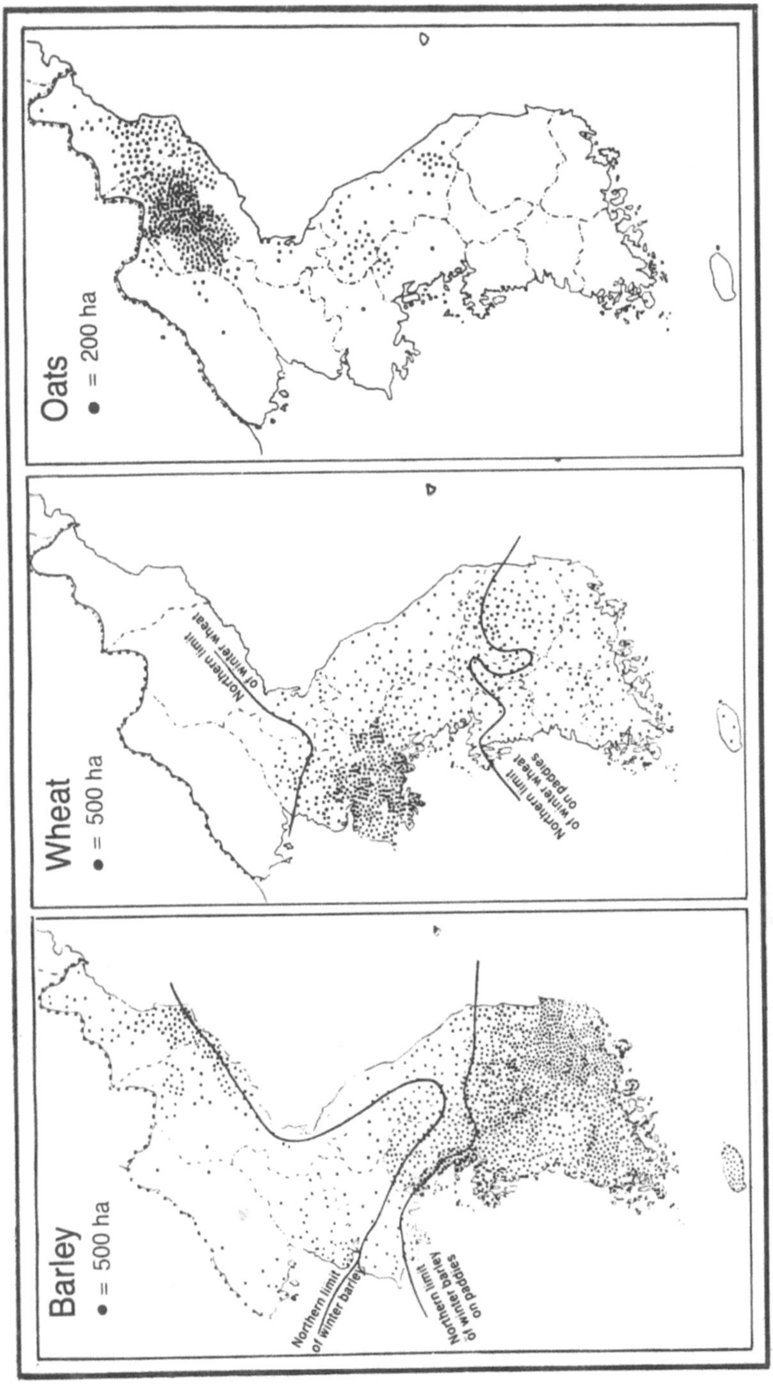

Fig. 88. Distribution of the areas cultivated with nine characteristic dry-field crops (1930 and 1934, after *Thriving Chosen* and Nagai, Nakagawa and Takasaki)

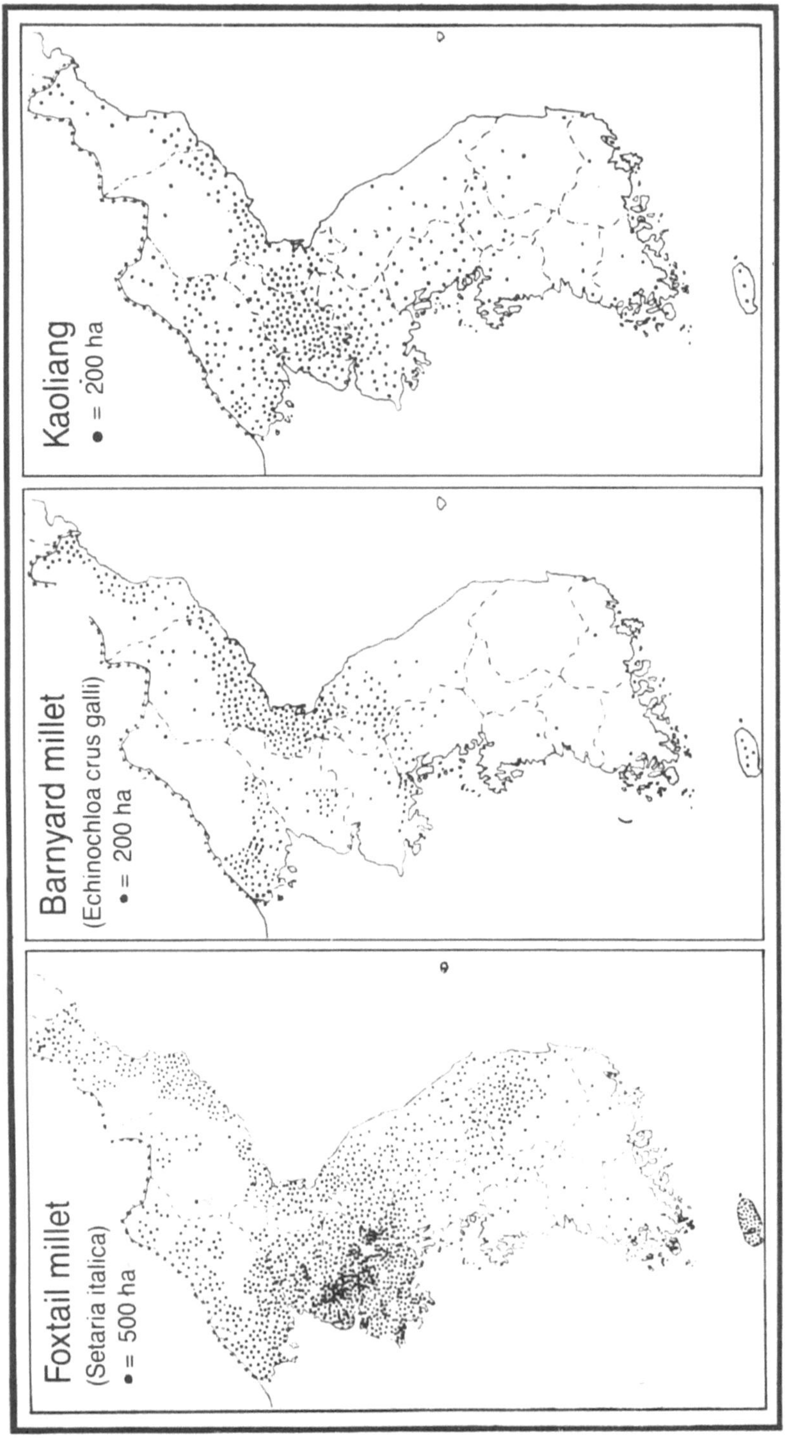
Foxtail millet
(Setaria italica)
● = 500 ha
Barnyard millet
(Echinochloa crus galli)
● = 200 ha
Kaoliang
● = 200 ha

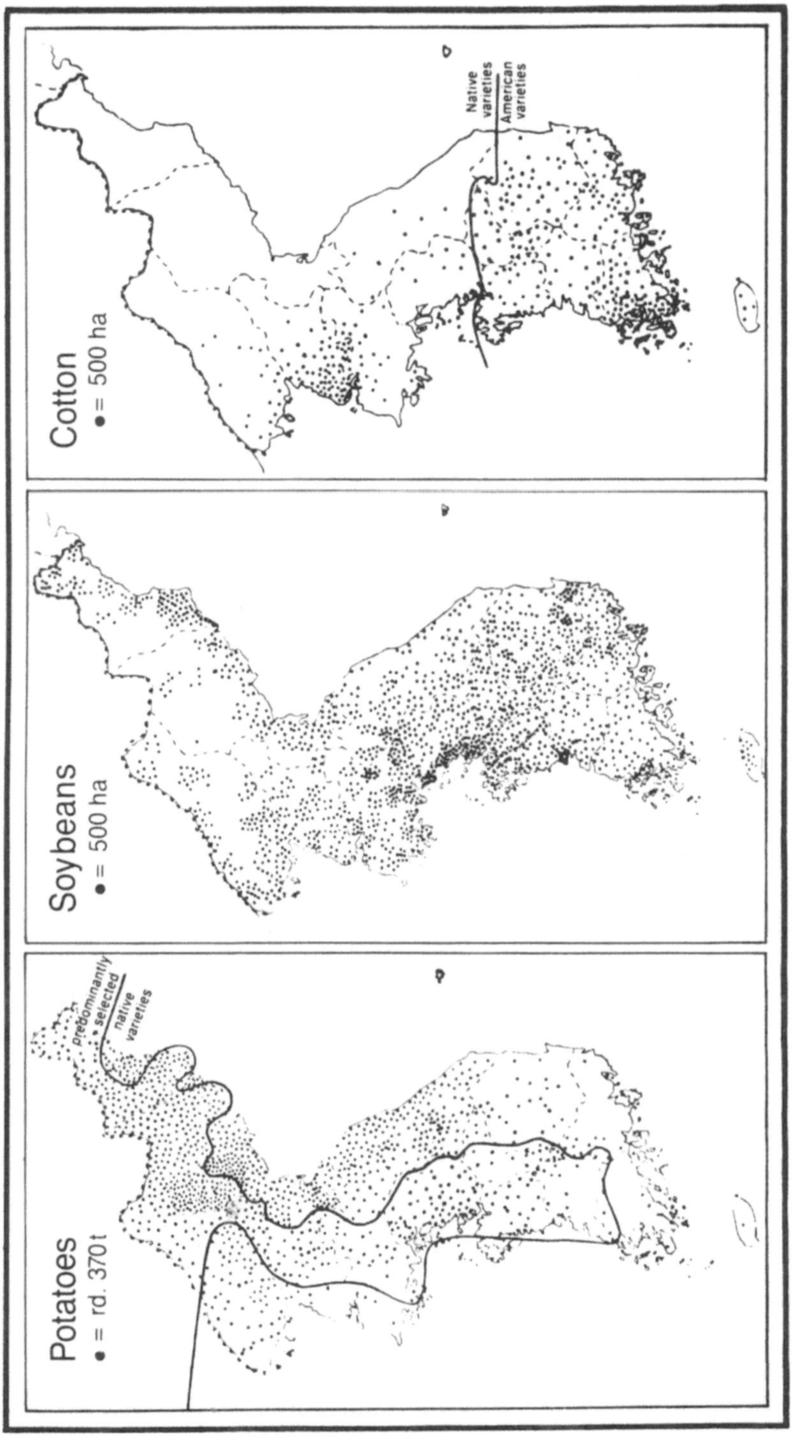
Cotton
● = 500 ha
Native varieties
American varieties
Soybeans
● = 500 ha
Potatoes
● = rd. 370 t
predominantly selected
native varieties

Winter cropping of naked barley, which brings a much higher price, is increasing strongly in southern Korea under Japanese guidance. The most resistant varieties of this plant, which is sensitive to cold, are preferred. On the national average, exactly one third of the area planted to naked barley lies on paddies, in Kyŏngsangnam-do even 56%. The cultivation is highly concentrated, for the reasons we have mentioned, in the four southernmost provinces, which contain 87.1% of the total acreage.

Only 8.4% of the wheat is grown on paddies, on the national average, and in connection with this it is not nearly as highly concentrated in the southern half of the country as the two preceding grains (Table 35 and Fig. 88). Rather, 49.4% of it is grown in P'yŏng-annam-do and Hwanghae-do alone. The northern limit of winter wheat now lies even farther north than that of winter barley, because the Japanese have developed varieties that are very resistant to cold. The acreage has decreased slightly since the government–general restricted the use of wheat for producing spirits in 1927.

In southern Korea barley has become the principal staple food under the influence of Japanese economic policy, whereas a large portion of the rice is sold. In Hwanghae-do and P'yŏng-annam-do, the second large center of cultivation, wheat plays the role of rice, and the millet varieties provide the most important staple food. A considerable amount is even imported from Manchuria [624]. The size of the area planted to the three millet varieties is extremely varied (Table 35.) Whereas the area of foxtail millet is not far behind that of barley, the area of barn-

1935	Cultivated area (1000 ha)	Harvest in: Million units of weight	Million yen	Yield per ha
Paddy rice	1642	31.8[a]	489.6	19.4[a]
Upland rice	38	0,5[a]		13.2[a]
Barley	870	15.8[a]	73.3	18.2[a]
Naked barley	162	2.9[a]	21.2	17.9[a]
Wheat	324	3.5[a]	24.5	10.8[a]
Foxtail millet	788	8.8[a]	58.0	11.2[a]
Barnyard millet	72	0.83[a]	3.60	11.5[a]
Broomcorn millet	15	0.13[a]	0.86	8.7[a]
Kaoliang	82	1.0	6.3	12.2[a]
Maize	118	1.29[a]	7.4	10.9[a]
Oats	116	0.75[a]	2.6	6.5[a]
Buckwheat	125	1.13[a]	6.1	9.0[a]
Potatoes	109	6.0[b]	19.8	55.0[b]
Sweet potatoes	24	2.1[b]	8.5	87.5[b]

1935	Cultivated area (1000 ha)	Harvest in: Million units of weight	Million yen	Yield per ha
Soybeans	785	7.9[a]	60.2	10.1[a]
Adzuki beans	233	1.7[a]	14.3	7.3[a]
Mung beans	38	0.24[a]	2.3	6.3[a]
Radishes and turnips	62	6.1[a]	12.7	98.4[b]
Cotton	208	128.2[a]	36.3	616[c]
Hemp	27	19.0[c]	6.2	704[c]
Ramie	2	0.57[c]	0.69	326[c]
Paper mulberry	7	6.9[c]	1.2	955[c]
Sesame	10	0.07[a]	0.96	7.0[a]
Perilla	13	0.11[a]	0.93	8.6[a]
Tobacco	17	0.22[b]	47.3	12.9[b]
Ginseng	12,5[d]	6.1[c]	2.6	
Total	4271		733.4	[a]

[a] Hectoliter, [b] Quintal (100 kg) [c] Kilogramm, [d] Are,

Table 37. Cultivated area, harvested amounts and yields per hectare of the most important crops, 1935

yard millet is not even one-tenth of the former, and broomcorn millet manages only one-third of the latter. Figure 88 shows that the cultivation of foxtail millet, similar to wheat, is concentrated in Hwanghae-do and P'yŏng-annam-do. Together they have 38.0% of the cultivated area, while in some southern Korean provinces the area planted to foxtail millet is negligible. The cheap barnyard millet is decidedly a grain of the northern Korean coastal fringes. The two Hamgyŏng provinces alone have 53.1% of its area.

In potato cultivation, too, Japanese research has brought substantial results. The "domestic" varieties, which probably derive from the tubers brought by Gützlaff (p. 47), have been pretty well replaced by "foreign" ones that are better adapted to the climate (Fig. 88). In 1930 52% of the acreage was still covered with "domestic" varieties. The fire-field province of Hamgyŏngnam-do leads, with 46% of the total area that is planted to potatoes.

The soybean vies with foxtail millet for third place in the order of crops, by cultivated area. It is a very profitable, undemanding crop that does not fail even in meteorologically abnormal years. Therefore its harvest curve does not have the ups and downs that are characteristic of many other crops, particularly paddy rice (Fig. 84). Acreage and yield per hectare have not increased nearly as much during the Japanese period, however, as for many other crops. Nevertheless, soybeans are the third most important item of agricultural export today, and the quality is supervised by the state. Approximately one-fourth of the harvest is exported. The most important export markets are Unggi, Ch'ŏngjin, Sŏngjin, Wonsan, Shinŭiju, P'yŏng-yang, Kŭmch'ŏn and Ch'ŏlsan [600]. They are located without exception in the northern half of the country.

The Korean adzuki bean, particularly that from the northeastern coast, is highly prized on the Japanese market because of its high quality and is exported to there from Ch'ŏngjin, though most of the production is for domestic consumption. By acreage it is the sixth most important crop. Of this area 59.4% lies in the three northern Yellow Sea provinces.

Another economically very important crop is cotton. Japan is known to have a very high demand for imports. It was therefore natural to partially transfer this raw material basis to the neighboring foreign possession, where a large portion of the domestic demand for textile raw materials was already met during the Korean dynastic period by growing cotton (p. 154). The Japanese agricultural experiment stations, particularly the one in Mokp'o, provided the scientific basis for the reformation of Korean cotton growing. Because their fibers are so short, the Korean varieties are poorly suited to factory processing by machines, and because of their coarseness they are not very useful for the production of quality merchandise. It turned out, however, that southern Korea was climatically well suited for growing American varieties of the group Kings Improved. In the six southern provinces, i.e., south of the boundary shown in Fig. 88, the native varieties have been completely supplanted by American ones. North of this line, which coincides approximately with the 20° C isotherm of the September mean temperature, the faster maturing native varieties have been retained. The acreage was systematically enlarged by the Japanese from 59,000 ha (1910) to 233,000 ha (1938), in particular by expanding cultivation in the southern provinces. Model cotton plantations and active propaganda in the newly founded cotton cooperatives served this purpose. The harvested amount slowly increased from 50,000 q of ginned cotton in 1910 to 295,000 in 1934. In 1933 a 10-year plan was launched to further increase cotton cultivation in all provinces, with the exception of the two Hamgyŏng provinces. The acreage is to be increased to 345,000 ha and production to 2.5 million q. According to Japanese statistics, there has actually been a strong increase since then. The data in the *Annuaire International de Statistique Agricole,* which were used for Fig. 84, indicate this rise only to a limited degree. The map in Fig. 88 shows that there are two centers of cultivation. The three northern Yellow Sea provinces had 27.4% of the acreage in 1930 and grew native varieties almost exclusively. The six southern provinces had simultaneously 68.6% of the acreage with 75.5% of the total production and worked almost entirely with American varieties. American cotton and its products are among Korea's biggest export articles (p. 476f). They go from Kunsan, Mokp'o, Yŏsu and other southern Korean ports to Japan.

The cultivation of tobacco is supervised completely by the state today, as the manufacture of tobacco products has been one of the four state monopolies since 1921 (p.31,179). Three varieties are grown, namely, Korean, Japanese and yellow, or American, tobacco. The first type outweighs the others by far in area and amount. The cultivation of American varieties is a specialty of Ch'ungch'ŏngbuk-do (p. 357). The Korean varieties are now grown separately in the back valleys of the mountains in P'yŏng-annam-do and Hwanghae-do and in the counties of Muju-gun, Yŏngdong-gun and Kŭmsan-gun. Only certain farmers are permitted to grow tobacco. They belong to tobacco cooperatives, which cooperate with the Office of Monopoly. The seeds are provided by branch offices, which draw them from a very limited number of farmers. The entire harvest must be delivered up after drying. In the tobacco-growing regions one often sees tobacco warehouses built of corrugated iron belonging to the Office of Monopoly. The total acreage reached a figure of 19,507 ha in 1938 with a production of 0.293 million q and a value of 61.9 million yen.

Another state monopoly is the production of red ginseng. Ginseng cultivation, as described on p. 182, is found especially in the environs of the old capital of Koryŏ, Kaesŏng. The name Koryŏ Ginseng is restricted to the products of the counties of Kaesŏng-gun, Kaep'ung-gun and Changdan-gun (Kyŏnggi-do) and Kŭmch'ŏn-gun, Sŏhŭng-gun, P'yŏngsan-gun and Pongsan-gun (Hwanghae-do, [632]). I also encountered a large number of ginseng plantations on the boundary between central and southern Korea in the counties of Ch'ungju-gun, Tanyang-gun, Yŏngju-gun and Andong-gun and in the county of Muju-gun on the upper Kŭmgang. The state monopoly developed out of the ancient crown rights (p. 33). The Japanese have investigated the pests and diseases of cultivated ginseng and given appropriate orders for its cultivation. For the cultivation of ginseng today, like tobacco, a special permit from the government–general is required. The producer hands over all roots to the state collecting center, which selects the roots that are suitable for producing red ginseng and returns the others for the production of white ginseng. The Japanese have developed ginseng production from small beginnings to a profitable branch of the economy. In 1911 the acreage was 473 ares with a harvest of 4631 kg, 1380 kg of which were used for producing red ginseng. The receipts from the ginseng monopoly amounted to only 0.1 million yen at that time. For 1931 the corresponding figures are: 12,511 are, 91,238 kg, 23,892 kg and 1.8 million yen. Since then a decline has set in, because the war with China began in 1937, and the majority of the exports went there (p. 182).

One of the agricultural activities that is particularly encouraged by the Japanese is the cultivation of mulberry bushes and sericulture. The area planted to mulberry bushes has been maintained at around 78,000 ha in recent years. Several of the agricultural experiment stations, such as those at Suwon and Ch'aryŏngwan [634], have special divisions for research and instruction in this branch of industry. Resarch has been devoted to the climatic resistance of Korean and Japanese varieties of mulberry, rate of growth of the leaves and how suited they are as food for silkworms. Further, the diseases of the silkworms and their causes have been studied, the sizes and qualities of the cocoons of the various species of silkworms have been determined, and considerable improvements have been made in quality and quantity by hybridization with Chinese and Japanese silkworms. The experiment station in Suwon has a very instructive museum, which shows these results, among other things. Since 1919 the government–general has passed legislation based on this research. Young girls are trained at silk schools for a year in the best methods of sericulture.

Sericulture is a domestic activity that is generally carried out on the side by farm households, and because of its profitability it is a suitable means of improving their lot. The eggs are laid by the moths on sheets of paper in 28 circular divisions per sheet and adhere firmly to the paper by themselves. The sheets with eggs are delivered to the raisers by state sericulture offices. In 1938 there were more than 1 million of them distributed to 817,000 families. The weight of the cocoons produced in the same year amounted to 26.3 million kg with a value of 21 million yen. The development began in 1910 with 76,000 families and a production valued at 0.4 million yen. There has thus been a fiftyfold increase. Figure 84 shows this development. Immediately after pupation the cocoons are delivered to the sericulture offices, which select the cocoons that are intended for breeding and have the rest of the pupae killed in a steam bath. Most of the cocoons then go to silk mills, which pass the raw silk on for export. The remaining cocoons are entrusted by the cooperatives to Korean families for hand reeling. This portion of the raw silk is for domestic use. The total weight of the raw silk was about 3 million kg in 1938, and the value of the exported raw silk was 15.6 million yen, that of the domestic silk 5.1 million yen. Sericulture is distributed across the entire country, everywhere where mulberry varieties find appropriate conditions for growth. It is very scarce in the two Hamgyŏng provinces, with the exception of the southern half of the North Korean Japan Sea borderland. On the whole it increases toward the south. The maximum production lies in the northwestern parts of Chŏllanam-do and Kyŏngsangbuk-do.

As we already pointed out on p. 185, the raising of wild silkworms is still common in P'yŏng-anbuk-do. In 1937 4257 ha of oak forest were used for this purpose, and the cocoon production amounted to 11.5 million. These are generally spring cocoons, which are exported to Manchuria. They produce tussah silk.

Whereas in traditional Korea cultivation of the tea plant was unknown, during the Japanese period small tea plantations have been established at <Ryōkurei> (for the significance of angle brackets cf. note in foreword) in Chŏllanam-do and on Chejudo. The obstacle to the expansion of these plantations is the sensitivity of the tea plant to cold winters.

The Japanese agricultural schools have also provided impulses for the cultivation of fruit trees, which have led to gratifying results. The yields of the Korean fruit trees have increased in both quality and productivity. Therefore exports increased from year to year until the beginning of the war with China. This is particularly true of the excellent Korean apples, whose yield amounted to 765,700 q in 1937; 150,000 q were exported in the same year. The most important areas of production lie in the southwestern part of P'yŏng-annam-do and the northern part of Hwanghae-do, in the North Korean Japan Sea borderland and in the boundary area between the two Kyŏngsang provinces. As a result of Japanese initiative, viticulture has flourished to a modest extent. In 1933 230,700 grapevines were counted with a yield of 24,500 q, used exclusively as fruit. The centers of production lie around the cities of P'yŏng-yang, Seoul, Taegu, P'ohang and Pusan. The total fruit harvest for 1937 is estimated at 1.02 million q with a value of 10.6 million yen.

From the outset the basis for the development of animal husbandry was somewhat broader than in the other East Asian countries, because the figures for some animals, especially cattle, and, compared with Japan, for pigs too, were already relatively high in ancient Korea (p. 186ff). Nevertheless, the Japanese authorities attached only secondary importance to increasing the number of livestock, and therefore this area is lacking in the enormous progress that characterizes many of the other branches of agriculture. This is shown by a comparison of the figures for 1914 and for 1937 in Table 38.

In all branches, with the exception of donkeys and mules, animal husbandry has made progress. But only in the case of pigs, sheep, goats and chickens is the increase greater than the population growth, so that the number of livestock in relation to population has increased. Otherwise there are fewer animals per 1000 inhabitants than in 1914, and by European standards the number of animals is still small for all species. But the quality has improved substantially as a result of Japanese efforts. This improvement took place primarily through the agency of the state-supported Chōsen Live Stock Association, which already had 213 individual branches in all parts of the country in 1923.

Even before the Japanese occupation cattle held an exceptional position, in that from the outset they were excellent both as draft animals and as meat producers, and their numbers were quite satisfactory even at that time. Japanese atten-

	In 1000 head		Number of livestock			
			per 1 km²		per 1000 inhabitants	
	1914	1937	1914	1937	1914	1937
Horses	52.50	52.6	0.24	0.24	3.50	2.40
Donkeys and mules	14.80	5.6	0.07	0.03	1.00	0.30
Cattle	1338.40	1710.0	6.10	7.80	89.80	76.30
Pigs	757.80	1625.1	3.40	7.40	50.90	72.50
Sheep	0.14	20.0	..	0.09	0.01	0.89
Goats	11.60	43.0	0.05	0.19	0.78	1.90
Chickens	4110.20	7221.1	18.60	32.70	276.00	322.00
Ducks	..	37.8	..	0.17	..	1.70

Table 38. Number of livestock in 1914 and in 1937

tion thus needed only turn to the maintenance of the high breeding standard and to an increase in the number of cattle. On the latter score the success has been rather incomplete because of the poverty of the Korean farmers thus far, although the government–general provided breeding bulls. Since 1938 a new 20-year plan has been in effect with the goal of increasing the number of cattle to at least 2.5 million. The demand for draft animals has grown with the increase in paddies, and beef consumption is increasing in the cities of Japan. The structure of the stock is completely different from that in Western countries, in that the number of female animals does not predominate to anything like the extent it does in Europe. The ratio between female and male animals is approximately 2:1 in Korea; in Switzerland, in contrast, it is 9:1 (Sang-uk Han [607]). The reason for this is that thus far very little milk is used in human nutrition. Recently there have been successful attempts at feeding dairy cattle with hay from lespedeza bushes, which grow rankly on fallow land [636, 637]. The density of the cattle population is about the same everywhere, with the exception of the Kaema upland. In southern Korea it is somewhat higher than in other parts of the country, because of the heavy demand for draft animals in rice farming. Since the north is much more sparsely populated, the number of animals per 1000 inhabitants is much greater there than in the south. In Chŏllabuk-do it is 36.7 and in Chŏllanam-do 60.1; in Hamgyŏngbuk-do, in contrast, it is 139.4, and in Py'ŏng-anbuk-do 139.9.

Japanese influence has been particularly beneficial in pig raising. Not only has it been possible to increase the number much more highly than the population has grown, so that the number of animals in relation to the population is now almost half again as large as in 1914. Above all, provision was made for the introduction of heavier, faster growing and fatter foreign breeds. In 1933 the number of these fat foreign pigs had already considerably outdistanced that of the small, uneconomical native animals. In Kyŏnggi-do, P'yŏng-annam-do and in the northern part of Hwanghae-do the native pigs have already been almost completely replaced by foreign ones; in the southern half of southern Korea, in contrast, the native animal still predominates by far (map of stock raising in [304]).

To promote the extremely scarce horse raising, the government– general has had Japanese stallions crossed with Mongolian mares on the state stud farms, particularly in Namgok (Kang-won-do) and Kyŏng-won (Hamgyŏngbuk-do). Since 1936 an 8-year plan has been in effect to cross good Japanese and Korean breeding mares with the best stallions. The expensive stallions all belong to the government–general itself. The hope is above all to be able to satisfy the demands of the army in Korea.

In 1914 the first farm with Mongolian sheep was opened in Sep'o-ri in the Ch'ugaryŏng furrow, thus instituting the beginning of sheep raising, which had been completely lacking in traditional Korea (p. 188). In 1929 a big plan for improving private sheep raising in the six northern provinces was published. "Sheep in the north, cotton in the south" became the slogan of the following years during the period of office of Governor–General Ugaki. The farmers of these provinces were encouraged to keep sheep, and for 5 years in a row 2500 sheep of the Corridale breed were imported annually from Australia and New Zealand. They were acclimatized on state farms, particularly in Myŏngch'ŏn-gun and then given to farmers for breeding purposes. The government–general grants these farmers a subsidy. This branch of industry is thus still quite new, and for the time being it plays only a slight role, even in the most favorable province, Hamgyŏngbuk-do.

Efforts to improve chicken farming are already of an older date, and the results are accordingly impressive. The Korean breeds are small and lay few eggs. They are kept only for their meat. The government–general provided for the introduction and breeding of foreign laying hens. In 1933 there were already 3.0 million foreign chickens, along with 3.8 million Korean chickens, and at the present date the former may already predominate, while the latter still dominate the field only in the interior of northern Korea.

Almost all of the individual elements of Korea's present-day agricultural geography existed in traditional Korea, as the remarks on pp. 173–182 show. Only in insignificant details, such as the introduction of the tea plant, has Japanese regional planning added completely new traits. Nevertheless, the character of all Korean landscapes has undergone profound changes because of it. From the land of ruins, as Ratzel described it (p. 29), Korea has become a country of vigorous agricultural productivity. The classification according to agricultural regions that have developed as a result of Japanese influence is shown in Fig. 38, and the agricultural types on which it is based were already pointed out on p. 186. Here our purpose is merely to subject these agricultural regions to a brief synthetic view from the standpoint of Japanese influence.

The Japanese influence is most intense in southern Korea, which affords the broadest possibilities for development, with its mild winter climate and its long summers. All of southern Korea, with the exception of the high mountains, has been made into a first-class rice-growing area. This is the area of types 10 and 11 in Fig. 38. The Japanese influence on the landscape is particularly massive in the southern Korean alluvial plains, which are crisscrossed by irrigation canals, and in the rice polders on the Yellow Sea coast (Figs. 74 and 76). The Naep'o Plain, which follows the lower Kŭmgang on the boundary between Ch'ungch'ŏngnam-do and Chŏllabuk-do, has been turned into a single tract of cropland by the Japanese. These central areas of rice export are simultaneously characterized by intensive cultivation of our native cereals and of naked barley on the paddies during the winter. Since from year to year alfalfa is being grown in winter on an increasing portion of the remaining paddies, the fields in this region are green then too. Winter grain crops grow on most of the dry fields as well. In summer the dry fields acquire a special character due to the intensive cultivation of American varieties of cotton. Fruit trees and mulberry plantations surround the settlements. Negative features are the diminishing role of millet and the absence of maize. In the southern half of this large area additional characteristics are the evergreen bamboo groves and the cultivation of plants that are sensitive to cold, such as ramie, paper mulberry and sweet potatoes, which, for their part, have increased substantially in extent during the Japanese period (type 11). The latter also characterize the remote islands, Ullŭngdo and Chejudo, with their mild winters. On them paddy rice cultivation recedes so completely, however, that they represent a special type (12). For the rest, Chejudo thus far has a stagnating, very ancient agricultural character, which is reminiscent of northern Korea with the intensive cultivation of foxtail millet and buckwheat (p. 378), while Ullŭngdo bears all the marks of a very recently settled agricultural region (p. 326).

The Kyŏnggi region and the Yŏngdong fringe of the T'aebaek Range are decidedly transitional areas (type 9). Paddy farming does still give their agricultural areas their basic character, to be sure, but double cropping on paddies is lacking or occurs only sporadically, and winter and summer dry-field crops are cultivated as intensively as rice.

An area with a very special character and intensive Japanese influence is again the Hwanghae peninsula and the adjoining southwestern part of the Taedong basin (type 6). Nowhere else is the dry-field area as extensive as here. Rice paddies are not completely lacking, but only locally, as in the Chaeryŏnggang basin and in the longitudinal valley of Chang-yŏn, do they form large continuous complexes, and nowhere does one find winter crops on them. The dry fields are characterized by intensive cultivation of barley, naked barley and wheat as winter crops and foxtail millet, kaoliang, buckwheat, soy, adzuki and mung beans and Korean varieties of cotton as spring crops.

Adjoining in the Yellow Sea fringe of the P'yŏngbuk region is another area with a pronounced character of its own (type 8). Here paddy rice farming has developed very intensively again, especially at the mouths of the Yalu and the Ch'ŏngch'ŏn-gang, due to the activities of the irrigation associations. Because of the extreme cold, however, winter crops are completely lacking, so that wheat growing stops here abruptly. Spring barley is grown sparsely. Barnyard millet and maize, however, make an appearance as important elements.

One last agricultural area that was substantially changed due to Japanese influence is the North Korean Japan Sea borderland (types 5 and 3). Here rice growing was practically completely lacking in traditional times. By providing rapidly maturing varieties the Japanese have made it into an important element of agricultural production, especially in the southern half. Cotton, in contrast, has to be completely ruled out, because of the cool summers. The characteristic plant of this area is barnyard millet, which has recently appeared in the northern part as a paddy crop, after rice growing had been pushed beyond the limit that promised permanent success. The other millet varieties and a large number of the bean varieties are also found. A good amount of barley is grown as well, but in the north it occurs only as a summer crop (type 3).

The three remaining types (7, 4 and 2) are restricted to the mountains of central and northern Korea, and the slight intensity of their agricultural production is therefore characteristic. In the areas where they predominate, the Japanese influence is also the slightest.

5 The Forests

With the exception of the fields almost all Korean land is under the supervision of the forestry division of the government–general. According to Table 32, these "forests" occupy no less than 73.2% of the national territory. The highest proportion of forest area in the total area is found in the Hamgyŏng provinces (83.5 and 82.8%), the lowest in Hwanghae-do (58.5%), which has few high mountains. Thus the regional differences are not nearly as great as in Central Europe.

In the chapter on plant geography we already pointed out, however, that only about one-third of the land that is under the supervision of the forestry authorities is covered by stands of tall timber. Of 16.2 million ha this is 5.4 million ha (1938). The remaining 10.8 million ha, i.e., one half of the entire country, consist of more or less degraded shrub formations, the composition of which was discussed in the chapter on plant geography. Indeed, even at the end of 1937 1.14 million ha, i.e., 5.2% of the area of the country, or an area as large as the province of Thuringia, was "without trees," i.e., practically totally barren and completely without use [XLIII]. This condition of the forests, which is the consequence of past and present sins, causes the terrible floods that severely damage the fields in the valley bottoms time and again (p. 119f).

The Japanese administration found most of the forests in a ravaged condition, the ownership rights in a state of confusion. At the recommendation of the resident–general, in 1908 the former Korean government promulgated a forest law, according to which all persons who believed they owned forest land were required to register it with the authorities within 3 years. The people assumed this meant a new tax, and most of them failed to register. When Korea then came under Japanese control, the government–general took over all unclaimed land as state land in 1911. Four fifths of the forests are supposed to have become state property in this manner, which, incidentally, they had been nominally for centuries, according to the laws of ancient Korea. This law did not meet with any sympathy among the Koreans, who were not accustomed to strictly defined laws. They felt they had been deprived of their property (Hoon K. Lee [605, p. 183]). The forestry law of 1911 simultaneously had a provision whereby this new state land could be leased to persons who had the intention of reforesting it. Once the leased land had been successfully reforested ownership was to be transferred to the persons who had accomplished the reforestation. In 1939 1.6 million ha of forest, i.e., a good tenth of the area, were leased for this purpose. The beneficial effects of the law are obvious everywhere today. Considering the tremendous size of the ravaged areas, rapid large-scale success would have been out of the question without widespread private assistance. Instruction in the schools, which were allocated land for planting trees, the stimulation to plant "memorial trees" and the nation-wide Arbor Day (p. 385) were all attempts to spread interest in reforestation among wide sections of the population.

The complaints of the former owners who had failed to register were so noisy, however, that the government–general was moved in 1917 and 1924 to make changes in the law of 1911. Persons who could demonstrate their former ownership rights to forest land were allowed to be "*enkosha*" (claimants) in this state-owned land. Since then there is claimed and unclaimed state land. The claimed land has been partially transferred to the ownership of the *enkosha*.

In 1936 58.4% of the forest area was again privately owned, 5.6% was owned by townships (*myŏn*), 1.1% was still owned by temples and only 35.0% was state-owned [XLIV]. The state-owned and temple lands are almost the only ones covered by tall stands of timber. These are restricted for the most part to the northern Korean plateaus and ranges, especially the Kaema upland (2.1 million ha) and the T'aebaek and Sobaek Ranges (2.8 million ha) (Map 4, cf. editors' note on p. 7). Of the privately owned forests, even today almost the only ones covered with tall trees are in the environs of yangban graves (p. 166). A small portion of the state forest is reserved as national parks and research stations for university purposes and is not exploited commercially. For instance, the lovely tall forest around Ch'ŏnwangbong in Chirisan has been entrusted to the University of Kyōtō. The private lands are owned not only by Koreans, but also by Japanese. Not only private persons, but also companies, such as the Chōsen Nobles Afforestation Co., the Oriental Development Co. (p. 384), the Kawasaki Plantation Co. and the Mitsui Gomei Kaisha participated in the reforestation, and, in accordance with the law of 1911, these companies now own the areas they reforested.

The first step taken by the Japanese forestry commission toward reforestation was the creation of model forests on the hills near Seoul and P'yŏng-yang, later followed by others in the vicinity of Suwon, Kaesŏng and Taegu. In Ch'ŏngnyang-ri near Seoul a central Forest Experiment Station was founded in 1913, and in 1922 it was expanded (638). Subsidiaries were set up in the state forest of Kwangnŭng (p. 289) and near Chinhae. It is the responsibility of the Forest Experiment Station to investigate the growing conditions of the trees, their distribution, diseases, insect pests and soils and to provide seedlings in nurseries. Such nurseries, producing red pine, oak, chestnut, poplar and larch seedlings, have been set up in increasing numbers since 1907, each province having today one, or several. Initially the seedlings were distributed gratis to interested persons. In addition, experimental nurseries were created. In 1913 there were already 319 of them.

A. Hofmann [494] reports vividly on the first attempts at reforestation, which were made in southern Korea in 1907. Five km north of Pusan he encountered an area 400 ha in size reforested with the broad-leaf oak, *Quercus dentata*, *Chamaecyparis*, *Cryptomeria japonica*, *Gingko biloba* and several species of poplar. Most of these species, which are all common in Japan, are no longer used in Korea for reforestation. The most important tree for reforestation today is the red pine, which was also used for this purpose in traditional Korea (p. 136). Usually alders and lespedezas are planted between the red pines. The lespedezas, which grow very fast in the beginning, are supposed to hold the soil. The alders influence the soil in a manner that has not yet been completely clarified, but which stimulates the growth of the pines. The Forest Experiment Station has done experiments along these lines. Red pines planted on a ridge with soil that was short on nutrients grew stunted at first. After alders were planted between them, they began to develop new, healthy, straight trunks, which had reached a height of 2 m when I was there. The alders that are most frequently used are *Alnus sibirica*, *Alnus firma* var. *multinervis* and *Alnus tinctoria* var. *glabra*. Other trees used for reforestation are *Quercus acutissima*, *Quercus serrata* and *Quercus dentata*, the hazel *Corylus heterophylla*, the juniper *Juniperus rigida* and the false acacia *Robinia pseudoacacia*. Naturally the trees used for reforestation change with the landscape. In southern Korea the black pine is used in preference to the red pine, because of its resistance to the pine bug (p. 136). Unfortunately, this lovely tree is very sensitive to cold. In regions whose climate even the red pine does not tolerate well, one finds the larch *Larix kaempferi* or a cross between the Wilson larch and a native variety. But the red pine is still the rule today. In most parts of the country once more or less completely barren slopes now covered by dense stands of red pine are a typical feature of the landscape. In many of its reports [e.g., VI, 304] the government–general has published pictures showing the same area before and a few decades after reforestation, and the change is indeed amazing. In regions where granite weathers to yellow soils, however, successful attempts at reforestation are not infrequently restricted to the ravines and small valleys and the parts of the slopes facing them. On the ridges separating them, in contrast, the saplings have remained small or died due to the lack of nutrients and the influence of wind and cold. The skeleton of the mountains stands out sharply even from a distance because of the color contrast green–yellow. Particularly striking is the successful reforestation along the Tumen and Yalu, which separate the still barren or exceedingly sparsely covered slopes in Manchuria from the copiously wooded and shrub-covered slopes in Korea. According to Table 32, areas with saplings covered 2.1 million ha (9.6% of the total area of the country) at the end of 1937, and most of this area, which is larger than Westphalia, owes its present-ay condition to reforestation. Since the occupation more than 4.4 billion trees have been planted. The direct activities of the forestry offices on state land account for 88,645 ha with 307.6 million seedlings. Thus 8000 ha are now being reforested annually by the state and 80,000 ha privately. Table 32 also shows the distribution of areas with saplings by province. The proportion varies between 2.1 and 17.3% of the total area of the province. This table simultaneously shows, likewise in percentages, the still barren areas of the provinces that are awaiting reforestation. They lie between 1.9 and 10.7%. The mountainous province of Kang-won-do has both the largest proportion of areas with saplings and that of still barren areas. Only in two provinces are the latter still more extensive than the former, and in both these areas are relatively small.

Often erosion is so intense on the barren slopes that it is impossible to plant seedlings directly on the slopes. Where granite weathers to yellow soils this is even the rule. The slopes are then divided with enormous effort into horizontal terraces about 1 m vertical distance from each other; the steep sides are plastered with sod, and then shrubs of the species mentioned above are planted on the level terraces. As long as the plants are still low, the slopes appear horizontally banded for miles around. Southwest of Chŏnju (Chŏllabuk-do) I observed a slightly altered system of slope terracing, which is shown in Fig 89. There are still 208,000 ha of barren land that require this "sand drift prevention" before they can be planted. The state is now terracing 3000 ha annually. It encourages private terracing by covering the complete costs of terracing and half of the costs of the

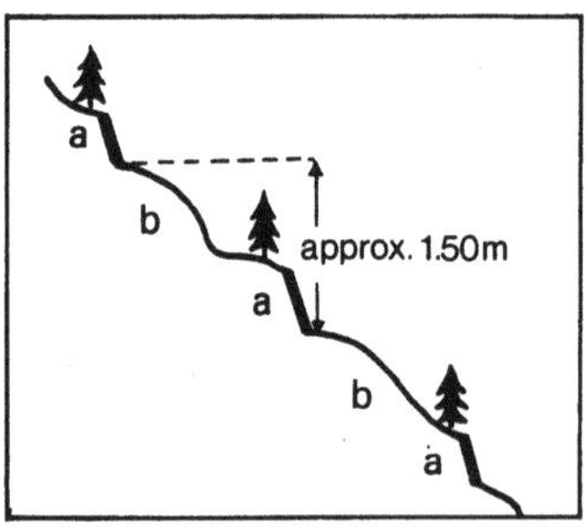

Fig. 89. System of slope terracing for reforestation in Chŏllabuk-do

trees planted on the terraces, with the provision that the trees may not be felled without its permission.

But reforestation alone will not suffice to cover Korea's mountains with trees again and to heal the severe wounds caused by the reckless exploitation of earlier times. The Koreans still resemble a man who once had a fair-sized capital that he has wasted without foresight except for a small remainder and even now does not want to be content with the meager interest alone. Their tremendous demand for fuel still leads them to use the remaining shrub formations much too intensively even today. Indeed one even sees not infrequently that an axe has been used surreptitiously on new plantings that are just beginning to grow up. Therefore, since January 1933 the government–general has been applying new guidelines stressing three important points: (1) Preservation of seed trees, to restore the productive forces of the country and achieve natural regeneration. (2) Efforts to obtain more wood for fuel from fully grown trees that are ready for felling, in order to use the good stands more completely and to set up new fields and pastures. (3) Protection for young trees and undergrowth and restriction of the use of grass and herbs, but encouragement to fell tall trees and increase the number of tree species. The third point is a reaction to the Korean habit, already criticized by Governor–General Ugaki, of valuing only the red pine and neglecting all other tree species. Since forestry measures always need a large number of years before their results are able to manifest themselves clearly, it will take at least another generation before Korea loses its barren appearance at lower elevations.

The tall stands of timber still covering one third of the forest land, which have been spared the same fate as the formerly luxuriant natural forests at lower elevations only because of their inaccessibility, have been increasingly in danger of being destroyed by fire-field farmers during the past decades. The Japanese forestry commission has taken over their management and is now enforcing the principle that no more wood may be felled than grows back. This forest management, developed according to sound forestry principles, whose directors were often trained in Germany and consult German literature on forestry, must naturally view the fire-field farmers as their main foes. According to a Japanese estimate, in 1916 50–60,000 ha of tall timber were destroyed by fire-field farming; 38,000 ha are still listed for 1930 [VI]. Immediately after the occupation, the Japanese administration therefore began the struggle against the fire-field farmers. In the same year (1910) a law prohibited all fire-field farming in state-owned forests, to violate it being an offense punishable by a high fine or imprisonment. It had practically no effect. The fire-field farmers did not find a prison sentence for violating it degrading, and they considered a fine to be a kind of fee authenticating their no longer clandestine ownership of their fire fields [560, p. 27ff].

In 1912 136 reserves were set up in the state-owned forests, each with a supervisor in charge. But even this measure was not effective enough, because the areas under the individual supervisors were much too large. Between 1932 and 1934, therefore, not only the number of state-owned forestry stations was raised from 53 to 64 and the number of supervisors by 11, but 242 jobs for assistant supervisors were also created. With these measures it has supposedly been possible to almost completely prevent new burning of virgin forest by fire-field farmers. The annual loss of forests has supposedly not exceeded 1000 ha since then. Thus far the struggle against the fire-field farmers has met with varying success in the different provinces. This stood out very clearly geographically, in that the parts of the provincial boundaries that ran across high drainage divides were occasionally also sharp boundaries in the intensity of fire-field farming (p. 255, 344f).

Years ago all kinds of attempts were considered to prevent further burning of virgin forests, to completely settle the fire-field farmers and to gradually have the property they had usurped recognized by the state. Since 1916 there are posters at the entrance to natural forests showing fiercely burning trees and huts to warn against setting fires, or tablets branding the destruction of trees in the virgin forests as a sin against the tree spirits (p. 267). Methods have also been used to develop in the fire-field farmers the desire for a settled form of life with property sanctioned by the state and thus to train them to become permanent farmers [561]: instruction in how to increase the yield of the fields, improvement of livestock, finding subsidiary occupations for the off-season, creation of marketing facilities as a stimulus to produce surpluses and to give up their previous meager and self-willed self-sufficiency. For this purpose the government-general has set up Mountain Agriculture Guidance Stations. All of these measures have of course not become superfluous with the prevention of further burning of natural forests. This has only ended the further expansion, but not the existence of fire-field farms. The renewed burning of secondary formations still continues for the time being. Now it is time to settle and intensify the existing farms and to let the forest take over unprofitable areas again, as much as possible by natural seeding. Japanese organizational talent has a long way to go, before this so interesting, but also detrimental, primitive form of economy in Korea is a thing of the past.

To supply the inhabitants of the lowlands with the necessary firewood and stop them from repeatedly overexploiting the shrub formations, arrangements have been made for systematic lumbering in the remaining natural forests and tall secondary forests. This is particularly true of the tremendous forests along the upper reaches of the Yalu and the Tumen. They contain reserves of about 150 million cubic meters of solid wood, most of which has decayed up till now. The forestry office in Shinŭiju was established to organized this lumbering. In 1937 it sold 464,000 cubic meters of wood that had been rafted down the Yalu for 10.2 million yen. The methods used in logging and transportation in the Kaema upland are described on p. 229f. Recently, mainly narrow-gauge railroads have been built to transport the wood to the rivers where rafting is possible. In 1938 there were 138 km of such railroads, which could transport 130,000 cubic meters per year.

In 1910 the value of the products of the forest lands amounted to 19.2 million yen. In 1930 it was 138.7 million yen. Of the latter 25.9% was for wood used for construction, cellulose and cut timber, 23.2% for firewood, 2.4% for charcoal and 0.3% for bamboo. The greatest part, by far, of the remainder of 58.2% is the estimated value of the masses of brush, twigs and grass cut and scraped off by the farmers. Forest production has increased constantly since 1910, but still much more slowly than other branches of the economy. The reason for this is that especially in this area, more strongly than in others, the aftereffects of past sins must be overcome, and this is impossible without the sympathetic cooperation of the broad masses of the population. For the time being Korea's forested areas are entitled to the name "forest" only inasmuch as they are under the control of a forestry office. Tall stands of timber that are rationally used and managed are found only to a very modest extent thus far. At present they form only a link in the chain beginning with the unused virgin forests with a decaying surplus of wood and ending with the barren areas recklessly exploited to the last stick on granitic yellow soils.

6 The Fishing Grounds

Fishery production has been increased more strongly still than agriculture and forestry by the Japanese. The value of the catches increased nine and a half times between 1911 and 1937. The seas around Korea are extraordinarily rich in the most varied economically useful-fauna. S. Tanaka counts 961 species of fish alone in Japanese waters [511], and only very few of them are not used by the Japanese. The Koreans, in contrast, did not know how to use this wealth adequately. Descriptions by early travelers, such as A. Hamilton (219, p. 241ff) clearly show that fishing was done only very casually by the Koreans at the turn of the century.

Korea's coasts, which reach a length of 18,203 km (p. 4), border on three seas with totally different oceanographical conditions (p. 122ff). A very deep sea with a small tidal range and a very narrow shelf zone spreads out to the east. The upper decameters of its water are warm in summer, particularly in the southern half, to which the Tsushima current brings the highly saline, warm water of the Kuroshio, while underneath extensive cold water areas follow. In winter the surface temperatures also are very low in the north. At that time the north Korean cold current extends far southward along the coast and dips under the warm current off Kyŏngsangbuk-do. All year round the two currents meet along a convergence that shifts its location seasonally. Eddies develop, in which the waters mix. The cold water area is highly aerated, enabling large amounts of plankton to exist. According to Uda (467), the intermediate layer at 25–50 m is most abundant in plankton. H. Aikawa, who devoted a special study to the plankton in Japanese waters (513), points out that the cold water off the northern Korean coast is characterized by diatom associations, and the warm water off the peninsula by phytoplankton associations.

To the west Korea borders on a shelf sea that is nowhere deeper than 100 m and has an enormous tidal range. Stretching out in front of the almost innumerable islands and peninsulas and the tidal flats surrounding them there is a broad zone of sand and mud banks. They are separated by deep tidal channels, in which the tidal currents develop enormous velocities. In summer the water of the Yellow Sea is warmed superficially by radiation even more than that of the Sea of Japan, and branches of the Kuroshio advance far northward along the Korean west coast as well, but already at a depth of 30 m a thermocline leads to constantly lower temperatures. In winter the Yellow Sea water is even colder on the surface than that of the Sea of Japan at the same latitude; in fact, along the coasts of the P'yŏng-an provinces ice occasionally begins to form. Close to the Korean coast a southward flowing current transports the cold, low saline water as far as the southwestern tip. In summer the large rivers pour large amounts of oxygen-rich, turbid water into the Yellow Sea from three sides. The coastal waters off northern and central Korea are again characterized by masses of phytoplankton, whereas otherwise diatoms predominate.

The sea off the Korean south coast, the Korean "South Sea," finally, is a shelf sea, which always has relatively warm water. The surface temperatures increase toward the east and the south in all seasons. In August they are around 24–27° C, and only off the southwestern tip are they lowered to 20° C by upwelling water. In February they remain at 8–15° C. The eastern half, through which the Tsushima current flows all year-round, is particularly warm and saline. The tidal range decreases sharply toward the east. The abundance of islands and peninsulas with many bays affords optimum conditions for setting up fishing harbors.

There are thus exceedingly strong regional and seasonal differences in the oceanographical conditions around Korea. Accordingly varied is the world of marine organisms that are used. They include not only fish, but also crabs, mollusks and cuttlefish, and even seaweed and algae, to a much greater extent than the seabeds exploited by Western peoples. The strongest of these regional and seasonal differences are caused by the change in surface temperatures, which is greater here than almost anywhere else at subtropical latitudes (p. 124ff). Warm-water, subtropical fish can be clearly distinguished from cold-water, boreal fish. The former come northward in summer into the Yellow Sea and the Sea of Japan and populate primarily the surface water layers. Many remain in the Korean south sea all year-round. Boreal fish come southward from the Bering Sea and the Sea of Okhotsk, following the Korean coast of the Sea of Japan southward with the north Korean cold current. In summer many of them retreat to the cold layers in the depths. In winter, however, they dominate the surface layers as well. Many of them are also found in the Yellow Sea. Since it is enclosed to the north by land, though, its fauna is much less varied than that of the Japan Sea. Tanaka rates 76.6% of the fish species listed by him as representatives of southern regions, 13.9% as ones of northern regions, while the small remainder is found in both.

The most important warm-water fish by far today is the sardine (*Amblygaster melanosticta* T. and S. [*Sardinops melanosticta*, Japanese Pilchard], Kor. *chŏng-ŏri*). It is a migratory fish that moves in swarms in the layers near the surface. Uda and Okamoto studied its living conditions in the Sea of Japan on the basis of observations for the years 1918–30 [517, 464]. In spring the sardines migrate from their spawning grounds off western Kyūshū northward along the east coast of Korea. They follow the Tsushima current at the same pace as the cold front retreats northward. In midsummer they also populate the surface layers of the cold current, which have warmed up strongly. From early autumn on the swarms migrate back. Sardine fishing begins accordingly along the coasts of the Kyŏngsang provinces in April. In May it has expanded as far as Yŏnghŭngman; in June it already extends past the mouth of the Tumen. In July its southern end has advanced to the northern edge of Kyŏngsangbuk-do, in August to Yŏnghŭngman. From October on the northern edge of the fishing area retreats southward, and in November it has reached the northern end of the Kyŏngsang provinces again. Along the coasts of Kyŏngsang-do and Kang-won-do there are thus two fishing periods, in spring and in autumn, and on the coast of Hamgyŏng-do only a single one, in midsummer and early autumn. The main catches are made off the latter between July and October. The fishing area extends here up to 30 nautical miles from the coast into the open sea. In accordance with the thickness of the upper warm-water layer, the depth at which the catches are made extends lower in the south than in the north. On the average the depths between 10 and 40 m are most abundant in sardines. The temperature span in the layers that are fished is narrow in the south and wider in the north. At a depth of 10 m it varies during the fishing season between 8° and 19° C, and during the main season between 12° and 16° C. It appears that 14° C is the optimum temperature for fishing. The annual catch varies, depending on how long these optimum conditions last during an individual year. Good and poor catches also depend on the growth conditions of the fish themselves, however. Therefore they each have the tendency to last for a few years, depending on the dominant age groups. The years 1926, 1927, 1929, 1930, 1936 and 1937 were years of peak catches along the Korean Sea of Japan coast; the catches in 1928, 1932 and 1938 were very small. In Korea's southern sea the catches are relatively small; the same is true in the Yellow Sea, where the fish are caught between July and September. In Kyŏnggiman sardine fishing is impossible, because of the extremely unfavorable conditions due to the currents. Off the coast of P'yŏng-anbuk-do sardine fishing is no longer worthwhile.

Sardine fishing is done today either with gill nets, in which the fish become entangled in the meshes with their gills, or with the sack-like purse seines introduced by the Japanese. Both types are pulled by fishing boats in the deep Sea of Japan, and are not anchored. According to F. Bartz [653], the principle catches there are made directly along the coast with the purse seine; the field of the gill net fishermen lies farther out to sea. Each ship with a purse seine is accompanied by a smaller auxiliary vessel that helps to lay out and draw in the net. The crew consists of approximately 40 men. During the main fishing season the fishermen work almost continously day and night. Special transport lighters keep up a shuttle service between the fishing grounds and the landing places on the coast of the Sea of Japan. The Shin Air Service Co. has set up an air service that searches for sardine swarms along the coasts of the Hamgyŏng provinces. Sten Bergman vividly describes sardine fishing with a gill net near Unggi in October [229, p. 102ff]. The Korean junk he went on appeared to be of a type that is at least 1000 years old (p. 439f). Far out at sea a gill net about 900 m long and 9 m high was set, its upper edge held by wooden swimmers at a depth of about 15 m. During this time the sails were reefed and the rudder was removed, so that the junk drifted with the current.

Mackerel are also pelagic fish from warm-water areas that move in swarms, like sardines. There are three types in Korean waters, the common mackerel (*Scomber japonicus* Houttuyn, Kor. *kodŭng-ŏ*), the larger Spanish mackerel ([*Scomberomorus nephonius* or *Scomberomorus semifasciatus*], Kor. *samch'i*), and the horse mackerel ([*Trachurus japonicus*] Kor. *chŏn-gaeng-i*). The latter gets to be as much as 1 m long with a weight of 15–20 kg. All three are voracious predators. As attempts at marking have shown, the common mackerel wanders northward along the coasts in spring and summer and back in autumn and winter [512]. Its optimum temperature lies around 14–16° C [516], whereas the horse mackerel prefers saltier and warmer water and therefore generally advances only as far north as the Korean south sea. There it is caught in summer and autumn, partly with nets and partly with fishing tackle. The area of distribution and fishing season of the common mackerel are very similar to those of the sardine. In the Yellow Sea it occurs only on the Korean side now, and even that only scarcely. Probably the turbidity of the water due to silt particles does not appeal to it. The Spanish mackerel is caught on all coasts, with the exception of Kyŏnggiman, primarily in spring and autumn. Common mackerel and Spanish mackerel were already popular food fish in traditional Korea and were eaten fried, cooked with soy sauce or salted and dried. Large amounts of horse mackerel are exported to Japan.

The shark (*Kor. sang-ŏ*) that comes into Korean waters is a long, slender animal that is fished with a line over a pulley in the summer half-year. Shark fins are sent to China as a delicacy.

In contrast to the warm-water fish mentioned thus far, the red sea bream (*Pagrus major* [*Chrysophrys major*, Japanese sea bream], Kor. *tomi*) is a fish that lives at the bottom of the sea. The tall, slender fish has a lovely coloring uniformly decorating the entire body. It is one of the most highly prized food fish and is eaten on special occasions, for instance Buddha's birthday (p. 168f). The season is limited to midsummer along the Hamgyŏng and P'yŏng-an coasts and to spring and autumn on the Ch'ungch'ŏng, Chŏlla and Kyŏngsang coasts. *Tomi* is often caught with a line over a pulley and lead weights.

Another benthic fish is the big, round corvenia, (*Sciaena schlegeli* Bleeker [*Sciaena* sp.], Kor. *chogi*) which is caught mainly in the Yellow Sea, partly with dragnets and partly with gill nets.

In contrast to the corvenia, the yellowtail (*Seriola quinqueradiata* T. and S., Kor. *pang-ŏ*) is rare in the Yellow Sea. This strapping fish is caught in the Korean southern sea in winter and on the Japan Sea coast in summer and autumn [512]. Fish that, conversely, are lacking in the Sea of Japan are the hairtail (Kor. *kalch'i*), whose shiny silver scales are used in making artificial pearls, and the croaker ([Nibea Croaker] *Nibea schlegeli* Blecker, Kor. *minŏ*), a medium-sized fish that is caught in large quantities in the Yellow Sea up to the mouth of the Yalu between May and October. The anchovy hauls (*Engraulis japonica* Schleg., Kor. *myŏlch'i*) are also very good-sized, but I do not know anything about the season or how they are fished.

Many fish are more highly restricted to warm-water areas than the ones we have just mentioned, and from Korea they are therefore caught only in the south sea. They include the varieties of tuna (*Thunnus* sp., Kor. *tarang-ŏ*), which reach a weight of 150–300 kg with a length of 2–3 m. Of the five species that come from the open sea into Japanese waters, two, the albacore (*Thunnus germo* [*Thunnus alalunga* Bonnaterre], Kor. *nalgae tarang-ŏ*) and the bluefin tuna (*Thunnus orientalis* [*Thunnus thynnus* L.], Kor. *hŭktarang-ŏ*) come as far as the Korean southern sea. The most favorable temperature for their grounds is 20–21° C [515]. They are caught with lines using live sardines as bait. The much smaller, likewise highly prized skipjack (*Katsuwonus pelamis* L., Kor. *kadaraeng-i*) reaches the northwestern boundary of its range here. It prefers temperatures of 20.5–26.5° C [514]. The rapidly moving swarms are caught with rods. The squid (Kor. *ojing-ŏ*), too, a soft-bodied invertebrate belonging to the cephalopods, occurs mainly in the Korean southern sea and is caught here at night using artificial lights as lures. It is sent to the market dried.

The warm-water fauna finally include many of the popular crustaceans and mollusks, such as the famous trepang (sea cucumber, Kor. *haesam*), of the genus *Holothuria*, which is considered a delicacy and was already one of ancient Korea's exports (p. 33), abalone (*Haliotis tuberculata*, Kor. *chŏnbok*), the short-necked clam (*Lima cardiformis* [*Venerupis japonica* Deshayes], Kor. *panjirak*), spiny lobsters and oysters (Kor. *kul*), the latter of which are cooked in soup or fried. Many of these are fauna of the mud flats and are thus common on the south coast and particularly on the Yellow Sea coast. They are gathered partly by hand by people walking across the tidal flats. Above all, however, cleverly constructed traps are used that can take the wear and tear of the strong and changing tidal currents [653]. They include the stow net (Kor. *an-gangmang*), an approximately 100-m-long net pouch that is used from an anchored vessel, and the *taesumang*, a smaller net pouch that is attached to posts rammed into the ground. In the Sea of Japan set nets are used correspondingly to catch benthic fish, migratory pelagic fish and all manner of other creatures (Fig. 90).

Not only animals, but also plants are marine cultures, particularly algae and seaweed. Algae grow pretty much only on the southern coasts. Above all *Ulva latissima* and the purple *Porphyra tenera* are used. They serve as glue or food (laver, Kor. *kim*, Jap. *nori*, p. 371). The seaweeds are much more widely distributed, particularly sea mustard (*Undaria pinnatifida* Suringar, Kor. *miyŏk*), which grows up to beyond the mouth of the Tumen and up to the mouth of the Taedonggang and tolerates the extremely low water temperatures there in winter, so that it can no longer be considered one of the subtropical forms. The giant greenish-brown leaves are torn loose from the bottom of the shallow sea in spring and early summer with whisks and small scythes attached to long, thin bamboo poles. They are scratched off the cliffs with sharp-edged seashells. On the beaches of the bars they are laid out to dry in long, narrow, rectangular strips. With soy sauce and other ingredients they are made into a soup that is a favorite among Koreans.

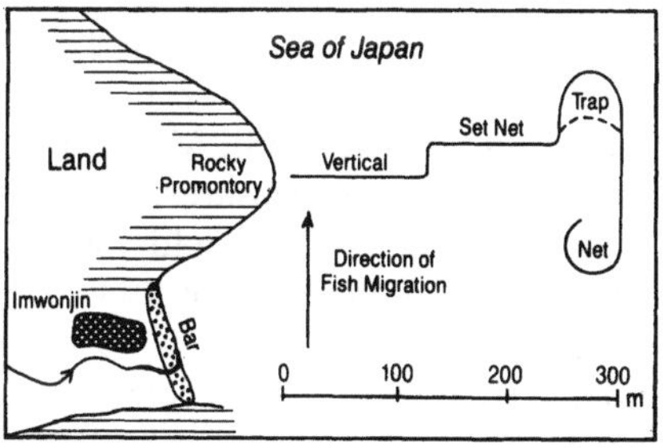

Fig. 90. Set net near Imwonjin (Kang-won-do). The net hangs vertically on posts in shallow water. Between the posts it is supported by thin beams of light wood. The sea creatures migrating northward are forced to the right by the net. The semicircular end prevents them from escaping, and thus they are finally pushed into the trap. It is closed off on its southern side by a sloping net that permits entry only. The trap is emptied three times a day.

Sea mustard and all kinds of creatures found in shallow waters are gathered by women divers on Chejudo. There are 8000 of them, who have now joined together in a cooperative, at Japanese instigation. They hire themselves out temporarily along the coasts of the mainland as well (p. 317).

Cold-ater fish advance more or less far southward along the coasts of the Sea of Japan. During the summer, the surface fauna are lacking, but the fauna of the cold depths are present even then. Most important is the Alaska pollack (*Theragra chalcogramma*, Kor. *myŏngt'ae*), a slender, medium-sized fish with magnificent dark stripes on its back. It is caught in the winter months on the Hamgyŏng and Kang-won coasts with four different types of nets and sent dried from there all over Korea.

The fishing grounds of the large Pacific herring (*Clupea harengus pallasi*, Kor. *ch'ŏng-ŏ*) extend farther southward, to Kyŏngsangnam-do. It is caught there in January and February, in Hamgyŏngbuk-do, in contrast, in late autumn and very early spring. Presumably the herring swarms migrate along the coast of the Sea of Japan opposite to the warm-water fish. The herring nets hardly differ from the European ones now.

The range of the Pacific cod (*Gadus* sp. [*Gadus macrocephalus*], Kor. *taegu*) extends around all of Korea. It is caught with trawls or long lines in the Korean south sea only from December to February, in the northern part of the Yellow Sea in March and April, on the Hamgyŏng coasts in spring and autumn.

Chum salmon (*Oncorhynchus keta*, Kor. *yŏnŏ*) appears to be limited to the area around the mouth of the Tumen. Its main range lies farther to the north. Other salmon species occur in both the Sea of Japan and the Yellow Sea [501a]. The medium-sized, slender common gray mullet (*Mugil cephalus*, Kor. *sung-ŏ*) is caught on all Korean coasts during the winter.

The cold-water fauna also include several flatfish, such as flounder (*Pleuronectes asperrimus* Schleg. [*Clidoderma asperrimum*], Kor. *kajami*) and Pacific halibut (Kor. *nŏpch'i*) and crustaceans, in particular crabs (*Paralithodes* sp., Kor. *ke*), shrimps and prawns (Kor. *saeu*). Flounders are caught everywhere on Korea's coasts, in the south from spring to autumn, on the Hamgyŏng coast with a break in midsummer. The common shrimp is rare in the Sea of Japan. The main fishing grounds are in the mud-flat zone of the Yellow Sea.

Finally, whales (Kor. *korae*) should be mentioned as cold-water fauna. Formerly whales were also common in Korean waters in the winter. In 1905 a Norwegian company caught 211 of them here [658]. Meanwhile, ruthless hunting has decimated their numbers, despite legal measures. Whaling is now done by the Japanese from special whaling stations. The yields, which still amounted to 1.2 million yen in 1930, had dropped to one-third by 1935; since then they have disappeared from the statistics.

In 1904 Japan concluded a treaty with the Korean emperor allowing Japanese to fish along Korean coasts. Japanese fishermen had already been operating in Korean waters for centuries, however. Immediately after the annexation, the Japanese began to organize operations. The fishery law of 1912 reserved the waters around Korea exclusively for fishing by inhabitants of the country. The limits of the area that is reserved for national use, shown on Figure 38, were set. They do not maintain a constant distance of a few kilometers from the mainland coasts, but instead include the wide expanses out to Ullŭngdo, south of Chejudo and

Sohŭksando and run as a straight line between the individual fixed points. The same law prohibited certain deeds that are detrimental to the catch in the national fishing grounds, regulated the issuing of fishing permits to local applicants and prevented the monopolization of any fishing grounds by individual persons. It also contained certain prohibitions regarding fishing equipment, seasons and grounds and stopped trawling, in particular, in some areas. Additionally it guaranteed measures for protecting the west coast against Chinese piracy. To be able to supervise these and other orders the government–general put the surveillance ship "Teru Kaze Maru" into service in 1936. In 1912 scientific investigations of the distribution of Korean fish, their migrations and their reproductive habits were also already begun, with a view toward developing new methods of fishing and preservation. Research to promote fish farming was also initiated. In 1922 the central Fishery Experiment Station, Pusan was founded for this research. Since 1924 it has issued annual publications ([XV, XVI] and p. 55). Recently branches have been established in Mokp'o, Inch'ŏn and some other ports. In addition schools for Korean fishermen were opened in Kunsan, Yŏsu, etc. To improve the quality of Korean canned fish for export purposes and thus to raise the reputation of Korean goods abroad, in 1918 it was ordered that all preserved goods, especially sardines in tomato sauce, seaweed and fish oil, were to be inspected prior to shipping. This led to a substantial increase in exports. The founding of fishing cooperatives was begun already in 1912. Their number increased to 197 with 136,000 members in 1937. Many are subsidized by the government–general. Their responsibilities consist in procuring fishing equipment, making arrangements for marketing, renting boats, providing capital, preparing anchorages, etc. In 1923, finally, the "Korean Fishing Association" was founded, with headquarters in Seoul and agencies in all provinces. It is concerned with lifeboat services, investigating the social conditions of fishermen, professional advice, etc. Above all, however, the government–general has promoted fishing by constructing a large number of fishing ports and marking the coasts with beacons (p. 469). In 1925 there were already 42 fishing ports equipped with breakwaters and berths. By 1937 the number had increased to 56. In all of these measures it was possible to take advantage of the Japanese homeland's many years of experience, it having already become the number one fishing nation in the world.

According to Japanese statistics, in 1911, shortly after the occupation, there were 226,208 persons who made their livelihood by fishing and fish farming in Korea [651]. This number increased to 522,108 in 1922 then sank to 333,882 at the end of 1938. By European standards these are very high figures. The number of fishing households rose from 72,030 (1911) to 108,624 (1925) and 120,000 (1937). They were distributed among 1800 fishing villages. The number of fishing vessels increased from 14,297 (1911) to 34,610 (1925) and 50,331 (1935). The latter included 1410 engine-powered vessels. Most of these are in Japanese hands and are stationed in the fishing ports constructed by the government–general. Gradually the number of Korean boats fitted with motors is increasing, however, especially along the Hamgyŏng coast. The Korean fishing boats are flat-bottomed vessels without a keel and with low sides, 15 m long at the most. Most of the types originated in traditional Korea (p. 190). To be able to move when it is calm they have one or several gigantic oars on which occasionally as many as ten men pull. The boats on the west coast, which have to reckon with strong tidal currents, extremely shallow water at times and surging seas and surf, are built more sturdily than the ones on the east coast and have two large, rectangular cotton sails with horizontal bamboo poles, instead of the one that predominates on the coast of the Sea of Japan. Korean boats can be distinguished from Japanese and Chinese boats at a distance by the way they are rigged [577, 578]. Where there are no fishing ports yet, these boats are still pulled up onto the beach on the Sea of Japan coast today by means of winches when bad weather approaches. In the Yellow Sea they use any tidal channel that extends up to the mainland as a mooring and usually lie aground at low tide.

Korean fishing was promoted most strongly by the experienced Japanese fishermen who settled in their midst. From them they learned an extraordinary amount, particularly about methods of fishing at greater depths, which were previously unknown in Korea. This was achieved despite the small number of Japanese fishermen who settled in Korea; in 1938 they numbered only 9537 persons, including family members, i.e., 2.9% of the total fishing population residing in Korea [XLI]. But many of these Japanese are entrepreneurs and own a large number of ships with Korean crews. They can thus make direct practical use of their experience by giving orders, and from these boats the ones in Korean hands learn the essentials.

In the course of a generation Korea has taken an equal place beside Japan in the production of marine foodstuffs. The Japanese empire's share of the world catch today amounts to 30–40%. Of the 5–6 million t of marine products with a value of 400–600 million yen now landed annually in Greater Japan, no less than a quarter is from Korea [657]. The per capita yield, related to the total population, was about 4.5 yen in Japan in 1937 and 4.0 yen in Korea. The latter has thus almost reached the intensity of the foremost world producer. The weight of the total Korean catch rose from 98,000 t (1911) to 651,000 t (1925), 825,000 t (1929), 1 million t (1933) and finally 1.6 million t (1937). A comparison of quantities is, however, not entirely appropriate, inasmuch as a unit of weight of the high-quality, expensive yellowtail is totaled up indiscriminately with a unit of weight of the mass product seaweed. The value of the catch increased from 9.4 million yen (1911) to 83.6 million (1925), then dropped to 47.8 million (1930) and rose again to 51.4 million (1933) and to the maximum value to date, 89.9 million (1937). The drop in the middle of the second decade, which we already encountered in the number of fishermen, is partially attributable to the boycott movement beginning in China against goods from the Japanese empire. The tremendous increase beginning anew thereafter is a result of the strong stimulation experienced by Korea's economy as a whole after Manchuria and heavily populated parts of China were incorporated into the Japanese economic sphere. And in particular it is the result of the mass fishing of sardines, which had previously been only incompletely exploited, this mass fishing having assumed an enormous volume since 1931. Since then the total yield has been determined to a great extent by the sardine catch in a given year, which depends on oceanographical and biological factors, and not economic conditions.

The overall structure of the Korean catch has changed fundamentally due to this development. Formerly fishing was done almost exclusively for food, and thus one year one food fish had a larger share in the total catch and the next year another one, depending on the oceanographical character of a given year, without a single one ever dominating overwhelmingly. Since 1931, however, from year to year an increasing portion of the catch is processed to fish oil, and as a result the sardine quickly became absolutely predominant. This can be seen from Table 39, which shows the percentage that the six top species of marine fauna, by value, have of the value of the total catch in five selected years.

This table shows that the sardine is the only fish that occurs among the six most important species of marine fauna in all years, and that it is always in first or second place. The overwhelming predominance of the sardine catch does not appear until 1937, however.

We should supplement these remarks on the role of the various fish species in the total Korean catch with a list of the other important catches, by value, in 1937. The proportions were: hairtail 2.8%, spiny lobsters, shrimps and prawns 2.4%, halibut 1.8%, Spanish mackerel 1.6%, cod 1.5%, sea bream 1.4%, horse mackerel 1.3%, croaker 1.0%, yellowtail 0.9%, shark 0.8%, oysters 0.6%, abalone 0.4%, crabs 0.4%, clams 0.3% [VI]. The remaining 19.3% was divided up among the large number of other marine fauna, some of which were described above, and none of which reached a proportion of 0.3%. Very important quantitatively, and even quite important by value, was the gathering of seaweed as well. Thus the overall picture is extremely varied and reflects the role of the three different seas and of cold and warm water. Subtropical fish are found along with boreal fish, crustaceans with mussels, oysters with squid, seaweed with algae. The Japanese source [651] contains statistics on the exploitation of no fewer than 79 different marine inhabitants.

The sardine was formerly a food fish that was not very highly prized and was therefore cheap. Quantitatively it thus had a greater share than it did by value. The tremendous increase in the sardine catch is expressed fully only in the figures on weight. The sardine catch amounted to: 9400 t in 1911 (9.5% of the total catch), 101,000 t in 1925 (15.5%), 330,000 t in 1930 (40.0%), 667,000 t in 1933 (66.7%), 1,387,000 t in 1937 (86.7%). In 1937 the Korean sardine catch exceeded the Japanese catch by 200,000 t, although it heads the list by far in Japan as well [657]. The sardine catch alone is two and one-half times as large, by weight, as the entire landings of the German sea fishery. Korea is the world's foremost sardine fishing country. Most of this enormous increase has been along the coasts of the two Hamgyŏng provinces, particularly that of Hamgyŏngbuk-do. It is a re-

1911	1925	1930	1933	1937
Alaska pollack 13.6	Sardine 13.6	Mackerel 12.4	Mackerel 12.4	Sardine 38.0
Sardine 11.7	Mackerel 11.3	Sardine 9.9	Sardine 11.2	Alaska pollack 7.2
Whale 10.2	Corvenia 6.9	Corvenia 7.4	Corvenia 7.2	Mackerel 6.7
Halibut 8.7	Whale 5.6	Laver 4.2	Alaska pollack 6.9	Corvenia 5.3
Sea bream 7.2	Alaska pollack 5.2	Herring 3.9	Laver 4.9	Anchovy 3.3
Spanish mackerel 6.7	Sea bream 4.0	Sea bream 3.7	Herring 3.7	Herring 3.0

Table 39. Percentage of the value of the total catch represented by each of the six most important species of marine fauna

sult of the fact that the sardine provides the most important raw material for producing fish oil today.

The Japanese have built up this kind of processing industry in Korea. Scattered everywhere in the sections of the Sea of Japan coast that have low beaches there are small fish oil pressing factories. The raw fish are cooked in large iron kettles directly after the catch has been landed. The stinking broth is either dumped into the sea or conducted onto the paddies as fertilizer. The cooked fish are crushed in cylindrical hand presses. The oil flows into wooden boxes, where it separates from the lighter water accompanying it. Through the cracks in the wooden boxes it drips into an adjacent wooden trough, from which it is filled into metal cans. The cylindrical pressed cakes that are the residue are chopped up with hammers, and the chunks are spread out on mats on the beach to dry. In the evening and in bad weather, the mats are folded together over the chunks. After drying the remains are packed in straw sacks and sent into the interior or to Japan as fertilizer. Almost all of the cans with fish oil go to Japan. Part of the herring catch is processed to oil in the same manner.

In the larger fishing ports in the Hamgyŏng provinces, which include primarily Ch'ŏngjin, Ŏdaejin, Unggi, Sŏsura, Tokchin, Sŏngjin, Shinch'ang, Sŏhojin and Wonsan, the Japanese have set up more economically functioning factories for processing oil in the last few years. The Chōsen Synthetic Oil Co., founded in 1934 in Ch'ŏngjin, is concerned with hardening the oil. In 1936 the fish oil production already amounted to 114,000 t and thus considerably surpassed that of Japan [653].

Part of the fish catch that is intended for eating purposes is sold fresh in the neighborhood, since the population is highly concentrated in the vicinity of the coasts. The consumption of slices of very fresh raw fish (Jap. *sashimi*) with soy sauce, introduced by the Japanese, has also spread among the Korean population. In the larger fishing towns Japanese companies, such as the Nippon Suisan Kabushiki Kaisha, have created ice and cold storage plants. From there the fish are transported in special ships to the Japanese consumers. Especially from Pusan a tremendous amount of fresh fish is sent to Japan. This port has become one of the foremost places for supplying western Japan. Most of the fish intended for Korean use is heavily salted and dried, as in olden times. Each fishing village has extensive racks for this purpose. Only a small portion of the catch goes to canning factories. According to F. Bartz [653] there are approximately 50 of them today, though mostly with a low production capacity. About ten, all of which lie on the Sea of Japan, produce sardines in tomato sauce, and eight can shrimps. The value of the processed fish products was 65 million yen in 1935, 45% of which was for food, 29% for fish oil and 24% for fertilizer.

For a regionalization of the fish grounds around Korea I have no more recent data on the amounts and values produced in the provinces than those of 1925. On the basis of these data and the previous remarks we arrive at the figures given in Table 40, to which the map in Fig. 38 refers.

The Yellow Sea zone had in 1925 and still has today by far the smallest catches, both by weight and by value. This is the result of its unfavorable oceanographical conditions. The smallest catches are made in Kyŏnggiman, which consists for the most part of sand banks, mud flats and tidal channels with raging tidal currents (Kyŏnggi section). In it only the gray mullet, corvenia and hairtail catches, along with crabs and eels, play a certain role. During the winter fishing stops almost completely in the Yellow Sea section, with the exception of cod and gray mullet. The catches include primarily corvenia, Spanish mackerel, shark, sea bream, flounder, croaker, abalone, shrimps and sea cucumber.

In the south coast zone oceanic and pelagic warm-water fauna dominate the catch: the three types of mackerel, sea bream, yellowtail, sardine, hairtail, shark, corvenia, skipjack, tuna, squid. In the bays with mud flats oysters, sea cucumber, laver, sea mustard, abalone and spiny lobsters are also caught. Cold-water fish are also hauled up from the lower layers, especially cod, mullet, flounder, and in the east even herring. Fishing is distributed more evenly throughout the year here. The warm-water fauna are caught primarily in summer, the cold-water fauna in winter. This oceanographically extraordinarily favored area was foremost in quantity in 1925, despite its relatively small size, and this was true of the value more highly than of the weight, because large amounts of particularly valuable fauna, such as sea bream and skipjack, are caught there. The Kyŏngsang section, whose center is Pusan, even surpasses the larger Chŏlla section.

Due to the development of large-scale sardine fishing in the past decade, the Japan Sea zone, which was second in 1925, has moved into first place meanwhile. In quantity it has even achieved an overwhelming predominance. This zone is completely dominated today by sardine fishing, which is done in spring and autumn in the Kyŏngsang – Kang-won section and in midsummer and autumn in the Hamgyŏng section in the manner described above. About one half of all Korean sardine catches are landed in the ports of Hamgyŏngbuk-do alone. As a result the focus within the Japan Sea zone, by value and most especially by quantity, has shifted much more heavily to the Hamgyŏng section than would be conjectured from the figures for 1925. The coast of Kang-won-do is still very far be-

Fish catches of the provinces	Amount (%)	Value (%)	of the fishery sections Amount (%)	Value (%)		of the fishery zones Amount (%)	Value (%)	
Hamgyŏngbuk-do	15.5	6.1	27.8	13.8	Hamgyŏng section	37.5	28.5	Japan Sea zone
Hamgyŏngnam-do	12.3	7.7						
Kang-won-do	2.1	4.9	9.7	14.7	Kyŏngsang-Kang-won-section			
Kyŏngsangbuk-do	7.6	9.8						
Kyŏngsangnam-do	26.6	35.6	26.6	35.6	Kyŏngsang section	45.2	52.9	South coast zone
Chŏllanam-do	18.6	17.3	18.6	17.3	Chŏlla section			
Chŏllabuk-do	1.6	2.7	2.8	6.0	Ch'ungch'ŏng-Chŏlla sec.	17.3	18.6	Yellow Sea zone
Ch'ungch'ŏngnam-do	1.2	3.3						
Kyŏnggi-do	1.0	1.3	1.0	1.3	Kyŏnggi section			
Hwanghae-do	4.8	4.6						
P'yŏng-annam-do	2.1	2.1	13.5	11.3	P'yŏng-an-Hwanghae section			
P'yŏng-anbuk-do	6.6	4.6						

Table 40. Regional distribution of amounts and values of marine production, 1925

hind in fishing, as in all other economic areas. Other warm-water fish, such as the common mackerel and the Spanish mackerel, yellowtail, sea bream and shark, are secondary to the sardine in the Japan Sea zone. Winter fishing for cold-water fauna, particularly Alaska pollack, herring, cod, flounder, chum salmon, mullet and shrimps, plays a much more important role. Large amounts of sea mustard are also gathered. This gives the Japan Sea zone a great deal of versatility along with the predominant sardine fishing. In its northern half it is Korea's most important and most modern fishing area today.

The Japanese introduced the cultivation of marine organisms. It employs a rather large number of persons, primarily on the side. The value of its products lay between 2.9 and 4.7 million yen between 1933 and 1936, and thus amounts to only about one twentieth of the fish catches. By far most important is the cultivation of algae for laver, followed by shellfish, especially oysters, whereas that of fish, crabs and mollusks plays only an unimportant role. The most important places of cultivation lie in the tidal flats of Chŏllanam-do (p. 371).

My travels showed me that river fishing in summer is an important subsidiary occupation for very many farmers. I found hardly any data in the literature on it, however. Fishing is done with casting nets, with fish traps set in obstructions, with lines and with harpoons, the latter of which occasionally have double flukes and are up to 6.5 m long. Repeatedly described in the literature, in contrast, is ice fishing on the frozen rivers, into which small holes are hacked [652]. It is more a sport that is done in the vicinity of larger cities, such as on the Han-gang near Seoul and on the Yalu near Shinŭiju. The Japanese founded a breeding station for freshwater fish in Chinhae (Kyŏngsangnam-do) in 1927, which distributes young fish to be released in rivers and irrigation ponds.

The marine products include, finally, sea salt. The primitive traditional Korean methods, which are still used on the Sea of Japan coast with its small tidal range, were already described on p. 189. Since rock salt deposits are lacking and salt springs are rare, the Japanese have created large modern saltworks on the Yellow Sea with its large tidal range. The production of sea salt is a state monopoly here. In 1907 the resident–general already began to build plants here that used the warmth of the sun instead of expensive fuel to separate the salt from the seawater. These plants have been methodically expanded more and more by the government–general. The most important lie near Chuan in the immediate vicinity of Inch'ŏn, near Kwangnyang-ri west of Chinnamp'o and near Namshi (Yong-amp'o-gun), that is, in the provinces Kyŏnggi-do and P'yŏng-an-do. In 1937 they had an area of 4285 ha and produced 240,000 t. The area was increased to 4605 ha by 1940 and now suffices to cover the household consumption. As a result of the industrialization of the country, however, the demand has increased by leaps and bounds, so that it has not been possible to reduce the consistently large imports.

The salt pans of Chuan, which I visited, stretch out with an area of 1644 ha in the end of a bay on former mud flats. They lie between a straight dike built of rectangular granite blocks and the naturally formed salt marsh. At high tide the seawater flows through the dike into narrow canals, out of which it is lifted by wooden treadmills, and then through a second system of canals into low, rectangular basins 50x40 m. The basins have a gray silt bottom and gray silt sides. In them the water is freed of sludge and concentrated by evaporation. Very slowly it flows from the higher into the lower of these basins. The concentrated salt solution is again raised by treadmills and diverted into the crystallization basins, measuring only 8x10 m, where it stands only around 5 cm high above a bottom lined with smooth stones. Here the salt (NaCl) crystallizes out. With wooden scrapers it is swept together into little heaps and immediately shoveled into baskets, which are brought into the storehouse. This work is done with great rapidity, to avoid oversaturation and keep the bittern salts from crystallizing out. The remaining salt solution flows back into the sea.

7. Mining and Industry

Korea is a country that is rich in the most varied mineral resources. This was shown by the overview of deposits (p. 65ff). Of the leading mineral products of the earth only petroleum and tin are missing. In ancient Korea this abundance was not used, however, to more than the most modest extent, for the scientific–technical view of the world was alien to the Koreans. The foreign companies, to whom the Korean government granted mining concessions from the end of the Sino–Japanese War (1894/95) on, were the first to begin real mining enterprises. It is due to their activities alone that production reached the still very modest value of 6.1 million yen in 1911. The Japanese had adopted the scientific–technical view during the Restoration Period, and this put them in the position to establish mines that could compete with the Western enterprises. It was natural that they soon began to consider these an alien element in their newly created economic system of Chōsen. Consequently, in 1916 a regulation was issued that in future mining rights would be conceded only to Japanese citizens or to legal corporations created under Japanese law. This regulation is still in force today. In addition, by 1939 the Japanese companies were able to buy up all foreign enterprises, so that all mines are in Japanese or Korean hands today. In 1936, of the 5518 mines in the country, 2944 were Japanese property, 2571 Korean, and only 3 still belonged to foreigners. The latter were among the most profitable enterprises, however, and the Korean ones were for the most part still primitive diggings carried out without any great amount of capital.

Mining began to flourish rapidly in the new Japanese foreign possession during the First World War with the lack of cargo space and the elimination of international competition. After the war, however, a severe depression soon set in, leading to the closure of a large number of mines. Slowly a new ascent began between 1925 and 1930. It increased when Governor–General Ugaki began to take an intense interest in mining in 1932 (p. 386f). Above all, the devaluation of the yen led to a rapid development of gold mining. Since the beginning of the conflict with China and the closing of the world markets, this upward movement has become tempestuous and has expanded to a large number of other mineral products. Today the mineral production is as intensive in all areas as in any way technically possible. In 1937 it had reached a value of 150.1 million yen and thus amounted to about 25 times as much as the production in 1911. Since then it has continued to rise. The most recent data, from 1936 and 1938, on the situation in the different branches of mining show the picture recorded in Table 41. I have no figures on mining of quicksilver, manganese, antimony, nickel and asbestos.

Gold mining is most important by far. Together with silver, with which it is very often directly connected, it provided 64.7% of the total value of mining production in 1936. To increase gold mining, for the past decade the government–general has been granting subsidies for prospecting activities and for the transport of low-quality ores to the nearest railroad or shipping route. In 1911 the gold production amounted to 4.5 million yen. After a peak year in 1916 it dropped to 3.4 million yen (2668 kg) in 1921; in 1928 it again amounted to 7.5 million yen (5553 kg). Then, however, under the influence of the circumstances described above, the tem-

	No. of establishments 1938		Production 1936			
	Operating	Suspended		Amount	Million Yen	(%)
Gold and silver mines	2774	1422	Lode gold	14,722 kg	49.9	45.2
			Silver	58,821 kg	2.8	2.5
			Gold and silver ore	79,029 t	9.4	8.5
Placer mining	304	162	Placer gold	2768 kg	9.4	8.5
Copper ore	11	9	Copper ore	2590 t	0.06	0.05
			Raw copper	3637 t	3.3	3.0
Iron ore	49	99	629,000 t with 283,000 t iron		1.9	1.7
Zinc ore	45	2			0.2	0.2
Lead ore	16	7		2738 t	0.8	0.7
Quicksilver ore	7	19				
Tungsten ore	17	16	Tungsten acid contents	1707 t	2.3	2.1
Molybdenum ore			Molybdenum contents	80 t	0.2	0.2
Manganese ore	0	3				
Arsenic ore			Arsenic ore contents	230 t	0.03	0.03
Bituminous coal and anthracite	149	257	Hard coal Anthracite Lignite	2,282,000 t	13.3	12.0
Graphite	89	94		40,914 t	1.0	0.9
Mica	14	16		70 t		
Fluorite	10	13		8740 t	0.1	0.1
Alunite	6	1		114,569 t	0.7	0.6
Barite	8	4		5113 t	0.07	0.06
Magnesite	4	3		14,258 t	0.04	0.04
Other minerals	399	484			15.0	13.6
Total	3902	2611			110.5	100.0

Table 41. Mining production, 1936 and 1938

pestuous upswing began to 31.3 million yen (12,130 kg, 1933), 48.5 million yen (16,296 kg, 1935) and 59.3 million yen (18,662 kg, 1937). In 1940 the 100 million yen limit was probably passed. The Korean gold production now contributes more than half of the total Japanese production. In 1937 it already amounted to 1.7% of the world production. Table 41 shows that this production was divided up among a very large number of enterprises, most of which must thus be quite modest in size. These small companies include all of the placer mining operations, which together still yielded 13.1% of the gold production even in 1936. During the present war they have even begun again to exploit the southern Korean alluvial deposits increasingly.

The gold mining areas are distributed across almost the entire country, because there are gray gneisses or "young granite" intrusions in almost all regions (p. 65f). Only the remote volcanic islands of Chejudo and Ullŭngdo and the Paektu-san area are without gold mines. In 1937 there were 13 mines that produced more than 1 million yen. The production was as follows: Taeyu-dong (P'yŏng-anbuk-do) 7.5 million yen; Unsan (near Pukchin, P'yŏng-anbuk-do) 5.5; Ongjin (Hwanghae-do) 3.3; Kŭmjŏng (Kyŏngsangbuk-do) 3.2; Sŏnghŭng (P'yŏng-annam-do) 3.0; Kimje (Chŏl-labuk-do) 2.6; Ŭiju (P'yŏng-anbuk-do) 2.0; Kwang-yang (Chŏllanam-do) 2.0; Suan (Hwanghae-do) 1.5; <Getsuden> (Chŏllanam-do, for the significance of angle brackets cf. note in foreword) 1.4; Haksan (Ch'ung-ch'ŏngbuk-do) 1.3; Sunan (P'yŏng-annam-do) 1.2; Chasŏng (P'yŏng-anbuk-do) 1.1 million yen. The province of P'yŏng-anbuk-do is thus foremost in gold production. Chŏllanam-do, Hwanghae-do and P'yŏng-annam-do follow. Some of the individual mines have been handled from a regional point of view in the regional section of this book (particularly p. 256ff). The larger gold mines now all belong to Japanese mining companies, such as the Nippon Kogyō K.K. (Japan Mining Co.) and the partly state-run Nippon Sankin Shinko K.K., founded in 1938 (K.K. = Kabushiki Kaisha = joint-stock company).

Silver production amounted to only 1702 kg in 1929, in 1937, in contrast, to 46,655 kg and thus 0.5% of the world production.

Coal mining takes second place by value and even first place by weight. The coal deposits were discovered and developed almost entirely by Japanese geologists and engineers (p. 188). In contrast to the gold deposits, they are restricted to much more limited areas. The anthracite and bituminous coal deposits occur in the strata of the P'yŏng-an and Taedong system (p. 66f). Mining was begun in the environs of P'yŏng-yang, where the conditions are very good for transportation and consumption, but the bedding is quite difficult for mining. In 1911 this "Southern Heian Coal-Field" was the only production center. Its yield was worth 0.4 million yen. The production here is in the hands of the Chōsen Anthracite Mining Co. and is still the most important today. The "Northern Heian Coal-Field," which lies in the counties of Kaech'ŏn and Tŏkch'ŏn, has been tackled only slightly thus far. The production is already stronger in the bituminous coal deposits of Sech'ŏn-dong in the vicinity of Hoeryŏng (Hamgyŏngbuk-do) and Munch'ŏn–Kowon (Hamgyŏngnam-do). A great future is in store for the deposits in the headwater region of the Namhan-gang, where a large electric power station was built near Yŏng-wol in 1937 to supply southern

Korea. Near Mun-gyŏng (Kyŏngsangbuk-do) and Hwasun (Chŏllanam-do) mining of bituminous coal has also begun. In 1934 Korea's production of bituminous coal and anthracite had reached 982,370 t (6.2 million yen).

Lignite is found, as mentioned on p. 66f, in the Tertiary strata in the Japan Sea fringe and those in Pongsan (Hwanghae-do) and Shinanju (P'yŏng-annam-do). Since the main railroad line Seoul – Shinŭiju runs past the last two deposits, they are being intensively mined. The production is much greater in the Japan Sea fringe, however. The most important mines lie near Kyŏng-won, Kyŏnghŭng, Hoeryŏng, Kyŏngsŏng, Chuŭl, Myŏngch'ŏn and Kilchu (Hamgyŏngbuk-do), Hamhŭng and Yŏnghŭng (Hamgyŏngnam-do) and T'ongch'ŏn (Kang-won-do). The mines of Aoji and Kogŏnwon here have reached a production of more than 100,000 t each. The total lignite production in 1934 was 706,277 t (3.8 million yen).

In the northern Yellow Sea fringe there are peat deposits at the mouths of the rivers, particularly of the Han-gang, Ch'ŏngch'ŏn-gang and Yalu. They are exploited by the farmers in the counties of Yŏnbaek-gun, Anju-gun, Pakch'ŏn-gun, Chŏngju-gun and Yong-amp'o-gun. Usually the peat is dug up from the bottoms of the rice paddies.

Very few figures are available to me on copper mines (p. 66). The most important mines appear to lie in northern Korea, for one in the Kaema upland (Kapsan) and for the other in the big bend of the Yalu (Tongjŏm). Although copper was already mined in ancient Korea (p. 29), the statistics for 1911 do not record any production. Even in 1921 the production was quite low, but during the past decade it then increased from 698 t of raw copper (1931) to 1434 t (1934) and to the amount mentioned in Table 41 and has increased considerably since then.

By value, the tungsten ores follow. They exist in Korea particularly in the form of scheelite ($CaWO_4$). The statistically recorded production did not begin until the past decade but soon increased to 368 t with a value of 0.7 million yen (1934) and then to the amount shown in Table 41. In 1939 further tungsten mines were opened, for this metal is indispensable for the steel and electric light bulb industries. The larger mines lie near Taehwa, Paengnyŏn, Kyŏngju, Koksan, Ch'ungju, <Kyurairei> and in Kŭmgangsan. The production for 1936 already amounted to no less than 7.5% of the world production.

The recovery of the iron ore deposits, in which Korea is rich, is just beginning. Particularly in this sphere Korea's mining still has a great future ahead. After all, the deposit at Musan on the Tumen is estimated to have reserves of 400 million t. The production amounted to 122,500 t in 1912 and had risen to 504,000 t by 1928. Since then the growth has slowed down, but in some years the production was already higher than Japan's. Magnetite ores are mined near Tanch'ŏn (Hamgyŏngbuk-do), Ch'ungju (Ch'ungch'ŏngbuk-do), Yang-yang (Kang-won-do) and Kimhae (Kyŏngsangnam-do). The enormous deposit at Musan consists of magnetite–hematite. Pure hematite ores are produced by the mines in Iwon (Hamgyŏngnam-do) and Anak (Hwanghae-do). Hematite– limonite is mined near Kaech'ŏn (P'yŏng-annam-do) and in the many mines in Hwanghae-do, particularly near Songnim, Chaeryŏng, Hasŏng, Naet'o, Ch'ŏlsan, Ŭnyul and Songhwa. Hwanghae-do, whose mines have the advantages of profitability and convenient locations, is at present far ahead of the other provinces in production. But it will probably soon be surpassed by the Hamgyŏng provinces.

Graphite production has assumed an important role. In 1911 and 1921 it lay at the low sum of 0.2 million yen. In 1934 it had increased to 0.5 million yen (31,294 t); by 1936 1 million yen was reached, and the rise continues. In 1935 Korea already had 22% of the world production and was the second largest graphite producer in the world. Crystalline graphite (p. 66) is recovered primarily from the deposits in the gray gneisses of northern Korea (Kanggye, <Shinkori>, Sŏngjin, <Fushiki>, <Jōkan>). Amorphous graphite is also mined in Hamgyŏngnam-do (Yŏnghŭng). The main mines lie in Ch'ungch'ŏngbuk-do and Kyŏngsangbuk-do, however, (Yamano Getsumei Mine, Komiya Graphite Mine).

The statistics for 1911 and 1921 do not show any lead production whatever. In 1929 lead production amounted to 333 t, and it increased to 2738 t in 1937. In the same year the raw lead content of the Korean production was already half that of the Japanese.

The still scanty mining of zinc ore is completely under the direction of the Sansei Mining Co. This company operates the mines at Somin-dong (Py'ŏng-anbuk-do), Sŏngch'ŏn (P'yŏng-annam-do) and <Ginseki> (Hwanghae-do).

Molybdenum ore is mined by the <Tenkei Mine> (Hwanghae-do), the Kŭmgang Mine (Kang-won-do) and the Chōsen Mine (Chŏllabuk-do). Its production has also increased strongly in recent years, particularly since 1939.

Korea is also one of the largest mica producers on earth (1934, 103 t, since then less). The most important mines lie in the gneisses of the Mach'ŏllyŏng Ridge (p. 219).

Governor–General Ugaki included the mining of light metals in his big mining program in 1934 [302]. It has begun meanwhile. Particularly worthy of mention are the mining of the aluminum mineral alunite on the Hwawon peninsula in the southern coastal fringe (p. 372f) and the mining of magnesite. The magnesite mines lie in the crystalline limestones of the Mach'ŏllyŏng Ridge in Sŏngjin–gun.

From the standpoint of world production Korean mining is still a modest enterprise in most of its branches. The exceptions have been pointed out above. In the Korean economy as a whole it also plays a rather minor role for the time being. At the end of 1938 only 2.3% of the Japanese, 1.2% of the Korean and 4.9% of the foreign, predominantly Chinese, population of the country was employed in mining, and the value in 1937 amounted to only 6.1% of the total economic production. But within the framework of the economy of the Japanese empire, the production of gold, iron ore, lead, tungsten, coal, graphite and magnesite is already an important factor today, and it is increasing in importance from year to year. The repercussions of the most recent immense increase in the Japanese territory can, however, in no way be assessed as yet.

In 1934 Governor–General Ugaki still spoke of Korean industry as an "infant in the cradle stage" [302]. Meanwhile the situation has changed quite fundamentally in this area as well. It is not easy to give figures on the degree of employment, because it is difficult to draw the line between industry and handicrafts. For this reason it is only with reservations that we can refer to the figures in "The Japan Year Book" [XLI], according to which there were 252 "factories" with 14,575 workers in 1911, but 6298 factories with 207,002 workers in 1937. This would be an increase in the number of factories by a factor of twenty-five and in the number of workers by fourteen and would mean that the number of factories increased more highly than the number of workers, and thus that the number of workers employed in a given factory has decreased. The reason for this is that in the more recent statistics "industry" includes every small family business with more than five employees, whereas in 1911 the statistical records were not yet advanced enough for such inclusions. More important for comprehending the growth is therefore the fact that in 1911 the value of industrial production was 19.6 million yen, and in 1938 1140.1 million yen. This is a 58-fold increase. At any rate, of the large-scale economic activities, industry is the one that has experienced by far the strongest growth during the Japanese period. And the major part of this increase has occurred during the past decade. In 1931 the value of industrial products amounted to only 252 million yen, in 1934 to 367 million, in 1937 to 959 million. In 1931 the value of industrial production amounted to 23% of the total production, in 1937 it was already 33%, and the proportion has continued to increase since then. Today it presumably already equals that of agricultural production [663]. At the end of 1938 16.5% of the Japanese employed in Korea were employed in industry, but only 2.9% of the Koreans. This comparison already shows how strongly the industrial enterprises, as leading economic establishments, are dominated by Japanese labor. With the lack of initiative among the Koreans, management is entirely in their hands.

Increasing industrialization is a feature of economic development that Korea shares in our times with the vast majority of countries that were previously oriented primarily toward agriculture. There are other, special reasons for the astonishing growth in Korea, though, which again are based on the country's affiliation with the Greater Japanese economic empire. The seclusion from the world markets has made it necessary to activate the native possibilities for development to the utmost. The present Governor–General Minami recognized this necessity, and since 1936 he has energetically tackled the industrialization of Chōsen. The Japanese homeland's capital turned to these tasks with great energy. Still unexploited sources of energy and raw material reserves, low wages and the lack of certain regulations restricting capitalistic development in the homeland attracted enterprising persons to the nearby colonial area. Therefore Korean industry is being developed for the most part with "island capital." No other branch of the economy has been developed so highly by Japanese companies. Most of these companies are organized in the form of joint-stock companies and have their seat, or at least their financial resources, in Japan. An exception is the Keijō Spinning Co., which is supported by "peninsular capitalists" [663]. To deal with these companies and their organizations individually would not be in keeping with the geographical purpose of this book. Most are concerned with a single branch of industry, such as, e.g., cotton spinning and weaving, hemp processing, magnesium production, production of enamelware, hardening of fish oil, or beer brewing. Some, however, are active in several very different fields simultaneously, such as one of the largest of them, the Chōsen Tisso Hiryo Kaisha (Chōsen Nitrogenous Fertilizer Co.) in Hŭngnam, which produces not only nitrogenous fertilizers, but also explosives, magnesium metal and soybean oil. We have already mentioned the extremely varied activities of the Tōyo Takushoku K. K. (p. 384). The Korean branches of almost all of these companies were not founded until after 1932. Korea's industrial development during the past decade was carried out from exclusively pan-Japanese standpoints, according to the imperative needs of the general situation. In no other area is the integration with Dai Nihon thus as great as in this youngest branch of the Korean economy.

Of the total industrial production, with a value of 1140 million yen (1938), 30.9% was chemicals, 24.3% food and beverages, 14.5% textiles, 8.1% metal products, 3.1% ceramic products, 2.4% machines and tools, 2.1% gas and electricity, 1.5% paper and printing, 1.3% wood products [XLI]. The chemical industry increased its production from 42 to 353 million yen between 1931 and 1938, the metal industry from 6 to 92, the textile industry from 32 to 165 million yen [663].

The colossal development of the chemical industry was possible only because of the large-scale development of hydroelectric plants, which was done predominantly with capital from the mighty Noguchi concern. The enormous power plants on the southeastern edge of the Kaema upland and the construction of the Yalu dams, which is still in progress, were described in detail in the regional section, as was the establishment of the thermal power plant near Yŏng-wol in Kang-won-do. At the end of 1936 there were already 147 hydroelectric and thermal power plants with a capacity of 2.4 million kW altogether. The high tension lines running from them are beginning to be a characteristic element of the Korean landscape. The investigations of the hydroelectric production capacity commissioned by the government–general showed 148 suitable sites for hydroelectric plants with a capacity of 5.7 million kW. Forty such plants with 2.8 million kW had been authorized for implementation by 1938 [VI]. The vast majority of them are based on the same principle as the plants on the Pujŏn-gang, the Changjin-gang and the Hwangsuwon-gang. Dams are built in the headwater regions of rivers that rise in the Hamgyŏng Range and the T'aebaek Range and flow northward or northwestward. The water is diverted through penstocks under the divide, and the intake conduits take advantage of the strong gradient on the Japan Sea side.

The main plant of the Chōsen Nitrogenous Fertilizer Co. is in Hŭngnam near Hamhŭng, in the vicinity of the Pujŏn-gang and Changjin-gang hydroelectric plants. A second one is near Kŭktong in Hamgyŏngbuk-do on the high tension line Hamhŭng – Ch'ŏngjin. These plants produced 418,000 t of ammonium sulfate, 109,000 t of ammonium phosphate and 42,000 t of calcium phosphate in 1937. The production of nitrogen compounds had already reached one-eighth of that of the German empire, the largest producer in the world, in 1936. Part is used in Korea itself to fertilize the rice paddies. Most goes to Japan. The lignite mining in Hamgyŏngbuk-do and the electrification have recently led to the construction of the first coal liquefaction plant near Nanam, which, however, has an annual consumption of only 100,000 t of coal.

Since Korea is primarily an agrarian country, industry has been concerned above all with the processing of agricultural products. Both the food and beverages industry and the textile industry use almost exclusively raw materials produced in the country. They are located primarily in the vicinity of the areas of production, but simultaneously where there is labor, transportation or consumption. The rice hulling and polishing plants are located above all around the centers of rice growing in the Ch'ungch'ŏng–Chŏlla region, but also in the consuming cities of Seoul, Inch'ŏn and P'yŏng-yang and the ports of exportation, Chinnamp'o and Kunsan. P'yŏng-yang is additionally known for its corn products industry, which was previously in American hands. At the same place is Korea's only sugar factory, which originally processed sugar beets grown in the vicinity and now has to depend on imported sugarcane from Taiwan, since the growing of sugar beets was abandoned. There are wheat mills in Seoul, Inch'ŏn and Sariwon. On the Japan Sea coast of Hamgyŏngbuk-do lie the soybean mills that process Manchurian beans. Fish canning factories are distributed along the entire coast of the Sea of Japan, from Ch'ŏngjin to P'ohang (p. 442). In Seoul and its suburb Yŏngdŭngp'o, P'yŏng-yang and a few other cities beer brewing has been developed, whereas until 1931 the entire beer consumption was covered by imports from Japan. The most common alcoholic beverage, rice wine (Jap. *sake*), reached a production value of 81 million yen in 1937. Despite this, sake for 2 million yen had to be imported from the homeland for the Japanese population. Cigarette making, which has been run by the state since 1921, is concentrated in Seoul, P'yŏng-yang, Taegu, Pusan and Chŏngju. In 1930 6.4 billion cigarettes were produced.

The production of cotton, hemp and silk fabrics reached a value of 68.4 million yen in 1937, but was nowhere near sufficient to meet the needs of the country. Cotton ginning plants and plants for producing cottonseed oil have been built in Kunsan and other cities in the Ch'ungch'ŏng–Chŏlla region. Cotton spinning and weaving are also localized in the parts of the country where a large amount of cotton is grown. Because they require trained workers, they are limited to the cities. The largest factories lie in P'yŏng-yang, Yŏngdŭngp'o, Anyang, Suwon, Inch'ŏn, Kunsan, Kwangju and Pusan. With the enormous increase in cocoon production (p. 423), silk reeling and weaving have also assumed considerable proportions. The most important center is Taegu. Other cities in southern Korea, where most of the cocoons are produced, such as the capitals of the Chŏlla provinces, Chŏnju and Kwangju, have silk factories as well. The same is true of some cities in P'yŏng-annam-do and of Seoul. Hemp processing is still done primarily by Koreans as a cottage industry. Factories have developed only in Pusan thus far. Linen weaving is also concentrated there. Due to its proximity to the raw material, a modest rayon industry has developed in Ch'ŏngjin. Wool processing, like sheep raising, is still in the beginning stages.

Along with the mining of the ores in question, the iron and metal industry is still modest. Songnim is for the time being the only large center of pig iron production. But the new plants in Ch'ŏngjin will soon surpass it. In 1936 the pig iron production still only amounted to 216,000 t (7.9 million yen). The Mitsubishi concern exports most of the iron ore that it mines in Hwanghae-do via Chinnamp'o to Japan. Songnim also has the only good-sized steel and cast iron plant. The steel production amounted to 97,000 t (6.5 million yen) in 1935 and has even dropped somewhat since then. The iron is processed in Chinnamp'o, Yŏngdŭngp'o, Pusan, Kunsan and other cities. The production of electric power on the interior edge of the Sea of Japan fringe has recently made it possible to produce light metals, in particular magnesium, as in Hŭngnam and Sŏngjin. As a result of electrification, gold refining is increasingly being done in the country itself. Works of this type have sprung up in Hŭngnam and in Changhang, across from Kunsan.

The ceramics industry includes above all pottery and tile and brick making, as far as this is done by large companies (p. 189). Such works have grown up near many cities. The largest lie in Yŏngdŭngp'o and P'yŏng-yang. The porcelain industry has developed particularly in Pusan and Kaesŏng. The Korean cement production is exclusively in the hands of the Onoda Company, which has a plant in Sŭngho near P'yŏng-yang and a second one in Ch'ŏnnae-ri near Wonsan, each with an annual capacity of 320,000 t. A third one, near Komusan in the valley of the Susŏngch'ŏn (Hamgyŏngbuk-do, 140,000 t), was added a few years ago. This expansion became necessary because of the constantly increasing consumption, which amounted to 288,000 t in 1932 and already to 519,000 t in 1935.

Since Korea is still in the first stages of industrialization, the production of machines and tools is quite limited to date. Large numbers of repair shops for machines and automobiles have sprung up in the large cities, however, particularly in Seoul and P'yŏng-yang. The plants opened by the administration of the government railways in Yongsan, Yŏngdŭngp'o and P'yŏng-yang are also primarily for repairs. The locomotives of the government railways were formerly from Germany or America. Now they are built in Japan, despite the differences in gauge.

Paper making was taken over as a cottage industry from traditional Korea (p. 188f) and is still done in a large volume in the same manner today. The mass production of printing paper from wood pulp is carried out especially by the Ōji Company. It has a large plant in Shinŭiju, to which the wood is rafted down the Yalu. Other productive paper factories have grown up in Seoul and in a few cities of the Ch'ungch'ŏng–Chŏlla region.

Aside from sugar manufacture, the making of rubber shoes is the only industry using a raw material that is exclusively imported. Rubber shoes are used directly to clothe the foot in Korea, and not as overshoes, and are very popular in this form. They are fabricated by the Senwa Gomu K.K., which has factories in Seoul, Iri, Kangggyŏng, Kwangju, Taegu, Mokp'o and Pusan. There is another large factory for rubber products in P'yŏng-yang.

Korea's industry does not dominate large areas as yet. One could most nearly say that industrial plants are extensively distributed in the Ch'ungch'ŏng–Chŏlla region, but they are small for the most part and do not dominate the overall economic character, which is agriculturally oriented here as well. A new economic area with a strong industrial element is, however, rapidly coming into being in the North Korean Japan Sea borderland. Otherwise industrial enterprises are concentrated like dots in the conveniently located large cities. Industrialization and population growth have mutually influenced each other here as well. Many of these industrial plants are simultaneously located where there are raw materials or fuels. Cities of such a character are above all Seoul with its suburbs, in particular Yongsan and Yŏngdŭngp'o, and nearby cities, P'yŏng-yang together with Chinnamp'o, Kunsan, Pusan, Taegu and Ch'ŏngjin. Some of the cities, in particular Songnim and Hŭngnam, actually only developed in connection with the construction of factories.

8 The Settlements

The effects of the great cultural geographical development that Korea has been going through for the last generation extend into almost every last village today.

The types of village settlement described earlier (p. 160f) have not changed basically, to be sure, but in connection with the rapid growth of the Korean population the villages have become larger and their number has increased. In almost every good-sized village there is a police station today, a simple, one-story wooden building, often with a roof of zinc sheeting. In its courtyard stands a tall, wide ladder with a fire bell hanging from its end. Occasionally more convenient wooden frames take the place of these fire alarm ladders. Often there is also a small post office or even a school. In the seats of the 2365 *myŏn* administrations, but also in other villages, the Japanese element is reinforced by a similarly constructed Japanese inn and a couple of Japanese shops.

The larger the settlement is, the more it generally has been altered by Japanese influence. The individual aspects of this development showed up in the description of the different cities in the regional section. Here we shall try to outline the typical features of city development under Japanese direction. Where statistical comparisons are made, the 61 settlements with *fu* (city) or *yu* (town) rights (p. 386) are considered "cities." On the whole this includes all settlements with more than 10,000 inhabitants. It is impossible to use another form of the term city as the basis for statistical comparisons, because the population figures are available not for individual settlements, but for the large townships ("*ho-men*"), and they include not only the main settlement, which is generally the seat of the *myŏn* administration, but also other settlements.

The Japanese who have settled in Korea are for the most part townspeople (p. 389ff). Their energies have been concentrated first and foremost on the cities, which therefore have generally grown incomparably faster than the total population. Urbanization of this type is a general characteristic of the economic development of the earth in the past decades, it is true, but in Korea it again has special features that are related to the country's incorporation into the Japanese sphere. On the national average the population growth during the period from 1910 to 1937 amounted to 67.9%, or 2.5% per year. In 1910 7.5% of the total population lived in the 61 cities, in 1937, in contrast, 13.1%. An investigation into the growth of the individual cities shows the following results.

A first group of cities did not even exist in 1910, or only in the form of tiny hamlets, so that it is not possible to obtain statistics on their population at that time. These newly founded cities include Najin and Hŭngnam. A second group was founded shortly before 1910. These cities were thus in the very first stages, so that the calculated growth rates are enormous. This group includes Ch'ŏngjin (124.8% growth per year on the average for the period 1910–1937), Songnim (124.8%) and Shinŭiju (50.8%). Although it was founded earlier, Mokp'o (66.4%) also belongs to this group.

The average growth in the third group is around 30% per year. This group includes Taejŏn (31.2), Chinju (31.2) and Ch'ŏngju (27.9). The two latter cities were provincial capitals that were still small at the end of the Korean dynastic period. Ch'ŏngju has retained this function to this day; Chinju did not lose it until quite late, so that the impulses arising from its central character had a long time to take effect. At the same time both have developed into important centers of industry and trade. Taejŏn did not become provincial capital until relatively late, but it gained importance shortly after 1910 as one of the most important railroad junctions.

The fourth group consists of the cities whose annual increase lies between 13 and 20%. They include four more provincial capitals (Ch'unch'ŏn 20.1, Nanam 15.2, P'yŏng-yang 14.1, Kwangju 13.6) and several ports that were opened by the Korean government around the turn of the century, so that by 1910 they had already finished part of their early development, such as Kunsan (17.7) in particular. In all of them industry contributed a great deal to the good-sized increase. This is particularly true of P'yŏng-yang, which was already Korea's third largest city in 1910. It is only due to the rapid development of industry and mining that it has experienced such strong absolute growth that it still belongs to this group even relatively.

The largest number of Korean cities belong to the fifth group, whose growth span encompasses the narrow scope between 7 and good 10% per year and is thus three to four times as large as the average population growth. The cities in this group are of quite different types. First there are several provincial capitals that already had a good-sized population during traditional Korean times, because they were located in a very fertile environment. They are Taegu (9.7), Hamhŭng (8.3), Haeju (7.9) and Chŏnju (7.5). Also belonging to this group are the four oldest treaty ports, which were already highly developed in 1910: Inch'ŏn (10.5), Wonsan (10.2), Chinnamp'o (7.7) and Pusan (7.4). It further includes some cities in the pioneer zone in the north, such as Hoeryŏng (9.3), which were only small county seats prior to 1910 and have participated particularly strongly in the general upward movement in northern Korea. Finally, there are many county seats in the center and the south in this group, which, being central places, have benefited especially from the general upswing, such as, e.g., Suwon (9.1).

These comments have shown that it is by no means the large cities with more than 100,000 inhabitants that are at the top. With their new development the Japanese started primarily with what already existed. The only settlements that have grown into large cities under their direction thus far are ones that had already reached a fair size during the Korean dynastic period, so that their relative growth remains within limited bounds. This holds particularly for the national capital, which already had 278,958 inhabitants in the census of 1910. It is only due to the incorporation of extensive surrounding areas in April 1936 that Seoul has reached a population of more than 700,000. The population data for these newly incorporated areas are not known for 1910, however, so that it is impossible to add them to the population figures for 1910. Thus one compares completely different city areas when one takes the statistical data for 1910 and 1937 as a basis and calculates an annual growth rate of 5.7%. Prior to the incorporations Seoul had only a good 400,000 inhabitants and thus a growth rate that, at 1.9%, even lay considerably below the national average.

The last group, finally, consists of the cities whose growth remained at the same rate as or even below the national average. It includes provincial capitals that were dethroned at an early period, such as Kongju (2.5%) and Ŭiju (2.3), the closed treaty port of Masan (2.7), the port of Ch'ungmu (2.9), which had already developed vigorously during the traditional Korean period due to fishery and trade with Japan, and the former capital Kaesŏng (2.5), which has hardly been Japanized in the modern period.

On the whole these remarks show that the individual groups include cities with very different types of origin, and that the classification into these groups is of no profound importance. It is therefore inappropriate to draw general conclusions from the growth rates without detailed knowledge, as Linke does [604, p. 104].

As far as the location of the 61 cities is concerned, it is important to stress how close to the sea the majority of them are (Fig. 83). Forty-five of them lie less than 50 km from the coast. Of the other 16, only three (Ch'ŏngju, Taejŏn and Taegu) have more than 30,000 inhabitants. The location of the cities is thus decidedly peripheral, indeed more so than the distribution of the total population (p. 397). The east coast of the peninsula is strikingly lacking in cities. Between 36° and 39° N it has only one, Kangnŭng, and its population is merely 18,000. In the western coastal fringe, in contrast, there are 14 cities between the same parallels, including three of the five large cities. In connection with the distribution of the population as a whole, the peripheral location of the cities is again much less pronounced in southern Korea than in the two other parts of the country. The six southern provinces contain 31, that is more than half of the cities, in an area that covers 31% of the country.

The location of the cities is closely related to the distribution of the population as a whole in more than one regard. The reason for this is that their economic and political function is governed by the same forces as those that determine the overall distribution. Nineteen of the cities are ports and fishing towns, and no single important port is missing among them. Likewise the 13 seats of the provincial governments are included. Twenty–five, all of which are located away from the coast, are central places, i.e., local centers of trade and administration, and not one of them has a population of more than 30,000. Many are primarily large farming towns with particularly productive environs. As characteristic examples we can mention Ch'ungju, Ch'ŏnan, Iri, Namwon, Sunch'ŏn, Andong, Sangju, Ulsan, Sariwon, Ch'ŏrwon, Pukch'ŏng and Chuŭl. Only 10 of the 61 cities are not yet connected to the railroad network. The great majority thus have an important function in transportation today. Those ten are connected with the closest railroad station by good streets. One city (Pukchin) is very predominantly a mining place, in two the importance is based solely on industry (Songnim, Hŭngnam), in three (Nanam, Kanggye, Chinhae) the military function is most important. Five (Taejŏn, Shinŭiju, Sunch'ŏn, Choch'iwon, Hoeryŏng) are important railroad junctions or border towns. In the five large cities (Seoul, P'yŏng–yang, Pusan, Taegu, Inch'ŏn) a large number of the just mentioned individual factors coincide.

As pointed out on p. 389, almost seven tenths of the Japanese in Korea live in the 61 cities and towns. Thus the new development is expressed more obviously and with greater geographical effect in them than anywhere else. To structure the growth of the cities rationally, the first planning law was issued in 1921. It turned out to be premature, however. In 1934 a second one followed dealing with the improvement and widening of the city streets and building of new ones. Thus far it has been applied in Seoul, Ch'ŏngjin, Najin and Sŏngjin, in other words, apart from the capital only in the extraordinarily rapidly growing ports of the North Korean Japan Sea borderland that started out as tiny settlements. But even before it was enacted much had already been done to adapt the street patterns of the cities to the needs of modern traffic and hygiene. This development had already begun during the last generation prior to the annexation, as described on p. 294, using Seoul as an example. The main streets, particularly the main thoroughfares, have been straightened and widened in almost all large cities meanwhile, partially in the form of relief works in the interest of unemployed Koreans. They stand out in the city plans at the first glance as special patterns in the maze of old alleyways. Along them run the rows of masts with the telephone and electric wires, which rise far above the level of the one-story houses and are a characteristic feature of East Asian cities. Only in the capital have they begun to lay these wires in cables underground. In the large cities the main streets often have a surface of macadam or even asphalt. They are followed by the lines of the electric streetcars, which already form extensive networks in Seoul, Pusan and P'yŏng-yang. This transformation has recently been accompanied by the construction of sewage and water systems. In the period prior to 1910 only Seoul, P'yŏng-yang, Pusan and Mokp'o were supplied by waterworks. Otherwise the water needed by the cities was supplied by wells, whose water occasionally was not sufficient during periods of drought and often was not pure, so that epidemics were frequent. At the end of 1937 74 large settlements already had an untainted and adequate water supply.

The construction of new city wards in the growing cities occurred more or less haphazardly at first. Korean suburbs sprang up with alleys just as narrow and crooked as in the old parts of town. The districts inhabited mainly by Japanese also grew without large-scale planning during the first decades of Japanese influence, as, e.g., the Japanese part of Seoul on the northern slope of Namsan clearly shows. Not until recently has the planning of the street patterns of such suburbs been prescribed by law, and this was applied for the first time in Najin. The government–general has also contemplated supervising the construction plans of the buildings that are to be erected in these street patterns. Hopefully in the future the style of new buildings, particularly in new city districts, will be more uniform because of this. Thus far individual caprice has predominated in this regard. In the straightened streets two-story Japanese shops and multistory buildings were set beside old, straw-roofed Korean huts or tile-roofed one-story Korean houses. This gave the streets a completely disharmonious character, which was pointed out in the regional section for many cities. Today it is still the rule in many cities. Just as the street pattern reveals the discord between the many old alleys and the few straightened, wide streets, the outline of the streets shows the juxtaposition of low huts and tall stone buildings lined up along the widened or newly constructed streets: provincial government and other administrative buildings, town halls, hospitals, schools, banks, post offices, hotels, restaurants, department stores, police stations, chambers of commerce, military and police barracks, factories, scientific research stations, etc.

The oldest public buildings erected by the Japanese in Korea are unassuming wooden buildings, usually not more than two stories high, with low gabled roofs and walls paneled with horizontal boards that overlap like tiles. The Japanese legation in Seoul was built on the northern slope of Namsan in 1885 in this style transmitted from the native country. It still serves as the residence of the governor– general today. In the same style is the adjacent old building of the government–general, which houses the government museum today. In the provinces very many old public buildings erected by the Japanese represent this style, in particular police and forestry stations, but, e.g., the older parts of the agricultural experiment station in Suwon as well. It is found particularly commonly in northern Korea, where wood is abundant and where even today public buildings are constructed in this style. In the past few decades, however, they have generally changed over to more durable stone buildings, for the size of the buildings has grown with the administrative requirements. Pure natural stone, particularly granite, is chosen just as well as cement and reinforced concrete. Recently the cheaper red brick has been preferred. With the change in the building material the style changes, and in this respect no uniform and original line has been achieved. One encounters the most varied Europeanized and Americanized styles slightly altered by Japanese influence. For prestigious buildings of the first rank a neo-Renaissance style has become popular. Seoul's central station was completed in this style in 1925 and the new palace of the government-general was constructed in the same from 1916–26. The new buildings of the provincial governments have repeatedly been patterned on these buildings in the capital. Schools, barracks and department stores, which are usually built of brick now, often have the ungainly box shape that is so easily connected with this material.

We already mentioned the Japanese residential areas on p. 391, but we need to describe the manner in which they are combined with the old and new Korean residential areas within the city as a whole. In this regard we can set up a whole series of forms. It begins with the cities in which Japanese influence is negligible, because only very few Japanese live in them. In Fig. 91, the plan of the old capital

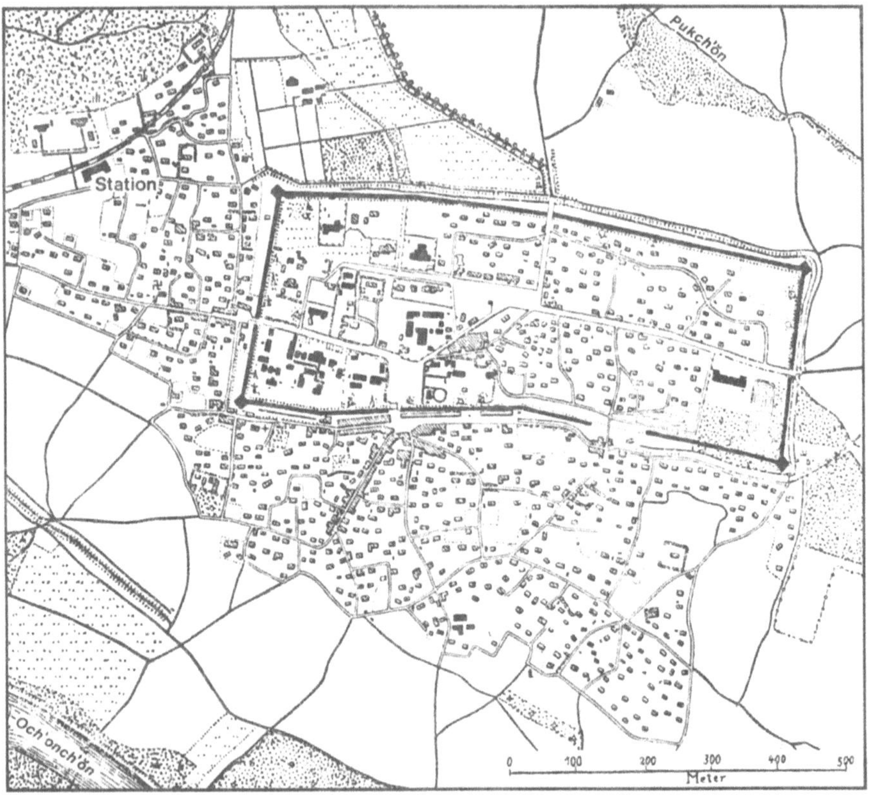

Fig. 91. Kyŏngsŏng, an example of a Korean city that has remained undeveloped (after an official plan)

of Hamgyŏngbuk-do, Kyŏngsŏng, is shown as an example. Here we still have a practically unaltered traditional Korean cityscape as described on p. 162f. The rectangular city wall is still well preserved. With its four corner towers and the gates lying primarily in the center of the sides of the rectangle, it surrounds a disorderly confusion of straw and tile-roofed houses. In the western half of the inner city, which probably represents the oldest town center, rise the – likewise low – buildings of the former provincial government. In the northwest and the south the built-up area has poured out of the walled enclosure in an equally irregular and disorderly fashion, and only in the extreme northwest, at the edge of the plain, where the main railroad line makes a curve around a range of hills projecting out from the west, do the train station and a few modern neighboring buildings rise as a symbol of the new age. Ŭiju, which is likewise a former provincial capital, represents the same type. Many county seats also belong to this type. Good examples are the Yalu cities of Sakchu, Pyŏktong and Ch'osan. Only the corrugated iron roofing that is gradually displacing the old straw, tile, shingle and slate roofs represents the new age.

The second type in the series already shows clearly recognizable Japanese influence. The main streets have been straightened. Often the straightening has gone so far that one could speak of a checkerboard pattern. Along these straight, wide streets stand a few tall new buildings with the above mentioned functions. With its Korean residential districts the city has grown beyond the area covered at the turn of the century, but there is no separate large Japanese residential district. Instead, the Japanese houses are predominantly on the straightened streets in the old city center. This type is represented particularly clearly by Kaesŏng. Ch'unch'ŏn, Kongju, Ch'ŏngju, Ch'ungju, Nonsan and Chŏnju also belong to it, but some of them have more Japanese houses interspersed. Kyŏngju is similar, but differs from the previous towns in that the built-up area is more dispersed. In Haeju the entire northern half of the town is already characterized by many Japanese houses.

In the third type good-sized Japanese residential areas of the type described on p. 391 have already developed. Depending on their location in relation to the old city, there are various subtypes. Most frequently the Japanese town is directly connected with the Korean town, which for its part has been remodeled in the same sense as in the second type. That is the case in P'yŏng-yang, where the new parts of the city constructed by the Japanese cover the area between the old South Gate and the central station and the western fringe of the city area. In Suwon new buildings have also begun to occupy the area between the southern end of the old town and the train station, which lies somewhat out of the way, but here many Koreans have also moved into the new part of town. The modern transformation has proceeded even farther in Hamhŭng, where more and more Japanese business houses are invading the center, while the extensive northeastern suburb is entirely dominated by Japanese barracks and other large buildings, along with Japanese residential houses. In Sŏngjin the Japanese town lies next to the harbor, the smaller Korean settlement just north of it. In the second subtype the Japanese settlement has been built right between the Korean residential areas. This is the case in Kwangju, the long, drawn-out capital of Chŏllanam-do. Finally, the Japanese town occasionally lies completely separate from the Korean settlement. A good example of this is Masan.

The same is true of Pusan. But here the Japanese town has long since become the main part of the entire establishment. The Korean settlement, which nestles in the northern end of the bay filled with tidal flats, is only an insignificant appendage. Pusan thus belongs to a fourth type for which the predominance of the Japanese settlement is typical. These Japanese towns are always characterized by methodically planned streets with right-angle crossings. Rectangular blocks of houses are more frequent than square ones. Occasionally the relief forces a division of the town plan, in which case the Japanese town consists of separate parts with variations in the direction and the width of the street grid. This is the case in Kunsan, Wonsan, Mokp'o and Chinnamp'o. This type is thus primarily represented by the large port cities. Inch'ŏn also belongs to it, but differs from the others in the European influence that still continues to have an effect on the cityscape. Taejŏn lies on a river plain, so that the terrain does not have any influence and there is only a single checkerboard system. Nanam is also similar to this type. Its primarily military character gives it a special touch, however. The Korean settlements affiliated with these towns, from which the names often came, have usually remained small fishing and farming villages. With their crooked streets they fit into the terrain better than the Japanese towns with their rigid street systems.

The end of the series is formed by the towns that are exclusively newly founded Japanese settlements. These are Shinŭiju, Shinanju, Songnim, Hŭngnam, Ch'ŏngjin and Najin. Here almost the entire city is dominated by the right angles of the new blocks. Shinŭiju, which lies in the alluvial plain of the Yalu, consists in its entirety of a single system of streets crossing at right angles. In Ch'ŏngjin, in contrast, it was necessary to conform as closely as possible to the mountainous terrain and the form of the coast, so that even the main streets have curves, and otherwise there are many individual rectangular blocks extending up the valleys. The plans included in the regional section demonstrate what we have said about the different types.

The capital is too extensive and the Japanese influence too varied to allow it to be classified according to a single one of the aforementioned types. Its Japanese city is a faithful copy of the native country in its unadulterated character. In the east there are, conversely, extensive areas that represent an untouched old Korea. The former legation district still shows strong European influence today. The center represents the second of the above types to a marked degree. In the northwest and northeast, in contrast, the palaces and parks of the old royal residence stand in glaring discord beside the prestigious modern buildings of the capital of Japan's oldest continental possession.

9 Traffic Routes and Transportation

For the economic development of the country the modernization of the transportation system was an absolutely necessary prerequisite. As the self-willed autarky of the old state as a whole and of the individual Korean family was broken, efficient transportation routes and harbors had to be built to handle the constantly increasing domestic and foreign trade and the correspondingly increasing passenger services, and there had to be a parallel development in means of communication. In the first decade of this century, certainly, military and political necessities took precedence in the modernization of the transportation system. Very soon, however, they were joined by economic considerations, and today Chōsen's transportation facilities have become an instrument of Japanese development and domination and of increasing Korean civilization that serves all purposes equally. No other sphere of the cultural landscape is as exclusively dominated by Japanese achievements as that of large-scale transportation.

The construction of a modern transportation network devoured very large funds in this country, which is extremely mountainous almost everywhere. Consequently the government-general was unable to develop the road and railroad networks simultaneously to the same degree. For strategic and political reasons railroads were given precedence. Road construction was therefore still in a poor state when I was there (1933), aside from isolated exceptions. Many roads constructed during the first years after the Japanese occupation, even first-class roads, had been abandoned in the meantime, because railroads had been built parallel to them. In regions with rice paddy plains they had occasionally been plowed under again by the residents. In mountainous terrain they were covered by grass and overgrown with shrubs. Boulders had tumbled down into the road cuts. These roads are shown in the topographical maps 1:50,000, however, and the maps include others that were no doubt part of the original plan but were never built, or at least not with the routing shown in the maps. The roads shown in the 1:50,000 maps have found their way into smaller scale maps. Anyone who, like Linke [604, p. 98ff], bases studies on the Korean road network on such maps is obviously going to come to false conclusions. Second and third-class roads are still improved today by compulsory enlistment of men who have not paid their taxes, this method of conscription having been retained from ancient Korea. These workers found it sufficient to pour layers of coarse gravel from nearby rivers and beaches or roughly hewn, sharp-edged hunks of rock onto the roads. No steam rollers or other such equipment were used. Instead, traffic had to pave the way for itself. Usually only two tracks in the width of the oxcarts or the automobiles were packed down into these beds. Quite generally the roads showed very proficient planning, often betraying Alpine engineering skill, but the surface was usually poor. Many river crossings were fraught with particular difficulties, because the concrete or wooden bridges had been built too narrow and too low, so that summer floods had washed them away. Also time-consuming were the frequent ferries along many roads. As the description by a more recent reporter shows ([XLVII], Vol. 1938), these road conditions, which meant that traveling entailed great difficulties, continued on after I was there. The feeder roads connecting important towns with the closest railroad station and the large thoroughfares that have to take over the role of still missing railroad lines were in better condition. Such roads cross the Hamgyŏng and T'aebaek Ranges, above all. It cannot be denied, therefore, that the construction of efficient roads has been very much retarded thus far, compared with railroad construction.

As an authority on the Korean road system, W.W. Taylor [667, p. 42], points out, the greatest period of road construction in Korea was that between 1907 and 1912. The same author stresses correctly that by no means all road construction was carried out for military reasons and that even those that were, were immediately utilized for economic purposes at the same time. The first road construction program was concluded in 1917, resulting in 2700 km. First-class roads were to have a width of 7 m and more, second-class ones 5.5 m and third-class ones only 4 m. Further programs followed, some of them planned from the standpoint of supporting poor and unemployed persons. As the annual report ([VI] 1938/39, p. 148f) shows, by the end of 1938 the government–general alone had spent a total of 120 million yen for constructing and improving roads. Further significant funds were provided by the provincial governments, which are responsible for the construction of second-class roads, and by the counties and townships, who build and maintain the third-class roads. Particularly high costs are connected with the construction of concrete bridges across the rivers, which are now built very long and solidly, with an eye to the summer floods. The irksome ferries are gradually being replaced by such bridges.

According to the report of the government–general, at the end of 1938 there were 38 first-class roads with a length of 3241 km, 97 second-class ones with 9966 km and 516 third-class ones with 14,524 km. In addition there were 278 km of roads leading to gold mines and to the most important forests. Thus there was a total of 28,009 km of roads in Korea 4 years ago according to this report, which corresponds to a road density of 12.7 km/100 km^2. This is not even half the figure for Portugal.

With the expansion and improvement of the road network, motorized public transportation has increased significantly. Such bus and freight lines run primarily from the county seats and other large towns to the nearest railroad stations. Other lines run for hundreds of kilometers through the country and handle the traffic that is not yet served by railroad lines in this mountainous country. There are scheduled and unscheduled freight and passenger lines. Their total length was 31,183 km in 1938.

Since, as we have said, the road network shown in the official maps is not in accordance with reality in many parts, Fig. 92 shows in its place the network of public transportation in 1930. In it one fact stands out clearly, namely, that along the main railroad lines there is no motorized public transportation. In this manner competition between rail and road is avoided from the onset. This is particularly true of the roads along the railroad lines Seoul – Taejŏn – Pusan, Taejŏn – Mokp'o, Seoul– Wonsan – Hamhŭng – Nanam – Namyang and Seoul – P'yŏng-yang – Shinŭiju. On the other hand, very many train stations along these four main railroad lines are served by feeder routes. In addition the map shows the large major routes, on which there is very busy truck traffic, by Korean standards. Apart from the short feeder routes serving the two main railroad lines, northern Korea has practically only such major routes. Particularly important are the lines: Shinbukch'ŏng – Hyesan or Shin-galp'ajin, Hamhŭng – Changjin and Shinanju – Hŭich'ŏn – Kanggye – Kŭmch'ang-dong or Chasŏng. Central Korea is crossed by the major routes P'yŏng-yang – Sŏngch'ŏn – Yangdŏk – Wonsan, Seoul – Ch'unch'ŏn – Inje – Kansŏng and Seoul – Hoengsŏng – Kangnŭng. They connect the large cities of the Yellow Sea fringe with the coast of the Sea of Japan. In southern Korea as well many major routes have a west–easterly direction. They include Ch'ŏngju – Mun-gyŏng – Andong – Yŏngdŏk, Kunsan – Chŏnju – Namwon – Chinju and Usuyŏng – Haenam – Pŏlgyo – Hadong – Chinju – Ch'ungmu. They are joined here, however, by several important north–south lines, such as Masan – Ch'angnyŏng – Taegu – Andong – Nogok, Chinju – Kŏch'ang – Kimch'ŏn and Mun-gyŏng – Namwon –

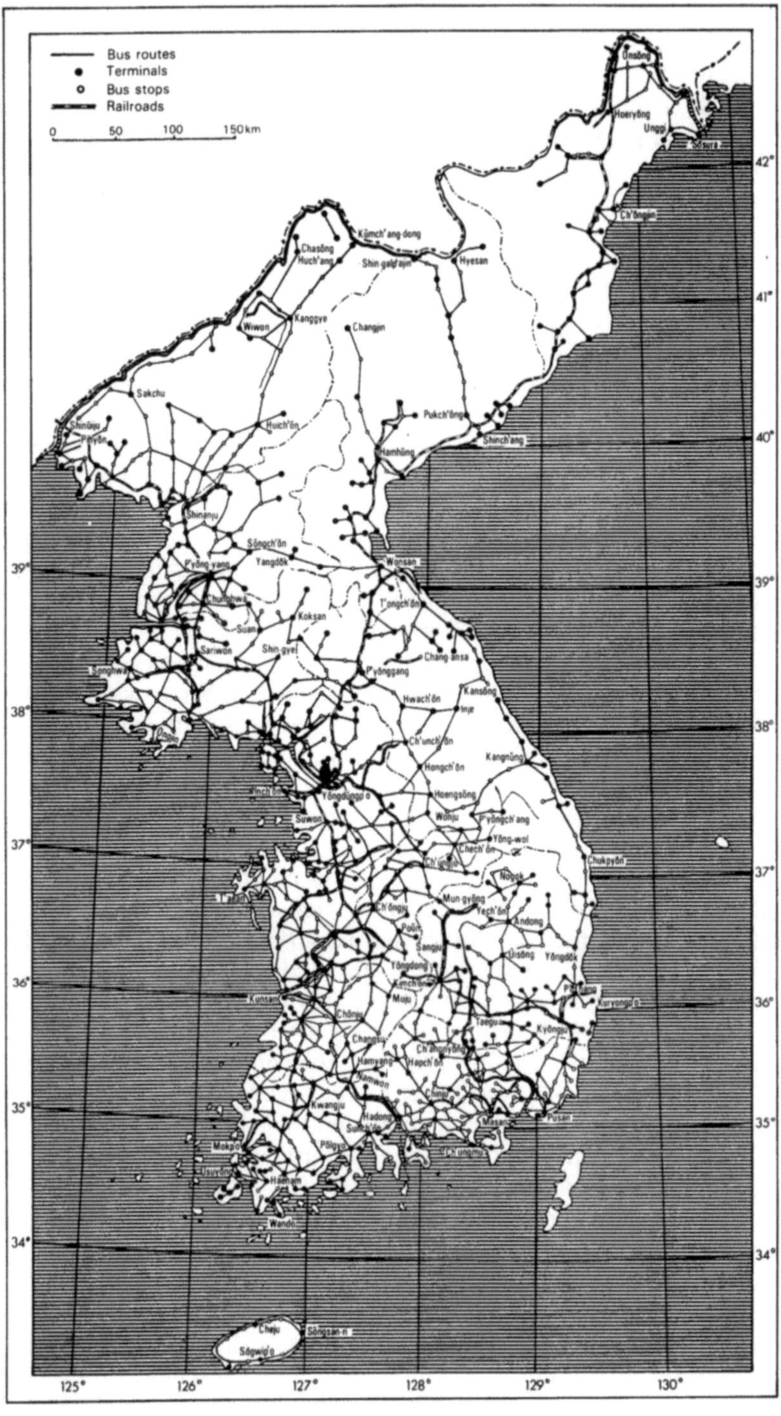

Fig. 92. Public transportation network in 1930 (after a map in the police department of the government-general)

Changsu – Yŏngdong – Poŭn – Ch'ŏngju. Of very special significance, finally, until the railroad construction is completed, is the line following the coast of the Sea of Japan, P'ohang – Yŏngdŏk – Kangnŭng – Kansŏng – Wonsan.

On the whole the network of motorized public transportation shows a close relationship to the population density and the density of cities. It is dense in southern Korea and along the entire Yellow Sea fringe, wide-meshed in the region of the T'aebaek Range and exceedingly full of gaps in the Kaema upland and the adjoining region to the west. It is densest in the southern half of the Naktong basin, on the one hand, and the southern Korean Yellow Sea fringe, on the other hand. The routing of the individual roads is naturally closely related to the terrain in mountainous country. The regional section provided several examples of this. The lowest passes represent the decisive criterion for crossing the main ranges, and very often structural valleys make construction easier for long distances.

Railroad construction began already during the Korean dynastic period shortly before the turn of the century. In 1896 an American businessman in Yokohama, J.R. Morse, obtained a concession from the Korean king for a section of track from Chemulp'o to Seoul. After the abandonment of the isolation policy, the traffic volume between the awakening capital and the nearby treaty port had increased greatly, but the waterway on the lower reaches of the Han-gang is not capable of handling very much traffic (p. 285). Railroad construction did not face any difficulties with the terrain, apart from the crossing of the Han-gang at Yongsan. As short as this stretch of railroad is (39 km), it was very important for the entire subsequent development. Morse used American tracks and cars, introducing the normal gauge to Korea in this manner. Most of the later railroad lines were built in this gauge, whereas Japan has the Cape gauge. Even before it was completed in 1900 the railroad passed over into the hands of a Japanese company. The Japanese thus knew in time to keep Western capital out of this politically so decisive sphere, transport and communications.

The king of Korea awarded two further concessions. Likewise in 1896 he entrusted a French company with the right to build a railroad Seoul – Ŭiju, which was intended to take the place of the old Imperial highway to China. Through their missionary activities, the French had gained a considerable degree of influence in Korea. This company began to use the rights invested in it late and hesitantly, however, for lack of capital, and finally gave them back to the Korean government. From the outset in 1898, though, the contract for construction of the stretch Seoul – Pusan was awarded to a Japanese company, the Keifu Railway Co. In the contract it was expressly provided that the line should have normal gauge. Preyer has given an excellent, well-informed survey of the history of railroad construction in Korea up to 1914 [673]. The official guide of the Railway Bureau of the Government – General [669] deals briefly with the entire history.

The Keifu Railway Co., for its part, did not begin constructing the railroad Seoul – Pusan (Jap. Keifu line, Kor. Kyŏngbu line) until 1901. When the Russo–Japanese War broke out in February 1904 three-quarters of the line had been completed. The remainder was constructed with the utmost speed. At the end of December 1904 the first train traveled the 450-km-long line. This line runs south from Seoul to Taejŏn without encountering any great difficulties with the terrain in the Yellow Sea fringe. Then the Sobaek Range has to be crossed, however and this occurs in its lowest pass, Ch'up'ungnyŏng (254 m). The most difficult stretch, technically, is that between Taegu and Miryang. In it a 1220-m-long tunnel had to be built under the divide. In the section from Samnangjin to Yangsan along the deeply entrenched lowest part of the Naktonggang, which is prone to flooding, a large number of expensive man-made structures were necessary. The line has a total of 24 tunnels with a combined length of 7300 m and two bridges, each more than 400 m long, across the Naktonggang and the Kŭmgang. In 1938 double tracking was begun and was to be finished by 1940, while all of Korea's other lines have remained single-track thus far. The Kyŏngbu line is the country's most important one. After all, it connects the capital diagonally across the highly developed southern half of the country with its most important port, from where the crossing to Japan is made.

After the outbreak of the Russo–Japanese War, the northern continuation of the Kyŏngbu line, the railroad Seoul – Shinŭiju (Kyŏng-ŭi line), was begun in great haste and initially in narrow gauge. The 501-km-long stretch was completed in the unbelievably short time from March 1904 to April 1905, an excellent feat by the young Japanese railroad troop. The Taedonggang and the Ch'ŏngch'ŏn-gang had to be crossed by ferry at first, though, and since the railroad runs exclusively through the Yellow Sea fringe, there were no real obstacles to overcome, aside from the numerous westward flowing rivers. Together with the line Dandong – Shenyang (Mukden), which was begun simultaneously and likewise originally in narrow gauge, this war construction handled a substantial portion of the supplies for the Japanese armies fighting on the Manchurian battlefields. After the war the Kyŏng-ŭi line was also rebuilt in standard gauge, and its routing was improved in many places. Simultaneously the 942-m-long bridge connecting Shinŭiju with Dandong across the Yalu was erected. This construction work was completed in 1911. Since then the 966 km long "Korean trunk line" Shinŭiju – Seoul – Pusan, which crosses the entire length of Korea, is a segment of the Eurasian railroad network. Until the outbreak of the current war with Russia, it was the last continental section of by far the most important connection between Europe and Japan. Twice a day it is traversed in each direction by the excellently equipped express train Pusan – Shenyang (– Changchun or Beijing), and for a time there was also a high-speed luxury train. In Pusan there is a direct connection to Shimonoseki. Still today the Korean trunk line is the most important connection between Japan and Manchuria.

The feeder line Masan – Samnangjin was likewise built during the Russo–Japanese war, and the 125 km stretch Ch'ŏngjin – Hoeryŏng was also begun by the military administration. It was at first narrow gauge, and its cars were pushed by coolies. With this we have listed the railroad lines that were built by the Japanese for purely military purposes. Only very recently have strategic viewpoints shown up again in the railroad construction in northeastern Korea. It is therefore false when Linke [604, p. 95ff] rates all Japanese railroad construction as strategic lines that neglect economic aspects. Rather, immediately after the annexation viewpoints related to the economic development of the country were put first, even in railroad construction. The administration of the lines changed from the Japanese Government Railways, which had taken charge in 1906, to the government–general. Only a few branch lines, above all the ones that run northeastward from the Kyŏngbu line and others that open up the Hwanghae peninsula have been left up to the initiative of private companies.

The first great period of Japanese railroad construction in Korea ends in 1914. In this year two more important lines were completed, the stretch Seoul – Wonsan (Kyŏng-won line, 223 km), which follows the Ch'ugaryŏng furrow in its entire length and connects the capital with the Sea of Japan (p. 300), and the stretch Taejŏn – Mokp'o (Honam line, 284 km), which crosses the best rice and cotton regions in the country.

The large-scale railroad construction projects of recent times have been devoted above all to opening up northern Korea, where the terrain presents by far the greatest difficulties. The most important ones are the stretch P'yŏng-yang – Sunch'ŏn – Kanggye – Manp'o (1931–39, 320 km), the Hamgyŏng line Wonsan – Hoeryŏng (1917–28, 625 km) and the Tumen line Hoeryŏng – Namyang – Unggi – Najin (1928–35, 240 km). In Manp'o and Namyang these lines are connected with the Manchurian railroad network by bridges across the Yalu and the Tumen. With the completion of the Manchurian line Changchun – Namyang in 1933 the railroad lines in the northeastern tip of Korea suddenly became a link in the connection between the Manchurian capital and the Japanese islands (p. 203), and consequently the government–general put the South Manchurian Railway in charge of running a total of 346 km of these lines.

In 1927 the Japanese Imperial Diet decided on a 12–year program for expanding the Korean railroad network and provided the necessary funds for this. It includes the completion of the diagonal line P'yŏng-yang – Wonsan (P'yŏng-won line, 214 km) and the line Anbyŏn – P'ohang (Tonghae line, 478 km). The same law provides for the purchase of 338 km of private Korean railroads. Additionally it calls for the construction of a second link between Seoul and Pusan. It is to cross the very mountainous interior of central and southern Korea. In 1941, under the influence of the war, the expansion of the railroad network was temporarily shelved, however, in favor of the development of mining and industry.

In 1939 the total length of the Korean railroad network measured 5411 km, 4177 km of which were in stretches run by the government railways and the South Manchurian Railway and 1234 km in private railroads. The average railroad density thus amounted to 2.5 km/100 km^2, which is quite a considerable figure, by Asian standards, and only marginally behind that of some European countries (e.g., Portugal 3.5 km/100 km^2). The present lines are still not sufficient for the great economic development plans of the government– general, however. Of the government-run railroads 235 km were narrow-gauge lines with a width of 0.762 m; the others are standard-gauge lines. The government railways had 558 stations. The average distance between stations was only 7.5 km and is thus well adapted to the diversity and population density of the country. They employ 33,000 persons, 14,000 of them Koreans. The Japanese officials are thus way in the majority (p. 388). Table 42 gives an overview of the development we have described [VI].

These figures provide a clear view of the absolute and relative growth of the government railways and their services in Korea. The latter have increased very rapidly in every respect, not only absolutely, but also per unit of length. Particularly impressive is the increase in the period 1930–1939. During this period the services far more than doubled, absolutely, and even in the figures per 1 km of length there was an increase of 50–90%. If we include the private railroads, the

Korean railroads transported 56.6 million passengers and 16.8 million t of goods in 1939 and earned 105.4 million yen. These figures do not yet include the northeastern Korean lines. They should cause the figures to rise by at least 10%. The majority of the passengers are Koreans. The signs with the names of the stations take this into account. In the middle of each sign the name is given in Chinese characters, to the left the Japanese pronunciation in the *hiragana* syllabary, to the right the Korean pronunciation in *han-gŭl* and underneath the Japanese pronunciation in Latin letters.

The Korean network consists of four main lines that converge in Seoul from the northwest, north, southeast and south. The two southern branches join already in Taejŏn, and the passenger and freight traffic of the Honam line is taken over by the Kyŏngbu line at this junction. These four lines were already finished for the most part in 1914, so that the centripetal character is the oldest trait in the general configuration of the railroad network. It will be reinforced even more by the second link between Pusan and Seoul, now under construction. Then the link from the middle of the east coast near Kangnŭng via Ch'unch'ŏn to Seoul, now handled by motorized vehicles, will be the only one lacking to make this centripetal character complete. In the later phases of development the construction of branch lines leading from the main lines into the interior increased in importance. They are joined by the branch lines to the ports of the peninsula. Particularly important are the connections P'yŏng-yang – Chinnamp'o, Iri – Kunsan, Sunch'ŏn – Yŏsu, Kyŏngju – P'ohang and the line that we have already described from Seoul – Inch'ŏn. There is not a single important harbor without a railroad connection today, indeed it was they that made the development of the harbors possible. The important lines that are under construction also take the need for peripheral connections into account, however. Such peripheral railroads are the east coast line Wonsan – P'ohang and the south coast line Songjŏng-ri – Posŏng – Sunch'ŏn – Chinju – Masan.

During the isolation period traditional Korea had only fishing ports without man-made structures. International ports were lacking, since the government did not allow any overseas trade. They began to be established in 1876 with the opening of the country. From this year on the number of treaty ports increased constantly. But the government did almost nothing to develop them and left everything up to foreigners, particularly the Japanese. After the annexation naturally all ports were open to Japanese shipping. Foreign shipping continued to be

		Absolute			Per km of length		
	Length in km	Passengers in mill.	Freight in mill. tons	Revenue in mill. yen	Passengers	Freight in tons	Revenue in yen
1911	1078	2.0	0.9	4.1	1877	824	3799
1920	1851	12.4	3.2	28.8	6705	1721	15,568
1930	2792	20.7	5.9	36.8	7396	2126	13,188
1939	3831	45.1	13.9	95.1	11,760	3634	24,832

Table 42. Development of the government railways

restricted to the treaty ports, however. Their number has not changed fundamentally since then. Masan was excluded; Najin was recently added. All large Korean harbors are treaty ports, for foreign shipping had discovered the most advantageous locations from the outset and had for the most part exhausted the possibilities in the generation prior to the annexation. There are 12 treaty ports today: Unggi, Najin, Ch'ŏngjin, Sŏngjin, Wonsan, Pusan, Mokp'o, Kunsan, Inch'ŏn, Chinnamp'o, Yong-amp'o and Shinŭiju.

Despite the enormous length of its coasts, Korea does not have many good locations for ports, although, in contrast to the adjacent coasts of the Ussuri region and Manchuria, they hardly freeze over in winter (p. 126). On the west coast the large tidal range, the rushing tidal currents and the dangerous shipping channels with their overabundance of reefs, shifting sandbanks, shoals, small islands and foggy days are obstacles. The east coast is cut off from the interior by the Hamgyŏng and T'aebaek Ranges, so that the harbors there find it difficult to gain a large enough hinterland, and the steep downwarped coast of the east side of the peninsula is also lacking in suitable bays. The tips of the bays of the south coast, in contrast, are filled with tidal flats, for the most part. The possibilities were thus limited from the start to a few suitable points, and solid and expensive constructions were often necessary to moderate the adversities of nature.

The most favorable stretch of coast by far lies on both sides of the mouth of the Naktonggang. We have called it the Naktonggang gateway (p. 27). Its advantages were pointed out on p. 395. A small tidal range is combined with deep bays, an extensive natural hinterland and the proximity of the Japanese coast. It is thus no accident that Pusan became the first port on the peninsula after the beginning of Japanese influence. The extreme northeastern part of the Korean coast has similar advantages. The bays there lacked the necessary hinterland, though, as long as the boundary of the Chinese empire lay only 55–75 km from this coast and Manchuria was undeveloped. Through the Ch'ugaryŏng furrow and the relatively low passes of the Wonsan arc of the Main Korean Range the port of Wonsan has the chance to develop its hinterland as far as Seoul and P'yŏng-yang and even into southern Manchuria. At the same time the spacious bay provides excellent opportunities for anchorage.

As southern Korea became a first-rate export country for products needed by Japan, an accordingly large volume of traffic developed between the south coast and the opposite Japanese coast. But due to the above described morphological and hydrographical conditions, it was difficult to build large-capacity ports. Ch'ungmu, Yŏsu and Mokp'o lie far out at the tips of their peninsulas.

The abundance of harbors on the west coast is the result of the fact that the regions of the country that are most populous, have the most cities and are most productive lie behind it and not that of any natural advantages. During the entire period prior to the opening of the country the Yellow Sea coast was dead, and even the so intensive cultural and martial contacts with the Middle Kingdom took place very predominantly via the roundabout way by land, as pointed out repeatedly in Part One. It was the encounters with foreigners and particularly with the Japanese that finally brought this dangerous coast to life. Shinŭiju–Yong-amp'o, Anju, Chinnamp'o and Kunsan–Changhang are its ports on river mouths. The mouth of the largest river of the peninsula, the Hangang is too shallow, the tidal range too great and the river floods too violent for the Yellow Sea port of the capital to develop at the mouth of the river. The capital's Yellow Sea port is instead Inch'ŏn, which, however, has the most difficult entrance and harbor conditions in all of Korea, and it is only due to the proximity of the capital that it became the country's second port chronologically and in rank.

The above named ports are primarily trading ports, serving overseas trade. Almost all are additionally fishing ports, some of them, such as, e.g., Ch'ŏngjin, of outstanding importance. We do not need to mention the fishing ports here again, however (p. 439). From the first years of their influence onward, the Japanese have done large-scale construction work on the trading ports. During the Protectorate work was already begun on 11 of them. After the annexation the work was taken over by the government–general and continued on a much larger scale. In all important harbors modern quays with loading facilities and railroad sidings were built. The shipping channels for many harbors, particularly those on the Yellow Sea coast, were deepened by constant dredging. Thus far, however, there are only very few ports where steamers with deep drafts can moor directly at the quay. They include Najin, Ch'ŏngjin, Pusan and Inch'ŏn. At most ports such steamers have to anchor in the roads, and transportation to and from land is carried out by lighters, which, considering the cheapness of labor in East Asia, does not have quite the same consequences as in Europe. In most cases the roads are protected against rough seas by long and expensive breakwaters. The difficulties presented by the enormous tidal range on the Yellow Sea coasts are surmounted in Inch'ŏn by docks, in Kunsan by pontoons.

The government–general has also put a lot of work into marking the coasts with beacons. The construction of lighthouses began in 1903 under the old government. In 1906 there were already 53 of them [682]. Considering how dangerous the south coast and especially the west coast is, this was still much too few. There was only one beacon per 330 km of the total length of the coast. Today there are 370 navigation aids, consisting of 195 lighthouses, 146 daylight signals and 29 fog sirens, so that now there is one lighthouse per 90 km of coastal length [VI]. Map 4 (cf. editors' note on p. 7) shows the location and range of the 81 most important lighthouses in Korea (after [683]). They have a range of up to 65 km. Understandably, the Yellow Sea coast and the southwestern corner are particularly well lighted. The east coast of the peninsula, which has no international ports, is not well marked. The northeast coast, whose importance for transport and communication is rapidly increasing these days, is again marked at closer intervals, although mostly with lighthouses that have a short range. In the coastal fringe of the Yellow Sea and along the south coast the lighthouses with their white paint often crown the tips of high rocky islands and are thus landmarks of these ingression landscapes (sketches in [459] and [460]).

The construction and the railroad connections of the ports have been carried out according to carefully considered plans, so that there is hardly any rivalry for hinterlands. Even in the three ports that lie so close together in the northeastern tip there is none, because in the past decade the volume of goods to be transported has increased so immensely that each of them has trouble handling its share. They are first of all the transit ports for freight and passenger traffic from southeastern Manchuria to Japan, and their hinterland extends beyond Harbin and Changchun. This is particularly true of Unggi and Najin. In Ch'ŏngjin exports of Korea's own products, in particular mineral, forest, fishing and industrial products, have a much greater share. The same holds for the ports in Hamgyŏngnam-do, particularly Sŏngjin, Shinch'ang and Sŏhojin, whose hinterland is the Kaema region only. Wonsan is the Japan Sea port of all of central Korea, and its hinterland will soon extend into southern Manchuria. P'ohang has only a small hinterland in Kyŏngsangbuk-do. Pusan's hinterland comprises the entire peninsula in freight traffic; in passenger services it includes not only the entire country, but most of Manchuria, indeed in times of peace it even extends as far as Europe. Masan has been restrained artificially. The other ports on the south coast serve only the rich direct hinterland. Mokp'o is the number one port of cotton exportation, as Kunsan is the most important port of rice exportation. With the lack of international ports on the opposite coast, Inch'ŏn's hinterland extends eastward as far as the T'aebaek Range. Above all, however, it is the passenger and freight services port for the capital. Chinnamp'o has the corresponding function for P'yŏng-yang. But it also has more extensive responsibilities; it exports most of the abundance of mineral, agricultural and industrial products of the Taedong and Chaeryŏnggang basin. Shinŭiju's main purpose is to transport on the wood that has been rafted down the Yalu. Its responsibilities and those of Dandong on the opposite bank will multiply, however, as soon as the Yalu has completely lost its character as a boundary river and the entrance from the sea has been improved.

By far the greatest portion of the maritime traffic of all ports is headed for Japan. Almost all are connected with Japan by regular freight and passenger ship lines. The northeastern ports also serve the very scanty traffic with Vladivostok, the Yellow Sea ports the traffic with the parts of China occupied by the Japanese, which has increased greatly in the past years. Prior to the current war European and American steamers occasionally called at the treaty ports, but these connections were not very frequent. A large portion of Korea's overseas trade with foreign countries was taken by Japanese ships to Japan or to Shanghai first. All ports simultaneously serve Korean coastal shipping, which also calls at the many small places, particularly on the east coast of the peninsula, where large ports are lacking, and thus performs feeder services for the major ports. Chejudo is connected to the peninsula via Mokp'o, and Ullŭngdo via Chukpyŏn (Map 1. cf. editors' note on p. 7).

The regular passenger and freight traffic crosses between Pusan and Shimonoseki twice daily on the two luxury steamers Kongō Maru and Koan Maru (7100 register tons) belonging to the Korean railways, which were put into service in 1936/37. In 1937 daily services between Yŏsu and Shimonoseki were also established. In addition there are eight Japanese shipping companies operating in Korean waters, the most important of which are the Nippon Yusen Kaisha and the Ōsaka Shosen Kaisha. Twenty-three shipping lines are supported by the government – general, 21 by the provincial governments, 8 by other public corporations. Finally, there are 203 purely private and mostly short lines serving coastal shipping only [305]. The number of ships registered in Korean ports was 87 in 1919 and, in contrast, 568 in 1937 with a net tonnage of 35,700 and 91,100, respectively. There has thus been a good-sized increase, although the total Korean tonnage is still very small. The tonnage of the individual ships is also still quite modest (410 and 160 register tons, respectively). The number of sailing ships serving maritime shipping has increased from 483 with 16,400 t to 1065 with 40,408 t [VI]. The proportion of junks in the total tonnage is thus still 30%. In 1938 13,062 steamers with a net tonnage of 14.2 million called at Korean ports; 89% of the ships and 86% of the tonnage were Japanese. In addition there were 26,275 junks with 0.51 million t. Of them only 45% were Japanese or Korean. However, 76% of the tonnage of the junks was national [XLIII]. The foreign junks were probably for the most part small Chinese vessels. I have no data available on the ships calling at the individual ports. The value of its trade is listed for each port in the regional section.

Quite unlike maritime shipping, river shipping has not been developed much by the Japanese. The reason for this lies in the nature of the Korean rivers, which we described previously. This makes them suited only to very limited junk traffic, which proceeds in the same manner as in traditional Korea (p. 190). Only on the longest Korean river, the Yalu, have the Japanese begun using the aircraft propeller-driven boats described on p. 248 over a 700-km-long stretch. In addition, with the rationalized utilization of the forest stands, rafting has assumed considerable proportions on the rivers flowing down the dip slope into the Yellow Sea. Japanese, Korean and Chinese rafts differ in their construction in the manner described on pp. 230, 247 and 258. Rafts of the same type as the Japanese ones are found in Yunnan as well, by the way (P. Deffontaines, *L'homme et la forêt.* Paris 1933[6]. Picture on p. 114/15).

Like maritime traffic, air traffic links Korea with connections going beyond the national boundaries. This is indeed generally the case. Very little leeway is left for inner-Korean air traffic. The entire length of the peninsula is covered by Japanese airways. Civil aviation was inaugurated by the Japanese in 1929. Today there are six civilian airports in Korea, in Ulsan, Taegu, Seoul–Yŏngdŭngp'o, P'yŏng-yang, Shinŭiju and Ch'ŏngjin. Beacons that emit revolving light rays at night stand along the line fixed by the five first mentioned places at Ulsan, Hwanggan, Taejŏn, Ch'ŏnan, Seoul, Sariwon, P'yŏng-yang, Chŏngju and Shinŭiju. According to the timetable for 1940 there was a scheduled daily flight from Tōkyō via Fukuoka, Ulsan, Seoul and Qingdao to Beijing, taking 11 h and 20 min. In Seoul there were connections to Shenyang – Changchun and to Lüda (Dalian). Between Fukuoka and Ulsan the route makes a detour around the fortified zones of Shimonoseki, Iki, Tsushima and Pusan–Masan. In addition there was an express connection between Tōkyō and Seoul via Yonago in 5 h. Finally, from 1937 on, long-range aircraft of German origin flew between Tōkyō and Changchun in 5–7 h with a stopover in Ch'ŏngjin (Map 4, cf. editors' note on p. 7).

For communications, too, there are no regionally differentiated data available, so that a geographical examination cannot be completely effectual. The first Japanese post office on Korean soil was established in Pusan in the year that the port was opened (1876). In 1896 the Korean government, for its part, proceeded to set up a postal service on a European model. In the year in which the Protectorate was set up (1905) there were 427 Korean and 89 Japanese post offices in Korea. In 1938 the number of post offices of the government–general had increased to 1183, and 444 million letters were carried.

Telegraph communications began in 1884 with the laying of the first submarine cable between Pusan and Shimonoseki. It was initially in foreign hands and was not acquired by the Japanese until 1910. Today a large number of telegraph cables run from the Korean south coast and from other places over to Japan. In 1894 the Korean government had the first overland telegraph route built from Seoul to Ŭiju along the old Imperial highway. In 1905 there were 44 telegraph offices in Korea. By 1938 the number had increased to 1019. On lines with a length of 51,624 km more than 10 million telegrams were sent.

In 1902 the first telephone line was established between Seoul and Inch'ŏn. In 1905 there were 16 such lines, in 1911 there were already 828, and in 1938 there were 12,969 suburban and long-distance lines. The total length of the telephone lines was 281,264 km. From 53,306 telephones 303 million calls were made. Direct telephone connections reach from Seoul to Tōkyō and Harbin [VI, XLIII]. The telephone and telegraph lines primarily follow the railroad lines, and very often the opening of new lines of communication was directly associated with that of railroads. Telephone poles along the roads, as is the rule in Central Europe, are an exception in Korea.

The first radio transmitter was set up in 1923 near Seoul. Others followed later in Ch'ŏngjin, Chinnamp'o, Ulsan, Pusan, Mokp'o and Cheju. There are radio stations in Seoul, P'yŏng-yang, Ch'ŏngjin and Pusan. The number of registered listeners is only 11,838 thus far, however, so that the possibilities for educating the Koreans by wireless are still quite limited.

All in all the Japanese have energetically pressed on with the development of transportation as well, and modern transportation facilities are of great significance for both the physiognomy and the physiology of the landscape, particularly in the cities and their environs. Naturally this development has not yet reached Central and Western European degrees of intensity.

10 Foreign Trade

Prior to the opening of the country, Korea's foreign trade was restricted to two or three points on the land boundary and to a few days per year, and it was very strictly controlled by the government. Its volume and value were infinitesimally slight. We already gave an impression of the exported goods on p. 32f. Overseas trade began with the end of the isolation period. Despite the great efforts of the Western powers to do business with Korea, the results were not very successful. In contrast, the trade relationships with nearby Japan developed favorably, and even some of the European, especially English, goods were brought to Korea via Japanese firms. On the whole, Korea's foreign trade remained slight, even though it had an upward trend. The total trade for 1890 is listed as 8.2 million yen, for 1900 as 20.6 million. In the year of the annexation it had grown to 59.7 million yen.

The development after that is shown in Fig. 93. Nothing demonstrates the enormous transformation that occurred in Korea from 1910 on more impressively than the immense rise in the export and import curves for specialized trade. In 1939 the total specialized trade reached the sum of 2395.2 million yen. From 1910 on there has thus been a fortyfold increase. This development is almost exclusively the product of Japanese initiative, which opened up the economic resources of the country in the above described manner so that a good-sized proportion of the production could be exported, enabling a corresponding amount of imports. The advantageous location of the country in East Asia and the fact that all important production centers are only a slight distance from the coast simplified the development of trade.

Figure 93 shows the development of total imports and total exports, imports to and exports from Japan and to and from all other countries in six curves. All six curves show essentially the same strong rise. It is particularly steep in the two total trade curves during the last war and the current one. The years after World War I, 1920–22 show a slight decrease, which is repeated during the Depression (1930–32). With the exception of the years 1924 and 1925 imports are always higher than exports, and the total specialized trade balance is thus always negative. This is a characteristic of recently colonized countries that are developing rapidly and therefore require a large amount of finished products for this development.

Japan developed the economy of its oldest continental possession primarily from the standpoint of its own needs. Thus the Japanese proportion of Korea's exports and imports has kept pace with the overall increase. Since 1910 it has always been very large. At that time it was 77% of the exports and 63% of the imports. In the former it sank to 71% in 1911, in the latter to 56% in 1913. Then the First World War, with the almost complete exclusion of the Western powers, brought a strong increase in the Japanese proportion. Exports to Japan grew to 90% in 1919, imports from there to 73% in 1918. From 1921 to 1932 the Japanese proportion of the exports was always above 90%, in imports the proportion rose gradually to 80% during this period. In 1933 the region in which the yen was currency began to expand considerably, and with it Korean trade with these other

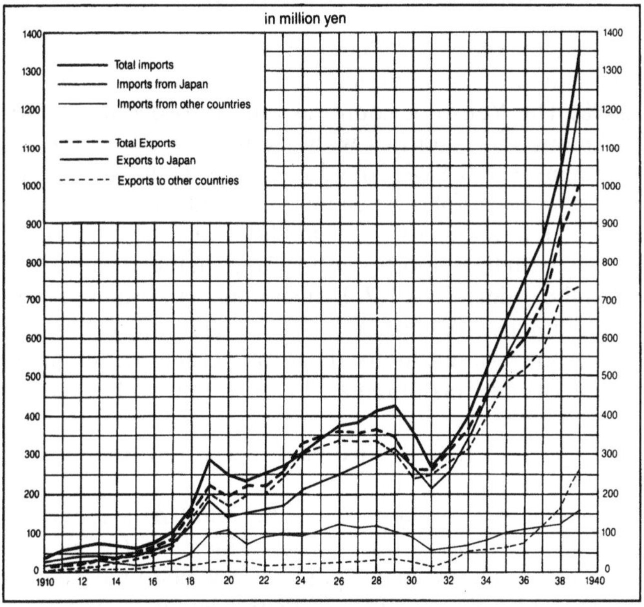

Fig. 93. Development of imports and exports 1910 to 1939

countries belonging to the Yen-Bloc increased substantially, particularly with the adjoining country of Manchuria, and since 1937 with the occupied parts of China. Above all Korean exports to these countries developed impressively. Imports from there did not increase to the same degree as imports from Japan proper, however, which rose by leaps and bounds, particularly in the years in which Korea's mining and industry were being developed. Consequently Japan's proportion of the exports dropped to 73% by 1939, but its proportion of imports rose further to 89%. In most of the years of the period from 1910 to 1939 Japan's share in Korea's exports was larger than its share in imports, and as a consequence in many years the absolute value of exports to Japan is greater than that of imports from there. The two Japan curves in Fig. 93 show this clearly.

In Korea's trade with "foreign countries", i.e., all countries aside from Japan, imports almost always exceed exports substantially. In 1922 and 1926 they even amounted to five times as much as the latter. Only in the two years 1938 and 1939 was the opposite the case. The increase in imports was interrupted already in 1914 by the beginning of the First World War. After the end of the war they at first suddenly increased and then stagnated. The period of large-scale railroad construction in Korea 1925–29 brought a renewed increase, which came to an end during the Depression. Only the last few years showed further growth. At this point it was not the big industrialized countries of the world that caused this increase, but the foreign countries of the Yen-Bloc. In exports the First World War does not show up as a drop in the curve, for the belligerent countries continued to have their needs. But the Depression later caused an incessant up and down. In 1931 exports to "foreign countries" again reached an extreme low; in this year they were only a bit more than twice as high as in 1910. With the incorporation of Manchuria, however, a steep rise set in in 1933, finally becoming so strong that the exports to these countries far exceeded the imports. Trade with the rest of the world sank to a minimum.

In 1938 the shares of the export markets were as follows (Fig. 94): Japanese islands 80.8, Manchukuo 13.9, occupied parts of China 2.5, Kwantung (Liaodong peninsula) 2.1%. The Yen-Bloc countries thus absorbed no less than 99.3% of the exports. Among the countries with foreign currencies the USA led with 0.15% and Germany with 0.12%. The shares of the countries exporting to Korea in the same year were as follows: Japanese islands 87.3, Manchukuo 5.5, occupied China 1.2, Kwantung 1.0%. Thus the Yen-Bloc countries did not have quite as large a share in the imports, with 95%, as in the exports. The other countries followed in the order: USA 1.7, British India 1.0, Indonesia 0.4, Straits Settlements 0.3, Phillipines 2.0, Germany 0.18%. The trade balance between Germany and Korea showed a surplus for the former (1.9 to 1.1 million yen). Korea's trade with Great Britain was minimal, likewise that with the adjoining Soviet Union.

The types of goods included in Korea's specialized trade can be seen from Table 43 (after [XLV]).

The detailed explanation of this table can be found in the remarks on economic geography in Chapters 4 to 7 of this section of the book (p. 398–453). Here we shall only need to stress a few major points, for which I have consulted sources [VI, XLI, XLIII and XLV] and additionally the data in the most recent issues of the *Annuaire International de Statistique Agricole.* The Japanese have made

Korea one of the foremost rice exporting countries in the world. Since 1934 it is always among the four countries with the largest rice export surplus, at times it has been number two, and for many years rice has led the list of exports. Most of the exported rice by far goes to Japan. Imports of low-quality rice, primarily from Japan itself, have almost ceased. In 1938 rice accounted for no less than 42% of the value of exports to Japan, because in 1937 Korea had the largest rice harvest ever before. In 1939, however, rice amounted to only 20% of the exports to Japan. Only a relatively modest quantity of rice went to other countries, primarily

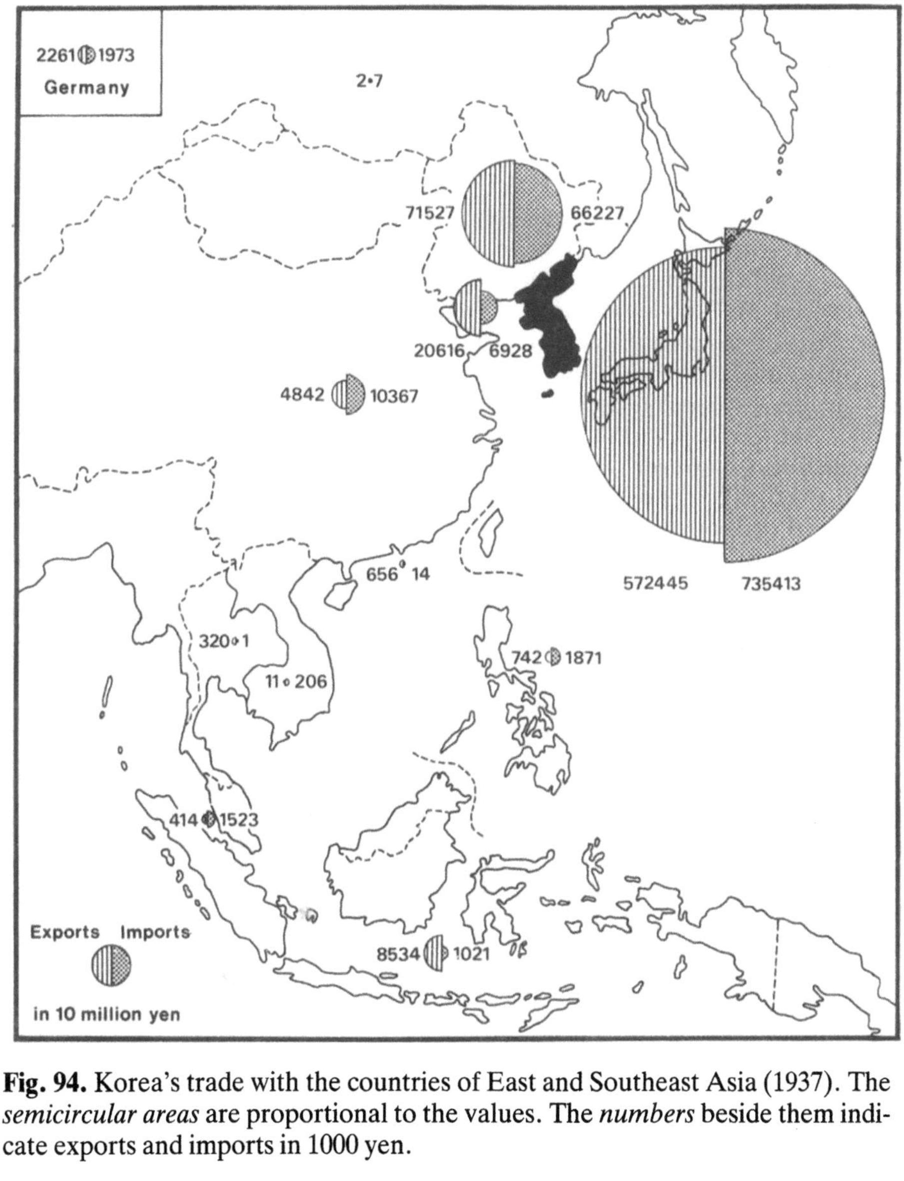

Fig. 94. Korea's trade with the countries of East and Southeast Asia (1937). The *semicircular areas* are proportional to the values. The *numbers* beside them indicate exports and imports in 1000 yen.

northern China. In the record export year it amounted to 0.47 million q, compared with 11.8 million q exported to Japan.

The second most important agricultural export product, by value, is raw silk. Exports to Japan lay between 2000 and 2750 t in the five years between 1934 and 1938. In addition there was a considerable export of cocoons. Part of these exports return from there in the form of silk fabrics and union silk.

Of the beans exported to Japan, according to Table 43, soybeans take first place. Soybean exports to Japan had a weight of 1.3–1.9 million q in the years mentioned above. In contrast, from the northeastern ports Manchurian beans were shipped to Germany and England. Additionally, not inconsiderable amounts of soybeans were imported by Korea itself from Manchuria (0.29 to 1.02 million q). Substantial amounts of the soybean products manufactured in the country were also shipped to Japan (soybean oil as much as 11,550 q, bean curd up to 1.2 million q). Accordingly, soybean oil (up to 3210 q) and bean curd (as much as 1.3 million q) were imported from Manchuria.

The last of the agricultural products listed among the goods exported to Japan is cotton. The exported amount lay between 82,600 and 135,100 q in the above mentioned years. Simultaneously fair amounts of raw cotton were imported via Japan, amounting at times to as much as 311,200 q, i.e., in some years exceeding the exported amounts. In addition, a considerable amount of cotton was imported from other countries directly, so that there was always a deficit in Korea's trade balance for ginned raw cotton. Korea's industry had already developed so strongly during the years included in the reports, however, that a large amount of cotton fabrics was exported to the Yen-Bloc countries, with the exception of Japan. The value of the cotton fabrics simultaneously imported from Japan was not much less, however. There is also a busy trade in other cotton products. As much as 123,000 q of cottonseeds were exported to Japan, and as much as half that weight imported from other countries. Cottonseed oil was an export item exclusively, to both Japan and other countries. As much as 92,000 q of cotton-

Exports 1939 in mill. yen				Imports 1939 in mill yen			
To Japan		To other countries		From Japan		From other countries	
Rice	149	Cotton fabrics	32	Ores, metals	138	Millet	20
Fertilizer	53	Machinery	19	Machinery	132	Beans	16
Copper	52	Fish etc.	18	Hardware	102	Coal	12
Fish etc.	32	Rayon staple fabrics	15	Vehicles	50	Cotton	10
Other ores	31	Vehicles	13	Silk fabrics	44	Fertilizer	7
Raw silk	24	Wood	10	Wood	39	Machinery	6
Beans	23	Rayon fabrics	9	Rayon staple	33		
Coal	14	Fruits, seeds	9	Cotton fabrics	32		
Fish oil	12	Hardware	9	Coal	28		
Cotton	12			Rayon fabrics	24		
				Paper	23		
				Stationery	23		
				Woolen fabrics	22		
				Rayon and union silk	22		
				Fertilizer	19		

Table 43. Korea's most important export and import goods, 1939

seed meal went to Japan. Thus both soybeans and cotton and their products have a highly varied role in Korea's foreign trade.

Of the fruits shipped to countries other than Japan apples head the list. Their exports reached 77,700 q. Apple exports to Japan were even more important. In 1938 81,100 head of cattle with a value of 8.1 million yen were exported, for the most part to Japan. Goodly amounts of hides were also exported (3.4 million yen).

Among the imported foodstuffs, millet, especially foxtail millet, tops the list. The significance of Manchurian millet for feeding the poorer strata of the Korean people was mentioned on p. 400. Since the production of sugar beets near P'yŏng-yang was discontinued all sugar used in the country must be imported. These imports reached a value of 10.5 million yen in 1938. They were partly in the form of refined sugar from Japan (up to 0.5 million q), partly as crude sugar from abroad (as much as 0.3 million q).

Trade with other agricultural products is secondary to the above mentioned ones in its importance for Korean economic life. In trade in wheat and wheat flour imports from Japan (up to 1.1 million q) and exports to other Yen-Bloc countries (up to 0.3 million q) are leading. By far most of the barley needed is produced in the country. There were also some imports from Japan (up to 0.8 million q), changing to considerable exports in the year of the surplus (1938). Despite the extensive government-supervised cultivation, substantial quantities of tobacco leaves were imported. They amounted to 48,190 q from Japan and even 68,010 q from other countries. Crude rubber amounting to a weight of up to 66,200 q was imported via Japan for processing in the country.

Of the nonagricultural trade items we should mention wood first. As Table 43 shows, exports from northern Korea to Manchuria and China contrast with much larger imports from Japan to southern Korea. With regard to manufactured wood products, a considerable amount of wood pulp is exported to Japan and cellulose and rayon fabrics to other countries. Much larger, by value, however, are the imports of cellulose, rayon, rayon fabrics, paper and paper products from Japan.

From what we have said about the high standing of Korean fishing, the large proportion of fishing products among the exports shown in Table 43 can be easily understood. Fresh fish, canned fish and fish oil go primarily to Japan, dried and salted fish to other countries. Considerable amounts of dried seaweed and laver are also exported to Japan, however. They had a value of 6.8 million yen in 1938. No marine products were imported, with the exception of salt.

Among the mineral products, the exports are led by gold, the total Korean production going to Japan. It is not included in the above figures, though, which refer only to specialized trade. Second place is now taken by copper. The role of iron ore and pig iron in foreign trade is extremely variable. The "other ores" in Table 43 otherwise include especially gold, silver and tungsten ores. Graphite exports to Japan are substantial. Goodly coal exports to Japan contrast with much larger imports from there and other countries. Petroleum and its derivatives are found exclusively in the list of imports.

Some industrial products, particularly textiles, have just been mentioned in connection with their agricultural raw materials. In certain areas of this sphere Korea has already become an important exporting country in the past years. Apart from cotton products, this is true especially of commercial fertilizers, which appear in second place in the above list of exports to Japan. This export of ammonium sulfate (as much as 110,000 q) and ammonium phosphate (up to 44,700 q) to Japan contrasts at times, however, with even larger imports from there and Germany. The third most important industrial export product is cement. Machines, vehicles and hardware are also already exported to the non-Japanese Yen-Bloc countries.

Imports of industrial products reach much higher values, however, most particularly from Japan, to a small extent from Western industrialized countries as well, most especially Germany and the USA. Crude metals, hardware and machines accounted for 30.3% of the imports from Japan in 1939. They also include vehicles (automobiles, railroad cars), woolen fabrics, which are not produced in the country, because of the minimal wool production, bicycles, matches, rice wine and beer. Korea's economic structure is currently in a stage of very rapid transformation. Whereas for decades after the end of the isolation policy it was primarily an area of agricultural production and only secondarily a mining region, now all forms of economic production are rapidly developing more or less equally. This development is clearly reflected in the character of foreign trade. Formerly agricultural products dominated exports by far, followed at a great distance by mineral products. The imports of the developing country consisted very predominantly of industrial products of all types. The latter continue to lead the imports, in fact, their proportion of the total imports has increased, but among the exports, mineral products have increased, and products of fishing, forestry and industry have been added to a remarkable degree. This development is reflected in Table 44.

11 Korea as a Part of Greater Japan

If we disregard the areas newly occupied during the current war, Greater Japan (Dai Nihon) covers 678,826 km^2 with 105.1 million inhabitants (1940); 32.4% of the area and 23.1% of the population are in Korea. Thus Japan's oldest continental possession forms a good-sized part of the Japanese empire. It represents its northwestern cornerstone. The domination of Korea is one of the most important prerequisites for Japan's control over the two adjoining marginal seas, which for

	Imports				Exports			
	Living animals	Food and beverages	Raw materials and semifin. products	Finished products	Living animals	Food and beverages	Raw materials and semifin. products	Finished products
1929	0.1	19.8	21.7	58.4	1.1	52.1	34.6	12.2
1937	0.2	11.5	22.8	65.5	0.8	42.3	33.6	23.3

Table 44. Percentages of Korean imports and exports according to groups of goods

their part form integrative elements of the Japanese empire. Thus a trait that is very characteristic of the Japanese empire, as K. Haushofer repeatedly stresses, namely, the interpenetration of Japanese–dominated land and sea areas, is shown to its full advantage in the northwestern sector of the Japanese sphere of influence as well, because Korea belongs to it.

As a result of the incorporation of Manchukuo into the Japanese continental area Korea has almost completely lost its peripheral location within the Japanese sphere of influence. Only at the mouth of the Tumen, along a 25-km-long stretch, does its land boundary still form the end of the same today. As a result its current strategic importance for the defense of the Japanese empire has decreased sharply. But the control of the two adjoining marginal seas and the Korea Strait, which provides access to the heartland, is safeguarded by a carefully considered system of "strategic zones" on the adjoining coasts, and fortified areas on Korean soil play a highly important role in this (p. 382 f, Map 4, cf. editors' note on p. 7). In a strategic sense, however, Korea itself does not need to play its historical role as a land bridge any more for the time being. The experience of the present war clearly shows, though, that passages even over very narrow straits that are completely under the control of one of the belligerent parties above water can be endangered under water by single submarines. For this reason Japan is in the process of considering a plan for a gigantic underwater tunnel that is to run from the Korean mainland via Kŏjedo, Tsushima and Iki to Kyūshū. In this connection they are thinking of building a standard-gauge express railroad to Tōkyō, so that a continuous standard-gauge connection from the mainland network of China, Manchukuo and Korea to the imperial capital is possible. For the time being it is not even worth putting a railroad ferry between Pusan and Shimonoseki into operation because of the different gauges of the railroad networks of island and mainland Japan.

Nevertheless Korea has been playing a more important role from year to year in the transportation system of the Japanese empire (Fig. 95). In this sphere the peninsula still forms the land bridge that routes the traffic from the Japanese islands into the northwestern periphery of the Japanese-dominated area of the mainland. This is above all the task of the Korean trunk line Pusan – Seoul – Shinŭiju – Shenyang (Mukden), which joins the South Manchurian Railway in the old capital of Manchuria. From here the overland traffic radiates as far as Changchun and Beijing. Parallel to this main line a second connection from Japan to Manchuria has been developed since 1933, the line Tōkyō – Niigata (Tsuruga) – Ch'ŏngjin (Najin) – Changchun. At the Korean– Manchurian border the railroad to Ning'an – Mudanjiang – Harbin branches off from this line. Whereas the access from Japan to the trunk line crosses 224 km of water, the northeastern stretch includes 860 or 900 km by water, respectively. This is the shortest connection from the imperial capital to Changchun, Harbin and the Transsiberian Railroad. It crosses Korea only on the the 215-km-long railroad stretch Ch'ŏngjin – Namyang, while the trunk line between Pusan and Shinŭiju is 966 km long. These pan-Japanese long-distance connections crossing Korea will soon be increased by others, particularly the line P'yŏng-yang – Manp'o – Changchun and the stretch (Tsuruga) – Wonsan – P'yŏng-yang – Shenyang. The latter will absorb some of the traffic from Japan to southern Manchuria.

As land to be settled by Japanese Korea is of no real significance, as pointed out in the second chapter of this section. The 0.6 million Japanese who live in Korea are only 0.85% of the total number of Japanese, which can be estimated as 74.5 million in 1938. The constantly increasing number of Koreans is economically of great value for Japanese purposes, but politically it is still a burden at the present and a task requiring much patience and effort.

All the greater is the role that Korea plays in the pan-Japanese economic system, and this role increases from year to year. In 1938 Japan's total imports amounted to 4.9 billion yen, and 46.2% of this sum came from the foreign possessions (Korea, Karafuto, Taiwan, Kwantung, South Pacific). Thus imports from the foreign possessions amounted to almost one-half of the total imports, and Korea supplied almost one-third (31.2%) of the former. Japan's total exports in the same year amounted to 5.2 billion yen. Moreover, 49.0% of these were to the foreign possessions, and of these exports 36.0% went to Korea. This integration of the island Japanese and pan-Japanese economy with that of Korea has increased extraordinarily in intensity since 1939 due to the war, the consequences

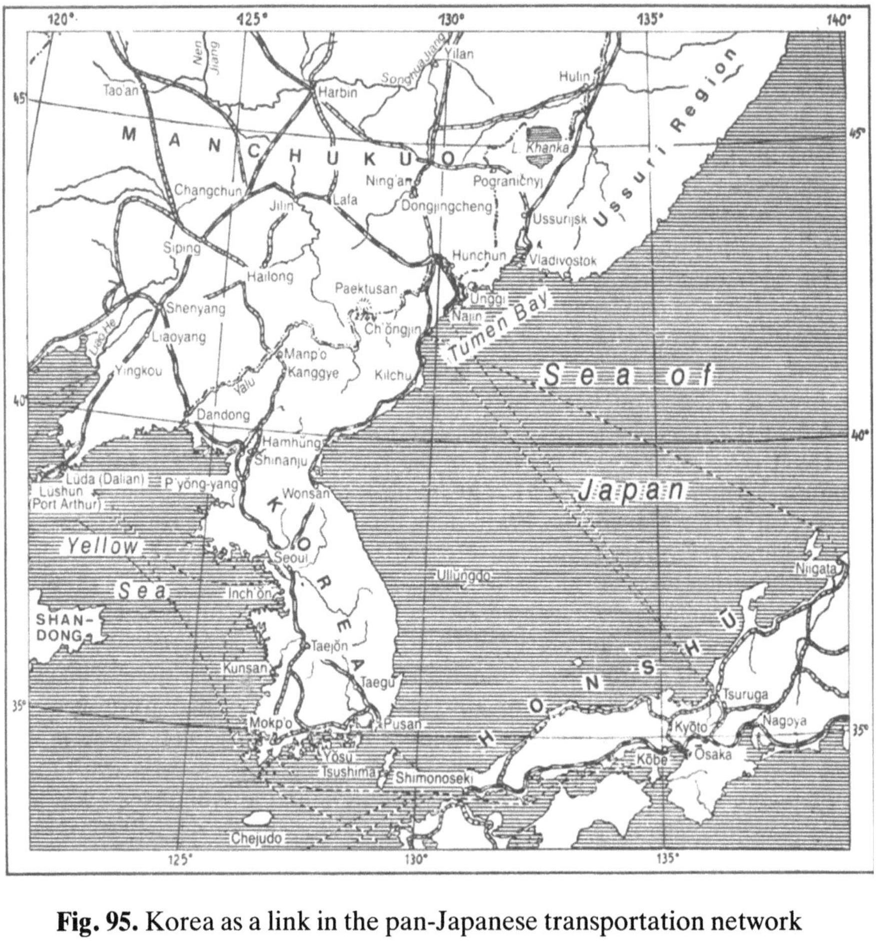

Fig. 95. Korea as a link in the pan-Japanese transportation network

of which are now affecting the entire globe most severely. Japanese trade is restricted almost entirely to the Japanese-dominated areas in East and Southeast Asia and in the Pacific at the moment. As a result the role of the foreign possessions has increased even more. The connections between the Korean and the pan-Japanese economy and their various features were described in great detail in the last chapters. They are the work of a generation of laborious development. The results of this development are summarized for the various branches of the economy dealt with above in Table 45.

The third column of this table shows how strongly Korean labor dominates the different occupations. The role played by the Japanese population element in them as lawmakers, managers, organizers, procurers of capital, industrial leaders, planners and research scientists has been mentioned repeatedly in the different chapters of this section. Four-fifths of the foreigners who appear in the table are Chinese. The columns showing how the nationalities are distributed among the different branches of industry demonstrate that among the Koreans farming predominates, while among the Japanese it is public services, commerce and manufacturing. The Chinese, for their part, are clever traders, but also industrious gardeners. The last columns give a summary of the immense increase in production values in all branches of industry since the occupation. At the same time they demonstrate the restructuring of the entire Korean economy that is in process. Whereas in 1910 agriculture dominated all other branches by production value, in the past years industry has come to equal agriculture, so that the share of agricultural production in percent has dropped to almost one half. Mining has also improved its position not only absolutely, but relatively as well. Absolutely speaking this is also true of fishing, but its relative growth has remained slight. Forestry production has not increased at the same rate as the economy as a whole, so that relatively speaking there has been a decline.

End of 1938	Number of employed persons including family members							Production value			
	Absolute	In percent of of nationalities			In percent of industrial branch			In mill. yen		In percent of total economy	
		Koreans	Japanese	Foreigners	Koreans	Japanese	Foreigners	1910	1937	1910	1937
Agriculture } Forestry	16,659,894	99.7	0.2	0.1	75.7	5.3	21.1	241.7	1134.1	81.,6	45.8
Forestry								19.2	138.7	6.5	5.6
Fishing and salt product.	338,882	97.1	2.9	..	1.5	1.5	0.1	9.6	04.8	3.2	3.8
Mining	274,188	93.8	5.3	0.9	1.2	2.3	4.8	6.1	150.1	2.1	6.1
Manufacturing	697,959	83.9	15.1	1.0	2.7	16.6	14.4	19.6	959.3	6.3	38.7
Commerce	1,585,271	89.7	9.4	0.9	6.5	23.4	29.6				
Transportation	235,279	83.6	15.9	0.5	0.9	5.9	2.5				
Civil service	892, 073	72.4	27.1	0.5	2.9	38.1	9.2				
Various occupations	1,559,218	98.2	1.2	0.6	7.0	2.9	18.0				
Unknown	395,987	93.6	6.4	..	1.6	4.0	0.3				
Total	22,633,751	97.0	2.8	0.2	100.0	100.0	100.0	296.2	2477.0	100.0	100.0

Table 45. The industrial branches in Korea

In order to characterize Korea's significance for the economy of Greater Japan in general and Japan in particular for the individual economic goods, I have calculated how large the Korean production was in 1938 in relation to the pan-Japanese production, on the one hand, and the island Japanese production, on the other hand. The pan-Japanese production includes that of Manchuria. The Korean rice crop in 1938 (44.8 million q) amounted to 23.4% of the pan-Japanese crop and 36.6% of that of Japan proper. The Korean population amounted at the same time to 15.8% of the pan-Japanese population and 31.8% of that of Japan proper. Because the per capita rice consumption in Korea was additionally much lower than in Japan proper, the enormous rice exports described in the last chapter were possible; 63.2% of Japan's rice imports came from Korea. The barley crop for the same year was 42.6% of the pan-Japanese crop and 79.5% of the island Japanese crop, so that 92.9% of the imports to Japan proper came from Korea. For soybeans the corresponding figures are: 6.4, 68.7 and 17.8%. In addition, 87.5% of the soybean oil imported by Japan came from Korea; 99.5% of the pan-Japanese cotton crop grew in Korea. But most of it was processed in Korea, so that only 1.2% of the total imports to Japan proper came from there. In contrast, 16.8% of the imports of cottonseed meal and even 96.3% of the cottonseed oil came from Korea. The Korean cocoon production amounted to 8.5% of the pan-apanese production and 9.3% of the island Japanese production, and 98.4% of the raw silk imported by Japan was of Korean origin. The corresponding figures for cattle raising are: 27.1, 93.9 and 100%. All of Japan's imported apples also came from Korea. The tungsten production was 96.8% of the pan-Japanese production, and all of it was shipped to the island country. The gold production accounted for 43.7% of the pan-Japanese production and 84.1% of the island Japanese production. The graphite mined in the Japanese sphere came almost entirely from Korea; 12.2% of the imported bituminous coal came from there and 20.5% of the imported ammonium sulfate. The value of the Korean fishing production was a good fourth of that of island Japan, which is known to be number one in the world. All of these figures impressively demonstrate the extremely important position that Korea occupies within the Japanese economic sphere. They intentionally include only products with substantial production and commercial values or weights.

In 1932 H. Schmitthenner gave a geographical classification of "the types of colonial areas" (Koloniale Rundschau p. 52–64) showing that to characterize the type of a given colony one must consider not only the political–economic standpoint, but also the location of the colony in relation to the homeland. He correctly considers "all areas that are in a relationship of dependency via organs of politics and power, of nationality and economics, to an area that is designated as the homeland and is culturally more highly organized" to be colonial areas. Colonies in the strict sense of the word additionally need the political title. According to this definition there can be no doubt that Korea is a Japanese colony, although this term is not used officially (p. 35, 381).

Since Korea and Japan lie on the same marginal seas, it belongs to the group of joint-sea or conmaritime colonies. Schmitthenner stresses that "such colonies can come into being only if the homeland is superior to the colonial area politically, militarily, economically and culturally or in the vigor and depth of its national traditions. The existence of an economic and cultural gradient between the homeland and the thalassic joint-sea colony is a necessary prerequisite." In our case this superiority has existed since the time of the Japanese Meiji Restoration, and only it enabled the island power to reach out to the nearby mainland bridge and to protect its new possession from the super powers China and Russia. Earlier attempts of the same type (p. 27, 31) failed, because at that time this economic and cultural superiority did not exist. Very many of the present-day relationships between the island nation and Korea in the spheres of strategy and transportation are comprehensible only in the context of their location as neighbors across the same sea. The role in the sphere of agriculture as well is due to the fact that Korea still lies in the same climatic realm and is thus able to supply the foodstuffs and raw materials that have long since become indispensable in the Japanese homeland, in particular rice and silk. The conmaritime location simplifies the shipping of these products quite considerably.

That and why Korea did not become a settlement colony, despite the neighboring location, has been pointed out repeatedly. Thus far it has been basically a domination colony. Schmitthenner defines this type as "a country that is dominated by the homeland. The native population is culturally advanced. It has retained its old economic system, which has merely been developed and modernized by measures taken by the homeland. The homeland's surplus population has no place here. In the economic sphere there are thus strong, though generally only indirect, relationships, brought about by the political domination, but between the population of the dominated colony and the homeland (in this case more appropriately called master country), there are actually no relationships at all. The character of the colony stands or falls with the political domination." In many respects Korea fits the thus described type. In two very basic points, however, there are deviations.

The old, purely agrarian–autarkic economic system was not retained. Instead it has been developed at the pace permitted by Korean comprehension and necessitated by the requirements of the Japanese homeland, and it is being supplemented by new economic forms. In the process the entire economic system has been changed fundamentally, and it has been done so according to pan-Japanese imperial ideas, as we have repeatedly stressed. In the course of the past generation Korea has thus made the transition from a pure domination colony to an economic colony.

Additionally, in the case of Korea there are quite close linguistic and racial relationships between the population of the homeland and that of the colony. The Koreans, however, have not attached any great importance to emphasizing this relationship thus far. They would have preferred to continue to form a family of their own, instead of being integrated into the family of their cousin, who has made himself at home in their house without invitation. But time has been on the side of Japan, and the Japanese methods today aim at the great goal of a spiritual assimilation of the Korean people. The economic–commercial assimilation is marching ahead with giant strides. In addition, the number of Koreans who understand and speak Japanese is growing from year to year, not only absolutely, but also relatively, due above all to the efforts of the schools. Optimistic estimates already include more than half of the Koreans in this group. The recent years of economic prosperity have temporarily lessened the Koreans' resistance to their membership in the Japanese empire, and Japan has thus slowly come closer to its goal of making Korea an "assimilation colony", i.e., of inducing the Koreans to accept the Japanese national polity and finally incorporating them into the Japanese nation. If no adverse conditions intervene, Korea will follow the same path as the countries around the western Mediterranean around 2000 years ago, which became "conmaritime assimilation colonies" of the Roman Empire.

12 Résumé: Essence and Change in Korea's Geographical Character

In its geographical location Korea is a transition zone on the Pacific rim of East Asia. Northern Korea is still a part of the uniform block that is the East Asian landmass, which is already slightly interrupted here, however, by the subsidence bay around the mouth of the Tumen and the ingression bay of the Gulf of Liaodong (Bo Hai). Central and southern Korea, in contrast, belong to the approx. 800 km wide zone of the marginal sea corridor, and Chejudo already extends out close to the zone of Pacific island arcs. The entire country is massively dominated by this transition from continental to maritime forms. Continental–maritime regional change is the primary formal principle. It takes place in a north-northwesterly to south-southeasterly direction, which is also the direction followed by the axis of the peninsula, and forms a transition from the purely continental forms of Manchuria to the purely maritime ones of western Honshū and Kyūshū. Its effect is amplified by the fact that everywhere in the world there is a tendency toward a change in forms in the direction of the meridians. This "planetarian" change is combined with the continental–maritime change here, because both run in approximately the same direction.

In its geological structure Korea is still basically a part of the East Asian landmass. The individual southwest–northeast, or west-southwest–east-northeast trending structural zones consisting of Precambrian to Jurassic rocks are closely related to those of Shandong, southeastern Manchuria and the Ussuri region. In the southeastern part of the peninsula, however, the "Tsushima basin," which is filled with Cretaceous sediments and eruptive rocks and extends as far as northwestern Kyūshū and western Honshū, already represents the transition to the Pacific rim, the Japanese islands. The recent volcanic extrusions are also typical phenomena of the continental rim. They are lacking in the west and northwest, which face the continent, but form important structural elements in the northeast, east and south. In some cases they form mighty stratovolcanoes (Paektusan, Ullŭngdo, Chejudo); in other cases they have welled up out of fissure systems belonging to the peri–Tunghai arc and then fill elongated structural valleys, especially the Ch'ugaryŏng furrow. The area in which they are distributed encircles the coast like an arc in the east and south and is closely related to the large-scale tectonic processes determining the basic features of the landforms. Their arrangement already brings west–east contrasts into the structural relationships of the country.

In the realm of landforms the continental–maritime regional change is most strongly pronounced in the northeast Korean block. This forms a part of Richthofen's Tungusic arc and thus represents a tilted fault block or an asymmetrical geanticline with its axis trending west-southwest–east-northeast. The dip slope, the Kaema upland, faces the continent and is thus completely exposed to its influences. Its drainageways run toward the continent, and because of the distance from the ultimate base level they have not produced recent erosion cuts to any great extent. There is a most powerful contrast between these old forms of the Kaema upland and those of the southeastern slope, the scarp face of the tilt block facing the sea. Here the influence of the recent downwarping of the Japan Sea area is noticeable all the way up to the drainage divide in the form of extraordinarily intense dissection. The peninsula itself, however, is predominantly under the influence of the east–west contrast. Only in the southeast, in the Naktong basin, do the forms of the continental–maritime change again appear, for the Naktonggang is the only river in the country that flows more or less parallel to the axis of the peninsula.

The continental–maritime change is expressed particularly strongly in the realm of climate and all geographical phenomena that are directly or indirectly dependent on it. The climate of midsummer is pretty much the same everywhere in Korea, however. The sea monsoon envelops the entire country in a mantle of tropical warmth at that time. Only the northeast Korean block with its strikingly small amount of precipitation is still on the side of the adjoining continental regions at that time of year. Most of the country is completely subject to the influence of cyclonic and typhoon downpours in midsummer, whose water masses are primarily oceanic in origin. Everywhere typhoons and cyclones are the main bringers of rain, not the summer monsoon itself. Despite its great elevation above sea level, the Kaema upland has continentally high temperatures at the same time. Thus the lines representing the reduced August isotherms do not reveal a planetary temperature gradient toward the north. In midwinter, in contrast, such a temperature gradient, and with it the continental-maritime change, is all the more pronounced. Then the north is much more often under the influence of those unmodified cold waves from the northern quadrant customarily known as the winter monsoon than the south. A decrease in the January mean temperature of 25° C within 9° of latitude, as is the case in Korea, is well-nigh unique on the earth. In connection with the immensely cold winter in the north, the annual temperature curve rises and falls very steeply, and the vegetation period is only half as long as in the extreme south.

Arising from these enormous climatic differences is a related continental–maritime change in the natural flora. Only the southern fringe of the peninsula and the remote islands of Chejudo and Ullŭngdo have a bottom zone with evergreen subtropical vegetation elements. Otherwise summergreen mixed forest predominates. In it the trees and shrubs that are sensitive to cold decrease in number of species and individuals the farther northward one proceeds and are replaced in this direction by plants that tolerate, or even require, low winter temperatures and are content with a short vegetation period. The zone of subtropical broadleaf evergreen forests is succeeded by the bamboo zone, the styrax zone and finally the oak zone, when one follows the coasts poleward.

The cultivated vegetation shows the same change. Certainly it consists for the most part of annual field crops, and as long as these are grown as summer crops, the degree of winter cold is immaterial to them. Subtropical crops such as rice and cotton can thus advance much farther north than the evergreen subtropical vegetation elements, especially since the Japanese have cultivated species that are adapted to the decreasing growing season in this direction. Thus the maritime–

continental change is not expressed fully until one takes the distribution of these varieties into account. But in general the intensity of rice cultivation as such does decrease toward the north. Cotton is not grown at all in northern Korea, with the exception of the Yellow Sea fringe, and the range of American varieties, which are sensitive to cold, already ends at the northern boundary of southern Korea. Our native crops have corresponding northern limits, which emphasize the maritime–continental change. Following each other from south to north are: the limits of winter naked barley, of winter wheat on paddies, of common barley on paddies, of winter barley on dry fields, of winter wheat on dry fields and finally of summer wheat, while the range of summer barley extends far beyond the northern boundary of Korea. Other plants, such as foxtail millet, barnyard millet and oats have their center of cultivation in the northern half of the country or are even limited to it. The ranges of tall, perennial cultivated plants show corresponding changes. Orange trees are limited to Chejudo, timber bamboo disappears at a latitude of 37–38°, the persimmon tree ends a bit farther north.

The proportion of the cultivated area in the total area of the counties decreases from south to north, and since the Koreans are first of all an agrarian people even today, it is not surprising that with the worsening climatic conditions, the population density also decreases toward the north. The northern fringe of the country is a pioneer zone, which has only begun to fill up with people under our eyes.

In contrast to this, the basic features of the traditional Korean cultural landscape are nearly the same all over the country. Differences between the south and the north like those existing around the time of the birth of Christ in the "period of the Three Hans" have disappeared during the last 2 millennia. Only in the racial structure are the migratory waves that came from the north on the one hand and the south on the other hand still recognizable today, in the form of certain regional differences. But the style of the Korean culture is surprisingly similar everywhere. The peninsula has been politically unified for 1300 years, and northern Korea has belonged to the same state for half a millennium, various states having previously extended far into the peninsula from the north Korean area. Everywhere one encounters the same white clothing on men and women, everywhere the same huts – differing only in their roofing – everywhere the same basic mentality with its merits and its disadvantages. In no other sphere does Korea appear more strongly as a unit than in that of the traditional Korean culture.

The regional features that are subject to the continental–maritime change in a north-northwest–south-southeasterly direction overlap with others that reveal a regular change in a west–easterly direction. Vast land areas spread out west of Korea too, separated from it only by the shallow Yellow Sea, which is subject to the full force of the seasonal climatic changes. East of the peninsula, in contrast, follows the wide and deep Sea of Japan, which is separated from the open Pacific Ocean only by the narrow ribbon of the Japanese island arc. As a result, in many of its features this west–east change is also climatically determined, in the sense of being more continental in the west and more maritime in the east. The eastern side is Korea's outer side. But this change includes features that have other roots. Since it is a regular change in forms leading over from the Yellow Sea (Kor. Hwanghae) to the Sea of Japan (Tonghae), we can speak of a Hwangdong change.

The Korean peninsula consists for its part of a tilted fault block. Its scarp slope faces eastward, the dip slope westward. The latter extends into northern Korea. From the Yalu on into southern Korea its crest is formed by the Main Korean Range. This tilt block is the meridional leg of the large peri–Tunghai arc, which extends across the western fringe of Kyūshū and the Ryūkyūs to Taiwan. In the southern two-thirds it clearly consists of elements arranged in an echelon pattern, in the sense meant by Tokuda. In southeastern Korea their place is taken by the structural valleys beginning at a very acute angle at the coast of the Sea of Japan and running straight into the body of the peninsula. The large rivers, from the Yalu to the Yŏngsan-gang follow the direction of the dip slope. The dip slope itself submerges gently under the level of the Yellow Sea, which is the result of a postglacial ingression. This ingression gave the Yellow Sea coast of Korea its immense wealth of forms contrasting in the strongest way conceivable with the smooth shape of the downwarped coast of the Sea of Japan. Only in the vicinity of the Yellow Sea coast do lowlands exist to any great extent, while behind the coast of the Sea of Japan the Main Range immediately rises up precipitously. The west side has thus been the civilized side of the country since prehistoric times. But the difficulties confronting navigation in the coastal waters, with their overabundance of broad tidal flats, sandbanks, reefs and rushing tidal currents kept this civilization from ever becoming a sea-going one.

Korea's narrow Sea of Japan fringe is cooler in midsummer and warmer in the other seasons than the wide Yellow Sea fringe. The annual range is thus much smaller in the former than in the latter. In spring and autumn the Sea of Japan fringe is distinguished by foehnlike katabatic winds, and in midwinter it receives relatively larger amounts of precipitation than the Yellow Sea fringe. The annual variation in precipitation is thus also somewhat more even, and it therefore has a more maritime character than the latter. For this reason the zones of natural vegetation mentioned above extend farther north in it, and the same is true of some cultivated plants, such as the timber bamboos.

Nevertheless, due to the relief, in its cultural geographical development the Sea of Japan fringe forms the retarded backside of the country. In the low-lying Yellow Sea fringe with its relative wealth of alluvial plains, the proportion of cultivated area in the total area of the counties is much larger, the number of crops greater. Accordingly the population crowds together extraordinarily more densely. Three of the five large cities and a large number of medium-sized towns lie in it, whereas the Japan Sea fringe between Wonsan and P'ohang contains only one single medium-sized town.

The third contrast that dominates Korea is that between the coastal fringes and the interior, the peripheral–central change. Towering mountain ranges fill the interior. They form the crests of the two tilt blocks and the dissected upper ends of the dip slopes. In addition there are the side ridges, which partially preserve older relief features and partially arose as the result of recent local uprise. The interior therefore also exhibits a very strong hypsometric change. This combines with the change that the increasing distance from the coast involves.

The climate of the interior parts of the country is much colder in winter than that of the coastal fringes, even when the isotherms are reduced. In reality this contrast is even greater, due to the influence of the high elevation. In summer the latter outweighs that of the more intense warming up, which is clearly shown in the broad southern half of the peninsula by the lines representing the reduced August isotherms. The interior is cooler than the coasts at all seasons, in midwinter even much more so than in midsummer. The natural vegetation is adapted to this decrease in temperature. The above mentioned vegetation zones each form bottom zones in the north and retreat in the south to the cooler heights. Only the subtropical zone, being the warmest one, occurs exclusively as a bottom zone. Above the coldest of these zones, the oak zone, there is a maple zone in central and northern Korea, and everywhere the forest ends with a coniferous forest zone, followed in the highest mountains by stunted subalpine shrub formations. The changes in vegetation with altitude thus create a very impressive contrast between the country's coastal fringes and its mountains.

With the increasing average steepness of the slopes and the increasing elevation the opportunities for agriculture grow more and more limited. Rice, cotton, kaoliang, wheat and soybean fields are left behind in the depths of the valleys. Millet, kidney beans, barley, potatoes, oats and buckwheat are the crops of the higher elevations, and only the last three reach the upper limit of cultivation. Where they are cultivated, the primitive economic form of migratory fire-field farming has been preserved in most of the higher mountains. The proportion of cultivated area decreases in counties lying at higher elevations, and the population density sinks accordingly. People and cities crowd together in the vicinity of the coasts, which offer an important subsidiary occupation in the form of fishing. In its principal features Korea's settlement is peripheral. Only in southern Korea with its more varied relief does dense settlement prevail throughout the country, with the exception of the Sobaek Range.

Korea's development persisted in the configuration arising out of the overlap of these three types of regional change until the Japanese began to shake the country out of its stagnation two generations ago. Then a profound transformation of the cultural landscape commenced. The country was set in motion. The regional change in forms was joined by a temporal change in structure, which gradually built up a more and more rapid tempo. Using carefully considered methods, they built on the existing conditions. Initially therefore the quantitative side of the change outweighed the qualitative side in many fields, particularly in agriculture. But already around the turn of the century, even before the actual occupation, the first railroads and roads brought new elements into the landscape. Their network became more and more dense from decade to decade. The Korean population grew constantly at a rapid pace, in the cities even much faster than in the country. The built-up city areas expanded, and Japanese districts developed alongside of the Korean ones. The harbors were fitted out with man-made structures. Forestry stations were put in charge of the remaining stands of tall timber, and reforestation of the barren hills and mountains that were the result of prior ruinous exploitation was begun. The first underground mines were opened, partly by the Japanese, and partly by representatives of the Western nations, and in productivity and in their influence on the landscape they soon far outdistanced the superficial burrowing done in traditional Korea. Fishing was intensified immensely on the basis of oceanographical and biological research and of the practical experience of the Japanese, so that it leaves its imprint everywhere along the coasts today to a greater or lesser degree. Finally, during the last decade the development of industry was begun, and today its production value has already caught up with that of agriculture.

Japanese influence has not affected the country and transformed it with the same intensity everywhere. Just as in the case of earthquakes zones of differing seismic activity are distinguished, zones of intensity of colonial development can be established. They are related in certain ways to the regular configuration of traditional Korea that we have just summarized from the standpoint of the threefold changes in forms. As we have said, the Japanese had to build on the conditions existing in traditional Korea. They also investigated the natural conditions scientifically and utilized them appropriately. Other directions of development, however, are linked to chance forms, such as the existence of minable mineral resources, or of particular features of the terrain, such as the junction of two natural traffic routes.

Generally speaking one can say that the coastal fringes with their varied possibilities for development have reached a higher degree of colonial transformation than the interior has. This is especially true of the Yellow Sea fringe. The southern coast lies so close to Japan that here, too, the development began early. After all, in Pusan, the southeastern entranceway, there has been a Japanese settlement almost continuously since 1443, and from here a vigorous stream of Japanese influence flows into southern and central Korea. The same can be said of the other harbors on the south coast to a much more limited degree only, for the relief makes it difficult to develop extensive hinterlands here. Not until the last decade was the North Korean Japan Sea borderland set in motion, but now it is developing fishing, mining, industry and transportation at a rate that is unprecedented in East Asia. Only the Japan Sea fringe of the peninsula still persists virtually under traditional Korean conditions today. It is after all, as we have just shown, the strip of Korea's coast that is least well endowed by nature.

The development of the coastal fringes started partially in the old Korean cities. The three capital cities, Seoul, P'yŏng-yang and Kaesŏng had already stood out among the other cities in ancient Korea on the basis of their population figures, magnificent buildings and fortifications, and this is also true to a lesser degree of other cities, such as Haeju, Chŏnju, Kwangju, Chinju and Kyŏngju. Most of them have now undergone profound changes in both their settlement pattern and their economic structure. Several important old Korean cities, such as Ŭiju, Anju and Kongju, have remained backward, because their functions were taken over by neighboring new cities on railroad lines (Shinŭiju, Shinanju, Taejŏn). Further starting points for Japanese colonization were the treaty ports, in which the Korean government granted free trade to the Japanese from 1876 on and to the Western nations from 1882 on. In the most advantageous locations for large ports they grew up out of nothing and were therefore retained and expanded by the Japanese after the occupation. Almost all of the large ports of the present day are among them. Finally, in the coastal fringes tiny Korean villages developed into industrial centers, such as Songnim and Hŭngnam, in locations where mining was being developed or where there were favorable conditions for power supply and transportation.

Toward the interior and higher elevations the intensity of the colonial development generally diminishes. It increases again in remote provincial capitals, such as Ch'unch'ŏn, military seats, such as Kanggye, county seats, at mines, reservoirs for producing power and along the railroad lines and roads. Along the land boundary on the Yalu and the Tumen Japanese influence also began early and vigorously, above all with a view to military security. The last outposts of the Japanese desire for transformation in the high mountains are the forestry stations with their logging roads, narrow-gauge railroads and rafting facilities and, in recent times, a few mountaineering huts. The intensive development of the Kaema upland did not begin until the past decade, for the terrain there confronted the builders of roads and railroads into it with virtually Alpine difficulties.

In general, the development in southern Korea began vigorously at an earlier date than in northern Korea, which is so much more poorly endowed by nature. In the south, in contrast to the north, one could say that colonization has permeated the entire country. Here, far from the coast, the old provincial capital of Taegu has developed into a large industrial and transportation center.

Taking all individual features of colonial transformation in economy, transportation, settlement and military security by land and by sea into account, one arrives at the following grades of intensity. (1) The five big cities Seoul, P'yŏngyang, Pusan, Taegu and Inch'ŏn. (2) The other large ports. (3) The other industrial centers, railroad junctions and main administrative seats. (4) Most of the county seats. (5) The rural counties of southern Korea with the exception of the island of Chejudo, which has thus far lagged far behind. (6) The rural counties of the central and northern Korean Yellow Sea coast. (7) The rural counties of the North Korean Japan Sea borderland. (8) The Japan Sea fringe of the peninsula. (9) The high mountains of central Korea. (10) The Kaema upland with the adjoining mountain ranges and Paektusan. Grades 1–4 are dotlike in their distribution, the others areal. Grade 1 is characterized by a comprehensive development of settlement, transportation, administration and, partially, mining and military security. In grade 2 the development is aimed more one-sidedly at overseas traffic and fishing. Grade 5 is characterized by the highest and most versatile state of agriculture, in particular by large irrigation works for the rice paddies. In grade 6 irrigation agriculture takes second place to dry-field farming. Grade 7 shows a diverse coexistence of agriculture, which is already quite limited by the climate, and new mining, electricity, transportation, industrial and fishing facilities. In grades 9 and 10 the Japanese reforms are limited to final emanations, particularly in the sphere of forestry, but also in the gradual settling of fire-field farmers.

Overlapping with the temporal change in structure accompanying the colonial development by Japan is the rhythm of seasonal change. Korea has a summer corresponding to that of the perennial climate of the tropical rainforests and a winter that is similar in the extreme south to Ireland and in the extreme north to the Pechora basin. Thus the seasonal changes in the climate assume tremendous dimensions, and they are much greater in the north even than in the south. The seasonal differences in the geographical character of the Korean landscape are accordingly great.

At the beginning of the year most of Korea lies under a blanket of snow. In northern Korea it extends all the way down to the coasts for months, but generally remains thin. The interior of the caldera of Paektusan is free of snow for only a few months each year. In southern Korea snow does not remain on the ground for very long in the vicinity of the sea, but in particularly cold winters with relatively large amounts of precipitation, such as that of 1932/33, the high mountains are sprinkled with white until March or April here as well, and on the summit of the Chejudo volcano one can run into snow patches as late as June. In northern and central Korea work on the fields ceases completely in midwinter. But in southern Korea's river valleys and plains the young winter crop sprouts pale green on paddies and dry fields, and other paddies are adorned by the red blossoms of alfalfa. On the slopes southern Korea's pine forests deck the ground with dark green even then, and in the extreme southern coastal fringe the evergreen trees stand out clearly against the gray of the trunks and branches of their bare neighbors. The floor heating system in the houses does not go out day or night anywhere in Korea, and the smoke and vapor from the fireholes and chimneys often completely envelop the villages. The Koreans wear thickly padded white clothing or several normal outfits on top of each other, and in northern Korea one sees many fur hats among the men.

The beginning of spring planting advances from south to north and from lower to higher elevations from March to April or May. The time also varies depending on the crop. Everywhere where rice is grown the seedbeds take preference over the neighboring fields in all types of work. The winter crops sprout vigorously from April on in the north as well, so that the fields come to life. Likewise from April on the summer crops are planted in rows between the winter crops in fields where intercropping is practiced. The deciduous trees and shrubs in the forests and shrub formations begin to bloom and turn green in the same month, one after another. South Korea's forests are adorned with the orange or yellowish–green catkins of the hornbeams; the thickets glow yellow with forsythias and pink, red and mauve with rhododendrons. Harvesting of the winter grain begins in May in southern Korea; in central Korea this is delayed until June. In southern Korea the rice plants are then immediately transplanted from the seedbeds to the paddies.

Between the middle of June and the middle of July the Koreans don their thin, yellowish–brown hempen clothing. Midsummer has arrived. The tall forests have developed their most luxuriant foliage; gigantic perennials and herbs have sprouted up in the clearings. Their multicolored blossoms are joined by those of many trees and lianas. In heat, hothouse humidity and driving rain nature displays an immense germinating power. In August hemp is ripe for fiber, and even before that harvesting begins on the melon fields. The watch towers erected on them are occupied day and night. In northern Korea the buckwheat fields, not sown until June, are covered by their white blossoms.

In September the hempen clothing is again exchanged for the usual white cotton apparel. Harvesting of the summer crops begins, generally starting earlier in northern than in southern Korea. Most of the dry-field crops are harvested prior to the rice, its reaping dragging on until early November in southern Korea. Occasionally the first frost already settles on the rich brown ears. The harvest round ends with soybeans, which remain on the fields with withered leaves into October. When they are harvested, the forests and thickets burst into a riot of colors unknown in Central Europe, gaudy red, pink, coppery brown, yellowish–green, dazzling yellow. Then the leaves begin to fall, and a dull gray descends upon the mountains. In September night frosts have already begun in northern Korea, and soon after the leaves have fallen the first snowfalls commence there. In southern Korea the first night frosts are delayed until November, the first snowfalls until December. Between the end of October and the middle of December, depending on the latitude, the heating system of the living rooms is again put into permanent operation, and winter is back again, with its frightful cold in the north.

By Asian standards, Korea is only a small country, a narrow peninsular bulge on its monsoon-swept continental Pacific coast. Its latitude at the transition from the subtropics to the temperate zone predestines it, however, to have a great wealth of forms. Its mountainous nature and the diversity of the seas surrounding it multiply this wealth, and the historical development of the past 60 years has increased it further. Nevertheless, on the whole this country gives the impression of homogeneity, for its plurality can be summarized from a few interrelated standpoints. Korea has been spared any profound long-range influences from a completely heterogeneous culture, such as are felt by most countries of colonial formation. The geographical features brought into the country from outside by man do not come from as far away as the winds that blow over it, with the exception of the elements that Japan itself originally adopted from the Western cultural sphere, but remodeled to suit its own purposes. As a peninsular country in the extratropical monsoon region of East Asia with a persevering culture of its own, Korea is a strong individuality that is truly worthy of such a detailed geographical treatment as the one presented here.

Table 46. Geological formations

Period	System	Series	Formation	Petrographical features	Thickness in m	Main area of occurrence	Remarks
Holocene				Alluvial plains along the rivers. Mud and sand flats. Sandbars. Most recent volcanism.			
Pleistocene				Extensive sheets of basalt, rhyolite and trachyte. Stratovolcanoes (Paektusan, Ullŭngdo, Chejudo). Marine Sŏgwip'o formation (Chejudo). Terraces in the valleys, deposits in limestone caves. Red Nangnang clays. Glacial deposits in the Kwanmobong group.			
Early Pliocene		Yŏnil	Yŏnil	Shale	400	west of Changgi peninsula	
			Chŏnbuk	Conglomerate	200		
Oligocene to Middle Miocene		Ch'ilbosan		Granite porphyry, alkali rhyolite, alkali basalt, rhyolite		Various, mostly small areas on the coast of the Sea of Japan	Stratigraphy and thickness of the formations listed for the Kilchu-Myŏngch'ŏn graben. In other smaller areas they may be quite different.
			period of erosion				
		Myŏngch'ŏn	Manhodong	Sandstone, conglomerate, coal	1000		
			Hamjindong	Shale with marl nodules	200		
			P'yŏng-yuktong	Conglomerate, sandstone shale (marine)	600		
		Yongdong	Yongdong Basalt				
			Yongdong	Shale, sandstone, coal	80		
Eocene			Period of erosion				
Late Cretaceous		Pulguksa		Granite porphyry, "young" granites, diorite, quartz porphyry, black felsitic porphyry		Around Kyŏngju in the Tsushima basin	The "young" granite of this period occurs everywhere in Korea as an intrusive rock. Pulguksa = Buddhist temple near Kyŏngju
			<Taishu>	Black shale, sandstone			

Period	System	Series	Formation	Petrographical features	Thickness in m	Main area of occurrence	Remarks
Middle Cretaceous	Kyŏngsang (Upper Taedong)	Shilla	Chusasan	Porphyry with tuff	1000	Tsushima basin	Stratigraphy and thickness of the formations listed for the Taegu region. In the other areas the same petrographical features. Shilla after the state of Shilla (1st millennium A.D.). Naktong after the Naktonggang. Kyŏngsang after the province.
			Kŏnch'ŏn-ri	Blackish shale, sandstone	800		
			Ch'aeyaksan	Porphyry with tuff	200–500		
			Taegu	Reddish and blackish shale	2000		
			Hakpong	Porphyry with tuff	0–250		
			Shilla	Conglomerate	200–600		
Early Cretaceous		Naktong	Chilgok	Reddish shale, sandstone	550–950	Tsushima basin	
			Chinju	Blackish shale, gray sandstone	1000		
			Hasandong	Red bituminous clay, shale, sandstone, conglomerate	1300		
			Naktong	Sandstone, shale, partly bituminous, conglomerate	700		
Late Jurassic			Great Taebo Orogeny				Taebo (Taebang) ef. p. 22
Lias and Dogger	(Lower) Taedong		Ongmasan	Sandstone and conglomerate alternating	800	In the fold belts on top of the Paleozoic and Triassic rocks, in the massifs sparsely found on the basement complex	Stratigraphy and thickness of the formations listed for Ch'ungch'ŏngnam-do. Near P'yŏng-yang, etc. slightly different stratigraphy, but same basic petrographical features. Taedong after the Taedonggang.
			Pyŏng-ri	Conglomerate	30		
			Paegunsa	Black shale	650		
			Kaehwa-ri	Conglomerate	70		
			Amisan	Shale and sandstone alternating, anthracite	750		
			Wolmyŏng-san	Conglomerate with interbedded sandstone	30–350		

Period	System	Series	Formation	Petrographical features	Thickness in m	Main area of occurrence	Remarks
Late Triassic			Songnim Disturbance: gentle folding, especially in the fold belts				
Middle Triassic	P'yŏng-an	Taejawon		Red conglomerates, devoid of fossils		P'yŏngnam zone, slightly different and even thicker in the Okch'ŏn zone. In the Tumen area only Hongjŏm strata.	P'yŏng-an after the province
Early Triassic		Green Series		Greenish shale and sandstone devoid of fossils	300–1000		
Permo-Triassic		Kobangsan		Sandstone, shale, anthracite seams	500		
Carbo-Permian		Sadong		Sandstone, shale, anthracite seams	100		
Middle Carboniferous		Hongjŏm		Marine limestone, red sandstone and shale	300		Base: red shale (P'yŏng-nam) or arcose (Okch'ŏn)
Late Ordovician to Early Carboniferous				Post-Nangnang iatus (parallel unconformity)			Nangnang (Lolang) cf. p 22
Middle Ordovician	Chosŏn	Mandal		Massive limestone and dolomite	600	P'yŏngnam zone. Also in Okch'ŏn zone, but slightly different stratigraphy	Chosŏn = Korea. Ŭnsan= town northeast of P'yŏng-yang, Sunch'ŏn ditto. Yangdŏk = county seat on the upper Namgang
Middle Cambrian to Early Ordovician		Ch'osan ("Great) limestone series")	Shinch'ang	Marl, limestone shale and limestone alternating	100		
			Ŭnsan	Dolomite and limestone, partly crystalline	800		
			Imch'ŏn	Black shale	80		
Early and Middle Cambrian		Yangdŏk	Sunch'ŏn	Shale, red slate	500–1000		
			Yangdŏk	Quartzite, micaceous shale	200		

Period	System	Series	Formation	Petrographical features	Thickness in m	Main area of occurrence	Remarks
Proterozoic	Sang-won	Sadang-u	<Madenri>	Limestone, light red dolomite, black shale	500	P'yŏngnam fold belt only	Sang-won = county in P'yŏng-annam-do. Sŏngch'ŏn = town in the same province. Chosŏn and Sang-won system together are also called Nangnang complex.
			Sŏngch'ŏn	Limestone, whitish gray crystalline dolomite	1600		
		Chik'yŏn	Sang-won mica schist formation (also phyllite)		660		
			Sang-won quartzite formation		100		
			Post-Kogurian erosional surface				
Archeozoic	"Basement complex"	"Gray Gneisses" (gray granite gneiss, augen gneiss), Kogurian granite. – Metamorphic sediments: Mach'ŏllyŏng group (metamorphic limestone, crystalline dolomite, mica schist, biotite gneiss, amphibolite). Yŏnch'ŏn group (amphibolite, mica schist, phyllite). Okch'ŏn group (ditto).			13,000 – 28,000	All massifs	Kogurian after the state of Koguryŏ. Mach'ŏllyŏng = mountain range on the border between the Hamgyŏng provinces. Yŏnch'ŏn = town north of Seaul. Okch'ŏn = county in Ch'ungch'ŏngbuk-do

1. Korean in a pavilion on the banks of the Taedonggang near P'yŏng-yang .

2. Korean potter with a soy sauce crock on his A-frame

3. Young Korean with firewood on his A-frame

4. The county seat of Musan, seen from Namsan. On the other side of the silt-filled bed of the Tumen, in Manchuria, a recent flood basalt forming an approx. 100-m-high terrace

5. Husking rice. In the background a typical house made of wickerwork plastered with mud, its hipped roof thatched with rice straw. Standing separately to the right of it the chimney of the floor heating system

6. Changsŭng posts at the southern entrance to a village 20 km north of Seoul

7. Traditional Korean farmhouse belonging to a wealthy farmer on the T'aean peninsula. In the foreground melons and hemp

8. Street in a northern Korean town. The roofs are covered with shingles

9. Mouth of the Tumen. In the background on the left the Russian shore. In the center a rocky island with cliffs that has become attached to the land. For the most part the mouth is blocked by a narrow bar

10. Landing the sardine catches

11. Drums of fish oil in the harbor of Unggi

12. Kwanmobong. View from the summit (2541 m) toward the jagged, steep forms on the side toward the Sea of Japan

13. Fire-field farming between P'ot'aeri and Taep'yŏng-ri (Paektusan). Of the primary forest not only stumps are still standing, but also a number of half-burned tree trunks. After the fields were abandoned the first time, brush grew up. The fields that are now being used were cleared out of this brush. But even when they were cleared the second time the stumps and trunks were not removed. In the background on the slope (white, in blossom) buckwheat, below oats and potatoes (30 August 1933)

14. Abandoned hut of a fire-field farmer between Hyesan and Poch'ŏnbo. In the background fire fields cut rectangularly into the brush. Forest is no longer visible; therefore the hut is not a log cabin. The walls are constructed of wickerwork and mud. The roof is covered with earth, on which weeds proliferate

15. Chejudo. Village street. The farmhouses, with their thatched roofs held down by straw ropes, hide behind walls of basalt rocks gathered from the fields

16. Japanese raft on the upper reaches of the Yalu. The logs are fastened together, forming a long wedge. The front section of the raft is skillfully steered by a Japanese raftsman by means of a steering beam

17. Dwarf pine and birch region on Maengbusan (2214 m) in the Nangnim Range.
Pinus pumila Regel and *Betula ermanii* Cham.

18. Very steep fire fields along the "Hemp Field Pass" between Wiwon and Ch'osan. The slopes covered with fire fields and brush have an inclination of approx. 60°

19. The Wonsan area viewed toward the north from the cape of Hapchin-ri. In the foreground a bar with a wide beach, behind it inactive cliffs. In between a coastal lake. In the background the bay, Yŏnghŭngman, with islands

20. View of the German abbey of Tŏgwon near Wonsan looking westward. In the center rice paddy plain, beyond it granitic hill country. On the horizon the more than 1000-m-high limestone ranges of the Wonsan arc of the Main Korean Range

21. The Valley of Ten Thousand Waterfalls (Manp'oktong) with the hermitage of Podŏgam in Kŭmgangsan

22. Temple of Chang-ansa in Kŭmgangsan

23. The valley of Ongnyudong in Kŭmgangsan, viewed from Sejonbong. Ridge with the boundary between rounded and jagged forms

24. Abraded and eroded forms in the "Sea Kŭmgang" (Haegŭmgang) near Kosŏng

25. Clusters of basalt pillars along the abrasion coast of the Sea of Japan near Kojŏ

26. Ullŭngdo. View from the highest peak (Sŏng-inbong, 984 m) toward the north. Looking into the subsidence caldera with the central eruption cone. In the background the Sea of Japan. The steep slopes are covered by luxuriant, dense, primarily summergreen forests. In the foreground the white blossoms of the ash *Sorbus commixta* Hedl. (5 June 1933)

27. Intercropping. Rice and sesame

28. Ullŭngdo. Town of To-dong

29. Bundling rice plants in the seedbed for transplanting. Near Yŏngju (Naktong basin). (2 June 1933)

30. Ch'ŏnwangbong (1915 m), the summit pyramid of Chirisan. High altitude forest, consisting of the Korean pine and birches, showing the influence of the weather. The highest peak (to the right of the more prominent mountain in front of it) is practically barren

31. The two "Horse Ear Mountains" (Maisan) near Chinan. Ch'ungch'ŏng–Chŏlla region. Floodplain with rice paddies and shrub-covered hills

32. Southwest coast near Mokp'o

33. The Taedong valley just above P'yŏng-yang. The river flows toward the right. On its left bank (background) alluvial plains; on the right bank hills composed of P'yŏng-an strata

36. Bay and harbor of Unggi

34. Pavilion northeast of the provincial capital of Haeju

35. Dome-shaped peak in Puk'ansan near Seoul. To the left the highest peak of the granite massif, Paegundae (836 m), with keystone, to the right "exfoliation from below"

37. View of the center of Seoul from Namsan (looking north). In the foreground: Japanese section. In the center: banks, department stores, central post office, city hall. At the foot of the mountain: palace of the government–general. In the background: Puk'ansan (836 m)

40. Street in a suburb of Seoul with straw-roofed Korean houses. Japanese flags on each house. Emperor's birthday

38. Namdaemunno (Great South Gate Street) in Seoul. Inorganic juxtaposing of old, low Korean buildings, two-story houses belonging to small-scale Japanese businessmen and a modern Japanese department store

41. South Gate of Suwon

39. Honmachi (Myŏngdong) in Seoul. Main shopping street in the Japanese section

42. Chejudo. Gorge in the basalt flow near Hahyo-ri east of Sŏgwip'o. The vegetation, consisting primarily of shiino trees, is evergreen

BIBLIOGRAPHY

Periodicals

(Abbreviations in parentheses)

I. Transactions of the Korea Branch of the Royal Asiatic Society. Vol. 1–30. Seoul 1900–1940. (Tr. Kor. Br. R. As. Soc.)
II. Korean Repository. Ed. F. and Mrs. F. Ohlinger. Vol. 1–5. Seoul 1892–1898. (Kor. Rep.)
III. The Korea Review. Ed. H. B. Hulbert. Vol. 1–6. Seoul 1901– 1906. (Kor. Rev.)
IV. The Korea Magazine. Vol. 1–3. Seoul 1917–1919. (Kor. Mag.)
V. The Korea Mission Field. Publ. in the Interests of all Evangelical Missions in Korea by the Fed. Council of Evang. Missions. Seoul since 1902, annually.
VI. Annual Report on Administration of Chōsen (Tyōsen). Since 1908, annually. Until 1923 under the title: Annual Report on Reforms and Progress in Chōsen. Keijō. Government-General of Chōsen (Eng.) – Jap. edition under the title: Chōsen Sōtokufu Shissei Nempō. (Ann. Rep. Adm. Chōs.)
VII. Chōsen Sōtokufu Tōkei Nempō (Statistical Yearbook of the Government–General of Chōsen). Keijō, annually (Jap.).
VIII. Chōsen Ihō (Chōsen Bulletin). Since 1916 (Jap.). (Chōs. Bull.)
IX. Chōsa Geppō. Research Monthly of the Government–General (Jap.).
X. Bunkyō no Chōsen. Scientific journal on Korean topics. (Jap.).
XI. Bulletin on the Geological Survey of Chōsen (Korea). Since 1919. Geol. Survey of the Gov.–Gen. of Chōsen (Jap. or Eng.). (B. Geol. S. Chōs.)
XII. Chōsen Tanden Chōsa Hōkoku (Report on the Geological Survey of the Coal–Fields in Chōsen). Since 1926 (Jap.). (Rep. Geol. Surv. Coal–Fields)
XIII. Bulletin on the Mineral Survey of Chōsen (Korea). Since 1912 (Jap., some with Eng. summary). (B. Min. S. Chōs.)
XIV. Annual Report of the Meteorological Observatory of the Government–General of Tyōsen. Since 1907, annually (Jap. and Eng.). (Ann. Rep. Met. Obs. Gov. Gen. Tyōsen)
XV. Annual Report of Hydrographical Observations. Suisan Shikenjō (Fishery Experiment Station). Chōsen Sōtokufu, Husan. Since 1926 (Jap. with Eng. summary). (Ann. Rep. Hydr. Obs. Husan)
XVI. Kayō Chōsa Hōkoku (Report of the Oceanographical Observations). Suisan Shikenjō (Fishery Experiment Station). Chōsen Sōtokufu, Husan. Since 1924 (Jap. with Eng. summary). (Rep. Oc. Obs. Husan)
XVII. Chōsen Hakubutsu Gakkai Zasshi (Journ. of the Chōsen Natural History Society). Since 1924 (Jap.). (Journ. Chōs. Nat. Hist. Soc.)
XVIII. Chōsen no Keisai Jijō (Economic Status in Korea). Statistical survey of the Government-General. Annually (Jap.).
XIX. Chōsen Sanrin Kaihō (Bulletin of the Forest Society of Korea) (Jap.). (Bull. For. Soc.)
XX. Ringyō Shikenjō Hōkoku (Bulletin of the Forest Experiment Station). Chōsen Sōtokufu. Since 1924 (Jap.). (Bull. For. Exp. Stat.)
XXI. Nōji Shinkenjō Ihō (Annals of the Agricultural Experiment Station). Suigen. Chōsen Sōtokufu. Since 1926, annually (Jap., some with Eng. summary). (Ann. Agr. Exp. Stat.)
XXII. Nōji Shinkenjō Kenkyū Hōkoku (Journal of the Agricultural Experiment Station). Suigen. Chōsen Sōtokufu. Since 1916 (Jap.). (Journ. Agr. Exp. Stat.)
XXIII. Chōsen Sōtokufu Kwangyō Mohanjō Ōbun Hōkoku (Bulletin of the Agricultural Experiment Station of the Government–General of Chōsen). Since 1922 (Jap.). (Bull. Agr. Exp. Stat. Chōs.)
XXIV. Jimbun Chiri (Anthropogeographical journal). Keijō (Jap.).
XXV. The Journal of the Keijō Imperial University Social Sciences Association. Keijō. Since 1928 (Jap. with Eur. summaries). (Journ. Keijō Imp. Univ. Soc. Sc. Ass.)

XXVI. The Oriental Development Company, Limited. A General Survey. Tōkyō, since 1924, semiannually (Eng.) (Or. Dev. Comp.)

XXVII. Japanese Journal of Geography and Geology. Ed. Nat. Research Council of Japan. Tōkyō. Since 1922 (primarily Eng.). (J. J. G. G.)

XXVIII. Chirigaku Hyōron (The Geographical Review of Japan). Ed. Association of Japanese Geographers. Since 1925 (Jap., some with Eng. summary). (Chir. Hyō.)

XXIX. Chigaku Zasshi (Geographical Journal of the Tōkyō Geographic Society). Since 1879 (Jap.). (Chi. Z.)

XXX. Chikyū (The Globe). Since 1913 (Jap.).

XXXI. Proceedings of the Imperial Academy of Tōkyō. Since 1925 (Eng.). (Pr. I. A. T.)

XXXII. Journal of the Geological Society of Tōkyō, Japan. Since 1894 (Jap.). (Journ. Geol. Soc. Tōkyō)

XXXIII. Journal of the Meteorological Society of Japan. Tōkyō. Since 1881 (Jap., some with Eng. summary). (Journ. Met. Soc. Japan.)

XXXIV. Suiro Yōhō (Hydrographic Bulletin). Tōkyō. Since 1922 (Jap.).

XXXV. Records on Oceanographic Works in Japan. Tōkyō. Since 1929 (Eng.). (Rec. Oc. Works.)

XXXVI. Journal of Oceanography. Tōkyō. Since 1930 (Eng.). (Journ. Oc.)

XXXVII. Bulletin of the Hydrographic Department of the Imperial Japanese Navy. Tōkyō. Since 1922 (Eng.). (Bull. Hydr. Dep. Imp. Jap. Navy)

XXXVIII. Memoirs of the Imperial Marine Observatory Kobe. Since 1930 (Eng.). (Mem. Imp. Mar. Obs. Kobe)

XXXIX. Journal of the Imperial Fishery Experiment Station Tōkyō. Since 1930 (Jap. with Eng. summary). (Journ. Imp. Fish. Exp. Stat. Tōkyō)

XL. Shizen Guku Zasshi (Journal of prehistory). Ed. K. Ōyama. Tōkyō. Since 1929 (Jap., some with Ger. summary). (Shiz. Z.)

XLI. The Japan Year Book. Ed. Foreign Affairs Association of Japan. Tōkyō. Since 1905. Korea: section XL. Chōsen (Korea). Most recent: 1940/41 (pp. 858–881).

XLII. The New Japan Year Book. Tōkyō annually. Since 1907/08.

XLIII. Japan–Manchoukuo Year Book. Tōkyō annually. Most recent: 1940.

XLIV. Résumé Statistique de l'Empire du Japon. Tōkyō. Since 1887 (Fr.).

XLV. The ...th Financial and Economic Annual of Japan. Tōkyō. Since 1901 (Eng.).

XLVI. The Statistical Abstract of the Ministry of Agriculture and Forestry. Tōkyō, annually.

XLVII. Economic Conditions in Japan. With Annexes on Economic Conditions in Formosa, Korea and Japanese Mandated Territory 1914/19. London 1920. Department of Overseas Trade. Stationary Office. Since then annually. (Dep. Ov. Tr.)

XLVIII. Mitteilungen der Deutschen Gesellschaft für Natur– und Völkerkunde Ostasiens. Tōkyō. Since 1873 (M. D. O. A. G.)

XLIX. Transactions of the Asiatic Society of Japan. Tōkyō. (Tr. As. Soc. Jap.)

L. Transactions of the North China Branch of the Royal Asiatic Society. (Tr. N. China Br. R. As. Soc.)

LI. T'oung Pao. Archives concernant l'histoire, les langues, la géographie, l'ethnographie et les arts de l'Asie orientale. Leyden.

LII. Memoirs of the National Research Institute of Meteorology. Nanking, now Chungking. Since 1928 (Chin. or Eng.). (Mem. Nat. Res. Inst. Met.)

LIII. Ostasiatische Rundschau. Since 1920. (O. A. R.)

LIV. Far Eastern Review. Since 1904. (F. E. R.)

LV. Proceedings of the ...th (Pan–)Pacific Science Congress (in Eur. languages). (Pr. P. S. C.)

P. M. = Petermanns Geographische Mitteilungen. Gotha, Justus Perthes
(L. B. = Literaturbericht)
Z. G. E. B. = Zeitschrift der Gesellschaft für Erdkunde zu Berlin

Bibliographies and Gazetteers

A. H. Lautensach, Korea (1926–36) mit Nachträgen aus älterer Zeit. Geogr. Jahrb. **53**, 1938. I, pp. 225–274.
B. H.H. Underwood, A Partial Bibliography of Occidental Literature on Korea, From Early Times to 1930. Tr. Kor. Br. R. As. Soc. **20**, 1931. Pp. 1–185.
C. E. and G. Gompertz, Supplement to "A Partial Bibliography of Occidental Literature on Korea" by H.H. Underwood. Tr. Kor. Br. R. As. Soc. **24**, 1935. Pp. 21–48.
D. O. Nachod, Bibliographie von Japan 1906–1926. 4 vols. Leipzig 1928. Korean titles esp. in vol. II, pp. 662–691, 757–759 and vol. IV, pp. 301–325. Continued by Praesent–Haenisch (vols. V and VI of O. Nachod's work). Leipzig 1937, 1940.
E. H. Praesent, Ostasien–Bibliographie der Ostasiatischen Rundschau. Since vol. **10**, 1929. – H. Praesent, Japan–Bibliographie der Zeitschrift Nippon. Since vol. **1**, 1935.
F. O. Nachod, Bibliographie über Bevölkerungsfragen. Ethnol.Anzeiger **1**, 1927, p. 66.
G. W. Gottschalck, Katalog der Handbibliothek der Orientalischen Abteilung der Preußischen Staatsbibliothek. Leipzig 1929. 573 p.
H. M. Courant, Bibliographie Coréenne. 3 vols. Paris 1894–96, Suppl. 1901. Geography: vol. II, pp. 474–535 and Suppl. pp. 52–58. – Introduction to Courant's "Bibliographie Coréenne". Translated from the French by Mrs. W. Massy Royds. Tr. Kor. Br. R. As. Soc. **25**, 1936. Pp. 1–100.
I. H.Cordier, Bibliótheca Sinica. Paris. Vol. IV, 1907/08², no. 2939– 3008, 3235–3236. Suppl. 1922², no. 4405–4428.
J. Publications of the Geological Survey. Government General of Chōsen, No.1. Keijō 1929. 8 p.
K. Kokusai Bunka Shinkōkai, Catalogue of the Kokusai Bunka Shinkōkai Library (Works in Western Languages). Tōkyō 1937. 203 p.
L. Kokusai Bunka Shinkōkai, Bibliographical Register of Important Works Written in Japanese on Japan and the Far East Published During the Year 1932. Tōkyō1937. 166 p.
M. Kokusai Bunka Shinkōkai, A Short Bibliography of English Books on Japan. Tōkyō 1936³. 44 p.
N. Kokusai Bunka Shinkōkai, Bibliographie abrégée des livres relatifs au Japon en français, italien, espagnol et portugais. Tōkyō 1936. 50 p.
O. K.B.S. Quarterly. Bulletin of the Kokusai Bunka Shinkōkai (The Society for International Cultural Relations). Tōkyō. Regularly contains: A List of Books on Japan in Western Languages.
P. E.D. Merrill and C.H. Walker, A Bibliography of Eastern Asiatic Botany. Jamaica Plains (Mass.) 1938. 715 p.
Q. R. Streit and J. Dindinger, Bibliotheca Missionum. Vol. X. Die Missionsliteratur Japans und Koreas von 1800 bis 1909. 32 + 565 p. Aachen (1937).
R. T. Kobayashi, Japan und Chōsen (Korea). Geol. Jahresberichte II B 1 (1937/38). Berlin 1940. Pp. 322–337. IV B (1939/40). Berlin 1943. Pp. 327–346.
S. S. Yoshimura, Classified List of Papers and Reports Bearing on Oceanic Geology Published in Japan During 1927–1937. Geologie d. Meere u. Binnengewässer **2**, 1938. Pp. 451–476.
T. C. Hasenclever, Bibliographie zur Industrialisierung Japans. Weltwirtsch. Archiv **46**, 1937 II. Pp. 332–366.
U. Manual of Corean Geographical and Other Proper Names Romanized. Yokohama 1883. English Legation.
V. E.M. Satow, List of Korean Geographical Names. Yokohama 1884. 70 p.
W. B. Kotō and S. Kanazawa, A Catalogue of the Romanized Geographical Names of Korea. Tōkyō (1903). 88 p.
X. S. Kanazawa, Untersuchungen über die japanischen und koreanischen Ortsnamen in alten Zeiten. Seoul, Government–General 1912. 18 p. German, 31 p. Japanese.

Map Series

a. Topographical Map of Korea 1:50,000. 723 sheets. Ed. Chōsen Sōtokufu (Government General of Korea). Cf. p. 53f.
b. Special maps 1:50,000: Environs of Seoul (monochrome), Kŏmgangsan (in five colors).
c. Topographical Map of Korea 1:200,000. 65 sheets. Ed. Chōsen Sōtokufu. Cf. p. 54.

d. Topographical map of parts of Korea 1:25,000. 97 sheets. Ed. Chōsen Sōtokufu. Cf. p. 54.
e. City maps 1:25,000, monochrome: Seoul (2 sheets), Kaesŏng, Kyŏngju etc. Ed. Chōsen Sōtokufu.
f. 45 City maps 1:10,000. In four colors. Ed. Chōsen Sōtokufu. Cf. p. 54.
g. Maps of the 13 Korean provinces 1:500,000. In 4 colors. Ed. Chōsen Sōtokufu. Cf. p. 54.
h. International Map of the World 1:1 Million. Sheets N.J.51 Ryōzyun (Lüshun), N.J.52 Keizyō (Seoul) and N.I.52 Nagasaki.
i. Kokusai Bunka Shinkōkai, Map of Japan and Adjacent Regions. 1:2 million. Tōkyō 1937.
j. General Geological Map of Chōsen (Korea). 1:1 million. Tōkyō 1926.
k. Geological Atlas of Eastern Asia 1:2 Million. Ed. Tōkyō Geographic Society. Tōkyō 1929. Especially sheet 8 (Wonsan) and 12 (Fukuoka).
l. Geological Atlas of Tyōsen 1:50,000. Issued to date: 1–19. Seoul 1924–1938. Ed. Geological Survey of the Government General of Tyōsen (Review of all installments by T. Kobayashi in J. J. G. G., especially in **16**, 1939. Pp. 33–42).
m. T. Okada, Climatic Atlas of Japan and Her Neighbouring Countries. Tōkyō 1929. 95 plates. Central Meteorological Observatory.
n. Hydrographic Department Tōkyō, The Current and Meteorological Charts of the Seas near Japan. Tōkyō 1925. 24 plates (Jap.).
o. Suisan Shikenjō (The Fisheries Experiment Station of the Government–General of Tyōsen, Husan), Oceanographical Charts of the Adjacent Seas of Tyōsen. Yearly since 1924. Husan.
p. Suisan Shikenjō (The Fisheries Experiment Station of the Government–General of Tyōsen, Husan), The Hydrographical Charts of the Neighbouring Seas of Tyōsen. Husan 1936.
q. British Admirality Charts No. 54, 104, 358, 882, 898, 913, 1257, 1258, 1270, 1271, 1316, 1558, 1656, 2365, 3365, 3366, 3480, 3642, 3666, 3699.
r. German charts No. 299, 552, 553, 554.
s. Kon. Nederlandsch Meteorol. Instituut, Oceanographische en Meteorologische Waarnemingen in de Chineesche Zeeen en in het Westelijk Deel van den Noord Stillen Ocean. Atlanten No. 115, 2 vols. 's Gravenhage 1935 and 1937.
t. Tozawa–Nakai, Atlas Illustrating Geographical Distribution of Korean Woody Plants and Bamboos. Part I. Forest Experiment Station. Government–General of Chōsen. Keijō 1929 (Jap. and Eng.).
u. The Map of Chōsen Showing Distribution of Staple and Special Products. Government–General of Chōsen. Commercial and Industrial Museum. Keijō 1933 (Jap.).
v. Forest and Economic Map of Korea 1:1.5 million. Keijō 1933 (with five inset maps, Jap.).
w. N. Takahashi, Map Illustrating the Distribution of Paddy Field and Field in Chōsen. 1:1.5 million. Keijō 1933.

Books and Articles

I. Prehistory

1. R. Torii, Yushi izen no Nippon (Prehistoric Japan). Tōkyō 1925 (Jap.).
2. R. Torii, Ancient Japan in the Light of Anthropology. Tōkyō 1937². 22 p., 12 plates, Kokusai Bunka Shinkōkai.
3. Z. Nakaya, Contributions à l'étude de la civilisation néolithique du Japon. Rev. des Arts Asiatiques **6**, 1929/30. Pp. 151–167.
4. Z. Nakaya, L'influence des civilisations continentales sur l'âge de pierre au Japon. Rev. des Arts Asiatiques **7**, 1931. Pp. 141–155.
5. M.C. Haguenauer, Notions d'archéologie japonaise (Bull. Maison Franco–Japonaise, Série Française, Vol. 3, 1931).
6. O. Menghin, Die Steinzeit Ostasiens. In: Festschrift P.W. Schmidt. Mödling 1928.
7. W. Eberhard, Die vorhistorische Entwicklung von Japan. O. A. R. 1936, p. 554.
8. W. Eberhard, Eine neue Arbeitshypothese über den Aufbau der chinesischen Frühkulturen. 2. Tagung d. Ges. f. Völkerkde. in Leipzig, 12.–15. Okt. 1936.
9. M. Ramming, Über die Herkunft der Japaner. Nippon **3**, 1937. Pp. 82–91.

10. K. Ohyama, Yayoi–Kultur. Eine prähistorische Kultur der japanischen Inseln. Jubiläumsbd. D. O. A. G. **I**, 1933. Pp. 127–134, Plates IX–XII.
11. S. Yokoyama, Keijō fugai Takonomine iseki hōkoku (Excavation report on Takonomine near Seoul). Shiz. Z. **2**, 1930. Pp. 289–300 (Jap.).
12. S. Yokoyama, Fusanfu Zetsuyeitō Tōsandō kaizuka hōkoku (Bericht über den Muschelhaufen Tōsandō auf Zetsuyeitō bei Fusan). Shiz. Z. **5**, 1933. 56 p. (Jap. with Ger. summary).
13. Taishō kyū nen do koseki chōsa hōkoku (Report on the Investigation of Antiquities in 1920): The Shell Mound of Kimhae. Chōsen Sōtokufu.
The Shell Mound of Kimhae. Chōsen Sōtokufu.
14. H. Lautensach, Die Urlandschaft in Portugal und in Korea, ein Vergleich. C. R. Congr. Intern. Géogr. Varsovie 1934, Vol. IV, Warszawa 1938. Pp. 161–172.
15. R. Torii, Les Dolmens de la Corée. Mem. Research Dept. Tōyō Bunko 1926. Pp. 93–100.
16. W. Gowland, Notes on the Dolmens and Other Antiquities of Korea. Journ. Anthropol. Inst. London **24**, 1895. Pp. 316–330.
17. E. Baelz, Dolmen und alte Königsgräber in Korea. Zeitschr. f. Ethnol. **42**, 1910. Pp. 776–781 (reprinted in Deutsche Japanpost **9**, 1910/11).
18. E. Baelz, Über Dolmen in Korea. Ostasiat. Lloyd. No. 50, 15 Dec. 1911.
19. F. Krüger, Koreanische Dolmen. Nordische Welt **2**, 1934, No. 9, pp. 1–12.
20. M. Colani, Mégalithes du Haut–Laos. 2 vols. Paris 1935. (Rev. in Z. G. E. B. 1936, p. 376.)
21. K. Hamada, Alte Gräber in Korea. Kokogaku Zasshi **14**, 1924. Pp. 915–932 (Jap.). (Cf. Ethnol. Anz. 1 B, 1928, p. 351.)

II. Archaeology

22. Taishō go nen Chōsen koseki chōsa hōkoku (Report on the investigation of Korean antiquities in 1916). Chōsen Sōtokufu 1917. – Taishō roku nen do koseki chōsa hōkoku (Report on the investigation of antiquities in 1917). Chōsen Sōtokufu 1920. This series has been continued regularly. Most recent: Reports on 1934 (Jap.). Cf. 13.
23. Koseki chōsa tokubetsu hōkoku (Special reports on antiquities). Vol. I. Heijō fukin ni okeru Rakurō jidai no fumbo (Tombs of the Lolang Period near P'yŏng-yang). Chōsen Sōtokufu 1919. 68 p. (Jap.). Supplement to 22. This series has also been continued. Vol. VI contains archaeological investigations in ancient Koguryŏ (two parts with 250 plates. Chōsen Sōtokufu 1933), vol. IV (1928) again reports on Lolang excavations.
24. Chōsen koseki zufu (An Illustrated Catalogue of Ancient Monuments and Historical Remains in Korea). Chōsen Sōtokufu. 14 vols., 1915–34. 6200 figures (text Jap.).
25. Chōsen koseki zufu kaisetsu (continuation of 24). Four issues to date. Chōsen Sōtokufu (Jap.).
26. O. Kümmel, Altchinesische Kunst auf koreanischem Boden. Forsch. u. Fortschr. **4**, 1928. P. 147f.
27. Y. Harada and K. Tazawa, Lo-Lang: A Report on the Excavation of Wang Hsü's Tomb in the "Lo Lang" Province, an Ancient Chinese Colony in Korea. Tōkyō 1930. 4°, 128 tables. Text partly Jap., partly Ger., partly Eng.
28. Rakurōgun jidai no iseki (Investigations on the period of the Lolang commandery). 1 vol. text, 2 vols. plates. Tōkyō 1932 (Jap.).
29. K. Hamada, On the Recent Excavations of the Han Tombs at the Ancient Lo-lang District, Korea. Pr. I. A. T. **10**, 1934, No. 4. Pp. 191–195.
30. Y. Harada, Excavation of the Mud-Castle at Lolang. Pr. I. A. T. **11**, 1935, No. 6. Pp. 207–209.
31. A. Koizumi and S. Sawa, Tomb of the Painted Basket and Other Two Tombs of Lo-lang. Detailed Report of the Excavations of the Tombs of the Lo-lang District, Korea, Ancient Chinese Colony of the Han Dynasty, Carried Out in the Season of 1931 by the Society for the Study of Korean Antiquities. New York 1935. 111 p., 132 maps.
32. H. Ikeuchi, A Study on Lo-lang and Tai-fang, Ancient Chinese Prefectures in Korean Peninsula. Mem. Research Dept. of the Tōyō Bunkō (The Oriental Library) no. 5. Tōkyō 1939. 17 p.
33. S. Umehara, Deux grandes découvertes archéologiques en Corée. Rev. des Arts Asiatiques **3**, 1926. Pp. 24–33.
34. H.E. Fernald, Rediscovered Glories of Korean Art. Asia 1931. Pp. 788–795, 799–802.

35. Studien zur Kunst der Han-Zeit. Die Ausgrabungen von Lo-lang in Korea. Festschrift J. Strzygowski (Wiener Beiträge zur Kunst– u. Kulturgeschichte Asiens VII). Wien 1933. 96 p., 4°.
36. Abbildungen von Rakurō-Gräbern. Atlantis 1929, p. 249 ff.
37. H.B. Hulbert, The Treasures of Kyong-ju. Kor. Rev. **2**, 1902. Pp. 385–392.
38. K. Hamada and S. Umehara, A Royal Tomb, "Kinkan tsuka" or the Gold Crown Tomb,at Keishū,and its Treasures. Chōsen Sōtokufu 1924. 1 vol. text (221 p.), 1 vol. plates (folio with 50 plates) (Jap.).
39. Memoir of Historic Remains in Keishū (Kyung-ju). Chōsen Sōtokufu 1926. 55 p., 27 plates (Jap. and Eng.).
40. The Oldest Relic in Korea (Tangun's Altar, Kangwha Island). Kor. Rev. **4**, 1904. Pp. 255–259.
41. T. de Lacouperie, The Miryeks or Stone-men of Corea. J. As. Soc. Jap. **22**, 1887.

III. History

42. H.B. Hulbert, The History of Korea. Seoul 1905. 2 vols. 409 and 374 p. Previously published in Kor. Rev. **1–4**, 1901–1904.
43. H.B. Hulbert, The Passing of Korea. New York 1906. 473 p. 44. J. Ross, Corea, Its History, Manners and Customs (also under the title: History of Corea, Ancient and Modern). London 1880 (1891). 404 p.
45. W.E. Griffis, Corea, the Hermit Nation. New York 1882. 462 p. 1907^2. 1911^3.
46. W.E.G. (Griffis), Corea Without and Within: Chapters on Corean History, Manners and Religion, with Hendrik Hamel's Narrative of Captivity and Travels in Corea. Philadelphia 1885. 315 p. (Rev. in P. M. 1886, L. B. 118).
47. W.E. Griffis, Korea and the Koreans: In the Mirror of Their Language and History. Am. Geogr. Soc. **27**, 1895. Pp. 1–20.
48. E.H. Parker, On Race Struggles in Corea. Tr. As. Soc. Jap. **18** II, 1890. Pp. 157–228.
49. J.S. Gale, The History of the Korean People. Seoul 1931. Also: Korean Mission Field 1924–27.
50. M. Courant, La Corée jusqu'au IXe siècle. T'oung Pao **9**, 1898. Pp. 1–27.
50a. Kei Won Chung and G.F. Hourani, Arab Geographers on Korea. Journ. Am. Oriental Soc. 1938, p. 658ff.
51. E. Baelz, Korea von seinen Anfängen bis zu seinem Ende. Frankfurt a. M. 1910. 24 p. (reprint from Frankf. Zeitung.)
52. W. Eberhard, Kultur und Siedlung der Randvölker Chinas. Leyden 1942. 506 p. Korea: Die koreanischen Völker, pp. 15–28.
53. J.H. Longford, The Story of Korea. London and Leipzig 1911. 400 p.
54. E.H. Parker, Corea. The Fortnightly Review N. S. **63**, 1898. Pp. 224–238.
55. A. Herrmann, Historical and Commercial Atlas of China. Cambridge, Mass. 1935.
56. O. Franke, Geschichte des Chinesischen Reiches. 3 vols. to date. Berlin and Leipzig 1930/37.
57. F.M. Trautz, Die Landbrücke von Korea. Z. F. Geopol. **1**, 1924. Pp. 485–496.
58. T. Hayashi, Chōsen tshushi (Comprehensive History of Korea). Tōkyō 1925. 608 p. (Jap.).
59. Chōsen rekishi chiri (Historical geography of Korea). Tōkyō 1913. 394 p. (Jap.).
60. K. Aoyanagi, Chōsen bunkashi taisei (Complete cultural history of Korea). Keijō 1924^4. Ges. z. Erf. Koreas. 1260 p. (Jap.).
61. J. Inaba, Chōsen bunkashi kenkyō (Studies on the cultural history of Korea). Tōkyō 1925. 378 p. (Jap.).
62. T. Shidehara, Chōsen shiwa (Historical Accounts of Korea). Tōkyō 1925. 531 p. (Jap.).
63. S. Shirabori, Manshū Chōsen rekishi chiri (Historical Geography of Manchuria and Korea). 8 vols., Tōkyō 1927 (Jap.).
64. T. Yoshida, Historical Atlas of Japan. Tōkyō 1935. 141 plates, 27 p. text (Jap.).
65. S. Umehara, Kodai Chōsen ni okeru hoppō-kei bumbutsu no konseki (The Traces of the Northern Civilisation in Ancient Korea). Seikyū gakusō 1932, no. 7 (Jap.).

66. C. Hentze, Schamanenkronen zur Han-Zeit in Korea. Ostasiat. Zeitschr. N. F. **9**, 1933. Pp. 156–163.
67. H. Koppelmann, Die Eurasische Sprachfamilie. Indogermanisch, Koreanisch and Verwandtes. Heidelberg 1933. 220 p.
68. H. Jensen, Indogermanisch und Koreanisch. Germanen und Indogermanen. Hermann Hirt-Festschrift vol. II. Heidelberg 1936. Pp. 159–170.
69. J.S. Gale, The Influence of China upon Korea. Tr. Kor. Br. R. As. Soc. **1**, 1900. Pp.1–24.
70. H.B. Hulbert, Korean Survivals. Tr. Kor. Br. R. As. Soc. **1**, 1900. Pp.25–50.
71. L.C. Goodrich, Korean Interference with Chinese Historical Records. Tr. N. China Br. R. As. Soc. **68**, 1937. Pp. 27–34 (cf. also Tr. Kor. Br. R. As. Soc. **30**, 1940. Pp. 33–45).
72. W.W. Rockhill, China's Intercourse with Korea from the XVth Century to 1905. London 1905. 60 p. (Rev. in P. M. 1906, L. B. 514.)
73. A. Herrmann, Das geographische Bild Chinas im Altertum. Sinica–Sonderheft 1937. Pp. 72–87.
74. F. Hirth and W.W. Rockhill, Chau Ju-Kua, His Work on the Chinese and Arab Trade in the 12th and 13th Centuries, Entitled Chu-Fan-Chi. Translated from the Chinese and annotated by... Petrograd 1911. Academy of Sciences. Korea: pp. 166–169.
75. J. Hackin, L'Oeuvre de la délégation archéologique française en Afghanistan (1922–32). I. Archéologie bouddhique. Tōkyō 1933. Maison Franco-Japonaise.
76. G. Haloun, Seit wann kannten die Chinesen die Tocharer oder Indogermanen überhaupt. 1. Teil. Leipzig 1926.
77. Z.N. Matveeff, The State Bohai. From the History of Eastern Asia from the 8th to the 10th Century. Vladivostok 1929 (Russ. with Eng. summary).
78. H.B. Hulbert, The Invention of the Ni-t'u. Kor. Rep. **5**, 1898. Feb.
79. M. Courant, Note sur les différents systèmes d'écriture employés en Corée. Tr. As. Soc. Jap. **23**, 1895. P. 5ff.
80. F.X. Biallas, Der Ursprung der koreanischen Schrift. Anthropos **25**, 1930. Pp. 731–736.
81. A. Eckardt, Ursprung der koreanischen Schrift. M. D. O. A. G. 1928, vol. B.
82. E. Satow, Further Notes on Movable Types in Korea and Early Japanese Printed Books. Tr. As. Soc. Jap. **10**, 1882. Part 2.
83. J. Daland, The Evolution of Modern Printing and the Discovery of Movable Metal Types by Chinese and Koreans in the 14th Century. Journ. Franclin Institute **212**, 1931. Pp. 209–234.
84. W.C. Rufus, Astronomy in Korea. Tr. Kor. Br. R. As. Soc. **26**, 1936. Pp. 1–52.
85. E. Satow, The Korean Potters in Satsuma. Tr. As. Soc. Jap. **6**, 1878. Part 2.
86. A. Wedemeyer, Japanische Frühgeschichte. Tōkyō 1930. M. D. O. A. G. Supplement XI, 346 p.
87. Y.S. Kuno, Japanese Expansion on the Asiatic Continent. Berkeley, Cal. Vol. I, 1937. 373 p. Vol. II, 1940. 352 p.
88. R.M. Cory, Some Notes on Father Gregorio de Cespedes, Korea's First European Visitor. Tr. Kor. Br. R. As. Soc. **27**, 1937. Pp. 1–55.
89. A. Pfizmaier, Der Feldzug der Japaner gegen Corea im Jahre 1597. Aus dem japan. Geschichtswerk Chōsen Monogatari übersetzt. 2 vols. Wien 1874.
90. J. Yamagata, Japanese–Korean Relations After the Japanese Invasion of Korea in the XVIth Century. Tr. Kor. Br. R. As. Soc. **4**, part 2, 1913. Pp. 1–11.
91. W. Krebs, Der Koreakrieg in seinen natürlichen Beziehungen zu den Witterungs– und Bevölkerungsverhältnissen Ostasiens. Samml. Wiss. Vortr. no. 232. Berlin 1895. 47 p., five maps.
92. E.M. Cable, The United States–Korean Relations 1866/1871. Tr. Kor. Br. R. As. Soc. **28**, 1938. Pp. 1–229.
93. C. Duncan, Corea and the Powers. A Review of the Far Eastern Question. Shanghai 1889. 96 p.
94. N. Weber, Gotteswege. Ein Blick in die koreanische Missionsgeschichte. St. Ottilien 1929.
95. L.G. Paik, History of Protestant Missions in Korea. Pyengyang 1929. 438 p.
96. J.D. van Buskirk, Korea, Land of the Dawn. New York 1931. 200 p.
97. H.A. Rhodes, History of the Korea Mission, Presbyterian Church, USA 1884–1934. Seoul 1934. 672 p.
98. C.A. Clark, The Korean Church and the Nevius Methods. New York etc. 1930. 278 p.
99. A. Eckardt, Eine deutsche Benediktinerabtei in Korea. O. A. R. **13**, 1932. Pp. 12–15.
100. H. Lautensach, Das Deutschtum in Korea. Neueste Zeitung 1935, no. 107.

IV. Ancient Korean Books and Maps

101. M.N. Trollope, Corean Books and Their Authors, Being an Introduction to Corean Literature. Tr. Kor. Br. R. As. Soc. **21**, 1932. Pp. 1–104.
102. Kim Pu-sik (1075–1151), Samguk sagi (Records of the Three Kingdoms). Completed around 1150. Eight books in 50 vols.
103. Sŏ Kŏ-jŏng (1420–1488), Tongguk t'onggam (Complete Mirror of the Eastern Country). Korean History from 2300 B.C. – 1392 A.D. Completed in 1484. 50 books in 28 vols.
104. J.S. Gale, Korean History (Translated from the Tong-gook T'ong-gam). Kor. Rep. **3**, 1895, pp. 321–327. **4**, 1896, pp.14–19, 95–100, 183–188.
105. No Sa-sin and Sŏ Kŏ-jŏng, Tongguk Yŏji Sŭngnam (Earth-loaded Cart-records). Completed 1481. Printed 1487. 11 books in 55 vols. One section is translated by H.B. Hulbert in Kor. Rep. 1898.
106. Sin Suk-chu and others, Kukcho pogam (Mirror of the Dynasty). History of the Chosŏn dynasty. Begun under King Sejo (1455–68).
107. Hong Pong-han, Tongguk munhŏn pigo (Korean Encyclopedia). First published in 1770. 100 books in 40 vols. Geography included.
108. J.S. Gale, A Shipwreck (Korean) in 1636 A.D. Translated from the Korean by ... Tr. Kor. Br. R. As. Soc. **15**, 1924. Pp. 1–22.
109. Ōtani, Chōsen kōchizu tenkan mokuroku (Catalogue of old Korean maps). Keijō 1933. 37 p. (Jap.).
110. H. Nakamura, Über die Entwicklung der Karten von China, Japan und Korea vom 8. bis zum 18. Jahrh. To be published in Imago Mundi IV.
111. H. Cordier, Description d'un atlas Sino–Coréen, manuscript du British Museum. Recueil de voyages et de documents p. s. (??). l'histoire de la géographie. Sect. Cart., no. 2. Folio. Paris 1896 (Rev. in P. M. 1896, L. B. 718).

V. Geographical Studies by Foreigners Unfamiliar with the Country
(Primarily from the Isolation Period)

112. G. Ferrand, Voyage du marchand arabe Sulaymân en Inde et en Chine rédigé en 851 suivi de remarques par Abû Zaid Hasan (vers 916). Traduit de l'arabe par... Paris 1922. Les Classiques de l'Orient vol. VII. 155 p.
113. Maçoudi, Les prairies d'or. Coll. d'ouvrages orientaux publ. par la Societé Asiatique. Paris 1859.
114. A. Jaubert, Géographie d'Édrisi. Traduite de l'arabe en français. Paris 1836 (Recueil de voyages et de mémoires). Vol. I, p. 93f, 500f.
115. K. Miller, Charta Rogeriana. Weltkarte des Idrisi vom Jahre 1154 n. Chr. Stuttgart 1926.– Mappae arabicae. Vol. VI, Stuttgart 1927. Plates 10 and 30.
116. Itinerarium Wilhelmi de Rubruk in: Receuil de voyages et de mémoires publié par la Soc. de Géogr. Vol. IV, Paris 1839. Pp. 205–396.
117. Marci Pauli Veneti...de regionibus orientalibus libri III.
118. Three Severall Testimonies Concerning the Mightie Kingdom of Coray, Tributary of the Kingdom of China...and Called by the Portugales Coria etc...Collected out of the Portugale Yearly Japonian Epistels, Dated 1590, 1592, 1594. Included in 119.
119. R. Hakluyt, The Principal Navigations, Voyages, Traffiques and Discoveries of the English Nation Made by Sea or Over Land to the Remote and Farther Distant Quarters of the Earth. 3 vols., London 1589–1600.
120. J. Huygen van Linschoten, Itinerario. Voyage oftte schipvaert van...naer oost ofte Portugaels Indien 1579–1592. Vierde deel. Werken uitgeg. door de Linschoten–Vereeniging XLIII, 1939.
121. M. Quadus, Fasciculus geographicus. Coloniae 1606.
122. P. Nicol. Trigautius, De christiana expeditione apud Sinas suscepta ab Societate Jesu, ex P. Matthei Riccii ejusdem Societatis Commentariis Libri V. Lugduni 1616.
123. M. Martini, Novus Atlas Sinensis. Amsterdam 1653. German edition Wien 1655.
124. C. Arnold, Wahrhaftige Beschreibungen dreyer mächtigen Königreiche, Japan, Siam und Corea...mit neuen Anmerkungen und schönen Kupferblättern vermehrt, verbessert und

geziert. Nürnberg 1672. On Korea: pp. 811–900, only a translation, however, of the text in Martini's atlas (no. 123) and Hamel's account (no. 171).
125. J.B. du Halde, Description géographique, historique, chronologique, politique et physique de l'empire de la Chine et de la Tartarie chinoise. 4 vols., Paris 1735. In vol. IV, pp. 529–537: Observations géographiques sur le royaume de Corée, tirées des mémoires du Pére Régis.
126. J.B.B. d'Anville, Nouvel Atlas de la Chine, de la Tartarie chinoise et du Thibet, contenant des cartes générales et particulières de ces pays ainsi que la carte du royaume de Corée. La Haye 1737.
127. J.B. du Halde, Ausführliche Beschreibung des chinesischen Reiches und der großen Tartarey, nebst Engelbr. Kämpfers Beschreibung des japonischen Reiches. 4 vols., Rostock 1747–1756.
128. J.B. Régis, Observations géographiques et histoire de la Corée. In: Prévost, Histoire générale des voyages vol. 8, 1749. La Haye. P. 395ff.
129. J. Klaproth, Mémoires relatifs à l'Asie. Vol. 1. Paris 1826. Korea: Voyage à la Montagne Blanche, traduit du Mandchou, pp. 455–460.
130. Timkowski, Travels of the Russian Mission Through Mongolia to China and Residence in Peking, in the Years 1820/21. London 1827. Vol. II, pp. 43–96.
131. J. Klaproth, San Kokf Tsou Ran To Setsu ou Aperçu Général des Trois Royaumes. Paris 1832. 288 p.
132. P. Fr. von Siebold, Nippon. Archiv zur Beschreibung von Japan und dessen Neben– und Schutzländern: Jezo mit den südlichen Kurilen, Krafto, Kooraï und den Liukiu-Inseln. Leyden 1832–1851. Especially: VII, Nachrichten über Kooraï. 163 p.
133. C. Ritter, Die Gebirgshalbinsel Korea. In: Die Erdkunde, 4. Teil. 2. Buch. Asien, 3, II. Berlin 1834. Pp. 573–647.
134. A. Pfizmaier, Nachrichten von den alten Bewohnern des heutigen Corea. Sitz.–Ber. Kais. Ak. Wiss. Wien. Phil.–hist. Cl. **57**, 1867. Pp. 461–523.
135. A. Williamson, Journeys in North China, Manchuria and Eastern Mongolia; with Some Account on Corea. London 1870. 2 vols. 444 and 442 p.
136. A. Pfizmaier, Darlegungen aus der Geschichte und Geographie Corea's. Sitz.–Ber. Kais. Ak. Wiss. Wien. Phil.–hist. Cl. **78**, 1874. Pp. 89–142.
137. C. Dallet, Histoire de l'Eglise de Corée, précédée d'une introduction sur l'histoire, les institutions, la langue, les moeurs et coutumes coréennes. 2 vols., Paris 1874. 1174 p.
138. P. Kempermann, Corea und dessen Einfluß auf die Bevölkerung Japans. Verhandl. Berl. Ges. f. Anthropol., Ethnol. u. Urgeschichte **8**, 1876. Pp. (78)–(83).
139. Der Markt am Thor von Korea. Das Ausland 1876, pp. 387–391.
140. F. v. Richthofen, Schreiben des Freiherrn Ferdinand von Richthofen über seine Reisen zur Grenze von Korea und in der Provinz Hu-nan. Z. G. E. B. **5**, 1870. Pp. 317–339.141. F. v. Richthofen, China. On Korea: vol. I, 1877, p. 575f. Vol. II, 1882, pp. 25, 52, 57, 63, 82–89, 131f, 141, 147f, 161–170, 244, 262.
141. F. v. Richthofen, China. On Korea: vol. I, 1877, p. 575f. Vol. II, 1882, pp. 25, 52, 57, 63, 82–89, 131f, 141, 147f, 161–170, 244, 262.
142. F. Scherzer, Journal d'une mission en Corée par Koei-Ling, ambassadeur de S. M. l'Empereur de la Chine près de la Cour de la Corée. Traduit du Chinois. Paris 1877. Recueil d'itinéraires dans l'Asie Centrale. Pp. 1–62, with map. Also in Revue de Géogr. 1877, pp. 129–134, 209–214, 287–292.
143. E. v. Kudriaffsky, Korea. Das Ausland 1877. Pp. 353–355, 502–504.
144. S. Mossmann, Corea. Geogr. Mag. **4**, 1877. Pp. 148–152.
145. F. Ratzel, Korea, die Liukiu-Inseln und die zwei ostasiatischen Großmächte. Österr. Monatsschr. f. d. Orient **5**, 1879. Pp. 189–196.
146. É. Reclus, Nouvelle Géographie Universelle. Vol. VII. L'Asie Orientale. Paris 1882. Korea: pp. 649–683.
147. F.G. Müller–Beeck, Unsere wissenschaftliche Kenntnis von Korea. 1. Jahresber. d. Geogr. Ges. Greifswald 1882. Pp. 35–57.
148. P.A. Dimitrewski, Facts on Korea. Compendium "General Geography," ed. R. Lenz. Vol. XII, section 4 (Russ.). Translated from the Japanese. (Rev. in P. M. 1884, p. 271).

149. P. Dimitrewski, Geografitscheskoe opisanie Korei. Perewod s kitaiskago (Geographical description of Korea. Translated from the Chinese). Chankou 1883 (Russ.).
150. Die Entwicklung der Kartographie von Korea. P. M. 1883, pp. 341–344. W. map 1:1.7 million (Plate 10).
151. W.J. Kenny, Account of a Secret Trip into the Interior of Korea. Tr. As. Soc. Jap. **11**, 1883. Pp. 141–148. Translated from the Japanese.
152. A. W. D., Notes on Corea. Shanghai 1884. Reprinted from "The Star in the East" (Rev. in P. M. 1885, p. 270).
153. M. Jametel, La Corée avant les traités. Souvenirs de voyages. Paris 1885. Also in Rev. de Géogr. **15**, 1884; **16**, 1885 and **17**, 1885.
154. M.F. Scherzer, Tchao-sien-tche, Mémoire sur la Corée, par un Coréen anonyme. Traduit pour la première fois du chinois, avec un commentaire perpétuel. Journ. Asiatique VIIIe Sér., V, Août–Oct. 1885, pp. 160–242. VII, Févr.–Avr. 1886, pp. 223–332. Also English translation.
155. L. de Rosny, Les Coréens. Aperçu ethnographique et historique. Paris 1886. 144 p.
156. M.A. Pogio, Korea. Wien and Leipzig 1895. Translated from the Russian by St. Ritter v. Ursyn-Pruszyński. 248 p. (rev. in P. M. 1896, L. B. 496).
157. L. Nocentini, Materiali per la storia degli antichi stati Coreanì. Rendiconti R. Ac. dei Lincei. Cl. di Sc. morali, stor. e fil. V, 5. 1896, pp. 12–36.– Materiali per la geografia della Corea, ibid. pp. 111–138.– Notizie generali della Corea, ibid. pp. 234–249 (Rev. in P. M. 1897, L. B. 360).
158. H. Cordier, De la situation du Japon et de la Corée. Manuscrit inédit du Père A. Gaubil, S. J. T'oung-Pao **9**, 1898. 16 p.
159. A. Pawlowski, Historique de la connaissance de la Corée d'aprés la cartographie. Bull. Soc. Gèogr. de Rochefort **26**, 1904. Pp. 216–225. (Geogr. Journ. **25**, 1905. P. 456).

VI. Western Naval Expeditions

160. Voyage de La Pérouse autour du monde, rédigé par M.L.A. Milet–Mureau. Paris, an V (1797). 4 vols. and atlas. Korea in vol. II, pp. 384–392.
161. W.R. Broughton, A Voyage of Discovery to the North Pacific Ocean: in which...the Coast of Corea (has) been examined and surveyed. Performed in His Majesty's Sloop "Providence" and her tender, in the years 1795, 1796, 1797, 1798. London 1804. 396 p.
162. Captain Basil Hall, Account of a Voyage of Discovery to the West Coast of Corea and the Great Loo-choo Island. London 1818. 222 p. App. 130 p. German translation: Weimar 1819. 2nd abbreviated edition: London 1820. 259 p.
163. Captain Basil Hall's Account of his Voyage to the West Coast of Korea in 1816. Tr. Kor. Br. R. As. Soc. **11**, 1920. Pp. 3–37. Excerpt from 162.
164. L.G. Paik, The Korean Record on Captain Basil Hall's Voyage of Discovery to the West Coast of Korea. Tr. Kor. Br. R. As. Soc. **24**, 1935. Pp. 15–19.
166. C. Gutzlaff, Journal of three Voyages along the Coast of China in 1831, 1832 and 1833, with Notices on Siam, Corea and the Loo-choo Islands. London 1834. 312 p.
167. Report of Proceedings on a Voyage to the Northern Ports of China, in the Ship "Lord Amherst". Extracted from papers printed by order of the House of Commons, relating to the trade with China. London 1833. A. Lindsay's Report pp. 215–259. B. Gützlaff's Report pp. 293–295.
168. Captain Sir Edward Belcher, Narrative of the Voyage of H. M. S. Samarang, During the Years 1843–46, Employed Surveying the Islands of the Eastern Archipelago. With Notes on the Natural History of the Islands, by A. Adams. London 1848, 2 vols.
169. E. Oppert, Ein verschlossenes Land. Reisen nach Corea. Leipzig 1880. 316 p. Also Engl. edition.
170. Captain Sir F.J.O. Evans, Report on Admirality Surveys for the Year 1882. Nautical Mag. **52**, 1883. Pp. 656–668 (British surveys on the west and south coast of Korea. (Cf. P. M. 1883, p. 387f).

VII. Western Travel Accounts

171. H. Hamel van Gorcum, Journael, Van de Ongeluckige Voyagie van't Jacht de Sperwer van Batavia gedestineert na Tayowan in't Jaer 1653, en van daer op Japan; hoe't selve Jacht door storm op't Quelpaerts Eylant is ghestrant... Als mede een pertinente Beschrijvinge der Landen Provintien Steden ende Forten leggende in't Coninghrijck Coeree:... Rotterdam 1668. Ger. translation in 124.
172. H. Hamel, Verhaal van het Vergaan van het jacht de Sperwer en van het wedervaren der schipbreukelingen op het eiland Quelpaert en het vasteland van Korea (1653–66), met eene beschrijving van dat rijk, door Hendrik Hamel. Uitgegeven door B. Hoetink. 's Gravenhage 1920. 220 p. Revised edition of 171 published as vol. 18 of Werken uitgegeven door de Linschoten–Vereeniging.
173. An Account of the Shipwreck of a Dutch Vessel on the Coast of the Isle of Quelpart, Together with the Description of the Kingdom of Korea. Engl. translation of 171. Tr. Kor. Br. R. As. Soc. **9**, 1918. Pp. 91–148.
174. H. Zuber, Une expedition en Corée. Le Tour du Monde **25**, 1873. Pp. 401–416.
175. J.C. Hall, A Visit to Corea, in October 1882. Proc. R. Geogr. Soc. N. S. **5**, 1883. Pp. 274–284.
176. J.C. Hall, A Visit to West Coast and Capital of Korea. Tr. As. Soc. Jap. **11**, part 2, 1883. Pp. 148–161.
177. H.A.C. Bonar, Notes on the Capital of Korea. Tr. As. Soc. Jap. **11**, part 2, 1883. Pp. 243–259.
178. W.G. Aston, A Journey from Söul to Songdo in August 1884. Comm. Reports by H. M. Consul-General in Corea, 1882–1883. London 1885.
179. P. Lowell, Chosön: the Land of the Morning Calm. Boston 1885. 402 p. (Rev. in P. M. 1886, L. B. 119).
180. P. Mayet, Ein Besuch in Korea im October 1883. M. D. O. A. G. **4**, 1884/88. Pp. 18–28, 146–152.
181. G.W. Morrison, Some Notes on a Trip to Corea,in July and August, 1883. Tr. N. China Br. R. As. Soc. **28**,1884. Pp. 141–157.
182. S.B. Bernerston, A Trip from Söul to Peng Yang. Journ. Am. Geogr. Soc. **16**, 1884. Pp. 234–241. (Cf. P. M. 1885, p. 182).
183. G.C. Foulk, Report on Four Important Towns of Corea. USA Foreign Relations 1885.
184. W. Gowland, Journey from Seoul to Pusan, rev. in Nature **31**, 1884/85. P. 397.
185. A. Glocker, Reise nach Korea. Dtsch. Rundschau f. Geogr. u. Stat. **9**, 1886/87. Pp. 108–116, 212–219, 358–366 (Dtsch. Kolonialztg. **3**, 1886. Pp. 729–733).
186. W.R. Carles Recent Journeys in Korea. Proc. R. Geogr. Soc. N. S. **8**, 1886. Pp. 289–312.
187. W.R. Carles, Life in Corea. London 1888. 317 p. (Rev. in P. M. 1888, L. B. 303).
188. C. Gottsche, Über Land und Leute in Korea. Verhandl. Ges. Erdk. Berlin **13**, 1886. Pp. 245–262. (Also as independent publication: Berlin 1886.)
189. C. Gottsche, Korea und seine Nachbarn. Mitt. Ver. Erdk. Leipzig 1886, p. XXXIIff.
190. H.E.M. James, The Long White Mountain. London 1888. 502 p. (Rev. in P. M. 1888, L. B. 304).
191. F. Kraus, Die Hauptstadt von Korea und ihre Bewohner. Ostasiat. Lloyd 1887. No. 218, 257, 263.
192. F. Kraus, Eine Reise nach Korea. Dtsch. Rundschau f. Geogr. u. Stat. **11**, 1889. Pp. 207–219.
193. O. Genest, Kapitän Jakobsen's Besuch bei den Koreanern. Globus **52**, 1887. Pp. 58–61, 71–75.
194. C. Varat, Voyage en Corée. Le Tour du Monde 1892, I. Pp. 289–368. (C. R. Soc. Géogr. Paris 1890, p. 67f).
195. Col. Chaillé–Long, De Séoul à Quelpaërt et Voyage de retour par Fou-San, Wôn-San et Vladivostok. Bull. Soc. Géogr. VIIe Série **10**, 1889. Pp. 425–444 (Eng. translation: From Corea to Quelpaert. Journ. Am. Geogr. Soc. **22**, 1890. Pp. 219–266).
196. Col. Chaillé–Long–Bey, La Corée ou Chösen. Ann. Mus. Guimet XXVI, 1. Paris 1894. Pp. 1–75 (Bull. Soc. Norm. Géogr. **14**, 1892, pp. 384–418. **15**, 1893, pp. 1–18). (Rev. in P. M. 1896, L. B. 499).
197. C.W. Campbell, A Journey through North Korea to the Ch'ang-pai Shan. Proc. R. Geogr. Soc. N. S. **14**, 1892. Pp. 141–161. (Scott. Geogr. Mag. **8**, 1892. Pp. 579–591).

198. J.S. Gale, To the Yaloo and Beyond. Kor. Rep. **1**, 1892, pp. 17–24, 37–51, 75–85.
199. L.O. Warner, A Journey of Exploration in Corea. Corea Mission Field **38**, 1893. Pp. 134–142 (Geogr. Journ. **9**, 1897. P. 661f. Österr. Monatsschr. f. d. Orient **23**, 1897. P. 82).
200. G.W. Gilmore, Korea from Its Capital. Philadelphia 1892. 328 p.
201. E. Brass, Reise durch Korea. Export **15**, 1893. Pp. 105f, 119–121, 136f, 167–170.
202. H. Goold–Adams, A Trip to the Mont Blanc of Corea. Kor. Rep. **1**, 1892. Pp. 237–244, 269–277, 300–307.
203. A.E.J. Cavendish, Korea and the Sacred White Mountain, Together with an Account of an Ascent of the White Mountain by Captain H.E. Goold–Adams. London 1894. 220 p. (Scott. Geogr. Mag. **10**, 1894. Pp. 561–574). (Rev. in P. M. 1896, L. B. 498).
204. E. v. Hesse–Wartegg, Korea. Eine Sommerreise nach dem Lande der Morgenruhe 1894. Dresden and Leipzig 1895. 220 p. (Rev. in P. M. 1896, L. B. 495).
205. A.H. Savage–Landor, Corea or Chosen. The Land of the Morning Calm. London 1895. 304 p. (Rev. in P. M. 1896, L. B. 497).
206. G.N. Curzon, Problems of the Far East. Japan–Korea–China. New York 1894. Korea: pp. 81–233.
207. I.B. Bishop, Korea and Her Neighbours. A Narrative of Travel, with an Account of the Recent Vicissitudes and Present Position of the Country. 2 vols. London 1898. 261 and 321 p. (Rev. in P. M. 1898, L. B. 799).
208. C.T. Gardner, Corea. Rep. VI. Meeting Australasian Ass. Brisbane 1895. Pp. 483–532 (P. M. 1896, p. 91).
209. J.S. Gale, A Trip Across Northern Korea. Kor. Rep. **4**, 1896. Pp. 81–89.
210. Frh. v. Grünau, Von Wönsan nach Soul. Globus **72**, 1896. Pp. 149–151. (P. M. 1897, p. 295).
211. J.H. Veitch, A Traveller's Notes; or Notes of a Tour Through India...Corea...During the Years 1891–1893. London 1896. 219 p.
212. D. Lièvre, Promenades japonaises et coréenes 1894–1896. Bull. Soc. Géogr. Commerc. 1898/99. Korea: pp. 195–211.
213. H. Webster, Korea – The Hermit Nation. Nat. Geogr. Mag. **11**, 1900. Pp. 145–155.
214. B. Knochenhauer, Korea. Dtsch. Kol. Ges. Abt. Berlin–Charlottenburg. Verh. 1900/01, no. 4. Pp. 74–124.
215. M. Monnier, A Travers la Corée. La Géogr. **1**, 1900. Pp. 35–50.
216. C. Rosetti, L'Impero di Corea. Bol. Soc. Geogr. Ital. Ser. IV, **5**, 1904. Pp. 550–574.
217. S. Genthe, Korea. Reiseschilderungen. Berlin 1905. 343 p.
218. E. Baelz, Das Leben eines deutschen Arztes im erwachenden Japan. Stuttgart 1931. Korea: pp. 219–246.
219. A. Hamilton, Korea, das Land des Morgenrots. Leipzig 1904. 296 p. (Also Eng. and Fr. editions).
220. W. Sieroszewski, Korea. Land und Volk nach eigener Anschauung gemeinverständlich geschildert. Berlin (1904). 302 p. (Rev. in P. M. 1906, L. B. 512).
221. V. Ehrencron–Müller, Korea og Koreanerne. Köbenhavn 1904. 79 p.
222. R. Zabel, Meine Hochzeitsreise durch Korea während des russisch–japanischen Krieges. Altenburg 1906. 462 p. (Rev. in P. M. 1907, L. B. 450).
223. A. v. Schmidt auf Altenstadt, Reisindrucken van en Hollandsche in Korea. Bull. Soc. Roy. Géogr. Anvers **29**, 1905. Pp. 75–142.
224. E. Kroebel, Wie ich an den Koreanischen Kaiserhof kam. Berlin–Schöneberg (1909). 185 p.
225. E.G. Kemp, The Face of Manchuria, Korea and Russian Turkestan. London 1910. 248 p. Korea: pp. 65–148.
226. N. Weber, Im Lande der Morgenstille. Reiseerinnerungen an Korea. St. Ottilien 1923^2. 467 p. (Rev. in P. M. 1920, p. 177).
227. S. Bergman, I Morgonstillhetens Land. Skildringar fran en Forskningsfärd till Korea. Stockholm 1937. 231 p.
228. S. Bergman, Koreanska Äventyr. Stockholm 1938. 281 p.
229. S. Bergman, In Korean Wilds and Villages. London 1938. 232 p. (Rev. in P. M. 1939, L. B. 119).
230. S. Bergman, När körsbärsträden blomma i Korea. Jorden Runt **9**, 1937. Vol. I, pp. 1–13.
231. H. Lautensach, Hauptergebnisse meiner Koreareise. P. M. 1934, pp. 172–175, 213–217, 256–259.

232. H. Lautensach, Forschungen in Korea. Verhandl. 25. Dtsch. Geogr.–Tag Bad Nauheim 1934, pp. 25–40.
233. H. Lautensach, Forschungsreise in Korea. Z. G. E. B. 1934, pp. 352–359.

VIII. Russian Literature

234. E. Bretschneider, Rußland und Korea. P. M. 1901, pp. 179–182.
235. K. v. Lysakowski, Die erloschenen Vulkane und die Bergketten der Halbinsel Korea. Das Weltall **6**, 1905/06. Pp. 133–139, 162–164.
236. J. Gontcharov, Fregat "Pallada". Otscherki puteschestwija (Frigate "Pallas." Brief accounts of the voyages.) Petrograd 1858 (Russ.).
237. P.M. Delotkevitch, Dnewnik po putji peschkom is Seula w Posjet, tscheres sew. Koreju. S. 6–go dekabrja 1885 g. 29 febr. 1886 g. (Diary written during the trip by foot from Seoul to Pos'yet, through northern Korea. From 6 Dec. 1885 to 29 Feb. 1886). Sbornik geogr. top. i stat. mater. po Asii **38**, 1998. P. 128–167 (Izvest. Imp. Russ. Geogr. Ob. **25**, 1889. Pp. 294–315) (Russ.).
238. Lieutenant Colonel Webel, Pojesdka w Koreju w 1889 g. (Travels through Korea in 1889). Sbornik geogr., top. i stat. mater po Asii **41**, pp. 143–255 (Russ.).
239. Lieutenant Colonel Alftan, Pojesdka w Koreju w dek. 1895 g. i w janw. 1896 g. (Travels through Korea in Dec. 1895 and Jan. 1896). Sbornik geogr., top. i stat. mater. po Asii **69** (Russ.).
240. M.A. Pogio, Otscherki Korei (Brief descriptions of Korea). Petrograd 1892. Original of 156 (Russ.).
241. A.G. Lubenzov, Chamkenskaja i pchienhanskaja prowinzii Korei (The Korean provinces of Hamgyŏng-do and P'yŏng-ando). Sapiski of the Priamur Section of the Imp. Russ. Geogr. Ob. II, 4. Khabarovsk 1897. 297 p. (Russ.).
242. A. Zweginzov, Pojesdka w sewernuju Koreju (Journeys in northern Korea). Izvest. Imp. Russ. Geogr. Ob. **36**, 1900. Pp. 502–518 (Russ.).
243. P.I. Schmidt, Westji iz Koreisko–Sachalinskoi ekspedizii 1900 (Investigations from an expedition through Korea and Sakhalin 1900). Izvest. Imp. Russ. Geogr. Ob. **36**, 1900. Pp. 463–501. Route map in **39**, 1903 (Russ.).
244. E. Ahnert, Puteschestwije po Manjurii (Travels through Manchuria). Sapiski Imp. Russ. Geogr. Ob., Sect. General Geography. vol. 35, 1904. Especially pp. 253–291, 300–308 (Russ.). (Rev. in P. M. 1905, L. B. 622).
245. Opissanie Koreai (S'Kartoi). (Description of Korea [with map].) Compiled in the Office of the Minister of Finance. Petrograd 1900. 3 vols. 448, 490 and 318 p. (Cf. P. M. 1901, pp. 179–182. La Géogr. **5**, 1902. Pp. 116–121) (Russ.).
246. D. Puschdnejeff, Description of Korea from the standpoint of physical, economic and ethnographical conditions. 4 vols. Petrograd 1910. Ministry of Finance. 1500 p. (Russ.).
247. N.W. Kjuner, Statistiko–geografitscherkij i jekonomitschjeskji otscherki Korei (Statistical-geographical and economic description of Korea). 2 vols. Vladivostok 1912. 376 and 380 p. Vostocnyi Institute (Rev. in P. M. 1913, I, p. 147. 1914, II, p. 37).

IX. Other Western Overviews

248. G. Baudens, La Corée. Paris 1884. 38 p.
249. J. Haas, Wirtschaftliche Zustände von Corea. Österr. Konsulatsbericht 1884. 30 p.
250. F. v. Hellwald, Korea. Österr. Monatsschr. f. d. Orient **13**, 1887. Pp. 17–21, 33–38.
251. C.W. Campbell, The Discovery of Korea: With a Brief Sketch of Korean Physiography. Tr. Ninth Internat. Congr. Or. London 1892, II. P. 423f.
252. C.W. Campbell, Report on a lecture on the earliest mention of the name Korea in the Occident. Tr. As. Soc. Jap. 1892.
253. F. v. Richthofen, Der Schauplatz des Krieges zwischen Japan und China. Verhandl. Ges. Erdk. Berlin **21**, 1894. Pp. 456–476 (Geogr. Journ. **4**, 1894).
254. F. v. Richthofen, Der Friede von Schimonoseki in seinen geographischen Beziehungen. Geogr. Zeitschr. **1**, 1895. Pp. 19–39.

255. D.L. Gifford, Places of Interest in Korea. Kor. Rep. **3**, 1895. Pp. 281–287.
256. Anonymous, Korea of to-day. London 1894. 128 p.
257. L. Jordan–Miln, Quaint Korea. London 1895. 306 p.
258. J.S. Gale, Korean Sketches. Chicago, etc. 1898. 256 p. (Scott. Geogr. Mag. **15**, p. 165).
259. C. Wolter, Korea, einst und jetzt. Mitt. Geogr. Ges. Hamburg **17**, 1901. Pp. 63–77.
260. M. de Lapeyrière, La Corée. Bull. Soc. Géogr. Commerc. **24**, 1902. Pp. 217–259.
261. É. Bourdaret, En Corée. Paris 1904. 363 p. (Rev. in P. M. 1905, L. B. 615).
262. C. Rosetti, Corea e Coreani. Impressioni e ricerche sull'Imperio del Gran Han. 2 vols. Bergamo 1904. 170 and 231 p. (Rev. in P. M. 1906, L. B. 513).
263. Generalmajor Krahmer, Die Beziehungen Rußlands zu Japan (mit besonderer Berücksichtigung Koreas). Rußland in Asien vol. VII, Leipzig 1904, 221 p.
264. A. von Schmidt auf Altenstadt, Van't land der morgenkalmte ent 't Hemelsche Rijk. Haarlem 1904.
265. H.N. Allen, Korea. Fact and Fancy. Seoul 1904. 285 p.
266. R. Villetard de Laguérie, La Corée, indépendante, russe, ou japonaise. Paris 1898, 304 p. 2nd edition Paris 1904. (Rev. in P. M. 1898, L. B. 800 and 1905, L. B. 617).
267. A. Legrand, La Corée. Bull. Soc. Bretonne de Géogr. (Lorient) 1905, pp. 315–351. 1906, pp. 391–413.
268. G. Trumbull Ladd, In Korea with Marquis Ito. London 1908. 477 p.
269. H.N. Allen, Things Korean. New York 1908. 256 p.
270. v. Ursyn–Pruszińki, Kleiner Führer durch das annektierte Korea. Wien 1910 (Rev. in P. M. 1911 I, p. 56).
271. M.A. Lushington and A. Hamilton, Korea, Its History, Its People and Its Commerce. Boston 1911. 326 p.
272. R.M. Keir, Modern Korea. Bull. Am. Geogr. Soc. **46**, 1914. Pp. 756–769, 817–830.
273. E.J. Urquhart, Glimpses of Korea. Mountain View, Cal. 1923. 103 p.
274. A. Ireland, The New Korea. New York 1926. 354 p. (Rev. in P. M. 1927, L. B. 526).
275. K. Haushofer, Korea (Chosen). In: Andree–Heiderich–Sieger, Geographie des Welthandels. Vol. II. Wien 1927. Pp. 574–583.
276. T.P. Terry, Terry's Guide to the Japanese Empire including Korea and Formosa. A Handbook for Travellers. London 1927[3]. Korea: pp. 693–756.
277. R. Goldschmidt, Neujapan. Reisebilder aus Formosa, den Ryukyu-Inseln, Bonininseln, Korea und dem südmandschurischen Pachtgebiet. Berlin 1927. Korea: pp. 235–290. (Rev. in P. M. 1927, L. B. 519, Z. G. E. B. 1927, p. 530f).
278. J. Sion, La Corée. In: Géographie Universelle. IX, Asie des Moussons. Part I. Paris 1928. Pp. 245–252.
279. B. Baráthosi–Balogh, Korea, a hajnalpir országa. Budapest 1929. 160 p.
280. F.M. Trautz, Japan mit Korea und Formosa. Berlin 1930. 54 p. text, 256 plates. 4°. (Rev. in P. M. 1933, L. B. 295).
281. H.B. Drake, Korea of the Japanese. London and New York 1930. 225 p.
282. E. Wagner, Korea: The Old and the New. New York etc. 1931. 160 p. (Rev. in P. M. 1933, L. B. 299).
283. J.D. van Buskirk, Korea and Manchuria. New York 1931. 200 p.
284. J. Wright Baylor, The Geography of Chosen. Econ. Geogr. **7**, 1931. Pp. 238–251.
285. M. Moncharville, Le Japon d'Outre mer. Paris 1931.
286. M. Croft Deering, Chosen – Land of Morning Calm. Nat. Geogr. Mag. **64**, 1933, pp. 421–448. Numerous black and white and color photos.
287. S. McCune, Chosen. Journ. of Geogr. **34**, 1935. Pp. 305–317.
288. G. Vacca, La Corea. In: Geografia Universale Illustrata. Vol. 4, 1936. Pp. 1230–1239.
289. H. Rosinski, Korea. In: Handb. Geogr. Wiss. Vol. Nordasien, Zentral– u. Ostasien. Potsdam 1937. Pp. 536–543.
290. W. Sieroszewski, Korea, klucz dalekiego Wschodu (Korea, key to the Far East). 2 vols., Warszawa 1937. 201 and 214 p.
291. W. Credner, In Japans Herrschaftsbereich. Die Auswahl **2**, 1940. Pp. 416–421.
292 H. Lautensach, Die amtlichen japanischen Kartenwerke. P. M. 1942, Pp. 33–38.

X. Japanese Literature on Korea as a Whole

293. An Official Guide to Eastern Asia. Vol. I. Manchuria and Chosen. Tōkyō 1913. Imp. Jap. Gov. Railways. 350 p. (Eng.).
294. Government General of Chosen, Chosen in Pictures. 58 Plates and Statistics. Seoul 1921 (Engl.).
295. Chosen (Korea). Keijo 1929. Illustrated guide publ. by the Government–General (Engl.).
296. S. Kiriyama, Chosen of to-day. Compiled in Commemoration of the Vicennial of the Government General of Chosen. Keijo 1929. 60 p. (Engl.).
297. F. Kobayashi, Dai Nihon Taikoku Chiri Seigi (Geography of the Japanese Empire). Tōkyō 1936. Korea: vol. II, pp. 260–373. Publ. by Geogr. Soc. Tōkyō (Jap.).
298. S. Nakamura, Nihon Chiri Taikei (Geography of Japan). Tōkyō 1930. Vol. XII. Chōsen-hen (Jap.).
299. T. Mukoyama, Nihon Chiri Fūzoku Taikei (Survey of Japan's geography and customs). Tōkyō 1931. Korea: vols. 16 and 17. 399 and 404 p. (Jap.).
300. Nihon Chiri Chizu (Geographical maps of Japan). Tōkyō 1934 (Jap.).
301. Governor–General Ugaki, Chosen and Among its People. Keijo 1934. 20 p. (Eng.).
302. Governor–General Ugaki, The Bright Future for Chosen. Keijo 1934. 34 p. (Eng.).
303. S. Uenoda, A Trip to Korea and Manchuria. Tōkyō 1934. 68 p. (Eng.).
304. Thriving Chosen. A Survey of Twenty-five Years' Administration. Compiled by the Government General of Chosen. Keijo 1935 (Eng.).
305. Chosen Is Land of Promise. Supplement to Osaka Mainichi, 11 Feb. 1936.
306. Shin Nihon Chiri (New geography of Japan). Vol. VI: Taiwan and Chōsen. Tōkyō 1936 (Jap.).
307. Shōwa Nihon Chizu (Japanese maps from the Shōwa period). Tōkyō 1937^2 (Jap.).
308. Kokumin hyakka daijiten. Art. Chōsen (Vol. IX, pp. 13578–13626) (Jap.).

XI. Geological Structure and Mineral Deposits

309. C. Gottsche, Geologische Skizze von Korea. Sitz.–Ber. Pr. Ak. d. Wiss. Berlin 1886. Vol. 2, pp. 857–873. With geol. map 1:4 million.
310. K. Inouye, Geology and Mineral Resources of Korea. Mem. Imp. Geol. Surv. Japan, Tōkyō 1907. 91 p., with colored geol. map 1:1.5 mill. (Geol. Zentralblatt **10**, 1908. Pp. 2–8).
311. N. Yabe, Geology of Korea. Journ. College of Sc. Imp. Univ. Tōkyō. **20**, 1905 (Jap.).
312. B. Kotō, Journeys Through Korea. Journ. College of Sc. Imp. Univ. Tōkyō. (a). **26** Art. **2**, 1909. 207 p., 36 plates. (b). **27**. Art. **12**, 1910, 32 p.
313. L. Leigh Fermor, On the Geology and Coal Resources of Korea State, Central Provinces. Mem. Geol. Survey of India **41**, II. 1914, pp. 148–245.
314. S. Kawasaki, Mesozoic Formations and Diastrophisms in Korea. Proc. 3rd Pan–Pac. Sc. Congr. Tōkyōo 1926. Vol. II, Tōkyō 1927, pp. 1705–1708.
315. S. Kawasaki, Geology and Mineral Resources of Korea. In: The Geology and Mineral Resources of the Japanese Empire. Tōkyō 1926. 20 p.
316. S. Nakamura, Chōsen chishitsu kōzoron chōsetsu (Treatise on the tectonics of Korea). Chikyū **8**, 1927 (Jap.).
317. G.B. Barbour, The Structural Evolution of Eastern Asia. In: J.W. Gregory, The Structure of Asia. London 1929, pp. 188–205.
318. I. Tateiwa, Igneous Activity During Cretaceous and Postcretaceous Times in Korea. Proc. 5th Pac. Sc. Congr. Canada 1933, vol. 3. Toronto 1934, pp. 2415–2419. (Rev. in J. J. G. G. 1939, p. 46).
319. T. Kobayashi, On the Ordovician Formation of North-Korea and South-Manchuria. Journ. Geol. Soc. Tōkyō **36**, 1929. **37**, 1930 (Jap.).
320. T. Kobayashi, Three Sedimentation Facies of the Korean and Sinian Systems in Korea and Manchuria and the Bearing of the Tsingling–Keijō Line on the Palaeogeography. Journ. Geol. Soc. Tōkyō **37**, 1930 (Jap.).
321. T. Kobayashi and S. Kin, Study on the Older Rock Formations of North Heian-do, Chōsen (Korea). Journ. Geol. Soc. Tōkyō **38**, 1931, pp. 349–355 (Rev. in J. J. G. G. 1931/32, p. 14f) (Jap.).

322. T. Kobayashi and S. Kin, Stratigraphical Study of the Early Palaeozoic Formation Found in the Southern Side of the River Oryokko. Chi. Z. **43**, 1931 (Jap.).
323. T. Kobayashi, Studies on the Ordovician Stratigraphy and Paleontology of North Korea with Notes on the Ordovician Fossils of Shantung and Liaotung. B. Geol. S. Chōs. **11**, 1931, no. 1. 60 p. (Jap.).
324. T. Kobayashi, A Sketch of Korean Geology. Am Journ. of Sc. 5th Ser. **26**, 1933. Pp. 585–606 (Eng.).
325. E. Takahashi, On the Geological Age of the Gishu Series. Chōsen Kogyō Kaishi **20**, 1937, no. 7, pp. 1–2 (Rev. in J. J. G. G. **15**, 1938. P. 8) (Jap.).
326. S. Nakamura and S. Matsushita, Pre-Cambrian in Manchuria and Corea. Proc. 6th Pac. Sc. Congr. 1939, pp. 311–319 (Eng.).
327. T. Kobayashi, The Sakawa Orogenic Cycle and its Bearing on the Origin of the Japanese Islands. Journ. Fac. of Sc. Imp. Univ. Tōkyō. Sect. II. Vol. V, part 7. Tōkyō 1941, pp. 230–578 (Eng.). Korea: pp. 234f, 440–443, 454–457, 467–469, 499–501, 516–522 (Cf. S. v. Bubnoff in: Die Naturwiss. 1942, pp. 569–577).
328. F.X. Schaffer, Lehrbuch der Geologie. 3. Teil Geologische Landeskunde. Wien 1941. Pp. 797–808: Die ostasiatischen Inselbogen.
329. J. Roth, Beiträge zur Petrographie von Korea. Sitz.-Ber. Pr. Ak. d. Wiss. Berlin 1886. Vol. 2. Pp. 875–881.
330. C. Gottsche, Über den Mineralreichtum von Korea. Mitt. Geogr. Ges. Jena **8**, 1890. Pp. 1–20.
331. K. Schulz, Zur Petrographie von Nordkorea. N. Jahrb. f. Min., Geol. u. Paläont. **29**, 1910. Suppl. pp. 1–52.
332. K. Inouye, On Gold Ore Deposits in Korea. Chi. Z. **23**, 1911. Pp. 97–109 (Jap.).
333. S. Kawasaki and E. Tamura, Coal in Chōsen. B. Min. S. Chōs. **1**, I, 1912 (Jap.).
334. M. Fukuchi, The Minerals of Chōsen. In: Contributions to the Mineralogy of Japan. Tōkyō 1915. Pp. 207–305 (Jap.).
335. Geological Survey of the Government General of Chōsen, Mineral Resources of Chōsen (Korea). Seoul 1913–29. 13 vols. (Jap.). Written primarily by S. Nakamura, E. Tamura and S. Kawasaki. The individual volumes deal with the mineral deposits of the 13 Korean provinces. (Rev. in J. J. G. G. **1**, 1922, p. 38ff, etc.).
336. H. Tanikawa, Minerals in Chōsen. B. Min. S. Chōs. **2**, 1923. 104 p. (Jap.) (Rev. in J. J. G. G. **2**, 1923. P. 20).
337. Y. Kamikishi and K. Seto, Chōsensan alkali ganseki ni tsuite (On alkali rocks in the Korean mountains). Chishitsugaku Zasshi **29**, 1923 (Jap.).
338. F. Yamanashi, Chōsen no alkali ganseki ni tsuite (On alkali rocks in Korea). Chōsen Hakubutsugaku Koenshu Dainishu 1924 (Jap.).
339. T. Kato, A Revised Summary of the Metallogenic Epochs of Japan, Including Korea, with Special Reference to the Periods of Orogeny and Igneous Activity. Proc. 3rd Pan-Pac. Sc. Congr. Tōkyō 1926. Vol. 1, Tōkyō 1928. Pp. 581–588.
340. M. Kamo, Korean Coals and Their Utilization. World Power Conf. London 1928. Vol. I, pp. 119–131.
341. T. Watanabe, On the Gold-Copper-Bismuth Ores of the New Ore Body, Holkol Goldmine, Suan, Korea. Journ. Geol. Soc. Tōkyō **40**, 1933. Pp. 70–85, 125–148, 188–209 (Jap.) (Summary in J. J. G. G. **11**, 1933/34, p. 4).
342. S. Kawasaki, Mineral Resources of Chōsen Appeared in Ancient Books. B. Min. S. Chōs. **9**, 1935 (Jap.).
343. T. Kato, Two Types of Nickeliferous Pyrrhotite Deposits Found in Korea. J. J. G. G. **13**, 1936. Pp. 269–275 (Eng., cf. also F. E. R. 1934, p. 19).
344. E. Ahnert, Zonale Verteilung der Erzlagerstätten in der Mandschurei. Intern. Geol. Congr. Rep. Sess. XVI, USA. 1933. Vol. 1. Washington 1936. Pp. 447–458.
345. I. Komada, Report on the Thermal Springs of Torai. B. Geol. S. Chōs. **2**, 1923 (Jap. with Eng. summary). – Report on the 3rd and 4th Examinations of the Thermal Springs of Torai. Ibid. **3**, 1925 (Jap.). – Report on the 5th and 6th Examinations of the Thermal Spring of Torai. Ibid. **7**, 1926 (Jap.).
346. I. Komada, Report on the Thermal Springs of Kaiundai, Jujo, Onyo, Shinsen, Angaku and Ryuko. B. Geol. S. Chōs. **3**, 1925 (Jap. with Eng. summary).

347. I. Komada, Thermal Springs in Onseiri, Shuotsu, Lower Shuotsu and Heisan and Cold Carbon-dioxidated Spring of Shōseiri. B. Geol. S. Chōs. **7**, 1926 (Jap. with Eng. summary).
348. S. Kawasaki, Report on the Thermal Springs of Masan. B. Geol. S. Chōs. **8**, 1928 (Jap.).
349. I. Tateiwa, Report on the Thermal Springs of Onyo, Unsan and Suianpo. B. Geol. S. Chōs. **8**, 1928 (Jap.).

XII. Geomorphology

350. W.G. Aston, Earthquakes in Corea. Transact. Seismol. Soc. Japan **12**, 1888. Pp. 77–79.
351. F. v. Richthofen, Geomorphologische Studien aus Ostasien. I–V. Sitz.–Ber. Pr. Ak. d. Wiss. 1900–1903, Phys.–math. Kl.
352. B. Kotō, An Orographic Sketch of Korea. Journ. College of Sc. Imp. Univ. Tōkyō **19**, Art. 1. 1903. 61 p. With map 1:2 mill.
353. L. Laloy, Orographie de la Corée. La Géogr. **8**, 1903. Pp. 377–381. Summary of [352].
354. W. Schröter, Korea und die riasverwandten Küsten dieser Halbinsel. Diss. Leipzig 1904.
355. L. Pervinquière, Constitution géologique et ressources minérales de la Manchourie et de la Corée. Rev. Scientif. Sér. V, vol. 1, 1904. P. 552ff.
356. T. Lorenz, Beiträge zur Geologie und Paläontologie von Ostasien. 1. Teil. Zeitschr. Dtsch. Geol. Ges. **57**, 1905. P. 495ff.
357. Orographical Features of Korea. Geogr. Journ. **22**, 1903. P. 567f. (Rev. of 352).
358. W. Volz, Der ostasiatische Landstufenbau als Ausdruck oberflächlicher Zerrung. P. M. 1914 II. Pp. 174–178.
359. B. Kotō, A Morphological Summary of Japan and Korea. Journ. Geol. Soc. Tōkyō **22**, 1915. Pp. 115–126, 147–157, 163–176.
360. W.H. Hobbs, The Asiatic Arcs. Bull. Geol. Soc. Am. **34**, 1923. P. 246ff.
361. S. Nakamura, Rakurō jun heigen (On the Nangnang plain). Chikyū **3**, 1925. P. 464.
362. S. Tokuda, On the Echelon Structure of the Japanese Archipelagos. J. J. G. G. **5**, 1926/27. Pp. 41–76 (Eng.).
363. W.H. Wong, Crustal Movements and Igneous Activities in Eastern China Since Mesozoic Time. Bull. Geol. Soc. China **6**, 1927. pp. 9–36. – The Mesozoic Orogenic Movement in Eastern China. Ibid. **8**, 1929. Pp. 33–44.
364. H. Schmitthenner, Die Oberflächengestaltung im außertropischen Monsunklima. Düsseldorfer Vortr. u. Erört. 3. Teil, Morphologie d. Klimazonen. Breslau 1927. Pp. 26–36.
365. S. Toyama, On the Marine Terraces Along the Coast of the Japan Sea. Chi. Z. **39**, 1927. Pp. 332–337, 393–400 (Jap.).
366. S. Hanai, Ryōtō hantō ni hattatsuru iwayuru tei ichi jun heigendo sono shoseishitsu ni tsuite (On the peneplain on the Liaodong peninsula and its characteristics). Chir. Hyō. **4**, 1928.
367. N. Mori, Chōsen san honyurui ka sekimokuroku (Overview of the mammalian fossils in Korea). Chōsen Hakubutsugaku Kaishi (Journ. of the Kor. Biol. Soc.) **9**, 1929.
368. S. Tokunaga, Mammalian Fossils Found in Limestone Caves in Korea. Proc. Imp. Ac. Tōkyō **5**, 1929. Pp. 139–141.
369. H. Yabe and R. Tayama, On Some Remarkable Examples of Drowned Valleys Found Around the Japanese Islands. Rec. Oc. Works II, 1, 1929. Pp. 11–17 (Eng.).
370. J. Ruud, Entwicklung der ostasiatischen Gebirgsbogen. P. M. 1929. Pp. 230–234.
371. M. Matuyama, Gravity Measurements in Tyōsen and Manchuria. Proc. 4th Pac. Sc. Congr. Java 1929. Vol. II B. Batavia–Bandoeng 1930. Pp. 745–748.
372. S. Togunaga, Kōseki jidai Hōnshu, Kyūshū, Chōsen ni okeru jurui kekiono iseki (Pleistocene finds from caves in Hōnshu, Kyūshū and Korea). Nippon Gakushutsu Hōkoku (Report of the Soc. for Jap. Art) **6**, 1930.
373. H. Yabe, The Latest Land Connection of the Japanese Islands to the Asiatic Continent. Proc. Imp. Ac. Tōkyō **5**, 1929. Pp. 167–169 (Rev. in J. J. G. G. **7**, 1929/30. P. 7. Cf. Journ. Geol. Soc. Tōkyō **36**, 1931. P. 253f).

374. H. Yabe, Geological Age of the Latest Continental Stage of the Japanese Islands. Proc. Imp. Ac. Tōkyō **5**, 1929. Pp. 430–433 (Rev. in J. J. G. G. **8**, 1930/31. P. 4. Cf. Journ. of Geogr. **42**, 1930. Pp. 324–329).
375. B. Kotō, The Rocky Mountain Arcs in Eastern Asia. Journ. Fac. Sc. Imp. Univ. Tōkyō. Sect. II, vol. III, part 3, 1931 (Eng.).
376. Y. Otuka, The Geological Age of Drowned Valleys in the Shelves Around the Japanese Islands and the Migration of Strandlines During the Early Alluvium. Chir. Hyō. **7**, 1931, pp. 447–458 (Jap.).
377. F. Ruellan, La décomposition et la désagré gation du granite à biotite au Japon et en Corée et les formes du modèle qui en resultent. C. R. Ac. Sc. Paris **193**, 1931. pp. 67–69.
378. W. Credner, Das Kräfteverhältnis morphogenetischer Faktoren und ihr Ausdruck im Formenbild Südost–Asiens. Bull. Geol. Soc. China **11**, no. 1, 1931.
379. H. Schmitthenner, Landformen im außertropischen Monsungebiet. Wiss. Veröff. Mus. f. Länderkunde, Leipzig, N. F. **1**, 1932. Pp. 81–101.
380. T. Kobayashi, Chōsen hantō hattatsushi to kindai chishi (Topography and Cenozoic Geology of the Korean Peninsula). Chir. Hyō. **7**, 1931. Pp. 523–550, 628–648, 708–733. (Jap., rev. in J. J. G. G. **12**, 1935. P. 19).
381. A.C. Lawson, Insular Arcs, Foredeeps and Geosynclinal Seas of the Asiatic Coast. Bull. Geol. Soc. Am. **43**, 1932.
382. S. Ebara, Geological Structure of the Median Line and the Origin of the Japan Sea. Journ. Assoc. Advanc. Sci. **7**, 1932. P. 346f (Jap.).
383. T. Kobayashi, Brief Notes on the Geomorphology of the Korean Peninsula and Its Relation to the Cenozoic History. Proc. 5th Pac. Sc. Congr. Canada 1933. Vol. II. Toronto 1934. Pp. 1605–1607 (Summary of [380]).
384. I. Tateiwa, On the Cycle of Erosion in Korea. Proc. 5th Pac. Sc. Congr. Canada 1933. Vol. II. Toronto 1934. Pp. 1609–1612 (Rev. in J. J. G. G. **16**, 1939. P. 46).
385. G. Imamura, Bottom Topography of the Japan Sea. Chikyū **20**, 1933. Pp. 270–275 (Jap.).
386. H. Yabe and R. Tayama, Bottom Relief of the Seas Bordering the Japanese Islands and Korean Peninsula. Bull. Earthquake Research Inst. Tōkyō Imp. Univ. **12**, 1934. Pp. 539–565 (Jap. with Eng. summary) (Rev. in J. J. G. G. **15**, 1938. Pp. 20–22).
387. P. Lejay, Exploration gravimétrique de l'Extrême–Orient. Comm. Nat. Franç. de Géodésie et de Géophysique. Paris 1936. 75 p.
388. J.S. Lee, The Framework of Eastern Asia. Intern. Geol. Congr. Rep. Sess. XVI, USA. 1933. Washington 1936.
389. S. Yehara, On the Echelon Structure of Shikoku and the Origin of Japanese Arcs. J. J. G. G. **13**, 1936. Pp. 1–24 (Jap.).
390. S. Nakai, Rotation of Chōsen Peninsula and the Origins of the Yellow Sea, Gulf of Liaotung and Japan Sea According to the Continental Drift Theory. Journ. Assoc. Advanc. Sci **11**, 1936. Pp. 308–313 (Jap.).
391. T. Kano, Chōsen tōhoku bu sanchi no hyōga chikei ni tsuite (On the glacial topography of the northeast Korean mountains). Chir. Hyō. **13**, 1937. pp. 1126–1145 (Jap. with Eng. summary).
392. Y. Sasa and T. Yamasaki, A Few Palaeoclimatological Facts in the Neogene and Later Periods Obtained by Pollen-Analysis in North Japan. Geol. Soc. Tōkyō **45**, 1938. P. 527f (Jap.) (Rev. in J. J. G. G. **15**, 1938. P. 62).
393. Y. Sasa and K. Tanaka, Glaciated Topography in the Kanbō Massif, Tyōsen (Korea). Journ. Fac. Sc. Hokkaidō Imp. Univ. Ser. IV, vol. 4, 1938. Pp. 193–212 (Jap.) (Rev. in J. J. G. G. **15**, 1938. P. 63f).
394. F. Machatschek, Das Relief der Erde. Vol. I. Berlin 1938. East Asia: Pp. 292–321.
395. K. Motizuki, A Study of Japanese Arcuate Mountains. J. J. G. G. **15**, 1938. Pp. 139–200.
396. K. Motizuki, On Some Features of the Meeting of Arcs in the Japanese Archipelago. J. J. G. G. **15**, 1938. Pp. 201–207.
397. K. Motizuki, Mountain Systems in Eastern Asia as a Peripheral Ring of Continent. Chir. Hyō. **15**, 1939. Pp. 343–372 (Jap., rev. in J. J. G. G. **16**, 1939. P. 81).
398. H. Lautensach, Unterirdischer Kaltluftstau in Korea. P. M. 1939, Pp. 353–355.
399. M. Schwind, Die eiszeitliche Vergletscherung Ostasiens. Geogr. Zeitschr. 1942, pp. 157–172.

XIII. Climatology

400. Y. Wada, The Amount of Precipitation in Keizyo During the Last 140 Years. Reports on the Korean Meteorological Observations in Former Times. Meteorol. Observ. in Korea. Jinsen 1917 (Jap.).
401. F. Tada, Über die periodische Änderung der Regenmenge in Chosen seit dem Jahre 1776. C. R. Congr. Intern. Géogr. Amsterdam 1938. Vol. II, Travaux Sections A–F. Leiden 1938. Pp. 305–308 (Ger.).
402. A. Woeikow, Zum Klima von Korea. Met. Zeitschr. **4**, 1887. Pp. 160–164.
403. L.E. Dinklage, Beitrag zur Kenntnis des Klimas von Korea. Ann. d. Hydr. etc. 1891, pp. 1–8.
404. C. Waeber, Notes on the Climate of Korea. Seoul 1891.
405. A. Supan, Regentafeln von China und Korea. P. M. 1896. Pp. 205–209.
406. J. Beicht, Das Koreanische Meteorologische Observatorium. Sturm–und Wettersignale an der Koreanischen Küste. Ann. d. Hydr. etc. **39**, 1911. P. 84f.
407. Meteorological Observatory of the Government–General of Chosen, Results of the Meteorological Observations Made at Chosen, for the Lustrum 1911–15. Chemulpo 1917 (Jap. and Eng.). Likewise for further lustra.
408. Meteorological Observatory of the Government–General of Chosen, Rainfall in Chosen (Korea). Zinsen (Chemulpo) 1925. 186 p.
409. Meteorological Observatory of the Government General of Tyōsen, Air Temperature in Tyōsen (Korea). Zinsen 1928. 343 p.
410. T. Hirata, Rainfall of Chosen (Korea). Ann. Rep. Met. Obs. Gov. Gen. Tyōsen 1915, pp. 133–175 (Eng.).
411. T. Hirata, The Climate of Korea. Jinsen 1916 (Jap.).
412. J.D. van Buskirk, The Climate of Korea and Its Probable Effect on Human Efficiency. Tr. Kor. Br. R. As. Soc. **10**, 1919. Pp. 1–58 (Cf. Geogr. Journ. **55**, 1920. P. 325f and Geogr. Rev. **10**, 1920. P. 45f).
412a P.I. Koloskoff, Klimatischerskoje oppisanie Korei. Leningrad and Moskva 1936. 46 p. (Russ.) (Climatological Description of Korea).
413. G. Schumacher, Das Klima Südkoreas auf Grund der japan. Wetterbeobachtungen der Jahre 1914–1933. Arch. Dtsch. Seewarte **59**, no. 2, 1939. Diss. Greifswald.
414. E. Trojan, Das Klima Nordkoreas auf Grund der japan. Wetterbeobachtungen der Jahre 1914–1933. Diss. Greifswald (in preparation).
414a S. McCune, Climatic Regions of Korea and Their Economy. Geogr. Rev. **31**, 1941. Pp. 95–99.
415. L. Froc, L'Atmosphère en Extrême–Orient. Paris 1920². 291 p.
416. T. Okada, The Climate of Japan. Bull. Central Met. Obs. Japan IV, 2. Tōkyō 1931. Pp. 89–416. 35 plates. Korea: pp. 156–160 (Eng.).
417. K. Tsukuda, The Mean Cloudiness in the Far East. Mem. Imp. Mar. Obs. Kobe **4**, 2. 1931. Pp. 75–98. 12 monthly charts.
418. E. Biel and O. Moese, Versuch einer Anwendung des Andauerprinzips auf das Klima von Japan. Gerlands Beitr. z. Geophys. **35**, 1932. Pp. 241–262.
419. H. v. Wissmann, Die Klima– und Vegetationsgebiete Eurasiens. Z. G. E. B. 1939, pp. 1–14.
420. H. v. Wissmann, Mean Variability of Annual Rainfall in Eurasia. Mem. Nat. Res. Inst. Met. XI, 1. 1937. 21 p.
421. T. W. F., The Aridity of North China. Scott. Geogr. Mag. **52**, 1936. Pp. 50f.
422. Y. Yamanouchi, Kishō shūsi (three cold four warm). Journ. Met. Soc. Japan **34**, 1914. P. 324.
423. T. Okada, On the Seven Day Period in the Winter Temperature of Korea and Manchuria. Pr. I. A. T. VIII, 4. 1932. Pp. 116–118 (Eng.).
424. K. Takahashi and S. Harada, On Sankansion (7 Days Periodicity in Air Temperature). Journ. Met. Soc. Japan, Ser. 2, **14**, 1936, Jap. pp. 370–376, Eng. p. (33).
425. T. Okada, On the Bai-u or Rainy Season in Japan. Quart. Journ. R. Met. Soc. 1911, p. 277.
426. J. Lee, Precipitation on the Islands Along the Chinese Coast. Monthly Weather Rev. **64**, 1936. Pp. 287–291.
427. J. Lee, The Cause of the Scarcity of Precipitations on the Islands Along the Chinese Coast. Mem. Nat. Res. Inst. Met. **9**, 1937. Pp. 43–53.

428. H. Arakawa, Die Luftmassen in den japanischen Gebieten. Met. Zeitschr. **54**, 1937. Pp. 169–174.
429. H. Arakawa, (The) Air Masses in Japan. Journ. Met. Soc. Japan, Ser. 2, **13**, 1935. Pp. 387–402. **14**, 1936. Pp. 328–338, p. (30). **15**, 1937. Pp. 185–189, p. (17f) (Jap. with Eng. summary).
430. T. Yamanaka and C. Ikeda, The 3 Principal Air Masses in Japan and Their Neighbourhood. Journ. Met. Soc. Japan, Ser. 2, **17**, 1939.
431. E. Gherzi, Air Masses Acting over China and the Adjoining Seas. Beitr. z. Phys. d. freien Atm. **24**, 1938. Pp. 45–52.
432. Coching Chu, The Climatic Factors of China. Journ. Geogr. Soc. China II, 2. 1935. P. 16.
433. Chang-Wang Tu, A Preliminary Study on the Mean Air Currents and Fronts of China. Mem. Nat. Res. Inst. Met. XI, 3. 1937. 12 p.
434. Chang-Wang Tu, The Air Masses of China. Mem. Nat. Res. Inst. Met. XII, 2. 1938. 50 p. (Cf. Quart. Journ. R. Met. Soc. **65**, 1939. Pp. 33–51).
435. R.W. Richardson, Winter Air-Mass Convergence over the North Pacific. Monthly Weather Rev. **64**, 1936. Pp. 199–203.
436. Sjan-zsi Li, Die Kaltlufteinbrüche in Ostasien. Diss. Berlin 1935. 80 p. (Met. Zeitschr. 1936, pp. 74–76).
437. A. Lu, The Cold Waves of China. Mem. Nat. Res. Inst. Met. **10**, 1937. 49 p.
438. Yellow Dust Haze at Mokuho, Tyosen (Corea). Journ. Met. Soc. Japan **43**, 1924. P. 76.
439. S.P. Peters, Note on a Cold Front at Chefoo, N. China. May 15, 1928. Quart. Journ. R. Met. Soc. **55**, 1929. Pp. 86–88.
440. K. Takahashi, Dynamic Climatological Consideration on the Alternation of Weather in Early Winter in the Far East. Journ. Met. Soc. Japan, Ser. 2, **18**, 1940, Jap. pp. 177–191, Eng. p. (21) (Cf. Das Wetter 1940, p. 304f).
441. Coching Chu, The Enigma of Southeast Monsoon in China. Journ. Geogr. Soc. China I, 1. 1934. P. 1.
442. L. Froc, Atlas of the Tracks of 620 Typhoons 1893–1918. Zikawei 1920.
443. S. Li, Untersuchungen über Taifune. Veröff. Met. Inst. Univ. Berlin I, 5. 38 p. Berlin 1936 (Met. Zeitschr. 1937, p. 349 f).
444. C.L. Mitchell, Cyclones and Anticyclones of the Northern Hemisphere, January to April, incl., 1925. Monthly Weather Rev. **58**, 1930. Pp. 1–22.
445. G. Tokuyama, On the Evolution of Cyclones in Japan Sea. Journ. Met. Soc. Japan **43**, 1924, Jap. pp. 184–189, Eng. p. 196 (J. J. G. G. **3**, 1924, p. [19]).
446. E. Gherzi, Note sur les dépressions extratropicales de Chine. Gerlands Beitr. z. Geophys. Köppen–Band I, 1931. Pp. 368–378.
447. Shio Wang Sung, The Extratropical Cyclones of Eastern China and Their Characteristics. Mem. Nat. Res. Inst. Met. **3**, 1931. 60 p.
448. Coching Chu, A Brief Survey on the Climate of China. Mem. Nat. Res. Inst. Met. **7**, 1936. 35 p.
449. S. Ooma, On the Development of Cyclones in the Neighbourhood of Japan (I). Journ. Met. Soc. Japan, Ser. 2, **17**, 1939, Jap. pp. 356–366, Eng. p. (33f).
450. H. Futi, S. Tazima and N. Murase, On the Structure of an Extratropical Cyclone in the Far East (I). Journ. Met. Soc. Japan, Ser. 2, **17**, 1939, Jap. pp. 343–355, Eng. p. (33).

XIV. Hydrography and Oceanography

451. Chisui oyobi suiri chōsa sho (Book on water regulation works and water usage). Chōsen Sōtokufu. 1920. 833 p. (Jap.).
452. Kinnen ni okeru Chōsen no fusuigai (Storm damage in Korea). Chōsen Sōtokufu, Meteorol. Observ., 1926. 264 S. (Jap.).
453. Chōsen kasen chōsa sho (Book on Korea's hydrographical conditions). Chōsen Sōtokufu 1929. 1 vol. text (902 p. and 80 p. appendix), 1 vol. tables (1163 p.) and 1 vol. with 343 maps and diagrams (Jap.).
454. S. Ohti, On the Freezing of Rivers in Korea and Southern Manchuria. Journ. Met. Soc. Japan, Ser. 2, **3**, 1925, Jap. pp. 211–221, Eng. p. (26f).

455. L. von Schrenck, Strömungsverhältnisse im Ochotskischen und Japanischen Meere und in den zunächst angrenzenden Gewässern. Mém. Ac. Imp. Sc. St. Pétersbourg, Ser. VII, XXI, 3. 1873.
456. A. Glocker, Die ostasiatischen Gewässer und der Korea-Archipel. Mitt. Geogr. Ges. Wien **30**, 1887. Pp. 482–489.
457. F.H. Mörsel, The Tidal Wave in the Yellow Sea. Kor. Rev. **1**, 1901. Pp. 202–206. (Cf. L.E. Dinklage in Ann. d. Hydr. **40**, 1891 and E. Brass in Export **15**, 1893. P. 169).
458. G. Schott, Oberflächen-Temperaturen und Strömungen der Ostasiatischen Gewässer. Arch. Dtsch. Seewarte **14**, 1898. No. 3. 46 p.
459. Hydrographic Department Admirality, China Sea Pilot, Vol. III. Comprising .... the Western Coast of Korea. London 1937. 652 p. Korea: pp. 402–540.
460. Hydrographic Department Admirality, South and East Coast of Korea, East Coast of Siberia, and Sea of Okhotsk Pilot (No. 2079). London 1937^3. 658 p. Korea: pp. 1–278.
461. V. Akhmatov, Oceanography. With map: The Probable Tracks of Bottles, Thrown for the Investigation of Currents. In: The Pacific Russ. Scient. Investigations. Leningrad 1926, pp. 95–112.
462. K. Nisida, On the Currents, Water Temperature and Salinity of the Adjacent Seas of Tyōsen (Chōsen). Rep. Oc. Obs. Husan **2**, 1927. 32 p.
463. L. Rudovitz, Oceanographical Researches in the Sea of Japan (1926–28). Bull. Pac. Comity Ac. of Sc. USSR. Leningrad 1929, no. 2.
464. M. Uda and G. Okamoto, Of the Monthly Oceanographical Charts of the Adjacent Seas of Japan Based on the Averages for the Eleven (Thirteen) Years from 1918 to 1929 (1930), with a Discussion of the Current-System Inferred from These Charts. Part I. From July to December. Part II. From January to June. Journ. Imp. Fish. Exp. Stat. Tōkyō 1930. I, 1. Pp. 39–55. I, 2. Pp. 59–81.
465. K. Nisida, Hydrographical Observations off the Shore Along the East Coast of Chōsen in October 1932. Charts and Tables. Oceanogr. Chart for the Year 1932. Suppl. B. Fish. Exp. Stat. Gov. Gen. Chōsen 1933.
466. M. Uda, Hydrographical Researches on the Normal Monthly Conditions in the Japan Sea, the Yellow Sea and the Okhotsk Sea. Journ. Imp. Fish. Exp. Stat. Tōkyō **5**, 1934. Pp. 191–236.
467. M. Uda, The Results of Simultaneous Oceanographical Investigations in the Japan Sea and its Adjacent Waters in May and June 1932. Journ. Imp. Fish. Exp. Stat. Tōkyō **5**, 1934. Pp. 57–190 (Jap. with Eng. summary). excerpt in Eng. in: Rec. Oc. Works VI, 1. 1934. Pp. 19–107.
468. M. Uda, Results of Simultaneous Oceanographical Investigations in the Japan Sea and its Adjacent Waters During October and November, 1933. Journ. Imp. Fish. Exp. Stat. Tōkyō **7**, 1936. Pp. 91–191 (Jap. with Eng. summary).
469. G. Wüst, Kuroshio und Golfstrom. Eine vergleichende hydrodynamische Untersuchung. Veröff. Inst. Meereskde. Univ. Berlin. N. F. Ser. A. No. 29. 1936. 69 p.
470. S. Ogura, On the Depths of the Adjacent Seas of Japan. Pr. 3rd P. S. C. Tōkyō 1926. Vol. I, 1928. Pp. 246–249. Also separately with large colored map.
471. S. Ogura, On Recent Soundings and Mapping of the Sea Floors in the Seas near Japan. Pr. 5th P. S. C. Canada 1933. Vol. I, 1934. Pp. 893–947.
472. T. Kumata, On the Temperature of Sea Water and Meteorological Elements Observed at the Coast of Korea. Journ. Met. Soc. Japan, Ser. 2, **4**, 1926. P. 1.
473. K. Suda and K. Seiki, Report on the Mean Surface Temperatures of the Neighbouring Seas of Japan. Journ. Oc. **1**, 1930. Pp. 57f, 132f, 379–382. **2**, 1930. Pp. 74–77, 266–275, 447f, 644–646.
474. S. Ogura, On the Currents in the Tung Hai. Suiro Yōhō **5**, 1926. Pp. 103–113 (Jap.).
475. K. Nisida, Results of the Current Measurements in the Adjacent Seas of Tyōsen, 1923–26. 1st Rep. Rep. Oc. Obs. Husan **2**, 1927. 150 p. (Jap. with Eng. summary).
476. S. Ogura, On the Tides in the Seas of Eastern Asia. Pr. 3rd P. S. C. Tōkyō 1926. Vol. I, Tōkyō 1928. Pp. 167–182.
477. T. Kubo, On the Surface Drift Currents in the Yellow and Chinese Eastern Sea, Determined by Drift Bottles. Journ. Oc. **2**, 1930. Pp. 289–296.
478. S. Ogura, The Discussion of the Results of Tidal and Current Observations Made by the Yodo in the Hwang Hai. Suiro Yōhō **11**, 1932. Pp. 185–203 (Jap.).
479. S. Ogura, The Tides in the Seas Adjacent to Japan. Bull. Hydr. Dep. Imp. Jap. Navy **7**, 1933. 189 p. (Eng.).

XV. Plant Geography

480. T. Nakai, Aperçu sur la flore de Corée. Bull. Acad. Intern. Géogr. Bot. **20**, 1910. P. 227f.
481. T. Nakai, Chōsen shokubutsu (Korean Plants) 1914. 430 p. (Jap.).
482. T. Nakai, Flora silvatica Koreana. To date 21 vols., 1906–36. The first two with the title: Flora Coreana. Journ. Coll. of Science Imp. Univ. Tōkyō 1909 and 1911, the following publ. by Government–General, Forest Experiment Station.
483. T. Nakai, Tōa shokubutsu kukei (The distribution of plants in East Asia). Iwanami seibutsugaku kōsa 1931 (Jap.).
484. T. Nakai, Lespedeza of Japan and Kora. Gov. Gen. Chōsen, For. Exp. Stat. Keijō 1927. 101 p. (Jap.).
485. T. Mori, Chōsen shokubutsu meii (An enumeration of the plants hitherto known from Corea). Seoul 1922. 372 p. (Jap. and Lat.).
486. I. Kokumen and T.T. Keng, Chōsen shinrin jimoku kanyō (Survey of the tree species in Korean forests). Chōsen Sōtokufu. For. Exp. Stat. 1923. 129 + 20 + 35 p. (Jap., Kor., Lat.).
487. B. Hayata, General Aspects of the Flora of Japan, Including ... Korea... In: Scientific Japan, Past and Present. Kyōtō 1926. Pp. 77–104.
488. H. Uyeki, Chōsen no rimboku (Corean timber trees). Vol. I. Gingkoales and Coniferae. Bull. For. Exp. Stat. **4**, 1926. 154 p. (Jap.).
489. H. Uyeki, On the Physiognomy of *Pinus densiflora* Growing in Corea and Sylvicultural Treatment for Its Improvement. Bull. Agr. For. Coll. Suigen **3**, 1928. 263 p. (Jap.).
490. H. Uyeki, Chōsen shinrin shokubutsu tai (Korean forest zones). Shokubutsu Bunrin Chiri II, 2. Keijō 1932 (Jap.).
491. H. Uyeki, On the Forest Zones of Korea. Acta Phytotaxonomica et Geobotanica **2**, 1933. Pp. 73–85.
492. H. Uyeki, The Regeneration of Conifers in Korea. Pr. 5th P. S. C. Canada 1933. Vol. 2, pp. 973–979.
493. Y. Takenaka, On the Botanical Natural Monuments in Chōsen. 47 p. (Jap.).
494. A. Hofmann, Die forstlichen Produktionsverhältnisse von Korea. M. D. O. A. G. XI, 1, 1906. Pp. 47–76.
495. A. Hofmann, Aus den Waldungen des Fernen Ostens. Wien and Leipzig 1913. 220 p.
496. E.H. Wilson, The Vegetation of Korea. Tr. Kor. Br. R. As. Soc. **9**, 1918. Pp. 1–68.
497. E.H. Wilson, A Phytogeographical Sketch of the Ligneous Flora of Korea. Journ. Arnold Arboretum **1**, 1919. Pp. 32–43.
498. E.H. Wilson, A Summary Report on the Forests, Forest Trees and Afforestation in Chosen (Korea). Tr. Scott. Arb. Soc. **33**, 1919. Pp. 44–51.
499. M.N. Trollope, Arboretum coreense. Tr. Kor. Br. R. As. Soc. **9**, 1918. Pp. 69–90. **11**, 1920. Pp. 39–100.
500. Wayside Flowers of Central Korea. Tr. Kor. Br. R. As. Soc. **18**, 1929. Pp. 22–82.
501. F. Heddleston Crane, Flowers and Folklore from Far Korea. Tōkyō and Osaka 1931. 92 + 4 p.
502. W. v. Burmeister, Die forsttechnisch wichtigen Bäume und Sträucher des Fernen Ostens. Zeitschr. f. Weltforsttwirtsch. **2**, 1935. Pp. 7–9, 385–417.
503. Korean Lespedeza. Kansas Agricultural Board. Biennial Report 1933, pp. 19–205.
504. J. Palibin, Conspectus Florae Koreanae. Trudy Imp. St. Petersb. Botan. Sada XII, 1. 117 p.
505. A. Eckardt, Ginseng. Die Wunderwurzel des Ostens. Festschr. P. Wilhelm Schmidt. Wien 1928. Pp. 220–230 (Cf. Heil– und Gewürzpflanzen **16**, 1935. Pp. 94–104).

XVI. Zoogeography

506. H.H. Underwood, Hunting and Hunters' Lore in Korea. Tr. Kor. Br. R. As. Soc. **VI**, 2. 1915. Pp. 23–43.
507. Y. Abe, On the Korean and Japanese Wolves. Journ. Sc. Hiroshima Univ. B 12, 1930. Pp. 33–37.
508. D.J. Cumming, Notes on Korean Birds. Tr. Kor. Br. R. As. Soc. **22**, 1933. Pp. 1–94.
509. D. Starr Jordan and C.M. Metz, Catalogue of the Fishes Known from the Waters of Korea. Carnegie Museum Memoirs. 1913.

510. T. Mori, A Catalogue of the Fishes of Korea. Journ. Pan–Pac. Res. Inst. (Honolulu) **3**, 1928. Pp. 3–8.
511. S. Tanaka, On the Distribution of Fishes in Japanese Waters. Journ. Fac. Sc. Imp. Univ. Tōkyō. Sect. IV, Zoology 3, 1931/34. Part 1. 90 p. – Figures and Descriptions of the Fishes of Japan, Including Riukiu Islands, Bonin Islands, Formosa, Kurile Islands, Korea and Southern Sakhalin. 48 vols. Tōkyō 1911–1933.
512. H. Marukawa and T. Kamiya, Some Results of the Migrations of Important Food Fish. Journ. Imp. Fish. Exp. Stat. Tōkyō **1**, 1930. Pp. 5–38 (Jap. with Eng. summary).
513. H. Aikawa, On the Quantitative Analysis of Plancton Associations in the Seas Surrounding Japan I and II. Journ. Imp. Fish. Exp. Stat. Tōkyō **2**, 1931. Pp. 37–57 and **5**, 1934. Pp. 237–247 Jap. with Eng. summary).
514. I. Takayama, N. Ikeda and S. Andō, A Study of the "Katsuwo" (*Katsuwonius pelamis*) Fishing in 1930. Journ. Imp. Fish. Exp. Stat. Tōkyō **5**, 1934. Pp. 23–56 (Jap. with Eng. summary).
515. I. Takayama and S. Andō, A Study of the "Maguro" (*Thunnus*) Fishing in 1930. Journ. Imp. Fish. Exp. Stat. Tōkyō **5**, 1934. Pp. 1–21 (Jap. with Eng. summary).
516. I. Takayama, M. Uda and Z. Tukusi, A Study of "Saba" (*Scomber japonicus* Houttuyn) Fishing in 1930 and 1931. Journ. Imp. Fish. Exp. Sta. Tōkyō **7**, 1936. Pp. 1–18 (Jap. with Eng. summary).
517. M. Uda and G. Okamoto, Effect of the Oceanographic Conditions on "Iwasi" (Sardine) Fishing in the Japan Sea. Journ. Imp. Fish. Exp. Stat. Tōkyō **7**, 1936. Pp. 19–49 (Jap. with Eng. summary).
517a T. Mori, On the Geographical Distribution of Korean Salmonoid Fishes. Bull. Biogeogr. Soc. Japan **6**, 1935. No. 8. Pp. 1–9.

XVII. Geographical Anthropology

518. H.B. Hulbert, The Origin of the Korean People. Kor. Rep. 1895, pp. 219–229, 255–264.
519. H.B. Hulbert, The Korean Physical Type. Kor. Rev. **3**, 1903. Pp. 55–59.
520. E. Baelz, Menschen-Rassen Ost-Asiens, mit specieller Rücksicht auf Japan. Zeitschr. f. Ethnol. (Verhandl. Berl. Ges. f. Anthropol. etc.) **33**, 1901. Pp. (166)–(189).
521. E. Baelz, Über die Rassenelemente in Ostasien. M. D. O. A. G. VIII, 2. 1900. Pp. 227–235.
522. E. Baelz, The Origin of the Korean People. Kor. Rev. **2**, 1902. Pp. 440–446, 453–458.
523. E. Baelz, Zur Rasse der Japaner und Koreaner. Verhandl. Ges. Dtsch. Naturf. u. Ärzte **78**, 1906. Part II, 1. P. 311f.
524. T. Koganei, Über Schädel und Skelette der Koreaner. Zeitschr. f. Ethnol. 1906. P. 523.
525. É. Chantre and É. Bourdaret, Esquisse anthropologique des Coréens. Bull. Soc. Anthropol. Lyon **21**, 1902. Pp. 22–38.
526. T. Kubo, Beiträge zur physischen Anthropologie der Koreaner. Part I. Mitt. Med. Fak. Kais. Univ. Tokyo **12**, 1911. 676 + 32 p. (Ger.). Part II. Mitt. Med. Fachschule Keijō 1917. 212 p. (Ger.).
527. R.G. Mills, Physical Anthropological Study of the Korean People. China Medical Journ. **31**, 1917. Pp. 523–549 (Excerpt of [526]).
528. R.G. Mills, Korean Stature and Weight, Compared According to Provinces. China Medical Journ. **32**, 1918. P. 566 (after Kubo).
529. R.G. Mills, A Contribution to the Nosogeography of Northern Korea. China Medical Journ. **25**, 1911. Pp. 277–293.
530. R.M. Wilson, Leprosy in Korea. Journ. Am. Med. Ass. **87**, 1926. P. 1211f.
531. H. Fowler, Leprosy in Korea. Leprosy Rev. **1**, 1930. Pp. 28–31.
532. R.M. Wilson, A Leper Colony. Its Management and Maintenance. China Medical Journ. **45**, 1931. Pp. 827–834.
533. H.C. de Souza–Aranys, Leprosy-Survey Made in 40 Countries 1924–27. Oswaldo Cruz Inst. Rio de Janeiro 1929. Korea: pp. 118–121.
534. R.M. Wilson, Pelagra in Korea. China Medical Journ. **38**, 1924. P. 705. **39**, 1925. P. 661f.
535. R.M. Wilson, Pelagra or Pelagroid in Leper Settlements in Korea. China Medical Journ. **47**, 1933. Pp. 287–289.

536. S.H. Martin, Hookworm Disease in Korea. China Medical Journ. **42**, 1928. Pp. 187–191 (Cf. Acta med. Keijō **12**, 1929. No. 2. Pp. 8–13).
537. W.R. Cate, Typhoid Fever in Korea. China Medical Journ. **40**, 1926. Pp. 554–558.
538. The Population of Korea. Kor. Rep. **5**, 1898. Pp. 28–31.
539. T. Imamura, Chōsen fūzoku shū (Collection of Korean traditions and customs). Keijō 1919[3]. 474 p. (Jap.).
540. Y. Murata, Chōsen no seikatsu to bunka (Life and culture in Korea). Tōkyō 1924. 358 p. (Jap.).
541. H.G. Arnous, Korea. Märchen und Legenden nebst einer Einleitung über Land und Leute, Sitten und Gebräuche Koreas. Leipzig o.J. 147 p.
542. H.B. Hulbert, Korean Folk-Tales. Tr. Kor. Br. R. As. Soc. **2**, 1902, II. Pp. 45–79.
543. W. Hough, Korean Clan Organisation. American Anthropologist **1**, 1893. Pp. 150–154.
544. D.L. Gifford, Every-Day Life in Korea. Chicago etc. 1898. 229 p.
545. C.I.D. Tayler, Koreans at Home. The Impressions of a Scots Woman. London 1904[2]. 80 p. (Rev. in P. M., L. B. 618).
546. J.R. Moose, Village Life in Korea. Nashville (Tenn.) 1911. 242 p. (Rev. in P. M. 1913, I, L. B. 276).
547. S.Y. Kwun, Korean Customs of the Year. Kor. Mag. **2**, 1918. Pp. 106–113).
548. P.L. Gillet, The Village Gilds of Old Korea. Tr. Kor. Br. R. As. Soc. **4**, 1913, II. Pp. 13–44.
549. Y. Kang, The Grass Roof. New York 1931 (also Ger.: Unter dem Grasdach. Leipzig 1933).
550. Mirok Li, Suam und Mirok. Kindheitsbilder aus Korea. Atlantis **7**, 1935. Pp. 349–360, 428–437.
551. A.C. Kim, Village Life in Korea. Journ. of Geogr. **36**, 1937. Pp. 315–320.
552. Essays on Korean Economic History (End of Li Dynasty). Journ. Keijō Imp. Univ. Soc. Sc. Ass. No. 6, 1933. 567 + 23 p. (Jap. with Eng., Ger. and Fr. summary).
553. Essays on Korean Economy. Journ. Keijō Imp. Univ. Soc. Sc. Ass. No. 2, 1929. 773 + 13 p. (Jap. with Eng. summary).
554. E. Werth, Zur Geographie und Geschichte der Hirsen. Angewandte Botanik **19**, 1937. Pp. 42–73.
555. S. Saigo, The History of Korean Agriculture. Suigen 1922 (Jap.).
556. M.C. Fenwick, Korean Farming. Kor. Rep. **5**, 1898. Pp. 288–293.
557. (H.B. Hulbert), The Products of Korea. Kor. Rev. **2**, 1902. Pp. 49, 108, 203, 300, 341, 393.
558. M. Heydrich, Koreanische Landwirtschaft. Beiträge zur Völkerkunde von Korea. I. Abhandl. u. Berichte d. Museen f. Tierkde. u. Völkerkde. Dresden **19**, 1931. 44 p.
559. H. Lautensach, Über den Brandrodungsfeldbau in Korea. P. M. 1941, Pp. 41–54.
560. U. Oda, Kadenmin raiju Shinajin (Fire-field farmers and Chinese immigration). Vol. 1. Chōsen buraku chōsa hōkoku (Research reports on Korean settlements). Keijō 1923. 65 p. and 33 plates (Jap.).
561. Kaden chōsa hōkoku shō (Investigation of fire-fields). Keijō 1928. 58 p. (Jap.).
562. T. Odauchi, Chōsen no kadenmin (Fire-field farmers in Korea). Jimbun Chiri **1**, 1926. Pp. 25–35 (Jap.) (Cf. A. Demangeon, La géographie humaine au Japon. Ann. de Géogr. **37**, 1928. Pp. 359–361).
563. Seihokusen-chihō ni okeru kaden ni kansuru chōsho (On the fire-fields in the northwest Korean provinces). Chōsen Sōtokufu Riuji Tochi Chōsa Kyoku. Keijō 1916 (Jap.).
564. Y. Niijima, Kaden, ein Röderland in Korea. Forstwissensch. Centralblatt **52**, 1938. Pp. 367–372.
565. D. Fairchild, Foods in Korea. Nat. Geogr. Mag. **36**, 1919. Pp. 57–77.
566. J.D. van Buskirk, Some Common Korean Foods. Tr. Kor. Br. R. As. Soc. **14**, 1923. Pp. 1–8.
567. J.D. van Buskirk, Studies in the Diet of the Korean People. China Medical Journ. **36**, 1922. Pp. 136–145.
568. R.L. Hobson, Corean Pottery. Burlington Mag. **56**, 1930. Pp. 54–64, 186–193.
569. E.W. Mills, Gold Mining in Korea. Tr. Kor. Br. R. As. Soc. **7**, 1916, I. Pp. 5–39, with map 1:1.5 million.
570. O. Kümmel, Die Kunst Chinas, Japans und Koreas. Handb. d. Kunstwissenschaft. Wildpark-Potsdam 1929. Korea: pp. 187–191.
571. A. Eckardt, Geschichte der koreanischen Kunst. Leipzig 1929. 227 p. text, 168 p. plates. Also Eng.: History of Korean Art. London 1929.

572. O. Kümmel, Koreanische Baukunst. Wasmuths Lexikon der Baukunst. Vol. 3, Berlin 1931. Pp. 416–419.
573. F. Tada, Morphology and Settlements in North Korea. Journ. Geol. Soc. Tōkyō 1935 (Jap.).
574. Chōsenfu raku chōsa hōkoku. Vol. I (House construction). Chōsen Sōtokufu 1924.
575. J.S. Gale, The Korean Pony. Kor. Rep. 1895, pp. 176–181.
576. H.H. Underwood, Korean Boats and Ships. Tr. Kor. Br. R. As. Soc. **23**, 1934, I. Pp. 1–99.
577. Fishing Boats on the East Coast of Chōsen. Publ. Fish. Exp. Stat. Gov. Gen. Chōsen **1**, 1924 (Jap.).
578. Fishing Boats on the South and West Coast of Chōsen. Publ. Fish. Exp. Stat. Gov. Gen. Chōsen **2**, 1928 (Jap.).
579. E.W. Koons, Beacon-Fires of Old Korea. Tr. Kor. Br. R. As. Soc. **16**, 1925. Pp. 46–52.
580. C.A. Clark, Religions of Old Korea. New York 1932. 295 p.
581. T. Ohm, Das religiöse Chaos in Korea. O. A. R. **10**, 1929. Pp. 223–227.
582. W. Gundert, Japanische Religionsgeschichte. Die Religionen der Japaner und Koreaner in geschichtlichem Abriß dargestellt. Stuttgart 1935. Korea: pp. 187–203.
583. H. Lautensach, Religion und Landschaft in Korea. Nippon **9**, 1943. Pp. 204–219.
584. M. Courant, Sommaire et histoire des cultes coréens. T'oung Pao II, 4. 1901, I. Pp. 37–58.
585. W.W. Rockhill, Notes on Some of the Laws, Customs and Superstitions of Korea. American Anthropologist **4**, 1896. P. 177.
586. G.H. Jones, The Spirit Worship of the Koreans. Tr. Kor. Br. R. As. Soc. **2**, 1901, I. Pp. 37–58.
587. G.H. J(ones), Hanging Rags on Trees. Kor. Rev. **1**, 1901. P. 21f.
588. T. Ohm, Naturgeisterverehrung in Korea. O. A. R. **11**, 1930. Pp. 742–744.
589. A. Eckardt, Vom Geisterglauben in Korea. Ostasiat. Beobachter **6**, 1938, Folge 60. P. 29f.
590. D.L. Gifford, Ancestral Worship as Practiced in Korea. Kor. Rep. 1892, pp. 169–176.
591. Y.K. Li, Ahnenkult in Korea. O. A. R. **14**, 1933. Pp. 255–258.
592. M.N. Trollope, Introduction to the Study of Buddhism in Corea. Tr. Kor. Br. R. As. Soc. **8**, 1917. Pp. 1–40.
593. F. Starr, Korean Buddhism: History – Condition – Art. Boston 1918. 19 + 104 p.
594. T. Ohm, Der koreanische Buddhismus. Zeitschr. f. Missionswiss. u. Religionswiss. **21**, 1931. Pp. 327–338.
595. H. Lautensach, Koreanische Klöster. Durch alle Welt. 1935, no. 11. P. 26f.
596. Picture Album of the 31 Main Korean Temples. Keijō 1931 (Jap.).
597. H.B. Hulbert, The Religion of the Heavenly Way. Kor. Rev. **6**, 1906. Pp. 418–424, 460–465.
598. Chuntokyo: Korea's New Religion. Review of Reviews **79**, 1929. Feb. Pp. 94–96.
599. T. Ohm, Neuere Religionsbildungen in Korea. O. A. R. **15**, 1934. Pp. 553–555.

XVIII. Economic and Settlement Geography

600. T. Tanaka, Chōsen no jimbun chishi gaku teki kenkyū (Anthropogeographical studies on Korea). Chiri Kyōiku **15**, 1931. Pp. 105–111, 226–230, 346–349. **16**, 1932. Pp. 587–592. **17**, 1933. Pp. 40–45, 256–263, 376–382.
601. H. Lautensach, Japan als kolonialer Gestalter der koreanischen Landschaft. (unpublished manuscript).
602. Y.S. Lee, Statistical Korea. Seoul 1931. 2 vols.
603. E. de Schweinitz Brunner, Rural Korea. A Preliminary Survey of Economic, Social and Religious Conditions. In: The Christian Mission in Relation to Rural Problems. Rep. Jerusalem Meeting Int. Missionary Council vol. VI. London etc. 1928, pp. 100–208.
604. J. Linke, Veränderung der wirtschaftsgeographischen Beziehungen in Korea unter dem Einfluß der Erschließung. Diss. Leipzig 1932. 134 S. Weltwirtschaftl. Abhandl. vol. 10, Stuttgart 1933.
605. H.K. Lee, Land Utilization and Rural Economy in Korea. The Institute of Pacific Relations, Honolulu. London 1936. 302 p. (Cf. Econ. Geogr. 1937).
606. J.O. Lévine, La situation économique du peuple coréen. L'Asie Française **38**, 1938. Pp. 217–222 (Excerpt of [605]).
607. Sang-uk Han, Die Landwirtschaft von Korea. Diss. Zürich 1927. 95 p.

608. J.W. Coulter and B.B. Hee Kim, Land Utilization Maps of Korea. Geogr. Rev. **24**, 1934. Pp. 418–422 (The eight maps are reproduced in: Krische, Mensch und Scholle, Berlin 1936, p. 119f).
609. R. Burnett Hall, Agricultural Regions of Asia. Part VII. The Japanese Empire. Econ. Geogr. **10**, 1934. Pp. 323–347. **11**, 1935. Pp. 33–52, 130–147.
610. Chōsen Sōtokufu nōji shikenjō nijūgo shūnen kinenshi (Publication on the occasion of the 25th anniversary of the Agricultural Experiment Station of the Government–General of Korea). Chōsen Sōtokufu 1931. 2 vols., 645 and 558 p. (Jap.).
611. Chōsen no nōgyō (Treatises on Korean agriculture). Chōsen Sōtokufu Nōrinkyōku. Keijō 1933. 211 p. (Jap.).
612. I. Nagai, Recent Progress of Agriculture in Tyosen. Pr. 3rd P. S. C. Tōkyō 1926. II, 1928. Pp. 2149–2154.
613. N. Mori, Tori to shokubutsu to nōringyō to no kankei (The relationships between birds, food and agriculture and forestry). Journ. Chōs. Nat. Hist. Soc. 1927, no. 5.
614. Chōsen no kosaku kwanku (Tenant customs in Korea). Chōsen Sōtokufu. Keijō 1930. Survey Material no. 26 (Jap.).
615. J. Lewin, Korea. Landwirtschaftliche Besitzverhältnisse. O. A. R. **13**, 1932. Pp. 26–28.
616. Development and Current Status of Agriculture in Korean Villages. Keijō 1935 (Jap.).
617. Bauernfrage und Selbsthilfeplan in Korea. Wirtschaftsdienst N. F. **23**, 1938 I. P. 265f.
618. H. Lautensach, Zur Geographie der künstlichen Bewässerung in Korea. P. M. 1940, pp. 289–303 (Jap. summary by T. Ootuki in Chir. Hyō. **17**, 1941. Pp. 152–158).
619. Korea and Irrigation. Compiled by the Oriental Development Co., Ltd., in: The Herald of Asia. Tōkyō 1927. 42 p. (cf. F. E. R. **21**, 1925. Pp. 207–210. **23**, 1927. Pp. 226–231).
620. I. Nagai and T. Nakagawa, Chōsen ni okeru suitō no shuyō hinshu to sono bumpu jōkyō (Important varieties of paddy rice in Korea and their distribution). Ann. Agr. Exp. Stat. **5**, 1931. Supplement 1. 42 p. and 31 maps (Jap.).
621. T. Ikeda, Rice in Chosen. F. E. R. **23**, 1927. Pp. 313, 316.
622. Chōsen no kome (Rice in Korea). Chōsen Shokusan Ginko Chōsaka. Keijō 1927 (Jap.).
623. S. Nakayama, The More Important Insect Enemies of the Rice Crop in Chosen. Pr. 4th P. S. C. Java 1929. Batavia–Bandoeng 1930. Vol. 4, pp. 181–183.
624. I. Nagai, T. Nakagawa and T. Takasaki, Chōsen ni okeru shuyō densakubutsu no bumpu oyobi saibai jōkyō (On the distribution and breeding of Korea's most important crops). Ann. Agr. Exp. Stat. **6**, 1932. No. 1 and 2. 87 p. and 27 maps (Jap.).
625. B. Laufer, Rye in the Far East and the Asiatic Origin of Our Word Series "Rye". T'oung Pao. **31**, 1935. Pp. 237–273.
626. Chōsen no daitsu (Soybeans in Korea). Chōsen Shokusan Ginko Chōsaka. Keijō 1924 (Jap.).
627. Chōsen no menka (Cotton in Korea). Chōsen Shokusan Ginko Chōsaka. Keijō 1924 (Jap.).
628. Tobacco in Korea. Kor. Mag. **1**, 1917. Pp. 248–254, 289–292.
629. Tobacco Industry of Chōsen. F. E. R. **21**, 1925. Pp. 31–34.
630. H.G. Appenzeller and G.H. Jones, Korean Ginseng. Kor. Rep. **5**, 1898. Pp. 161–170.
631. C.T. Collyer, The Culture and Preparation of Ginseng in Korea. Tr. Kor. Br. R. As. Soc. **3**, 1903, I. Pp. 18–30.
632. The Chosen Government Monopoly Bureau, Do You Know Koryo Ginseng? Keijō 1937. 11 p.
633. Chōsen no mayu ni tsuite (On sericulture in Korea). Chōsen Sōtokufu Shokusan Kyōku. Keijō 1927 (Jap.).
634. Sericulture Office in Ch'aryŏn-wan, Facts on the Sericulture School 1931. 45 p. (Jap.).
635. Chōsen no chikusan (Livestock raising in Korea). Chōsen Sōtokufu Shokusan Kyōku. Keijō 1927 (Jap.).
636. W.C. Etheridge and C.A. Helm, Korean Lespedeza in Rotation of Crops and Pastures. Univ. of Missouri. Agricult. Exp. Stat. (Columbia). Bull. no. 360, 1936. Pp. 1–22.
637. C.W. Holdaway, Korean Lespedeza and *Lespedeza sericea* Hays for Producing Milk. Virginia Polytechnic Inst. Agricult. Exp. Stat. (Blacksburg). Bull. no. 305, 1936. Pp. 1–7.
638. Chōsen Sōtokufu ringyō shikenjō yō ran (Facts on the forestry experiment stations of the government–general of Korea). Keijō 1932. 30 p. and 29 col. maps (Jap.).
639. E.W. Koons, Afforestation in Korea. Tr. Kor. Br. R. As. Soc. **6**, 1915, I. Pp. 35–42.
640. P.M. Roxby, Afforestation in Korea. Scott. Geogr. Mag. 1923, pp. 3–6.

641. P.M. Roxby, Forestry in Korea. F. E. R. **22**, 1926. Pp. 9–12.
642. Chōsen no ringyō (Forestry in Korea). Chōsen Sōtokufu Sanrinbu. Keijō 1929 (Jap.).
643. Chōsen no kogyō (Mining in Korea). Chōsen Sōtokufu Sanrinbu. Keijō 1922 (Jap.).
644. A.A. Friedman, Gold Mining in Korea. Eng. and Mining Journ. **132**, 1931. P. 453f.
645. B.P. Torgašev, The Mineral Industry of the Far East: Economic and Geological Report on the Mineral Resources and Mineral Industries of the Far Eastern Countries:....Japan with Its Dependencies Korea and Formosa...Shanghai 1930. 510 p.
646. Salt Industry of Korea. F. E. R. **21**, 1925. P. 81.
647. K. Tago, The Marine Production of Korea. Chi. Z. **33**, 1921. Pp. 595–606 (Jap.).
648. Chōsen no jūdai gyogyō (The ten most important branches of fishing in Korea). Chōsen Sōtokufu Shokusan Kyōku. Keijō 1921 (Jap.).
649. Chōsen no jūyo gyogyō (The most important branches of fishing in Korea). Chōsen Sōtokufu Shokusan Kyōku. Keijō 1923 (Jap.).
650. Chōsen no suisan shōken (Trade in marine products in Korea). Chōsen Sōtokufu 1938.
651. Chōsen no suisan gyō (Marine products of Korea). Chōsen Sōtokufu 1927. 44 p. with map (Jap.).
652. A.M. Karlin, Fischfang in Korea. Der Erdball **4**, 1930. P. 332f.
653. F. Bartz, Die Seefischerei Koreas. Zeitschr. f. Erdk. 1940, Pp. 296–303.
654. W.H. de Roos, De Visscherij in Japan. In: Economische Verslagen van Nederlandsche Diplomatieke en Consulaire Ambtenaren. 's Gravenhage **24**, 1930. Pp. 121–184.
655. H. Schepers, Japans Seefischerei. Breslau 1935. 228 p.
656. S. Yoshimura, Fishing in Japanese Waters. Chir. Hyō. **15**, 1939. Pp.57–65, 76–84 (Jap.).
657. F. Bartz, Japans Seefischereien. P. M. 1940, pp. 145–160.
658. C. Rabot, La chasse à la baleine sur les côtes du Japon et de la Corée. La Géogr. **15**, 1907. P. 126.
659. Chōsen no mentai (The Korean Alaska pollack). Chōsen shokusan Ginko Chōsaka. Keijō 1925 (Jap.).
660. Production and Industry in Korea. F. E. R. **24**, 1928. Pp. 469–474 (Excerpt of [VI]).
661. Chōsen no shokogyō (Trade and industry in Korea). Chōsen Sōtokufu Shokusan Kyōku. Keijō 1928 (Jap.).
662. Y. Muto, New Economic Trends in Korea. Contemporary Japan **5**, 1936/37. Pp. 204–210.
663. K. Moritani, The Present and Future of Chosen Industry and Culture. Bull. South Seas Ass. **3**, 1940. Pp. 1–9 (Eng.).
664. K. Ishiyama, Industrialization of Korea. Contemporary Japan **10**, 1941. Pp. 1160–1165.
665. Chōsen no sanshigyō (Silk reeling in Korea). Chōsen Sōtokufu Shokusan Kyōku. Keijō 1929 (Jap.).
666. Katei kogyō chōsa (Investigation on cottage industry). Keijō Shogyō Kaigisho. 1927 (Jap.).

XIX. Geography of Transport and Commerce

667. W.W. Taylor, Korean Roads, Past and Present. Tr. Kor. Br. R. As. Soc. **15**, 1924. Pp. 35–82.
668. Improvements in Chosen. F. E. R. **30**, 1934. Pp. 425–428 (Cf. **24**, 1928. P. 270f. Excerpt of [VI]).
669. Chosen and its Railways. Railway Bureau, Gov. Gen. of Chosen. New editions each 2–3 years.
670. The Seoul – Chemulpo Railroad. Kor. Rep. **5**, 1898. P. 272f.
671. The Seoul Fusan Railroad. Kor. Rev. **5**, 1905. Pp. 16–21, 183–187.
672. K. Haushofer, Der Ausbau des Eisenbahnnetzes in Korea. P. M. 1914, II. P. 56.
673. Dr. Preyer, Die Eisenbahnen in Korea (Chosen). Arch. f. Eisenbahnwesen **37**, 1914. Pp. 402–418, 720–743.
674. L. Mayer, Die japanischen Eisenbahnen und ihre Entwicklung. Mitt. Geogr. Ges. Wien **70**, 1927. Pp. 327–337.
675. T. Ohmura, Present Railway Conditions of Korea. F. E. R. **24**, 1928. Pp. 111–114.
676. Railway Development in Chosen. F. E. R. 1929, pp. 25–27, 380, 570. 1931, pp. 250–253. 1932, pp. 321, 329. 1933, pp. 331–335.
677. Das Eisenbahnwesen Koreas. Die Lokomotive **27**, 1930. P. 37f.

678. W. Paschen, Die Eisenbahnen in Chosen (früher Korea). Arch. f. Eisenbahnwesen 1930 II, pp. 520–524.
679. M. Arikawa, Communications in Chosen. Japan Mag. **24**, 1934. Pp. 124–126.
680. Government Railways of Chosen. Japan Mag. **24**, 1934. Pp. 127–129.
681. W. Paschen, Die Eisenbahnen in Japan. Arch. f. Eisenbahnwesen 1940, pp. 849–876.
682. J.R. Harding, Memorandum on the Lighthouse Department of the Korean Customs Service in November, 1905. Kor. Rev. **5**, 1905. Pp. 414–423.
683. Oberkommando der Kriegsmarine, Verzeichnis der Leuchtfeuer und Signalstellen aller Meere 1938. Teil X, Stiller Ozean, W-Seite. Berlin 1938, 66th edition. Pp. 60–89: Tyōsen (Korea) mit naheliegenden Inseln.
684. Chōsen shijō (The Korean markets). Chōsen Sōtokufu Chōsa Shiryō. Keijō 1923 (Jap.).
685. Shigaichi no shōken (Street trading). Chōsen Sōtokufu Chōsa Shiryō. Keijō 1928 (Jap.).
686. Chōsen shijō keizai (Markets in Korea). Chōsen Sōtokufu Chōsa Shiryō. Keijō 1929 (Jap.).
687. S. Miyake, Economic Relations Between Chosen and Japan Proper. Journ. Keijō Imp. Univ. Soc. Sc. Ass. **1**, 1928, IV (Jap.).
688. Y. Yagi, The Relation Between Japan Proper and Korea as Seen from the Standpoint of Rice-upply. Kyōtō Univ. Econ. Rev. **6**, 1931, II. Pp. 86–112.
689. Le Clerc, Rice Trade in the Far East. Washington 1927. 73 p.

XX. Population Distribution

690. Chōsen no jinko gensho do fuzu (Population distribution in Korea and its cartographic representation). Chōsen Sōtokufu Chōsa Shiryō. Keijō 1927 (Jap.).
691. H. Sawada, Chōsen, Karafuto, Kantoshū, Nanyoshōtō. Chimei no yomi kata oyobi jinko hyō (Results of the population census in Korea, Karafuto, Kwantung und the South Seas territories). Tōkyō 1938 (Jap., with transcription of names).
692. Chōsen ni okeru Naichijin (The Japanese in Korea). Chōsen Sōtokufu. Keijō 1923.
693. T. Ogishima, L'émigration japonaise. Rev. intern. Travail, Genève **34**, 1936. pp. 663–698.
694. K.J. Pelzer, Japanese Migration and Colonization. In: I. Bowman, Limits of Land Settlement. New York 1937, Pp. 155–194. Korea: p. 165.
695. M. Schwind, Der japanische Bevölkerungszuwachs im Rückgang. O. A. R. **22**, 1941. Pp. 216–220. Included: Map of the distribution of Japanese in Korea.
696. M. Schwind, Japanische Bauern jenseits des Stammlandes. 3. Japanische Bauern in Chōsen und auf Taiwan. Nippon **7**, 1941. Pp. 163–168.
697. H. Lautensach, Das japanische Bevölkerungselement in Korea. Geogr. Anz. **43**, 1942. Pp. 371–375.
698. Chōsen ni okeru Shinajin (The Chinese in Korea). Chōsen Sōtokufu. Keijō 1924.
699. H.K. Lee, The Pioneer Belt in Manchuria with Special Emphasis on Korean Settlements. Am. Geogr. Soc. 1931.
700. H.K. Lee, Korean Immigrants in Manchuria. P'yŏng-yang (1932). 270 p. (Kor.).
701. H.K. Lee, Korean Migrants in Manchuria. Geogr. Rev. **22**, 1932. Pp. 196–204.
702. G. Dubarrier, Un aspect du problème mandchourien: Les populations coréenes de Mandchourie. Rev. Franco–Chinoise 1932, Pp. 87–91.
703. G. Dubarrier, La question des Coréens en Mandchourie. Rev. du Pacifique 1932, pp. 687–691.

XXI. Japan in Korea

704. A.J. Brown, The Mastery of the Far East. The Story of Korea's Transformation and Japan's Rise to Supremacy in the Orient. New York 1919. 671 p.
705. A.J. Brown, Japan, Aufstieg zur Weltmacht. Zürich 1931. Japan and Korea: pp. 109–127.
706. H. Schmitthenner, Die japanische Expansion und Kolonisation in Ostasien. Geogr. Zeitschr. 1928, pp. 1–22.

707. H. Schmitthenner, Kulturgeographische und politisch-geographische Entwicklungen in Ostasien. Geogr. Zeitschr. 1935, pp. 56–78.
708. L. Mecking, Japans räumliche und wirtschaftliche Entwicklung und ihre geographischen Grundlagen. Jahrb. Ak. d. Wiss. Göttingen 1940/41, pp. 37–51.
709. R. S., Die japanische Expansion. Zeitschr. f. Geopol. 1939, pp. 617–622.
710. M. Schwind, Japanische Raumnot und Kolonisation. M. D. O. A. G. **32**, 1940. 23 p.
711. T. Gollier, Le Japon colonisateur: Les Japonais en Corée. Rev. Économ. Intern. **4**, 1909. Pp. 242–282.
712. A. Backhausen, Die japanische Verwaltung in Korea und ihre Tätigkeit. Berlin 1910. 79 p.
713. M. Komatsu, The Old People and the New Government. Tr. Kor. Br. R. As. Soc. **4**, 1912, I. Pp. 1–12.
714. A. Halot, La colonisation japonaise en Corée. Bull. Soc. Franco–Japonaise de Paris **26/27**, 1912. Pp. 61–74.
715. K. Haushofer, Dai Nihon. Betrachtungen über Groß-Japans Wehrkraft, Weltstellung und Zukunft. Berlin 1913. 377 p. Korea: pp. 187–202.
716. T. Hoshino, Economic History of Chosen. Seoul 1920. 266 p.
717. S. Motono, Le rôle de la Corée dans la politique japonaise. La Revue Mondiale. Sér. 11, année 27, vol. 172, 1926. Pp. 119–126.
718. K.K. Kawakami, Korea Under Japanese Rule. The Review of Nations 1927, pp. 65–75.
719. M. Moncharville, Études sur la colonisation japonaise. La Corée. L'Asie Française 1929, pp. 385–394.
720. O. Richter, Die Kolonien Korea und Formosa. O. A. R. **11**, 1930. Pp. 182–184.
721. O. Mossdorf, Der Weg zur koreanischen Verwaltungsreform. O. A. R. **11**, 1930. P. 278f. (Cf. ibid. p. 244).
722. Twenty-five Years of Progress in Chosen. Peninsula's Resources and Development. Japan Mag. **24**, 1934. Pp. 112–119.
723. C. Porter, Korea and Formosa as Colonies of Japan. Far Eastern Survey **4**, 1935. Pp. 81–88.
724. B. Nikitine, Où en est la Corée? Notes politico– économiques. Rev. Économ. Intern. **27**, 1935, I. Pp. 489–522.
725. A.J. Grajdanzev, Korea Under Changing Orders. Far Eastern Survey **8**, 1939. Pp. 291–297.
726. H. Thierbach, 30 Jahre japanische Kolonialpolitik in Korea. Deutscher Kolonialdienst **5**, 1940. Pp. 136–139.
727. A. Eckardt, Der Shintoismus in Korea. Die kathol. Mission **61**, 1933. Pp. 249–251, 277–278, 306–307.
728. C.W. Kim, Die Volksschule in Korea. Die japanische Assimilationserziehung. Diss. München 1934. 69 p.

XXII. North Korean Japan Sea Borderland

729. Chōsen Sōtokufu, Tetsudōkyōku Eigyōka, Tomankō ryuiki keizai jijō (The economic situation in the drainage area of the Tumen River). Keijō 1927.
730. T. Ichimura, On the Basaltic Rocks in Kainei-District, Kankyō-Hokudō, Chōsen. Journ. Geol. Soc. Tōkyō **30**, 1923. Pp. 413–428 (Jap., rev. in J. J. G. G. **3**, 1924. P. [1]).
731. T. Ichimura, Kainei Coal Field. Rep. Geol. Surv. Coal–Fields **1**, 1927. Pp. 1–76 (Jap. rev. in J. J. G. G. 1939, p. [45]).
732. S. Ushimaru, Geology of the Kainei–Chientao Coal-Fields. Bull. Geol. Inst. South Manchurian Railway Comp. Dairen 1936. No. 87, pp. 27–37 (Jap.).
733. H. Lautensach, Japanische Eisenbahnen und Kriegshäfen in Korea. P. M. 1935, p. 151f.
734. A.J. Avdoščenkov, Severo-koreiskie žel dorogi, kak novyi etap transportnoj svazi Japonii s materikom. Vestnik Mańčžurii (Harbin) 1934, pp. 102–111, 141–149 (Russ. with Eng. summary: North Korean Railways as a New Stage of Transport Connections Between Japan and the Mainland).
735. A.J. Avdoščenkov, Stroitel' stvo porta Rašin. Vestnik Mańčžurii (Harbin) 1934, no. 6 (Russ. and Eng.). 17 + 15 p.
736. E. Schultze, Rashin, ein neuer Welthafen in Ostasien. Verkehrstechn. Woche **29**, 1935. P. 283f.

737. G.B. Cressey, Rashin, New Gateway to Asia. Asia **35**, 1935. Pp. 587–589.
738. Rashin Kō (Rasin Hafen), Rashin Wan, Thōsen (Korea), O-Küste. Der Seewart **7**, 1938. Pp. 248–256.
739. S. Shimamura, Geological Atlas of Chōsen no. 16. Renshin and Seishin Sheets. Keijō 1933 (Jap.).
740. W. Muroi, On the Geology and Groundwater of the Vicinities of Seisinfu, Korea. Chi. Z. **50**, 1938. Pp. 442–456 (Jap. with Eng. summary).
741. T. Shiraki, Coal-Fields in Kyōjō-gun. Rep. Geol. Surv. Coal-Fields **5**, 1930. 105 p. (Jap., rev. in J. J. G. G. **19**, 1940. P. [8f]).
742. F. Yamanari, Alcaline Rock Region in the Kisshū–Meisen District, North-East Korea. Journ. Geol. Soc. Tōkyō **30**, 1923. Pp. 212–226 (Jap., rev. in J. J. G. G. **2**, 1923. P. [27f]).
743. F. Yamanari, Geological Atlas of Chōsen no. 3. Kayōho–Kisshū–Shiho and Rinmei Sheets. Keijō 1925 (Jap. w. Eng. summary).
744. I. Tateiwa, Geological Atlas of Chōsen no. 4. Kyokudō–Meisen–Shichihōzan and Kotendō Sheets. Keijō 1925 (Jap. with Eng. summary).
745. A. Lacroix, Les rhyolites et les trachytes hyperalcalins quartzifères á propos de ceux de la Corée. C. R. Ac. Sc. Paris **185**, 1927. Pp. 1410–1415.
746. J. Makiyama, The Meisen Miocene of North Korea. Mem. College Sc. Kyōtō Imp. Univ. Ser. B **11**, no. 4. 1936. Pp. 193–288 (Jap., rev. in J. J. G. G. **16**, 1937. P. [14f]).
747. Phook-Chyung Guidance (Hamnam Chosun) (Illustrated guide to Pukch'ŏng-gun, Hamgyŏngnam-do). Hokusei 1933.
748. K. Morita and Y. Kubota, The Fusenko Hydro-electric Power Plant. Tr. Tōkyō Meeting World Power Conference 1929, vol. II. Pp. 473–531 (Cf. F. E. R. 1930, pp. 131–138. 1932, pp. 475–477).
749. Das Kraftwerk Fusenko auf Korea. Handel u. Industrie **41**, 1932. Pp. 497–499.
750. I. Tateiwa, Geological Atlas of Chōsen no. 6. Shinkō, Kodosui, Genpeijo, Gorori, Kankō and Seikoshin Sheets. Keijō 1926 (Jap. with Eng. summary).

XXIII. Kaema Upland

751. S. Nakamura, Kankyō-Nandō hokubu kōsho chōsa hōkokubun (Research report on the conditions in the northern part of Hamgyŏngnam-do). Chōsen Kōsho Chōsa Hōkoku **5**, 1927.
752. T. Watanabe and K. Nagazawa, Chōsen, Kanhoku no yama yama (Kambōhō, Hizanhō, Hakutōzan) (The mountains of Hamgyŏngbuk-do in Korea). Sangaku **27**, 1932. No. 2.
753. S. Nakamura, Kōzan Chōshin kōgen no nan yen (The southern limit of the P'ungsan–Changjin plain). Chikyū **8**, 1919.
754. T. Nakai, Roho no shokubutsu chōsa (Report on vegetation of Mt. Waigalbong, northern Chosen). Chōs. Bull. 1916. 39 p. (Jap.).
755. Y. Kishinami, Über die alpinen Pflanzen der Tōrohōkette in Nordkorea. Bot. Mag. Tōkyō **46**, 1932. P. 257 (Ger.).
756. C.H. Archer, Some Climbs in Korea. Alpine Journ. 1931, pp. 103–113.
757. N. Mori, Hakutōzan oyobi Kankyō-Nanhoku kōchitai no dobutsu ni tsuite (On the fauna of Paektusan and the high regions of Hamgyŏngbuk-do and Hamgyŏngnam-do). Bunkyō no Chōsen 1926, no. 11.
758. Geological Map of Manchuria with Explanatory Text (Scale 1:400,000). Tou-Man-Chiang. Geol. Inst. South Manch. Railway Comp. 1932 (Jap. with Eng. summary).
759. T. Ichimura, Geological Notes on the Mozan Iron-Bearing District, N. Kankyōdō, Korea. Mem. Fac. Sc. and Agriculture, Taihoku Imp. Univ. **VI**, 6 (Geology). 1933 (Jap.).
760. T. Kato, On the Origin of the Mozan Iron Ore Deposits, North Korea. J. J. G. G. **16**, 1939. Pp. 233–238.
761. T. Ichimura, Geological Consideration of Some Basalt Flows on the Frontier Region of North Korea. Pr. 5th P. S. C. Canada 1933. Vol. 3, 1934. Pp. 2421–2427 (Rev. in J. J. G. G. **16**, 1939. P. [45]).
762. Y. Kinosaki, Geological Atlas of Chōsen no. 14. Saitoku, Shimpokujō, Kōho and Gosokuri Sheets. Keijō 1932 (Jap. with Eng. summary).

763. Y. Kinosaki, Geological Atlas of Chōsen no. 19. Gyohyōri, Kozyōri, Tyokodō and Zyōnōri Sheets. Keijō 1938 (Jap. with Eng. summary).
764. Y. Nakai, Chōsen Kankyō-Hokudō no meizan Kambōhō to sono shokubutsu (Kambōhō, a noted mountain in the province of Kankyō-Hokudō, Korea, and its vegetation). Jissai engei (Practical Horticulture) 1929, pp. 25–32 (Jap.).
765. T. Nakai, List of Plants, Collected at Mt. Matinryöng, Korea. Bot. Mag. Tōkyō **22**, 1908. Pp. 179–182 (Lat.).

XXIV. Paektusan Area

766. T. Nakai, Hakutōzan shokubutsu chōsa sho (Florula of Mt. Paik-tu-san). Gov. Gen. Chōsen. Seoul 1918. 77 p.
767. G. Fenzel, Report on Forest Regions of Fengtien and Kirin Provinces, Manchuria. Lingnan Sc. Journ. (Canton) **7**, 1932. P. 539ff.
768. S. Kawasaki, Hakutōzan myaku (On the Hakutō volcanic chain). Journ. Chōsen Nat. Hist. Soc. **4**, 1927, no. 4 (Jap.).
769. T. Mori, On the Flora of Mt. Hakutō. Journ. Chōsen Nat. Hist. Soc. **4**, 1927. Pp. 25–54 (Jap.).
770. Report on the District of the Forestry Station Hyesan. 1928. 30 p. text. 28 plates. 2 maps 1:200,000 (Jap.).
771. Survey of the Forests of the Forestry Station Musan (Hamgyŏngbuk-do). Map 1:200,000 (Jap.).
772. M.B. Ordun, Mountain-Water Adventure in Korea. Asia **33**, 1933. Pp. 344–350.
773. W. Fuchs, The Long White Mountain in Manchuria. China Journal **18**, 1933 I. Pp. 196–198.
774. A. de C. Sowerby, The Heavenly Lake. China Journal **19**, 1933 II. P. 54.
775. H. Lautensach, Eine Räuberreise zum Weißkopfberg an der koreanisch–mandschurischen Grenze. Die Umschau 1935, pp. 221–227.
776. Kyōtō teikohu daigaku Hakutōzan enseitai hōkoku (Report on the climbing of Paektusan by researchers from the University of Kyōtō). Kyōtō 1935. 148 p.
777. K. Takahashi, Hakuto-san im Winter (Korea). Deutsche Alpenzeitung **31**, 1936. Pp. 123–126.
778. H. Lautensach, Der Hakutōzan, eine vulkanische Landschaft im koreanisch–mandschurischen Grenzbereich. Geogr. Zeitschr. 1941, pp. 354–371, 409–431.

XXV. P'yŏngbuk Region

779. T. Ōzaki, Ōryokukō. Tōkyō 1910. 258 p. (Jap.).
780. R. Farquharson, The Yalu-River to Linkiang. China Journal **8**, 1928 I. Pp. 22–27.
781. R.G. Mills, Ecological Studies in the Tong-nai River Basin, Northern Korea. Tr. Kor. Br. R. As. Soc. **12**, 1921. Pp. 1–78.
782. Map of the Forestry District Kanggye, as of 1 April 1930. 1:200,000 (Jap.).
783. Survey of the Activities of the Forestry Station Ch'osan. 1932. 33 p. (Jap.).
784. Survey of the Production of Ch'osan-gun (P'yŏng-anbuk-do) 1932 (Jap.).
785. S. Shimamura, Geological Atlas of Chōsen no. 17. Seijochin, Tenmadō and Eizanshi Sheets. Keijō 1936 (Jap.).
786. Y. Kinosaki, Geological Atlas of Chōsen no. 15. Hokuchin and Gyukenchin Sheets. Keijō 1933 (Jap. with Eng. summary).
787. S. Amanuma, The Third Account of my Travels in Korea. 1. The Buddhist Temple Myōkōsan Hokenji. Tōyō Bijutsu 1932, no. 15 (Jap.).
788. The Cave of Dong Long Kul. Picture album (Jap.).
789. C.D. Morris, The City of Yung-byun (Neihen). Kor. Rev. **5**, 1905. P. 134.
790. Kisengun no kaden-jōkyō (The situation of fire-field farming in Hŭich'ŏn-gun). Chōsa Geppō 1928 (Jap.).

XXVI. Taedong Basin

791. H. Yabe, Report on the Anthracite-Bearing Formation of S. Heian-Do. B. Geol. S. Chōs. **I**, 1. Seoul 1919 (Jap.).
792. Ichimura, Kohira and Soki, Heian-Nandō hokubu tanden chishitsu zu (On the bituminous coal deposits in the northern part of P'yŏng-annam-do). Seoul 1927 (Jap.).
793. T. Shiraki, The Northern Coal-field in the Tokusen District in Heian-Nandō. Rep. Geol. Surv. Coal-fields **10**, 1931. 133 p. (Jap., rev. in J. J. G. G. 1940, p. [8]).
794. I. Tateiwa, Geological Atlas of Chōsen no. 12. Tōsō, Inzan, Bessori and Seisen Sheets. Keijō 1931 (Jap. with Eng. summary).
795. S. Mizugori, Geography and Geology of Tenseiri in Chosen. Journ. Geol. Soc. Tōkyō **43**, 1931. Pp. 377–383 (Jap., rev. in J. J. G. G. 1931/32, p. [15]).
796. S. Matsushita, The Stratigraphy and Structure of the District North-East of Heijō, Korea. Chikyū **20**, 1933. Pp. 1–26 (Jap., rev. in J. J. G. G. 1933/34, p. [10]).
797. S. Matsushita, Geological Structure of Kōto Area and Northeastern Area of Heijō, Chōsen. Journ. Geol. Soc. Tōkyō **45**, 1938. P. 487f (Jap., rev. in J. J. G. G. **17**, 1938. P. [63]).
798. N. Ikebe, Geological Structure of Santō and Syōgen Districts, Heizyō Coal-Field, Korea. Journ. Geol. Soc. Tōkyō **46**, 1939. Pp. 433–440 (Jap., rev. in J. J. G. G. 1940, p. [62]).
799. N. Kōbatake, Thrust-tructure in the Santō Area in Heijō Coal–field. Journ. Geol. Soc. Tōkyō **44**, 1937. P. 616f (Jap., rev. in J. J. G. G. 1937, p. [37]).
800. N. Ikebe, On the Geology of Taisei Mine and its Vicinity. Journ. Geol. Soc. Tōkyō **44**, 1937. P. 614f (Jap., rev. in J. J. G. G. 1937, p. [37]).
801. S. Maejima, Geology of the Vicinity of the City of Heijō, S. Heian-dō, Korea. Chikyū **24**, 1935. Pp. 81–88 (Jap., rev. in J. J. G. G. 1936, p. [1]).
802. E. Konnō, The Geology and Geological Structure of the Western Marginal Region of the Heijō-Coal-Field, Korea. Journ. Geol. Soc. Tōkyō **35**, 1928. pp. 65–104 (Jap., rev. in J. J. G. G. 1928/29, p. [12]).
803. S. Nakamura, The Geological Structure of the Western Heijō Coal Field, Korea. Chikyū **25**, 1936. Pp. 159–171 (Jap., rev. in J. J. G. G. 1936, p. [23]).
804. S. Matsushita, The Heian-Nandō Southern Coal Field. Rep. Geol. Surv. Coal-Fields **13**, 1938. 164 p. (Jap., detailed rev. in J. J. G. G. 1939, p. [43f]).
805. K. Saitō, Cambrian Formations of the Chunghwa District, Western North Korea. Journ. Geol. Soc. Tōkyō **40**, 1932. Pp. 242–260 (Jap., rev. in J. J. G. G. 1933/34, p. [3]).
806. D.F. Higgins, Geology and Ore Deposits of the Collbran Contact of the Suan Mining Concession, Korea. Economic Geology **13**, 1918. Pp. 1–34.
807. T. Watanabe, The Geology of the Suian Gold Mining District. Journ. Fac. Sc. Hokkaidō Imp. Univ. Sapporo III, 1935, no. 1. Pp. 49–59.
808. (H.B. Hulbert), The Caves of Kasa. Kor. Rev. **5**, 1905. Pp. 292–298.
809. J.H. Wells, Pyengyang. Kor. Rep. **4**, 1897. Pp. 57–59.
810. L. Nocentini, P'ieng-iang (città della Corea). Giornale Soc. Asiat. Ital. XIV, 1901. Pp. 217–240.
811. (H.B. Hulbert), A Visit to Pyeng-yang. Kor. Rev. **5**, 1905. Pp. 287–292.
812. Oh Seung Keun, What to See at Pyeng-yang. Kor. Rev. **6**, 1906. Pp. 321–324.
813. Brief History of P'yŏng-yang. 57 and 50 p. 1932 (Jap.).
814. F.H. Morsel, Chinnampo and Mokpo. Kor. Rep. **4**, 1897. Pp. 334–338.
815. Das Klima von Chinnampo. Ann. Hydr. u. Mar. Met. **40**, 1912. Pp. 489–498, 567.
816. K. Saitō and M. Matsuno, The Coal Shipping Plant at Chinnampo Harbour, Korea. Journ. Civil Engineering Soc. (Tōkyō) **20**, 1934. Pp. 977–1021.

XXVII. Hwanghae Peninsula

817. S. Shimamura, Geological Atlas of Chōsen no. 8. Kenjiho, Shariin and Sainei Sheets. Keijō 1929 (Jap. with Eng. summary).
818. Matsushita, Onoyama and Maejima, Geology and Fossil of Hōzan Coal-Field, Kōkaidō, Chōsen. Chikyū **23**, 1935. Pp. 403–430 (Jap., rev. in J. J. G. G. 1938, p. [63]).

819. Y. Kinosaki, Angaku Iron Ore Deposit, Kōkaidō, Chōsen. Journ. Geol. Soc. Tōkyō **33**, 1926. pp. 142–154, 175–186, 232–242 (Jap., rev. in J. J. G. G. 1925, p. [26]).
820. F. Takai, A Tooth of Elephant from the Seabottom near the Island of Hakurei, Kōkaidō, Korea. Journ. Geol. Soc. Tōkyō **44**, 1937. P. 304f (Jap.).
821. S.F. Moore, Note on Hai-ju. Kor. Rev. **4**, 1904. Pp. 207–211.
822. Excursion to Songdo. Kor. Mag. **2**, 1918. Pp. 290–292.
823. J.W. Hirst, A Trip to Sorai Beach. Kor. Rev. **6**, 1906. Pp. 27–34.

XXVIII. Kyŏnggi Region

824. Tateiwa, Shimamura, Tamura and Yamazawa, Report on the Flood of the Kankō in 1925. B. Geol. S. Chōs. **V**, 1. 1925 (Jap.).
825. T. Ishidoya and P.S. To, Florula Seoulensis. Journ. Chōs. Nat. Hist. Soc. **14**, 1932. Pp. 1–48 (Jap. with Lat. plant lists).
826. M. Tozawa and T. Nakai, Kōryō shikenrin no ippan (Brief description of the experimental forest of Kwangnŭng). Forest Experiment Station of the Government–General 1932. 171 p. (Jap.).
827. H. Uyeki, Vegetations in the Kwa-San Hill and the Environs of Suigen. Bull. Agr. For. Coll. Suigen **5**, 1936. 155 p. (Jap.).
828. E.B. Landis, Notable Dates of Kang-wha. Kor. Rep. **4**, 1897. Pp. 245–248.
829. M.N. Trollope, Kang-wha. Tr. Kor. Br. R. As. Soc. **2**, I, 1901. Pp. 1–36.
830. F.H. Mörsel, Climatical Records for Chemulpo. Kor. Rep. **3**, 1896. P. 460f. **4**, 1897. Pp. 76f, 156f, 318f.
831. G.H. Jones, Chemulpo. Kor. Rep. **4**, 1897. Pp. 374–384.
832. Beiträge zur Küstenkunde: Chemulpo–Chemuropo (Westküste Koreas). Der Seewart **3**, 1934. Pp. 180–195.

XXIX. Seoul

833. H.G. Appenzeller, The Beginnings of Seoul. Kor. Rep. **1**, 1892. May.
834. H.N. Allen, Places of Interest in Seoul with History and Legend. Kor. Rep. **2**, 1895. Pp. 127–133, 182–187, 209–214.
835. M. Courant, Souvenir du Séoul, Corée. 8 p., 24 plates. Paris 1900.
836. J.S. Gale, Han-Yang (Seoul). Tr. Kor. Br. R. As. Soc. **2**, II, 1902. Pp. 1–43.
837. Oh Seung-Keung, The Fortress of Puk-han. Kor. Rev. **3**, 1903. Pp. 44–51.
838. J. D., Die Gründung der Hauptstadt Koreas. Deutsche Rundschau f. Geogr. u. Stat. **26**, 1903/04. P. 337f.
839. Places of Interest About Seoul. Kor. Mag. **2**, 1918. Pp. 252, 302, 409, 443, 492.
840. Keijō. 42 plates. Text Jap. and Eng.
841. Japan Tourist Bureau, Chōsen Branch, City plan and geographical and historical description of Seoul (Eng.).
842. Japan Tourist Bureau, Chōsen Branch, How to See Keijō. Annually.
843. H. Lautensach, Koreas Hauptstadt und ihre Umwelt. Geogr. Zeitschr. **46**, 1940. Pp. 1–17.

XXX. Upper Han-gang Region

844. T. Kobayashi, Geological History of Kōgen-dō, Chōsen. Journ. Geol. Soc. Tōkyō **34**, 1927 (Jap.).
845. F. Yamanari, Imbricated Structure of Kōgen-dō, Chōsen. Chir. Hyō. **2**, 1926 (Jap.).
846. I. Yoshimura, On the Schuppenstruktur in the Environs of Neietsu, Kōgendō. Journ. Geol. Soc. Tōkyō **46**, 1939. P. 348f (Jap., rev. in J. J. G. G. 1940, p. [8]).
847. I. Yoshimura, Geology of the Neietsu District, Kōgendō, Tyōsen. Journ. Geol. Soc. Tōkyō **47**, 1940. Pp. 112–122 (Jap.).

848. G. Ozawa, Akiyoshidai no chishi to chikei to chikasui (On the geographical history, the geographical forms and the underground water courses of Akiyoshidai). Chir. Hyō. **1**, 1925 (Jap.).
849. S. Nakamura, Chiri kyōsai to shite no chikeizu daisanjugo Chōsen shimbaku fuki no karsto (On karst in Kang-won-do). Chikyū **14**, 1926 (Jap.).
850. J. R. Moose, The Iron Mines of Kang-won Province. Kor. Rev. **5**, 1905. Pp. 8–11.

XXXI. Wonsan Area

851. R. Kodaira, The Kogen Anthracite Field. Rep. Geol. Surv. Coal-Fields **5**, 1930. 91 p. (Jap., rev. in J. J. G. G. 1940, p. [8]).
852. R. Kodaira, The Bunsen Coal-Field. Rep. Geol. Surv. Coal-Fields **5**, 1930 (Jap., rev. in J. J. G. G. 1940, p. [8]).
853. T. Okada, Foehnwind at Wonsan in Korea. Journ. Met. Soc. Japan **26**, 1907. Pp. 1–8 (detailed rev. by A[lbert] D[efant] in Met. Zeitschr. 1908, p. 88).
854. F. Immanuel, Die Broughton-Bai (Ostküste von Korea). Globus **66**, 1894. Pp. 357–360.

XXXII. T'aebaek Range, Including Kŭmgangsan

855. T. Ichimura, The Tsūsen Coal-Field. Rep. Geol. Surv. Coal-Fields **3**, 1938^2. 23 p. (Jap., rev. in J. J. G. G. 1940, p. [8]).
856. J. Suzuki, On the Metamorphic Rocks in the Southeastern Part of Kōryōgun, Kōgendō, Chōsen. Journ. Geol. Soc. Tōkyō **42**, 1935. P. 309f (Jap., rev. in J. J. G. G. 1939, p. [43]).
857. C. Hunt, Diary of a Trip to Sul-Ak San (via the Diamond Mountains) 1923. Tr. Kor. Br. R. As. Soc. **24**, 1935. Pp. 1–14.
858. W.B. Scranton, The Fifty Three Buddhas and the Nine Dragons. Kor. Rep. **4**, 1897. Pp. 321–324.
859. Government–General Seoul, Keum-Gang-San (Diamond Mountain). Tourist map in five colors. 1:50,000. 1917.
860. T. Nakai, Kongōsan shokubutsu chōsa sho (Report on the vegetation of Diamond Mountains, Corea). Gov. Gen. Seoul 1918. 204 p. (Jap.).
861. Cho-Sung-Ha, The Diamond Mountains. Kor. Mag. **2**, 1918. Pp. 449–452.
862. E.J. Coatsworth, Through the Diamond Mountains of Korea with the Ancients. Asia **19**, 1919. Pp. 16–23.
863. H.C. Bowser, The Kongo-San. Geogr. Journ. **55**, 1920. Pp. 48–50.
864. J.S. Gale, The Diamond Mountains. Tr. Kor. Br. R. As. Soc. **13**, 1922. Pp. 1–74.
865. Marquess Curzon of Keddleston, In the Diamond Mountains: Adventures Among the Buddhist Monasteries of Eastern Korea. Nat. Geogr. Mag. **46**, 1924. Pp. 353–374.
866. P. Klautke, Keum Gang San, die Diamantberge Koreas. M. D. O. A. G. **21**, part C. 1926. 60 p.
867. N. Weber, In den Diamantbergen Koreas. St. Ottilien 1927. 109 p.
868. E.S. Fischer, Die "Kongo-Shan" oder Diamanten-Bergkette in Korea. Mitt. Geogr. Ges. Wien **71**, 1928. Pp. 240–248.
869. C.S. Deming, A Guide to the Diamond Mountains. Seoul 1929. 47 p.
870. P. Chizallet, Le Keum-kang-san. Bull. Soc. Missions–Étrangères de Paris **9**, 1930. Pp. 633–638.
871. Chōsen, Kongōsan. 1930. Album with 88 plates. Captions in Jap. and Eng.
872. P. Klautke, Beitrag zur Pflanzenwelt der Diamantberge. Jubil. Bd. D. O. A. G. **I**, 1933. Pp. 57–75.
873. D. Weber, In den Diamantbergen Koreas. Chinadienst **3**, 1934. Pp. 392–398.
874. K. Scott, The Diamond Mountains of Korea. Canad. Geogr. Journ. **12**, 1936. Pp. 183–192.
875. The Government Railways of Chosen, Kongosan (Chosen). Seoul, repeated editions, last in 1937. Travel guide, 27 p.
876. H. Lautensach, Das Diamantgebirge in Korea. Zeitschr. f. Erdk. **10**, 1942. Pp. 442–452.

XXXIII. Ullŭngdo

877. F.H. Mörsel, Wölung Do (Dagelet Island). Kor. Rep. **2**, 1895. P. 412f.
878. (H.B. Hulbert), Ul-leung Do (Dagelet Island). Kor. Rev. **6**, 1906. Pp. 281–285.
879. T. Nakai, Utsuryōtō shokubutsu chōsa sho (Report on the vegetation of the island Ooryongto or Dagelet Island, Corea). Gov. Gen. Chōsen, Seoul. 1919. 87 p.
880. S. Tsuboi, On a Leucite Rock from Utsuryōtō Island in the Sea of Japan. Journ. Geol. Soc. Tōkyō **27**, 1920 (Jap.).
881. T. Nakai, The Vegetation of Dagelet Island: its Formations and Floral Relationship with Korea and Japan. Pr. 3rd P. S. C. Tōkyō 1926. P. 911–914.
882. T. Ishidoya, On the Flora of the Island Ooryongto or Dagelet Island I. Journ. Chōs. Nat. Hist. Soc. **7**, 1928. Pp. 21–43 (Jap.).
883. H. Lautensach, Quelpart und Dagelet. Vergleichende Landeskunde zweier koreanischer Inseln. Wiss. Veröff. Mus. f. Länderkunde Leipzig 1935, pp. 177–206.
884. F. Tada, Saishūto oyobi Utsuryōtō. Chikyū **24**, 1936. 11 p. (Jap. summary of 883).

XXXIV. Naktong Basin

885. F. Yamanari, Geological Atlas of Chōsen no. 1. Mitsuyō and Yusen Sheets. Keijō 1924 (Jap. with Eng. summary).
886. I. Tateiwa, Geological Atlas of Chōsen no. 2. Ennichi–Kyūryūho and Choyō Sheets. Keijō 1924 (Jap. with Eng. summary).
887. I. Tateiwa, Geological Atlas of Chōsen no. 10. Keishū, Eisen, Taikyū and Waikan Sheets. Keijō 1929 (Jap. with Eng. summary).
888. N. Hatae, Geological Atlas of Chōsen no. 18. Neikai and Eitoku Sheets. Keijō 1937 (Jap. with Eng. summary).
889. K. Kanehara, The Geology of the Northern Part of Geizitu District, North Keisyō-dō, Korea. Journ. Geol. Soc. Tōkyō **43**, 1936 (Jap.).
890. T. Katō, The Alaskitic Gold Vein of the Kinsei-Mine, Korea. J. J. G. G. **13**, 1936. Pp. 103–105.
891. Statistical Yearbook of the Province of Kyŏngsangbuk-do (Jap.). Annually.
892. Agriculture in the Province of Kyŏngsangbuk-do (Jap.). Annual statistics.
893. Commerce, Industry and Products of the Province of Kyŏngsangbuk-do (Jap.). Annual statistics.
894. W.C. Rufus, Trip to Kyungju. Kor. Mag. **1**, 1917. Pp. 532–541.
895. B. Gottsche, Sok-kul-am, das Steinhöhlenkloster. Ostasiat. Zeitschr. **7**, 1918/19. Pp. 161–167.
896. Sekkutsuan and Bukkokuji. Gen. Surv. Korean Art no. 1. Nara 1928. 100 plates, Jap. text.
897. The Old Capital of the Shilla Dynasty, Kyŏngju. 44 plates, Jap. text.
898. R.H. Sidebotham, Unknown Land. Kor. Rev. **5**, 1905. pp. 223–227.
899. K. Nisida, Result of Current Observations Made in the Vicinity of Husan. Umi to sora (Sea and Sky) **8**, 1928. Pp. 184–186 (Jap.).
900. M. Courant, Unétablissement japonais en Coré. Pou-san depuis le XVe siècle. Bibliothéque de la France Coloniale Moderne 1904. 24 p.

XXXV. Sobaek Range

901. I. Tateiwa, Report on the Groundwater in Kan-an District. B. Geol. S. Chōs., **V**, 2, 1926 (Jap.).
902. T. Nakai, Chirisan shokubutsu chōsa hōkoku sho (Report on the Vegetation on Chirisan). Chōsen Sōtokufu 1915. 98 p. (Jap.).

XXXVI. Ch'ungch'ŏng–Chŏlla Region

903. S. Shimamura, Geological Atlas of Chōsen no. 5. Chinan and Zenshū Sheets. Keijō 1925 (Jap. with Eng. summary).
904. S. Shimamura, Geological Atlas of Chōsen no. 7. Seizan and Eidō Sheets. Keijō 1927 (Jap. with Eng. summary).
905. S. Shimamura, Geological Atlas of Chōsen no. 13. Daisenri, Fuyo and Ranpo Sheets. Keijō 1931 (Jap. with Eng. summary).
906. F. Yamanari, Report on the Geology of the Reservoir of Tōshin Irrigation Association. N. Zenra-Do. B. Geol. S. Chōs. **I**, 2. 1923 (Jap.).
907. I. Komada, On the Pre-cambrian Formation of the Environs of Chūshū, N. Chūsei-Do, Korea, with Special Reference to the Geology and Ore-deposits of the Chūshū Iron Mine. Chōsen Kōgyō Kai-Shi (Journ. Korean Assoc. Min. Eng.) **5**, 1922, no. 8 (Jap., rev. in J. J. G. G. 1923, p. [7f]).
908. I. Tateiwa, Report on the Water Supply for Dried Laver Industry and for Fishing Boats in S. Zenra-Do. B. Geol. Soc. Chōs. **VIII**, 2. 1929 (Jap.).
909. S. Shimamura, Report on the Groundwater in Shinho-men, N. Zenra-Do. B. Geol. S. Chōs. **VIII**, 2. 1929 (Jap.).
910. S. Shimamura, Report on the Groundwater in Riri, N. Zenra-Do. B. Geol. S. Chōs. **VIII**, 2. 1929 (Jap.).
911. Statistical Yearbook of the Province of Chŏllabuk-do (Jap.). Annually.
912. Asan-gun, Ch'ungch'ŏngbuk-do. 1929. 221 p. (Jap.).
913. Introduction to the Fuji-Nōson-Farm. (Jap.)
914. Fuji-Nōson sangyō kumiai (General Report on the Fuji Agricultural Company) (Jap.).
915. D.L. Gifford, A Visit to a Famous Mountain (Mt. Kay-Riong-San). Kor. Rep. **1**, 1892. Pp. 41–45.
916. G.H. Jones, Korea's Colossal Image of Buddha. Tr. Kor. Br. R. As. Soc. **1**, 1900. Pp. 51–79.

XXXVII. Southern Coast Borderland

917. Y. Kinosaki, Geological Atlas of Chōsen no. 9. Kainei and Usuiei Sheets. Keijō 1929 (Jap. with Eng. summary).
918. S. Shimamura, Geological Atlas of Chōsen no. 11. Kwantō–Kokwatō–Seizantō and Tarotō and Shoantō Sheets. Keijō 1930 (Jap. with Eng. summary).
919. T. Katō, Geology and Ore Deposits of the Tōei Gold Mine, South Keishō-dō, Korea J. J. G. G. **2**, 1923. Pp. 85–93.
920. T. Ichimura, The Wajun Coal Field. Rep. Geol. Surv. Coal-Fields **2**, 1927. Pp. 1–23 (Jap., rev. in J. J. G. G. 1939, p. [44]).
921. T. Nakai, A Comparison of the Florulas of the Dagelet Island and the Daikoku-zantō, Which Are Isolated in the East and West of Korean Peninsula. Journ. Orient. Art. Sci. **43**, 1927, pp. 214–227 (Jap.).

XXXVIII. Chejudo

922. L. Nocentini, Port Hamilton e Quaelpart. Boll. Soc. Geogr. Ital. II, **12**, 1887. Pp. 389–398.
923. B.S. Ohlinger, The Island of Quelpart. Kor. Rep. **1**, 1892. Pp. 94–96.
924. A.A. Pieters, A Visit to Quelpart. Kor. Rev. **5**, 1905. Pp. 172–179, 215–219.
925. H.B. Hulbert, The Island of Quelpart. Bull. Am. Geogr. Soc. **37**, 1905. Pp. 396–408.
926. M.P. Anderson, Forty Days in Quelpart Island. Overland (San Francisco). N. S. **63**, 1914. Pp. 392–401.
927. R. Burnett Hall, Quelpart Island and Its People. Geogr. Rev. **16**, 1926. Pp. 60–72.
928. W. Stötzner, Eine Reise nach Quelpart. Forsch. u. Fortschr. 1930. P. 425f.
929. W. Stötzner, Have You Been to Quelpart? Asia **33**, 1933. Pp. 412–417.
930. W. Stötzner, Steingötter wachen über Quelpart. Die Umschau **39**, 1935. P. 778f.

931. A. Eckardt, Quelpart, die merkwürdigste Insel Ostasiens. Zeitschr. f. Erdk. **9**, 1941. Pp. 742–750.
932. S. Julien, Note sur une montagne volcanique sortie du fond de la mer au sud de la Corée, l'an 1007 de J.–C. Extrait de l'Encyclopédie japonaise. C. R. Ac. Sc. Paris **10**, 1840. P. 835.
933. K. Haraguchi, Saishūtō no chishi zu (Saishū Volcano). B. Geol. S. Chōs. **X**, I. Keijō 1931. 46 p. (Jap.).
934. H. Takeda and T. Nakai, Plantae ex insula Tschedschu (Quelpart). Bot. Mag. Tōkyō **23**, 1909. Pp. 46–58 (Ger.).
935. T. Nakai, Saishūtō oyobi Kwantō ni okeru shokubutsu chōsa hōkoku sho (Report on the botanical investigation of Chejudo and Kwantō). Chōsen Sōtokufu 1916 (Jap.).
936. T. Nakai, The Floras of Tsuma and Quelpart as Related to Those of Japan and Korea. Pr. 3rd P. S. C. Tōkyō 1926. Pp. 893–912.
Cf. 883 (Lautensach, Quelpart und Dagelet).

INDEX OF TOPICS

INDEX OF PERSONS

INDEX OF PLACE NAMES

The index of place names lists all places mentioned in this book and provides the following information:

1. Korean pronunciation romanized according to the McCune-Reischauer system, with the modifications announced by the Ministry of Education, Republic of Korea, in 1984; place names in China romanized according to the Pinyin system;
2. Chinese characters;
3. Japanese pronunciation in angle brackets, romanized according to the Romaji-Kai, as used by H. Lautensach;
4. Geographical coordinates in whole degrees of latitude and longitude;
5. Coordinates in the UTM (Universal Transverse Mercator) grid system, consisting of two letters designating the 100 km square and two digits designating the 10 km square; the Grid Zone Designations pertaining to Korea (51S, 52S and 51T, 52T) have been omitted;
6. Reference to figures (in italics) and pages on which the name occurs.

MIX
Papier aus verantwortungsvollen Quellen
Paper from responsible sources
FSC® C105338

If you have any concerns about our products,
you can contact us on
ProductSafety@springernature.com

In case Publisher is established outside the EU,
the EU authorized representative is:
Springer Nature Customer Service Center GmbH
Europaplatz 3, 69115 Heidelberg, Germany

Printed by Libri Plureos GmbH
in Hamburg, Germany